Basin Inversion

Geological Society Special Publications

Series Editor A.J. FLEET

GEOLOGICAL SOCIETY SPECIAL PUBLICATION NO. 88

Basin Inversion

EDITED BY

JAMES G. BUCHANAN
British Gas Exploration and Production Ltd.,
Reading, UK

AND

PETER G. BUCHANAN
Oil Search Ltd.,
Port Moresby, Papua New Guinea

1995
Published by
The Geological Society
London

THE GEOLOGICAL SOCIETY

The Society was founded in 1807 as the Geological Society of London and is the oldest geological society in the world. It received its Royal Charter in 1825 for the purpose of 'investigating the mineral structure of the Earth'. The Society is Britain's national society for geology with a Membership of 7500 (1993). It has countrywide coverage and approximately 1000 members reside overseas. The Society is responsible for all aspects of the geological sciences including professional matters. The Society has its own publishing house which produces the Society's international journals, books and maps, and which acts as the European distributor for publications of the American Association of Petroleum Geologists and the Geological Society of America.

Fellowship is open to those holding a recognized honours degree in geology or cognate subject and who have at least two years relevant postgraduate experience, or who have not less than six years relevant experience in geology or a cognate subject. A Fellow who has not less than five years relevant postgraduate experience in the practice of geology may apply for validation and subject to approval, may be able to use the designatory letters C. Geol (Chartered Geologist).

Further information about the Society is available from the Membership Manager, The Geological Society, Burlington House, Piccadilly, London W1V 0JU, UK.

Published by The Geological Society from:
The Geological Society Publishing House
Unit 7
Brassmill Enterprise Centre
Brassmill Lane
Bath BA1 3JN
UK
(*Orders*: Tel. 01225 445046
         Fax 01225 442836)

First published 1995

British Library Cataloguing in Publication Data

A catalogue record for this book is available from the British Library

ISBN 1–897799–29–2

Typeset by Type Study, Scarborough, UK

Printed in Great Britain by
Alden Press, Oxford

Distributors

USA
  AAPG Bookstore
  PO Box 979
  Tulsa
  OK 74101–0979
  USA
  (*Orders*: Tel. (918)584–2555
            Fax (918)584–0469)

Australia
  Australian Mineral Foundation
  63 Conyngham Street
  Glenside
  South Australia 5065
  Australia
  (*Orders*: Tel. (08)379–0444
            Fax (08)379–4634)

India
  Affiliated East-West Press PVT Ltd
  G-1/16 Ansari Road
  New Delhi 110 002
  India
  (*Orders*: Tel. (11)327–9113
            Fax (11)326-0538)

Japan
  Kanda Book Trading Company
  Tanikawa Building
  3–2 Kanda Surugadai
  Chiyoda-Ku
  Tokyo 101
  Japan
  (*Orders*: Tel. (03) 3255–3497
            Fax (03) 3255–3495)

Contents

Case studies: Asia

Case studies: Australasia

Index — 589

Introduction

This volume presents recent work on the causes and effects of inversion tectonics; it is not intended to be a comprehensive review of basin inversion and associated topics. It brings together studies from both academia and the petroleum industry and attempts to forge links between workers in these areas.

Inversion tectonics has previously been reviewed in Cooper & Williams (1989) and more recently by Coward (1994). The Cooper & Williams volume was the first detailed work on inversion tectonics and it acted as a catalyst for focusing future studies in the subject.

The volume highlights the following issues:
(a) there is no concise, universal definition of basin inversion;
(b) there are numerous examples from around the world which attest to the increasing recognition and importance of inversion tectonics;
(c) the understanding of the causes of inversion is still limited;
(d) the precise mechanics of fault reactivation are clearly important yet they are still unconstrained in detail;
(e) the 3D geometries of inversion related structures are relatively poorly described and not fully understood.

As stated above there is no agreed or universally used definition of basin inversion or inversion tectonics. The term is very widely used and in the most general sense well understood. As evidenced by the papers in this volume, the process occurs in a wide range of tectonic setting thus a universal definition would be extremely cumbersome to frame. It is recommended that every author defines their use of the term in their paper. In most cases the author(s) in this volume have taken this approach.

As inversion occurs globally and in a wide range of tectonic settings, there is a need to differentiate between minor fault reactivation and major uplift of sediments within fault bounded basins. It is very important to study inversion in its regional context in order to distinguish between regional and local tectonics. The term 'inversion' is often used to cover all compressional tectonics which have occurred post-basin formation. There is a need to differentiate where basin inversion ceases and orogenesis begins.

The papers in the volume have been subdivided into seven sections. Many papers, however, could have been placed in several different sections owing to the breadth of their content.

Mechanics, dynamics and geometry

In this first section, Sibson presents a study on the relationship of fluid flow, fault reactivation and inversion tectonics. He examines the nature of the controls on selective fault reactivation during inversion. Brodie & White detail an interesting study on how uplift can be produced by underplating igneous material. The paper uses both geochemical and geophysical data to understand further the crustal processes which may produce regional uplift and denudation. In a wide-ranging review paper on inversion tectonics from several basins across the world, Lowell presents a discussion on the control which pre-existing basin configuration and tectonic setting can have on inversion. A detailed field study by Knott *et al.* in the Gulf of Suez focuses on fault geometries during compressional tectonics. This work highlights unusual fault geometries which may be produced during basin inversion and which have not been previously recognized. Macgregor reviews the effect tectonic inversion can have on rift basins and on hydrocarbon prospectivity. He presents a classification of inverted rift basins and their relative prospectivity for gas/oil exploration.

Modelling

In the first paper of this section McClay reviews a series of analogue sandbox model experiments. The kinematics and dynamics of both the extensional and compressional tectonics are examined. A detailed analysis of the geometries produced in the models is also presented and the implications for other inversion studies discussed. In another paper based on sandbox modelling, Eisenstadt & Withjack document and discuss the underestimation of tectonic inversion. This study is of major importance in the measurement of inversion using purely geometric criteria, for example, in the analysis of seismic data. Keller & McClay present a brief study on the modelling of 3D inversion tectonics using large scale sandbox models.

Recognition and measurement

Section three deals with papers on the recognition and measurement of basin inversion. In this section, Green *et al.* present a review of the measurement of uplift in inverted basins using fission track (apatite) and related data. The data are integrated and analysed within basin modelling software and the results discussed with reference to the basin uplift history. In two papers, Hillis and Menpes present their work on the use of sonic velocity data in the study of uplifted basins in the United Kingdom Continental Shelf (UKCS). This method is a powerful way of analysing uplift across a large geological area using well data which are often readily available.

American case studies

The last four sections deal with Case Studies of inverted basins across the world. Section four deals with three studies from the Americas. Uliana and his co-workers review inversion tectonics within several basins in Argentina using both field and seismic data to constrain their interpretations. They also review the mechanical controls on basin inversion in their area of study. Homovc *et al.* present a well constrained case study of the inversion tectonics in the San Jorge basin in Argentina. Sinclair uses a detailed grid of seismic data and associated wells to review the compressional or transpressional tectonics within the Jeanne d'Arc basin, offshore Eastern Canada.

European case studies

Case Studies from Europe are presented in Section Five. The first three studies are largely based on the interpretation of seismic data from the North Sea. Thomas & Coward use good quality 2D seismic data to analyse the evidence for small scale compressional tectonics and their implications for regional structural evolution in the East Shetland Basin of the North Sea. Hooper *et al.* studied the basin evolution of the Broad Fourteens Basin in the Southern North Sea, utilizing a high resolution 3D seismic survey. In contrast to the two preceding studies, Deeks *et al.* interpreted deep seismic data, in addition to conventional seismic and other geophysical data, to constrain the basement control on inversion tectonics in the Southern Baltic Sea. A comparison of several inverted Western European basins which have different rifting and post-rifting histories is presented by Huyghe & Mugnier. Their paper is concluded by a valuable analysis of the control which the amount of rifting and length of post-rifting interval has on the nature of basin inversion.

The rest of the papers in this section rely in the main on detailed fieldwork studies. Nemock *et al.* studied the onshore/offshore portion of the Bristol Channel Basin, south Wales and western England looking at the relationship of inversion tectonics to fracture permeability within the basin. Dart & McClay present a detailed field study of the North Somerset coast off western England looking at the inversion of mesoscopic scale extensional fault systems.

The final two papers are field-based studies of inverted basins in northern Spain. Bond & McClay present a well documented study of syn-sedimentary fault reactivation and folding related to basin inversion in the Southern Pyrenees of Spain. Guimerá *et al.*, in a study of the Cameros Basin, document and discuss the inversion of the basin by a newly formed thrust fault.

An abstract by Turner & Corbin (at the end of the book) discusses studies which focus on inversion tectonics within the Cardigan Bay Basin, offshore Wales.

Asian case studies

In section six basin inversion studies in Asia from three contrasting tectonic settings are presented. In the first paper Lambaise & Bosworth discuss their study of the inversion of the Kyokpo Basin, South Korea. The paper concludes that the inversion of the basin is related to regional strike-slip tectonics. Samuel *et al.* present a detailed, integrated study from the Sunda Forearc, Indonesia, looking at inversion in an outer arc ridge context. They discuss the alternative deformation scenarios prior to outlining preferred inversion model. Wang and his co-workers present a case study of basin inversion within an East China Sea basin based on 2D seismic data interpretation. They document the initiation and evolution of a growth fold system within the inverted basin.

Australasian case studies

The last section deals with papers from Australasia. Hill *et al.* present a detailed, integrated analysis of Bass Strait Basins, Australia. This well documented paper studies the inversion histories of several basins and focuses on their contrasting tectonic evolution. They conclude with stimulating comments on both thermal and tectonic inversion processes based on their study. Bishop & Buchanan use seismic and outcrop data in conjunction with restored cross-sections to analyse basins within the South Island of New Zealand.

Overall the papers in this volume highlight that research in recent years has increased our understanding of the inversion processes. The control which pre-existing structures have on both tectonic and thermal inversion is now more clearly understood. The study of 2D/3D seismic surveys, field work and analogue modelling has documented the range of geometries produced in inversion tectonics. In many cases inversion tectonics is the controlling factor on hydrocarbon prospectivity. In particular the timing, extent, amount and controls on basin inversion will need to be understood in order to successfully and cost-effectively explore sedimentary basins.

More research is required into the mechanics and kinematics of 3D basin scale inversion tectonics in order to establish how long distance transfer of stress and strain occurs (for example in the Alpine Foreland of the Southern North Sea). A focused effort is also needed in order to differentiate the uplift of sediments produced by tectonic compression and fault reactivation from regional thermal processes. Hill *et al*. and Brodie & White (both this volume) are addressing these complex problems in a meaningful way.

One strength of the papers in this volume is that their conclusions are often based on numerous data types and both surface mapping and sub-surface seismic interpretation. In order to understand fully the 3D evolution of an inverted basin every available dataset requires detailed analysis to produce a well constrained, integrated study.

The papers in this volume are based on presentations made at the Basin Inversion Conference held at the University Museum, Oxford in October 1993. The conference was sponsored solely by British Gas Exploration and Production Limited (BGE&P) on a non-profit basis and their support is gratefully acknowledged. The Conference was held in association with the Petroleum Group of the Geological Society whose assistance is much appreciated. Thanks also go to Professor John Dewey of Oxford University and the other members of the conference committee for their assistance with conference logistics at Oxford and for their help with technical aspects of the meeting.

In conjunction with the meeting, two fieldtrips were organized to study inversion tectonics. Our thanks go to the leaders of these trips. The first, prior to the conference, was led by Alastair Beach to the north Somerset coast. The trip focused on excellent coastal exposures of the mesoscopic extension and inverted faults. The second trip to the south coast of England was led by Steve Hesselbo and Alastair Ruffell, following the technical sessions of the Conference. The trip studied the classical Tertiary inversion structures in Dorset and the Isle of Wight.

All authors are thanked for their support of the conference and their diligent efforts in the preparation of technical papers. We thank all the technical referees for their hard work reviewing the contributions submitted for publication. The staff at the Geological Society Publishing House are thanked for all their efforts during the publication of this volume.

We would also like to thank Peter Schwarz and John Field of BGE&P for their consistent support and encouragement of this venture. JGB and PGB acknowledge the support of the management and colleagues within BGE&P and CogniSeis Development Inc respectively. The Graham James Consultancy are thanked for the high quality of their organisational skills during the preparation and running of the conference. We would like to thank our families and friends especially our wives, Ruth and Aarti, for their constant help and support during this project.

References

COWARD, M. P. 1994. Inversion Tectonics. *In*: HANCOCK, P. L. (ed.) *Continental Deformation*. Pergammon Press, 280–304.

COOPER, M. A. & WILLIAMS, G. D. 1989. *Inversion Tectonics*. Geological Society, London, Special Publication, **44**.

James G. Buchanan
Peter G. Buchanan

Mechanics, dynamics and geometry of basin inversion

Selective fault reactivation during basin inversion: potential for fluid redistribution through fault-valve action

RICHARD H. SIBSON

Department of Geology, University of Otago, PO Box 56, Dunedin, New Zealand

Abstract: Inversion structures, associated with the compressional reactivation of moderate to steeply dipping normal faults inherited from earlier crustal extension, form important structural traps for hydrocarbons. Migration into these traps must occur syn- or post-inversion. Seismic reflection profiles show that inversion is frequently highly selective, with only some of an existing normal fault set being reactivated. Neglecting marked stress-field heterogeneity, frictional mechanics suggests three possible explanations for this selective reactivation: (1) preferential reactivation of shallowest-dipping normal faults in a region that previously underwent the greatest extensional 'dominoing' of fault blocks; (2) the presence of anomalously low friction material along particular faults; and (3) a heterogeneous distribution of fluid overpressures (P_f > hydrostatic) with preferential reactivation occurring in the area of most intense overpressuring. The last possibility is favoured by the likelihood that fluid overpressures develop during inversion as a consequence of the dramatic increase in mean stress that accompanies the transition from an extensional stress regime, with high capacity to store fluids, to a compressional regime with comparatively low storage capacity.

Compressional reactivation of moderately to steeply dipping faults likely requires significant overpressures; in the case of faults with dips in excess of the frictional 'lock-up' angle (typically 50–60°), supralithostatic fluid pressures (P_f > σ_3 = σ_v) are a necessary prefailure condition in rupture nucleation sites. Extreme fault-valve action then becomes possible, with postseismic flushing of fluids upwards along faults from overpressured compartments. Evidence for such activity comes from mesothermal Au-quartz veins hosted in steep reverse faults, where repeated attainment of supralithostatic pressures alternated with discharge episodes along the faults. Episodes of vertical hydrocarbon migration along reverse faults are therefore a likely accompaniment to basin inversion; there are, for instance, historical records of postseismic discharge of aqueous and hydrocarbon fluids from the overpressured basins of the Western Transverse Ranges, California, where steep reverse faults remain active today. Careful assessment of the relative timing of fault reactivation during inversion and hydrocarbon migration is needed to evaluate the hypothesis for ancient inverted basins.

During positive basin inversion, normal fault structures inherited from an earlier phase of crustal extension are reactivated in compression. The importance of such inversion structures as traps for hydrocarbons has become very apparent over the past decade or so (e.g. Bally 1983; Fraser & Gawthorpe 1990). Seismic reflection profiling has revealed the diagnostics of positive basin inversion and a great deal of attention has been paid to the geometrical, stratigraphic and kinematic characteristics of the inversion structures (Ziegler 1987, 1989; Williams *et al.* 1989). Replication of the general geometrical form of inversion structures from analogue modelling (Koopman *et al.* 1987; McClay 1989; Buchanan & McClay 1992; Sassi *et al.* 1993) has yielded additional insights. However, important questions remain related to the mechanics of inversion and the timing of hydrocarbon migration into the inversion structures, which must occur either during or after the phase of tectonic inversion.

In this paper, the frictional mechanics of fault reactivation are explored to investigate the particular stress and fluid pressure conditions under which inherited normal faults may be reactivated in compression. Only the simplest case of coaxial inversion (extensional, followed by contractional pure dip-slip faulting) is considered to keep the analysis tractable. However, in a qualitative sense the results are extendable to more general situations of oblique inversion under triaxial stress. The analysis yields some insights into processes of fluid redistribution likely to accompany inversion. An important aspect is the peculiar ability of high-angle reverse faults of comparatively low displacement to act as major conduits for episodic

From BUCHANAN, J. G. & BUCHANAN, P. G. (eds), 1995, *Basin Inversion*,
Geological Society Special Publication No. 88, 3–19.

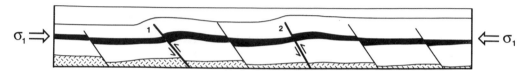

Fig. 1. Cartoon summarizing observations of selective fault reactivation during mild positive basin inversion beneath an unconformity: (1) preferential reverse reactivation of a relatively low-dipping fault; (2) selective reverse reactivation of a fault dipping steeper than its neighbours.

passage of overpressured fluids through fault-valve action, for which evidence exists throughout the geological record (Sibson 1990a).

Observations of fault reactivation during inversion

Much of the structural evidence for positive tectonic inversion comes from the detailed seismic reflection profiles available for areas such as the North Sea and related areas in NW Europe, the Taranaki Basin in New Zealand and the Gippsland Basin in southeastern Australia (e.g. Bally 1983; Williams *et al.* 1989). Several important observations can be made from these different studies (Fig. 1).

(i)   While it is clear that contraction during the inversion phase is sometimes oblique to the former direction of extension, there are areas such as the Taranaki Basin in New Zealand (King & Thrasher 1993), and the Wessex and Weald Basins of southern England (Stonely 1982; Simpson *et al.* 1989; Butler & Pullan 1990) where extension and contraction appear to have been close to coaxial (but, see also Chadwick 1986).

(ii) In many instances fault reactivation during inversion is highly selective, with only a few structures within an extensive set of pre-existing normal faults, or only individual segments of a normal fault system undergoing compressional reactivation as reverse faults (Badley *et al.* 1989; Hayward & Graham 1989; Williams *et al.* 1989). Nor is it always the apparently more optimally orientated faults (see later) that are selectively reactivated.

(iii) In areas of low to moderate inversion, the reactivated faults have moderate to steep dips (40–60°) and low-angle short-cut thrusts through the footwall are comparatively rare (Hayward & Graham 1989; Simpson *et al.* 1989).

We are therefore faced with a situation where it is apparently easier to reactivate moderate to steeply dipping faults that are not well

orientated for reactivation under compression in preference to the formation of new, favourably oriented thrusts. Moreover, such reactivation as occurs is highly selective.

Mechanics of inversion

Topography is unlikely to be extreme in areas undergoing incipient inversion and, while competent strata may act as stress guides to some extent, they are unlikely to deviate significantly from the subhorizontal until inversion is well advanced. The standard 'Andersonian' assumption of horizontal and vertical principal stress trajectories (Anderson 1951) is therefore probably warranted. Under triaxial stress (principal compressive stresses, $\sigma_1 > \sigma_2 > \sigma_3$) with the vertical stress, σ_v (equivalent to the overburden pressure) staying constant, the simplest case of coaxial inversion involves a progressive increase in the horizontal stress from an extensional stress state with $\sigma_v = \sigma_1$ to a compressional state with $\sigma_v = \sigma_3$ (Fig. 2). Normal faults formed in the extensional phase likely initiated as Coulomb shears in planes containing the σ_2 axis with dips of $c.60°$ (Anderson 1951), but may since have rotated to lower dips.

Frictional reactivation of faults

In fluid saturated crust, the effective principal compressive stresses are $\sigma_1' = (\sigma_1 - P_f) > \sigma_2' = (\sigma_2 - P_f) > \sigma_3' = (\sigma_3 - P_f)$, and the condition for frictional reactivation of existing cohesionless faults may be represented by the equivalent of Amontons's Law:

$$\tau = \mu_s \sigma_n' = \mu_s(\sigma_n - P_f) \qquad (1)$$

where μ_s is the static coefficient of friction, P_f is the fluid pressure within the rock mass, and τ and σ_n are respectively the resolved shear and normal stress components on the fault. From an extensive series of low-temperature experiments, Byerlee (1978) found the static coefficient of rock friction to be largely independent of rock type and restricted to the range, $0.6 < \mu_s < 0.85$, prominent exceptions being material rich in montmorillonite, allied

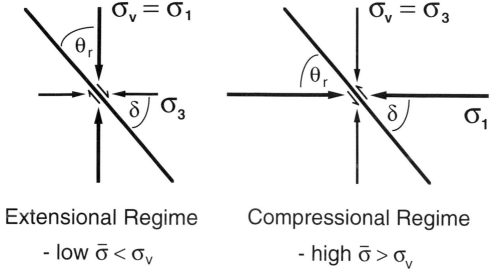

Fig. 2. Coaxial inversion from extensional to compressional stress regimes for a fault with dip, δ where θ_r is the reactivation angle defined with respect to the σ_1 direction.

clay minerals, or evaporites. Elevated temperatures have little effect on rock friction coefficients up to $c.350°C$ (Stesky *et al.* 1974). Note that the higher bound ($\mu_s = 0.85$) was derived from comparatively low normal stress experiments ($\sigma_n' < 200\,MPa$) most likely to be applicable in the upper few kilometres of the crust.

For the situation where the pole to existing faults lies within the σ_1/σ_3 plane (so that σ_2 has no effect), the reactivation criterion (1) may be recast in terms of the ratio of effective principal stresses as:

$$\frac{\sigma_1'}{\sigma_3'} = \frac{(\sigma_1 - P_f)}{(\sigma_3 - P_f)} = \frac{(1 + \mu_s\cot\theta_r)}{(1 - \mu_s\tan\theta_r)} \quad (2)$$

where θ_r is the angle of reactivation as defined in Figs 2 & 3 (Sibson 1985, 1990*b*). This expression can be rewritten in terms of the fault dip, δ, by substituting $\theta_r = \delta$ for compressional stress regimes and $\theta_r = (90° - \delta)$ for extensional regimes. The expression is a useful measure of the relative ease of reactivation of differently oriented faults (Fig. 3a). The optimal angle for reactivation, where σ_1'/σ_3' is a positive minimum, occurs when $\theta_r^* = 0.5\tan^{-1}(1/\mu_s)$. For the Byerlee range of coefficients, $25° < \theta_r^* < 30°$, not dissimilar to the orientation of faults during their initiation as Coulomb shears. As θ_r deviates increasingly from θ_r^*, reactivation becomes progressively more difficult, requiring a higher stress ratio. Frictional 'lock-up' with $\sigma_1'/\sigma_3' \to \infty$

occurs as $\theta_r \to 0$ or $2\theta_r^*$, so that either σ_1' has to become very large, or σ_3' has to become very small. The lock-up angle corresponding to $2\theta_r^*$ is given by $\theta_l = \tan^{-1}(1/\mu_s)$, and for Byerlee friction coefficients should occur in the range, $50° < \theta_l < 59°$ (Fig. 3b).

The physical basis of frictional lock-up is that at large reactivation angles further boosting of σ_1 adds more to the normal stress 'gluing' the walls of the fault together than to the resolved shear stress promoting instability. As lock-up is approached, the rapid increase in the stress ratio for continued reactivation may lead to the formation of new, favourably orientated faults in accordance with the Coulomb criterion unless $\sigma_3' = (\sigma_3 - P_f)$ is made very low through fluid overpressuring. Reactivation at $\theta_r > 2\theta_r^*$ (the field of severe misorientation) is only possible in special circumstances when the least stress becomes effectively tensile (i.e. $\sigma_3' < 0$, or $P_f > \sigma_3$).

Frictional constraints on inversion

Extensional faulting phase During crustal extension with $\sigma_v = \sigma_1$, normal faults generally initiate with steep dips (typically 55–65°) in accordance with Coulomb failure theory (Anderson 1951). With continued extension, sets of normal faults dipping in the same direction (and their intervening fault-blocks) may rotate progressively to lower dips in a domino-like manner. As domino rotation

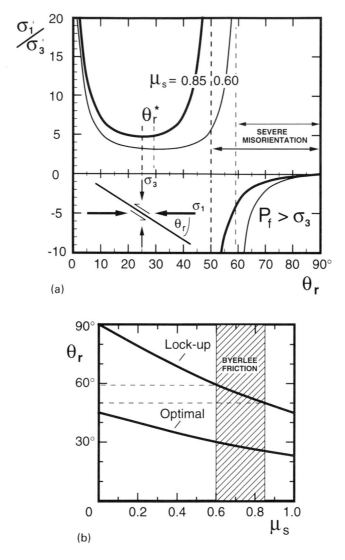

(a)

(b)

Fig. 3. (a) Stress ratio (σ_1'/σ_3') for the frictional reactivation of a cohesionless fault, whose pole lies in the σ_1/σ_3 plane, plotted against the reactivation angle, θ_r, for $\mu_s = 0.6$ (light line) and $\mu_s = 0.85$ (bold line) (after Sibson 1985). (b) Values of the optimal angle for reactivation and the frictional lock-up angle for cohesionless faults plotted against the friction coefficient, μ_s.

proceeds, the fault reactivation angle increases so that for Byerlee friction coefficients, lock-up is expected in the dip range 30–40°. The observation that modern, seismically active planar normal faults only remain active down to this dip range (Jackson 1987; Doser & Smith 1989; Roberts & Jackson 1991) supports the application of Byerlee-type friction to natural systems of planar faults.

Compressional faulting during inversion At the onset of a subsequent phase of horizontal compression with $\sigma_v = \sigma_3$, dip angles of faults inherited from the extensional phase (now corresponding to the reactivation angle, Fig. 2) could therefore range from around 30–65°. The ease with which these existing faults are reactivated in compression as reverse faults clearly decreases with increasing dip. In fact, the range of dip values for which fault reactivation can occur within the bounds of frictional lock-up for both extension and compression is quite sensitive to the friction coefficient (Fig. 4). At the high end of the Byerlee range ($\mu_s = 0.85$) only

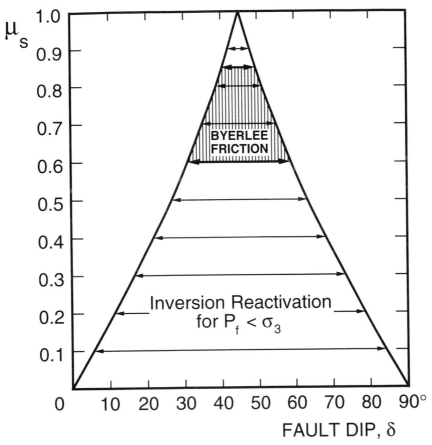

Fig. 4. Permissible range of dips allowing reactivation of cohesionless faults in both extension and compression (provided $\sigma_3' = (\sigma_3 - P_f) > 0$), plotted against static friction coefficient, μ_s. The Byerlee (1978) range of rock friction coefficients is shaded.

faults dipping in the range $40° < \delta < 50°$ can be reactivated in both compression and extension when $P_f < \sigma_3$. At the lower end ($\mu_s = 0.6$), the dip range permitting reactivation in both extension and compression broadens to about $30° < \delta < 60°$ for horizontal σ_3 and σ_1 stress trajectories, respectively.

The magnitude of differential stress required for compressional reactivation varies with overburden pressure and the fluid pressure within the rock mass. At a depth, z, in the crust, the effective overburden pressure is:

$$\sigma_v' = (\sigma_v - P_f) = \rho gz(1 - \lambda_v) \qquad (3)$$

where ρ is the average rock density, g is gravitational acceleration, and the pore-fluid factor, $\lambda_v = P_f/\rho gz$. Equation (2) can then be rewritten as:

$$(\sigma_1 - \sigma_3) = \frac{\mu_s(\tan\delta + \cot\delta)}{(1 - \mu_s\tan\delta)}\rho gz(1 - \lambda_v) \qquad (4)$$

an expression giving the differential stress required for compressional reactivation of a fault with dip, δ, at a depth, z, for particular values of μ_s, ρ, and λ_v (Sibson 1990b). For ease of scaling, the differential stress required to reactivate a cohesionless fault in compression at 1 km depth at different values of λ_v is plotted against dip angle in Fig. 5, for crust with average density, $\rho = 2350$ kg/m^3, and with $\mu_s = 0.85$, the most appropriate friction coefficient for the uppermost few kilometres of the crust. For this rock density, the curve with $\lambda_v = 0.4$ corresponds approximately to hydrostatic fluid pressure conditions, higher values representing different degrees of overpressuring. Note how rapidly the required stress level for reactivation rises as the lock-up angle is approached. It is in these fields that new, favourably oriented thrusts are likely to form in preference to continued reactivation of the existing structures. The comparative scarcity of such low-angle thrusts in

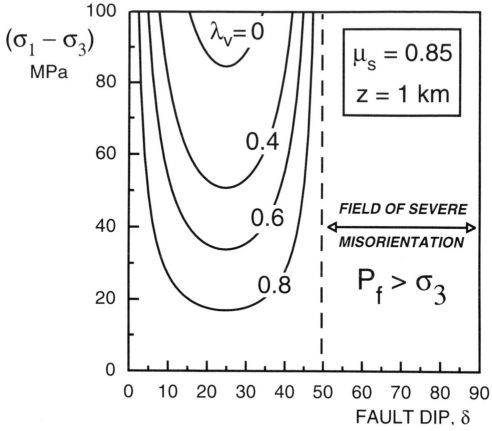

Fig. 5. Differential stress required for compressional reactivation of a cohesionless fault at a depth of 1 km plotted against dip angle, δ, for $\rho = 2350 \, kg/m^3$, $\mu_s = 0.85$, and varying λ_v. The field where $P_f > \sigma_3$ is required for reactivation is the field of severe misorientation.

regions of mild inversion where moderate to steep faults have been reactivated in compression has, therefore, the implication that the reactivated faults are extremely weak.

Reactivation of listric faults Because of decreasing dip with depth, originally listric normal faults should experience varying ease of compressional reactivation at different structural levels. Low fault dips at depth allow easy reactivation, but steep dips at higher levels may lead to lock-up and the accommodation of shortening through wallrock strain. This distributed strain at high levels during inversion may contribute to the development of major antiformal buckles in hangingwall strata, perhaps amplifying earlier 'roll-over' structures formed during the extensional phase (Fig. 6). Alternatively, 'short-cut' thrusts may develop in the footwall to accommodate continued shortening.

Analogue modelling Sandbox modelling of inversion tectonics involving the compressional reactivation of earlier extensional faults (McClay 1989; Buchanan & McClay 1992; Sassi *et al.* 1993) lends support to several aspects of the simple frictional mechanics analysis presented above. Lock-up of reverse faults is observed to occur in the dip range 50–60° and is often followed by the formation of new, more favourably oriented low-dipping thrusts as footwall 'short-cuts'.

Mechanisms for selective reactivation

From analogue modelling, Sassi *et al.* (1993) demonstrated that selective reaction could occur within a set of parallel, favourably oriented faults when the faults were closely spaced. However, this form of selective fault reactivation likely arises from stress-field

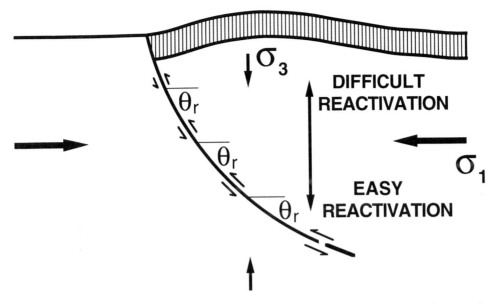

Fig. 6. Cartoon showing how the changing dip of a listric normal fault with depth leads to varying ease of reactivation under compression.

heterogeneity caused by the experimental boundary conditions. On the basis of the 2-D reactivation theory outlined above, and neglecting the possibility of such stress inhomogeneities, three other mechanisms may account for the selective fault reactivation that is observed during inversion.

(i) An obvious expectation is that the lowest dipping faults in a rotated set would be preferentially reactivated in compression at lower stress ratios than steeper, less favourably oriented structures. Inversion should then initiate in the areas of greatest former extension and spread to surrounding areas as the earliest reactivated faults steepen by reverse dominoing during regional contraction. If inversion does not proceed beyond a certain stage, fault reactivation will appear highly selective.

(ii) The presence of anomalously low-friction material on a fault, such as montmorillonite-rich gouge, could lead to its selective reactivation.

(iii) For faults of comparable dip, preferential fault reactivation would occur where fluid pressure levels were locally elevated. This last mechanism takes on special significance given the now widespread recognition of both vertical and lateral compartmentalization of fluid pressures in sedimentary basins (Hunt 1990).

While the first mechanism is undoubtedly important in some areas, there are clear instances where it is not the shallowest dipping fault that has undergone compressional reactivation (e.g. Hayward & Graham 1989; Simpson *et al.* 1989). The second mechanism is difficult to discount but requires rather special pleading because selective reactivation often occurs within a set of faults cutting essentially the same stratified sequence. The third mechanism, invoking locally elevated fluid pressures as a cause of selective reactivation, is of particular interest in view of the fluid overpressuring that is likely to develop during inversion (see below).

Fluid storage in compressional vs. extensional regimes

Both porosity and permeability characteristics of a rock mass are likely to vary with the state of stress, either through the influence of stress-controlled features such as faults, microcracks, hydrofractures and stylolites, or through changes in the level of mean stress, $\bar{\sigma} = (\sigma_1 + \sigma_2 + \sigma_3)/3$, which affects the volumetric elastic strain. Stress-controlled structures may enhance or counteract existing anisotropic permeability depending on the relative attitude of bedding

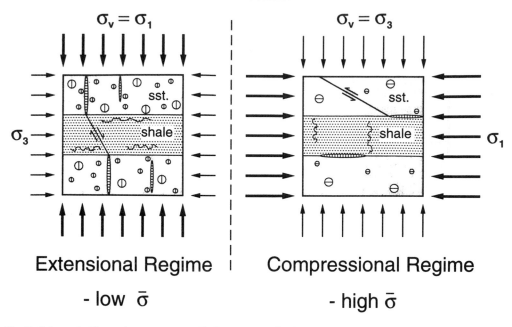

Fig. 7. Schematic illustrating stress-controlled structures affecting the porosity and permeability of the rock mass, and the contrasting fluid storage capacity in extensional and compressional stress regimes (split circles = transgranular microfractures; squiggly lines = stylolites; cross-hatched ellipses = hydraulic extension fractures).

and the principal stresses (e.g. du Rouchet 1980; Sibson 1994).

As illustrated in Fig. 7, vertical permeability is likely to be enhanced in extensional stress regimes ($\sigma_v = \sigma_1$) through the development of subvertical transgranular microcracks and extensional hydraulic fractures (both forming perpendicular to the least compressive stress, σ_3), and by steep normal faults. This may be countered to some extent by the development of flat-lying stylolites forming perpendicular to σ_1 in the finer grained sediments. By contrast, bedding-parallel permeability may be enhanced in compressional settings by microfractures, by hydraulic extension fractures and by low-angle thrusts. The relative development of these various structural features in different stress fields affects the capacity of the rock mass to store fluids.

Conditions for hydraulic extension fracturing

Because of their large aperture and continuity, macroscopic extension fractures formed by hydraulic fracturing are particularly important from the viewpoints of both fluid storage and

rock mass permeability. In a rock mass with tensile strength, T, hydrofractures form when:

$$P_f = \sigma_3 + T \qquad (5)$$

provided the differential stress, $(\sigma_1 - \sigma_3) < 4T$, so that shear failure is inhibited (Secor 1965). Development of transgranular microcracks through grain impingement is also likely to be enhanced as $\sigma_3' = (\sigma_3 - P_f) \to 0$ (Palciauskas & Domenico 1980).

Following Secor (1965), the fluid pressure conditions under which hydraulic extension fractures can develop at different depths in extensional (normal faulting) and compressional (reverse faulting) regimes can be represented on a plot of the pore-fluid factor ($\lambda_v = P_f/\rho gz$) against depth (Fig. 8). The curves are plotted for T = 1 MPa and T = 10 MPa, bracketing the common range for long-term rock tensile strength (Etheridge 1983), and represent the λ_v conditions for hydraulic fracturing when differential stress is at the limiting value. In compressional settings, supralithostatic ($\lambda_v > 1$) fluid pressures are required at all depths to induce hydraulic fracturing. In extensional settings, however, hydraulic fracturing can potentially develop under hydrostatic levels of fluid pressure close to the Earth's surface, and at greater

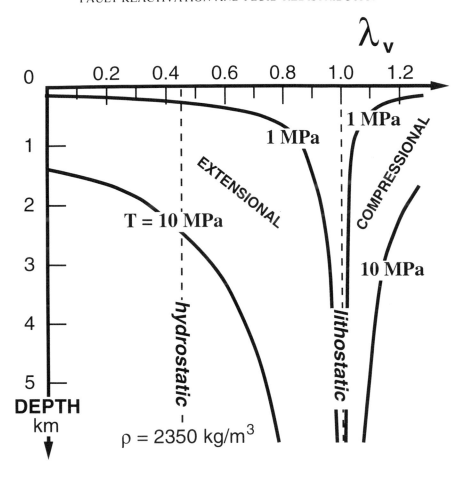

Fig. 8. λ_v plots defining the conditions for hydraulic fracturing at maximum allowable differential stress in extensional and compressional stress regimes for rock tensile strengths, T = 1 MPa and 10 MPa and an average rock density, $\rho = 2350\,\text{kg/m}^3$ (after Secor 1965).

depths requires a relatively small degree of overpressuring above hydrostatic. As a consequence, in comparison with a compressional stress regime, the capacity of a rock mass in an extensional tectonic regime to store fluids may be significantly enhanced, especially in the near-surface, through the development of subvertical hydrofractures and transgranular microcracks.

Mean stress effects

Elastic volumetric strain within a rock mass is related to the level of mean stress, $\bar{\sigma}$. In fluid saturated crust, the effective mean stress becomes $\bar{\sigma}' = (\bar{\sigma} - P_f)$ through the principle of effective stress. Increasing the level of compressive mean stress therefore either decreases the volume of the solid framework within the rock mass, causing fluid to be expelled from existing pore space in a manner analogous to squeezing a sponge,

or, if fluid cannot escape, boosts the level of fluid pressure. Consider the stress states illustrated in Fig. 2 for coaxial inversion. Clearly, the level of mean stress in an extensional regime is always less than the vertical stress (overburden pressure) while in a compressional regime it must exceed the vertical stress. The question to be addressed, therefore, is the extent to which the mean stress level increases during the transition from an extensional regime with active normal faulting to a compressional regime with the same faults reactivated in reverse mode.

For the 2-D reactivation analysis, where mean stress reduces to $\bar{\sigma} = (\sigma_1 + \sigma_3)/2$, (2) can be rearranged in terms of the effective mean stress, giving:

$$\frac{\bar{\sigma}'}{\sigma_3'} = \frac{2 + \mu_s(\cot\theta_r - \tan\theta_r)}{2(1 - \mu_s\tan\theta_r)} \qquad (6)$$

an expression which gives the ratio of the

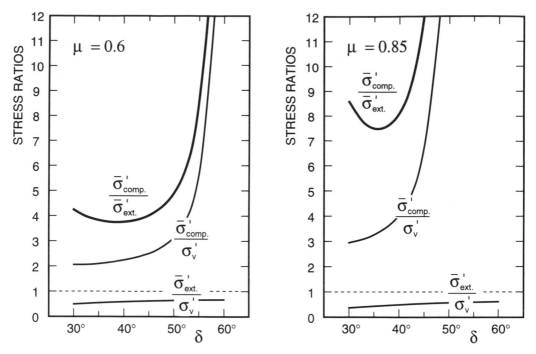

Fig. 9. Mean stress ratios as defined in text plotted against dip angle for $30° < δ < 60°$, and for $μ_s = 0.6$ and 0.85, for existing faults at the point of frictional shear failure.

effective mean stress to the least compressive stress at frictional failure for a cohesionless fault oriented at a reactivation angle, $θ_r$. For a compressional regime, where $σ_v = σ_3$ and $δ = θ_r$, the expression can be written:

$$\frac{\bar{σ}_{comp}'}{σ_v'} = \frac{2 + μ_s(\cot δ - \tan δ)}{2(1 - μ_s \tan δ)} \quad (7)$$

while for an extensional regime with $σ_v = σ_1$ and $δ = (θ_r - 90°)$, it becomes:

$$\frac{\bar{σ}_{ext}'}{σ_v'} = \frac{2 + μ_s(\tan δ - \cot δ)}{2(1 + μ_s \tan δ)} \quad (8)$$

These two equations give the ratio of effective mean stress to effective vertical stress for faults with dip, $δ$, at the point of frictional failure in compressional and extensional regimes. They can be combined to yield $\bar{σ}_{comp}'/\bar{σ}_{ext}'$, the ratio of the levels of effective mean stress for a fault with dip, $δ$, at frictional failure in compression and extension at the same level of fluid pressure. Values from these expressions are plotted for the dip range 30–60° for $μ_s = 0.6$ and 0.85 (Fig. 9). Clearly, the increase in mean stress during inversion can be quite dramatic; for example, for $δ = 45°$, $\bar{σ}_{comp}'/\bar{σ}_{ext}'$ has values of 12.3 and 4.0 for $μ_s = 0.85$ and 0.6, respectively. Note, however, that these plots presuppose the same level of

fluid pressure to be maintained in compressional as in extensional regimes, whereas what is in fact likely to happen is a significant boosting of fluid pressure as mean stress progressively increases.

Combined effect on fluid pressure level

Both the higher fluid storage capacity of extensional stress regimes and the increase in mean stress accompanying the transition from extension to compression are likely to contribute to increased fluid pressures during inversion. The onset of compression will lead to closure of the subvertical microcracks and extension fractures formed during the extensional phase (Fig. 7), inhibiting vertical migration. Coupled with the increasing mean stress, this reduction in void space will either force fluids out of the rock mass, or boost the level of fluid pressure if drainage is inhibited by low permeability caps such as shale horizons. The magnitude of the pressure increase will depend on the permeability characteristics of the rock mass and the length of time over which the transition from an extensional to a compressional regime takes place, shorter transition intervals being more likely to lead to substantial increases in fluid pressure. In some respects, the conclusions reached here are similar to those arrived at by Berry (1973) in his consideration of the development of high fluid pressures in the California Coast ranges.

A good example of the structural features diagnostic of boosted fluid pressures during inversion is provided by a reverse-reactivated fault at Ogof Gynfor on the north coast of Anglesey, Wales (Sibson 1981). The fault, with a finite reverse slip in excess of 20 m, dips moderately to the north and disrupts a sequence of cleaved black shale overlying a quartzite breccia-conglomerate. Two incrementally developed sets of quartz veins document a history of inversion reactivation. Prominent subvertical extension veins striking parallel to the fault were apparently developed during an early phase of extensional normal faulting, but are cut across by a sparsely developed second set of subhorizontal extension veins associated with the reverse reactivation of the fault. The vein-sets demonstrate that overpressuring ($P_f > \sigma_3$) accompanied both extensional normal faulting and compressional reverse reactivation, with fluid pressures intermittently boosted to supralithostatic values ($P_f > \sigma_3 = \sigma_v$) during the phase of reverse reactivation.

Fluid redistribution during fault reactivation

The boosting of fluid pressures during inversion allows existing faults that are not favourably oriented to become reactivated under compression, and provides a plausible mechanism for the selective reactivation that is observed. Any overpressures that develop within a basin undergoing inversion are likely to be very heterogeneous because of varying fluid storage and permcability characteristics. In this connection, it may be significant that Tertiary inversion within the Wessex Basin, for example, tends to be concentrated within the sub-basins which accommodated the thickest accumulations of Late Jurassic–Early Cretaceous sediments (Simpson et al. 1989), where fluid content is likely to have been greatest.

Fault-valve activity

Fault-valve behaviour becomes possible wherever active faulting occurs in overpressured portions of the crust and ruptures transect suprahydrostatic vertical gradients in fluid pressure (Sibson 1981). Valving action depends on the ability of faults to behave as impermeable seals through the interseismic period from the presence of clay-rich or cataclastic gouge and/or

hydrothermal cementation, but to become highly permeable channelways for fluid discharge immediately postfailure as a consequence of the intrinsic roughness of natural rupture surfaces (Power et al. 1987). Breaching of low-permeability seals to overpressured zones by fault rupture thus leads to upwards discharge of fluids along the rupture zone and local reversion towards a hydrostatic gradient (Fig. 10). While valving leading to minor postfailure discharge may occur in any tectonic setting where overpressuring has developed, there are mechanical reasons why extreme fault-valve action involving substantial postfailure discharge tends to be associated with steep reverse faults (Sibson 1990a).

Ordinary versus extreme valving action Progressively increasing fluid pressure may trigger movement on existing faults in accordance with the reactivation criterion (1), thereby promoting valve action. It is apparent from the 2-D analysis of reactivation (Fig. 3) that favourably oriented faults would be the first to be reactivated, and that reactivation of unfavourably orientated faults in areas of increasing fluid pressure depends on the absence of favourably orientated structures. Provided fault orientation lies in the range, $0° < \theta_r < \theta_1 = 2\theta_r^*$, fault reactivation will occur before the condition for hydrofracturing ($P_f > \sigma_3$) can be achieved with fluid pressures then dropping postfailure through valve action. In such circumstances, the volume of fluid discharged is likely to be comparatively small.

For faults that are in the field of severe misorientation, however, the condition $P_f > \sigma_3$ is a prerequisite for reactivation, and hydraulic extension fractures are likely to develop prior to failure. The significance of this is that an array of hydrofractures in the rupture nucleation site provides a reservoir of overpressured fluids with ready access to the fault postfailure. Fluid volumes available for postfailure discharge are thus likely to be much greater in the case of severely misorientated faults; such structures are therefore capable of extreme fault-valve behaviour. In compressional settings, supralithostatic fluid pressures ($P_f > \sigma_3 = \sigma_v$) are required to meet the prefailure condition for steep, severely misorientated reverse faults (Fig. 8), but the prefailure hydrofractures are then subhorizontal and unlikely to breach impermeable sealing horizons themselves. It is this special combination of circumstances that gives steep reverse faults their peculiar ability for extreme fault-valve action, with the discharge of large fluid volumes postfailure (Sibson 1990a).

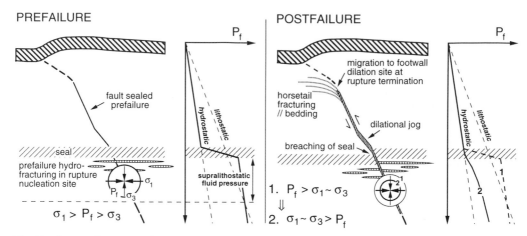

Fig. 10. Cartoon (not to scale) illustrating stress and fluid-pressure states prefailure and postfailure associated with extreme fault-valve action on a steep reverse fault. Only the simplest situation is shown where one impermeable seal separates hydrostatic and overpressured fluid regimes; reality is likely to be more complex.

The stress and fluid pressure cycling associated with extreme valve action on steep reverse faults is illustrated in Figs 10 and 11. In the immediate prefailure state, the condition $\sigma_1 > P_f > \sigma_3$ prevails in the overpressured rupture nucleation site, with formation of gaping subhorizontal hydrofractures. If relief of shear stress along the fault during rupturing is near-complete (and there is textural evidence from vein systems to support this, see Boullier & Robert 1992), the condition immediately post-failure is $P_f > \sigma_1 \approx \sigma_3$. Under these conditions, fluids from the hydrofracture array, and from any grain-scale microfracture porosity developed at low effective stress, are expelled into the rupture zone which becomes the principal avenue for fluid discharge. For total stress release, or if slight 'overshoot' occurs, actual dilation of the rupture zone may take place. Discharge continues until the pressure gradient reverts to hydrostatic or the rupture zone self-seals through hydrothermal precipitation ($\sigma_1 \approx \sigma_3 > P_f$). Differential stress and fluid pressure then start to rebuild towards the next failure episode ($\sigma_1 > \sigma_3 > P_f$). Valving activity thus depends on the competition between creation and destruction of permeability in fault zones (Sibson 1992), for which textural evidence is widespread (e.g. Roberts 1991; Hippler 1993).

Analogy with mesothermal Au-quartz lodes
Mesothermal gold-quartz vein systems developed in sub- to low-greenschist metamorphic environments provide evidence that steep reverse faults of comparatively low displacement can serve as conduits for focused flow of

aqueous fluids through extreme fault-valve action (Sibson *et al.* 1988; Cox *et al.* 1991), and demonstrate the massive scale of flow that can occur. Typical vein assemblages comprise a mixture of flat-lying extensional veins in mutual cross-cutting relationships with steep fault-veins hosted by reverse faults. Hydrothermal textures of both vein-sets record histories of incremental deposition (Boullier & Robert 1992), their mutual cross-cutting relationships suggesting that the veins within the two sets formed at different stages of a repeating cycle. The flat-lying extension veins demonstrate that the condition $P_f > \sigma_3 = \sigma_v$ was repeatedly attained and are interpreted as representing prefailure hydrofractures, while the fault-veins are inferred to have developed incrementally during episodes of postfailure discharge (see Figs 10 & 11).

As an example, consider the Mother Lode vein system of early Cretaceous age in the western Sierra Nevada foothills of California (Knopf 1929). The principal quartz veins are hosted on individual reverse faults within the Melones fault zone. In places, continuous veins averaging over a metre in thickness can be traced for kilometres along strike and have been mined to depths in excess of 1 km. Per kilometre of fault strike-length, the volume of fault-hosted quartz therefore commonly exceeds $10^6 \, m^3$. Established reverse displacements on the hosting faults range up to 100 m or so, with ribbon vein textures recording hundreds of episodes of hydrothermal deposition. Given the low solubility of quartz (Fyfe *et al.* 1978), something like $10^9 - 10^{10} \, m^3$ of aqueous fluid (more than the

(a) IMMEDIATELY PREFAILURE   **(b) IMMEDIATELY POSTFAILURE**

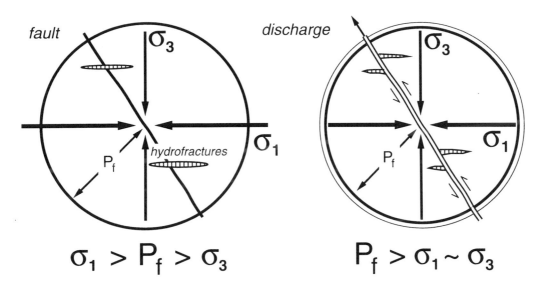

(c) POST DISCHARGE

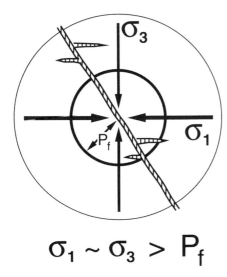

(d) REACCUMULATION OF DIFFERENTIAL STRESS AND FLUID PRESSURE

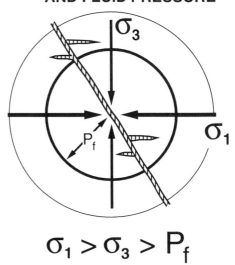

Fig. 11. Cycle of stress and fluid-pressure states accompanying extreme valve action on a steep reverse fault. The radius of the bold circle is a measure of the fluid pressure level in relation to the principal stresses.

fluid volume of a supergiant oil-field!) would have to be flushed through the fault per kilometre strike-length to deposit the hydro- thermal infilling. And the Mother Lode vein system can be traced along strike for at least 200 km!

Extent of vertical migration through valving action There are historical instances of surface hydrocarbon discharge, potentially attributable to valving action on steep reverse faults, following shallow crustal earthquakes in areas such as the Western Transverse Ranges of California where overpressuring is widespread (Hamilton *et al.* 1969; Sibson 1990*b*). In general, however, only the larger earthquake ruptures extend to the ground surface. Moreover, in the case of inversion tectonics, normal growth faults developed during the extensional phase may not continue through the upper levels of the stratigraphic succession. This helps to explain the tendency of some steep faults reactivated in compression to shallow into optimally oriented low-angle thrusts at higher levels (e.g. Badley *et al.* 1989).

Such geometrical heterogeneities may play a role in the termination of reverse ruptures, as an upwardly propagating reverse rupture will induce dilation in the footwall near the rupture tip (Pollard & Segall 1987). Termination of upwardly propagating ruptures into 'horsetail' extensional fracturing in the footwall would also be facilitated by bedding anisotropy. Other rheological factors may also play a role. The expectation, therefore, is that individual reverse ruptures are likely to terminate at dilational sites involving horsetail fracturing in the footwall, and it is to these sites that fluids discharged by valve action from overpressured zones at depth may migrate (Fig. 10). Subsequent migration into hangingwall anticlinal structures would depend on the permeability of the fault at high levels (see Knipe 1992).

As a possible example of an area where these processes are likely to be active today, consider the Ventura Avenue anticline in the Western Transverse Ranges of California. Oil is produced from a thick Plio–Pleistocene turbidite sequence in the tight core of the anticline which is disrupted by reverse faults and is strongly overpressured (λ_v c.0.8 at 3.8 km depth) (Yeats 1983). The anticline is also flanked by the Red Mountain Fault, a pure reverse structure which dips 60–65° to at least 6 km depth and remains microseismically active (Yeats *et al.* 1987). It seems probable that fault-assisted migration into the upper levels of the structure is still ongoing as the anticline has evolved as an oil-hosting structure within only the last 0.2 Ma!

Diagnostics of fault-valve action Recognition of former fault-valve action (especially extreme valving action) on reverse reactivated faults is of some importance given its potential role in fluid migration during inversion tectonics. While no single item is fully diagnostic by itself, the following characteristics would be expected on or in the vicinity of a fault that has undergone extreme valving action (Fig. 10): (1) evidence of reverse reactivation on a fault dipping steeper than 50–60°; (2) the presence of subhorizontal extension veins (parallel to wallrock bedding?) adjoining faults in formerly overpressured sites of rupture nucleation; (3) traces of hydrothermal and hydrocarbon material along the fault zone, especially in sites such as dilational fault jogs; (4) the presence of 'horsetail' fracturing with hydrothermal/hydrocarbon infilling at former rupture termination sites in the footwall; and possibly (5) the presence of multiple generations of cement in fault-bounded reservoirs, as described by Burley *et al.* (1989) from sandstones of the Tartan Reservoir in the North Sea, and attributed to episodic flushing of hot fluids up fault conduits.

The previously described fault at Ogof Gynfor on the north coast of Anglesey displays some of these structural characteristics. Several generations of lensing quartz veins with surfaces covered with slickenfibres developed along the fault during the phase of reverse reactivation and may result from intermittent valving discharge. Structurally analogous assemblages of gypsum veins occur west of Watchet on the Bristol Channel coast of West Somerset in the vicinity of E–W striking faults that have undergone reverse reactivation juxtaposing crumpled Liassic strata in the hangingwall against Triassic red beds in the footwall (Whittaker 1972). Multiple generations of predominantly flat-lying fibrous gypsum veins are concentrated in the footwall and the fault is coated with several sets of gypsum slickenfibres. The concentration of veining in the footwall, with some local semblance of horsetail geometry with respect to the fault, suggests that the veining may have developed in a footwall dilational site which acted as a migratory sink for fluids transported by valving action during the reverse reactivation.

Discussion

A circumstantial case has been presented that extreme fluid overpressures are likely to develop during the transition from an extensional to a compressional stress regime accompanying basin inversion, and that it is the development of localized overpressuring that allows selective reactivation of faults that would otherwise be unfavourably orientated for reactivation. In such circumstances, extreme valve action on severely misorientated faults may allow episodic vertical migration of large fluid volumes along the fault zone. The association of active high-angle reverse faulting with overpressured basins as in the Western Transverse Ranges of California, coupled with historical records of post-seismic hydrocarbon discharge in such areas (Hamilton *et al.* 1969), lends support to the hypothesis. Judging by the hydrothermal vein systems associated with steep reverse faults of comparatively low displacement in other settings, the fluid volumes transported in this manner may be very large. The case is based on a 2-D analysis of the mechanics of coaxial inversion where pure reverse faulting follows earlier normal slip. Frictional lock-up may also, however, occur in more general situations of oblique slip under triaxial stress, but the fields of severe misorientation where extreme fault-valve action is likely to operate are then critically influenced by the value of the intermediate principal stress, σ_2, whose value is not easily determinable.

A range of other mechanisms may also contribute to the generation of fluid overpressures in sedimentary basins. They include the smectite to illite transition and other diagenetic processes, aquathermal pressuring, disequilibrium compaction, chemical osmosis and the generation of hydrocarbons, especially gas, from kerogen (e.g. Shi & Wang 1986; Buhrig 1989; Williamson & Smyth 1992). No account has been taken of these additional mechanisms in this analysis, but it may prove interesting to integrate these processes, and thermal maturation analyses, with the stress effects considered in this paper to evaluate the optimal conditions under which overpressuring during basin inversion may lead to fault reactivation and hydrocarbon migration. What does seem clear is that the length of time for the transition to take place from an extensional to a compressional stress regime is likely to be a crucial factor.

Direct evidence for the role of faults as migratory conduits for hydrocarbons is accumulating through detailed studies of fault rock assemblages. Hippler (1993) recognizes evidence for hydrocarbon migration through cataclastic material along extensional faults in the Orcadian Basin, while Roberts (1991) describes an association between bitumen and ferroan calcite in the gouge zone of an alpine thrust. Additional evidence comes from comparative studies of incrementally developed cements in reservoir sandstones from the Tartan Field in the North Sea and those developed within adjacent faults (Burley *et al.* 1989).

Other indirect evidence of faults as migratory paths is also coming to hand. In the Sable Basin off Nova Scotia, it appears that overpressured gas has migrated into hydrostatically pressured reservoirs at higher levels only where large listric growth faults are available as conduits (Williamson & Smyth 1992). Likewise, Barnard & Bastow (1991) have emphasized the role of faults in the North Sea basins as conduits for migration into the higher Palaeocene and Eocene reservoirs. Butler & Pullan (1990) note that the second phase of hydrocarbon migration within the Weald Basin of southern England took place during the early to mid-Tertiary accompanying the development of inversion structures, and that there is a clear association between major reactivated faults cutting the entire section, the occurrence of multiple hydrocarbon plays at both deep and shallow levels, and the presence of oil and gas shows along the faults. However, they note that no overpressuring has been encountered within the basin during exploration, suggesting that faulting has been an efficient pressure release mechanism.

In order to evaluate the primary hypothesis developed here – that major episodes of vertical migration are likely to be associated with compressional reactivation of steep faults during inversion – it will clearly be necessary to assess very carefully the relative timing of hydrocarbon migration and reverse reactivation of fault structures in areas that have undergone tectonic inversion.

Many of the ideas expressed here arose through discussions with colleagues at Imperial College, the University of California at Santa Barbara and the University of Otago. I thank referees Tim Buddin and Dick Nieuwland for helpful comments, and Alastair Beach especially, for introducing me to the thought-provoking exposures of fault-related veining along the Somerset coast.

References

ANDERSON, E. M. 1951. *The Dynamics of Faulting and Dyke Formation with Application to Britain.* 2nd edition. Oliver and Boyd, Edinburgh.

BADLEY, M. E., PRICE, J. D. & BACKSHALL, L. C. 1989. Inversion, reactivated faults and related structures: seismic examples from the southern North Sea. *In*: COOPER, M. A. & WILLIAMS, G. D. (eds) *Inversion Tectonics*. Geological Society, London, Special Publication, **44**, 201–219.

BALLY, A. W. (ed.) 1983. *Seismic expression of structural styles*. American Association of Petroleum Geologists, Studies in Geology, **15**, vol. 1–3.

BARNARD, P. C. & BASTOW, M. A. 1991. Hydrocarbon generation, migration, alteration, entrapment and mixing in the Central and Northern North

Sea. *In*: ENGLAND, W. A. & FLEET, A. J. (eds) *Petroleum Migration*. Geological Society, London, Special Publication, **59**, 167–190.

BERRY, F. 1973. High fluid potentials in California Coast Ranges and their tectonic significance. *American Association of Petroleum Geologists Bulletin*, **57**, 1219–1249.

BOULLIER, A.-M. & ROBERT, F. 1992. Paloseismic events recorded in Archean gold-quartz vein networks. *Journal of Structural Geology*, **14**, 161–179.

BUCHANAN, P. & McCLAY, K. R. 1992. Experiments on basin inversion above reactivated domino faults. *Marine and Petroleum Geology*, **9**, 486–500.

BUHRIG, C. 1989. Geopressured Jurassic reservoirs in the Viking Graben: modelling and geological significance. *Marine and Petroleum Geology*, **6**, 31–48.

BURLEY, S. D., MULLIS, J. & MATTER, A. 1989. Timing diagenesis in the Tartan Reservoir (UK North Sea): constraints from combined cathodoluminescence microscopy and fluid inclusion studies. *Marine & Petrolum Geology*, **6**, 98–120.

BUTLER, M. & PULLAN, C. P. 1990. Tertiary structures and hydrocarbon entrapment in the Weald Basin of southern England. *In*: HARDMAN, R. F. P. & BROOKS, J. (eds) *Tectonic Events Responsible for Britain's Oil and Gas Reserves*. Geological Society, London, Special Publication, **55**, 371–391.

BYERLEE, J. D. 1978. Friction of rocks. *Pure & Applied Geophysics*, **116**, 615–626.

CHADWICK, R. A. 1986. Extension tectonics in the Wessex Basin, southern England. *Journal of the Geological Society*, London, **143**, 465–486.

COX, S. F., WALL, V. J., ETHERIDGE, M. A. & POTTER, T. F. 1991. Deformational and metamorphic processes in the formation of mesothermal vein-hosted gold deposits – examples from the Lachlan Fold Belt in central Victoria, Australia. *Ore Geology Reviews*, **6**, 391–423.

DOSER, D. I. & SMITH, R. B. 1989. An assessment of source parameters of earthquakes in the cordillera of the western United States. *Bulletin of the Seismological Society of America*, **79**, 1383–1409.

DU ROUCHET, J. 1980. Stress fields, a key to oil migration. *American Association of Petroleum Geologists Bulletin*, **65**, 74–85.

ETHERIDGE, M. A. 1983. Differential stress magnitudes during regional deformation and metamorphism: upper bound imposed by tensile fracturing. *Geology*, **11**, 231–234.

FRASER, A. J. & GAWTHORPE, R. L. 1990. Tectono-stratigraphic development and hydrocarbon habitat of the Carboniferous in northern England. *In*: HARDMAN, R. F. P. & BROOKS, J. (eds) *Tectonic Events Responsible for Britain's Oil and Gas Reserves*. Geological Society, London, Special Publication, **55**, 49–86.

FYFE, W. S., PRICE, N. J. & THOMPSON, A. B. 1978. *Fluids in the Earth's Crust*. Elsevier, Amsterdam.

HAMILTON, R. M., YERKES, R. F., BROWN, R. D., BURFORD, R. O & DE NOYER, J. M. 1969. Seismicity and associated effects, Santa Barbara region. *U.S. Geological Survey Professional Paper*, **679**, 47–72.

HAYWARD, A. B. & GRAHAM, R. H. 1989. Some geometrical characteristics of inversion. *In*: COOPER, M. A. & WILLIAMS, G. D. (eds) *Inversion Tectonics*. Geological Society, London, Special Publication, **44**, 17–39.

HIPPLER, S. J. 1993. Deformation microstructures and diagenesis in sandstone adjacent to an extensional fault: implications for the flow and entrapment of hydrocarbons. *American Association of Petroleum Geologists Bulletin*, **77**, 625–637.

HUNT, J. M. 1990. Generation and migration of petroleum from abnormally pressured fluid compartments. *American Association of Petroleum Geologists Bulletin*, **74**, 1–12.

JACKSON, J. A. 1987. Active normal faulting and crustal extension. *In*: COWARD, M. P., DEWEY, J. F. & HANCOCK, P. L. (eds) *Continental Extensional Tectonics*. Geological Society, London, Special Publication, **28**, 3–17.

KING, P. R. & THRASHER, G. P. 1993. Post-Eocene development of the Taranaki Basin, New Zealand: convergent overprint of a passive margin. *In*: WATKINS, J. S., ZHIQIANG, F. & McMILLEN, K. G. (eds) *Geology and Geophysics of Continental Margins*. American Association of Petroleum Geologists Memoir, **53**, 93–118.

KNIPE, R. J. 1992. Faulting processes and fault seal. *In*: LARSEN, R. M. (ed) *Structural and Tectonic Modelling and it's Application to Petroleum Geology*. NPF Special Publication, **1**, 325–343.

KNOPF, A. 1929. *The Mother Lode System of California*. U.S. Geological Survey Professional Paper, **157**, 88pp.

KOOPMAN, A., SPEKSNIJDER, A. & HORSFIELD, W. T. 1987. Sandbox model studies of inversion tectonics. *Tectonophysics*, **137**, 379–388.

McCLAY, K. R. 1989. Analogue models of inversion tectonics. *In*: COOPER, M. A. & WILLIAMS, G. D. (eds) *Inversion Tectonics*. Geological Society, London, Special Publication, **44**, 41–59.

PALCIAUSKAS, V. V. & DOMENICO, P. A. 1980. Microfracture development in compacting sediments: relation to hydrocarbon-maturation kinetics. *American Association of Petroleum Geologists Bulletin*, **64**, 927–937.

POLLARD, D. D. & SEGALL, P. 1987. Theoretical displacements and stresses near fractures in rocks: with application to faults, joints, veins, dikes, and solution surfaces. *In*: ATKINSON, B. K. (ed.) *Fracture Mechanics of Rock*. Academic Press, New York, 277–349.

POWER, W. L., TULLIS, T. E., BROWN, S., BOITNOTT, G. N. & SCHOLZ, C. H. 1987. Roughness of natural fault surfaces. *Geophysical Research Letters*, **14**, 29–32.

ROBERTS, G. 1991. Structural controls on fluid migration through the Rencurel thrust zone, Vercors, French Sub-Alpine Chains. *In*: ENGLAND, W. A. & FLEET, A. J. (eds) *Petroleum Migration*. Geological Society, London, Special Publication, **59**, 245–262.

ROBERTS, S. & JACKSON, J. A. 1991. Active normal

faulting in central Greece: an overview. *In*: ROBERTS, A. M., YIELDING, G. & FREEMAN, B. (eds) *The Geometry of Normal Faults*. Geological Society, London, Special Publication, **56**, 125–142.

SECOR, D. T. 1965. Role of fluid pressure in jointing. *American Journal of Science*, **263**, 633–646.

SHI, Y. & WANG, C-Y. 1986. Pore pressure generation in sedimentary basins: overloading versus aquathermal. *Journal of Geophysical Research*, **91**, 2153–2162.

SASSI, W., COLLETTA, B., BALE, P. & PAQUEREAU, T. 1993. Modelling of structural complexity in sedimentary basins: the role of pre-existing faults in thrust tectonics. *Tectonophysics*, **226**, 97–112.

SIBSON, R. H. 1981. Fluid flow accompanying faulting: field evidence and models. *In*: SIMPSON, D. W. & RICHARDS, P. G. (eds) *Earthquake Prediction: an International Review*. American Geophysical Union, Maurice Ewing Series, **4**, 593–603.

—— 1985. A note on fault reactivation. *Journal of Structural Geology*, **7**, 751–754.

—— 1990*a*. Conditions for fault-valve behaviour. *In*: KNIPE, R. J. & RUTTER, E. H. (eds) *Deformation Mechanisms, Rheology and Tectonics*. Geological Society, London, Special Publication, **54**, 15–28.

—— 1990*b*. Rupture nucleation on unfavourably oriented faults. *Bulletin of the Seismological Society of America*, **80**, 1580–1604.

—— 1992. Implications of fault-valve behaviour for rupture nucleation and recurrence. *Tectonophysics*, **211**, 283–293.

—— 1994. Crustal stress, faulting, and fluid flow. *In*: PARNELL, J. (ed.) *Geofluids: Origin, Migration and Evolution of Fluids in Sedimentary Basins*. Geological Society, London, Special Publication, **78**, 69–84.

——, ROBERT, F. & POULSEN, K. H. 1988. High-angle reverse faults, fluid pressure cycling and meso-thermal gold-quartz deposits. *Geology*, **16**, 551–555.

SIMPSON, I. R., GRAVESTOCK, M., HAM, D., LEACH, H.

& THOMPSON, S. D. 1989. Notes and cross-sections illustrating inversion tectonics in the Wessex Basin. *In*: COOPER, M. A. & WILLIAMS, G. D. (eds) *Inversion Tectonics*. Geological Society, London, Special Publication, **44**, 123––129.

STESKY, R., BRACE, W., RILEY, D. & ROBIN, P-Y. 1974. Friction in faulted rock at high temperature and pressure. *Tectonophysics*, **23**, 177–203.

STONELY, R. 1982. The structural development of the Wessex Basin. *Journal of the Geological Society*, London, **139**, 545–554.

WHITTAKER, A. 1972. The Watchet fault – a post-Liassic reverse fault. *Bulletin of the Geological Survey of Great Britain*, **41**, 75–80.

WILLIAMS, G. D., POWELL, C. M. & COOPER, M. A. 1989. Geometry and kinematics of inversion tectonics. *In*: COOPER, M. A. & WILLIAMS, G. D. (eds) *Inversion Tectonics*. Geological Society, London, Special Publication, **44**, 3–15.

WILLIAMSON, M. A. & SMYTH, C. 1992. Timing of gas and overpressure generation in the Sable Basin offshore Nova Scotia: implications for gas migration dynamics. *Bulletin of Canadian Petroleum Geology*, **40**, 151–169.

YEATS, R. S. 1983. Large-scale Quaternary detachments in Ventura Basin, southern California. *Journal of Geophysical Research*, **88**, 569–583.

——, LEE, W. H. K. & YERKES, R. F. 1987. Geology and seismicity of the eastern Red Mountain fault, Ventura County. *U.S. Geological Survey Professional Paper*, **1339**, 161–167.

ZIEGLER, P. A. 1987. Late Cretaceous and Cenozoic intra-plate compressional deformations in the Alpine foreland – a geodynamic model. *Tectonophysics*, **137**, 389–420.

—— 1989. Geodynamic model for Alpine intra-plate compressional deformation in Western and Central Europe. *In*: COOPER, M. A. & WILLIAMS, G. D. (eds) *Inversion Tectonics*. Geological Society, London, Special Publication, **44**, 63–85.

The link between sedimentary basin inversion and igneous underplating

JAMES BRODIE & NICKY WHITE

Bullard Laboratories, Madingley Rise, Madingley Road, Cambridge CB3 0EZ, UK

Abstract: There is now general agreement that many sedimentary basins on the northwest continental shelf of Europe underwent permanent exhumation during the Tertiary. The most dramatic indicator of this process is the present-day absence of up to 3 km of anticipated post-rift subsidence in the midlands of Britain and in the East Irish Sea. Any explanation must take into account the fact that the entire shelf has very small, long wavelength, free-air gravity anomalies. This constraint is of fundamental importance and implies either that the crust has been thickened, that phase changes have occurred within the lithosphere, or that low density material has been added to the lithosphere. Tertiary epeirogenic uplift and exhumation is often attributed to horizontal shortening which is assumed to be related in a general sense to Alpine mountain building. However, the removal of 3 km of sediment from a basin, which was originally 100 km wide, requires 20–30 km of shortening. Whilst minor Tertiary shortening is observed all over the continental shelf, nowhere is it sufficient to account for the inferred amount of denudation. More significantly, exhumation is thought to have commenced in the Early Tertiary and dramatically increases from south to north. Shortening is generally younger (mid-Tertiary) and decreases in intensity from south to north. Here we argue that Tertiary uplift and denudation are a consequence of regional igneous underplating. At the beginning of the Tertiary, rifting associated with the initiation of the Iceland plume generated substantial volumes of melt. Petrological arguments and the results from inversion of rare earth element concentrations of MgO-rich igneous rocks suggest that a minimum of 2–5 km of melt were produced beneath a substantial part of the continental shelf. We infer that much of this melt was trapped within the lithosphere, presumably close to the Moho, which would have acted as a density filter. Such underplating will have caused rapid surface uplift whilst maintaining isostatic equilibrium. Simple calculations based on deep seismic reflection data and the high P-wave velocities observed beneath Scotland are consistent with the petrological arguments.

A substantial part of the British Isles suffered rapid exhumation at the beginning of the Tertiary (Ziegler 1982; Johns & Andrews 1985; Roberts 1989; Lewis *et al.* 1992*a*). This conclusion is based primarily on subsidence analyses, on vitrinite reflectance profiles, and on fission track analyses. Exhumation is also manifest in the onshore and offshore Mesozoic outcrop pattern (Fig. 1). Nonetheless, the extent and magnitude of exhumation are disputed and it is often assumed that most of the British Isles remained at or close to sea-level from the end of the Carboniferous until at least the Cretaceous (see, for example, Ziegler 1982; Glennie 1990; Naylor 1992). Here we briefly review the evidence in favour of Tertiary uplift and consequent erosion. We point out that there is a considerable body of data which suggests that rapid and permanent exhumation must have been caused by regional igneous underplating associated in a general way with the opening of the North Atlantic ocean over the Iceland plume.

The clearest evidence in favour of Tertiary uplift and erosion depends upon our understanding of the evolution of extensional sedimentary basins. According to the lithospheric stretching model (McKenzie 1978; Jarvis & McKenzie 1980), syn-rift subsidence is followed by an exponentially decreasing post-rift or thermal subsidence (Fig. 2). The most recent episodes of extension in northwest Europe took place during the Permo–Triassic and in the Late Jurassic (Ziegler 1982; Glennie 1990). Upper Cretaceous and Tertiary extension events have also occurred but they are largely confined to the continental margin farther north and west of the area of interest and so are not considered here (Brooks & Glennie 1987). In the North Sea basin, where both Permo–Triassic and Late Jurassic extension have been documented (Glennie 1990), the predictions of the stretching model and the observations are in agreement (Barton & Wood 1984). Similar agreement has been documented for many extensional basins throughout northwest Europe (Ziegler 1982). In

From BUCHANAN, J. G. & BUCHANAN, P. G. (eds), 1995, *Basin Inversion*, Geological Society Special Publication No. 88, 21–38.

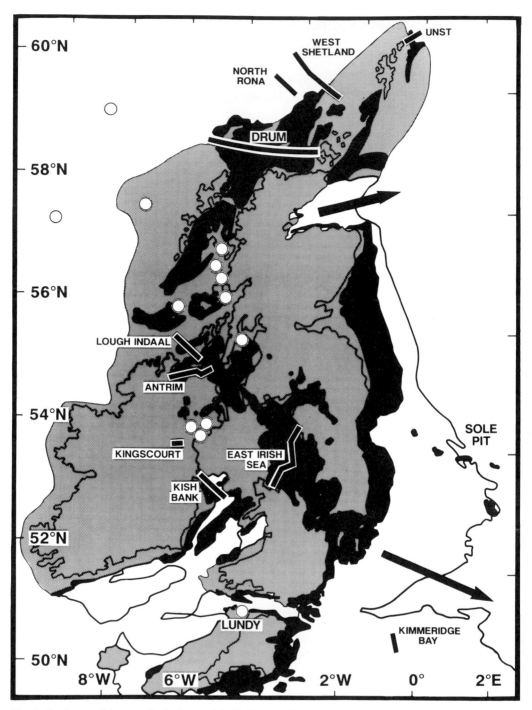

Fig. 1. Onshore–offshore geological map of the British Isles (sub-Pleistocene strata only) showing location of principal sedimentary rocks (redrawn from Woodland 1979). Note that coastlines of southern England and France have been omitted for clarity. Thin black line is boundary between pre-Mesozoic and Cretaceous rocks; thick black line is top of Cretaceous; outcrop pattern of Mesozoic and Cenozoic sedimentary rocks omitted for clarity; black areas show sub-crop of Permian–Triassic rocks, including some Jurassic and Cretaceous rocks in places; shading indicates Palaeozoic and older rocks, each representing basement in this paper. Arrows point in direction of progressively younger Mesozoic and Cenozoic rocks. Circles = major Tertiary igneous centres (note position of Lundy). Thick black lines with white margins and labels = location of profiles shown in Figs 4–6.

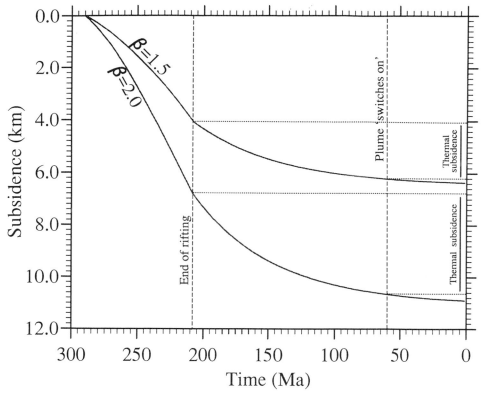

Fig. 2. Theoretical sediment-loaded subsidence curves for stretching factors of β = 1.5, 2. Rifting lasts from 290 to 210 Ma (i.e. Permian–Triassic) assuming a constant strain rate. Sediment density = 2.2 Mg m^{-3}. Vertical dashed lines indicate end of rifting and 60 Ma. Horizontal dotted lines indicate thicknesses of thermal subsidence from end of rifting until 60 Ma for each subsidence curve.

the vicinity of the British Isles, stretching factors for the Permo–Triassic extension are generally small (β = 1.1–1.3), although higher values (c. 1.5) have been proposed for basins north of Scotland (Cheadle et al. 1987; Enfield & Coward 1987). Late Jurassic extension is often substantial, especially within the Viking and Central Graben of the North Sea where β reaches values of 1.5 (Barton & Wood 1984). In the Porcupine basin west of Ireland, Jurassic stretching factors vary from 1.3–6.0 (Tate et al. 1993).

If rifting lasts for 80 Ma, stretching factors of 1.5 can give rise to sediment-loaded post-rift subsidence of c. 2 km (Jarvis & McKenzie 1980; Fig. 2). Shorter rifting periods give rise to greater amounts of post-rift subsidence. However, on mainland Britain, in the Irish Sea, and north of Scotland, fault-controlled (i.e. syn-rift) Permo–Triassic sedimentary rocks crop out close to the surface and the predicted post-rift subsidence is absent (Jenner 1981; Evans et al. 1982; Enfield & Coward 1987; Kirton & Hitchen

1987; Roberts 1989). Our understanding of extensional sedimentary basin formation suggests that these basins are likely to have subsided thermally after Permo–Triassic rifting. We conclude that post-rift sediments were deposited and have been subsequently removed. This inference is indirectly corroborated by anomalous porosity-depth variation and by high vitrinite reflectance in the Irish Sea, Inner Moray Firth and elsewhere (Roberts 1989; Andrews et al. 1990; Hillis et al. 1994). There is also excellent evidence that the products of large-scale erosion were deposited during the Palaeocene in rapidly subsiding basins to the north, east and west of the British Isles (Glennie 1990).

Finally, apatite fission track analyses from Mesozoic sedimentary rocks also appear to favour substantial Tertiary denudation (Lewis et al. 1992a; Green et al. 1993). These data can provide estimates of the time at which sedimentary rocks began to cool from maximum palaeotemperatures. Using analyses from the East Midland shelf and from northern England,

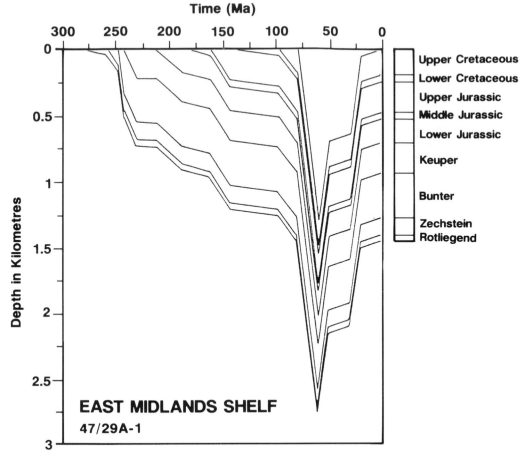

Fig. 3. Reconstructed burial history for post-Carboniferous stratigraphic section in off-shore East Midlands Shelf, based on apatite fission track analyses and vitrinite reflectance data (redrawn from Green *et al.* 1993). See also Roberts (1989) for slightly different interpretation of vertical movements in this area.

Green (1986, 1989), and Bray *et al.* (1992) infer that maximum palaeotemperatures occurred at about 60 Ma and that up to 2 km of section have been removed since that time (Fig. 3). Offshore, in the Irish Sea, Lewis *et al.* (1992a) argue that between 2.7 and 3.3 km of overburden was removed. In northern Scotland some fission track results are complicated by local heating effects associated with the Tertiary intrusive centres. Away from these centres, pre-Tertiary fission track ages suggest that a maximum of 1–1.5 km of overburden could have been removed from Scotland (Lewis *et al.* 1992b). Inferences based upon fission track analyses must be treated with caution since they rely upon the geothermal gradient remaining relatively constant during the last 60 Ma (Brown 1991). Modelling results are also critically dependent upon important assumptions about the kinetics

of fission track growth and annealing (Green *et al.* 1993).

Sedimentary basin geometry

The outcrop pattern of the British Isles attests to exhumation: along the length of Britain, Mesozoic and Tertiary sedimentary rocks offlap to the southeast (Fig. 1). Farther north, eastward-directed offlap is observed in the Moray Firth graben. This general pattern, coupled with our understanding of the evolution of extensional sedimentary basins, can be used to delineate a region within which significant amounts of denudation have occurred (Fig. 1). This region extends from the Unst basin (north east of the Shetland Islands) to the Bristol Channel basin (southwest of Wales) to the west coast of Ireland.

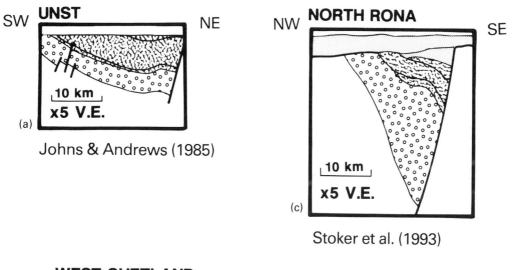

Johns & Andrews (1985)

Stoker et al. (1993)

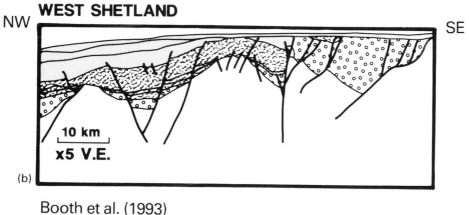

Booth et al. (1993)

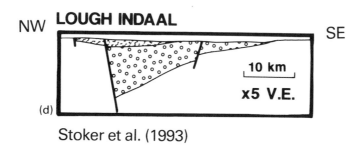

Stoker et al. (1993)

Fig. 4. Depth-converted and interpreted seismic reflection profiles from vicinity of British Isles (see Fig. 1 for locations). Thin solid lines = interpreted strata; thick solid lines = normal faults; scalloped shading = pre-Permian sedimentary rocks and basement; open circle shading = Permian–Triassic sedimentary rocks; irregular shading = Jurassic–Cretaceous sedimentary rocks; fine shading = Tertiary and Quaternary sedimentary rocks. Solid black = Tertiary igneous rocks. Scale, vertical exaggeration (V.E.) and original source indicated in each case. Where appropriate, depth conversion was carried out using a uniform velocity of $3\,km\,s^{-1}$. *continued overleaf*

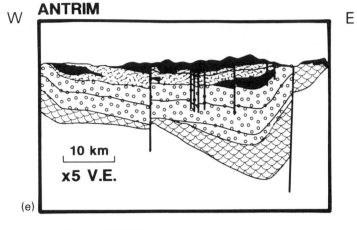

McCann (1988)

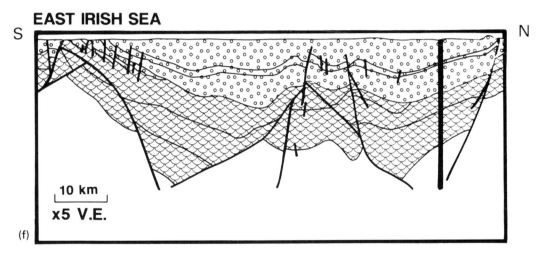

British Geological Survey (1994)

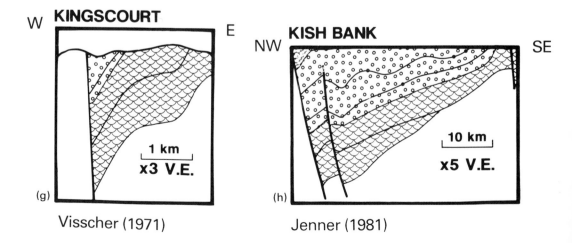

Visscher (1971)   Jenner (1981)

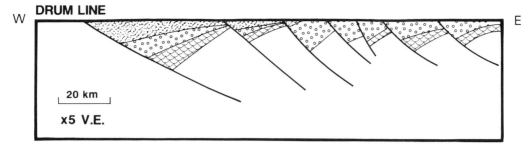

Cheadle et al. (1987)

Fig. 5. Depth-converted interpretation of upper 7 seconds (two-way travel time) of DRUM deep seismic reflection profile. Note absence of thick pile of unfaulted post-rift sedimentary rocks younger than Lower Jurassic. Redrawn from Cheadle *et al.* (1987). See Fig. 1 for location and Fig. 4 for key to shading.

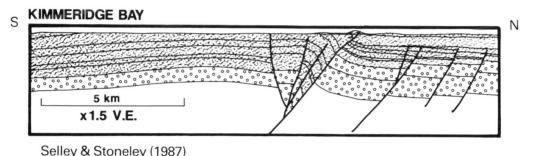

Selley & Stoneley (1987)

Fig. 6. North–south cross-section through Kimmeridge Bay and Wareham oil fields (see Fig. 1 for location and Fig. 4 for key). Redrawn from Selley & Stoneley (1987).

In Fig. 4, a series of simplified cross-sections through selected sedimentary basins located within the exhumed region are shown. In all cases, it is clear that syn-rift sedimentary rocks (i.e. strata which have been rotated and displaced by normal faulting) crop out close to the Earth's surface or beneath a thin cover of late Tertiary and Quaternary sedimentary rocks (e.g. Fig. 4b & c). It is especially clear that mappable syn-rift horizons are truncated close to the surface and that the youngest syn-rift sedimentary rocks occur within the hangingwall, against large normal faults. There is little evidence for regional horizontal shortening. The syn-rift sedimentary rocks are generally 2–5 km thick. Comparison with Fig. 2 suggests that the stretching factor, β, reaches values of about 1.5, which implies that 2 or more kilometres of post-rift sedimentary rocks are missing.

The discrepancy between predicted and observed subsidence is most clearly illustrated by the DRUM seismic reflection profile, which is located 30 km off the north coast of Scotland (Figs 1 & 5). An array of fault-bounded blocks is discernable in the upper crust where individual

half graben are characterized by differentially rotated syn-rift sedimentary rocks. As before, unfaulted post-rift sedimentary rocks are absent, despite the fact that estimates of β, determined from the normal faulting geometry and from subsidence analysis, vary between 1.5–2.0 (Enfield & Coward 1987). Whilst acknowledging that the thickness of the Permo–Triassic fill is disputed, we can use these rough estimates of β to infer that 2–4 km of post-rift sedimentary rocks are missing. Cheadle *et al.* (1987) suggested that the absence of post-rift subsidence could be accounted for by stretching crust which was originally 45 km thick by a factor of 1.5–2.0. This scheme is unlikely to account for the absence of post-rift subsidence because it disregards other evidence, cited above, in favour of regional Tertiary exhumation.

Published hypotheses

Despite the fact that most of the observations summarized here are not new, little attention has been paid to the causes of Tertiary epeirogeny. Usually uplift is attributed in a general and rather unspecified way either to flexural and thermal effects associated with the opening of the North Atlantic ocean (Glennie 1990; Lewis *et al.* 1992*a*) or to horizontal shortening associated with the Alpine orogeny (Roberts 1989).

We argue that any explanation must take into account the absence of significant gravity anomalies on wavelengths of greater than 50 km. This important observation means that the entire continental shelf is in approximately local (Airy) isostatic equilibrium. Bearing this important constraint in mind, we will now examine the two main categories of hypotheses used to account for Tertiary epeirogeny.

Flexural and thermal uplift

Admittance studies of the free-air gravity anomalies and load topography in basins like the North Sea (Barton & Wood 1984) together with flexural cantilever modelling of syn-rift topography (Marsden *et al.* 1990) have demonstrated that the elastic thickness of the lithosphere beneath these offshore basins is very small ($\tau_e <$ 5 km). Barton (1992) has shown that the elastic thickness beneath Scotland is equally small. Hence Early Tertiary uplift and denudation of the British Isles cannot be accounted for by lithospheric flexure which, in any case, would generate large amplitude gravity anomalies.

Other models based on transient thermal uplift associated with the opening of the North Atlantic and the initiation of the Iceland plume, would seem to be more plausible (Ziegler 1982). However, the exhumed sedimentary basins around the British Isles have generally subsided very little since the beginning of the Tertiary (Figs 4 & 5). This observation means that thermal uplift would have to have been maintained right through to the present day. The only exceptions occur north of Scotland, where Tertiary subsidence has been important (Fig. 4b & c). The British Isles are now *c.* 1000 km away from the centre of the Iceland plume. Thermal uplift must have decreased by a considerable amount over this distance with a time constant of 40–60 Ma. Once the causes of uplift are removed then sedimentary basins must subside back to the level they would have attained in the absence of an uplift event.

Lithospheric shortening

Permanent, isostatically compensated, uplift can only be generated either by thickening the crust, or by adding material which is less dense than asthenosphere (3.2 Mg m^{-3}), to the lithospheric column. Crustal thickening caused by

horizontal shortening associated with Alpine mountain building would initially appear to be a satisfactory way of producing permanent exhumation. There is evidence for slight Tertiary shortening all over the northwest European shelf (Ziegler 1982; Roberts 1989; Evans 1990; Chadwick 1993). The classical 'inversion structures' seen along the southern coast of Britain (e.g. Chadwick 1993) and in the Sole Pit basin of the southern North Sea (Glennie & Boegner 1981) are generally absent further north. An important exception is the Wyville–Thomson ridge in the Faeroe–Shetland basin. This ridge formed by local shortening, which started in the early Eocene and continued until the Miocene (Boldreel & Andersen 1993). However, horizontal shortening reverses the throw across major normal faults, resulting in hangingwall anticlines (Fig. 6). Coupled with erosion, 'structural inversion' produces a complicated Mesozoic outcrop pattern with length scales of tens of kilometres rather than wholesale exhumation with length scales of hundreds of kilometres (Figs 1 & 4). Thus, basins which have undergone even quite modest horizontal shortening should have easily identifiable geometries both in plan view and in cross-section. It is clear that such geometries do not generally occur within the region dominated by Tertiary exhumation (compare Figs 4 & 6).

It is straightforward to quantify the amount of regional shortening that is required to produce a given level of denudation. If a standard column of continental lithosphere is shortened uniformly by a factor, f, then the amount of uplift, U, is given by

$$U = \frac{a\left[(\rho_m - \rho_c)\frac{t_c}{a}\left(1 - \frac{\alpha T_1}{2}\frac{t_c}{a}\right) - \frac{\alpha T_1}{2}\rho_m\right]}{\rho_m(1 - \alpha T_1)}(f - 1) \quad (1)$$

where the symbols are defined in Table 1. Substitution of the values given in Table 1 yields

$$U = 2.5(f - 1) \quad (2)$$

Table 1 *Symbols and values of parameters used in text*

a	lithospheric thickness	125 km
t_c	crustal thickness	30 km
ρ_w	sea water density	1.0 g cm^{-3}
ρ_c	crust density (at 0° C)	2.8 g cm^{-3}
ρ_m	mantle density (at 0° C)	3.33 g cm^{-3}
ρ_a	asthenosphere density (at 1333° C)	3.2 g cm^{-3}
α	thermal expansion coefficient	$3.28 \times 10^{-5 \circ}$ C^{-1}
T_1	asthenospheric temperature	1333° C

The amplification effect of erosion can result in a maximum amount of denudation, D, which is given by

$$D = \left(\frac{\rho_a}{\rho_a - \rho_s} \right) U \qquad (3)$$

where ρ_s is the density of sediment and ρ_a is the density of the asthenosphere. If $\rho_s = 2.4 \, \mathrm{Mg \, m^{-3}}$ and $\rho_a = 3.2 \, \mathrm{Mg \, m^{-3}}$, horizontal shortening of $f = 1.3$ is required to achieve 3 km of denudation. If the lithospheric mantle does not thicken during shortening then

$$U = t_c \left[1 - \frac{\rho_c}{\rho_m} \left(\frac{2a - \alpha T_1 t_c}{2a(1 - \alpha T_1)} \right) \right] (f - 1) \qquad (4)$$

Substitution yields

$$U = 4.1(f - 1) \qquad (5)$$

In this case, $f = 1.2$ if $D = 3$ km.

These simple calculations show that regional shortening is relatively inefficient at generating substantial amounts of denudation. The level of denudation observed in the vicinity of the British Isles would require a sedimentary basin which was originally 100 km wide to be shortened by 20–30 km. There is no evidence for such large amounts of shortening, either onshore or off-shore, on any of the cross-sections shown in Figs 4 & 5. Other simple observations bear out this conclusion. For example, the most significant amounts of Tertiary shortening occur along the south coast of Britain ($f c. 1.05$; Fig. 6). Despite these signs of horizontal shortening, the level of denudation is modest: most of the Cretaceous sedimentary rocks are still preserved (Glennie 1990). In contrast, the shortened passive continental margin exposed in southeast France has been exhumed to Permo–Triassic and Lower Jurassic stratigraphic levels (i.e. syn-rift). Here, several kilometres of denudation have only been achieved by predictably large amounts of short-ening: there is excellent evidence for fault throw reversal, extensive folding, and widespread cleavage development (Lemoine et al. 1986).

The only other geologically plausible means of thickening the crust is by igneous underplating (Cox 1980; McKenzie 1984; Brown 1991). We now show that the large volumes of basaltic melt produced at the beginning of the Tertiary must have resulted in rapid uplift and denudation of a considerable portion of the continental shelf.

Tertiary magmatism

The Tertiary igneous rocks of the northwest European continental shelf form part of the North Atlantic Tertiary Igneous Province, which also encompasses the flood basalts of East Greenland and dipping reflector sequences located along the continental margins of north-west Europe and East Greenland (Upton 1988; White & McKenzie 1989). Radiometric dating of the province, combined with biostratigraphic and palaeomagnetic constraints, shows that igneous activity commenced at magnetochron C24R just prior to the initiation of sea-floor spreading in the North Atlantic (Harland et al. 1989; Larsen et al. 1992). Recent models suggest that rapid generation of the great volumes of igneous material observed in the North Atlantic province resulted from continental rifting in the vicinity of a proto-Iceland plume (e.g. White & McKenzie 1989; Skogseid et al. 1992; White 1992; Eldholm & Grue 1994).

The distribution of Tertiary igneous rocks around the British Isles is shown in Fig. 7. Major intrusive centres occur throughout the region, from the west of Ireland to the Erlend centre north of the Shetland Isles. Flood basalts are associated with some of these centres but are more widely distributed in the Rockall Trough, Antrim (Northern Ireland) and in the vicinity of the Faeroe Islands. Minor intrusive bodies and dyke swarms are common both onshore and offshore, as far south as Lundy Island in the Bristol Channel (Thorpe et al. 1990), as far east as the western margin of the North Sea basin, and as far west as the Porcupine basin (Tate & Dobson 1988). Further northwest, along the continental margin, the results of wide-angle seismic experiments have indicated the presence of substantial volumes of underplated igneous rock (up to 15 km in thickness) within the lower crust of the Hatton–Rockall margin and the Vøring Plateau (Hinz et al. 1987; White et al. 1987). In the British Isles, the oldest Tertiary igneous rocks have been dated at c. 63 Ma, with the most intense magmatic activity concentrated at c. 59 Ma (Mussett et al. 1988).

Here we are principally concerned with estimating the volume and distribution of melt generated in the vicinity of the British Isles. Recent work by McKenzie & O'Nions (1991) and White et al. (1992) suggests that rare earth element (REE) concentrations in MgO-rich igneous rocks ($\geqslant 6 \mathrm{wt\%} \ MgO$) can be used to constrain the partial melt distribution with depth at the time of melt generation. Estimates of the one-dimensional melt thickness may then be obtained by integration over depth. This inversion technique has been successfully ex-ploited in the ocean basins, where predicted crustal thicknesses are in good agreement with those determined independently by seismic refraction experiments. We shall not describe

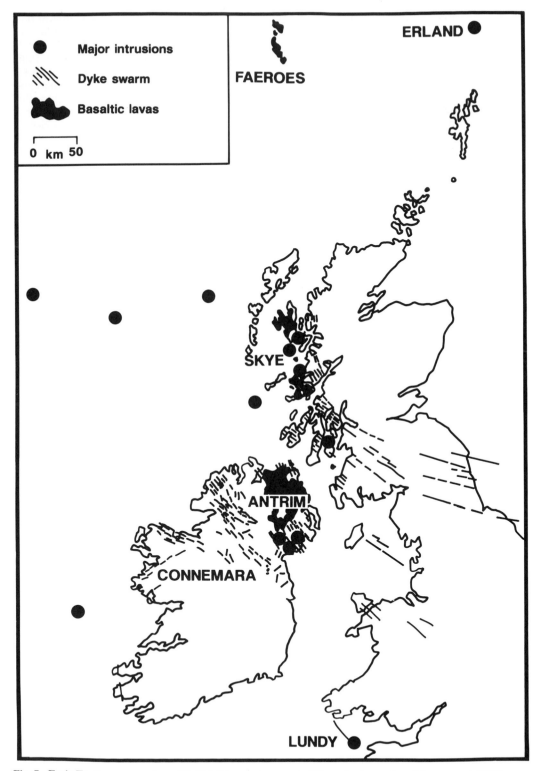

Fig. 7. Early Tertiary magmatism of British Isles, onshore and offshore (redrawn from Mussett *et al.* 1988).

the inverse model here, but refer the interested reader to McKenzie & O'Nions (1991) and to Brodie *et al.* (1994).

There are several important difficulties associated with applying the inversion technique to the British Tertiary Igneous Province. First, there is substantial evidence to suggest that MgO-rich basalts exposed on, for example, the Isle of Skye, have been contaminated with material derived from the continental crust and/or lithospheric mantle (e.g. Dickin 1981; Thompson *et al.* 1982; Thirlwall & Jones 1983; Menzies *et al.* 1987; Thompson & Morrison 1988). The effects of contamination on the REE are difficult to quantify, although selective enrichment in lighter REE is anticipated. This enrichment implies that melt thicknesses calculated using the REE inversion technique are likely to be minimum estimates of the true melt thickness generated prior to contamination (Brodie *et al.* 1994). Kerr (1993) records modest light REE enrichment in crustally contaminated Mull basalts compared to uncontaminated basalts. The degree of enrichment is not sufficient to affect significantly the results of the inversion technique. Similarly, inversion of REE abundances in selected Skye rocks, which are thought to be virtually uncontaminated by lithospheric material (the Beinn Dearg Mhor dykes: Thompson & Morrison 1988), yield partial melt distributions which are similar to those determined for the variably contaminated Skye Main Lava Series. We conclude that the inversion technique is relatively insensitive to the modest amounts of contamination by lithospheric material inferred by previous workers.

The second difficulty is that within a buoyancy-driven flow such as a mantle plume, asthenospheric material is continuously fluxed through the melting region, and in the case of a stationary lithospheric plate, one-dimensional melt thicknesses calculated using the inversion technique are likely to underestimate the true thickness of melt generated (Latin *et al.* 1993). It is not yet obvious how plume fluxes during the Tertiary can be estimated and we have not attempted to correct our predicted melt thicknesses for this effect. These estimates are therefore minima.

Rare earth element abundances have been measured in Tertiary MgO-rich igneous rocks collected from onshore and offshore Ireland and Scotland, and from Lundy Island. We have supplemented our own database with suitable data drawn from the literature. Modelling using the inversion technique suggests that typically 2–5 km of melt were produced after correction for fractional crystallisation of observed melt compositions. Two examples are shown in Fig. 8 for flows sampled from the Skye Main Lava Series and for dykes collected in the west of Ireland. Before correction for fractional crystallisation, the predicted melt thickness typically exceeds *c.* 1.8 km. Calculated melt thicknesses are strongly dependent on composition of the source mantle, which must be specified a priori using the inversion technique. Greater melt thicknesses are predicted from more primitive source compositions (Brodie *et al.* 1994). The cited range of 2–5 km was determined for depleted mantle source (McKenzie & O'Nions 1991), since isotopic constraints suggest that the source of melts in the British province was depleted relative to bulk earth (e.g. Thirlwall & Jones 1983; Kerr 1993). For this reason, and from the arguments outlined in the preceding paragraphs, we suggest that a minimum of *c.* 2–5 km melt was produced beneath a substantial part of the continental shelf encompassing the British Isles.

Although Fig. 7 shows that Tertiary melts are widespread around the British Isles, generally less than 1 km of igneous rock is observed near the surface at any one location (e.g. Upton 1988). The discrepancy between predicted (fractionated) and observed melt thicknesses could arise in several ways. First, melt may have been redistributed by lateral flow prior to solidification. Secondly, erosion could have removed large amounts of igneous material over the past 60 Ma. For example, there is some evidence to suggest that of the order of 1 km of basalt has been eroded from Skye (Anderson & Dunham 1966; Craig 1983). Alternatively, and more probably, much of the predicted melt could have been trapped at depth, as inferred for the Hatton–Rockall margin (White *et al.* 1987). This inference is reasonable since the Moho will act as a density filter for melts which have a density of *c.* 2.8 Mg m^{-3} (Stolper & Walker 1980; Cox 1980).

The addition of material which is less dense than asthenosphere to the lithospheric column must cause uplift, regardless of the precise distribution of added material within that column. Assuming Airy isostasy, the amount of uplift, U, is given by

$$U = (1 - \rho_\chi/\rho_a)\chi \qquad (6)$$

where χ and ρ_χ are the thickness and density of the added material respectively and ρ_a is the density of the asthenosphere (*c.* 3.2 Mg m^{-3}). Thus 5 km of basalt or gabbro with a density of 2.8 Mg m^{-3} causes an initial uplift of 625 m (Fig. 9). As before, the amplification effect of erosion

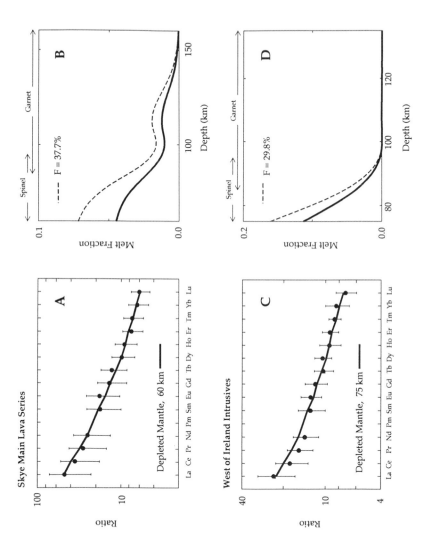

Fig. 8. Inversion of rare earth element concentrations from Isle of Skye, Scotland and from the west of Ireland. (a) and (b) Skye Main Lava Series. (A) Rare earth element concentrations normalised with respect to depleted mantle, circle denotes mean of dataset, bar is 1 standard deviation; thick line shows values calculated from the melt distribution obtained by inversion. (B) Corresponding distribution of melt fraction with depth obtained by inversion (thick line); dashed line shows resultant melt distribution when corrected for fractional crystallisation of olivine, based on misfit of modelled Fe and Mg concentrations to observed Fe and Mg concentrations (see McKenzie & O'Nions (1991) for further details). Mantle mineralogy and composition taken from McKenzie & O'Nions (1991): 'spinel' = spinel peridotite; 'garnet' = garnet peridotite. (C) and (D) Dykes from western Ireland (Brodie, unpublished data). Source composition and depth to top of melting column are indicated on (A) and (C). (B) produce 1.8 km of melt (2.9 km after correction for fraction crystallisation); (D) also produces 1.8 km of melt (2.5 km after correction for fraction crystallization).

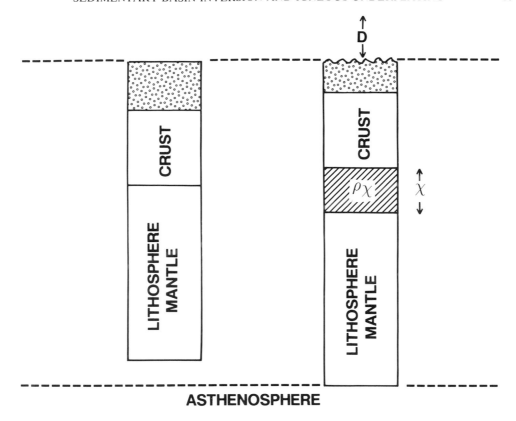

Fig. 9. Cartoons illustrating the effect of adding a thickness of melt, χ, with density ρ_x to the lithosphere. If melt is less dense than asthenosphere ($3.2\,\mathrm{Mg\,m^{-3}}$), surface uplift will result. Erosion can result in a maximum amount of denudation, D, whose value can be determined from Equations (6) and (7). Note that precise distribution of underplating does not affect isostatic balance.

will increase U so that the total amount of material removed, D, is

$$D = \left(\frac{\rho_a - \rho_x}{\rho_a - \rho_s}\right) \chi \qquad (7)$$

where ρ_s is the density of sediment. If $\rho_s = 2.4\,\mathrm{Mg\,m^{-3}}$ and if $U = 625\,\mathrm{m}$, a total of $2.5\,\mathrm{km}$ will be removed. Picritic melt, which have densities of $c.\ 3.0\,\mathrm{Mg\,m^{-3}}$, will produce less denudation.

Geophysical evidence for igneous underplating

We have already remarked that about $2\,\mathrm{km}$ of post-rift subsidence-related sedimentary rocks are missing from the DRUM deep seismic reflection profile located north of Scotland (Figs 1 & 5). Assuming a crustal velocity of $6\,\mathrm{km\,s^{-1}}$, the average crustal thickness along the profile excluding syn-rift sedimentary rocks is $c.\ 26\,\mathrm{km}$

(Cheadle *et al.* 1987). This thickness is in broad agreement with the results of the LISPB wide-angle line, which was positioned north–south along the length of Scotland (Bamford *et al.* 1976; Barton 1992). However, the average stretching factor along DRUM is 1.5–2.0, which implies that the crust should only be 15–20 km thick. If we assume that 5 km of Tertiary magma were generated beneath Skye and trapped within the lower crust then this discrepancy can be explained.

Deep seismic reflection profiles collected around the British Isles (e.g. DRUM profile; McGeary & Warner 1985), indicate that the lower crust is strongly reflective (e.g. Hall *et al.* 1984; Cheadle *et al.* 1987; Freeman *et al.* 1988). It has already been suggested that this reflectivity could be explained by gabbroic intrusions which underplate the lower crust (Warner 1990; Nelson 1991). If so, we suggest that underplating is Early Tertiary in age since this period is the only time since the Late Carboniferous–Early Permian when large quantities of basaltic melt have been generated over such a wide region (e.g. Craig 1983).

Barton's (1992) recent reinterpretation of the LISPB dataset has yielded a detailed cross-section of crustal structure north–south along the length of Scotland. The crust has a relatively uniform thickness (*c.* 30 km) and there is no evidence for crustal roots. Thus any significant topography, such as the Grampian mountains, is likely to be supported by variations in crustal density. Velocities within the upper 20 km of the crust vary from 6.2–6.45 km/s. At 20 km depth, the velocity gradient increases sharply so that the velocity within 5 km of the Moho is 7.1 km/s. Lower crustal igneous rocks are expected to have seismic velocities of *c.* 7.2 km/s (White & McKenzie 1989) and so Barton's (1992) results may be consistent with underplating. However, high velocities are a necessary but not sufficient condition for the existence of igneous underplating. Furthermore, there is no age constraint on timing of formation of lower crustal layering.

The strong correlation between Bouguer gravity anomalies and topography over Scotland show that this region must have an elastic thickness of less than 5 km (Barton 1992). The most significant anomalies are quite clearly associated with granitic and gabbroic bodies within the upper crust. Thus there is little evidence supporting Watson's (1985) argument that Scotland has had a much higher rigidity than the surrounding continental shelf and so was unaffected by Mesozoic extension. Mesozoic fault patterns clearly show that the Scottish massif did not act as a horst block (Fig. 10). With the exception of the northern margin of the Moray Firth, the Scottish coastline is cut at a high angle by numerous faults which bound offshore basins. This observation also applies to the rest of the British Isles landmass.

Discussion & conclusions

Extensional sedimentary basins in the Irish Sea and immediately to the northwest of Scotland formed by lithospheric stretching during the Permo–Triassic and, in some cases, during the Late Jurassic. The general acceptance of the stretching model together with the fact that it is testable has focussed attention on the striking absence of post-rift thermal subsidence in these basins. In many cases, compacted

syn-rift sediments occur at the sea-bed, masked only by a thin veneer of Late Tertiary and Quaternary sediment. Subsidence calculations show that 2–4 km of sediment are missing. Knox *et al.* (1981) and Mudge & Bliss (1983) have independently suggested that a period of widespread uplift and erosion must have occurred west of the Moray Firth graben after deposition of Danian chalk and marl since Danian sediments have since been removed from all but the deepest parts of graben. Resedimented chalk deposits, slumping, and debris flows have also been documented in this area (Morton 1982).

Any explanation of these observations must take into account the fact that the entire region is characterized by small gravity anomalies. This fundamental and often overlooked constraint implies that models which appeal to flexural affects arising from the opening of the North Atlantic ocean are incorrect. Furthermore, any thermal effects must decay with time and with increasing distance from the Iceland plume. We argue that the crust has been thickened permanently, either by shortening or by igneous underplating. Basin inversion caused by shortening is observed throughout the region but simple calculations show that the observed shortening cannot account for the magnitude, location, or timing of denudation.

We have shown that it is much easier to explain Tertiary uplift and erosion along a NNE–SSW axis by igneous underplating. Inverse modelling of rare earth element concentrations from MgO-rich igneous rocks collected from the British Isles and continental shelf suggest that at least 2–5 km of melt was generated at the beginning of the Tertiary. Much of this melt has probably been trapped within the lower crust. Simple calculations show that the addition of 5 km of igneous rock can give rise to 2.5 km of exhumation. Since magmatism occurred rapidly and over a wide region, regional epeirogenic uplift would also have been rapid. The consequent tilting about a NNE-SSW trending horizontal axis eventually produced the familiar outcrop pattern of these islands.

Finally, it is often assumed that substantial portions of the British Isles were unaffected by the widespread Permo–Triassic and Upper Jurassic extension documented throughout Western Europe and beyond (Anderton *et al.* 1979; Holland 1981; Ziegler 1982; Glennie 1990). However, the Scottish, Welsh, Cornish, and Irish massifs are completely surrounded by

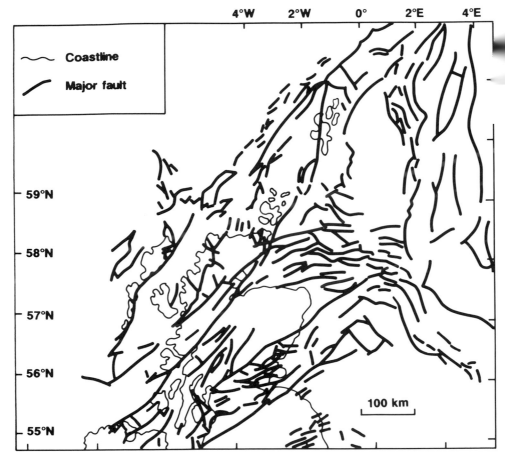

Fig. 10. Map of northern British Isles showing principal post-Palaeozoic normal and strike-slip faulting (onshore and offshore). Redrawn from Cook (1991).

Permo–Triassic and Jurassic outcrop and have dimensions comparable with the thickness of the lithosphere. In all cases, significant Mesozoic faults can be traced onshore and coasts are not bounded, except locally, by normal faults (e.g. Fig. 10). More importantly, outliers of Permo–Triassic sedimentary rock (Craig 1983; Holland 1981) together with minor Early Permian volcanic centres (Rock 1983; Penn *et al*. 1983; Mitchell & Mohr 1987) are recorded from the highlands of northern Scotland and from Ireland. We infer from these observations that both landmasses underwent at least modest Permo–Triassic and/or Upper Jurassic stretching. Depending upon location, 1–4 km of Cretaceous post-rift sediments are assumed to have been deposited, followed by extensive denudation in the Tertiary.

JAB gratefully acknowledges receipt of a Shell studentship. We thank Paul Mohr, Paul Lyle and Joanne Wallace for their generous help in the field in Ireland. Major and trace element analyses were carried out on the x-ray fluorescence spectrometer at the Department of Geology and Geophysics, University of Edinburgh. Rare earth element analyses were carried out at the Natural Environment Research Council ICP-MS facility at the Department of Geology, Royal Holloway and Bedford New College, University of London and at CARE, Silwood Park, Ascot. Dan McKenzie kindly provided us with a copy of his rare earth element inversion program. We are grateful to G. Budd, S. Capon, R. England, G. Fitton, B. Hall, H. Iwamori, D. James, K. Jarvis, D. Johnston, A. Kerr, H. Kerr, D. Latin, L. Lonergan, D. Lyness, K. Martyn, R. Newman, G. Sevastopulo, and J. Wills for their help. M. Bamford and R. Sibson provided thorough reviews.

References

ANDERSON, F. W. & DUNHAM, K. C. 1966. The geology of northern Skye. *Memoir of the Geological Survey of Great Britain.*

ANDERTON, R., BRIDGES, P. H., LEEDER, M. R. & SELLWOOD, B. W. 1979. *A dynamic stratigraphy of the British Isles.* George Allen & Unwin, London, 301pp.

ANDREWS, I. J., LONG, D., RICHARDS, P. C., THOMSON, A. R., BROWN, S., CHESHER, J. A. & McCORMAC, M. 1990. *The geology of the Moray Firth*, British Geological Survey, United Kingdom Offshore Regional Report, HMSO, London.

BAMFORD, D., NUNN, K., PRODEHL, C. & JACOB, B. 1976. LISPB-IV. Crustal structure of Northern Britain. *Geophysical Journal of the Royal Astronomical Society,* **54**, 43–60.

BARTON, P. J. 1992. LISPB revisited: a new look under the Caledonides of northern Britain. *Geophysical Journal International,* **110**, 371–391.

BARTON, P. & WOOD, R. 1984. Tectonic evolution of the North Sea basin: Crustal stretching and subsidence. *Geophysical Journal of the Royal Astronomical Society,* **79**, 987–1022.

BOLDREEL, L. O. & ANDERSEN, M. S. 1993. Late Palaeocene to Miocene compression in the Faeroe-Rocall area. *In:* PARKER, J. R. (ed.) *Petroleum geology of Northwest Europe: Proceedings of the 4th conference.* Geological Society, London, 1025–1034.

BOOTH, J., SWIECICKI, T. & WILCOCKSON, P. 1993. The tectono-stratigraphy of the Solan basin, west of Shetland. *In:* PARKER, J. R. (ed.) *Petroleum geology of Northwest Europe: Proceedings of the 4th conference.* Geological Society, London, 987–998.

BRAY, R. J., GREEN, P. F. & DUDDY, I. R. 1992. Thermal history reconstruction using apatite fission track analysis and vitrinite reflectance: a case study from the UK East Midlands and Southern North Sea. *In:* HARDMAN, R. F. P. (ed.) *Exploration Britain: geological insights for the next decade.* Geological Society, Special Publication, **67**, 3–25.

BRITISH GEOLOGICAL SURVEY 1994. *East Irish Sea (Special Sheet Edition).* 1 : 250000. Edinburgh, Scotland.

BRODIE, J., LATIN, D. & WHITE, N. 1994. Rare earth element inversion of forward-modelled melt compositions. *Journal of Petrology*, **35**, 1155–1174.

BROOKS, J. & GLENNIE, K. W. (eds) 1987. *Petroleum geology of northwest Europe.* Graham & Trotman, London, 623–632.

BROWN, R. W. 1991. Discussion on thermal and tectonic history of the East Midlands shelf (onshore UK) and surrounding regions assessed by apatite fission track analysis. *Journal of the Geological Society, London,* **148**, 785–787.

CHADWICK, R. A. 1993. Aspects of basin inversion in southern Britain. *Journal of the Geological Society, London,* **150**, 311–322.

CHEADLE, M. J., McGEARY, S., WARNER, M. R. &

MATTHEWS, D. H. 1987. Extensional structures on the western UK continental shelf: a review of evidence from deep seismic profiling. *In:* COWARD, M. P., DEWEY, J. F. & HANCOCK, P. L. (eds) *Continental extensional tectonics.* Geological Society, Special Publication, **28**, 445–465.

COOK, P. J. 1991. *Geology of the United Kingdom, Ireland and the adjacent continental shelf (North Sheet).* British Geological Survey, Natural Environment Research Council.

COX, K. G. 1980. A model for flood basalt volcanism. *Journal of Petrology,* **21**, 629–650.

CRAIG, G. Y. (ed.) 1983. *Geology of Scotland.* Scottish Academic Press, Edinburgh. Second edition.

DICKIN, A. P. 1981. Isotope geochemistry of Tertiary igneous rocks from the Isle of Skye, *Journal of Petrology,* **22**, 155–190.

ELDHOLM, O. & GRUE, K. 1994. North Atlantic volcanic margins: dimensions and production rates. *Journal of Geophysical Research,* **99**, 2955–2968.

ENFIELD, M. A. & COWARD, M. P. 1987. The structure of the West Orkney basin, northern Scotland. *Journal of the Geological Society, London,* **144**, 871–884.

EVANS, C. D. R. 1990. *The geology of the Western English Channel and its western approaches.* British Geological Survey, United Kingdom Offshore Regional Report. HMSO, London.

EVANS, D., CHESHER, J. A., DEEGAN, C. E. & FANNIN, N. G. T. 1982. *The offshore geology of Scotland in relation to the IGS shallow drilling programme, 1970–1978.* Institute of Geological Sciences, Report 81/12. HMSO, London.

FREEMAN, B., KLEMPERER, S. L. & HOBBS, R. W. 1988. The deep structure of northern England and the Iapetus Suture zone from BIRPS deep seismic reflection profiles. *Journal of the Geological Society of London,* **145**, 727–740.

GLENNIE, K. W. (ed.) 1990. *Introduction to the petroleum geology of the North Sea.* Blackwell Scientific Publications, London. Third edition. 402pp.

—— & BOEGNER, P. 1981. Sole Pit inversion tectonics. *In:* ILLING, L. V. & HOBSON, G. D. (eds) *Petroleum geology of the continental shelf of Northwest Europe.* Heyden & Son, London, 110–120.

GREEN, P. F. 1986. On the thermo-tectonic evolution of Northern England: evidence from fission track analysis. *Geological Magazine,* **123**, 493–506.

—— 1989. Thermal and tectonic history of the East Midlands shelf (onshore UK) and surrounding regions assessed by apatite fission track analysis. *Journal of the Geological Society, London,* **146**, 755–773.

——, DUDDY, I. R., BRAY, R. J. & LEWIS, C. L. E. 1993. Elevated palaeotemperatures prior to Early Tertiary cooling throughout the UK region: implications for hydrocarbon generation. *In:* PARKER, J. R. (ed.) *Petroleum geology of Northwest Europe: Proceedings of the 4th conference.* Geological Society, London, 1067–1074.

HALL, J., BREWER, J. A., MATTHEWS, D. H. &

WARNER, M. R. 1984. Crustal structure across the Caledonides from the WINCH seismic reflection profile: influences on the evolution of the Midland Valley of Scotland. *Transactions of the Royal Society of Edinburgh: Earth Sciences*, **75**, 97–109.

HARLAND, W. B., ARMSTRONG, R. C., CRAIG, L., SMITH, A. G. & SMITH, D. G. 1989. *A Geologic Time Scale*. Cambridge University Press.

HILLIS, R. R., THOMSON, K. & UNDERHILL, J. R. 1994. Quantification of Tertiary erosion in the Inner Moray Firth by sonic velocity data from the Chalk and Kimmeridge Clay. *Marine and Petroleum Geology*, **11**, 283–293.

HINZ, K., MUTTER, J. C., ZEHNDER, C. M. & NGT STUDY GROUP. 1987. Symmetric conjugation of continent-ocean boundary structures along the Norwegian and East Greenland margins. *Marine and Petroleum Geology*, **3**, 166–187.

HOLLAND, C. H. (ed.) 1981. *A geology of Ireland*. Scottish Academic Press, Edinburgh. 335pp.

JARVIS, G. T. & McKENZIE, D. P. 1980. Sedimentary basin formation with finite extension rates. *Earth & Planetary Science Letters*, **48**, 42–52.

JENNER, J. K. 1981. The structure and stratigraphy of the Kish Bank basin. *In*: ILLING, L. V. & HOBSON, G. D. (eds) *Petroleum geology of Northwest Europe*. Institute of Petroleum, London. 426–431.

JOHNS, C. R. & ANDREWS, I. J. 1985. The petroleum geology of the Unst basin, North Sea. *Marine and petroleum geology*, **2**, 361–372.

KERR, A. C. 1993. The geochemistry and petrogenesis of the Mull and Morvern Tertiary lava succession, Argyll, Scotland. PhD thesis, University of Durham.

KIRTON, S. R. & HITCHEN, K. 1987. Timing and style of crustal extension N of the Scottish Mainland. *In*: COWARD, M. P., DEWEY, J. F. & HANCOCK, P. L. (eds) *Continental extensional tectonics*. Geological Society, Special Publication, **28**, 501–510.

KNOX, R. W. O' B., MORTON, A. C. & HARLAND, R. 1981. Stratigraphic relationships of Palaeocene sands in the UK sector of the central North Sea. *In*: ILLING, L. V. & HOBSON, G. D. (eds) *Petroleum geology of the continental shelf of Northwest Europe*. Heyden & Son, London, 521pp.

LARSEN, L. M., PEDERSEN, A. K., PEDERSEN, G. K. & PIASECKI, S. 1992. Timing and duration of Early Tertiary volcanism in the North Atlantic: new evidence from West Greenland. *In*: STOREY, B. C., ALABASTER, T. & PANKHURST, R. J. (eds) *Magmatism and the Causes of Continental Break-up*, Geological Society Special Publication, London, **68**, 321–333.

LATIN, D., NORRY, M. J. & TARZEY, J. R. E. 1993. Magmatism in the Gregory Rift, East Africa: evidence for melt generation by a plume. *Journal of Petrology*, **34**, 1007–1027.

LEMOINE, M., BAS, T., ARNAUD-VARNEAU, A., ARNAUD, H., DUMONT, T., GIDON, M., BOURBON, M., DE GRACIANSKY, P.-C., RUDKIEWICZ, J.-L., MEGARD-GALLI, J. & TRICART, P. 1986. The continental margin of the Mesozoic Tethys in the western Alps. *Marine and Petroleum Geology*, **3**, 179–199.

LEWIS, C. L. E., GREEN, P. F., CARTER, A. & HURFORD, A. J. 1992a. Elevated K/T palaeotemperatures throughout Northwest England: three kilometres of Tertiary erosion? *Earth and Planetary Science Letters*, **112**, 131–145.

——, CARTER, A. & HURFORD, A. J. 1992b. Low-temperature effects of the Skye Tertiary intrusions on Mesozoic sediments in the Sea of Hebrides basin. *In*: PARNELL, J. (ed.) *Basins on the North Atlantic Seaboard: petroleum geology, sedimentology, and basin evolution*. Geological Society, Special Publication, **62**, 175–188.

McCANN, N. 1988. An assessment of the sub-surface geology between Magilligan Point and Fair Head, northern Ireland. *Irish Journal of Earth Sciences*, **9**, 71–78.

McGEARY, S. & WARNER, M. R. 1985. Seismic profiling of the continental lithosphere. *Nature*, **317**, 795–797.

McKENZIE, D. 1978. Some remarks on the development of sedimentary basins. *Earth & Planetary Science Letters*, **40**, 25–32.

—— 1984. A possible mechanism for epeirogenic uplift. *Nature*, **307**, 616–618.

—— & O'NIONS, R. K. 1991. Partial melt distributions from inversion of rare earth element concentrations. *Journal of Petrology*, **32**, 1021–1091.

MARSDEN, G., YIELDING, G., ROBERTS, A. M. & KUSZNIR, N. J. 1990. Application of a flexural cantilever simple-shear/pure-shear model of continental lithosphere extension to the formation of the northern North Sea basin. *In*: BLUNDELL, D. J. & GIBBS, A. D. (eds) *Tectonic evolution of the North Sea rifts*. Oxford Science Publications, 240–261.

MENZIES, M. A., HALLIDAY, A. N., PALACZ, Z., HUNTER, R. H., UPTON, B. G. J., ASPEN, P. & HAWKESWORTH, C. J. 1987. Evidence from mantle xenoliths for an enriched lithospheric keel under the Outer Hebrides. *Nature*, **325**, 44–47.

MITCHELL, J. G. & MOHR, P. 1987. Carboniferous dikes of West Connacht, Ireland. *Transactions of the Royal Society of Edinburgh: Earth Sciences*, **78**, 133–151.

MORTON, A. C. 1982. Lower Tertiary sand development in the Viking Graben, North Sea. *American Association of Petroleum Geologists Bulletin*, **66**, 1542–1559.

MUDGE, D. C. & BLISS, G. M. 1983. Stratigraphy and sedimentation of the Palaeocene sands in the Northern North Sea. *In*: BROOKS, J. (ed.) *Petroleum geochemistry and exploration of Europe*. Geological Society, Special Publication, **12**, 95–111.

MUSSETT, A. E., DAGLEY, P. & SKELHORN, R. R. 1988. Time and duration of activity in the British Tertiary Igneous Province. *Geological Society, Special Publication*, **39**, 337–348.

NAYLOR, D. 1992. The post-Variscan history of Ireland. *In*: PARNELL, J. (ed.) *Basins on the North Atlantic Seaboard: petroleum geology, sedimentology, and basin evolution.* Geological Society, Special Publication, **62**, 255–275.

NELSON, K. D. 1991. A unified view of craton evolution motivated by recent deep seismic reflection and refraction results. *Geophysical Journal International,* **105**, 25–35.

PENN, I. E., HOLLIDAY, D. W., KIRBY, G. A., KUBALA, M., SOBEY, R. A., MITCHELL, W. I., HARRISON, R. K. & BECKINSALE, R. D. 1983. The Larne No. 2 borehole: discovery of a new Permian volcanic centre. *Scottish Journal of Geology,* **19**, 333–346.

ROCK, N. M. S. 1983. *The Permo–Carboniferous camptonite–monchiquite dyke-suite of the Scottish Highlands and Islands: distribution, field and petrological aspects.* Institute of geological sciences, Report 82/14, HMSO, London.

ROBERTS, D. G. 1989. Basin inversion in and around the British Isles. *In:* COOPER, M. A. & WILLIAMS, G. D. (eds) *Inversion tectonics.* Geological Society, Special Publication, **44**, 131–150.

SELLEY, R. C. & STONELEY, R. 1987. Petroleum habitat in south Dorset. *In:* BROOKS, J. & GLENNIE, K. W. (eds) *Petroleum geology of northwest Europe.* Graham & Trotman, London, 623–632.

SKOGSEID, J., PEDERSEN, T., ELDHOLM, O. & LARSEN, B. T. 1992. *Tectonism and magmatism during NE Atlantic continental break-up: the Vøring Margin.* Geological Society, Special Publication, **68**, 305–320.

STOKER, M. S. HITCHEN, K. & GRAHAM, C. C. 1993. *United Kingdom offshore regional report: The geology of the Hebrides and West Shetland shelves and adjacent deep-water areas.* London: HMSO for the British Geological Survey.

STOLPER, E. & WALKER, D. 1980. Melt density and the average composition of basalt. *Contributions to Mineralogy and Petrology,* **74**, 7–12.

TATE, M. P. & DOBSON, M. R. 1988. Syn- and post-rift igneous activity in the Porcupine Seabight Basin and adjacent continental margin W of Ireland. *In:* MORTON, A. C. & PARSON, L. M. (eds) *Early Tertiary volcanism and the opening of the N. E. Atlantic.* Geological Society, Special Publication, **39**, 309–334.

——, White, N. & CONROY, J.-J. 1993. Lithospheric extension and magmatism in the Porcupine basin west of Ireland. *Journal of Geophysical Research,* **98**, 905–13,923.

THIRLWALL, M. F. & JONES, N. W. 1983. Isotope geochemistry and contamination mechanisms of Tertiary lavas from Skye, NW Scotland. *In:* HAWKESWORTH, C. J. & NORRY, M. J. (eds) *Continental basalts and mantle nodules.* Shiva. 186–208.

THOMPSON, R. N. & MORRISON, M. A. 1988. Asthenospheric and lower lithospheric mantle contributions to continental extensional magmatism: an example from the British Tertiary Province. *Chemical Geology,* **68**, 1–15.

——, DICKIN, A. P., GIBSON, I. L. & MORRISON, M. A. 1982. Elemental fingerprints of isotopic contamination of Hebridean Palaeocene mantle-derived magmas by Archaean Sial. *Contributions to Mineralogy and Petrology,* **79**, 159–168.

THORPE, R. S., TINDLE, A. G. & GLEDHILL, A. 1990. The petrology and origin of the Tertiary Lundy Granite. *Journal of Petrology,* **31**, 1379–1406.

UPTON, B. G. J. 1988. History of Tertiary igneous activity on the North Atlantic borders. *Geological Society, Special Publication,* **39**, 429–453.

VISSCHER, H. 1971. The Permian ans Triassic of the Kingscourt outlier, Ireland. *Geological Survey of Ireland Special Paper* **1**.

WARNER, M. 1990. Basalts, water, or shear zones in the lower crust? *Tectonophysics,* **173**, 163–174.

WATSON, J. 1985. Northern Scotland as an Atlantic-North Sea divide. *Journal of the Geological Society, London,* **142**, 221–243.

WHITE, R. S. 1992. Crustal structure and magmatism of North Atlantic continental margins. *Journal of the Geological Society, London,* **149**, 841–854.

—— & MCKENZIE, D. 1989. Magmatism at rift zones: The generation of volcanic continental margins and flood basalts. *Journal of Geophysical Research,* **94**, 7685–7729.

——, SPENCE, G. D., FOWLER, S. R., MCKENZIE, D. P., WESTBROOK, G. K. & BOWEN, A. N. 1987. Magmatism at rifted continental margins. *Nature,* **330**, 439–444.

——, MCKENZIE, D. & O'NIONS, R. K. 1992. Oceanic crustal thickness from seismic measurements and rare earth element inversions. *Journal of Geophysical Research,* **97**, 19,683–19,715.

WOODLAND, A. W. 1979. *Sub-Pleistocene geology of the British Isles and the adjacent continental shelf.* Institute of Geological Sciences, Second edition.

ZIEGLER, P. A. 1982. *Geological atlas of Western and Central Europe.* Shell Internationale Maatschappij B. V., The Hague.

Mechanics of basin inversion from worldwide examples

JAMES D. LOWELL

2200 W. Berry Ave., Littleton, CO 80120, USA

Abstract: Inversion is dependent on pre-existing basin configuration in the initial subsidence, usually extensional phase, and the resolution of compressional forces in the later shortening phase. External horizontal, rather than isostatic vertical forces are required for inversion because many deep sediment-filled basins around the world have never been inverted, but should have been, if isostatic rebound were the driving mechanism.

Inversion can range in scale from basins to sub-basins to selected structures within basins. Typically, rift basins can be later inverted. Mainly by reactivation of older normal faults, inversion selects rift basins where, in pure shear, weakening because of necking or thinning of lithosphere has occurred; and where, in simple shear, mechanical detachment surfaces are available for subsequent movement. Some pre-existing lows or sags can apparently be inverted in the absence of reactivated normal faults, as in the southern Altiplano of Bolivia and offshore Sabah, Borneo.

Basins can be inverted by dominantly strike-slip with some convergent component, e.g. offshore northeast Brazil, and by almost direct compression, Atlas Mountains, Morocco. A prime difference is that convergent strike-slip can reactivate relatively high-angle normal faults as reverse faults, whereas these faults tend to lock when subjected to more direct compression which then creates lower angle contractional faults. Usually, inversion is caused by a combination of compression and strike-slip (transpression) because the azimuth of maximum principal compressional stress to the direction of original basin trend vectors into an oblique-slip component. This is illustrated in northern Argentina where compression derived from Andean deformation is resolved obliquely (about 45°) against an older rift basin in the foreland nearby. Interestingly, for this area at least, the 45° angle which affords an equal contribution of strike slip and compression results only in reactivation of older normal faults and does not create younger contractional features.

Apparently compressive forces can be transmitted backward from the lead edge of an underthrusted foreland plate to invert rather remote regions and basins that are carried on that plate; this may be a significant part of the mechanism for basin inversion in northwest Europe and the southern North Sea as that foreland region underthrusted along the Alpine system.

Inverted basins[1] and their associated structures have been recognized on every continent that has been explored for petroleum. Recognition has most often been by reflection seismic profiling which is ideally suited to provide the regional scope and subsurface detail needed to document these features. While occurrences of inversion are universally accepted, the mechanisms, whether ultimately vertical or horizontal forces were responsible, have been debated.

The idea that structural inversion could be caused by more or less vertical forces related to isostatic rebound of depressed relatively light sediments (Voigt 1962) is not tenable because many deep sediment-filled basins around the world have never been inverted. This is nowhere better demonstrated than in the Dneiper–Donets basin in Russia where up to 18 km of sediment that accumulated in Devonian, Carboniferous and Early Permian times remain depressed (Ulmishek *et al.* 1994). Only where it is crossed at right angles by the Donbas fold belt (Fig. 1), has inversion occurred; otherwise rocks of the NW-trending rift basin remain undeformed, likely because compression is dominantly parallel to the trends of the boundary faults of the rift (see discussion of Transpressive-driven Inversion).

It seems inescapable that external horizontal or far-field forces along and transmitted within tectonic plates are required for inversion. This paper emphasizes that the most common horizontal driving force of inversion is transpression, or the combination of compression and strike-

---

[1] Basin inversion can be defined as 'a basin controlled by a fault system that has been subsequently compressed–transpressed producing uplift and partial extrusion of the basin fill' (Cooper *et al.* 1989). Original normal faults are often, but not always, reactivated as reverse faults.

From BUCHANAN, J. G. & BUCHANAN, P. G. (eds), 1995, *Basin Inversion*,
Geological Society Special Publication No. 88, 39–57.

Fig. 1. Map showing relationship of Donbas fold belt to Dnieper–Donets basin, Ukraine. Only where fold belt crosses basin does inversion occur (Ulmishek *et al.* 1994, reprinted by permission).

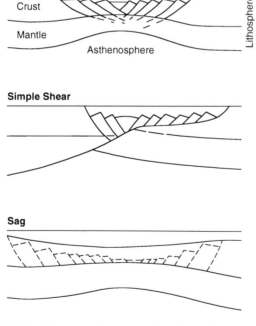

Fig. 2. Diagrammatic cross-sections illustrating pre-inversion basins formed by pure shear and simple shear that can be subsequently inverted. The observation that some areas are inverted in the absence of earlier normal faults (see text discussion) suggests that sag basins or sub-basins can also be inverted.

slip. Deep-seated transpression can convert to substantial uplift of the region above.

Pre-existing basins and scale of deformation

Typically, earlier extensional or rift basins with bounding normal faults are subject to later inversion (Fig. 2). In such basins formed by pure shear, coincident thinning of the lithosphere creates areas of weakness that are selectively shortened and inverted. In simple shear lithospheric thinning also occurs, but the locus of thinning is offset from the overlying basin and it is probable that a mechanical detachment surface guides basin inversion (Gibbs 1987). In some cases, the detachment may be a thrust surface related to a still earlier period of compression. In both pure and simple shear, listric normal faults are ideally suited to be reactivated as thrusts, along the lower, flatter portion of the listric fault, with associated uplift in the higher, steeper part (Letouzey *et al.* 1990).

In basins formed by transtension, those segments formed by listric normal faults are also prone to reactivation during inversion. However, the steep, strike-slip segments are more difficult to invert, especially by direct compression.

Some local or sub-basinal areas are inverted even though no recognizable earlier normal

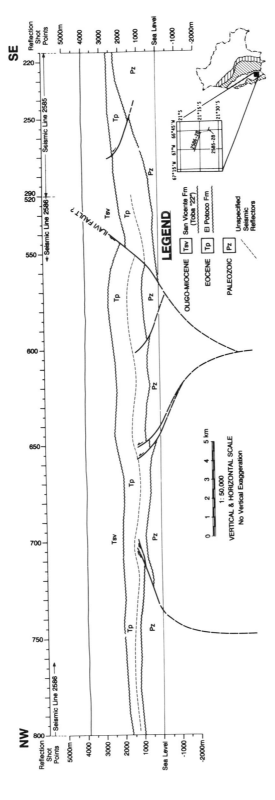

Fig. 3. Cross-section controlled by seismic lines showing inversion of Eocene section (El Potoco Formation) in former sag(?) basin of southern Altiplano of Bolivia on faults paralleling and approximately 30 km east of the NNE-trending Uyuni–Keniani wrench fault.

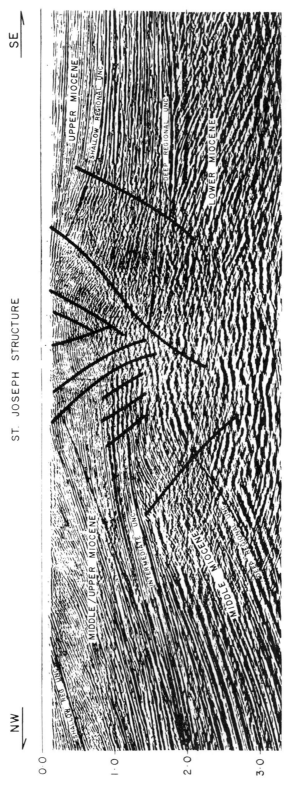

Fig. 4. Seismic section (migrated) across St. Joseph structure, offshore Sabah, demonstrating inversion of mainly Middle Miocene section interpreted to have originally been deposited in a sag area. Line is c. 18 km across. (Section from Bol & Van Hoorn 1978.)

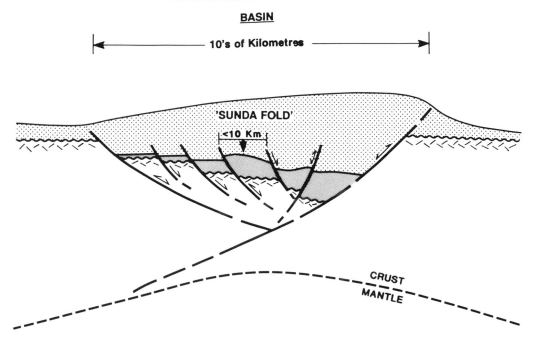

Fig. 5. Diagrammatic cross-section showing coincident inversion at both individual structure and basinal scales. Note that all pre-existing normal faults are not shown to be reactivated.

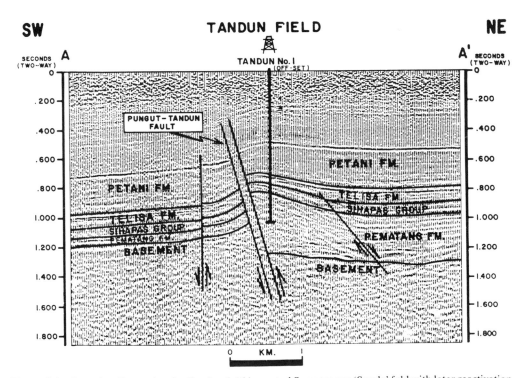

Fig. 6. Seismic section illustrating the Tandun field in central Sumatra as a 'Sunda' fold with later reactivation of the Pungut–Tandun fault that earlier effected drastic thickening of the Pematang Formation (after Mertosono 1975). Inversion is considered to be strike-slip dominant as reactivation is concentrated on normal faults and no later contractional faults are present.

Fig. 7. Seismic line from offshore Gippsland basin, southeastern Australia showing that during inversion of Latrobe group (Upper Cretaceous–Upper Eocene) only faults designated by number '2' were reactivated; faults with number '1' remain normal faults (Davis 1983).

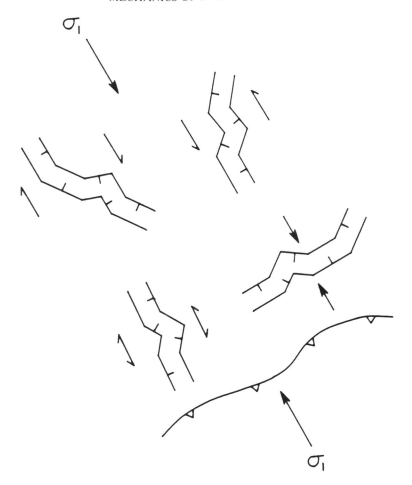

Fig. 8. Map demonstrating range of possible angles of incidence of compression (during later inversion stage) to original rift-bounding faults. σ_1 is maximum principal compressive stress (see text for additional discussion).

faults are available for reactivation. This can be observed in the southern Altiplano region of Bolivia (Fig. 3) and in offshore Sabah, Borneo (Fig. 4) where original sedimentary thickening into the presently inverted area appears to have taken place independently of any normal faults. Subsequent transpression through nearly the centre of the region caused inversion that was characterized by broad positive flower structures and most, if not all faults were created at the time of inversion and had no earlier history. It is possible that this phenomenon can be extrapolated to the scale of sag basins where normal faults did not play a part in original thickening (Fig. 2).

Inversion at a basinal scale, on the order of tens of kilometres in width, probably requires prior crustal thinning (Fig. 5). Inversion of individual anticlines (Sunda folds of Eubank &

Makki 1981; Fig. 6), generally much less than 10 km wide, can take place by reactivation of individual faults within a basin (Fig. 5). It is one of the enigmas of inversion, however, that some older normal faults are reactivated and others having the same attitude and presumably the same orientation are not. This is rather common behaviour of faults during inversion (Fig. 7) that has received very little study.

Transpressive-driven inversion

Deformational forces responsible for inversion are oriented from 0–90° to the pre-existing basin-bounding faults. The zero direction is pure strike-slip that is not favourable in effecting inversion. Clay model experiments (Lowell 1985, figs 10–13, 10–14, 10–15) show that pure strike-slip superposed on an earlier normal fault

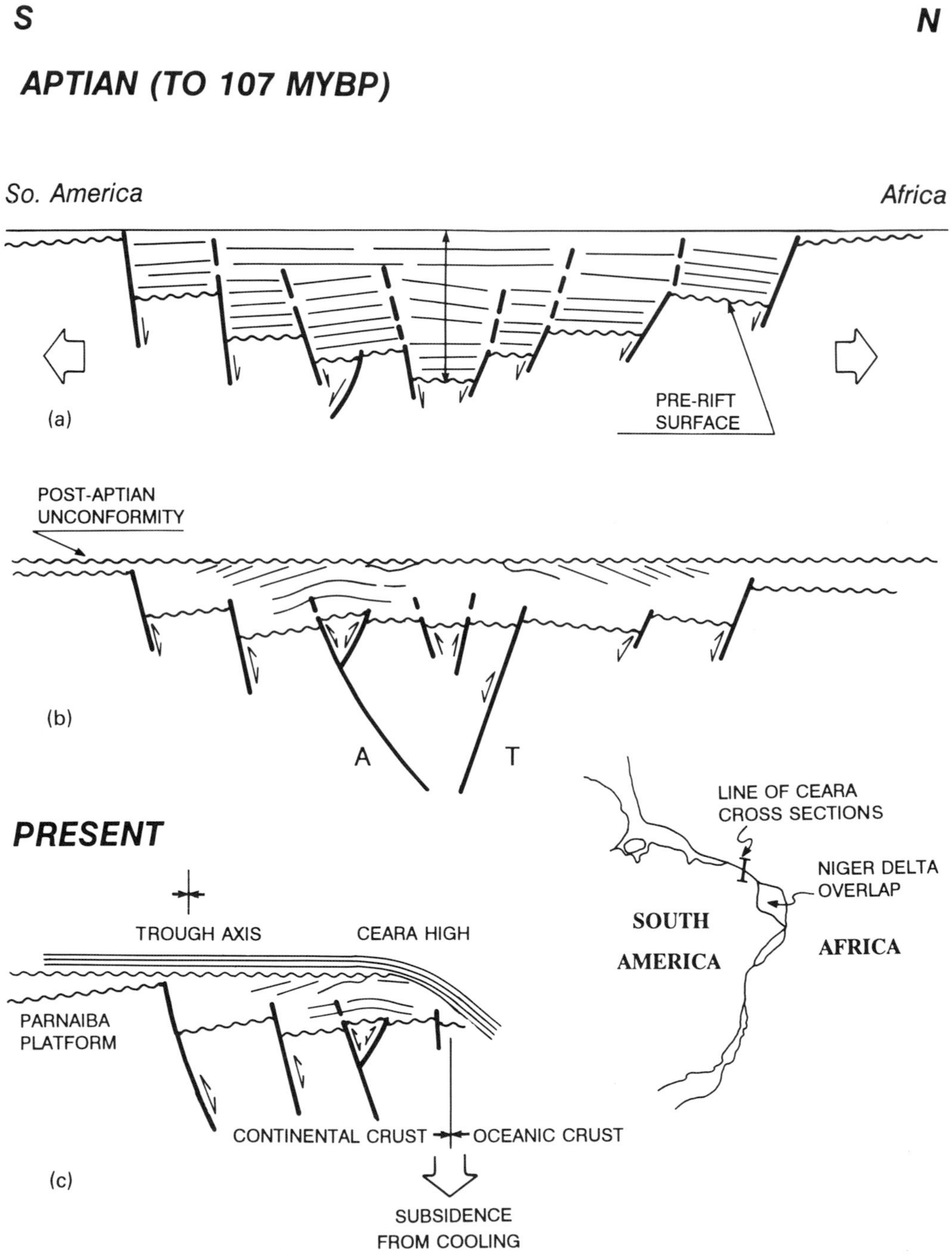

S   **N**

APTIAN (TO 107 MYBP)

So. America   *Africa*

(a)

PRE-RIFT
SURFACE

POST-APTIAN
UNCONFORMITY

(b)

A   T

PRESENT

TROUGH AXIS   CEARA HIGH

PARNAIBA
PLATFORM

LINE OF CEARA
CROSS SECTIONS

NIGER DELTA
OVERLAP

SOUTH

AMERICA

AFRICA

CONTINENTAL CRUST ⟷ OCEANIC CRUST

(c)

SUBSIDENCE
FROM COOLING

Fig. 9. Cross-sections (based on reflection seismic lines) and pre-drift index map showing evolution of northeastern Brazilian continental margin as strike slip-dominant in a transpressional regime. (**a**) Aptian rift between South America and Africa; (**b**) inversion and erosion of upper Aptian section below a post-Aptian unconformity; (**c**) present continental margin with post-Aptian upbuilding and prograding section related to thermal subsidence as Africa moved out of plane of cross-section.

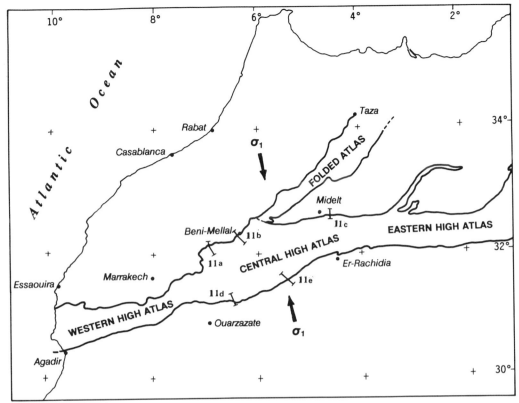

Fig. 10. Map of Morocco showing Tertiary Atlas Mountain trends that reflect earlier, Mesozoic pattern of rifting. Direction of maximum principal compressive stress indicated by σ_1.

fabric is simply resolved as horizontal slip on the old normal fault surfaces with little to no evidence manifested of the compressional component of strike-slip. Compression at 90° to existing structures should be highly effective in inversion (Letouzey *et al.* 1990), but instances in which later compression is precisely normal to the original multi-trended boundary faults are probably rare (Fig. 8). Thus, transpression is usually the most common cause of inversion.

Difficulty arises, however, in determining the relative contribution of strike-slip and compression. Strain measurements in outcrop studies could determine azimuth of slip, but the precise resolution of slip direction from 2D reflection seismic sections is not possible. However, it is reasonable to speculate that when steep pre-existing faults are reactivated during inversion, a strong strike-slip component of motion must be operative in as much as these faults would tend to lock when subjected to direct compression. Conversely, inversion structures marked by little or no reactivation of earlier normal faults and creation of younger

low-angle thrusts must reflect a dominant element of compression. Occurrences of strike slip-dominant and compression-dominant inversion are illustrated by the following examples.

Strike slip-dominant inversion

The offshore Ceara Piaui basin in northeast Brazil is an area of inversion in which plate kinematics indicate that strike slip must have been dominant in the transpressive process. A very thick rift section accumulated between the northeastern Brazilian portion of South America and adjacent Africa during Aptian time (Ponte & Asmus 1978). Then, as South America and Africa separated, the offshore area underwent convergent right-lateral movement that strongly inverted the earlier formed rift section stripping off much of the Aptian section beneath the post-Aptian unconformity (Fig. 9). Dominance of right-lateral motion was required as South America and Africa moved past one

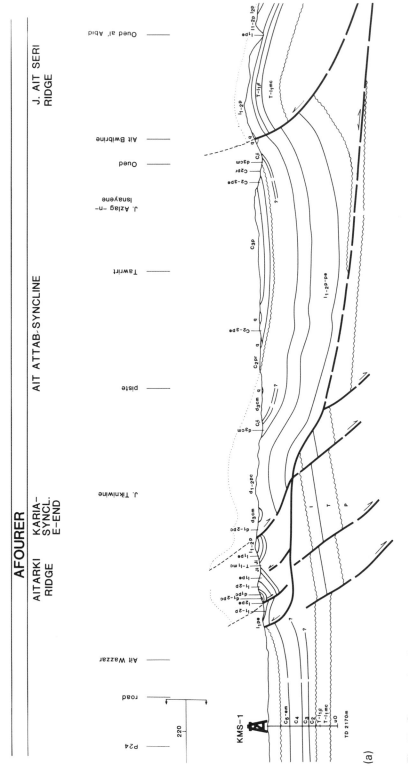

Fig. 11. Cross-sections interpreting low-angle thrusts on north and south sides of inverted compression-dominant High Atlas Mountains, Morocco. North: (**a**) Afourer, (**b**) Beni Mellal, (**c**) Tarhemt Pass; South: (**d**) Toundout–Boumalne, (**e**) Est de Tinerhir. Note that normal faults are not reactivated in 11(a) and 11(c). Available seismic and well control shown (Bennett *et al.* 1992). See Fig. 10 for location of sections.

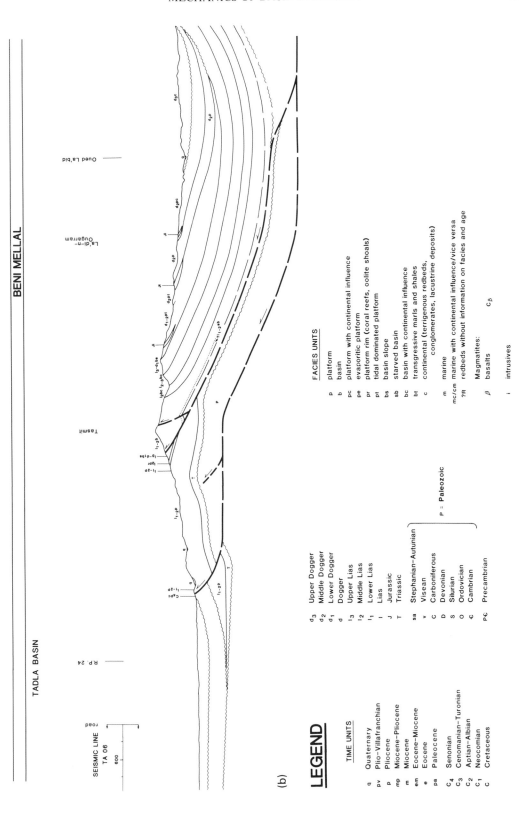

(b)

LEGEND

TIME UNITS

q	Quaternary
pv	Plio–Villafranchian
p	Pliocene
mp	Miocene–Pliocene
m	Miocene
em	Eocene–Miocene
e	Eocene
pa	Paleocene
c₄	Senonian
c₃	Cenomanian–Turonian
c₂	Aptian–Albian
c₁	Neocomian
c	Cretaceous

d₃	Upper Dogger
d₂	Middle Dogger
d₁	Lower Dogger
d	Dogger
l₃	Upper Lias
l₂	Middle Lias
l₁	Lower Lias
l	Lias
J	Jurassic
T	Triassic
sa	Stephanian–Autunian
v	Visean
C	Carboniferous
D	Devonian
S	Silurian
O	Ordovician
€	Cambrian
P€	Precambrian

P = Paleozoic

FACIES UNITS

p	platform
b	basin
pc	platform with continental influence
pe	evaporitic platform
pr	platform rim (coral reefs, oolite shoals)
pt	tidal dominated platform
bs	basin slope
sb	starved basin
bc	basin with continental influence
bt	transgressive marls and shales
c	continental (terrigenous redbeds, conglomerates, lacustrine deposits)
m	marine
mc/cm	marine with continental influence/vice versa
?R	redbeds without information on facies and age

Magmatites:
β	basalts Cβ
i	intrusives

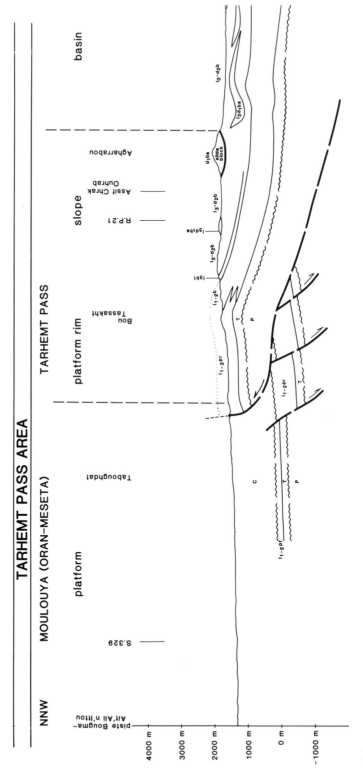

(c)

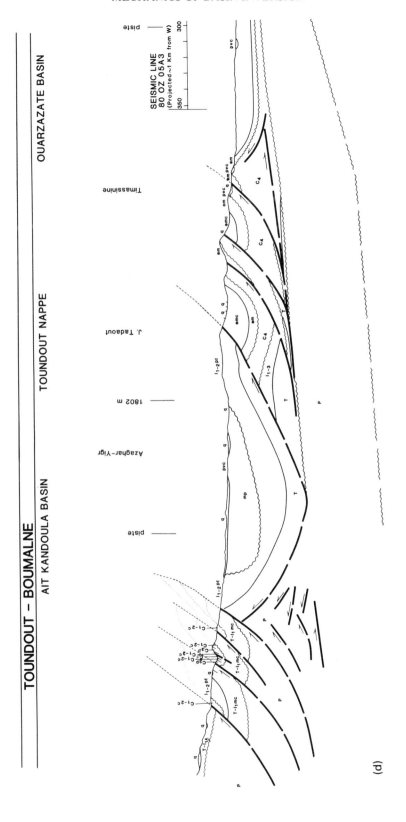

(d)

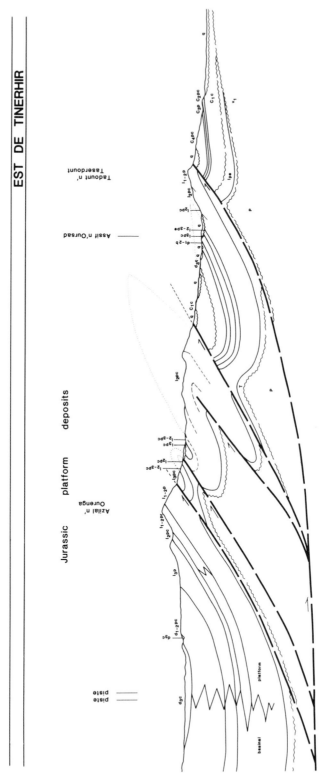

(e)

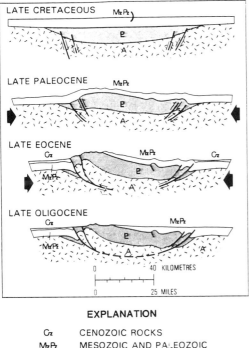

EXPLANATION

Cz	CENOZOIC ROCKS
MzPz	MESOZOIC AND PALEOZOIC ROCKS
P	PROTEROZOIC ROCKS
A	ARCHEAN ROCKS

——————— CONTACT

——————— FAULT – Queried where inferred

Fig. 12. North–south cross-section (Hansen 1986) showing inversion of the Uinta Mountains of northeast Utah as compression dominant (see text for additional discussion).

another. Inversion was concentrated on relatively high-angle, pre-existing normal faults further pointing to the importance of a strike-slip component in transpression.

Compression-dominant inversion

The High Atlas Mountains of Morocco are a prime example of compression-dominant transpression. During final convergence of Africa and Iberia mainly in Miocene time, NNW–SSE compression (Brede *et al.* 1992) inverted the ENE–WSW-trending Triassic–Jurassic Atlas rift creating low-angle thrusts present on both the north and south sides of the High Atlas (Bennett *et al.* 1992; Figs 10, 11). Some of the original rift-bounding normal faults were not apparently reactivated (Fig. 11a, c), as thrusts, more typical of compression were formed.

The Unita Mountains of northeast Utah also were dominantly compressed during inversion. Situated in the foreland of the Wyoming–Utah–Idaho thrust-fold belt, the Uintas have been considered a Proterozoic aulacogen that was inverted in late Laramide (early–middle Eocene) time (Fig. 12; Hansen 1986; Stone 1989). The Unitas have an E–W trend that was no doubt established during Proterozoic rifting. The much later Laramide inversion has been attributed to a maximum principal compressive stress oriented either N–S (Gries 1983) or NE–SW (Stone 1989). Whether a product of direct N–S compression or regional NE–SW compression that by strain partitioning is resolved as local N–S compression (see Varga 1993), the inversion of the Unitas has been almost solely by compression. This is further evidenced by: (1) extreme reactivation of the older normal faults and possibly creation of new thrusts so that several kilometres of thrust overlap occurred; and (2) outcrops of Palaeozoic and Mesozoic rocks continuous and not offset around the western end of the Uintas demonstrating that later strike slip parallel to the original rift basin was negligible (first noted by Kaspar Arbenz, pers. comm.).

Strike slip and compression approximately equal

The Upper Cretaceous Lomos de Olmedo rift basin in Salta province in northern Argentina appears to have undergone roughly equal strike slip and compression, i.e. the azimuth of compression was about 45° to the trend of the rift basin. Figure 13 shows Upper Cretaceous (F. Yacoraite) thickening to the northwest across two former normal faults that were reactivated during Andean orogenesis. The NE–SW striking rift basin trends beneath the NNE–SSW trending Andean thrust front at about 45°, suggesting that strike-slip and compressional components are essentially equal. Of interest here is that all inversion has taken place on the normal faults; no additional compressional features were formed implying that, at least for this area if not others, a compressional component exceeding 45° is required for the creation of new contractional structures.

The close proximity of the older rift to the Andes and the very late inversion that is consistent with the young age of Andean deformation demonstrates conclusively a cause and effect relation between Andean compression and basin inversion. In the examples to follow, the distance of rift or structural features

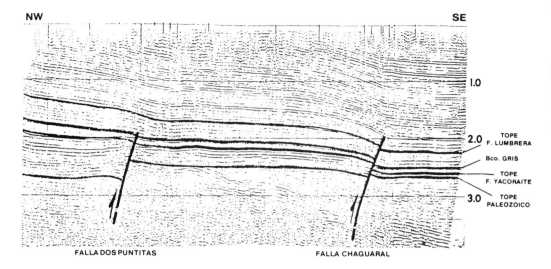

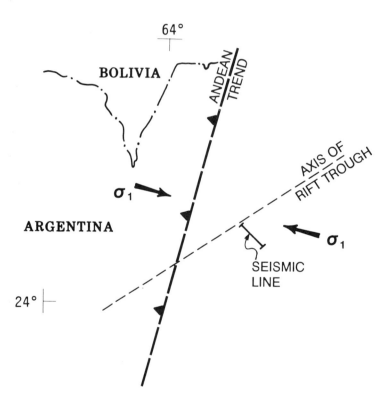

Fig. 13. Seismic line from Salta Province, northern Argentina showing inversion along SE part of Lomos de Olmedo rift basin. Tope F. Lumbrera is approximate top of Palaeocene. Tope F. Yacoraite is approximate top of Cretaceous (Bianucci *et al.* 1982). Dos Puntitas and Chaguaral faults are believed to be listric at depth. Location map shows relation of rift and Andean trends, probable σ_1, and approximate position of seismic section.

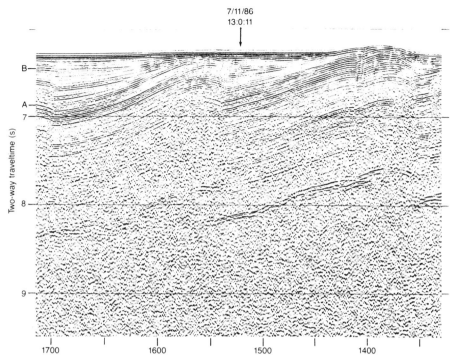

Fig. 14. Single-channel migrated seismic line showing compressive blocks on the Indian Ocean floor about 800 km south of Sri Lanka. Basement is prominent reflector on either side of 8 sec. Unconformity A is Late Miocene, B within the Pleistocene. Approximate 3 : 1 squeeze. South to the right. Line *c*. 10 naut. mi. (17.6 km) across. (Cochran *et al*. 1989, fig. 12, reprinted by permission.)

to the locus of deformation for inversion is much greater, yet a cause and effect relationship no doubt exists.

Possible role of underthrusting in inversion

As no criteria exist to distinguish underthrusting from overthrusting, partiality for the former is a matter of personal preference. Underthrusting, however, can be easier to conceptualize. Rather than deformation occurring by a 'push from behind' with resultant transmission of stress for great distances, it is easier to conceive of stresses accumulating in an underthrust plate as it encounters resistance at its lead edge. Hence, although they are not inversion structures, compressive blocks (Fig. 14) on the Indian Ocean plate more than 3000 km from the Himalayas have been interpreted as being caused by resistance of the Indian Ocean plate as it underthrusted Asia (Cochran *et al.* 1989). It is almost impossible to conceive of compressive stresses being transmitted this distance by push from Asia along the line of plate interaction; accumulation of compressive stress backward

(southward) in the underthrust Indian Ocean plate seems easier to visualize.

Similarly in northwest Europe and the southern part of the North Sea, inversion-causing compressive stresses may have been stored back (to the NW) in the European foreland that underthrusted along the Alpine system (Fig. 15). It would follow that the amount and intensity of inversion should generally decrease northwestward from the Alpine deformation front and this appears to be the case. Recent maximum principal compressive stress in northwestern Europe is known to be horizontal and aligned NNW–SSE (Griener & Lohr 1979; Froidevaux *et al.* 1979; Betz *et al.* 1987) so rift troughs normal to this direction should receive maximum inversion. Figure 8 was designed for the northwest European foreland to show how differently orientated rifts and various fault segments comprising the rifts present different angles of incidence to compression. Some rifts would experience dominant compression and others dominant strike-slip. Owing to changes in rift trend and variations of rift fault patterns, however, transpression is the most common

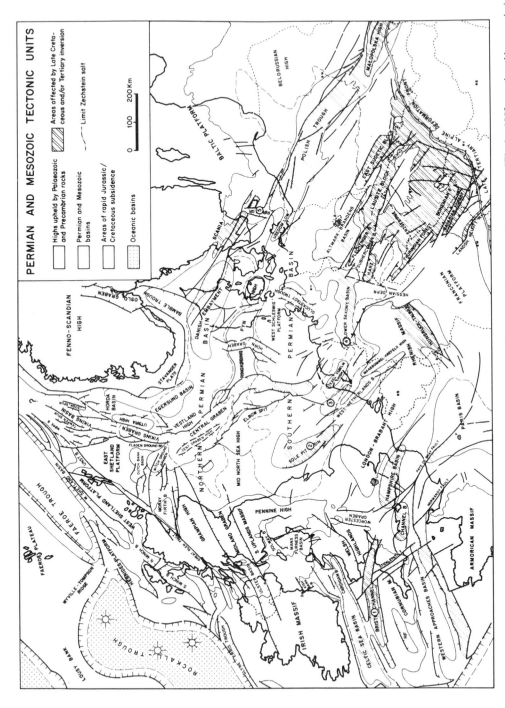

Fig. 15. Map showing extent of inversion in Alpine foreland with Late Tertiary Alpine deformation front in southeast corner (Ziegler 1983, reprinted by permission).

agent for inversion. Even in the example of direct (90°) compression against the rift, some transpression is inevitable where dog-leg bends are present in the rift. Transpression must also occur along dog-legs in the dominantly strike-slip example.

Conclusions

As suspected intuitively and suggested by clay models, pure strike-slip operating parallel to an earlier rift basin or a smaller normal-faulted associated structure is not effective in inversion; compression normal to such features very effectively inverts them. Given the variety of pre-existing basinal trends to the azimuth of later compression, it is likely that most inversion is caused by components of both compression and strike slip, i.e. transpression. Empirical evidence shows that where the strike-slip component is dominant, relatively high-angle normal faults can be reactivated and new structures need not be created. Where compression is dominant, parts of older normal faults may be rejuvenated, but more importantly, contractional structures are formed.

References

BENNETT, R. L., SCHUEPBACH, M. A., HAUPTMANN, M. K. & LOWELL, J. D. 1992. The Moroccan High Atlas interpreted as a low angle thrust belt and its relation to structural traps. *First Conference on Petroleum Exploration, ONAREP, Rabat.* 37.

BETZ, D., FUHRER, F. GRIENER, G. & PLEIN, E. 1987. Evolution of the Lower Saxony basin. *Tectonophysics,* **137**, 127–170.

BIANUCCI, H., HOMOVC, J. F. & ACEVEDO, O. M. 1982. Inversion tectonica y plegamientos resultantes en la comarca Puesto Guardian-Dos Puntitos, Dpto. Oran, Provincia de Salta. *5 Congreso Latinoamericano de Geologia.* Instituto Argentino del Petróleo. 23–30.

BOL, A. J. & VAN HOORN, B. 1978. Structural styles in western Sabah offshore. *Geological Society Malaysia, 2nd Petroleum Seminar, Kuala Lumpur.*

BREDE, R., HAUPTMANN, M. & HERBIG, H. B. 1992. Plate tectonics and intracratonic mountain ranges in Morocco – The Mesozoic–Cenozoic development of the Central High Atlas and the Middle Atlas. *Geologische Rundschau,* **81**, 127–141.

COCHRAN, J. R., STOW, D. A. V., *et al.* 1989. *Proceedings of the ODP, Initial Reports, 116.* College Station, TX (Ocean Drilling Program), 197–210.

COOPER, M. A. *et al.* 1989. Inversion tectonics – a discussion. *In*: COOPER, M. A. & WILLIAMS, G. D. (eds) *Inversion Tectonics.* Geological Society, London, Special Publication, **44**, 335–347.

DAVIS, P. N. 1983. Gippsland Basin, Southeastern Australia. *In*: BALLY, A. W. (ed.) *Seismic expression of structural styles.* American Association Petroleum Geologists, Studies in Geology, **15**, vol. 3.

EUBANK, R. T. & MAKKI, A. C. 1981. Structural geology of the central Sumatra back-arc basin. *Indonesian Petroleum Association, 10th Anniversary Convention.*

FROIDEVAUX, C., PAQUIN, C. & SOURIAU, M. 1979. Tectonic stresses in France. *American Geophysical Transactions,* **60, 32**, 607.

GIBBS, A. D. 1987. Basin development, examples from the United Kingdom and comments on hydrocarbon prospectivity. *Tectonophysics,* **133**, 189–198.

GRIENER, G. & LOHR, J. 1979. Tectonic stresses in the northern foreland of the Alpine system measurements and interpretation. *American Geophysical Union Transactions,* **60, 32**, 607.

GRIES, R. 1983. North–South compression of Rocky Mountain foreland structures. *In*: LOWELL, J. D. (ed.) *Rocky Mountain foreland basins and uplifts.* Rocky Mountain Association Geologists, 9–32.

HANSEN, W. R. 1986. *Neogene tectonics and geomorphology of the eastern Uinta Mountains in Utah, Colorado and Wyoming.* United States Geological Survey, Professional Paper 1356, 78.

LETOUZEY, J., WERNER, P. & MARTY, A. 1990. Fault reactivation and structural inversion. Backarc and intraplate compressive deformations. Example of the eastern Sunda shelf (Indonesia). *Tectonophysics,* **183**, 341–362.

LOWELL, J. D. 1985. *Structural styles in petroleum exploration.* Oil and Gas Consultants International Inc., Tulsa, 460.

MERTOSONO, S. 1975. Geology of the Pungut and Tandun oil fields. *Indonesian Petroleum Association, Proceedings of the 4th Annual Convention* I, 165–179.

PONTE, F. C. & ASMUS, H. E. 1978. Geological framework of the Brazilian continental margin. *Geologische Rundschau,* **67, 1**, 201–235.

STONE, D. S. 1989. Uinta Mountains: An inverted Proterozoic aulacogen. *Rocky Mountain Association Geologists Outcrop*, 5 (October).

ULMISHEK, G. F., BOGINO, V. A., KELLER, M. B. & POZNYAKEVICH, Z. L. 1994. Structure, stratigraphy, and petroleum geology of the Pripyat and Dnieper–Donets basins, Byelarus and Ukraine. *In*: LANDON, S. M. (ed.) *Interior rift basins.* American Association Petroleum Geologists, Memoir, **59**, 125–156.

VOIGT, E. 1962. Über Randtroge vor Schollenrander und ihre Bedeutung im Gebiet der Mitteleuropaischen Senke und Angrenzender Gebiete. *Zeitschrift Deutsche Geologische Gesellschaft,* **114**, 378–418.

VARGA, R. J. 1993. Rocky Mountain foreland uplifts: Products of rotating stress field or strain partitioning. *Geology,* **21**, 1115–1118.

ZIEGLER, P. A. 1983. Inverted basins in the Alpine foreland. *In*: BALLY, A. W. (ed.) *Seismic expression of structural styles.* American Association Petroleum Geologists, Studies in Geology, **15, 3**.

Basin inversion in the Gulf of Suez: implications for exploration and development in failed rifts

S. D. KNOTT, A. BEACH, A. I. WELBON & P. J. BROCKBANK

Alastair Beach Associates, 11 Royal Exchange Square, Glasgow G1 3AJ, UK

Abstract: The Gulf of Suez (Clysmic) rift is a classic and well documented extensional province and has many similarities to the North Sea rift to which it is often compared. New structural mapping and fault analysis in western Sinai have shown that post-rift crustal shortening and inversion occurred across the Gulf of Suez, demonstrated by the presence of reverse faults formed as a result of overturning of normal faults, widespread sub-horizontal hydraulic fractures due to vertical stretching, low-angle thrusting and associated folding in the hangingwalls of normal faults, and steepening of bedding dips adjacent to basement stress risers, in some cases overturning stratigraphy. These features show the effect that superposition of inversion on extensional structures has on the geometry of the bedding in footwalls and hangingwalls. Observations and data from outcrop can be used to assist in the interpretation of similar structural relations in the North Sea, where post-rift contraction is also seen. Recognition of inversion-related structures is important in exploration, appraisal and development of North Sea reservoirs, for example, in estimating palaeobathymetry of Late Jurassic depocentres, in constructing cross-sections for reservoir zonation, and in evaluating fault seal and reservoir compartmentalization. Shortening in the Gulf of Suez occurred immediately after rift failure and was probably a gravity-driven process. Similar processes may have occurred in other failed rifts.

The Gulf of Suez (Clysmic) rift (Fig. 1) (Robson 1971) lies between the Eastern Desert of Egypt and the Sinai peninsula. The rift trends NW–SE and terminates at a triple junction between the Red Sea to the south and the Gulf of Aqaba transform to the east. In the north, the rift is buried beneath sediments of the Nile delta and the continuation of the rift into the Mediterranean is therefore difficult to image on seismic data. The regional structure of the rift is an elongate ribbon of thinned and subsided lithosphere bounded on either side by massifs of relatively thick crust (Steckler 1985). The across strike structure of the rift is characterized by large, widely spaced (10 to 20 km) tilted fault blocks with eroded crests and deep hangingwall basins. Along strike the polarity of the rift changes with eastward dipping faults in the north, westward dipping faults in the central part (around Hammam Faraun) and faults dipping eastward in the south (around Gebel Zeit) (Fig. 1) (Colletta *et al.* 1988). Faults in the Clysmic rift have two main trends; a NW trending set of faults – the Clysmic trend, and a NE to NNE trending set – the Aqaba trend. Both trends are present in the study area and the majority of structures studied have the Clysmic trend.

Field work was carried out along the western margin of the Sinai peninsula (Fig. 1) stretching from Wadi Tayiba in the north to Gebel Ekma in the south, some 80 km, and from Abu Zenima in

the west to Wadi Feiran in the east, roughly 40 km. Reconnaissance mapping was carried out over a larger area. The stratigraphical succession of this part of the Gulf of Suez rift is summarized in Fig. 2. The Nubian Sandstone Formation (?Cambrian to Early Cretaceous, divisions D to A; Robson 1971) lies unconformably upon Precambrian basement. The lower part of the Nubian stratigraphy contains distinctive manganiferous beds interbedded with limestone and red and black shale. These beds proved useful in across fault correlation and the estimation of fault throws. Cenomanian to Maastrichtian limestone and chalk are overlain conformably by the Esna Shale Formation (Palaeocene) and the Thebes Limestone Formation (Eocene).

A major regional erosional unconformity overlies the preceding succession which is essentially pre-rift. The unconformity is of probable Oligocene age. The Abu Zenima Formation (Oligocene) has a highly localized distribution immediately above the unconformity and is locally volcaniclastic, particularly in the northern part of the study area. Also above the unconformity, locally developed, lie basic igneous volcanics with associated intrusives. The conglomeratic Nukhul Formation (Early Miocene) overlies the Oligocene unconformity representing the early stages of the syn-rift sedimentary fill.

The main episode of rifting occurred during

From BUCHANAN, J. G. & BUCHANAN, P. G. (eds), 1995, *Basin Inversion*, Geological Society Special Publication No. 88, 59–81.

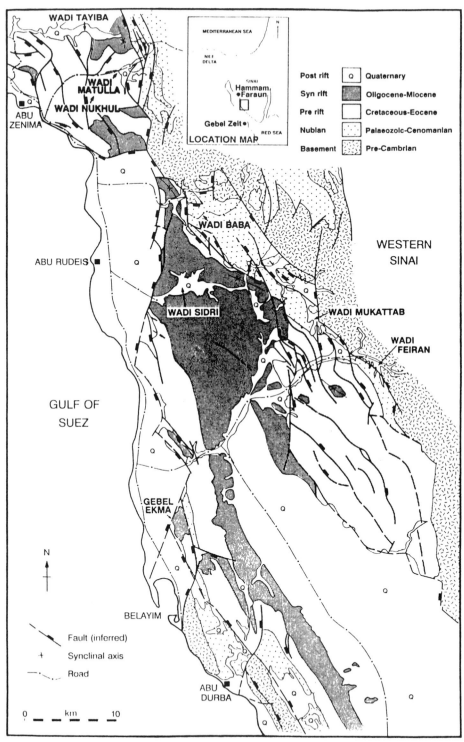

Fig. 1. Geological map of western Sinai.

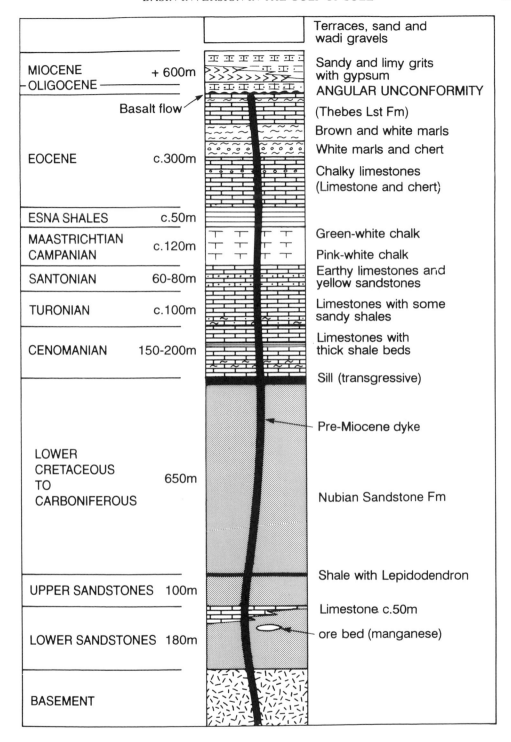

Fig. 2. Stratigraphical succession of the Gulf of Suez (Clysmic) rift, after Robson (1971).

deposition of the Rudeis Formation (Steckler *et al*. 1988). The upper and lower Rudeis are separated by an angular unconformity associated with the mid-Clysmic event during which a reorganization of active fault bocks occurred (Garfunkel & Bartov 1977). The upper Rudeis, Kareem and Belayim Formations all form the Abu Alaqa Group (Steckler *et al*. 1988, Gawthorpe *et al*. 1990). Subsidence in the Gulf of Suez slowed during deposition of the Kareem Formation and at this time the Gulf of Aqaba was the main northern plate boundary (Steckler *et al*. 1988).

The Abu Alaqa Group is overlain conformably by hundreds of metres of post-rift succession, comprising Miocene evaporite interbedded with siliciclastics, which thickens to several kilometres in the centre of the rift to the west. Another marked erosional unconformity occurs above the tilted Miocene succession which is overlain by Pliocene to Pleistocene sands and gravels which are themselves locally tilted and incised by Recent uplift and erosion.

Regional structural history of the Gulf of Suez

As mentioned above the Gulf of Suez (Clysmic) rift underwent a relatively complex tectonic history, which may be indicative of the type of history experienced by other, similar rifts, such as the North Sea. The Gulf of Suez is not necessarily a special case, and many of the phenomena, including immediate post rift shortening, may be present in other failed rifts. The deformation sequence started with uplift and erosion in Oligocene time, accompanied by volcanism, followed by extension in early Miocene time. Rifting in the Gulf of Suez was followed, very closely in time, by an episode of shortening. A more detailed review of the evidence for shortening is given later.

Although the sediments older than Miocene are considered to be pre rift, due to their relatively constant thickness over much of Egypt (Robson 1971), there is some evidence for active faulting during the deposition of the Nubian Sandstone Formation – in Wadi Matulla, Wadi Baba, Wadi Sidri and Wadi Mukattab. The evidence for this is soft sediment deformation of beds over large areas, the most common structures being slump folds. The environment of deposition was fluvial in the lower portion of the Nubian Sandstone Formation (?Carboniferous) in western Sinai, so slope instability is unlikely to have caused the slumping. Instead, deformation due to active seismicity is the most likely explanation.

The main rifting took place after Eocene deposition (Robson 1971). The rift initially formed an elongate arch with pre-rift sediments dipping gently away from the rift in both directions. The post rift succession is dominated by evaporite (Miocene Evaporitic Group; Robson 1971) interbedded with sandstone and shale. Thickness and facies variations are common, particularly in the lower part of the succession.

Clear evidence for post-rift shortening across the region is manifested by major reverse faults due to overturning of normal faults, asymmetric contractional folds in the sedimentary cover, calcite-filled sub-horizontal hydraulic fractures indicating vertical stretching, low-angle thrusts and associated folds in the hangingwalls to normal faults, shortening at stress risers, including steepening of bedding dips at normal faults which, in some cases, has overturned stratigraphy. These features are described in detail below.

Late extension is also seen in the Gulf of Suez where normal faults offset the Pliocene and Pleistocene sands and gravels, which are particularly well exposed along the coast. Present day hot springs adjacent to faults also attest to active normal faulting, and recent instrumental seismicity has been detected in the area (Daggett *et al*. 1986). Incision of Pliocene to Pleistocene sands and gravels is common across the entire Sinai peninsula. Alluvial fans and braid plains have been eroded during uplift, and the present day topography is the result of this relatively recent phase of roughly 3 km uplift (including erosion and isostatic rebound) in the past 9 Ma (Steckler 1985).

Evidence for inversion

A series of maps and geological cross sections will be used to describe the characteristic geometries cited as evidence for post rift contraction and these can be located in Fig. 3. Extensional faults that formed during rifting and were subsequently modified during inversion are well exposed in the following localities: Wadi Tayiba, Gebel Musaba Salama, Wadi Baba, Wadi Mukattab, and Wadi Feiran (Fig. 1). These areas were mapped at 1:50,000 regional scale (Fig. 1) and in detail at 1:10,000 scale (Fig. 3) based on commercial interpreted satellite images (ERA Dublin, unpublished data).

Wadi Tayiba

In Wadi Tayiba (Fig. 4) the main basement-involved normal fault (the Gamal Fault, Robson 1971) has a throw of roughly 2 km and juxtaposes

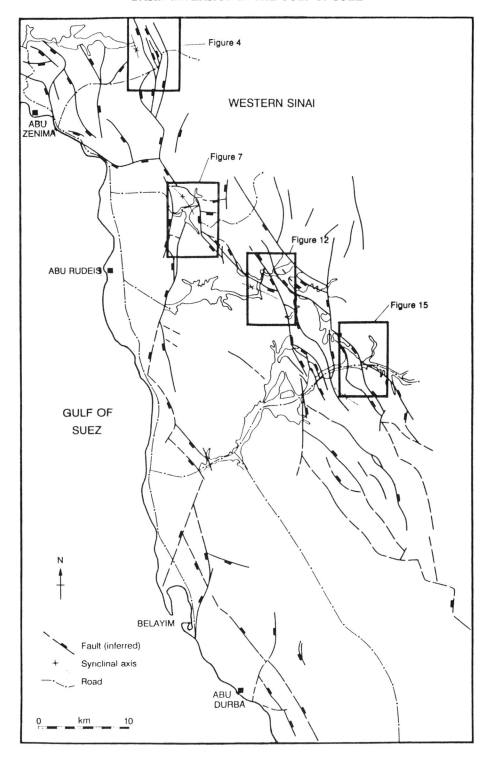

Fig. 3. Geological map and cross-section locations.

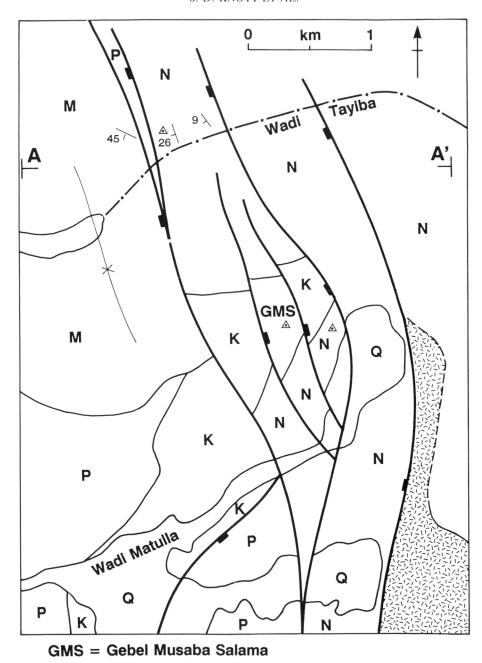

GMS = Gebel Musaba Salama

Fig. 4. Wadi Tayiba/Gebel Musaba Salama geological map. N – Nubian Sandstone Formation, K – Cretaceous, P – Palaeocene, M – Miocene, Q – Quaternary, GMS – Gebel Musaba Salama.

Nubian Sandstone Formation against Miocene Abu Alaqa Group. The beds in the footwall and the hangingwall to the fault indicate reverse reactivation or shortening across it with folding and steepening of bedding (Fig. 5). The smaller normal faults in the footwall to the Gamal Fault appear to have been less affected by shortening.

Some of the change in dips of bedding adjacent to the Gamal Fault have been interpreted to be due to deposition of Miocene fan

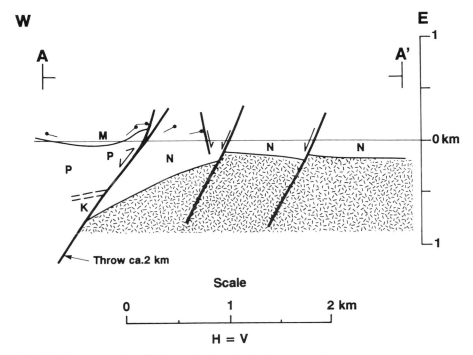

Fig. 5. Wadi Tayiba cross-section. Horizontal and vertical scales are equal.

deltas during active extension (Gawthorpe *et al.* 1990). The geometry of the Miocene fan delta top and fore sets in Wadi Tayiba and in Wadi Sidri (Fig. 6a) indicate increased tilting of older fan deltas during faulting with bedding dipping synthetically to the dip of the fault. This geometry is also characteristic of uplift associated with reverse faulting (e.g. Fig 6b) although fan delta top sets could have been rotated to dip synthetically to the fault during extension and then bedding dips steepened during contraction. In general, and as exemplified in other rift settings (e.g. Fig. 6c), during extension faulting, top sets of older fan deltas that were originally horizontal have a regional dip towards a normal fault. In contrast, the bedding and fault geometry across the Gamal Fault (Fig. 6a) indicate an origin, at least in part, due to contraction with a sense of overthrust shear to the east. A component of the deformation may also be due to compaction but the wavelength (kilometres) and vergence indicate that shortening was the dominant folding mechanism.

Wadi Baba

In Wadi Baba (Fig. 7), beds in the hangingwall of a basement involved normal fault are steep to overturned (Fig. 8). Close to the footwall small low angle thrusts occur indicating shortening up against a rigid buttress (Fig. 8). Also, a minor antithetic fault has been reactivated in a reverse sense. In the footwall, beds of Nubian Sandstone Formation dip towards the hangingwall and may indicate rotation of the footwall block in a counterclockwise sense (as viewed from the south in Fig. 8) during contraction.

Along strike, in Wadi Sidri (see below), bedding dips towards the footwall, presumably preserving the original extensional geometry. Small scale folds and thrusts (Fig. 9) also indicate a sense of overthrust shear to the east. Locally extension faults cut through earlier thrust faults which are then re-imbricated by later thrust faults (Fig. 10).

The overall geometry of the Wadi Baba section evolved from a tilted fault block during the extensional episode (Fig. 11a) to a fault block tilted in the opposite sense during contraction with footwall beds dipping synthetically to a steepened normal fault, and a tightened hangingwall syncline containing overturned normal faults (Fig. 11b). Shortening was probably episodic indicated by the cross cutting relations and temporal sequence of early thrusts, extension faults and later thrusts.

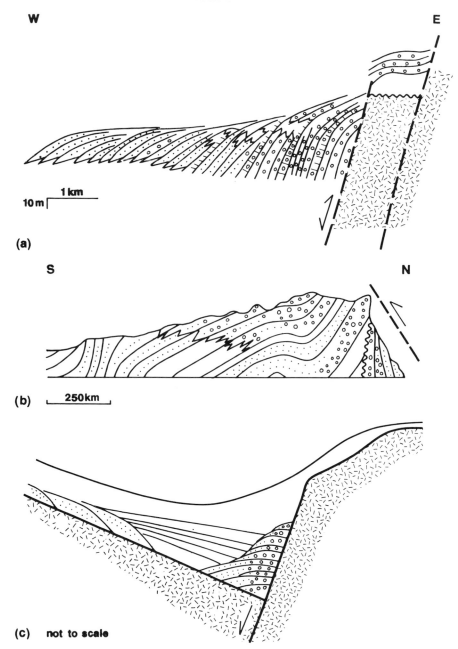

Fig. 6. Fan geometries associated with active faulting. (a) Schematic cross section based on composite sections from several exposures around Wadi Sidri, western Sinai (after Gawthorpe *et al.* 1990). (b) Growth folds in alluvial fanglomerates at the emergent thrust-front margin of the Ebro Basin, Spain (after Nichols 1984). (c) Possible reflector configurations associated with active extension faulting. Wedge shaped bodies near the footwall are likely to be fanglomerates and talus (after Prosser 1993).

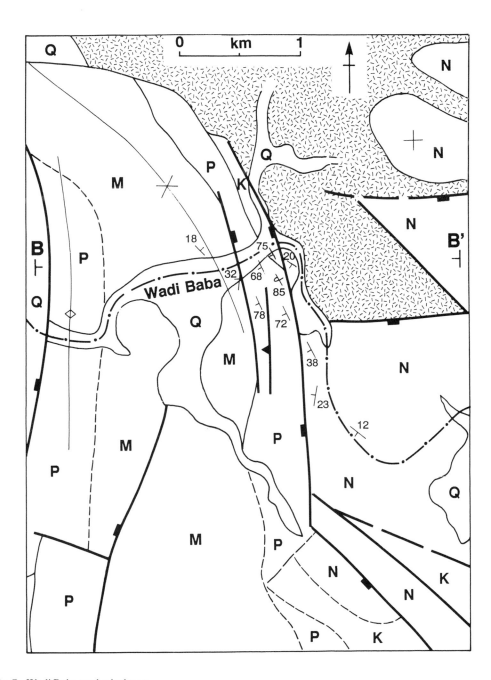

Fig. 7. Wadi Baba geological map.

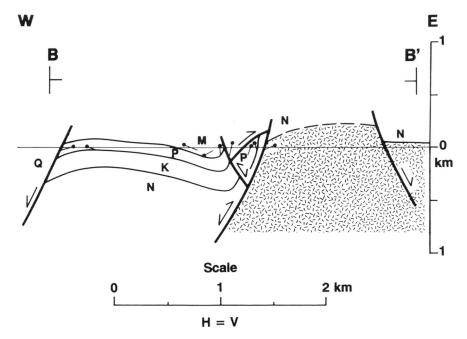

Fig. 8. Wadi Baba cross-section.

Wadi Sidri

Wadi Sidri (Fig. 12) displays the clearest example of an overturned normal fault in the field area. The eastern bounding fault to a basement horst in Wadi Sidri changes dip from eastward on the north side of the wadi to westward on the south side (Fig. 12). In cross section (Fig. 13) the overturning of the normal fault is depicted and in the field Nubian Sandstone Formation can be seen in the footwall lying beneath basement rocks in the hangingwall (Fig. 14).

Wadi Feiran

In Wadi Feiran (Fig. 15) a major fault juxtaposes basement against Nubian Sandstone Formation. This fault has been overturned and was initially a normal fault. The fault was overturned during shortening but the last phase of movement was normal, with the basement block on the down-thrown side (Fig. 16). The present day down-throw to the north is confirmed by kinematic indicators in the fault zone such as minor synthetic Reidel shears and shear gouge fabric. This fault retains a reverse fault geometry at the present day and, although poorly exposed, the fault trace and dip can be mapped based on topographic relief and by drawing structural contours on the fault surface. Indeed the entire

structural history of the Gulf of Suez (Clysmic) rift is represented in the geometry and kinematics of this major basin margin fault (Fig. 17), starting with Early Miocene extension, then mid-Miocene contraction followed by Pliocene to Recent extension.

Fault kinematics

Rifting in late Oligocene to early Miocene time gave rise to the Clysmic trend of extensional faults (NNW) that lay within a zone of stretching which trended WNW. The regional extension direction is considered to be NE–SW (Steckler *et al.* 1988) assuming no rotation of fault blocks in map view about a vertical axis. This extension direction may be significantly different if block rotations about a vertical axis had occurred. The fault kinematic data shown in Fig. 18a indicate that the faults within the extending zone are oriented obliquely to the main rift trend and have a left slip component to the extension across them (Fig. 18b). During inversion, faults that were selectively reactivated in a reverse sense would have had a right slip component due to the NE–SW trending shortening direction (Fig. 18c). The above kinematic observations are important in that the structures in the Clysmic rift formed during predominantly dip slip faulting, and evidence for strike slip during any phase of the deformation history is rare, and

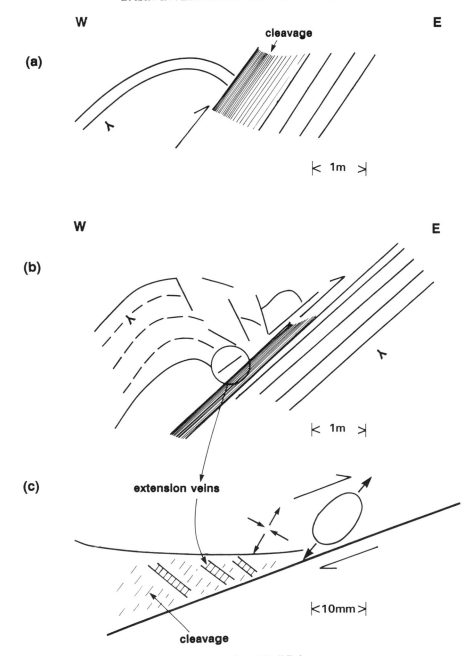

Fig. 9. Sketch of small scale fold and thrust structures from Wadi Baba.

only found on minor, second order faults formed during mid-Miocene shortening (Fig. 18c).

Discussion

Timing of the contractional event is based on syn-deposition contractional folding of the Abu Alaqa Group (early–mid Miocene) seen, for example, in Wadi Tayiba and Wadi Sidri. This deformation post-dates deposition of the Thebes Limestone Formation (Eocene), and pre-dates deposition of the Miocene Evaporitic Group. This indicates that soon after rifting – the first syn-rift deposits are the Nukhul Formation

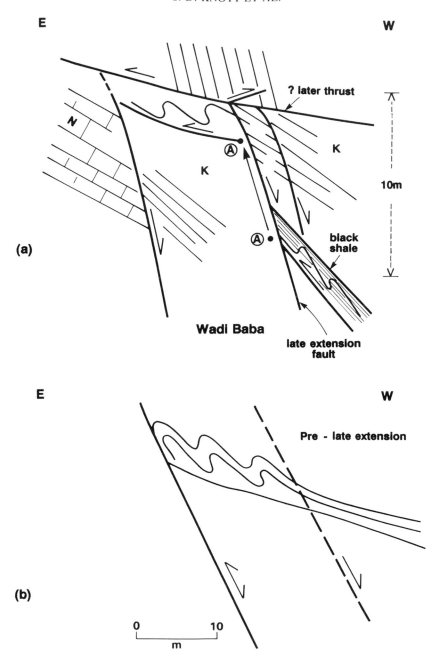

Fig. 10. Wadi Baba. (**a**) Sketch showing structures related to episodic shortening and extension during inversion. The fault on the left-hand side is the main bounding fault juxtaposing Nubian Sandstone Formation and Cretaceous rocks. A minor set of fold and thrust structures are offset by a small normal fault which is itself cut by a later thrust. (**b**) Cartoon showing folds developing in the hangingwall during inversion cut by normal faults which are in turn cut by later thrust faults (not shown). N – Nubian Sandstone Formation, K – Cretaceous.

clastics and the lower part of the Rudeis Formations (Steckler *et al.* 1988; Gawthorpe *et al.* 1990) – the Gulf of Suez and environs experienced compression, producing crustal shortening of between 2 and 5%, measured from cross sections shown in this paper, or even

E   W

(a)

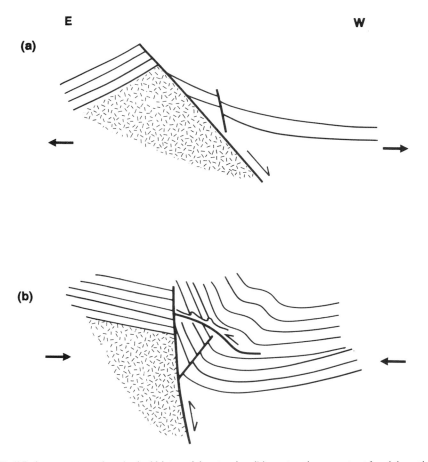

(b)

Fig. 11. Wadi Baba – cartoon of geological history (**a**) extension (**b**) contraction – see text for elaboration. Not to scale.

greater values of shortening immediately adjacent to basement stress risers.

An explanation for the contraction event can be related to changes in intraplate stress due to the failure of rifting in the Gulf of Suez and the locus of rifting transferring to the Gulf of Aqaba (mid Clysmic event; Steckler *et al.* 1988). Regional plate boundary conditions do not indicate compression at this time (?Burdigalian–Langhian). Earlier Miocene compression oriented NE–SW and associated with the Syrian arc (Garfunkel & Bartov 1977) could not have produced the contractional structures seen in the Gulf of Suez (Clysmic) rift although deformation in the Syrian arc was active during deposition of the Nukhul Formation (Steckler *et al.* 1988).

It is possible that the initiation of left oblique slip across the Gulf of Aqaba rift in Burdigalian to mid-Langhian time (Steckler *et al.* 1988 and references therein) caused inversion in the Gulf of Suez. However, extension across the Aqaba rift was active prior to and during the shortening documented from western Sinai and no evidence for compression due to plate motions has been identified in the region of the Sinai triple junction (Steckler *et al.* 1988).

The first evidence for regional shortening in the Afro-Arabian region, following extension in the Gulf of Suez and Red Sea, occurred in the Bitlis collision zone during Serravallian time (Hempton 1987). This collision appears to post date the shortening documented here for the Gulf of Suez. Therefore, from the above discussion, it is unlikely that plate boundary forces caused the contractional deformation seen in the Clysmic rift (Burdigalian–Langhian). It must be made clear, however, that the dating of these events is not very well constrained and further work is required, particularly on the age of the syn-contraction sediments, to confirm this.

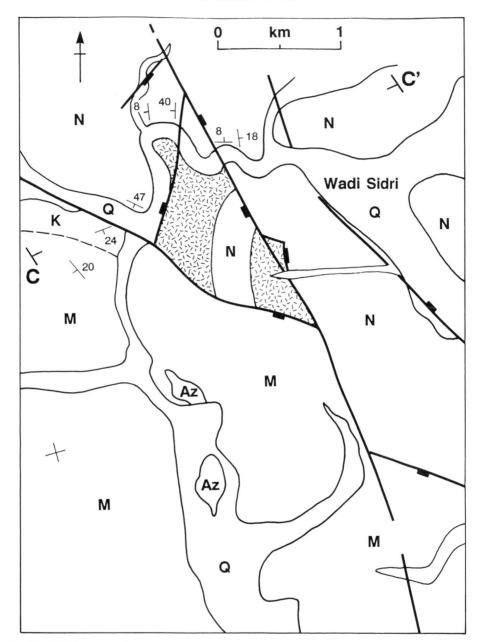

Fig. 12. Wadi Sidri geological map.

Instead of plate boundary forces, contractional deformation could have been the response of the crust to body forces. The gravitational potential of the massifs on either side of the failed rift could have provided the necessary compressional stress for shortening. Artyushkov (1973) demonstrated that deviatoric stresses can arise within an isostatically compensated crustal layer of variable thickness floating on a fluid (Fig. 19). The potential energy of an uplift produces tensional stresses which can lead to gravity-driven crustal extension. The extension is balanced by contraction in the surrounding regions. The magnitude of the tensional deviatoric stresses within the uplift (a few MPa) are sufficient to overcome the strength of the upper

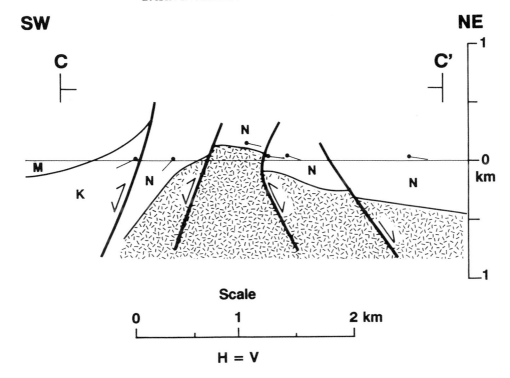

Fig. 13. Wadi Sidri cross-section.

crust, particularly in areas of high heat flow such as the Gulf of Suez. These stresses could induce contractional strains of the order of 2 to 5% given typical strength values for upper crustal rocks (Hobbs *et al.* 1976). This physical model can explain the origin of contraction in the Gulf of Suez, in the absence of suitable plate boundary forces, and may also be applicable to other failed rifts such as the North Sea (Fig. 19).

In the Gulf of Suez other studies suggest that, at the start of the main rifting in the Gulf of Aqaba, perhaps at the mid Clysmic event (Steckler *et al.* 1988), the Gulf of Suez (Clysmic) rift had either failed completely and was undergoing thermal subsidence (Courtillot *et al.* 1987), or very minor extension followed by thermally induced uplift on the rift flanks (Steckler 1985). This combination of factors and the collective field evidence for crustal shortening point to gravity-driven contraction in the Gulf of Suez (Fig. 19).

Application to North Sea exploration and development

At this point it is convenient to make comparisons between the Gulf of Suez and the North Sea. The succession of events is similar in both rifts, where uplift and erosion during the development of a thermal dome was followed by crustal extension. In the central North Sea thermal doming occurred in Bathonian time (Underhill & Partington 1993) and was followed by episodes of extension in the Callovian and Oxfordian/Kimmeridgian (Rattey & Hayward 1993). The North Sea experienced uplift along the Norway margin during syn to post rift times (latest Jurassic to earliest Cretaceous). This uplift is documented by progradation of clastics westward off the Norwegian landmass (Spencer *et al.* 1993). The origin of the uplift is uncertain. However, uplift in Norway could have induced contraction due to the gravitational potential of the massif and the topographic relief of greater than 2 km (Rattey & Hayward 1993) between the top of the massif and the floor of the rift (Fig. 19).

Contractional structures of Late Jurassic to earliest Cretaceous age have been documented from the North Sea by Fossen (1989), Wood & Hall (in press), Cherry (1993), Bartholemew *et al.* (1993), Lee & Hwang (1993), Thomas & Coward (this volume) and we have recognized contractional structures that developed soon after rifting in various places from the Haltenbanken area of Norway to the central North Sea.

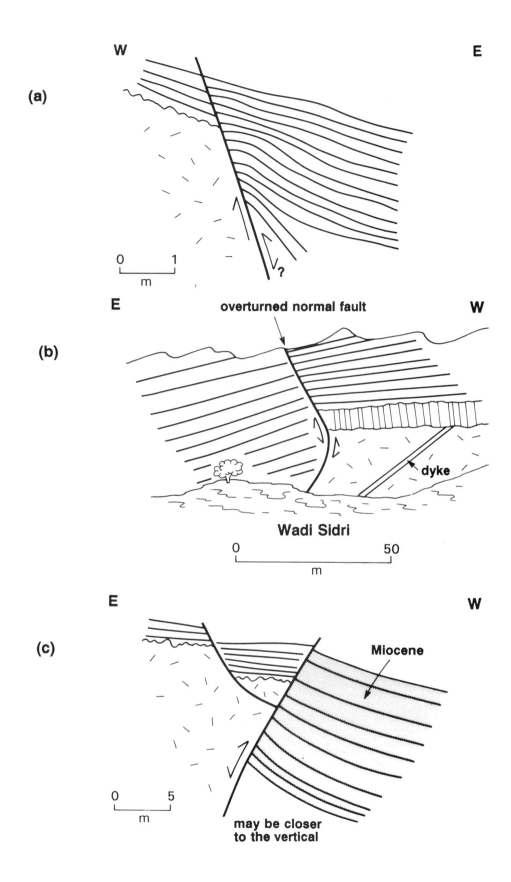

(a)

W   E

0   1

m

?

(b)

E   overturned normal fault   W

dyke

Wadi Sidri

0   50

m

(c)

E   W

Miocene

0   5

m

may be closer
to the vertical

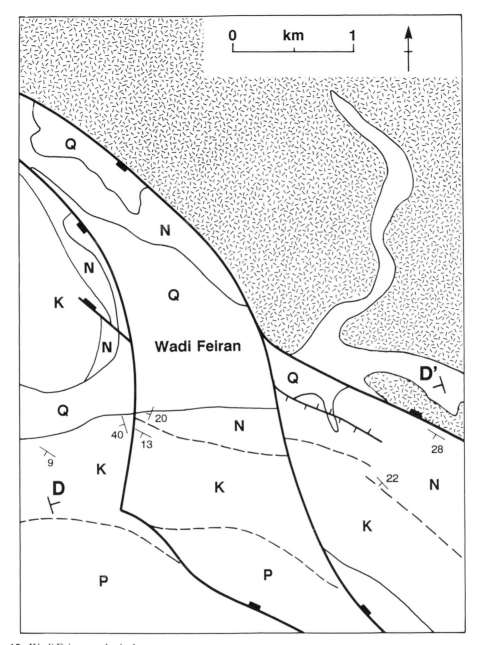

Fig. 15. Wadi Feiran geological map.

Fig. 14. Wadi Sidri detailed field sketches (**a**) normal fault (**b**) overturned normal fault on the north east side of the basement horst (**c**) overturned normal fault on the south west side of the basement horst. Stipple is basement; parallel lines on white background are Nubian Sandstone Formation; parallel lines on shaded background are Miocene syn-rift clastics.

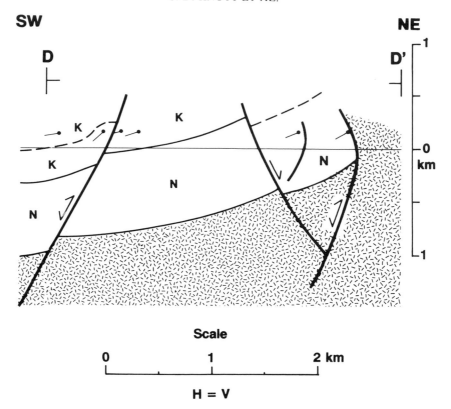

Fig. 16. Wadi Feiran cross-section.

The age of formation of the contractional structures varies along the length of the North Sea rift. Rifting is earlier in the Viking Graben (Oxfordian/Kimmeridgian) than in parts of the Central Graben (Volgian) and post-rift contraction is similarly diachronous, immediately post-dating the cessation of rifting in these two areas.

Although transpression is often invoked to explain these structures (e.g. Bartholomew *et al.* 1993), it is also possible that these structures are due to the mechanism of post rift gravity-driven contraction based on the model of Artyushkov (1973) described above. Shortening would be characterised by predominantly dip slip movements in the North Sea, with transpression a localized phenomenon on favourably oriented faults. The presence of contractional structures of Late Jurassic to earliest Cretaceous age in various places along the entire length of the North Sea and mid Norway rifts indicates that the causal mechanism for the structures is on a much larger scale than local rift kinematics (Bartholemew *et al.* 1993, Thomas & Coward this volume).

These observations have some important implications for exploration and development of hydrocarbons in the North Sea. Modifications to sea floor bathymetry during post-rift contraction controlled the deposition of some late Jurassic deep marine sandstones. Syn-rift deposition of the Brae trend submarine fan apron (Cherry 1993) was followed by compaction and contractional fold growth (Wood & Hall in press). The site of later deep marine turbidite deposition was basinward and the Miller field reservoir sandstones bypassed the topographic high of the Brae trend which lay adjacent to the south Viking Graben bounding fault (Rattey & Hayward 1993).

Modifications to fault and bedding geometry during post rift contraction can have important implications for structural interpretation of three-dimensional seismic data during field development. Contraction may give rise to detailed structural patterns that are not usually associated with a rift setting. The cross sections from the Gulf of Suez indicate some of the geometries likely to be encountered. Fault reactivation during contraction and basin inversion may also modify the fault rock within

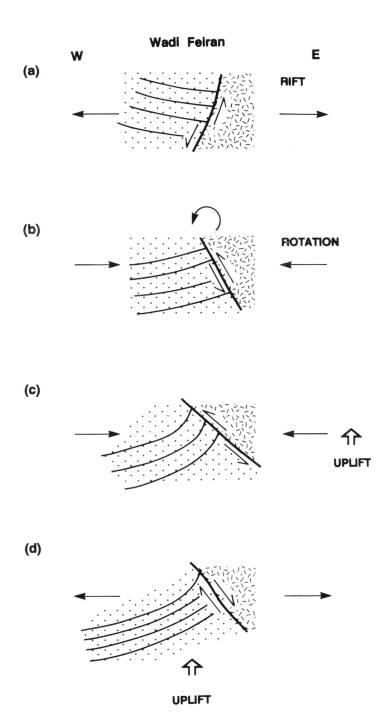

Fig. 17. Wadi Feiran geological history (**a**) extension (**b**) contraction (**c**) uplift and further contraction (**d**) Pleistocene uplift and late-stage extension – see text for elaboration.

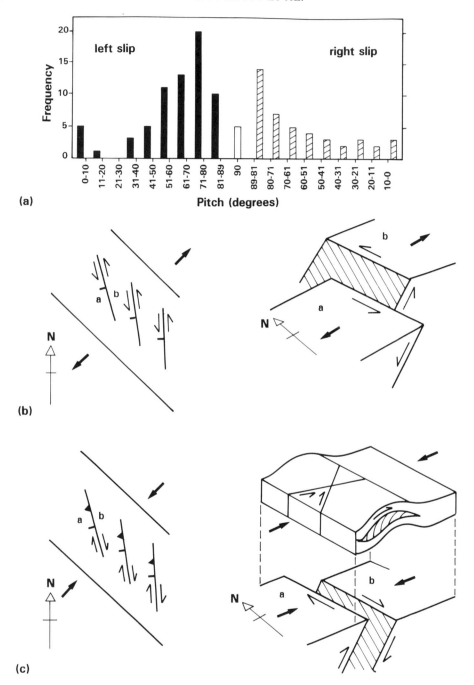

Fig. 18. Kinematics of Gulf of Suez (Clysmic) rift: (**a**) histogram of fault striae pitch data, note the predominance of oblique slip and the paucity of pure strike slip and dip slip measurements; (**b**) left oblique slip during extension, heavy arrow – extensional component, half arrow – left slip component; (**c**) right oblique slip reverse reactivation of normal faults during contraction with deformation in the sedimentary cover including vertical stretching, contractional folding and strike slip faulting. The main Clysmic trend faults strike roughly 140 degrees east of north – see text for elaboration.

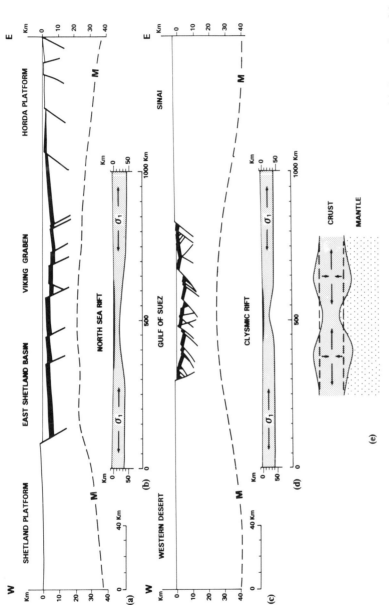

Fig. 19. Gravity-driven contraction in rifts, a suggested mechanism for the post rift shortening in the North Sea and Gulf of Suez. (**a**) Cross-section of the immediate post-rift geometry of the North Sea rift. Crustal thicknesses and geometry modified after Beach *et al.* (1987) and Cheadle *et al.* (1987). The cause of syn to post-rift uplift in Norway is speculatively related to a deep mantle extensional shear zone dipping eastwards beneath Norway (not shown). Shaded area – pre and syn-rift sediments, M – Moho. (**b**) Simplified true scale cross-section of the North Sea rift showing horizontal compressive stresses applied across the rift due to the potential energy of the adjacent uplifts of Norway and Scotland. Black – sea, shaded area – crust, unshaded – mantle. (**c**) Cross section of the immediate post rift geometry of the Clysmic rift. Crustal thicknesses and geometry modified after Colletta *et al.* (1988) and Steckler *et al.* (1988). The cause of syn to post-rift uplift of Sinai is related to induced mantle convection associated with the Sinai triple junction to the south (Steckler 1985). Shaded area – pre-rift sediments, M – Moho. (**d**) Simplified true scale cross section of the Clysmic rift showing horizontal compressive stresses applied across the rift due to the potential energy of the adjacent uplifts of Sinai and the Western Desert. Black – sea, shaded area – crust, unshaded – mantle. (**e**) Dynamic model for gravity driven contraction in rifts modified after Artyushkov (1973). Contraction is generated by the gravitational potential of the uplift adjacent to the rift with a topographic relief of over 2 km between the top of the uplift and the rift floor – see text for elaboration. Not to scale: crustal thicknesses are 40 km and 20 km for the uplift and rift respectively.

fault zones and hence their seal capacity (Sibson this volume). It is therefore important to recognize whether faults have been reactivated when assessing the compartmentalization of a hydrocarbon trap by sealing faults.

Conclusions

The regional structural history of the western Sinai area has been updated by this field work study through the recognition of a phase of shortening immediately post-dating the cessation of rifting. The sequence of tectonic events can be summarized as follows: 1) deposition of the pre rift stratigraphy, although there is some evidence for active faulting during Nubian Sandstone deposition, from the base of the Nubian Sandstone Formation (?Cambrian to Early Cretaceous) to the top of the Thebes Limestone Formation (Eocene), 2) thermal doming, uplift, erosion and minor associated igneous activity during Oligocene time, followed by 3) early Miocene crustal stretching, tilting of fault blocks, and deposition of syn-rift clastics in subsiding hangingwalls, then 4) a brief period (early to mid Miocene) of pulsed, gravity-driven shortening documented by overturning of large normal faults, thrusting, steepening of bedding dips, and finally 5) regional uplift, erosion and increased heat flow in Pliocene to Recent time.

Identification of a phase of contraction is important, not only for the regional understanding of the Gulf of Suez, but also for other failed rifts such as the North Sea. These observations will have implications for seismic and structural interpretation, and reservoir distribution, particularly in the Late Jurassic successions of the North Sea. The timing of uplift with respect to faulting in both the Gulf of Suez and the North Sea is well constrained (Steckler 1985, Spencer *et al.* 1993) and lends some support to the inference that post-rift topographic elevation drove shortening deformation due to the gravitational potential of uplifts several hundreds of kilometres wide. Once the tensional stresses are removed on rift failure, probably by rifting in the Levant transform in the case of the Gulf of Suez and by transfer of stretching to the proto-North Atlantic in the case of the North Sea (Knott *et al.* 1993), contraction of the rift would have occurred as thermal relaxation takes over. The Gulf of Suez probably reflects the complexities of many rift provinces, and this study therefore has applications not just in the Gulf of Suez but in other failed rifts around the world.

This work was funded by Norsk Hydro a.s and Statoil a.s. We thank them for giving permission to publish. Logistical and geological help from Mosleh Salah of EGPC is greatly appreciated. Initial ideas for gravity-driven contraction were originally perceived as the converse of extensional collapse in overthickened accretionary wedges. These ideas stemmed from discussions with John Platt and Joe Cartwright and SDK thanks them for their stimulating thoughts. Sincere thanks to Mike Coward, David Thomas and particularly John Walsh for helpful reviews which improved the layout and text of this article. Barbara Bruce is thanked for skillfully drafting the figures.

References

ARTYUSHKOV, E. V. 1973. Stresses in the lithosphere caused by crustal thickness inhomogeneities. *Journal of Geophysical Research* **78**, 7675–7708.

BARTHOLEMEW, I. D., PETERS, J. M. & POWELL, C. M. 1993. Regional structural evolution of the North Sea: oblique slip and the reactivation of basement lineaments. *In*: PARKER, J. R. (ed.) *Petroleum Geology of Northwest Europe: Proceedings of the 4th Conference*. Geological Society, London, 1109–122.

BEACH, A., BIRD, T. & GIBBS, A. 1987. Extensional tectonics and crustal structure: deep seismic reflection data from the northern North Sea Viking Graben. *In*: COWARD, M. P., DEWEY, J. F. & HANCOCK, P. L. (eds) *Continental Extensional Tectonics*. Geological Society, London, Special Publication, **28**, 467–476.

CHEADLE, M. J., MCGEARY, S., WARNER, M. R. & MATTHEWS, D. H. 1987. Extensional structures on the western UK continental shelf: a review of evidence from deep seismic profiling. *In*: COWARD, M. P., DEWEY, J. F. & HANCOCK, P. L. (eds) *Continental Extensional Tectonics*. Geological Society, London, Special Publication, **28**, 445–466.

CHERRY, S. T. J. 1993. The interaction of structure and sedimentary processes controlling deposition of the Upper Jurassic Brae Formation conglomerates, Block 16/17, North Sea. *In*: PARKER, J. R. (ed.) *Petroleum Geology of Northwest Europe: Proceedings of the 4th Conference*. Geological Society, London, 387–400.

COLLETTA, B., LEQUELLEC, P., LETOUZEY, J. & MORETTI, I. 1988. Longitudinal evolution of the Suez rift structure (Egypt). *Tectonophysics*. **153**, 221–233.

COURTILLOT, V., ARMIJO, A. & TAPPONIER, P. 1987. Kinematics of the Sinai triple junction and a two phase model of Arabia–Africa rifting. *In*: COWARD, M. P., DEWEY, J. F. & HANCOCK, P. L. (eds.) *Continental Extensional Tectonics*. Geological Society, London, Special Publication, **28**, 559–573.

DAGGET, P. H., MORGAN, P., BOULOS, F. K., HENNIN, S. F., EL-SHERIF, A. A., EL-SAYED, A. A., BASTA, N. Z. & MELEK, Y. S. 1986. Seismicity and active tectonics of the Egyptian Red Sea margin and the Northern Red Sea. *Tectonophysics,* **125**, 313–324.

FOSSEN, H. 1989. Indication of transpressional

tectonics in the Gullfaks oil-field, northern North Sea. *Marine and Petroleum Geology,* **6**, 22–30.

GARFUNKEL, Z. & BARTOV, Y. 1977. The tectonics of the Suez rift. *Geological Survey of Israel Bulletin,* **71**, 44.

GAWTHORPE, R. L., HURST, J. M. & SLADEN, C. P. 1990. Evolution of Miocene footwall-derived coarse-grained deltas, Gulf of Suez, Egypt: implications for exploration. *AAPG Bulletin,* **74**, 1077–1086.

HEMPTON, M. R. 1987. Constraints on Arabian plate motion and extensional history of the Red Sea. *Tectonics,* **6**, 687–705.

HOBBS, B. E., MEANS, W. D. & WILLIAMS, P. F. 1976. *An outline of structural geology.* Wiley and Sons, 571 pp.

KNOTT, S. D., BURCHELL, M. T., JOLLEY, E. J. & FRASER, A. J. 1993. Mesozoic to Cenozoic plate reconstructions of the North Atlantic and hydrocarbon plays of the Atlantic margins. *In*: PARKER, J. R. (ed.) *Petroleum Geology of Northwest Europe: Proceedings of the 4th Conference.* Geological Society, London, 953–974.

LEE, M. J. & HWANG, Y. J. 1993. Tectonic evolution and structural styles of the East Shetland Basin. *In*: PARKER, J. R. (ed.) *Petroleum Geology of Northwest Europe: Proceedings of the 4th Conference.* Geological Society, London, 1137–1149.

NICHOLS, G. J. 1984. *Thrust tectonics and alluvial sedimentation, Aragon, Spain.* PhD thesis, University of Cambridge, 243 pp.

PROSSER, S. 1993. Rift-related linked depositional systems and their seismic expression. *In*: WILLIAMS, G. D. & DOBB, A. (eds) *Tectonics and Seismic Sequence Stratigraphy.* Geological Society, Special Publication, **71**, 35–66.

RATTEY, R. P. & HAYWARD, A. W. 1993. Tectono-stratigraphic evolution of a failed rift system – the

MIDDLE JURASSIC TO EARLY CRETACEOUS BASIN EVOLUTION OF THE CENTRAL AND NORTHERN NORTH SEA. *In*: PARKER, J. R. (ed.) *Petroleum Geology of Northwest Europe: Proceedings of the 4th Conference.* Geological Society, London, 215–249.

ROBSON, D. A. 1971. The structure of the Gulf of Suez (Clysmic) rift, with special reference to the eastern side. *Journal of the Geological Society, London,* **127**, 247–276.

SIBSON, R. H. 1995. Selective fault reactivation during basin inversion: potential for fluid redistribution through fault-valve action. *This volume.*

SPENCER, A. M., BIRKELAND, O. & KOCH, J-O. 1993. Petroleum geology of the proven hydrocarbon basins, offshore Norway. *First Break.* **11**, 161–176.

STECKLER, M. S. 1985. Uplift and extension of the Gulf of Suez, indications of induced mantle convection. *Nature,* **317**, 135–139.

——, BERTHELOT, F., LYBERIS, N. & LE PICHON, X. 1988. Subsidence in the Gulf of Suez: implications for rifting and plate kinematics. *Tectonophysics,* **152**, 249–270.

THOMAS, D. & COWARD, M. P. 1995. Anomalous structures within the East Shetland Basin: implications for Late Jurassic/Early Cretaceous inversion. *This volume.*

UNDERHILL, J. R. & PARTINGTON, M. A. 1993. Jurassic thermal doming and deflation in the North Sea: implications of the sequence stratigraphic evidence. *In*: PARKER, J. R. (ed.) *Petroleum Geology of Northwest Europe: Proceedings of the 4th Conference.* Geological Society, London, 337–386.

WOOD, J. L. & HALL, S. in press. A tectono-stratigraphic model for the development of submarine slope apron fans and basin floor fans during the Late Jurassic-Early Cretaceous of the South Viking Graben. *Basin Research.*

Hydrocarbon habitat and classification of inverted rift basins

DUNCAN S. MACGREGOR

*BP Exploration Operating Company, 4/5 Long Walk, Stockley Park, Uxbridge
UB11 1BP, UK*

*Present address: BP Exploration, Kuningan Plaza South Tower, PO Box 2749 Jakarta
Selatan, Indonesia*

Abstract: This shortened paper presents a brief analysis of the relationship between hydrocarbon distribution and the intensity of inversion in rift basins. For this purpose, three intergradational classes of rift were identified according to the intensity and areal extent of inversion-related uplift. Simple rifts, showing no significant inversion, are characterised by high success rates and dispersed hydrocarbon distributions. Locally inverted rifts, such as those showing restricted areas of uplift along faults, also show a high degree of petroleum retention and success, with the key success factor often being the relationship between timing of trap formation and petroleum charge. Regionally inverted rifts show lower success rates, failures being often attributable to a redistribution of hydrocarbons during inversion, with losses having occurred through biodegradation, surface erosion or seepage on surface penetrating faults. Both types of inverted rift frequently show a concentration of reserves in one large field.

The aim of this brief study was to identify the effects of differing styles and intensities of inversion on hydrocarbon distribution and exploration success rates in rift basins. A three fold classification of rift basins was attempted, as is summarized below. Data were collected from published literature and internal BP sources on hydrocarbon resource size, field size distribution and seep occurrence for over 100 basins, including 175 giant fields (>250 MMBOE), while more detailed research was conducted on a selection of 16 Old World major petroleum provinces (Table 1). The accuracy of the views presented in this paper is reliant on the accuracy of the interpretations given in the published data sources on the basins concerned. A full listing of references for the key basins discussed is given in Table 1. The use of references is consequently reduced in the following shortened text.

Attempted classification of rifts

For the purposes of this paper, three categories of rift are identified. The classification is based on the areal extent of inversion related uplift in a basin, with the aim of discerning petroleum systems with differing relationships between source rock generation and inversion-related structuring. Type examples are illustrated schematically in Fig. 1 and listed in Table 1.

Simple rifts have not undergone significant inversion and thus show a largely unaltered extensional structural regime. Source rocks have thus undergone largely uninterrupted and progressively deeper burial through time.

Locally inverted rifts are those in which sections of the basin have been uplifted in post-rift times while others have continued to subside. The original extensional topography is altered or destroyed, with the creation of a series of broad compressional anticlines, often caused by the inversion of former lows. Source rocks have continued to subside and mature during and/or following the inversion event in the subsiding areas. This type of inversion is frequently associated with transpression, as is the case in both the examples shown in Fig. 1.

Regionally inverted rifts are those in which the majority of the original basin area has undergone late stage uplift accompanied by deep regional scale erosion. The original basin centres usually show the greatest degree of uplift. The extensional topography is lost in the basin centres, though may be preserved on less inverted basin rims. Most of the examples of regional inversion discussed in this paper are the product of large-scale compression, although similar effects may be produced by thermal/isostatic uplift (e.g. the late stage uplift in the Timan–Pechora Basin, see Table 1). These basins show complex source rock histories with at least one phase of petroleum generation usually preceding inversion. Generating kitchens have thus usually been regionally uplifted and at least temporarily shut off.

While this classification can easily be applied to the majority of basins studied, there is a minority of basins which do not readily fall into the defined classes. Some basins show varying

From BUCHANAN, J. G. & BUCHANAN, P. G. (eds), 1995, *Basin Inversion*,
Geological Society Special Publication No. 88, 83–93.

83

Table 1 *Hydrocarbon habitat characteristics of identified rift types*

Basin	Country	Total Reserves Oil & Gas (Billion Barrels Oil Equivalent)	Largest Field (% of basin reserves)	Predominant Trap Types	Seepage/ Palaeoseepage	Notes on Inversion Style	Notes on Source Burial/Generation History	References
(a) SIMPLE RIFTS								
West Siberia	Russia	c. 200	Samotlor (22%)	Drape Anticlines	Apparently none	No significant inversion	Continuous burial and generation over long period	Rigassi, 1986, Kontorovich 1984, Lopatin 1993
North Sea	UK, Norway	c. 60	Troll (13%)	Fault Blocks, Drape Ant., Strat. (various)	Some indications of leakage from high pressure traps only	Very minor transpressional structures	Continuous burial and generation over long period	Harding 1983, Goff 1984
Sirte	Libya	c. 35	Sarir (13%)	Drape Ant., Fault Blocks, Reefs	None known	Some regional uplift to west	Continuous burial and generation over long period	Harding 1983, Parsons et al. 1980
Huabei	China	c. 17	Renqui (14%)	Fault Blocks, Strat., Hanging Wall Closures	Minor	None known	Continuous burial and generation over long period	Li 1991, Zhai 1988
Gulf of Suez	Egypt	c. 7	Morgan (23%)	Fault Blocks	Seeps on surface penetrating normal faults bounding basin	Very minor inversion structures	Continuous burial and generation over long period	Brown 1980, Harding 1983

(b) LOCALLY INVERTED RIFTS

1) without regionally inverted basin margins

Basin	Location		Field	Traps/Structures	Seepage	Inversion style	Generation timing	References
Central Sumatra	Indonesia	c. 12	Minas (35%)	Uplifted/wrench-altered basement highs (major), inversion structures (minor)	None in main petroleum system, minor on flanks	Local narrow inversions on wrench faults ('Sunda Folds')	Generation in mid Miocene may have preceded latest phase of wrenching, largest fields are in old traps	Eubank & Matti 1981, Macgregor 1993, Hudbay Oil & BP Internal Data
Malay	Malaysia, Indonisia	c. 6	Seligi (10%)	Inversion Anticlines	None known	Narrow inversions of original half-grabens	Source generating during and after inversion phase	ASCOPE 1981
SE Triassic (Rhourde Nouss)	Algeria	c. 3	Rhourde Nouss (27%)	Inversion Anticlines	None evident	Narrow inversions along transpressional faults	Source generating during and after inversion phase (gas charge postdates inversion)	Boudjema 1987, Attar 1994

2) with regionally inverted basin margins

Basin	Location		Field	Traps/Structures	Seepage	Inversion style	Generation timing	References
Songliao	China	c. 16	Daqing (80%)	Basin Centre Uplift, Strat. Pinchouts	Seeps on uplifted basin margins	Large basin centre uplift	Intimate near-synchronous relationship between generation and structuring	Yang 1985, Li 1991
South Sumatra	Indonesia	c. 2.5	Talang Akar (c. 15%)	Inversion Anticlines, Diapiric Anticlines, Reefs	Major seepage on basin flanks and over many small inversion/diapiric oilfields	Local Miocene inversion followed by more regional uplift associated with Plio-Pleistocene Barisan uplift to SW. Northeastern areas less or not affected	Not well documented. Much of generation may be broadly synchronous with Miocene inversions, and precedes Plio-Pleistocene uplift	BP Internal Data, Macgregor 1993

Table 1 *continued*

Basin	Country	Total Reserves Oil & Gas (Billion Barrels Oil Equivalent)	Largest Field (% of basin reserves)	Predominant Trap Types	Seepage/ Palaeoseepage	Notes on Inversion Style	Notes on Source Burial/Generation History	References
(c) REGIONALLY INVERTED RIFTS								
Timan–Pechora	Russia	c. 6	Usa (c. 30%)	Inversion Anticlines	Large numbers of seeps known	Numerous local inversions in Palaeozoic followed by broad regional epeirogenic/ isostatic uplifts in Tertiary/ Quaternary	Generation roughly contemporaneous with local Permo-Triassic inversions – most hydrocarbons sourced around this time. Phases of tertiary migration and leakage are evident during later regional uplifts	Bogatsky et al. 1993, Navilkin et al. 1984, Pairazain 1993, Ulmishek 1982
Sole Pit	UK	c. 5	Leman (36%)	Fault Blocks	Gas plumes on sonar largely outside area of evaporite devt.	Basin–scale inversion of compressional origin	Generation prior to inversion. Re-migration into present highs during inversion: evaporite seal aided preservation at this time	Hillier & Williams 1991, Walker & Cooper 1986, BP Internal Data
Ahnet	Algeria	c. 2?	??	Inversion Anticlines	None evident at present day	Local Inversions in late Carb. immediately followed by regional uplift and erosion of compressional origin. Basin transitional between locally and regionally inverted classes	Poorly understood. Much of oil charge may have preceded inversion and was lost. Dry gas charge roughly contemporaneous with early inversion	Attar 1994, Macgregor 1994

Basin	Country		Field	Trap type	Seepage	Inversion	Charge/timing	References
Lower Saxony	Germany, Netherlands	1.2	Schoonebeek (30%)	Inversion Anticlines	None evident at present day, but large losses interpreted during inversion	Large scale uplift and structuring of compressional origin in late Cret/early Tertiary	Much of charge preceded Late Cret. inversion and was lost through erosion/seepage at that time. Kitchen 'switched on' again by Tertiary burial	Binot et al. 1993
Wessex	UK	0.6	Wytch Farm (c. 98%)	Fault Block	Many impregnations and some flowing gas seeps in inverted basin area	Large scale compressional regional inversion of most of basin. Northern flank (Wytch Farm area) unaffected – can be considered as area of simple rift tectonics	Generation preceded basin inversion and much of hydrocarbon resource possibly lost through seepage	Colter and Havard 1981, Ebukanson & Kingston 1986, Selley & Stoneley 1987, Miles et al. 1993
West Netherlands	Netherlands	c. 0.3	Ijsselmonde (50%)	Inversion Anticlines	None evident at present day, but large losses interpreted during inversion	Large scale uplift and structuring in late Cret/early Tertiary	Much of charge preceded Late Cret. inversion and was lost through erosion/seepage at that time. Part of kitchen 'switched on' again by Tertiary burial	Bodenhausen & Ott 1981, De Jager et al. 1993

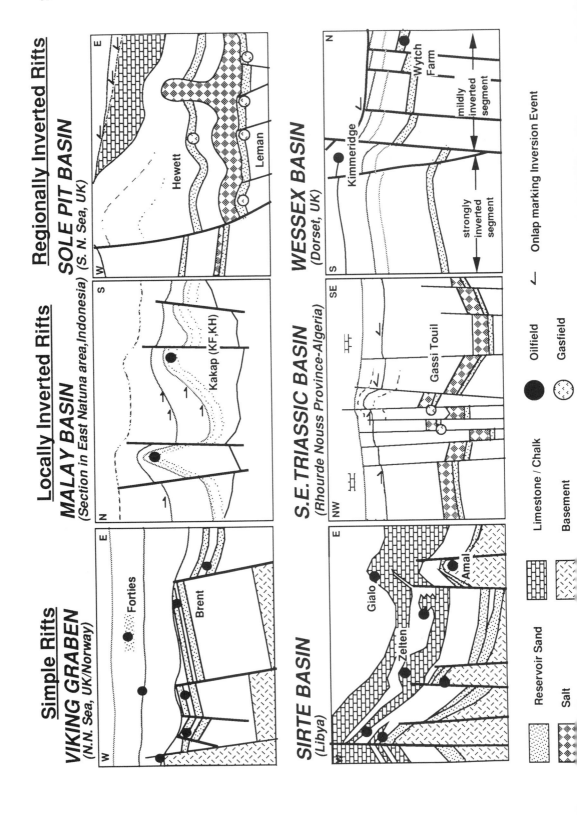

Simple Rifts

VIKING GRABEN
(N.N. Sea, UK/Norway)

Locally Inverted Rifts

MALAY BASIN
(Section in East Natuna area, Indonesia)

Regionally Inverted Rifts

SOLE PIT BASIN
(S. N. Sea, UK)

SIRTE BASIN
(Libya)

S.E. TRIASSIC BASIN
(Rhourde Nouss Province-Algeria)

WESSEX BASIN
(Dorset, UK)

Reservoir Sand

Salt

Limestone / Chalk

Basement

Oilfield

Gasfield

Onlap marking Inversion Event

inversion styles in different areas of the basin (e.g. South Sumatra, Table 1), while others show complex structural histories involving multiple inversions of often different styles (e.g. Timan–Pechora). It is also accepted that regional inversions may be caused by various different mechanisms and there is a case for distinguishing regional inversions of thermal/isostatic origin from those caused by compressional activity. Further development of the classification is undoubtedly required.

Results

Results of the analyses of hydrocarbon resource size and distribution and of seep occurrence in 105 Mesozoic rift basins are summarized in Figs 2 & 3. The more detailed literature review of 16 major petroleum provinces is summarized in Table 1.

Exploration success and resource size (Fig. 2)

From this database the frequency of exploration success is measured, using a number of different criteria. The frequency of large reserves is relatively high for simple and locally inverted rifts, but low for regionally inverted rifts. A comparison is also shown against thrust belts, some of which can be considered as severely inverted rifts (e.g. Papua New Guinea), and therefore as a continuation of the same trend.

The analysis presented indicates that local inversion is frequently a favourable factor for the development of major hydrocarbon provinces, with the key factor in many cases being the relationship between trap growth and timing of generation and migration (e.g. Songliao Basin). It is apparent that many locally inverted rifts, particularly those without pre-rift reservoirs (Malay and Songliao Basins), would contain no or few large traps if it was not for the effects of local inversions.

Cases of successful regionally inverted rifts are less frequent and often tied to special criteria. Of the four well documented European examples reviewed in Table 1, two (southern West Netherlands, Lower Saxony) can be explained by the reactivation of kitchens by post-inversion burial, with it being significant in both cases that the authors on these basins

consider that the pre-inversion petroleum systems have been completely destroyed. The other two European examples are explicable by the occurrence of an unbroken evaporite seal which preserved hydrocarbons during inversion (Sole Pit) and by the preservation of a single large palaeotrap in an uninverted basin flank (Wessex). The Timan–Pechora example may be important in illustrating significant preservation in a basin inverted by epeirogenic/isostatic/thermal effects. The high proportion of low reserves and failure cases amongst regionally inverted rifts is often explained in the literature by the destruction of pre-existing hydrocarbon systems during inversion, e.g. northern part of West Netherlands, (Bodenhausen & Ott, 1981), the Atlas rifts of Morocco and Algeria (Macgregor et al. 1994), inverted Nigerian rifts (Whiteman 1982).

Field size distribution (Table 1)

Simple rifts are typified by dispersed field size distributions with rarely more than 25% of basin reserves in the largest field. This can be attributed to the high degree of structural segmentation of these basins, affecting both trap and kitchen geometries (e.g. Viking Graben, Goff 1984). The locally and regionally inverted rifts studied tend to show a higher degree of concentration of reserves in the largest field (Table 1). Two hypotheses can be proposed for this: either one inversion structure has formed a migration focus accessing the majority of the basin's kitchen area, or a more dispersed field size distribution once existed, but has been altered as a result of the destruction of many of the original fields. The former hypothesis seems most applicable to the cases of high reserves concentration in a single field in locally inverted rifts, particularly those cases in which uplift of a basin centre has formed one predominant structure (e.g. Songliao Basin). The latter hypothesis is more applicable to regionally inverted rifts such as Wessex, where there is abundant evidence for trap destruction in the form of seeps and surface impregnations, and where reserves can be seen to be concentrated in a surviving palaeostructure.

Seepage

The lower reserves and success rates in regionally inverted rifts are often accompanied,

---

Fig. 1. Schematic cross-sections of basins discussed in this paper based on published data. The key difference between locally inverted rifts, which show continued subsidence of adjoining synclines, and regionally inverted rifts, which do not, is highlighted. Note the significance of evaporite seals and palaeostructures in controlling hydrocarbon habitat in cases of severe inversion. For sources see Table 1.

FREQUENCY OF SUCCESS

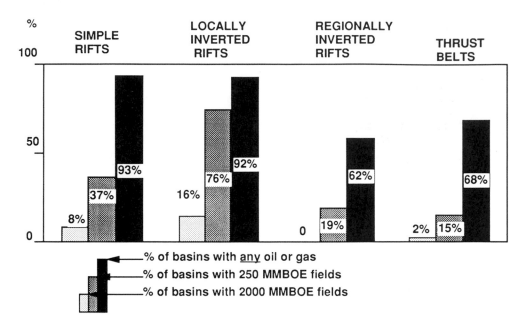

MAXIMUM FIELD SIZE

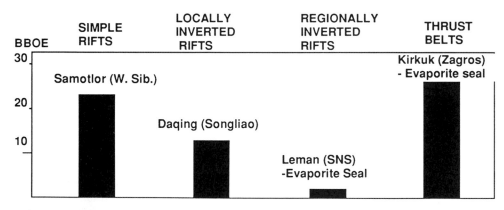

Fig. 2. Exploration success criteria on the three categories identified as measured from a database of 105 Mesozoic rift basins (BP internal data). Thrust Belts are shown for the purposes of comparison.

particularly for those which have suffered recent inversions, by considerable direct evidence for loss of petroleum at surface (Fig. 3). This may be either in the form of flowing seeps, generally concentrated along reverse fault planes (Macgregor 1993), or as exposed oil impregnations, representing eroded oil accumulations or migration pathways. Generation in an inverted basin is interpreted by most authors (e.g. Bodenhausen & Ott 1981) to have been shut off during uplift, and these shows must therefore

represent a partial and continuing destruction of a petroleum system that existed prior to inversion.

It is suggested, based on the discussions of seepage and other destructive processes in Navilkin *et al.* (1984) and Macgregor (1993), that the critical point at which inversion begins to have a negative impact on petroleum productivity is when previously existing accumulations are uplifted into zones where shallow degradational processes such as biodegradation

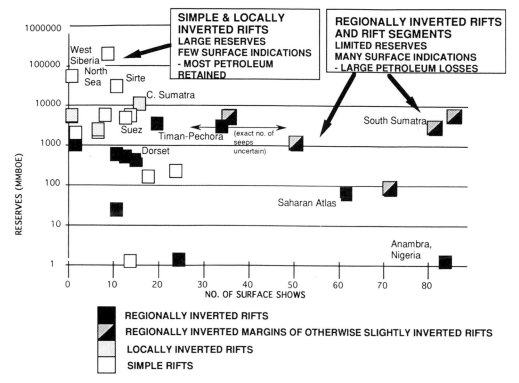

Fig. 3. Relationship between reserves and evidence for seepage, as measured by number of reported surface shows. See Macgregor (1993) for further details of the databases used. Note the apparent negative relationship seen, which contrasts with positive relationships between reserves and seeps in other basin types. This indicates simple rifts to be unusually retentive, with seepage only occurring from rifts if they are subject to later inversions that cause fault connections of pre-existing fields with surface.

may occur and/or fault connections are established to surface, thereby enabling leakage to surface in the form of seeps. The significance of faulting would infer that regional inversions of compressional origin are likely to be more destructive than those of isostatic and thermal origin.

Conclusions

Inversion is a process which, dependent on its severity and its relationship to hydrocarbon generative phases, may have a positive or negative effect on the hydrocarbon systems of a rift basin. Mild inversions clearly play a beneficial role in creating large simple anticlinal traps. Continued subsidence and generation in surrounding kitchens is a critical factor, with migration often occurring synchronously with, or shortly after, formation of inversion structures and traps.

Regional inversions, particularly those associated with compressional surface-penetrating faults and non-evaporite seals, frequently cause

significant losses from pre-existing petroleum systems. These losses result in lower exploration success rates and field sizes for regionally inverted rifts than for other basin types. However, the successful examples of regionally inverted rifts listed in Table 1 demonstrate that, under the right conditions, success can be obtained. Areas of surviving palaeotraps or traps near kitchens that may have been reactivated post-inversion are clearly areas where exploration effort is better concentrated. A generalized model, to aid the prediction of hydrocarbon habitat in unexplored inverted rifts, is shown in Fig. 4.

Further work

This brief literature study has highlighted some relationships between inversion style/intensity and hydrocarbon habitat. However, the scope of the study has been limited and it is accepted that more detailed study of a wider range of basins is required to develop the classification into a more

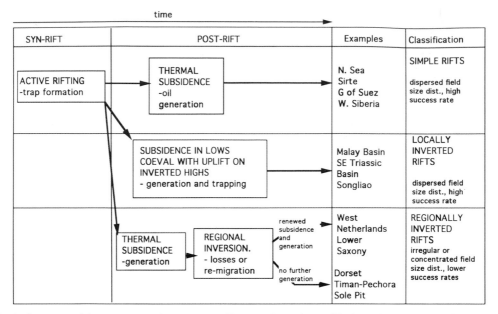

Fig. 4. Summary of the sequences of events controlling trap formation and hydrocarbon generation in the identified categories of rift basins. The implications of these relationships on field size distribution and success rates are summarized in the right hand column.

reliable predictive tool. The locally inverted and regionally inverted classes will need refinement and subdivision, possibly into categories of different inversion style and mechanism.

References

ASCOPE 1981. *Tertiary Sedimentary Basins of the Gulf of Thailand and South China Sea, Stratigraphy, Structure and Hydrocarbon Occurrences.* ASEAN Council on Petroleum, ASCOPE Sekretariat, Jakarta, Indonesia.

BINOT, F., GERLING, P., HILTMANN, W., KOCKEL, F. & WEHNER, H. 1993. The Petroleum System in the Lower Saxony Basin. *In:* SPENCER, A. M. (ed.) *Generation, Accumulation, and Production of Europe's Hydrocarbons, III.* Special Publication of European Association of European Petroleum Geoscientists, Springer-Verlag, Chapter 11.

BODENHAUSEN, J. W. A. & OTT, W. F. 1981. Habitat of the Rijswijk Oil Province, Onshore, the Netherlands. *In:* ILLING, L. V. & HOBSON, G. D. *Petroleum Geology of the Continental Shelf of Northwest Europe, 2nd Conference.* Graham & Trotman, London, 301–309.

BOGATSKY, V. & Pankratov, I. 1993. Controls on the Distribution of Oil, Gas & Condensate Pools in the Timan–Pechora Basin (abs). *American Association of Petroleum Geologists Bulletin,* **77**, 1608.

BOUDJEMA, A. 1987. *Evolution Structurale du Bassin Petrolier Triasique du Sahara Nord-Oriental (Algerie).* PhD thesis, Universite Paris-Sud.

BROWN, R. N. 1980. History of Exploration and Discovery of Morgan, Ramadan and July Oilfields, Gulf of Suez, Egypt. *In:* MAILL, A. D. (ed.) *Facts and Principles of World Petroleum Occurrence.* Canadian Society of Petroleum Geologists, Memoir **6**, 733–764.

COLTER, V. S. & HAVARD, D. J. 1981. The Wytch Farm Oil Field, Dorset. *In:* ILLING, L. V. & HOBSON, G. D. *Petroleum Geology of the Continental Shelf of Northwest Europe, 2nd Conference.* Graham & Trotman, London, 139–148.

DE JAGER, J., DOYLE, M., GRANTHAM, P. & MABILLARD, J. 1993. Hydrocarbon Habitat of the West Netherlands Basin (abs). *American Association of Petroleum Geologists Bulletin,* **77**, 1618.

EBUKANSON, E. J. & KINGSTON, R. R. F. 1986. Maturity of Organic matter in the Jurassic of Southern England and its Relation to the Burial History of the Sediments. *Journal of Petroleum Geology,* **9**, 259–280.

EUBANK, R. T. & MATTI, A. C. 1981. Structural Geology of the Central Sumatra Backarc Basin. *Proceedings of the Indonesian Petroleum Association,* 10, Jakarta.

GOFF, J. C. 1984. Hydrocarbon Generation and Migration from Jurassic Source Rocks in the East Shetland Basin and Viking Graben of the Northern North Sea. *In:* DEMAISON, G. & MURRIS, R. (eds) *Petroleum Geochemistry and Basin Evaluation.* American Association of Petroleum Geologists, Memoir **35**, 273–302.

HARDING, T. P. 1983. Graben Hydrocarbon Plays and Structural Styles. *In:* KAASCHIETER, J. P. H. & REIJER, T. J. A. (eds) *Petroleum geology of the Southeastern North Sea and the Adjacent Onshore Areas.* Geologie en Mijnbouw, **62**, 003–023.

HILLIER, A. P. & WILLIAMS, B. P. J. 1991. The Leman Field, Blocks 49/26, 49/27, 49/28, 53/1, 53/2, UK North Sea. *In*: ABBOTTS, I. L. (ed.) *United Kingdom Oil and Gas Fields, 25 Years Commemorative Volume*. Geological Society, Memoir **14**, 451–458.

LI DESHENG 1991. Tectonic Types of Oil & Gas Basins in China. *China Petroleum Industry Press, Beijing*.

LOPATIN, N. 1993. Occurrence of Oil & Gas Fields and Source Rock Transformation in the West Siberia Basin and Barents Sea Platform (abs). American Association of Petroleum Geologists Bulletin, **77**(9), 1643.

MACGREGOR, D. S. 1993. Relationships between Seepage, Tectonics and Subsurface Petroleum Reserves. *Marine & Petroleum Geology*, **10**, 606–619.

——, ROBERTS, D. G. & DALTON, D. G. 1994. Hydrocarbon Systems of North Africa. *Proceedings of the Mediterranean Oil & Gas Conference, Malta, January 1994*.

MILES, J. A., DOWNES, C. J. & COOK, S. E. 1993. The Fossil Oil Seep in Mupe Bay, Dorset, a Myth Investigated. *Marine & Petroleum Geology*, **10**, 58–70.

NAVILKIN, V. D., GOLDBERG, I. S., KRUGLIKOV, N. M., LAZAREV, V. S., SAKHIBGARAYEV, G. P., SVERCHKOV, G. P. & SIMAKOV, S. N. 1984. Destructive Processes affecting Oil and Gas Pools and Estimation of the Hydrocarbon Loss. *International Geology Review*, **26**(12), 1185–1198.

PAIRAZIAN, V. V. 1993. Petroleum Geochemistry of the Timano–Pechora Basin. *First Break*, **11**, 279–286.

PARSONS, M. G., ZAGAAR, A. M. & CURRY, J. J. 1980. Hydrocarbon Occurrences in the Sirte Basin, Libya. *In*: MAILL, A. D. (ed.) *Facts and Principles of World Petroleum Occurrence*. Canadian Society of Petroleum Geologists, Memoir **6**, 723–732.

RIGASSI, D. A. 1986. Wrench Faults as a Factor Controlling Petroleum Occurrences in West Siberia. *In*: HALBOUTY, M. T. (ed.) *Future Petroleum Provinces of the World*. American Association of Petroleum Geologists, Memoir **40**, 529–544.

SELLEY, R. C. & STONELEY, R. 1987. Petroleum Habitat in South Dorset. *In*: BROOKS, J. & GLENNIE, K. (eds) *Petroleum Geology of Northwest Europe*. Institute of Petroleum, London, 494–503.

ULMISHEK, G. 1982. *Petroleum Geology and Resource Assessment of the Timan–Pechora Basin, USSR, and the Adjacent Barents-Northern Kara Shelf*. Argonne National Laboratory, Report ANL/EES/-TM-199.

WALKER, I. M. & COOPER, W. G. 1986. The Structural and Stratigraphic Evolution of the N. E. Margin of the Sole Pit Basin. *In*: BROOKS, J. & GLENNIE, K. (eds) *Petroleum Geology of Northwest Europe*. Graham & Trotman, London, 263–275.

WHITEMAN, A. J. 1982. *Nigeria – its Petroleum Geology, Resources and Potential, volumes 1 and 2*. Graham & Trotman, London.

YANG WANLI 1985. Daqing Oil Field, People's Republic of China, A Giant Field with Oil of Non-Marine Origin. American Association of Petroleum Geologists Bulletin, **69**, 1101–1111.

ZHAI GUANGMING, WANG SHENYAN & LI GANSHENG 1988. Characteristics and Oil and Gas Potential of Sedimentary Basins in China. *In*: WAGNER, H. C. *et al*. (eds) *Petroleum Potential of China and Related Subjects*. Circum Pacific Council for Energy and Mineral Resources, Houston, Earth Science Series, **10**, 1–22.

Modelling of basin inversion

The geometries and kinematics of inverted fault systems: a review of analogue model studies

K. R. MCCLAY

Fault Dynamics Project, Department of Geology, Royal Holloway, University of London, Egham, Surrey, TW20 0EX, UK

Abstract: The geometries and kinematics of 2D inverted extensional fault systems are reviewed using the results of scaled physical models together with case histories from inverted basin fault systems. 2D analogue models of detached terranes, listric, planar, ramp/flat listric and domino arrays of planar faults were constructed from homogeneous sandpacks and from anisotropic sand/mica layers. The models were first extended and then subjected to horizontal compression in order to reactivate the extensional fault systems. Upon extension simple listric faults produce a characteristic roll-over anticline and crestal collapse graben. Inversion of this system produces reactivation of the main detachment with the development of a fault-bounded wedge of syn-extensional strata elevated above regional together with tightening of the crestal collapse graben. New thrust faults initiate from the tips of the crestal collapse extensional faults. Characteristic 'harpoon' structures develop associated with the reactivation of the main detachment. Similar inversion architectures were also produced for the planar fault systems. Extension of ramp-flat listric fault systems produces an upper roll-over and crestal collapse graben together with a ramp syncline and lower roll-over and crestal collapse graben. Inversion of this system only reactivates a part of the main detachment producing a hangingwall shortcut fault that bypasses the upper roll-over system. Inversion of domino fault arrays produced characteristic harpoon geometries of the syn-rift wedge together with shortcut faults in the footwalls of the main domino faults.

Analysis of the progressive deformation during both extension and inversion has been carried out using marker points embedded in the models. In the analogue models of detached terranes hangingwall collapse during extension occurs along planar to curved shear surfaces, whereas upon inversion, different, lower angle shear trajectories are utilized. In the domino fault arrays more complicated shear paths are observed as a result of the footwall shortcut faults developed during inversion. These results have important implications for the analysis of extensional and inverted terranes and for fault reconstruction and section balancing techniques.

The architecture of inverted fault systems developed in the analogue models show striking similarities to the inverted basin geometries found in natural fault systems such as those in the North Sea and South East Asia. Examples from these terranes are compared and contrasted with the results of the analogue models. Conceptual models for inversion kinematics and fault system architectures are presented.

Inversion tectonics involves a switch in tectonic mode from extension to contraction such that extensional basins are contracted and become regions of positive structural relief. This process is generally accepted to involve the reactivation of pre-existing extensional faults such that they undergo reverse slip (Cooper & Williams 1989). The term positive inversion refers to a switch in tectonic mode from extension to contraction whereas the term negative inversion refers to a switch from contraction to extension (Williams *et al.* 1989). This paper reviews the results of an extensive programme of analogue modelling of positive inversion tectonics carried out in the modelling laboratory at Royal Holloway, University of London (McClay 1989; Buchanan & McClay 1991, 1992; Buchanan 1991; Simmons 1991). The focus of this review is upon the geometries and kinematics of 2D dip-slip extension and dip-slip inversion. The results of some 3D analogue models of inversion structures are presented in Keller & McClay (this volume).

Inversion geometries have commonly been drawn conceptually as involving complete reactivation of the basin forming extensional faults such that the syn-extensional ('syn-rift') sedimentary wedge is pushed up – 'inverted' above the pre-extensional 'regional elevation' (Fig. 1) (Bally 1984). Such conceptual models of inversion ignore the mechanics of the inversion process and do not necessarily resemble natural geometries (Fig. 2). A wide range of natural

From BUCHANAN, J. G. & BUCHANAN, P. G. (eds), 1995, *Basin Inversion,*
Geological Society Special Publication No. 88, 97–118.

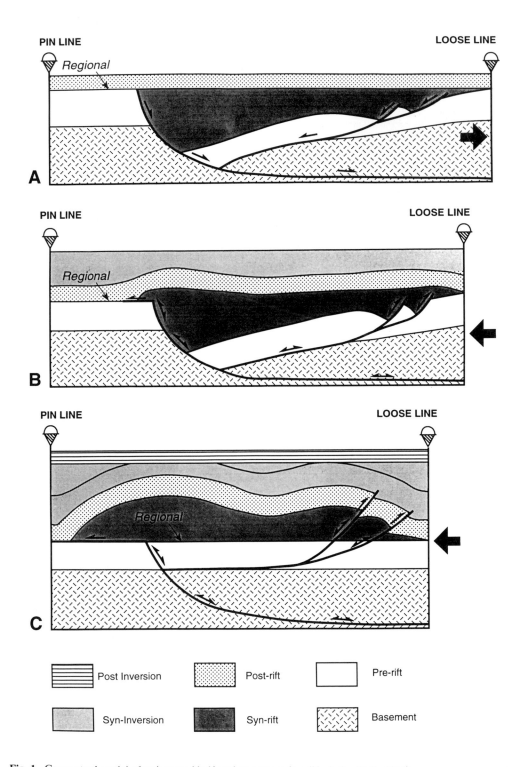

Fig. 1. Conceptual model of an inverted half-graben system (modified after Bally 1984).

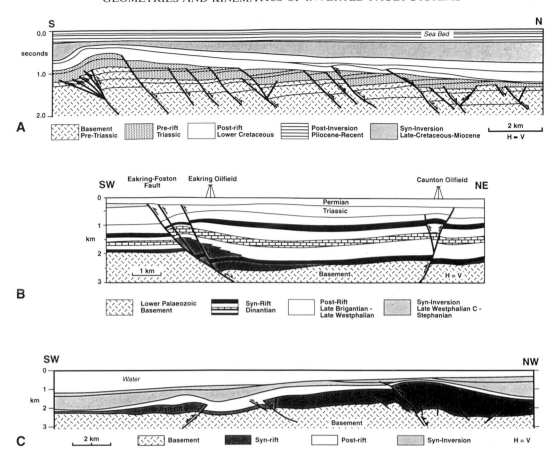

Fig. 2. Examples of inverted extensional fault systems: (**A**) geoseismic section across the South Hewett fault zone, southern North Sea. The inverted domino fault array shows a major growth anticline at the left hand end of the section and high angle thrust faults have produced arrowhead or 'harpoon' structures at the right hand end of the section; (**B**) cross-section through the Eakring oilfield, England (modified after Fraser *et al*. 1990). The extensional half-graben system has undergone partial inversion with contractional displacement on the main extension fault and a 'pop-up' at the location of the Caunton oil field; (**C**) cross-section through part of the Sunda Arc (modified after Letouzey 1990) showing inversion of half-graben fault systems to produce characteristic 'Sunda folds' ('arrowhead' or 'harpoon' geometries).

inversion structures have been identified from basins such as the southern North Sea, the Sunda Arc and the East Java Sea, Indonesia (e.g. Fig. 2; and other papers in this volume). This paper aims to illustrate the inversion geometries in scaled analogue models of simple extensional fault systems and analyse the kinematics of dip-slip inversion. The results of the analogue models are compared and contrasted with natural examples of inverted extensional fault systems.

Geometries of inverted extensional fault systems

Inverted extensional fault systems in sedimentary basins vary from reactivation of isolated extensional faults within the basin to reactivation of all major faults and elevation of the basin to form significant anticlinal fold structures. Syn-inversion sedimentation over reactivated faults produces growth anticlines that are characteristic of positive structural inversion. Extensional fault systems which have undergone relatively small amounts of positive inversion commonly show reactivation of only a few of the faults (Fig. 2a). The inversion structures across the South Hewett fault zone in the southern North Sea show reactivation of the main basin bounding fault and development of a fault-propagation growth anticline together with a number of small footwall shortcut thrusts (Fig. 2a). There is only localized reactivation of

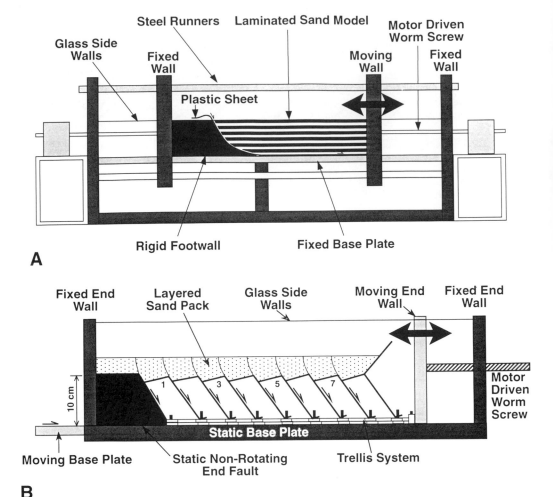

Fig. 3. Experimental apparatus: (**A**) deformation rig for the modelling of extension and inversion of listric and planar faults; (**B**) deformation rig for modelling the extension and inversion of a domino fault array.

extensional faults in the main domino array within the basin. The growth fold displays thinning over the fold crest and thickening away from the reactivated basin margin fault (Fig. 2a). In examples where listric extensional faults had been reactivated during inversion, hangingwall anticlines are developed and 'pop-up' structures are characteristic (Fig. 2b). At greater values of contraction reactivated extensional faults propagate and flatten upwards into the post-rift and syn-inversion sequences (Fig. 2c). Complex 'arrowhead' or 'harpoon' geometries result from the wedge of 'syn-rift' sediments being elevated above regional. The examples in Fig. 2 illustrate some of the common geometries of inverted extensional fault systems. These will be compared with the

results of scaled analogue model experiments described below.

2D analogue models of inverted extensional fault systems

Experimental method

The analogue modelling method for inverted extensional fault systems has been described by Buchanan & McClay (1991) and McClay & Buchanan (1992) and only a brief review will be given here. Two types of sandbox experiment have been carried out. The first type, the simple listric and the ramp-flat listric fault shapes, consisted of fixed footwall geometries defined by preformed wooden blocks over which a deformable hangingwall was first extended and then contracted by horizontal compression. The second type of experiment simulated a series of rigid, domino-style, basement fault blocks that firstly underwent extension and rotation and then were inverted by reversing the motion such that they returned to their pre-extensional configuration.

The simple listric and ramp-flat listric experiments were carried out in a glass-walled deformation rig (Fig. 3a) that was 150 cm long, 20 cm wide and up to 20 cm deep. Initial model dimensions were typically 30 cm long, 20 cm wide and 10 cm deep. The glass side-walls of the rig were coated with a low-friction polymer to minimize frictional edge effects which are minor, and the deformation patterns viewed through the glass walls are representative of the internal deformation in the models. The models were constructed between the end walls of the deformation rig and consisted of alternating layers of white and coloured sand (average grain size 225 μm). Deformation was achieved by moving one of the end walls at a constant displacement rate of 4.16×10^{-3} cm sec^{-1}. A plastic sheet between the sand model and the footwall block/base of the deformation rig permitted constant displacement of the hangingwall block over the rigid footwall block during extension. For extension the hangingwall was pulled down and along the basal detachment, whereas for inversion, the plastic sheet was fixed to the baseplate and the hangingwall was pushed back up the detachment. This was a mechanical limitation of the modelling method in order to prevent folding and rucking up of the plastic sheet during contraction. During extension syn-rift layers were incrementally added to the model in order to simulate syn-rift sedimentation. At the end of the extensional phase, a post-rift sequence was added to the models and during the contraction, syn-inversion layers were introduced in order to simulate sedimentation during inversion and to prevent the formation of unstable surface slopes where the thrust faults breached the surface of the model.

The second type of experiment simulated domino extensional fault blocks and utilized the glass-sided deformation rig shown in Fig. 3b (Buchanan 1991). This apparatus consists of eight rigid domino-shaped plates linked by a trellis system such that they can be simultaneously extended by moving the end wall to the right and inverted by moving the end wall back (Fig. 3b). In this series of experiments a 3 cm thick pre-extensional sequence was placed

in the apparatus and syn-rift sediments were added incrementally during extension. A 1.5 cm thick post-rift sequence was added at the end of extension and then the models were inverted by pushing the end wall back such that the domino fault blocks slid back to their pre-extensional configuration.

The experiments were scaled to simulate the brittle deformation of sedimentary rocks in the upper crust. Previous studies (Horsfield 1977; McClay 1990) have shown that dry, cohesionless sand is a good analogue model material for experiments carried out under conditions of 1 g. Under the laboratory conditions used in this research 1 cm of sandpack in the model scales to approximately 750 m of sedimentary section in the natural prototype. The sand used in the experiments described in this paper have (at normal stress levels greater than 200 Pa) an internal angle of friction of 28.6° for unfaulted sand (Fig. 4a) and an angle of friction of 31.3° for faulted sand (i.e. reactivation of pre-existing faults) (Fig. 4b). During extension the normal stress levels from gravity within the models are 164 Pa per 1 cm thickness of sand, i.e. 1640 Pa for the 10 cm deep models during extension. The stress needed to shear faulted sand is on average 16.7% lower than that required to shear unfaulted sand (Fig. 4).

In both series of experiments, progressive deformation during both the extension and inversion phases was monitored using time lapse photography. The completed models were impregnated and serially sectioned to study the final deformation geometry. In addition to the experimental procedure outlined above additional models were run in which marker triangles were placed in the sandpacks in order to monitor displacement of individual points within the model during both the extension and inversion phases. The marker points are formed by black sand with identical grain sizes and material properties as that utilized in the rest of the model. All of the experiments have been repeated several times (in most cases at least four). Reproducible results were obtained for these repeat experiments and the results described below are typical examples of the inversion geometries produced.

Results of 2D analogue modelling of dip-slip inversion

Listric faults

Extension above a listric detachment fault produces a characteristic roll-over structure with

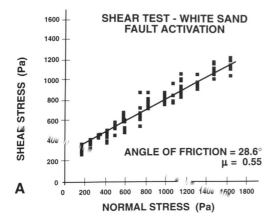

**SHEAR TEST - WHITE SAND
FAULT ACTIVATION**

ANGLE OF FRICTION = 28.6°
μ = 0.55

A

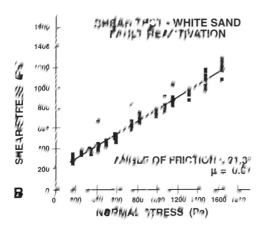

SHEAR TEST - WHITE SAND
FAULT REACTIVATION

ANGLE OF FRICTION = 21.3°
μ = 0.61

B

Fig. 4. Material properties of the quartz sand used in the experiments. (A) shear strength versus normal stress data for inflating sand; (B) shear strength versus normal stress for faulting sand.

an associated crestal collapse graben (Fig. 5a). These geometries have been described in detail by Ellis & McClay (1988) and McClay (1990). The antithetic faults are planar and bound the right hand side of the crestal collapse graben whereas the synthetic faults in the roll-over are sigmoidal in shape (Fig. 5a). The amount of bed rotation in the roll-over is proportional to the amount of extension and the shape of the detachment (Ellis & McClay 1988). The crestal collapse graben is generated by arc stretching as the hangingwall accommodates to the underlying detachment geometry.

Inversion of the listric extensional fault model

was achieved by horizontal recompression. Upon inversion the hangingwall was pushed upwards and rotated back up with reactivation of the main detachment surface to produce a fault-bounded wedge above the main extensional fault breakaway (Fig. 5b). The main detachment fault propagated into the overlying post-rift and syn-inversion strata with the same geometry as the listric extensional segment (Fig. 5b). Small displacement short-cut faults formed in the post-rift and syn-rift section in the footwall of the main fault. These short-cut thrust faults flatten upwards (i.e. become convex upwards) in the post-rift and syn-inversion strata. During inversion the crestal collapse graben was tightened and small thrust faults nucleated from the tips of the pre-existing crestal-collapse graben extensional faults. This produces a characteristic 'pop-up' structure above the inverted crestal collapse graben (Fig. 5b). Syn-inversion strata thin over the crest of the main fault-bounded wedge and thicken away to either side (Fig. 5b).

Particle displacement paths Marker triangles placed within the models were used to monitor the progressive displacement of individual points within the models during both extension and inversion. The corner points of the triangular triangles were digitized for each centimetre of extension and inversion and then analysed with reference to fixed points within the footwall and within the hangingwall. Displacement calculated with reference to a fixed point in the footwall define the actual displacement trajectories of the hangingwall whereas displacement calculated with respect to a fixed point only undergoing horizontal translation in the hangingwall (both in extension and inversion define the relative movement (apparent trajectories) during hangingwall on the fault surface.

For the simple listric fault the trajectories relative to a fixed footwall show particle paths parallel to the main detachment (Fig. 6a). Measured relative to a fixed point in the hangingwall displacements for hangingwall extension show curved trajectories that dip shallowly in the model and become steep at the detachment surface (Fig. 6b) the particle displacement relative to a point in the footwall parallel the main detachment. The particle displacement relative to a fixed point ...

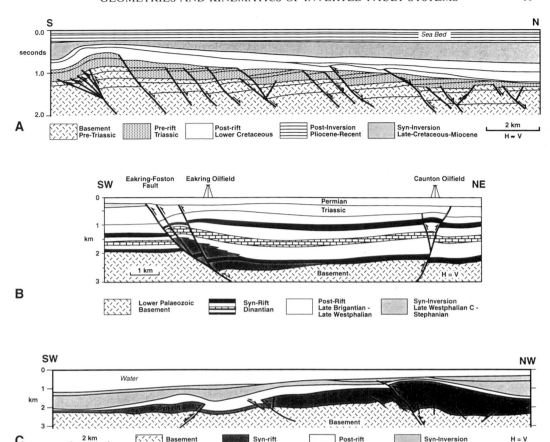

Fig. 2. Examples of inverted extensional fault systems: (**A**) geoseismic section across the South Hewett fault zone, southern North Sea. The inverted domino fault array shows a major growth anticline at the left hand end of the section and high angle thrust faults have produced arrowhead or 'harpoon' structures at the right hand end of the section; (**B**) cross-section through the Eakring oilfield, England (modified after Fraser *et al.* 1990). The extensional half-graben system has undergone partial inversion with contractional displacement on the main extension fault and a 'pop-up' at the location of the Caunton oil field; (**C**) cross-section through part of the Sunda Arc (modified after Letouzey 1990) showing inversion of half-graben fault systems to produce characteristic 'Sunda folds' ('arrowhead' or 'harpoon' geometries).

inversion structures have been identified from basins such as the southern North Sea, the Sunda Arc and the East Java Sea, Indonesia (e.g. Fig. 2; and other papers in this volume). This paper aims to illustrate the inversion geometries in scaled analogue models of simple extensional fault systems and analyse the kinematics of dip-slip inversion. The results of the analogue models are compared and contrasted with natural examples of inverted extensional fault systems.

Geometries of inverted extensional fault systems

Inverted extensional fault systems in sedimentary basins vary from reactivation of isolated extensional faults within the basin to reactivation of all major faults and elevation of the basin to form significant anticlinal fold structures. Syn-inversion sedimentation over reactivated faults produces growth anticlines that are characteristic of positive structural inversion. Extensional fault systems which have undergone relatively small amounts of positive inversion commonly show reactivation of only a few of the faults (Fig. 2a). The inversion structures across the South Hewett fault zone in the southern North Sea show reactivation of the main basin bounding fault and development of a fault-propagation growth anticline together with a number of small footwall shortcut thrusts (Fig. 2a). There is only localized reactivation of

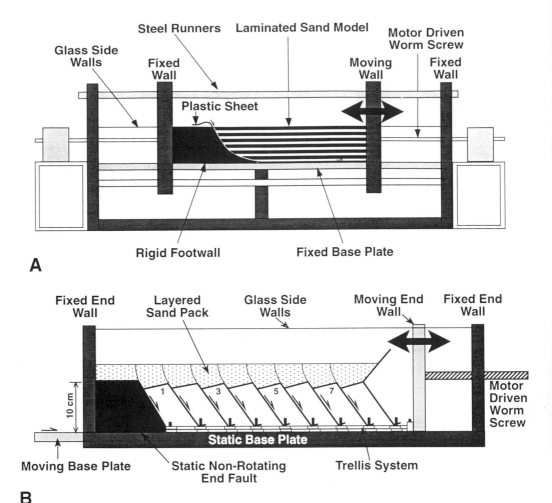

Fig. 3. Experimental apparatus: (**A**) deformation rig for the modelling of extension and inversion of listric and planar faults; (**B**) deformation rig for modelling the extension and inversion of a domino fault array.

extensional faults in the main domino array within the basin. The growth fold displays thinning over the fold crest and thickening away from the reactivated basin margin fault (Fig. 2a). In examples where listric extensional faults had been reactivated during inversion, hangingwall anticlines are developed and 'pop-up' structures are characteristic (Fig. 2b). At greater values of contraction reactivated extensional faults propagate and flatten upwards into the post-rift and syn-inversion sequences (Fig. 2c). Complex 'arrowhead' or 'harpoon' geometries result from the wedge of 'syn-rift' sediments being elevated above regional. The examples in Fig. 2 illustrate some of the common geometries of inverted extensional fault systems. These will be compared with the

results of scaled analogue model experiments described below.

2D analogue models of inverted extensional fault systems

Experimental method

The analogue modelling method for inverted extensional fault systems has been described by Buchanan & McClay (1991) and McClay & Buchanan (1992) and only a brief review will be given here. Two types of sandbox experiment have been carried out. The first type, the simple listric and the ramp-flat listric fault shapes, consisted of fixed footwall geometries defined by preformed wooden blocks over which a deformable hangingwall was first extended and then contracted by horizontal compression. The second type of experiment simulated a series of rigid, domino-style, basement fault blocks that firstly underwent extension and rotation and then were inverted by reversing the motion such that they returned to their pre-extensional configuration.

The simple listric and ramp-flat listric experiments were carried out in a glass-walled deformation rig (Fig. 3a) that was 150 cm long, 20 cm wide and up to 20 cm deep. Initial model dimensions were typically 30 cm long, 20 cm wide and 10 cm deep. The glass side-walls of the rig were coated with a low-friction polymer to minimize frictional edge effects which are minor, and the deformation patterns viewed through the glass walls are representative of the internal deformation in the models. The models were constructed between the end walls of the deformation rig and consisted of alternating layers of white and coloured sand (average grain size 225 μm). Deformation was achieved by moving one of the end walls at a constant displacement rate of 4.16×10^{-3} cm sec^{-1}. A plastic sheet between the sand model and the footwall block/base of the deformation rig permitted constant displacement of the hangingwall block over the rigid footwall block during extension. For extension the hangingwall was pulled down and along the basal detachment, whereas for inversion, the plastic sheet was fixed to the baseplate and the hangingwall was pushed back up the detachment. This was a mechanical limitation of the modelling method in order to prevent folding and rucking up of the plastic sheet during contraction. During extension syn-rift layers were incrementally added to the model in order to simulate syn-rift sedimentation. At the end of the extensional phase, a post-rift sequence was added to the models and during the contraction, syn-inversion layers were introduced in order to simulate sedimentation during inversion and to prevent the formation of unstable surface slopes where the thrust faults breached the surface of the model.

The second type of experiment simulated domino extensional fault blocks and utilized the glass-sided deformation rig shown in Fig. 3b (Buchanan 1991). This apparatus consists of eight rigid domino-shaped plates linked by a trellis system such that they can be simultaneously extended by moving the end wall to the right and inverted by moving the end wall back (Fig. 3b). In this series of experiments a 3 cm thick pre-extensional sequence was placed

in the apparatus and syn-rift sediments were added incrementally during extension. A 1.5 cm thick post-rift sequence was added at the end of extension and then the models were inverted by pushing the end wall back such that the domino fault blocks slid back to their pre-extensional configuration.

The experiments were scaled to simulate the brittle deformation of sedimentary rocks in the upper crust. Previous studies (Horsfield 1977; McClay 1990) have shown that dry, cohesionless sand is a good analogue model material for experiments carried out under conditions of 1 g. Under the laboratory conditions used in this research 1 cm of sandpack in the model scales to approximately 750 m of sedimentary section in the natural prototype. The sand used in the experiments described in this paper have (at normal stress levels greater than 200 Pa) an internal angle of friction of 28.6° for unfaulted sand (Fig. 4a) and an angle of friction of 31.3° for faulted sand (i.e. reactivation of pre-existing faults) (Fig. 4b). During extension the normal stress levels from gravity within the models are 164 Pa per 1 cm thickness of sand, i.e. 1640 Pa for the 10 cm deep models during extension. The stress needed to shear faulted sand is on average 16.7% lower than that required to shear unfaulted sand (Fig. 4).

In both series of experiments, progressive deformation during both the extension and inversion phases was monitored using time lapse photography. The completed models were impregnated and serially sectioned to study the final deformation geometry. In addition to the experimental procedure outlined above additional models were run in which marker triangles were placed in the sandpacks in order to monitor displacement of individual points within the model during both the extension and inversion phases. The marker points are formed by black sand with identical grain sizes and material properties as that utilized in the rest of the model. All of the experiments have been repeated several times (in most cases at least four). Reproducible results were obtained for these repeat experiments and the results described below are typical examples of the inversion geometries produced.

Results of 2D analogue modelling of dip-slip inversion

Listric faults

Extension above a listric detachment fault produces a characteristic roll-over structure with

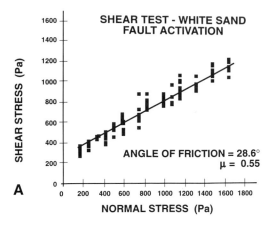

A

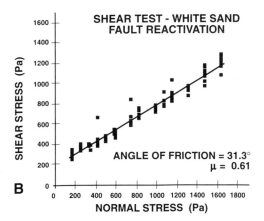

B

Fig. 4. Material properties of the quartz sand used in the experiments: (**A**) shear strength versus normal stress data for unfaulted sand; (**B**) shear strength versus normal stress for faulted sand.

an associated crestal collapse graben (Fig. 5a). These geometries have been described in detail by Ellis & McClay (1988) and McClay (1990). The antithetic faults are planar and bound the right hand side of the crestal collapse graben whereas the synthetic faults in the roll-over are sigmoidal in shape (Fig. 5a). The amount of bed rotation in the roll-over is proportional to the amount of extension and the shape of the detachment (Ellis & McClay 1988). The crestal collapse graben is generated by arc stretching as the hangingwall accommodates to the underlying detachment geometry.

Inversion of the listric extensional fault model

was achieved by horizontal recompression. Upon inversion the hangingwall was pushed upwards and rotated back up with reactivation of the main detachment surface to produce a fault-bounded wedge above the main extensional fault breakaway (Fig. 5b). The main detachment fault propagated into the overlying post-rift and syn-inversion strata with the same geometry as the listric extensional segment (Fig. 5b). Small displacement short-cut faults formed in the post-rift and syn-rift section in the footwall of the main fault. These short-cut thrust faults flatten upwards (i.e. become convex upwards) in the post-rift and syn-inversion strata. During inversion the crestal collapse graben was tightened and small thrust faults nucleated from the tips of the pre-existing crestal-collapse graben extensional faults. This produces a characteristic 'pop-up' structure above the inverted crestal collapse graben (Fig. 5b). Syn-inversion strata thin over the crest of the main fault-bounded wedge and thicken away to either side (Fig. 5b).

Particle displacement paths Marker triangles placed within the models were used to monitor the progressive displacement of individual points within the models during both extension and inversion. The corner points of the triangular markers were digitized for each centimetre of extension and inversion and then analysed both with reference to fixed points within the footwall and within the hangingwall. Displacement paths calculated with reference to a fixed point in the footwall define the actual displacement trajectories of the hangingwall whereas displacements calculated with respect to a fixed point that has only undergone horizontal translation in the hangingwall (both in extension and in inversion) define the relative movement (apparent shear trajectories) during hangingwall collapse above the fault surface.

For the simple listric fault the displacement trajectories relative to a fixed point in the footwall show particle paths that parallel the main detachment (Fig. 6a). Measured relative to a fixed point in the hangingwall the particle displacements for hangingwall collapse during extension show curved apparent shear trajectories that dip shallowly in the upper part of the model and become steep towards the main detachment surface (Fig. 6b). During inversion the particle displacement paths relative to a fixed point in the footwall show trajectories that parallel the main detachment surface (Fig. 6c). The particle displacement paths calculated relative to a fixed point in the hangingwall show

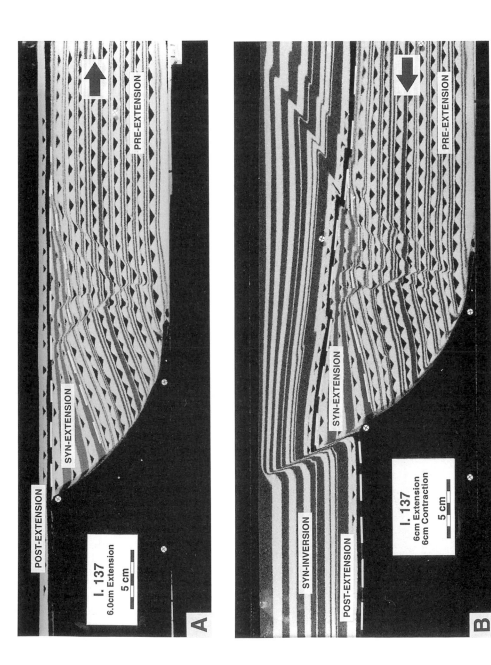

Fig. 5. Extension and inversion of a simple 60° listric fault: (**A**) photograph of Experiment I 137 at the end of 6 cm of extension; (**B**) photograph of Experiment I 137 at the end of 6 cm of contraction.

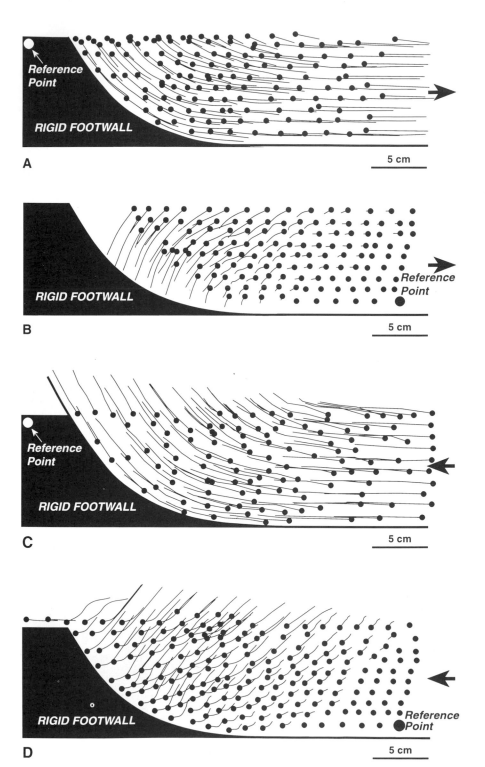

Fig. 6. Analysis of particle displacements for extension and inversion above a simple 60° listric fault: (**A**) particle displacement paths for extension above a 60° listric fault, measured relative to a fixed point in the footwall fault block; (**B**) particle displacement paths for extension above a 60° listric fault, measured relative to a fixed point in the hangingwall fault block. These trajectories indicate the apparent shear collapse of the hangingwall; (**C**) particle displacement paths for inversion above a 60° listric fault, measured relative to a fixed point in the footwall fault block; (**D**) particle displacement paths for inversion above a 60° listric fault, measured relative to a fixed point in the hangingwall fault block. These trajectories indicate the apparent shear for inversion of the hangingwall.

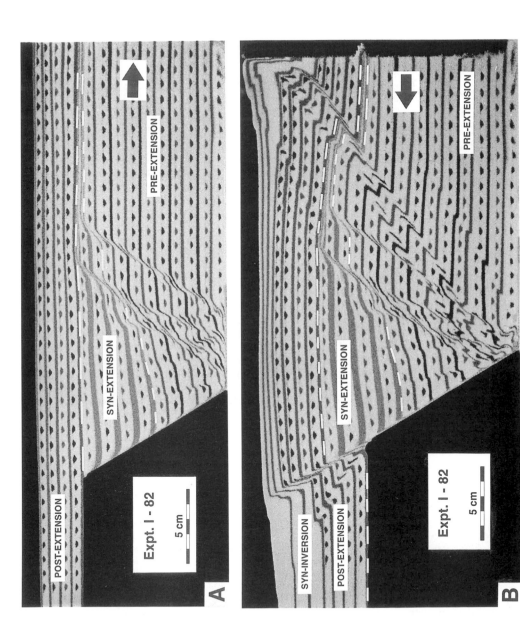

Fig. 7. Extension and inversion of a simple 60° planar fault: (**A**) photograph of Experiment I 82 at the end of 6 cm of extension; (**B**) photograph of Experiment I 82 at the end of 6 cm of contraction.

apparent shear trajectories for inversion that are
initially shallow and then steepen upwards with
increased contraction (Fig. 6d). The initial flat or
shallow apparent shear trajectories reflect an
important component of dilation (volume contraction) at the early stages of inversion. It is also
important to note that the apparent shear
trajectories during inversion are less steep than
those during extension thus indicating a significant change in the angle of shear failure between
extension and inversion.

Planar faults

Extension above a 60° planar detachment fault
produces a characteristic planar roll-over structure bounded by two planar antithetic faults
(Fig. 7a). The amount of bed rotation in the
roll-over is proportional to the amount of
extension and the kinked shape at the base of the
detachment. The roll-over has horizontal hangingwall strata adjacent to the steep section of the
main detachment and gently tilted and faulted
beds near the antithetic faults (Fig. 7a) on the
right hand side of the roll-over. The displacement on the planar antithetic faults dies out
downwards towards the main detachment surface (Fig. 7a). In the pre-extension strata, near
the steep portion of the main detachment,
complex, convex-up and curved synthetic faults
are cut by planar antithetic faults (Fig. 7a). As
these are carried down onto the flat section of
the main detachment they become highly rotated and form a complex structure that is
difficult to interpret. As a result particle displacement paths are obscured within this part of
the model (Figs 8 a, b).

Inversion of the planar extensional fault
model was achieved by horizontal recompression in the same manner as for the simple listric
fault. The main detachment surface was reactivated to produce a fault-bounded wedge above
the steep planar extensional fault breakaway
(Fig. 7b). The main detachment fault propagated as a 60° planar fault into the overlying
post-rift and syn-inversion strata (Fig. 7b).
Upon inversion the hangingwall was moved
back up the main detachment surface. A small
displacement footwall short-cut thrust fault
formed in the post-rift and syn-rift section in
front of the main fault. A small displacement
thrust fault nucleated from the tip of one of the
antithetic faults (Fig. 7b). With increased contraction new hangingwall-vergent (McClay
1992) thrust faults developed on the right hand
side of the model. These link to the kink point in
the basal detachment and appear to be the result
of buttressing against the steep section of the

planar fault. The overall inversion geometry is
that of a broad, asymmetric 'pop-up' bounded
on the left-hand side by a thrust fault in the
post-extension   and   syn-inversion   strata
(Fig. 7b). The inversion structure shows relatively   horizontal   post-extension   and   syn-
inversion strata in contrast to the architecture of
the inverted listric fault described above.

Particle displacement paths For the simple
planar fault the displacement trajectories relative to a fixed point in the footwall show particle
paths that parallel the main detachment (Fig. 8a)
similar to those described for the simple listric
fault model. Measured relative to a fixed point in
the hangingwall the particle displacements for
hangingwall collapse during extension show
straight, planar apparent shear trajectories that
dip uniformly at 56–58° in an antithetic sense
towards the main detachment surface (Fig. 8b).
During inversion the particle displacement paths
relative to a fixed point in the footwall show
trajectories that parallel the main detachment
surface (Fig. 6c). The particle displacement
paths calculated relative to a fixed point in the
hangingwall show apparent shear trajectories for
inversion that are initially shallow and then
steepen upwards and become planar at a
constant dip of 41–42° with increased contraction (Fig. 6d). As for the listric fault model the
initial flat or shallow apparent shear trajectories
reflect an important component of dilation
(volume contraction) at the early stages of
inversion. The apparent shear trajectories during inversion for the planar fault model are also
less steep (41–42°) than those during extension
(56–58°).

Ramp-flat listric fault

The structural architecture which evolves above
an extensional ramp-flat detachment geometry
has been discussed in detail by Ellis & McClay
(1988) and McClay & Scott (1991). The basic
structural elements are an upper roll-over and
associated crestal collapse graben developed
above the upper listric portion of the detachment (Fig. 9a) together with a hangingwall
syncline overlying the convex-up ramp section of
the main detachment, and a larger roll-over and
crestal collapse graben structure formed above
the lower concave-up section of the main
detachment (Fig. 9a). Several small displacement 'reverse' faults develop during extension
above the convex upwards ramp segment of the
main detachment (Fig. 9a).

Upon inversion, reactivation of the lower
listric ramp resulted in the propagation of a

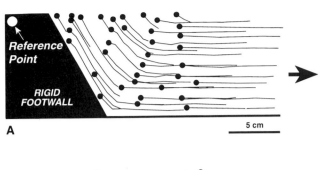

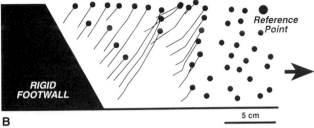

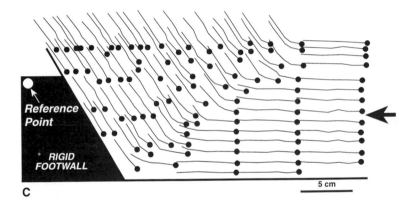

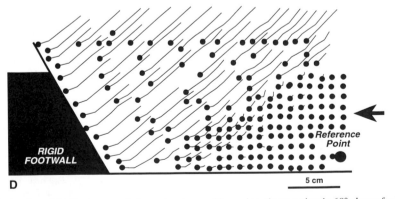

Fig. 8. Analysis of particle displacements for extension and inversion above a simple 60° planar fault: (**A**) particle displacement paths for extension above a 60° planar fault, measured relative to a fixed point in the footwall fault block; (**B**) particle displacement paths for extension above a 60° planar fault, measured relative to a fixed point in the hangingwall fault block. These trajectories indicate the apparent shear collapse of the hangingwall; (**C**) particle displacement paths for inversion above a 60° planar fault, measured relative to a fixed point in the footwall fault block; (**D**) particle displacement paths for inversion above a 60° planar fault, measured relative to a fixed point in the hangingwall fault block. These trajectories indicate the apparent shear for inversion of the hangingwall.

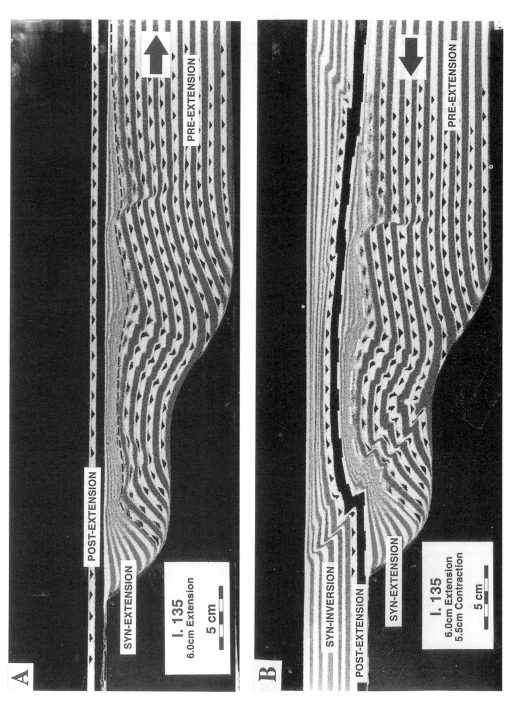

Fig. 9. Extension and inversion of a 60° ramp-flat listric fault: (**A**) photograph of Experiment I 135 at the end of 6 cm of extension; (**B**) photograph of Experiment I 135 at the end of 5.5 cm of contraction.

major, footwall-vergent thrust fault upwards through the upper crestal collapse graben (Fig. 9b). The greater footwall vergent displacement along this thrust produced a broad 'pop-up' structure and a marked thickness contrast across the fault-bounded wedge in the syn-inversion sequence. Minor amounts of thrusting parallel to this major thrust were also initiated in its hangingwall. This footwall-vergent, 'hangingwall-bypass thrust' is relatively planar and the upper extensional roll-over structure, in the footwall to this by-pass thrust, is preserved unaffected by the contraction. Fixing of the plastic detachment during inversion permits hangingwall bypass faults to form whereas in the extensional deformation mode the whole hangingwall is detached by the moving plastic sheet and bypass faults cannot form. As in the case of the listric fault model (Fig. 5b) buttressing against the lower ramp in the detachment results in the development of a hangingwall-vergent 'backthrust' that nucleates from the crestal collapse graben in this region (Fig. 9b). The synthetic (right-dipping) extensional fault to this crestal collapse graben is reactivated but also undergoes significant curvature during inversion, in contrast to the antithetic faults (left-dipping), giving a highly asymmetric profile to this 'pop-up'. The 'pop-up' structure is centred above the soling-out point of the lower ramp of the original extensional detachment. The maximum uplift in the hangingwall during inversion is located above the top of the ramp in the main detachment (Fig. 9b). As a consequence of the relatively planar nature of the major footwall vergent thrust, little rotation is observed in the hangingwall which is simply displaced over the undeformed footwall.

Particle displacement paths For the ramp/flat listric fault the displacement trajectories relative to a fixed point in the footwall show particle paths that parallel the main detachment (Fig. 10a) similar to the simple listric and planar fault models described above. Measured relative to a fixed point in the hangingwall the particle displacements for hangingwall collapse during extension show more complex paths than the simple models described above. Above the simple listric geometry of the upper section of the main detachment the apparent shear trajectories display steepening downwards curved paths (Fig. 10b) as in the simple listric fault model (Fig. 6b). Above the convex-up section of the detachment complex paths and steep trajectories reflect the interference of the upper and lower roll-over structures. Above the lower listric section of the main detachment steepening

downwards curved shear trajectories are found (Fig. 10b).

During inversion the particle displacement paths relative to a fixed point in the footwall show trajectories that parallel the lower section of the main detachment surface but because of the development of the hangingwall bypass thrust they parallel this structure and bypass the upper listric section of the detachment (Fig. 10c). The particle displacement paths calculated relative to a fixed point in the hangingwall show apparent shear trajectories for inversion that are initially shallow and then steepen upwards (Fig. 10d). As for the listric fault model the initial flat or shallow apparent shear trajectories reflect an important component of dilation (volume contraction) at the early stages of inversion. The apparent shear trajectories during inversion of the ramp-flat fault model are also less steep than those during extension.

Domino fault array

The extensional architecture produced above a series of domino faults is illustrated in Fig. 11a and is similar to that described in detail in Buchanan (1991) and in Buchanan & McClay (1992). The structure is similar for each domino fault and consists of a half graben filled with a syn-extension wedge in which the layering is planar. The final profile geometry of the domino faults has a slight listricity acquired during extension as a result of the interaction between sedimentation and rotation (see Vendeville & Cobbold 1988). Minor antithetic planar faults form in the hangingwall to each domino fault (Fig. 11a). The displacements on these antithetic faults die out upwards, from the pre-extension strata into the syn-extension strata.

Inversion of this domino fault array and associated half-grabens was achieved by pushing the individual domino fault blocks back to their pre-extension configuration such that the top of the rigid basement was returned to horizontal (Fig. 11c). Contractional strain in the sand pack above was accommodated by reactivation of the extensional faults. With increased contractional displacement and back rotation of the domino fault blocks, footwall shortcut faults developed in the post-rift and also syn-rift and pre-rift strata (Fig. 11b). These shortcut faults are planar to slightly convex upwards in profile and define downward tapering wedge shaped blocks at the margin of the basins. Where much of the displacement is transferred on to the footwall shortcut faults, the upper portion of the domino

K. R. McCLAY

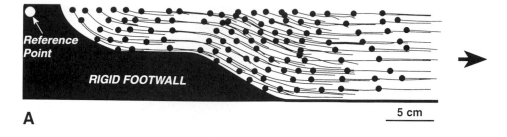

A   5 cm

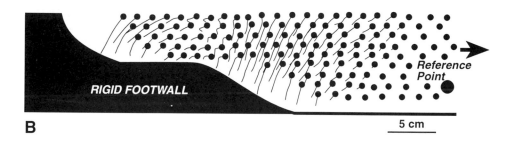

B   5 cm

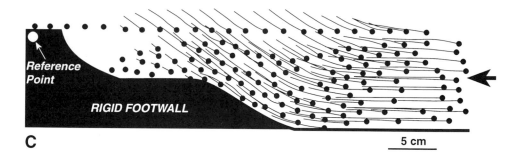

C   5 cm

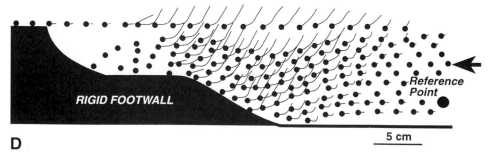

D   5 cm

Fig. 10. Analysis of particle displacements for extension and inversion above a 60° ramp-flat listric fault: (**A**) particle displacement paths for extension above a 60° ramp-flat listric fault, measured relative to a fixed point in the footwall fault block; (**B**) particle displacement paths for extension above a 60° ramp-flat listric fault, measured relative to a fixed point in the hangingwall fault block. These trajectories indicate the apparent shear collapse of the hangingwall; (**C**) particle displacement paths for inversion above a 60° ramp-flat listric fault, measured relative to a fixed point in the footwall fault block; (**D**) particle displacement paths for inversion above a 60° ramp-flat listric fault, measured relative to a fixed point in the hangingwall fault block. These trajectories indicate the apparent shear for inversion of the hangingwall.

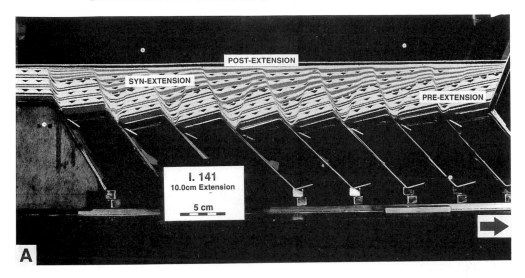

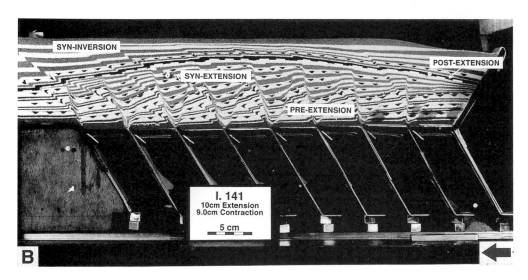

Fig. 11. Extension and inversion of a 60° domino fault array: (**A**) photograph of Experiment I 141 at the end of 10 cm of extension; (**B**) photograph of Experiment I 141 at the end of 9 cm of contraction.

fault remained unreactivated. All the extensional and contractional faulting in the cover above each domino block links into the domino master fault in the rigid basement below. Reactivation of individual domino faults causes post-rift and upper syn-rift layers to be put into net contraction whilst below, the syn-rift and pre-rift layering remains in net extension. As a result, null points where net displacement across the fault is zero and classic 'arrowhead' or 'harpoon' structures are formed. Reactivation of the antithetic faults during inversion did not generally occur in these experiments. The style

of the contractional deformation is strongly asymmetric, all the reactivated and shortcut faults verging in the same direction (i.e. footwall-vergent).

Discussion

The dip-slip extension and dip-slip inversion analogue models reviewed in this paper illustrate many of the fundamental features on inversion tectonics that are found in natural examples (e.g. Fig. 2). A number of characteristics are found in all four types of models.

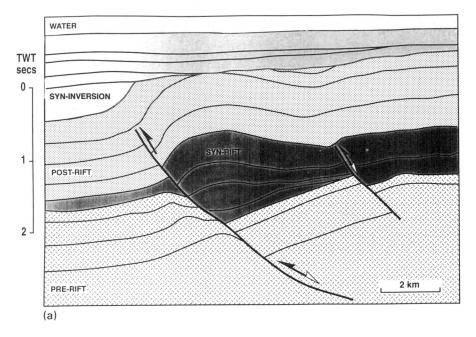

(a)

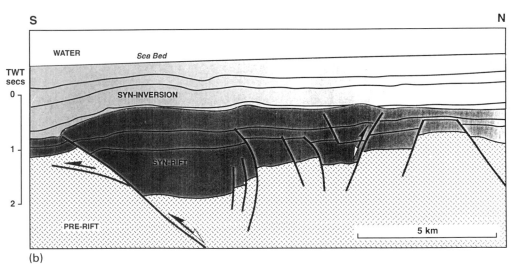

(b)

Fig. 12. Examples of inverted fault systems: (**a**) geoseismic section of an inverted fault with characteristic harpoon structure (line drawing of a seismic section in Goudswaard & Jenyon 1988). The white arrows indicate early extensional faulting whereas the black arrows indicate later contractional faulting; (**b**) geoseismic section across an inverted planar fault from the East Java Sea basin, Indonesia. The white arrows indicate early extensional faulting whereas the black arrows indicate later contractional faulting; (**c**) geoseismic section across an inverted domino fault array from the Hatton Bank (line drawing of a seismic section in Boldreel & Andersen 1993). The white arrows indicate early extensional faulting whereas the black arrows indicate later contractional faulting.

Inversion of the wedge shaped syn-extensional strata produces typical 'harpoon' or 'arrowhead' geometries, the shape of which depends upon the geometry of the underlying extensional fault (e.g. Figs 5, 7, 9 & 11).

Elements of such features have been also discussed by Buchanan & McClay (1991, 1992) and by Koopman et al. (1987). Syn-inversion strata thin on top of this wedge and thicken away from it.

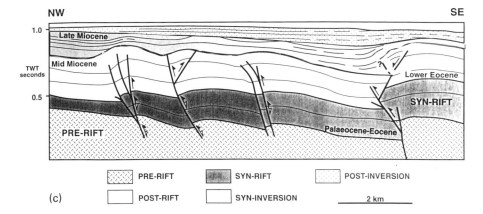

(c)

During inversion of the detachment models, reactivation results in propagation of the main fault into the post-extension and syn-inversion strata (e.g. Figs 5b, 7b & 9b). In modelling materials such as clay, which appears to be more competent than sand, propagation of the main detachment during inversion produces a fault-propagation growth anticline (e.g. Mitra 1993). These anticlines typically show asymmetric growth stratal relationships with a tight growth syncline in the footwall and thin growth geometries on the anticline crest.

In the detachment models (Figs 5, 7 & 9), where crestal collapse grabens and strong antithetic faulting is developed, inversion compresses these features such that thrust faults propagate from this region producing 'pop-up' structures. In cross-section the superposition of contractional structures on pre-existing extensional structures in these regions produces a geometry that has the appearance of a 'flower structure' but is generated simply by dip-slip extension followed by dip-slip inversion (see also Harding 1985; Buchanan & McClay 1991; Mitra 1993).

One of the most characteristic features of all of the experimental models carried out in this programme and in other experiments (Buchanan 1991; Buchanan & McClay 1991, 1992) is the development of shortcut thrusts in the footwall to the reactivated extensional faults (see Fig. 11b). These are commonly typically convex upwards and form a small fault-propagation fold at their tips. At high values of contraction the footwall shortcut thrusts are responsible for generating a lower angle, more smoothly varying thrust trajectory and may result in the incorporation of exotic basement slices within the inversion structure. In the detachment models (Figs 5, 7 & 9) the rigid footwall block precludes the development of

significant footwall shortcut thrusts whereas in the inverted domino arrays (Fig. 11) the deformable sand footwalls in the pre-extension strata readily develop footwall shortcut faults.

The experiments presented in this paper have been repeated many times and consistent inversion architectures have been obtained. In particular similar results have been obtained in experiments where the footwall block was moved out for extension and pushed back into the model to produce inversion. As noted in the section on modelling method above, during inversion the plastic sheet usually remained stationary and the sandpack was pushed back up the detachment. This procedure slightly changes the frictional constraints at the base of the model but comparisons with models in which the plastic sheet was translated up the detachment during inversion, however, show only minor differences in the inversion geometries. In these models in which the plastic sheet was moved during inversion more rotation of the syn-extensional wedge occurred but buttressing was also developed.

The analysis of particle displacement paths indicates significant changes in the angles of apparent hangingwall shear between extension and inversion. Importantly the shear angles for inversion are generally 15° lower than for extension, a factor that needs to be borne in mind for numerical modelling and construction of balanced sections. The particle path analysis also shows that the early stages of inversion involve significant dilation and compaction as the hangingwall is compressed against the rigid footwall of the main detachment. Further detailed testing with different fault geometries and differing boundary conditions is warranted in order to determine fully the deformation mechanisms and particle displacement paths in a variety of inversion models.

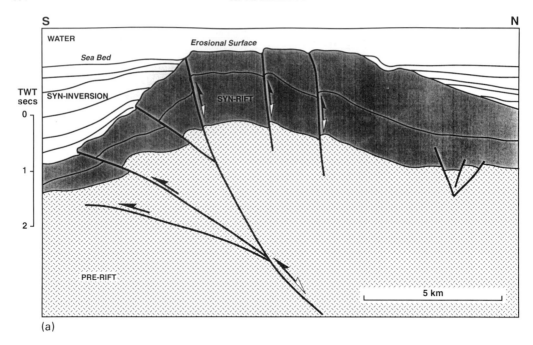

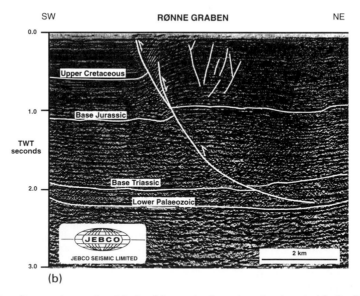

Fig. 13. Examples of inverted extensional faults: (**a**) geoseismic section of an inverted fault with development of an emergent growth fold, East Java Sea basin, Indonesia. The white arrows indicate early extensional faulting whereas the black arrows indicate later contractional faulting; (**b**) seismic section across an inverted fault with a fault propagation growth anticline, Ronne graben, Bornholm, Baltic Sea. Jurassic to Lower Cretaceous is the syn-extensional sequence and inversion occurred in the Late Cretaceous and Tertiary (Deeks & Thomas, this volume). Seismic data courtesy of JEBCO Seismic Limited; (**c**) seismic section across an inverted and buttressed fault at the southeast margin of the Ronne graben, Bornholm, Baltic Sea. Pre-extensional strata (Lower Palaeozoic and Triassic) and syn-extensional strata (Jurassic and Lower Cretaceous) are buttressed and folded in the hangingwall of the basis margin fault. Only post-extension Upper Cretaceous and syn-inversion Tertiary strata occur above crystalline basement on the footwall side of the fault (Deeks & Thomas, this volume). Seismic data courtesy of JEBCO Seismic Limited.

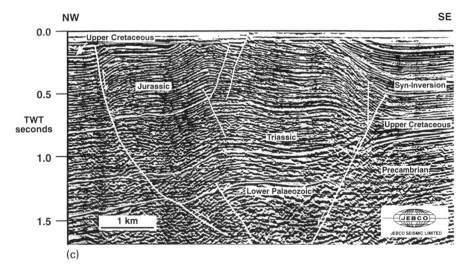

(c)

Natural examples of inverted extensional structures show many similarities to the models described above. Harpoon structures formed by wedges of syn-extension strata are formed by reactivation of both planar and listric extensional faults (e.g. Fig. 12a,b). Fault-propagation growth anticlines are commonly formed at the tips of the reactivated extensional faults (e.g. Mitra 1993). In the example shown in Fig. 12a the fault-propagation fold that is developed in the post-extensional sequence has become exposed and eroded during the inversion. Figure 12b shows a reactivated planar fault system from the East Java Sea, Indonesia. Syn-inversion strata thin onto the uplifted syn-extensional sedimentary wedge and thicken away from it (Fig. 12b). These features clearly show striking similarities to the inversion architectures found in the analogue models (e.g. Fig. 7b). A striking example of an inverted domino fault array is shown in Fig. 12c near the Hatton Bank, west of Shetland UK (Boldreel & Andersen 1993). Inversion occurred in the Mid to Late Miocene (Boldreel & Andersen 1993) with the domino faults being reactivated and developing footwall shortcut faults. The resultant structure is very similar to the domino fault models shown in Fig. 11b and also in Buchanan & McClay (1992).

Fault-propagation growth anticlines are characteristic of many natural inversion structures (e.g. Mitra 1993) but are only poorly developed in the sandbox models due to the homogeneous nature of the sandpack and the lack of strong, competent bedding units that could undergo folding during inversion. Two outstanding examples of natural fault-propagation anticlines are shown in Figs 13a, b. Here asymmetric growth anticlines are developed above reactivated extensional faults. Figure 13a shows an emergent growth anticline in the East Java Sea. Syn-inversion strata onlap onto the anticline whose crest has undergone erosion and truncation. Figure 13b shows a well developed fault-propagation fold formed by inversion of the Ronne graben, southern Baltic Sea (see Deeks & Thomas this volume). This structure has undergone strong inversion with the entire fault now in net contraction (Fig. 13b) and the syn-extensional Jurassic to Lower Cretaceous strata are folded into a hangingwall anticline by the inversion. Figure 13c shows buttressed folding against the inverted basin margin fault of the Ronne graben (Deeks & Thomas this volume). Buttressing, broad arching and back-thrusting are commonly developed in the inversion models (Figs 5b, 7b & 9b) and in outcrop examples of reactivated faults but is not usually shown on seismic interpretations.

A series of simple synoptic models for each of the four styles of extension and inversion models described in this paper are shown in Fig. 14. These illustrate the differences both in extension and inversion between the simple listric and simple planar faults (Fig. 14a,b) and the fundamental architectures found in the ramp/flat and domino fault arrays (Fig. 14c, d). In all of the four model styles the inverted syn-extensional stratal wedge develops a characteristic 'harpoon' or 'arrowhead' geometry that shows many of the aspects of the 'Sunda folds' described from inverted rift basins in Indonesia.

A INVERSION OF A SIMPLE LISTRIC FAULT

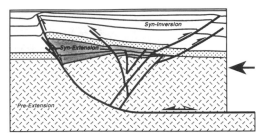

B INVERSION OF A SIMPLE PLANAR FAULT

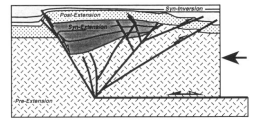

C INVERSION OF A RAMP/FLAT LISTRIC FAULT

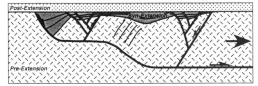
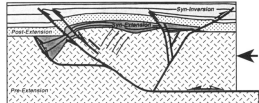

D INVERSION OF DOMINO FAULT BLOCKS

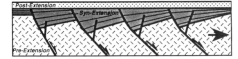
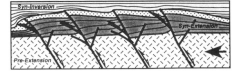

Fig. 14. Synoptic models for inversion of simple extensional fault systems: (**A**) inversion of a simple listric fault; (**B**) inversion of a simple planar fault; (**C**) inversion of a ramp-flat listric fault; (**D**) inversion of a domino fault array.

Conclusions

The experimental models and examples cited in this paper illustrate some of the complexities of the structures that are found when extensional basin systems are subjected to dip-slip inversion. The analogue models show many striking similarities to natural examples of inverted basin fault systems. Harpoon and arrowhead syn-extension stratal wedge geometries are charac-teristic of inversion tectonics. Fault-bounded wedges typically develop above reactivated sections of the main detachments or basin bounding faults. 'Pseudo flower structures' can be formed by dip-slip extension followed by dip-slip inversion indicating that flower ge-ometries must be interpreted with caution and do not necessarily indicate strike-slip tectonics. Footwall shortcut faults and backthrusts are common features of both the experimental models and are also found in nature. Buttressing effects against steep sections of bounding faults produce compactional features and folding. These characteristics are found both in the analogue models and in natural examples of inverted extensional faults. The models re-viewed in this paper combined with the detailed analysis of the particle displacement paths indicate that scaled analogue modelling is a powerful technique in helping to understand the geometries and kinematics of basin inversion.

The research presented in this paper is part of the Fault Dynamics Project sponsored by ARCO British Ltd., BP Exploration, BRASOIL UK Ltd., CONOCO UK Ltd., MOBIL North Sea Ltd. and Sun Oil Britain Ltd. B. Adams, L. Thompson and P. Hollings are thanked for assistance in the analogue modelling laboratory. Aspects of the modelling research presented in this paper are based upon MSc research by M. Simmons and PhD research by P. Buchanan and their collabor-ation is gratefully acknowledged. D. Waltham and J. Clark wrote the digitising and analysis programs for particle displacement calculations. The Bornholm seismic data was provided courtesy of JEBCO Seismic Limited who are thanked for permission to publish. N. Deeks is thanked for advice on the seismic interpre-tations. C. Powell and M. Withjack are thanked for critical reviews of the manuscript. Fault Dynamics Publication No. 43.

References

BALLY, A. W. B. 1984. Tectogenese et seismique reflexion. *Bulletin de la Société Géologique de France*, **7**, 279–285.

BOLDREEL, L. O. & ANDERSEN, M. S. 1993. Late Paleocene to Miocene compression in the Faeroe–Rockall area. *In*: PARKER, R. J. (ed.) *Petroleum Geology of Northwest Europe: Pro-ceedings of the 4th Conference*, The Geological Society, London, 1025–1034.

BUCHANAN, P. G. 1991. *Geometries and kinematic analysis of inversion tectonics from analogue model studies*. PhD thesis, University of London.

—— & McCLAY, K. R. 1991. Sandbox experiments of inverted listric and planar faults systems. *Tec-tonophysics*, **188**, 97–115.

—— & —— 1992. Experiments on basin inversion above reactivated domino faults. *Marine and Petroleum Geology*, **9**, 486–500.

COOPER, M. A. & WILLIAMS, G. D. (eds) 1989. *Inversion Tectonics*. Geological Society, London, Special Publication, **44**.

DEEKS, N. R. & THOMAS, S. A. 1995. Basin inversion in a strike-slip regime: the Tornquist Zone, southern Baltic Sea. *This volume*.

ELLIS, P. G. & McCLAY, K. R. 1988. Listric extensional fault systems: results from analogue model experiments. *Basin Research*, **1**, 55–71.

FRASER, A. J., NASH, D. F., STEELE, R. P. & EBDON, C. C. 1990. A regional assessment of the intra-Carboniferous play of Northern England. *In*: BROOKS, J. (ed.) *Classic Petroleum Provinces*. Geological Society, London, Special Publication, **50**, 417–440.

GOUDSWAARD, W. & JENYON, M. K. (eds) 1988. *Siesmic atlas of structural and stratigraphic fea-tures*. European Association of Exploration Geo-physicists.

HARDING, T. P. 1985. Seismic characteristics and identification of negative flower structures, posi-tive flower structures, and positive structural inversion. *American Association of Petroleum Geologists*, Bulletin, **69**, 582–600.

HORSFIELD, W. T. 1977. An experimental approach to basement controlled faulting. *Geologie en Mijnbouw*, **56**, 363–370.

KELLER, J. & McCLAY, K. R. 1995. 3D sandbox models of positive inversion. *This volume*.

KOOPMAN, A., SPEKSNIJDER, A. & HORSFIELD, W. T. 1987. Sandbox model studies of inversion tec-tonics. *Tectonophysics*, **137**, 379–388.

LETOUZEY, J. 1990. Fault reactivation, inversion and fold-thrust belt. *In*: LETOUZEY, J. (ed.) *Petroleum and Tectonics in Mobile Belts*. IFP Editions Technip, Paris, 101–128.

McCLAY, K. R. 1989. Analogue models of inversion tectonics. *In*: COOPER, M. A. & WILLIAMS, G. D. (eds) *Inversion Tectonics*. Geological Society, London, Special Publication, **44**, 41–59.

—— 1990. Extensional fault systems in sedimentary basins. A review of analogue model studies. *Marine and Petroleum Geology*, **7**, 206–233.

—— 1992. Glossary of thrust tectonics terms. *In*: McCLAY, K. R. (ed.) *Thrust Tectonics*, Chapman & Hall, London, 419–434.

—— & BUCHANAN, P. G. 1992. Thrust faults in inverted extensional basins. *In*: McCLAY, K. R. (ed.) *Thrust Tectonics*, Chapman & Hall, Lon-don, 419–434.

—— & SCOTT, A. D. 1991. Experimental models of hangingwall deformation in ramp-flat listric ex-tensional fault systems. *Tectonophysics*, **188**, 85–96.

MITRA, S. 1993. Geometry and kinematic evolution of inversion structures. *American Association of Petroleum Geologists*, Bulletin, **77**, 1159–1191.

SIMMONS, M. 1991. *An experimental study of the geometries and kinematics of inversion and growth structures*. MSc thesis, Royal Holloway, Univer-sity of London.

VENDEVILLE, B. & COBBOLD, P. R. 1988. How normal

faulting and sedimentation interact to produce
listric fault profiles and stratigraphic wedges.
Journal of Structural Geology, **10**, 649–659.
WILLIAMS, G. D., POWELL, C. M. & COOPER, M. A.
1989. Geometry and kinematics of inversion
tectonics. *In*: COOPER, M. A. & WILLIAMS, G. D.
(eds) *Inversion Tectonics*. Geological Society,
London, Special Publication, **44**, 3–15.

Estimating inversion: results from clay models

GLORIA EISENSTADT & MARTHA OLIVER WITHJACK

*Mobil Research and Development Corporation, PO Box 650232, Dallas, TX 75265,
USA*

Abstract: Physical models using wet clay show that, although both fold and fault geometries change with increasing amounts of inversion, it is difficult to estimate the magnitude of inversion. During extension, a listric main normal fault and a rollover fold cut by secondary normal faults developed above two diverging, basal plates. With 50% inversion (4 cm extension, 2 cm shortening), the main normal fault underwent reverse displacement, and an asymmetric syncline formed in the pre-growth layers in the hangingwall. Secondary antithetic and synthetic normal faults passively rotated. Faults on the synclinal limb near the main normal fault rotated away from the fault, whereas those on the opposing limb rotated towards the fault. A broad anticline formed at shallow levels above the original half graben. Compressional reactivation of the main normal fault ceased between 50% and 100% inversion (4 cm extension, 4 cm shortening). Instead, low-angle thrust faults with small displacements accommodated most shortening. The shallow anticline above the original half graben expanded laterally and vertically. With 200% inversion (4 cm extension, 8 cm shortening), low-angle thrust faults with large displacements deformed the hangingwall and footwall of the main normal fault. Thrust faults cut the main normal fault and many of the secondary normal faults. The geometry of the hangingwall fold varied with depth. A broad anticline deformed the upper growth layers, changing to a syncline in the lower growth layers and an anticline in the pre-growth layers.

Estimates of inversion from null point analysis, amount of uplift above regional datum, or line length balancing, grossly underestimate the amount of shortening in the models. In addition, both the 50% and 100% inversion models lack the large-scale deformation features typically associated with compression. These results suggest that both quantitative and qualitative techniques used to calculate inversion magnitude can significantly underestimate the amount of shortening. Furthermore, if the inversion uplift is eroded or if many small-scale deformation features are not imaged seismically, then inversion may go unrecognized.

Quantifying the magnitude of shortening during inversion is a fundamental problem in basin reconstruction and modelling. The most widely cited technique for estimating inversion magnitude involves identifying a null point, which is a level of no stratigraphic offset between normal and reverse motion along a fault (De Paor & Eisenstadt 1987; Williams *et al.* 1989). This method assumes that all shortening occurs along a single normal fault. Therefore, the null point should reflect the relative importance of extension versus contraction (Mitra 1993). Other cited methods of estimating inversion magnitude involve measuring the amount of uplift above a regional datum (Stoneley 1982; Mitra 1993) or estimating shortening by line-length restoration techniques (Badley *et al.* 1989; Chadwick 1993), effectively treating inversion structures as compressional structures. Chadwick (1993) has tried to quantify the former technique by suggesting that shortening can be calculated as a function of the uplift amplitude above the regional datum and the dip of the planar, controlling fault.

In practice, most geologists make qualitative judgments to estimate inversion magnitude. These judgments are based primarily on how much the inversion geometry differs from typical extensional structures, or resembles typical compressional structures. Figure 1 shows two examples of compressional inversion structures. Cartwright (1989) suggested that the structure in the Danish Central Graben (Fig. 1a) represents mild inversion because of its similarity to typical extensional structures. The structure in Fig. 1b, however, appears to represent greater inversion because of its resemblance to a compressional structure.

In this paper we address the questions of how fault and fold geometries vary with increasing inversion and how well quantitative and qualitative methods determine the magnitude of inversion. We approached the problem experimentally with a series of inversion models of listric normal faults using wet, layered clay as the modelling material. For comparison we also performed pure extensional and contractional

From BUCHANAN, J. G. & BUCHANAN, P. G. (eds), 1995, *Basin Inversion,*
Geological Society Special Publication No. 88, 119–136.

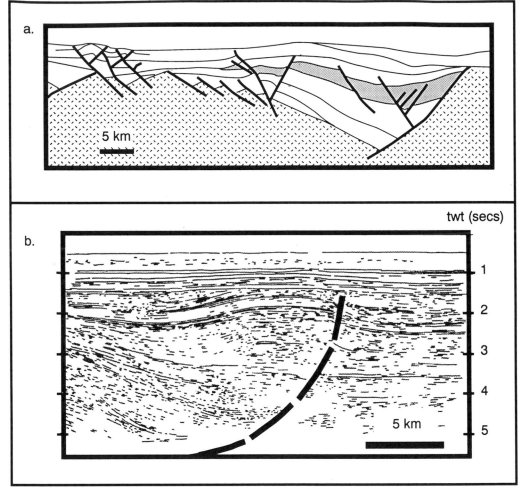

Fig. 1. Examples of inversion structures. (a) Inversion structure from Danish Central Graben, considered to be mild (after Cartwright 1989). Vertical exaggeration is approximately 1 : 1. (b) Line drawing of proprietary time-migrated seismic line showing inversion structure believed to have undergone moderate amount of inversion. Vertical exaggeration is approximately 1 : 1 at 2 seconds TWT.

experiments. Although other physical models of inversion structures above listric normal faults have been reported, these models focused on hangingwall deformation using a rigid footwall. Some of those models used sand (McClay 1989; Buchanan 1991), and other models used un-layered clay that extruded sideways during inversion, causing loss of area in cross sections (Mitra 1993). In our clay models both the hangingwall and footwall could deform, the model's side surfaces were constrained to pre-vent extrusion and, because the clay was layered, we were able to examine cross sections free of edge effects.

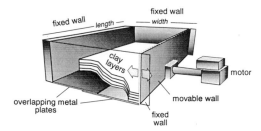

Fig. 2. Diagram of experimental apparatus showing cut-away view of clay layers above thin, overlapping metal plates. The upper metal plate attaches to a fixed wall. The bottom plate is attached to a movable wall.

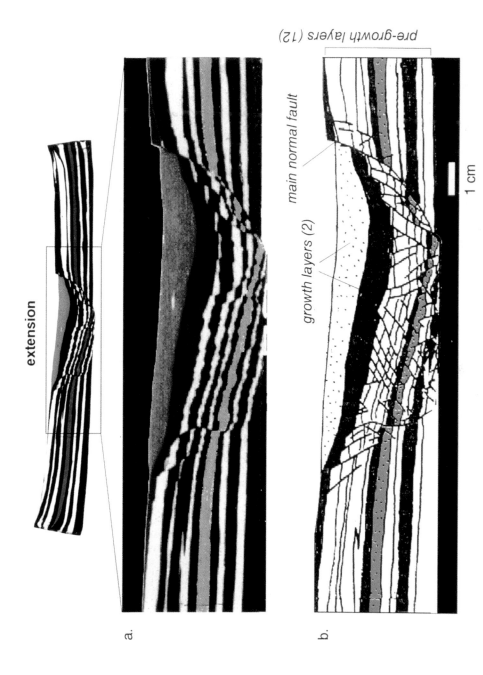

Fig. 3. Extensional model after 4 cm of displacement of the movable wall. (a) Photograph of section through model. Only central part of model is shown. (b) Interpreted line drawing of same cross-section showing main normal fault, rollover and secondary antithetic and synthetic normal faults. The sixth layer from the bottom is highlighted (purple) for comparison with the other models.

Experimental procedure

The experimental apparatus had three fixed walls, a movable wall, and two thin, overlapping basal metal plates (Fig. 2). The lower basal plate was attached to the movable wall and the upper basal plate was attached to the opposing, fixed wall. Clay, consisting of water (50% by weight), kaolin, powdered nepheline-syenite and powdered flint, covered the metal plates. Its cohesive strength was about 40 Pa (Sims 1993). The clay, initially 4 cm thick, 60 cm long and 24.5 to 32 cm wide, was composed of twelve thin, coloured layers of identical composition.

During modelling, displacement of the movable wall and the attached metal plate produced extension and/or contraction in the overlying clay. The model length remained constant while the width changed (Fig. 2). The edge of the upper plate determined the location of the main normal fault during extension and localized contractional deformation during shortening. In each model, the displacement rate of the movable wall was 4 cm/hr (0.001 cm/s). Additional coloured clay layers were added during extension to simulate deposition and record the timing of normal faulting. Each growth layer was added such that its upper surface was flat and horizontal. After each model run, the clay was left to dry for approximately one week. It was then serially sectioned, lacquered and photographed. Although the extent of edge effects (as evidenced by curved faults along the fixed walls) appear to only extend 2 to 4 cm from the walls, measurements of strain markers on the surface showed that the edge effects actually extended 8 to 10 cm from the fixed walls. Consequently only sections through the central part of the models were used for analysis.

Extensional experiments

The initial width of the extensional models was 24.5 cm, and they were extended to 28.5 cm. In these models a main normal fault formed above the edge of the upper metal plate and propagated upward through the clay (Fig. 3). The main normal fault initially formed as discontinuous fault segments along strike. With greater extension, some normal faults coalesced along strike to form a through-going main normal fault. Other normal faults, bypassed by the main normal fault, became inactive. As the experiment continued, secondary synthetic and antithetic normal faults and a rollover fold formed in the hangingwall of the main normal fault. The main normal fault and most secondary normal

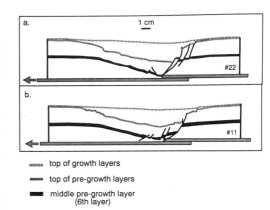

  top of growth layers
  top of pre-growth layers
  middle pre-growth layer
     (6th layer)

Fig. 4. Simplified line drawings of cross-sections from extensional model showing variation in fold geometry of pre-growth layers along stike. (a) Cross-section where all displacement occurred along the main normal fault, creating a rollover into the fault. (b) Cross-section where displacement occurred along several fault splays, creating a syncline in the hangingwall next to the main normal fault.

faults dipped 60° to 70°. The geometry of the hangingwall fold varied along strike. A hangingwall rollover formed in the lower pre-growth layers where most of the displacement occurred on the main normal fault (Fig. 4a). Where displacement was accommodated amongst the main normal fault and several fault splays, a hangingwall syncline formed instead (Fig. 4b).

The extensional models resemble those published by H. Cloos (1928, 1930) and E. Cloos (1968). Unlike these published experiments, however, our models were internally layered, permitting us to analyse changes in fold and fault geometries laterally through the models. Additionally, the layers added during extension allowed us to compare structural geometries in the pre-growth and growth sections. Surface and side observation showed no apparent difference between the behaviour of the clay in layered and unlayered models, suggesting that the layering exerted no influence on the deformation.

Withjack *et al.* (1995) demonstrated that some boundary conditions have little effect on fold and fault geometries, whereas other boundary conditions profoundly affect resulting geometries. For example, they showed that extensional deformation patterns that form over overlapping basal plates are similar whether the footwall is rigid or deformable. In contrast, extensional physical models with a mylar sheet along the main fault surface have very different secondary fault patterns. One of the major differences is the absence of a crestal collapse graben in models without a mylar sheet.

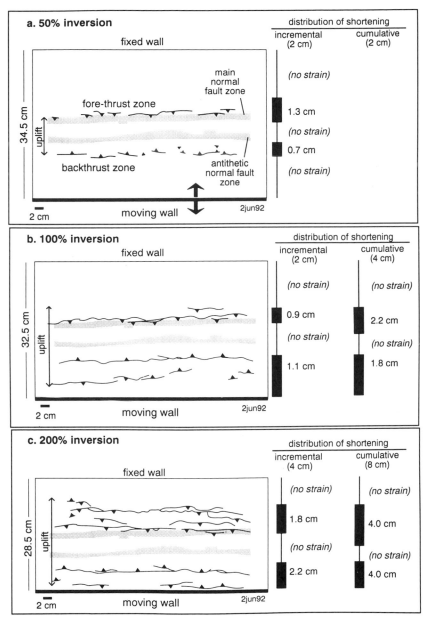

Fig. 5. Map views of progressive deformation in 200% inversion model. Original main normal fault and antithetic normal fault zones shown in gray. (a) At 50% inversion, numerous small fore- and backthrusts are visible. The fore-thrust zone overlies the old main normal fault zone and accommodates twice as much shortening as the backthrust zone. An uplift develops between the fore- and backthrust zones. (b) At 100% inversion, thrusts link along strike, new thrusts develop further away from the half graben, and the uplift grows laterally and vertically. The distribution of shortening along the fore- and backthrust zones has almost equalized. (c) At 200% inversion, new thrusts form. The proximity of the moving wall hinders the development of backthrusts.

Cloos (1968) showed that the overall ge-ometry of extensional sand and clay models with overlapping basal plates is very similar; the different materials cause only minor variations. Our preliminary tests indicate that this con-clusion is not true for inversion models; clay and

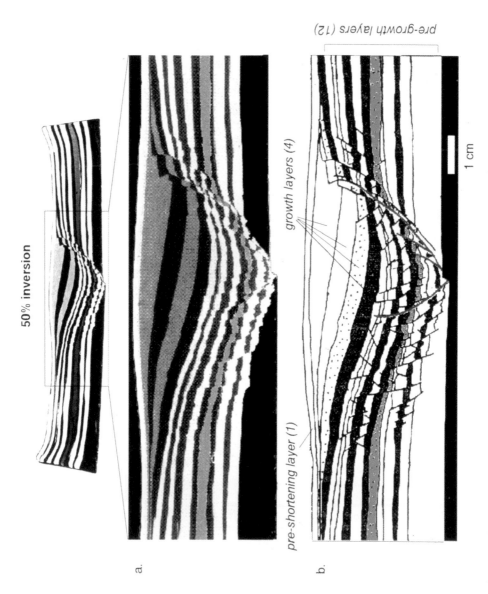

Fig. 6. (a) Photograph of cross-section through 50% inversion model. Only central part of model is shown. (b) Interpreted line drawing; black lines are old normal faults; red lines are normal faults reactivated as reverse faults. The sixth layer from the bottom is highlighted (purple) for comparison with the other models.

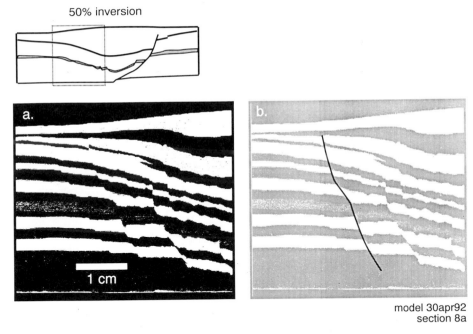

Fig. 7. (a) Close-up photograph of 50% inversion model. Sketch indicates location. (b) Same photograph highlighting old normal fault reactivated as reverse fault. Note normal displacement of lower layers and reverse displacement of upper layers.

sand inversion models formed over overlapping basal plates have very different fault geometries. Sand inversion models show only a minor amount of reactivation of older normal faults and develop thrust faults much sooner than clay models. Faults in clay seem to both reactivate more readily and to accommodate much more displacement than those in sand.

Inversion experiments

Each inversion experiment started with an extensional growth model 28.5 cm wide before shortening (except for the 200% inversion model that was 36.5 cm wide). A thin layer of clay (0.25 cm thick) covered the model after extension and before shortening. Each model was initially extended 4 cm, and the amount of shortening varied: the 50% inversion model was shortened 2 cm; the 100% inversion model was shortened 4 cm; and 200% inversion model was shortened 8 cm. The surface of the inversion models was photographed after every 0.25 cm of displacement to record fold, fault and strain patterns.

Map view

Figure 5 shows the progression of deformation in map view of the 200% inversion model. The first thrust faults were visible after 1.75 cm of displacement of the movable wall (44% inversion). At 2 cm of displacement (50% inversion), fore-thrusts and backthrusts appeared on either side of an inversion uplift (Fig. 5a). The fore-thrusts developed above the main normal fault zone of the extensional models. The backthrusts, however, were not spatially associated with the zone of antithetic normal faults. Shortening was not equally partitioned between the fore-thrust and backthrust zones, demonstrating that preferential movement occurred along the original main normal fault zone.

By 4 cm of displacement (100% inversion), the early-formed thrust faults had coalesced along strike and new fore-thrusts and back-thrusts began to form away from the inversion uplift, causing the uplift to broaden (Fig. 5b). Shortening was confined to the fore- and backthrust zones and was almost equally partitioned between the two zones. By 8 cm of displacement (200% inversion), additional fore- and backthrusts developed away from the inversion uplift, causing the uplift to further broaden. At 200% inversion, shortening was equally partitioned between the fore- and backthrust zones (Fig. 5c).

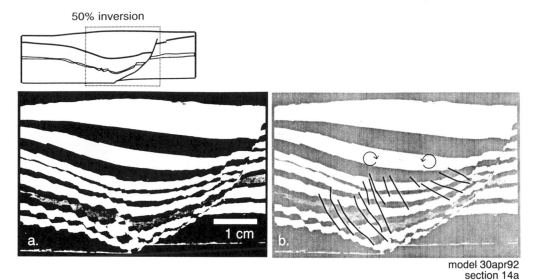

model 30apr92
section 14a

Fig. 8. (a) Close-up photograph of 50% inversion model. Sketch indicates location. (b) Same photograph highlighting rotated antithetic normal faults. Antithetic faults on the synclinal limb near the original main normal fault rotate counter-clockwise (away from the fault) and become more gently dipping, whereas antithetic faults on the opposing limb rotate clockwise (towards the fault) and become more steeply dipping as the normal fault reverses movement.

50% inversion – cross-sectional view

In cross-sectional view, the 50% inversion model lacked deformation features typically associated with compression (Fig. 6). Shortening reactivated the main normal fault and other secondary normal faults as reverse faults (compare position of highlighted layer in the extensional model (Fig. 3) and 50% inversion model (Fig. 6)). Most faults maintained a normal sense of offset, although some faults showed reverse offset in the uppermost layers (Fig. 7). Inversion produced an uplift in the upper growth and post-growth layers above the original half graben. At deeper levels, in the pre-growth layers, an asymmetric syncline formed in the hangingwall of the main normal fault. This hangingwall syncline developed in all cross sections, regardless of whether the original extensional geometry was a rollover monocline into the fault or a gentle hangingwall syncline. The development of this hangingwall syncline during inversion caused secondary normal faults on the limb near the main normal fault to rotate counter-clockwise (away from the fault). As a result the dip of synthetic faults increased and the dip of antithetic faults decreased (Fig. 8). Similarly, secondary normal faults on the opposing limb of the syncline rotated clockwise (towards the fault), causing the dip of synthetic

faults to decrease and the dip of antithetic faults to increase.

Only a few thrust faults can be seen in cross-sections of the 50% inversion model, all with minute displacements. Although the model surface showed two discrete zones of thrusts (Fig. 5a), these zones are not evident in cross-sectional view. The lack of correlation between surface and cross-sectional deformation may be due to few pre-existing normal faults in the uppermost growth layers. The upper growth layers accommodated the shortening along new thrusts while the interior of the original half graben accommodated the shortening by reverse motion along old normal faults.

100% inversion – cross-sectional view

Reverse motion along the original main normal fault stopped between 50% and 100% inversion (Fig. 9), leaving most normal faults with a normal sense of offset. The hangingwall syncline in the pre-growth layers became very gentle and those layers almost resumed their pre-extensional geometry. The syncline in the growth layers became more pronounced with increased shortening. The inversion uplift expanded laterally and vertically, and pervasive low-angle thrust faults cut the hangingwall and footwall of the original main normal fault (Figs 10 & 11). These

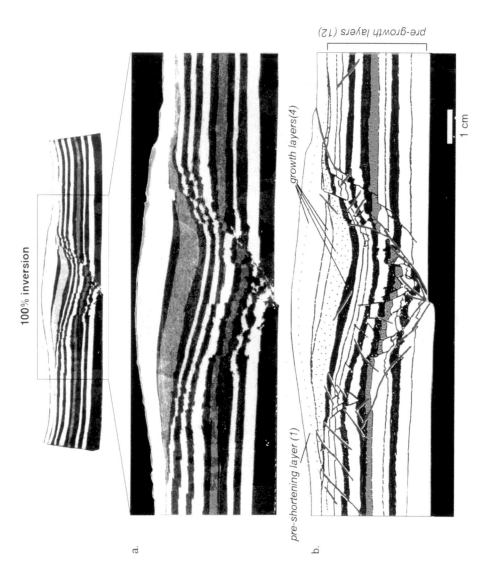

Fig. 9. (a) Photograph of cross-section through 100% inversion model. Only central part of model is shown. (b) Interpreted line drawing: black lines are old normal faults; red lines are normal faults reactivated as reverse faults and newly formed thrust faults. The sixth layer from the bottom is highlighted (purple) for comparison with the other models.

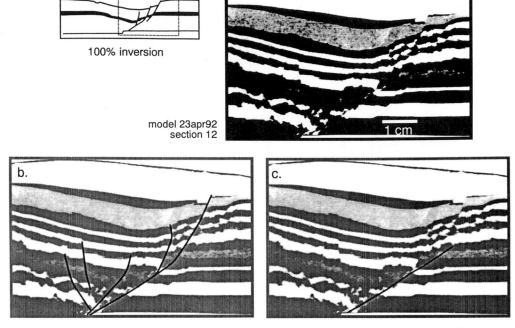

100% inversion

model 23apr92
section 12

Fig. 10. (a) Close-up photograph of 100% inversion model. Sketch shows location of photograph. (b) Same photograph highlighting old normal faults, and (c) same photograph showing one of several newly formed thrust faults cutting the old main normal fault zone.

thrusts verge toward the centre of the half graben. The thrust faults mapped at the surface of the 100% inversion model cannot be traced at depth in cross-sectional view. These surface thrust faults appear to acommodate the shortening in the upper growth layers (Fig. 5b).

200% inversion – cross-sectional view

By 200% inversion, low-angle thrust faults accommodated the shortening throughout the model (Fig. 12). Several thrust faults cut the main normal fault, giving it the appearance of a folded fault with a lower overall dip of about 45° (Fig. 13). With increasing shortening, the inversion uplift expanded laterally and vertically. The fold geometry changed dramatically with depth: a broad anticline developed in the uppermost layers, the syncline in the lower growth layers became more asymmetric, and an asymmetric anticline formed in the lower pre-growth layers. The central part of the half graben was uplifted and rotated towards the main normal fault. Many unusual structures exist in the 200% inversion model, although they do not all appear on every cross section. For example, on some sections small inverted grabens in the footwall of the original main normal

fault have structural characteristics of flower structures (Fig. 14), and on other sections fault-propagation folds at depth appear to link upward into normal faults (Fig. 15).

Influence of pre-existing structures

To investigate the influence of pre-existing zones of weakness on inversion geometries, we compared two models shortened the same amount, a 100% inversion model and a contractional model. The inversion model was initially 24.5 cm wide. It was then extended 4 cm, and then shortened 4 cm. The contractional model was initially 28.5 cm wide, and then shortened 4 cm. Comparisons of the 100% inversion model with the contractional model show that the pre-existing zones of weakness associated with extension profoundly affected the fault patterns during shortenings (Figs 9 & 16). Cross-sections through the contractional model show that faulting occurred in two discrete zones on either side of a symmetrical pop-up structure (Fig. 16). In the 100% inversion experiment, however, deformation was more evenly distributed throughout the model as reactivated normal faults and newly formed thrust faults (Fig. 9). The strain distribution in map view also differed

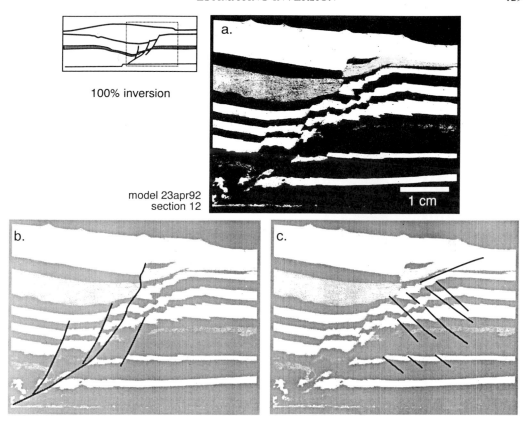

Fig. 11. (a) Close-up photograph of 100% inversion model. Sketch shows location of photograph. (b) Same photograph highlighting old normal faults, and (c) same photograph showing a few of the newly formed thrust faults deforming the footwall.

markedly in the inversion and contractional models. In the 100% inversion model, shortening occurred unequally in two discrete zones (Fig. 5b), with most initial shortening occurring over the original main normal fault. In the contractional model, strain was initially evenly distributed throughout the model (Fig. 17a). With increasing shortening the strain then became partitioned equally into two discrete zones (Fig. 17b). It is clear from a comparison of these maps that the original extensional faults control the early distribution of shortening in the inversion models.

Discussion of modelling results

The layered clay models show that fold geometries change with increasing amounts of inversion (Fig. 18). In the extensional models, either a rollover or a syncline forms in the lower pre-growth layers in the hangingwall of the main normal fault (Fig. 18a). With up to 50%

inversion, the main normal fault experiences reverse movement, and an asymmetric syncline develops in the pre-growth layers in the hangingwall, regardless of the original extensional geometry (Fig. 18b). A broad anticline forms at shallow levels above the original half graben. With 50% inversion, fold geometries generally resemble those associated with extension rather than compression (compare Figs 4b & 18b). Compressional reactivation of the main normal fault ceases between 50% and 100% inversion. The anticline above the original half graben becomes higher and broader, whereas the hangingwall syncline in the pre-growth layers becomes very gentle (Fig. 18c). At 200% inversion, the geometry of the hangingwall fold varies with depth (Fig. 18d). An asymmetric anticline exists in the lower pre-growth layers, a syncline affects the lower growth layers, and a broad anticline deforms the uppermost growth and post-growth layers.

The models also show that fault geometries

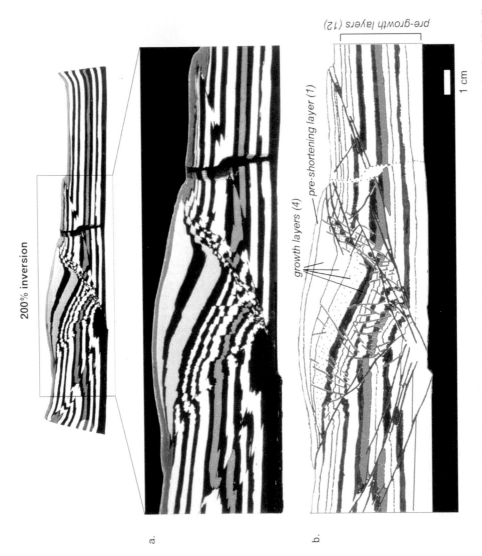

Fig. 12. (a) Photograph of cross-section through 200% inversion model. Only central part of model is shown. (b) Interpreted line drawing: black lines are old normal faults; red lines are normal faults reactivated as reverse faults and newly formed thrust faults. The sixth layer from the bottom is highlighted (purple) for comparison with the other models.

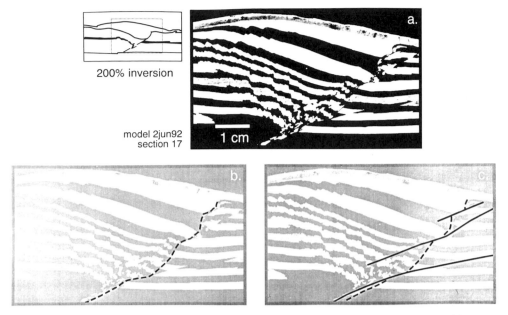

Fig. 13. (a) Close-up photograph of 200% inversion model. Sketch shows location of photograph. (b) Same photograph showing erroneous interpretation of main normal fault being folded or having ramp/flat geometry. (c) Same photograph showing correct interpretation with old main normal fault being cut by several low-angle thrusts.

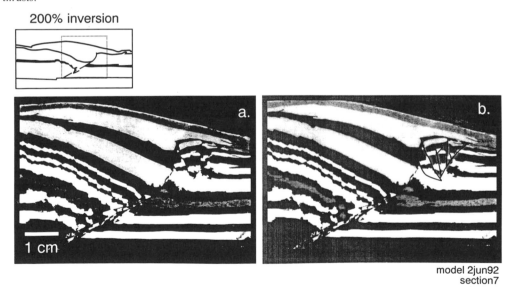

Fig. 14. (a) Close-up photograph of 200% inversion model. Sketch shows location of photograph. (b) Same photograph highlighting inverted graben in the footwall resembling a flower structure.

change with increasing amounts of inversion (Fig. 19). A main normal fault and secondary normal faults, having dips of 60° to 70°, develop in the extensional models. With 50% inversion, secondary antithetic and synthetic normal faults

passively rotate as the hangingwall reverses displacement along the main normal fault and forms a syncline in the pre-growth layers (Fig. 19b). Faults on the synclinal limb near the main normal fault rotate counter-clockwise (away

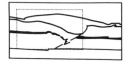

200% inversion

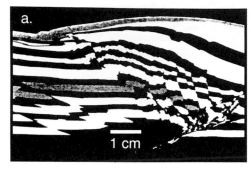

a.

1 cm

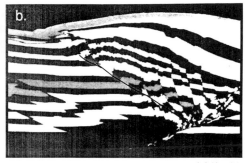

b.

model 2jun92
section11

Fig. 15. (a) Close-up photograph of 200% inversion model. Sketch shows location. (b) Same photograph showing newly formed fault-propagation fold (solid line) linking upsection to older normal fault (dashed line).

from the fault), whereas those on the opposing limb rotate clockwise (towards the fault). This rotation produces anomalous high and low fault dips. By 100% inversion, reverse motion has ceased along the main normal fault. Instead, low-angle thrust faults with small displacements accommodate most of the shortening and deform both the hangingwall and footwall (Fig. 19c). These thrusts faults form on both sides of the original half graben, far from the end walls, and have a vergence towards the half graben. With 200% inversion, the hangingwall and footwall of the main normal fault are cut by low-angle thrust faults with large displacements, these faults have a vergence away from the half graben. These later thrust faults have the same dip and vergence as the thrusts formed in the pure shortening model. Some thrust faults cut the main normal fault, causing it to appear folded and more gently dipping (Fig. 19d).

Techniques for determining inversion magnitude (i.e. null point analysis, line-length restoration and uplift calculations) underestimate the amount of shortening in these inversion models. Null-point analysis along the main normal fault suggested that the amount of extension was greater than the amount of shortening in the three inversion models. This prediction is correct for the 50% inversion model, but erroneous for the 100% and 200% inversion models. Null-point analysis fails because shortening during inversion is accommodated by newly formed thrust faults and reactivated secondary normal faults that are distributed throughout the

model, not just by displacement along the reactivated main normal fault. In cases where the main normal fault has been cut by low angle thrust faults (200% inversion), a null point analysis becomes meaningless.

Line-length restoration assumes that layer thickness remains constant during deformation and all the shortening occurs along visible thrust faults or folds. Neither assumption is valid for these inversion models and therefore the technique grossly underestimates the amount of shortening. The calculated amount of shortening for the upper layers in the 50%, 100% and 200% inversion models is less than 0.1 cm, 0.2 cm, and 1.6 cm, respectively. The actual amounts are 2 cm, 4 cm and 8 cm respectively. The difference between the measured and actual amount of shortening suggests that much of the deformation is occurring at a scale too small for observation.

Measuring the area of uplift provided the best estimate, although still too low, of the amount of inversion. The technique assumes that the area of uplift above the regional datum divided by the detachment depth is equivalent to the amount of shortening associated with inversion. This method suggests that the amount of shortening in the 50%, 100% and 200% inversion models was about 0.6 cm, 1.6 cm and 4.8 cm respectively. Although area is preserved in the cross-sections during inversion, uplift calculations fail to accurately predict the amounts of shortening in the physical models because both the footwall and hangingwall are elevated during inversion.

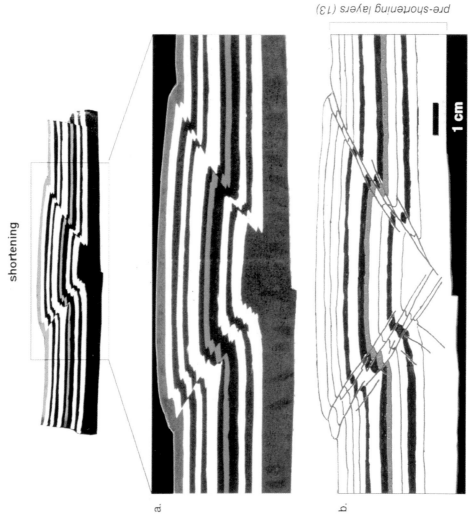

Fig. 16. Contractional model after 4 cm of displacement of the movable wall. (a) Photograph of centre of cross-section through model. (b) Interpreted line drawing of same cross-section showing thrust faults (red). The sixth layer from the bottom is highlighted (purple) for comparison with the other models.

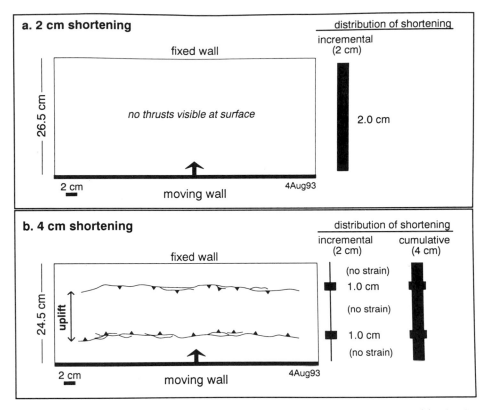

Fig. 17. Map view of shortening model showing thrusts, uplift and distribution of shortening. (a) After 2 cm of displacement of the movable wall, no thrusts are visible. Shortening is distributed evenly throughout the model. (b) After 4 cm of displacement, fore- and backthrusts bound a well-developed uplift. Shortening is distributed equally along the two fault zones. Continued shortening causes new thrusts to form further away from the centre of the model, increasing both the height and width of the uplift.

Thus, the interpreted regional datum is too high and the measured amount of uplift above the regional datum is too low. In real examples there is also the danger that a seemingly undeformed regional datum has been uplifted during inversion.

Qualitative judgements based on how much the geometry of an inversion structure resembles either an extensional or compressional structure would also underestimate the magnitude of inversion. For example, the structural features in the 50% and 100% inversion models more closely resemble those in the extensional model than those in the contractional model. Consequently subtle inversion structures, like the feature in the Danish Central Graben (Fig. 1a), might actually have undergone significant amounts of shortening. Erosion of the uplift associated with mild inversion would leave only structures with extensional geometries.

Conclusions

The clay models show that although quantitative techniques fail to estimate correctly amounts of inversion, there are systematic changes in fold and fault patterns with increasing magnitudes of inversion. These changes in structural geometries, along with the recognition of small-scale deformation, may help define the magnitude of inversion in real examples. Initially shortening during inversion is accommodated along pre-existing normal faults and small-scale thrust faults. For less than 100% inversion, the main normal fault experiences reverse displacement, and an asymmetric hangingwall syncline forms in the pre-growth layers. Secondary antithetic and synthetic normal faults passively rotate. Faults on the synclinal limb near the main normal fault rotate away from the main normal fault, whereas those on the opposing limb rotate

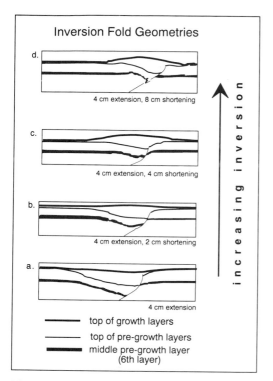

Fig. 18. Diagram summarizing changes in fold geometries with increasing inversion. (a) A rollover into the main normal fault is developed in the pre-growth layers during extension. (b) Characteristics of the 50% inversion model include reactivation of the main normal fault as a reverse fault creating a hangingwall syncline in pre-growth layers, and the development of a broad surface uplift. (c) Characteristics of the 100% inversion model include continued reactivation of the main normal fault as a reverse fault, continued development of the surface uplift and flattening of pre-growth layers. (d) Characteristics of the 200% inversion model include continued development of the surface uplift, development of a hangingwall anticline in the pre-growth layers, and development of a hangingwall syncline in the growth layers.

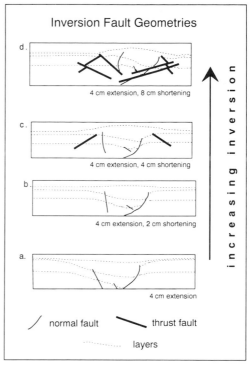

Fig. 19. Diagram summarizing changes in fault geometries with increasing inversion. (a) Extensional antithetic and synthetic faults have dips of 60° to 70°. (b) Characteristics of 50% inversion include reactivation of main normal fault as a reverse fault, causing rotation of secondary normal faults with anomalous high and low fault dips. (c) Characteristics of 100% inversion include continued rotation of secondary normal faults and pervasive development of fore- and backthrusts (verging towards centre of model) in both hangingwall and footwall. (d) Characteristics of 200% inversion include offset of main normal fault by low-angle thrusts (causing it to appear more gently dipping), continued uplift and rotation of secondary normal faults, and new fore- and backthrusts verging away from the centre of the model.

towards the fault. A broad anticline forms at shallow levels above the original half graben. Compressional reactivation of the main normal fault ceases between 50% and 100% inversion, and low-angle thrust faults with small displacements accommodate most shortening. The broad anticline above the original half graben expands laterally and grows vertically. With 200% inversion, the hangingwall and footwall of the main normal fault are cut by low-angle thrust faults with large displacements. This causes the main normal fault to appear to be more gently dipping with a ramp/flat or folded trajectory.

Thrust faults cut the main normal fault and many of the secondary normal faults. The geometry of the hangingwall fold varies with depth. A tight anticline in the pre-growth layers changes to a syncline in the lower growth layers and a broad anticline in the upper growth and post-growth layers.

The clay models suggest that both quantitative and qualitative methods would consistently underestimate inversion magnitude. If the footwall and hangingwall are elevated during inversion, then techniques using the amount of uplift above the regional datum would underestimate inversion magnitude. Similarly, if pervasive small-scale thrust faults and reactivated normal faults accommodate much of the shortening, then techniques using line-length restoration or null-point analysis would also underestimate the amount of shortening associated with inversion. A broad, gentle anticline and small-scale thrust faults are the only evidence of shortening in both the 50% and 100% inversion models. Amounts of inversion for either model would be underestimated because the large-scale faults and folds in those models are more typically associated with extension rather than compression.

We thank Kris Meisling, Bruno Vendeville and Chris Banks for their thoughtful and critical reviews. Peter Hennings, Eric Peterson and Joana Vizgirda provided helpful discussions, and Charlie Wall assisted with the physical modelling. We also thank Mobil Research and Development Corporation and their management for support of this research and permission to publish.

References

BADLEY, M. E., PRICE, J. D. & BACKSHALL, L. C. 1989. Inversion, reactivated faults and related structures: seismic examples from the southern North Sea. *In*: COOPER, M. A. & WILLIAMS, G. D. (eds) *Inversion Tectonics*. Geological Society, London, Special Publication, **44**, 201–219.

BUCHANAN, P. 1991. *Geometries and Kinematic Analysis of Inversion Tectonics from Analogue Model Studies*. PhD thesis, Royal Holloway and Bedford College, University of London.

CARTWRIGHT, J. A. 1989. The kinematics of inversion in the Danish Central Graben. *In*: COOPER, M. A. & WILLIAMS, G. D. (eds) *Inversion Tectonics*, Geological Society, London, Special Publication, **44**, 153–175.

CHADWICK, R. A. 1993. Aspects of basin inversion in southern Britain. *Journal of the Geological Society*, **150**, 311–322.

CLOOS, E. 1968. Experimental analysis of Gulf Coast fracture patterns. *American Association of Petroleum Geologists Bulletin*, **52**, 420–444.

CLOOS, H. 1928. Experimente zur inneren Tektonik. *Centralblatt für Mineralogies*, **Abt B**, 609–621.

—— 1930. Kunstliche Gebirge, II. *Natur und Museum*, **60**, 258–269.

DE PAOR, D. G. & EISENSTADT, G. 1987. Stratigraphic and structural consequences of fault reversal: An example from the Franklinian Basin, Ellesmere Island. *Geology*, **15**, 948–949.

McCLAY, K. R. 1989. Analogue models of inversion tectonics. *In*: COOPER, M. A. & WILLIAMS, G. D. (eds) *Inversion Tectonics*. Geological Society, London, Special Publication, **44**, 41–59.

MITRA, S. 1993. Geometry and kinematic evolution of inversion structures. *American Association of Petroleum Geologists Bulletin*, **77**, 1159–1191.

SIMS, D. 1993. The rheology of clay: a modeling material for geologic structures. *Eos, Transactions, American Geophysical Union*, 569.

STONELEY, R. 1982. The structural development of the Wessex Basin. *Journal of the Geological Society*, **139**, 543–554.

WILLIAMS, G. D., POWELL, C. M. & COOPER, M. A. 1989. Geometry and kinematics of inversion tectonics. *In*: COOPER, M. A. & WILLIAMS, G. D. (eds) *Inversion Tectonics*, Geological Society, London, Special Publication, **44**, 3–15.

WITHJACK, M. O., ISLAM, Q. & LA POINTE, P. 1995. Normal faults and their hangingwall deformation: an experimental study. *American Association of Petroleum Geologists Bulletin*, **79**, 1–18.

3D sandbox models of positive inversion

J. V. A. KELLER & K. R. McCLAY

Fault Dynamics Project, Department of Geology, Royal Holloway, University of London,
Egham, Surrey, TW20 0EX, UK

Abstract: 3D scaled physical models of positive inversion involving idealized, simple cylindrical listric and ramp-flat listric fault systems were carried out. In both types of models extension and inversion produced strongly segmented faults with fault displacement varying along-strike. Displacement transfer was achieved through relay ramp structures between overlapping faults. No transfer faults developed during extension or inversion. Inversion produced four distinct responses, basal décollement reactivation, folding, faulting and back-rotation of faults associated with localized layer-parallel shortening. Upon inversion reverse faults nucleated at the tips of pre-existing extensional faults. Reverse faults formed longer segments and accommodated less displacement than extensional faults of the same length. Maximum fault displacement/length ratios (γ) for extensional and reverse faults in the models varied between 2.5×10^{-1} and 7.0×10^{-3} (best fit values range from 1.1×10^{-2} to 6.4×10^{-2}) and are within the range measured for natural faults. γ values for reverse faults are consistently lower than for extensional faults in both listric or ramp-flat experiments. The results of these analogue models are comparable to natural examples of inverted fault systems.

Tectonic inversion is a switch in tectonic mode such that a region which has been extended subsequently undergoes contraction or vice versa. A region affected by extension and subsequently subjected to contractional deformation is said to have undergone positive inversion, whilst a region which suffered contraction and later underwent extension is said to have experienced negative inversion (Harding 1985; Gillcrist *et al.* 1987).

Physical analogue model studies of inversion have been described by Koopman *et al.* (1987), McClay (1989), Buchanan & McClay (1991), Roure *et al.* (1992), Mitra (1993) and Krantz (1991). All of these studies, however, have focused on the 2D geometries of inversion structures. This paper presents and analyses the results of a series of 3D sandbox positive inversion experiments carried out in the Fault Dynamics modelling laboratory at Royal Holloway, University of London. These experiments were aimed at a better understanding of the along-strike geometry of inverted extensional fault systems and their progressive evolution, with particular focus being placed upon the study of fault propagation.

Experimental set-up

The experiments were carried out in a 150 cm × 60 cm × 10 cm deformation rig and typical initial model dimensions were 55 cm × 60 cm × 10 cm (Fig. 1). Deformation was basically plane strain

as the rig has constant width, with out of the plane movement and oblique-slip faults beyond the scope of this paper. Dry quartz sand was the main modelling material used (average grain size 212 μm; average density 157 g/cm^3; angle of internal friction ø = 30–31° (McClay 1990a)). Dry cohesionless sand is a good analogue for upper crust brittle rocks in the laboratory (see Horsfield 1977; McClay 1990b) and deforms independently of strain rate. Models were built using 0.5 cm thick layers of different coloured sand interbedded with thin films of mica flakes in order to enhance bed-parallel slip (McClay 1990b). A rigid wooden footwall block acted as basement fault with a pre-defined geometry. For simplicity the footwalls used in the experiments had cylindrical listric and ramp-flat listric geometries, with fault cut-off angles of 60°. The hangingwall was decoupled from the footwall by using a plastic sheet as a low friction décollement horizon (Fig. 1). The plastic sheet was attached to the moving wall during the experiments and transfers the extensional and/or contractional displacements to the sand-pack in the hangingwall. The moving wall was connected via a worm screw to a motor that generated a constant displacement velocity $v = 4.2 \times 10^{-3}$ cm/s.

Syn-rift layers were added incrementally during extension to maintain a horizontal upper surface to the model. At the end of extension a 1.0 cm post-rift layer was deposited. Inversion was achieved by reversing the motion of the end-wall and syn-inversion layers were added

From BUCHANAN, J. G. & BUCHANAN, P. G. (eds), 1995, *Basin Inversion,*
Geological Society Special Publication No. 88, 137–146.

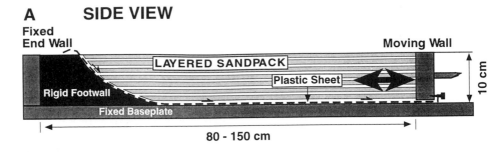

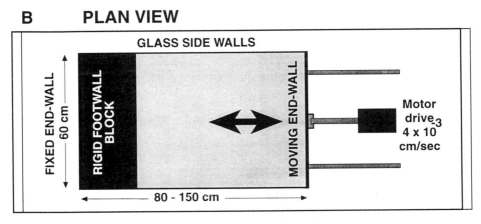

Fig. 1. Schematic diagram of deformation rig. (**A**) side view; (**B**) top view.

after every 1.0 cm increment of contraction. The top surface in each model was monitored using 35 mm still photography during extension and contraction. All the models were extended to a total of 10.0 cm followed by 5.0 cm contraction. At the end of the experiments the models were impregnated and serially sectioned, either vertically or horizontally, to analyse the finite 3D fault geometries of the inverted fault systems. Each experiment was repeated under identical conditions to assure the reproducibility of the results.

Experimental results

Two series of 4 3D inversion experiments were carried out and representative model results are presented below. The two different systems studied involved deformation above a 60° cylindrical listric detachment and above a 60° cylindrical ramp-flat listric detachment. The

ramp-flat detachment consisted of an upper cylindrical listric ramp followed by a flat segment and a lower cylindrical listric ramp.

60° listric fault

Extension Extension above the 3D listric décollement produced a roll-over anticline and a long 15.0 cm wide, crestal-collapse graben. The surface faulting pattern was defined by straight, steeply-dipping (60–65°) overlapping fault segments that extended across the width of the model (Fig. 2a). The graben was formed by prominent graben-bounding faults and shorter intra-graben faults, both sets forming perpendicular or at high angles to the extension direction (parallel to the footwall breakaway). During the early stages of extension the graben-bounding faults developed propagating straight segments that overlapped, forming relay ramps.

Fig. 2. (**a**) – (**d**) Sequential diagrams showing the top surface of the model Inv–3 at 10.0 cm extension and the evolution of the inversion fault system after 1.0, 3.0 and 5.0 cm of contraction. Note how the main displacement transfer zones observed during extension coincide with discontinuous displacement zones during inversion.

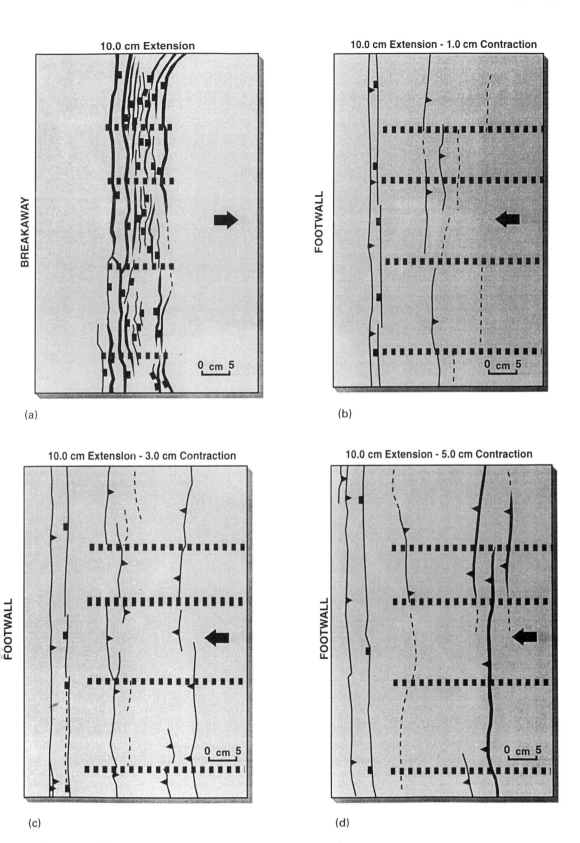

10.0 cm Extension

BREAKAWAY

0 cm 5

(a)

10.0 cm Extension - 1.0 cm Contraction

FOOTWALL

0 cm 5

(b)

10.0 cm Extension - 3.0 cm Contraction

FOOTWALL

0 cm 5

(c)

10.0 cm Extension - 5.0 cm Contraction

FOOTWALL

0 cm 5

(d)

■■■■■■■ Displacement transfer zone

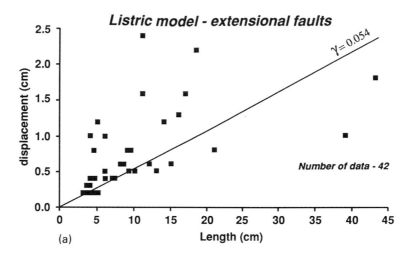

(a)

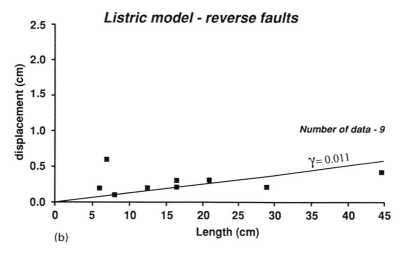

(b)

Fig. 3. (a) Maximum fault displacement d vs. fault length L for extensional faults developed in the 60° listric model. (b) Maximum fault displacement d vs. fault length L for reverse faults developed during inversion in the 60° listric model.

Intra-graben faults, with less displacement than the graben-bounding faults, also developed laterally overlapping segments. With continuation of the extension the graben-bounding faults, as well as intra-graben faults, propagated towards each other breaching the relay zones and coalescing to form longer faults. Some faults developed highly sinuous convergent segments that exhibit fault polarity reversal along-strike (Fig. 2a). No strike-slip transfer faults were observed.

Inversion Inversion was characterized by the

reverse reactivation of the low-friction basal décollement and the formation of new reverse faults. Reverse reactivation of the main décollement generated a fault propagation growth anticline above the main detachment breakaway, thinning of the syn-inversion strata over the growth anticline and an upward-propagating thrust. Long and straight, shallowly dipping (30–40°), overlapping reverse faults formed in the region above the crestal-collapse graben (Fig. 2b). The reverse faults nucleated at the tips of synthetic crestal-collapse graben extensional faults and propagated upwards, with the same dip sense as the extensional faults. The thrusts propagated along-strike with increasing horizontal shortening forming a well developed system of straight and overlapping thrust faults (Fig. 2c). At about 25% contraction hangingwall vergent thrusts formed as a result of buttressing against the main detachment and a 19.0 cm wide 'pop-up' structure developed (Fig. 2c). Further contraction caused some of the segments to grow laterally and coalesce forming longer faults (the pop-up's left-hand side thrust stretches across the entire width of the model; Fig. 2d). As some fault tips curved during growth, some thrusts exhibited more sinous traces. No strike-slip transfer faults were observed to develop during inversion.

Fault analysis Analysis of the growth of faults during extension shows that initial surface traces at 2.0 cm of extension had lengths between 2.0 and 8.5 cm. At 10.0 cm of extension fault traces varied between 2.0 and 45.0 cm. Analysis of maximum displacement/length ratios for 42 extensional faults show a wide range of values; $\gamma(d/L)$ varied from 2.6×10^{-2} to 2.2×10^{-1} (Fig. 3a). A best fit straight line through the data set gives a γ value of 5.4×10^{-2} (Fig. 3a).

During inversion thrust lengths varied between 6.5 and 19.5 cm in length, at 2.0 cm of contraction, and reached a maximum of 45.0 cm at the end of inversion. The thrusts were typically longer along-strike than the extensional faults. Fewer faults developed than during extension. For the thrust faults γ varied from 7.0×10^{-3} to a maximum of 8.6×10^{-2} with a typical value of 1.5×10^{-2} (Fig. 3b). These results highlight geometric differences between the extensional faults and the reverse faults formed during inversion.

60° ramp-flat listric fault

Extension Extension above the 3D ramp-flat listric décollement produced two complex rollover/crestal-collapse graben (upper and lower crestal-collapse graben), separated by a ramp syncline associated with reverse faults (Fig. 4a) (see also Ellis & McClay 1988; McClay 1990a). Both crestal-collapse graben developed parallel to the basal breakaway. The left-hand border faults consisted of an en echelon array of segments (3.0–14.5 cm in length), whereas the right-hand border faults were made up of co-planar segments (3.0–24.0 cm in length). A complex array of intra-graben faults also formed, but with shorter fault segments than the graben-bounding faults. Some of the intra-graben faults showed marked overlaps (Fig. 4a).

At the end of extension the basic architecture persisted, but both graben increased in width (8.0 and 12.0 cm respectively) and complexity (Fig. 4a). No strike-slip transfer faults were observed to develop during extension.

Inversion During inversion the low-friction basal décollement was reactivated, producing a fault propagation anticline above the breakaway, thinning of the syn-inversion strata above it and a second growth anticline above the ramp in the basal décollement. Reverse faults nucleated at the tips of extensional faults and propagated upwards forming long, straight segments (up to 50.0 cm long) (Fig. 4b). A 'pop-up' structure developed in the central part of the model, bound by a system of co-planar thrusts on the left-hand side and single back-thrust on the right-hand side (Fig. 4b). As shortening accumulated segments propagated laterally (the fault tips propagated in both directions) overlapping each other forming relay ramps (Fig. 4c). Tightening of the crestal-collapse graben by back-rotation of extensional faults occurred as the horizontal shortening increased. With further shortening propagating fault tips intercepted and the segments coalesced into longer faults. The thrust faults continued to propagate laterally, sometimes until a single fault developed across the entire width of the model (Figs 4c,d). No strike-slip transfer faults developed during inversion.

Fault analysis The progressive evolution of faults during extension in the ramp-flat experiments showed that the fault segments developed from 3.0–8.0 cm in length, at the beginning of extension, to a maximum of 29.5 cm at 10.0 cm of extension. d/L (γ) ratios for 34 extensional faults vary between 3.4×10^{-2} and 1.5×10^{-1} (Fig. 5a). A best fit curve, in the form of a linear fit constrained to pass through the origin, gives a γ value of 6.4×10^{-2} (Fig. 5a).

During inversion thrust segment lengths varied between 6.0 and 34.0 cm at 1.0 cm of contraction, and developed to a maximum length of 60.0 cm at the end of inversion. The thrust faults were usually longer than the extensional faults, with considerably fewer faults developing during inversion than during extension. γ values for the thrusts vary from 1.5×10^{-2} to a maximum of 7.0×10^{-2} (Fig. 5b), with the majority of faults having values between 1.5×10^{-2} to 2.8×10^{-2} and an average value of 2.1×10^{-2}. This indicates that within the modelling medium it was easier for thrust faults to propagate along-strike than extensional faults.

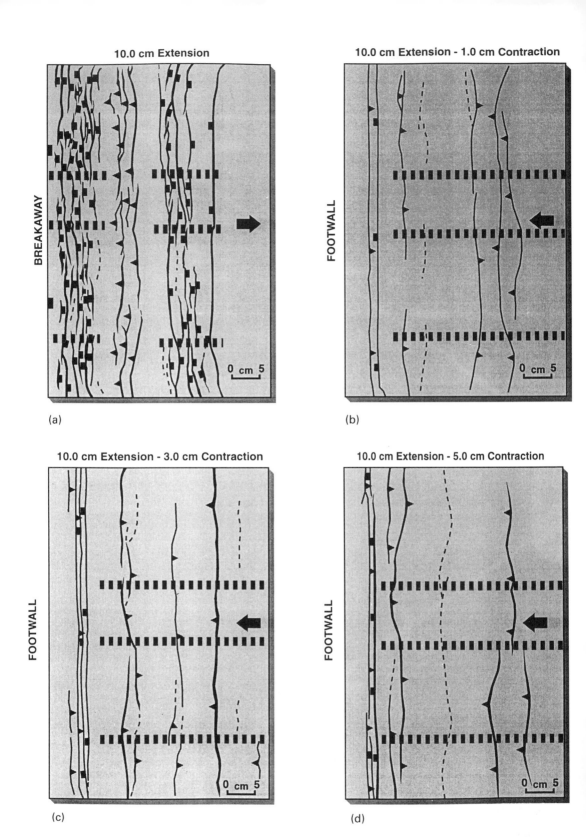

(a) 10.0 cm Extension — BREAKAWAY

(b) 10.0 cm Extension - 1.0 cm Contraction — FOOTWALL

(c) 10.0 cm Extension - 3.0 cm Contraction — FOOTWALL

(d) 10.0 cm Extension - 5.0 cm Contraction — FOOTWALL

0 cm 5

▮▮▮▮▮▮▮ Displacement transfer zone

Discussion

In these inversion experiments horizontal shortening was partitioned in different amounts between four deformation mechanisms; (i) reactivation of the low-friction basal décollement, (ii) folding, (iii) faulting and (iv) back-rotation of faults associated with localized layer-parallel shortening. Boundary conditions in the models were such that the extensional and inverted fault systems developed and grew, as a response to the translation of the hangingwall above a rigid footwall. The rigid footwall acts as basal décollement and controls the geometry of the faults in the hangingwall. As the experiments were carried out using cylindrical footwall detachments under plane strain conditions the faults could only exhibit dip-slip motion.

Stresses generated during horizontal shortening were close to perpendicular to the steep pre-existing extensional faults. In a few cases in the ramp-flat listric experiments extensional faults rotated to shallower dips and were able to be reactivated as the horizontal stress increased. However, in most cases fault orientations were such that it was mechanically easier for new reverse faults to form. Nevertheless, the pre-existing extensional faults define the nucleation sites of the new thrusts as their fault planes, and especially tip zones, are areas of local strain softening. This is due to shear dilatancy developed during the faulting (Mandl *et al.* 1977; McClay 1990*a*).

Along-strike geometries

The 3D geometries of the faults were investigated using serial parallel sections. The main characteristic found was along-strike displacement variation which was responsible for the high degree of along-strike segmentation found in the models.

The extensional faults were first observed as isolated, small amplitude perturbations or deformation bands on the upper surface of the model. These perturbations were usually 1.5–2.5 cm in length and were arranged in a narrow band across the model surface. With increasing extension the deformation bands showed localized propagation along-strike and developed into small faults. The faults extended along-strike with both fault tips undergoing contemporaneous lateral propagation. As individual faults grew two or more propagating co-planar faults merged and formed larger faults (non co-planar faults produced displacement transfer zones in the form of relay ramps). En echelon geometries were commonly developed. Coalescence of non co-planar fault segments produced 'kinks' in the fault traces. In all these experiments no strike-slip transfer faults were observed either on the upper surface or at depth in the sections.

During inversion isolated, small amplitude perturbations, in the form of monoclines, started to form with increasing shortening, and developed into small reverse faults. The granular nature of the modelling material precluded the formation of buckle style folds. As horizontal shortening increased these isolated faults propagated along-strike and coalesced or formed arrays of overlapping thrusts separated by relay ramps. This indicates that the faults produced in the models exhibit variable displacement along-strike (displacement transfer zones are indicated by isolated fault tips, relay ramps and kinks in the fault trace). Moreover, during inversion the displacement transfer zones observed on the upper surface of the models appeared to be located along vertical planes that also contained the overlap zones formed during extension. With increasing strain, however, the thrusts propagated across these transfer zones. This indicates that the main transfer zones formed by the extensional faults also remain as zones of discontinuous displacement during the inversion phase. The pre-existing extensional architecture appears to influence the inversion geometry by establishing zones of easy fault nucleation at the tips of pre-existing extensional faults and zones of discontinuous displacement along-strike.

Fewer faults developed during inversion than during extension. Moreover, the reverse faults formed longer segments that accommodated less displacement than the equivalent extensional faults. Such differences may be a limitation of the analogue modelling – the dimensions of faults in the models only vary over two orders of magnitude. Alternatively, this may reflect real differences in the propagation mechanisms of reverse and extensional faults. Further research both in the laboratory and on natural faults in homogeneous and anisotropic lithologies is warranted.

Fig. 4. (**a**) – (**d**) Sequential diagrams showing the top surface of the model Inv–4 at 10.0 cm extension and the evolution of the inversion fault system after 1.0, 3.0 and 5.0 cm of contraction. Note how the main displacement transfer zones observed during extension correspond in most cases to discontinuous displacement zones during inversion.

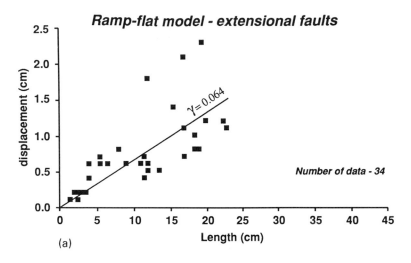

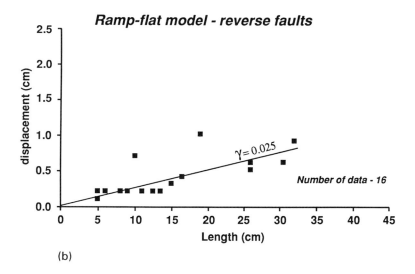

Fig. 5. (**a**) Maximum fault displacement *d* vs. fault length *L* for extensional faults developed in the 60°
ramp-flat listric model. (**b**) Maximum fault displacement *d* vs. fault length *L* for reverse faults developed during
inversion in the 60° ramp-flat listric model.

Cross-sectional geometries

In 2D (cross-section) the geometry of the
inverted listric fault system shows a fault
propagation growth anticline above the reactivated main detachment. The syn-extension
strata was pushed up along the detachment
surface forming an 'arrowhead' wedge geometry. The extensional crestal-collapse graben
was tightened and a broad 'pop-up' structure,
bounded by low-angle thrusts, developed. This
affected mainly the syn-inversion and, to a lesser

extent, the syn-extensional strata. These features are similar to those described in 2D
experiments by Buchanan & McClay (1991) and
by Roure *et al.* (1992).
    The cross-sectional geometry of the inverted
ramp-flat listric system exhibits growth anticlines developed above the reactivated detachment breakaway and above the ramp in the
detachment surface. A small 'arrowhead' structure, disrupted by syn-inversion extensional
faults, developed in the syn-extensional sequence. The extensional crestal-collapse graben

have been tightened and some faults in the upper crestal-collapse graben showed signs of reactivation. Faulting in the inverted upper crestal-collapse graben formed a complex flower-like structure, with superposed extensional and reverse faults. Similar structures were described by Buchanan & McClay (1991) in 2D inverted ramp-flat experiments.

Fault analysis

The analysis of fault displacement versus fault length in the sandbox models gives γ values which range from 0.25–0.007 with best fit values between 0.011 and 0.064 (Figs 3 & 5). These values are comparable with those measured for natural faults (e.g. Barnett *et al.* 1987; Cowie & Scholz 1992). A significant scatter has been observed in the data with faults of similar type and length showing displacements that can vary by a factor of 3 and in some isolated cases by a factor of 5 (Figs 3 & 5). As the experiments were always carried out using the same material (sand and mica flakes) the results obtained for the *d*/*L* analysis were not affected by different material properties. Moreover, significant differences exist between the plots for extensional and thrust faults. Typically, reverse faults formed longer segments and accommodated less displacement than extensional faults of equivalent length (Figs 3 & 5), with the reverse faults showing consistently lower values of γ than extensional faults in both the listric and ramp-flat experiments.

Natural fault systems

The inversion architectures obtained in these 3D experiments show many similarities to natural examples of inversion geometries as described by Buchanan & McClay (1991) and also those shown in Cooper & Williams (1989). The patterns of faulting and the nature of the accommodation zones between fault systems in the models show similarities to inverted fault systems at a variety of scales.

Conclusions

The following conclusion can be drawn from this study.
(i) The above experiments showed four distinct deformation mechanisms to operate during inversion. These are reactivation of the low-friction basal décollement, folding, faulting and fault back-rotation associated with localized layer-parallel shortening.
(ii) The main feature of the 3D models is the

variation in fault displacements along-strike. This variation produced along-strike fault segmentation, with faults dying out laterally or overlapping across displacement transfer zones in the form of relay ramp structures. The faults produced in the experiments appear to behave in a similar fashion to extensional and reverse faults in nature.
(iii) The pre-existing extensional fault architecture appears to exert a distinct control on the inversion geometry. This is achieved by creating zones of easy fault nucleation at their tips and by establishing the location of discontinuous displacement zones, that remain 'active' and affect the evolution of faults during inversion. No strike-slip transfer faults were observed to develop in the models either in extension or upon inversion.
(iv) Reverse faults were observed to form longer segments and to accommodate less displacement than extensional faults. γ values for reverse faults are consistently lower than for extensional faults in both listric and ramp-flat experiments.

The research presented in this paper is part of the Fault Dynamics Project sponsored by ARCO British Ltd., BP Exploration, BRASOIL UK Ltd., CONOCO UK Ltd., MOBIL North Sea Ltd. and SUN Oil Britain Ltd. B. Adams, L. Thompson and P. Hollings are thanked for assistance in the analogue modelling laboratory. Peter Ellis and an anonymous referee are gratefully acknowledged for reviews and suggestions that greatly improved this manuscript. Fault Dynamics Publication No. 41.

References

BARNETT, J. A. M, MORTIMER, J., RIPPON, J. H., WALSH, J. J. & WATTERSON, J. 1987. Displacement geometry in the volume containing a single normal fault. *American Association of Petroleum Geologists Bulletin*, **71**, 925–937.

BUCHANAN, P. G. & McCLAY, K. R. 1991. Sandbox experiments of inverted listric and planar fault systems. *Tectonophysics*, **188**, 97–115.

COOPER, M. A. & WILLIAMS, G. D. (eds). 1989. *Inversion Tectonics*. Geological Society, London, Special Publication, **44**.

COWIE, P. A. & SCHOLZ, C. H. 1992. Displacement-length scaling relationship for faults: data synthesis and discussion. *Journal of Structural Geology*, **14**, 1149–1156.

ELLIS, P. G. & McCLAY, K. R. 1988. Listric extensional fault systems: results of analogue model experiments. *Basin Research*, **1**, 55–71.

GILLCRIST, R., COWARD, M. & MUGNIER, J.-L. 1987. Structural inversion and its controls: examples from the Alpine foreland and the French Alps. *Geodinamica Acta*, **1**, 5–34.

HARDING, T. P. 1985. Seismic characteristics and

identification of negative flower structures, positive flower structures and positive structural inversion. *American Association of Petroleum Geologists Bulletin,* **69**, 582–600.

HORSFIELD, W. T. 1977. An experimental approach to basement-controlled faulting. *In*: FROST, R. C. T. & DIKKERS, A. J. (eds) *Fault Tectonics in NW Europe*. Geologie en Mijnbouw, **56**, 363–370.

KOOPMAN, A., SPEKSNIJDER, A. & HORSFIELD, W. T. 1987. Sandbox models of inversion tectonics. *Tectonophysics,* **137**, 379–388.

KRANTZ, W. K. 1991. Normal fault geometry and fault reactivation in tectonic inversion experiments. *In*: ROBERTS, A. M., YIELDING, G. & FREEMAN, B. (eds) *The Geometry of Normal Faults*. Geological Society, London, Special Publication, **56**, 219–229.

MANDL, G., DE JONG, L. N. J. & MALTHA, A. 1977. Shear zones in granular material. *Rock Mechanics,* **9**, 95–144.

MITRA, S. 1993. Geometry and kinematic evolution of inversion structures. *American Association of Petroleum Geologists Bulletin,* **77**, 1159–1191.

McCLAY, K. R. 1989. Analogue models of inversion tectonics. *In*: COOPER, M. A. & WILLIAMS, G. D. (eds) *Inversion Tectonics*. Geological Society, London, Special Publication, **44**, 41–59.

—— 1990*a*. Deformation mechanics in analogue models of extensional fault systems. *In*: RUTTER, E. H. & KNIPE, R. J. (eds) *Deformation Mechanisms, Rheology and Tectonics*. Geological Society, London, Special Publication, **54**, 445–454.

—— 1990*b*. Extensional fault systems in sedimentary basins. A review of analogue model studies. *Marine and Petroleum Geology,* **7**, 206–233.

ROURE, F., BRUN, J.-P., COLLETA, B. & VAN DEN DRIESSCHE, J. 1992. Geometry and kinematics of extensional structures in the Alpine Foreland Basin of south-eastern France. *Journal of Structural Geology,* **14**, 503–519.

Recognition and measurement of basin inversion

Applications of Thermal History Reconstruction in inverted basins

PAUL F. GREEN[1], IAN R. DUDDY[1] & RICHARD J. BRAY[2]

[1]*Geotrack International Pty Ltd, PO Box 4120, Melbourne University, Victoria 3052, Australia*

[2]*Geotrack International UK Office, 30 Upper High Street, Thame, Oxfordshire OX9 3EX, UK*

Abstract: Studies of inverted basins based solely on the preserved section provide only partial insight into the thermal and tectonic history of basin development. Equally important is that part of the history represented by section removed by erosion during inversion. Thermal History Reconstruction, involving application of Apatite Fission Track Analysis (AFTA™) and Vitrinite Reflectance (VR) to define the timing and magnitude of key erosional and/or thermal episodes, and integration of this information with data from the preserved section, provides a more complete description. Case histories from the UK, New Zealand and Australia are presented to demonstrate this approach. Much of northern and eastern England, the Irish Sea and the Southern North Sea was affected by heating apparently due largely to additional burial, followed by early Tertiary cooling due to regional uplift and erosion, synchronous with recognized basin inversion events. Heating associated with this episode caused generation of hydrocarbons throughout the region. Areas conventionally regarded as not inverted appear to have been subjected to km-scale uplift and erosion, which was not restricted to classic 'inversion axes'. Failure to allow for these effects can lead to a significant underestimation of regional maturity trends. Thermal History Reconstruction is also essential in areas that have undergone more than one episode of inversion. For example in Inner Moray Firth well 12/16—1 maximum palaeotemperatures in the Carboniferous section were reached prior to Hercynian inversion and the chances of preserving any hydrocarbons generated from Carboniferous or older source rocks appear poor. In southeastern Australia, early Cretaceous rift basins underwent mid-Cretaceous inversion at a time of high palaeogeothermal gradients (c. 60°C km^{-1}), and basement margins were also inverted at that time. Again, early generation implies that the chances of preservation of hydrocarbons are small. By integrating stratigraphic and structural relationships within the preserved section with data on the palaeo-thermal history of a section derived from direct measurements of palaeotemperature profiles, the complete history of hydrocarbon generation may be reconstructed with confidence.

In inverted basins, the timing of oil generation in relation to that of trap formation is a critical factor in determining hydrocarbon prospectivity. It is therefore essential to reconstruct the thermal and tectonic histories of inverted basins as accurately and as precisely as possible. Specifically, it is important to define the time at which hydrocarbon generation effectively ceases (which we can take to be synchronous with the onset of cooling from maximum palaeotemperatures), in relation to episodes of uplift and erosion (e.g. due to inversion) which might have produced potential trapping structures, particularly where multiple episodes may have occurred.

Studies of inverted basins which are based only on the preserved section, using techniques such as backstripping, structural analysis, section balancing etc., provide only partial insight into the thermal and tectonic history of basin development, although this approach is quite common (e.g. Chadwick 1986; Karner *et al.* 1987). In this paper we highlight the importance of considering that part of the history represented by the section which may have been removed by erosion, and integrating this with information derived from the preserved section.

Thermal History Reconstruction

If inversion is accompanied by erosion of the inverted sequence, then the preserved section cools. Methods of quantitative palaeotemperature assessment, such as Vitrinite Reflectance (VR) and Apatite Fission Track Analysis (AFTA), may then be used to reconstruct profiles of maximum palaeotemperature as a function of depth, from which the palaeogeothermal gradient prior to inversion can be estimated (to within ±5°C km^{-1} in favourable cases). Extrapolating this gradient to an

From BUCHANAN, J. G. & BUCHANAN, P. G. (eds), 1995, *Basin Inversion*, Geological Society Special Publication No. 88, 149–165.

149

assumed palaeo-surface temperature then provides an estimate of the amount of section eroded (Bray *et al.* 1992).

This estimate relies on the assumption that extrapolation of a linear gradient through the eroded section is valid. This assumption can never be verified, since the section is no longer present, and Holliday (1993) has recently challenged the validity of this assumption. However good agreement between estimates of section removed using the above approach and non-thermal techniques (discussed in later sections of this paper) suggests that the results are generally reliable. Importantly, the approach outlined here provides an internally consistent framework for reconstructing patterns of palaeotemperature variation.

Estimates of the amount of section removed also rely on an assumed palaeo-surface temperature. This parameter is very difficult to control quantitatively, and so in the absence of more reliable values the present value is usually used. Estimates of section removed derived in this way can easily be converted to relate to other palaeo-surface temperature values by subtracting or adding the depth equivalent to the difference in palaeo-surface temperature for the appropriate palaeogeothermal gradient.

AFTA also provides an independent estimate of the time at which the section began to cool from maximum palaeotemperatures (for more details see e.g. Green 1989, Kamp & Green 1990). These data often provide information that cannot be obtained from any other source, particularly in basins with complex histories and in basins where deep erosion has removed much of the stratigraphic record.

Determination of palaeogeothermal gradients at the time of maximum palaeotemperatures can provide insight into the cause of heating and subsequent cooling. For example, if heating was due solely to a period of high heat flow, the geothermal gradient at the time of maximum temperature would be higher than the present-day gradient, whereas if heating was caused solely by deep burial followed by uplift and erosion with no change in heat flow, the palaeogeothermal gradient should be similar to the present gradient.

Thermal History Reconstruction, involving application of AFTA and VR to define the timing and magnitude of key erosional and/or thermal episodes, and integration of this information with data from the preserved section therefore provide a more complete description than that obtained in the absence of measured palaeotemperature constraints. Further details of Thermal History Reconstruction using AFTA

and VR are given in Bray *et al.* (1992).

In inverted basins, integration of such information with results obtained from the preserved section is essential in order to reconstruct the complete history. Situations where Thermal History Reconstruction may be particularly useful include:

● areas which have undergone regional episodes of uplift and erosion, such that no complete reference section exists to serve as a comparison;
● basins where stratigraphic and structural relationships allow only broad constraints on the timing of inversion (e.g. where an unconformity spans a broad range of time);
● basins subject to multiple episodes of inversion;
● areas where seismic data are either unavailable or allow only limited resolution, or where inversion involves basement.

In addition, Thermal History Reconstruction may reveal aspects of palaeothermal history which are not related to burial and are not evident from the preserved section (e.g. episodes of elevated heat flow; hot fluid movements).

In this paper, we review a series of case histories involving inverted basins where Thermal History Reconstruction provides important constraints on thermal and tectonic history which are not evident from study of the preserved section, and where possible we discuss the implications for hydrocarbon prospectivity.

Fresne-1, Taranaki Basin, New Zealand

This example is included to demonstrate a case where information derived from Thermal History Reconstruction is consistent with that obtained from reconstructions based on the preserved section. This provides strong support for the approach used and the assumptions on which it is based. This example is also used to demonstrate some of the methodology used in Thermal History Reconstruction.

Kamp & Green (1990) reported the results of an AFTA study of four wells in the Southern Taranaki Graben. Discussion here is restricted to the Fresne-1 well, which was drilled on a late Miocene inversion structure (Fig. 1), dated by the presence of truncated mid-Miocene strata overlain by Pliocene to Recent sediments. By reconstructing the original section by comparison with that preserved off structure in a non-inverted setting, Knox (1982) suggested that at least 2 km of section had been eroded

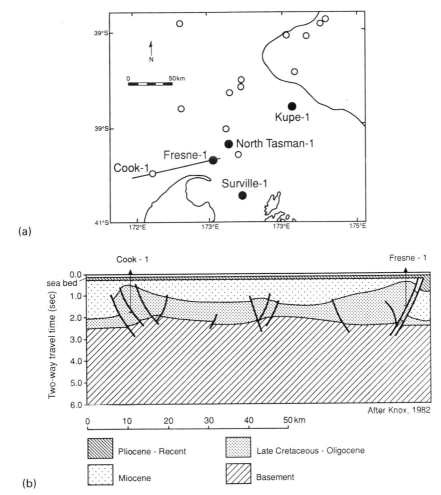

Fig. 1. (a) Location map of the Southern Taranaki Basin showing locations (solid symbols) of wells studies by Kamp & Green (1990) and other wells in the region (open symbols), plus the location of the section shown in (b). (b) Simplified section from Cook-1 to Fresne-1 (location shown in (a)), showing late Miocene inversion structures (after Knox 1982).

from the structure during late Miocene inversion, while Ellyard & Beattie (1990) suggested that 3.5 km of section had been eroded, using a similar approach. The late Cretaceous to early Tertiary coal measures within the inverted sequence contain hydrocarbon source rocks but no significant hydrocarbon accumulation was found.

AFTA data in samples from the inverted section over a depth range of around 2 km show consistent evidence of maximum palaeotemperatures around 80°C higher than present temperatures, and suggest that cooling from these palaeotemperatures began at c. 10 Ma. This suggests that the palaeogeothermal gradient at the time of maximum palaeotemperatures was close to the present value of 28°C km^{-1}

(Kamp & Green 1990), and that cooling was due almost solely to uplift and erosion during late Miocene inversion. Allowing for reburial by c. 100 m of Pliocene to Recent sediment, the difference between maximum palaeotemperatures and present temperatures indicates around 3 km of section was removed by uplift and erosion. Kamp & Green (1990) also reported that VR data were consistent with that conclusion.

Applying the methodology outlined by Bray et al. (1992), we can now obtain more rigorous estimates of palaeogeothermal gradient and section removed in Fresne-1. Maximum palaeotemperatures from AFTA and VR data in Fresne-1 are shown plotted against depth in

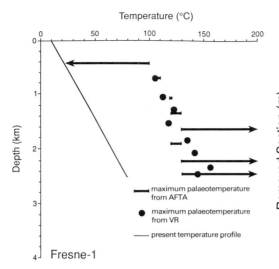

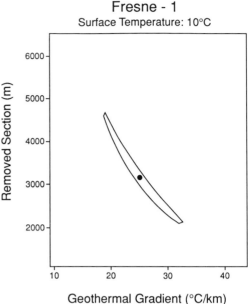

Fig. 2. Palaeotemperature estimates from AFTA and VR in Fresne-1 (from Kamp & Green 1990), plotted against depth. Arrows indicate mimimum limits to maximum palaeotemperatures due to total annealing of all fission tracks prior to cooling (see Kamp & Green 1990 for more details).

Fig. 3. Crossplot of section removed against palaeogeothermal gradient, showing the range of allowed values (within the contoured region) consistent with the palaeotemperature constraints in Fresne-1, shown in Fig. 2. The shape of the contoured region results from the negative correlation between values of palaeogeothermal gradient and section removed which are compatible with the data. The maximum likelihood values of removed section and palaeogeothermal gradient are indicated by the spot.

Fig. 2, while Fig. 3 illustrates the range of palaeogeothermal gradients and section removed which are compatible with the palaeotemperature estimates. The data are consistent (at ±95% confidence limits) with gradients between 19 and 32°C km^{-1} and sections removed between 2.0 and 4.5 km, with maximum likelihood values of 25°C km^{-1} and 3.1 km, respectively.

Figure 4 shows the reconstructed burial and uplift history for Fresne-1 using the maximum likelihood values of removed section and palaeogeothermal gradient. Biostratigraphical evidence shows that the youngest sediment preserved below the sub-Pliocene unconformity is mid-Miocene, deposited at *c.* 15 Ma, which means that the 3 km or more of section that was eroded must have been deposited between 15 and 10 Ma, as shown. Section of this age is preserved off-structure, where erosion has been much less pronounced, validating the overall form of the burial history shown here.

The form of the reconstructed burial history in Fig. 4 emphasizes that the burial phase responsible for producing the observed palaeotemperatures is equally as important as the subsequent uplift and erosion, in terms of the tectonic evolution of the basin. Such episodes of accelerated burial prior to uplift and erosion appear to be a common feature of many inverted sedimentary basins (see e.g Figs 10 and 13, and

fig. 4 of Green *et al.* 1993). This aspect of inverted basins and basins affected by regional uplift and erosion is often overlooked, although the amount of section that has been eroded is sometimes close to or greater than that still preserved above basement.

The preserved section in Fresne-1 also shows an early Eocene to mid-Oligocene unconformity (52 to 29 Ma). Since the effects of early heating episodes in both AFTA and VR data are overprinted by subsequent heating to higher temperatures, no constraints are possible on the palaeotemperatures reached during the time span represented by this earlier unconformity, except that they were less than the maximum palaeotemperatures reached during the Miocene.

Thus maximum palaeotemperatures were reached immediately before late Miocene inversion, and maximum hydrocarbon generation would have occurred during the phase of burial between 15 and 10 Ma prior to late Miocene inversion. This is an important factor with

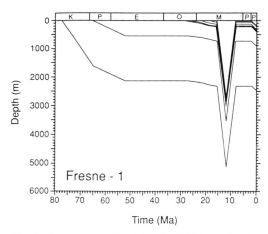

Fig. 4. Reconstructed burial and uplift history for the Fresne-1 well, using the maximum likelihood values of removed section and palaeogeothermal gradient shown in Fig. 3. Note that the c. 3 km of section removed during inversion must have been deposited between c. 15 Ma (when the youngest sediment preserved below the unconformity as deposited), and the onset of inversion at c. 10 Ma. This section is preserved off structure, and validates the form of the burial history shown here.

respect to hydrocarbon prospectivity, as although source rocks within the Late Cretaceous to Early Tertiary coal measures sequence entered the oil window, any oil generated cannot have been trapped in late Miocene inversion-related structures. A much more favourable situation exists to the east, where late Miocene inversion was less severe, and where Pliocene to Recent burial (up to 2 km or more) has been sufficient to take the source rock sequence to higher temperatures than reached prior to late Miocene inversion, such that peak maturity was reached well after formation of structures. Discovery of a significant oil accumulation in this region at Kupe South (Schmidt & Robinson 1990) provides strong support for this model.

In summary, application of Thermal History Reconstruction to the Fresne-1 well provides a best estimate of palaeogeothermal gradient ($25°C\,km^{-1}$) which is close to the present gradient ($28°C\,km^{-1}$) and a best estimate of section removed (3.1 km) which is consistent with estimates derived from extrapolation of seismic sections (3.5 km). On this basis, we propose that the techniques employed to reconstruct a more complete history can be used with confidence in less well controlled situations.

Onshore East Midlands Shelf, UK

The onshore East Midlands Shelf of eastern England (Fig. 5) contains a number of small oil accumulations, generally trapped in late Carboniferous inversion structures, overlain by Permo–Triassic red beds. Fraser *et al.* (1990) described the province in detail. Stratigraphic and structural relationships within the preserved section show that inversion occurred between Westphalian and early Permian times, but since the youngest sediments preserved in the immediate area are Triassic, the post-Triassic history cannot be reconstructed from the preserved section alone.

Application of Thermal History Reconstruction using AFTA and VR (Green 1989; Bray *et al.* 1992) shows that across the onshore East Midlands Shelf the entire section reached maximum palaeotemperatures during the late Cretaceous to early Tertiary, and began to cool at c. 60 Ma. Application of the methods described above shows that palaeogeothermal gradients at the time of maximum palaeotemperatures were close to present values, suggesting that the observed heating was due to additional burial by between 1 and 2 km of post-Triassic section, subsequently removed by Tertiary uplift and erosion.

The information revealed by Thermal History Reconstruction is again crucial to understanding hydrocarbon prospectivity, as it reveals that maximum hydrocarbon generation occurred well after the late Carboniferous structures were formed, making them suitable for preservation of oil generated from Carboniferous source rocks during Cretaceous and early Tertiary burial, and explaining the presence of numerous accumulations of oil and gas in the region.

East Midlands Shelf and Sole Pit axis, UK Southern North Sea

Application of AFTA and VR data around many of the classic inversion axes in Northwest Europe has shown that exposure to higher palaeotemperatures prior to inversion is often not restricted to the recognized inversion axes, but is also observed in areas well away from the axes, classically regarded as being non-inverted.

Data from the UK sector of the Southern North Sea (Fig. 5) demonstrate this clearly. Figure 6 summarizes the conventional interpretation of the region in terms of Cretaceous/Tertiary inversion, as depicted for example in the palaeogeographical reconstructions of Ziegler (1990). The Sole Pit axis has long been

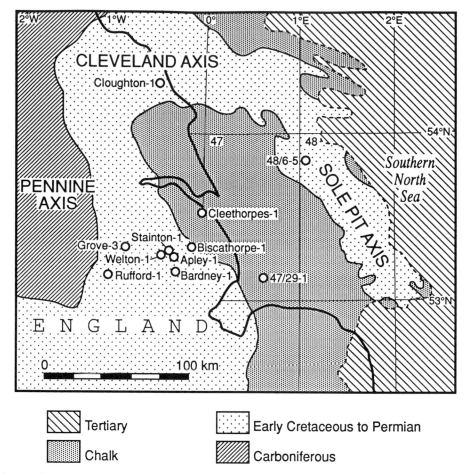

Fig. 5. Location map showing wells in the Southern North Sea and adjacent areas of eastern England referred to in the text, and in papers referenced herein.

recognized as a major axis of inversion (Marie 1975; Glennie & Boegner 1981; Van Hoorn 1987l; Bulat & Stoker 1987), but the East Midlands Shelf has been widely regarded as not having been buried more deeply in the past (e.g. Glennie & Boegner 1981; Cope 1986).

Bray *et al.* (1992) reported that Thermal History Reconstruction in well 47/29a-1 from the offshore shelf region (Fig. 5) suggests that the preserved section in this well was also more deeply buried in the past, with a best estimate of 1.3 km (between 0.7 and 2.19 km at ±95% confidence limits) of post-Campanian sediment removed by Tertiary uplift and erosion beginning at *c.* 60 Ma. This is consistent with results from the neighbouring onshore shelf region, which also show evidence of km-scale section removed by Tertiary uplift and erosion. In particular, independent estimates of the amount of section removed, derived from sonic velocity

data in Chalk in the Cleethorpes-1 well and other onshore East Midlands Shelf wells (Hillis 1991, 1993) are in good agreement with those derived from Thermal History Reconstruction. This provides more support for the validity of the assumptions implicit in this approach.

As discussed in detail by Green *et al.* (1993), the palaeothermal effects identified in well 47/29a-1 appear to be part of a regional episode of deeper burial followed by uplift and erosion beginning in the early Tertiary which affected most of the northern half of England and much of the UK sector of the Southern North Sea. Within this regional effect, the recognized axes of inversion appear to represent local maxima of erosional removal, where deep incision has produced obvious structural and stratigraphic evidence.

Van Hoorn (1987, 1989) suggested that the Sole Pit axis (Fig. 6) underwent an early episode

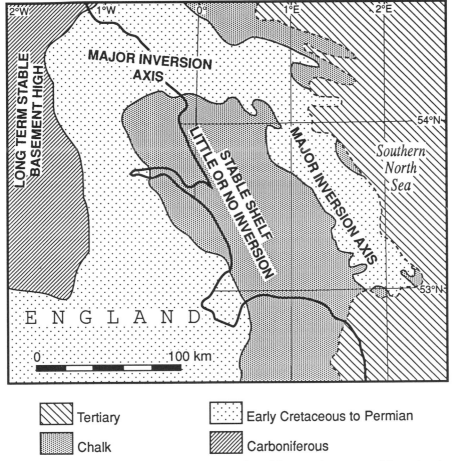

Fig. 6. Simplified summary of the conventional understanding of the tectonic structure of the region shown in Fig. 5. As discussed in the text, application of Thermal History Reconstruction across this region suggests a much more complex history. Recognized inversion axes appear simply to represent local maxima in the magnitude of section removed, while areas previously considered to represent stable shelves or highs have undergone regional burial by between 1 and 2 km of additional section which was removed by Tertiary uplift and erosion.

of inversion in the Turonian to Campanian and a later episode in the Oligocene. Ziegler (1990) followed this interpretation. This differs significantly from the thermal history of the shelf outlined above, dominated by an episode of regional uplift and erosion beginning in the early Tertiary. However AFTA data from Sole Pit axis well 48/6–5 (Fig. 7) show that the section preserved in this well also began to cool from maximum palaeotemperatures in the early Tertiary, synchronous with that in other wells from the East Midlands Shelf, onshore and offshore (discussed above). Thermal History Reconstruction shows that this event dominates the thermal history of most of the Sole Pit axis as well as the shelf area to the west, although both

Cretaceous and mid-Tertiary events have also been observed in different regions of the axis. A complex picture emerges, highlighting the necessity of determining the timing of maximum temperature directly wherever possible, rather than relying on extrapolation of regional structural or stratigraphic trends.

Irish Sea

The East Irish Sea Basin (EISB) provides an example of the ability of AFTA to identify independently the timing of a major phase of inversion. Ziegler (1990) interpreted the EISB as a region of Eo-Oligocene tectonics on the

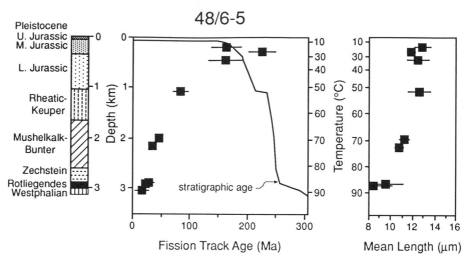

Fig. 7. AFTA parameters in Sole Pit axis well 48/6-5, plotted against depth and present temperature. The fission track age data shown the classic pattern of a section which has been hotter in the past. The 'break in slope' between 1000 and 200 m where the rate of decrease of fission track age with depth becomes much less pronounced, marks the transition from partial to total annealing of fission tracks formed prior to the onset of cooling. Fission track ages below this depth relate only to tracks formed after the onset of cooling, which these data suggest occurred at around 60 Ma. Thus the thermal history of this well is dominated by the same event that dominates the history of hydrocarbon generation in the East Midlands Shelf offshore and onshore, in northern England and eastwards into the Irish Sea.

basis of geological evidence from surrounding regions. However, the youngest sediments preserved in the EISB are Triassic (except for local early Jurassic deposits and Tertiary intrusives) and the preserved evidence provides no direct constraint on the timing of inversion within the EISB. Lewis *et al.* (1992) showed, using AFTA, that the EISB and neighbouring parts of northern England underwent a major episode of cooling at 65 ± 5 Ma, and these regions were affected by the same regional episode of uplift and erosion described in preceding sections. Therefore it seems clear on the basis of Thermal History Reconstruction that the major inversion event in the Irish Sea occurred in the early Tertiary.

This is consistent with the published model for the history of hydrocarbon generation in the Morecambe Bay gas field, which is based primarily on study of diagenetic patterns within the reservoir section (Bushell 1986; Woodward & Curtis 1987; Macchi *et al.* 1990; Stuart & Cowan 1991). The model suggests that hydrocarbon generation probably began in the early Jurassic, with liquids and gas migrating into a low-relief structure formed by early Cimmerian movements (late Triassic to early Jurassic). Generation of hydrocarbons continued through the Jurassic until late Cimmerian (late Jurassic to early Cretaceous) uplift and erosion at which

time the structure was breached and the hydrocarbons were lost. Renewed burial in the Cretaceous then led to further generation of gas, which accumulated in structures resealed by salt movement in the overlying Keuper Saliferous Beds. Gas generation was terminated by uplift and erosion in the early Tertiary, which led to the present configuration and caused some expansion of the accumulated gas due to the pressure reduction caused by uplift and erosion.

The heating revealed by AFTA and the subsequent cooling beginning in the late Cretaceous/early Tertiary therefore played a critical role in the evolution of the Morecambe Bay gas field. Integration with regional data show these events to be part of a major episode which exerted similar influence over the hydrocarbon prospectivity of much of the onshore region to the east and further eastwards into the Southern North Sea.

The Pennine Axis

The Pennine Axis of northern England provides an example of basement uplift recognized from Thermal History Reconstruction, coincident in time with inversion of adjacent basins. The Pennine Axis (Fig. 5) is often described as a long-term stable high throughout much of post-Carboniferous time (e.g. Fraser &

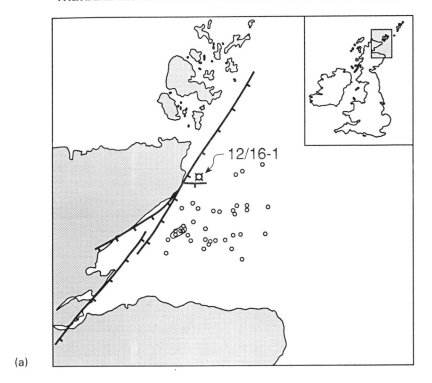

(a)

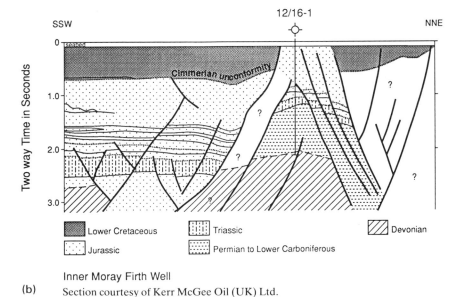

(b)

Inner Moray Firth Well
Section courtesy of Kerr McGee Oil (UK) Ltd.

Fig. 8. (a) Location map of the Inner Moray Firth showing the location of well 12/16-1 and the location of the section shown in (b). (b) Simplified section through the '12/16-1 horst' (provided by Kerr McGee Oil (UK) Ltd.).

Gawthorpe 1990; Ziegler 1990; Cope *et al.* 1992). However, AFTA data (Lewis *et al.* 1992; Green *et al.* 1993) clearly show that samples from outcrop in this area cooled from palaeotemperatures of around 90 to 100°C or higher in the early Tertiary, suggesting that this area was

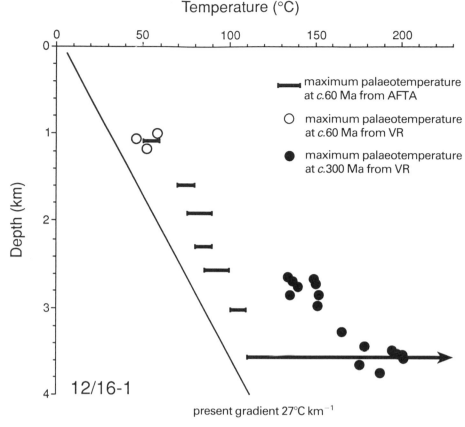

Fig. 9. Palaeotemperature estimates from AFTA and VR in 12/16-1, plotted against depth. The difference between palaeotemperatures derived from AFTA and VR in the deeper (Carboniferous) section reflects an earlier heating episode, in which the Carboniferous section reached maximum palaeotemperatures, while the AFTA data represent a later heating episode prior to early Tertiary cooling.

also buried by as much as 2 km or more of section prior to the onset of regional uplift and erosion in the early Tertiary. Estimates of removed section based on sonic velocities (Whittaker *et al.* 1985) also show an increase towards the Pennine Axis, and support the conclusions derived from AFTA.

While not related directly to basin inversion *sensu stricto*, this is worth noting in the context of the previous sections, as it emphasizes the link between uplift and erosion of basement blocks such as the Pennines and the Lake District to the west (Green 1986; Green *et al.* 1993) and inversion in the Irish Sea and Sole Pit Basins, as well as in other neighbouring basins (Cleveland, Solway, Vale of Eden) which underwent inversion at the same time. This link between basement uplift and basin inversion has been largely ignored to date, but remains a key aspect

of the late Cretaceous to early Tertiary tectonics of the UK region.

Inner Moray Firth well 12/16-1

In areas with more complex histories involving two or more periods of inversion, seismic sections may be complex and it may not be possible to determine which event caused maximum hydrocarbon generation from the preserved section alone. The Inner Moray Firth well 12/16-1 (Fig. 8) illustrates this situation. AFTA data from samples throughout the well show clear evidence of cooling in the early Tertiary from palaeotemperatures around 20°C higher than present temperatures. Figure 9 shows palaeotemperatures estimated from AFTA plotted against depth. The VR data from the Carboniferous section indicate higher

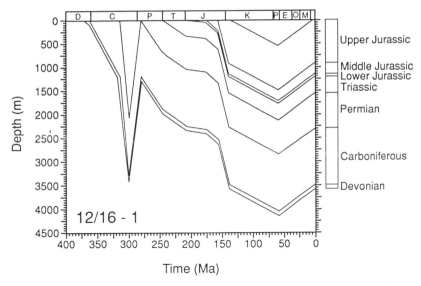

Fig. 10. Reconstructed burial and uplift history for the 12/16-1 well, using the maximum likelihood values of removed section and palaeogeothermal gradient discussed in the text, derived from the palaeotemperature data shown in Fig. 9.

palaeotemperatures than the early Tertiary palaeotemperatures derived from AFTA, showing that maximum palaeotemperatures as reflected in the VR data were achieved during an earlier event. As the VR data provide no independent evidence of the timing of this earlier event, and since the section contains a major unconformity from Namurian to Permian, we have assumed that cooling was due to Hercynian tectonism at the end of the Carboniferous. Maximum palaeotemperatures have been calculated from the VR data on this basis and are also plotted in Fig. 9.

Palaeotemperatures during the early event give a maximum likelihood estimate of $57.4°C\,km^{-1}$ for the end-Carboniferous palaeogeothermal gradient (with upper and lower 95% confidence limits of 71 and $44°C\,km^{-1}$, respectively). Extrapolating this gradient to a palaeo-surface temperature of 6°C gives a maximum likelihood estimate of 2.0 km for the amount of late Namurian to Westphalian section removed during Hercynian inversion (with upper and lower 95% confidence limits of 2.8 and 1.5 km, respectively). Similar analysis of the early Tertiary palaeotemperatures gives maximum likelihood estimates of $28°C\,km^{-1}$ (34 to $22°C\,km^{-1}$) and 0.6 km (1.2 to 0.2 km) for the early Tertiary palaeogeothermal gradient and section removed during Tertiary uplift and erosion, respectively.

Figure 10 shows the full reconstruction of the burial and uplift history for this well. Note that the higher gradient during the earlier event has allowed this event to be recognized although maximum burial depths were reached in the early Tertiary.

As noted previously, estimates of the amount of section removed during each event depend critically on the assumed value of palaeo-surface temperature. In the absence of reliable estimates we have assumed a value similar to the present-day value, which may well underestimate the actual values. Therefore the estimates of section removed are best viewed as maximum limits in this case. If higher palaeo-surface temperatures are preferred the estimated section removed can be easily corrected, as discussed earlier. Whatever the actual values of palaeo-surface temperature, the parameters presented here and used in constructing Fig. 10 provide an internally consistent scheme for predicting patterns of palaeotemperature variation.

Because of the early occurrence of maximum palaeotemperatures in the vicinity of well 12/16-1, maximum hydrocarbon generation would have occurred prior to formation of Hercynian structures and well before more recent phases of structuring in the Inner Moray Firth Basin. Therefore the chances of discovering significant accumulations of hydrocarbons sourced from the Carboniferous and older section would appear to be small and would depend strongly on preservation through several episodes of remigration.

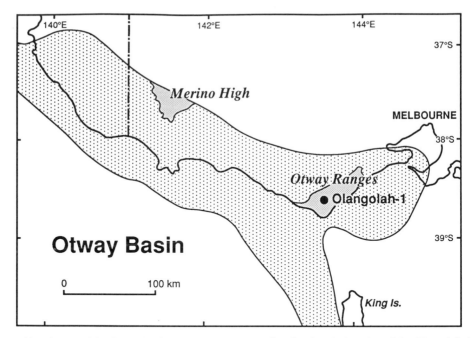

Fig. 11. Sketch map of the Otway Basin, south eastern Australia, showing the location of the Olangolah-1 well, drilled in the Otway Ranges.

Olangolah-1, Otway Basin, SE Australia

This example illustrates a basin with an extremely high palaeogeothermal gradient, which can only be revealed using Thermal History Reconstruction, and also a situation where until recently it has been difficult to obtain useful seismic data because of a lack of resolution within a thick pile of volcanogenic sediments.

The Olangolah-1 well was drilled on the crest of the Otway Ranges, a northeast–southwest trending topographic high in the Eastern Otway Basin, southeastern Australia (Fig. 11). Early Cretaceous Otway Group sediments outcrop across most of the Otway Ranges, and the well penetrated over 2 km of these volcanogenic sediments. VR data throughout the well reveal extremely high palaeotemperatures, as shown in Fig. 12, while AFTA data show that cooling took place around 95 Ma.

The VR data define a palaeogeothermal gradient between 45 and 74°C km^{-1} (maximum likelihood value of 59°C km^{-1}), with between 2.1 and 4.3 km of section removed (maximum likelihood value of 3.0 km). Stratigraphic information provides little or no constraints on the late Cretaceous to Recent history, but AFTA data suggest that since initial rapid cooling, only very limited amounts of re-burial and sub-

sequent uplift have affected the sequence. Fig. 13 shows the reconstructed burial and uplift history. Mid-Cretaceous inversion in the Otway Basin, together with synchronous events around many of the basin margins in southeastern Australia (see next section), appears to have been due to a major plate rearrangement at that time.

Note that the high gradient prior to inversion is detectable only because inversion was early, and was followed by relatively little thermal disturbance so that the palaeotemperature evidence for this early event could be preserved. In other parts of the basin, unaffected by the inversion although early Cretaceous gradients may well have been of similar magnitude to that in Olangolah-1, the section is now at maximum temperature since deposition due to the cumulative effects of late Cretaceous to Recent burial. In this case, no evidence remains in either AFTA or VR data of the early episode of elevated gradients.

Southeastern Australia basin margins

Inversion involving basement can readily be studied using AFTA, although since vertical sections are generally not available, it is not

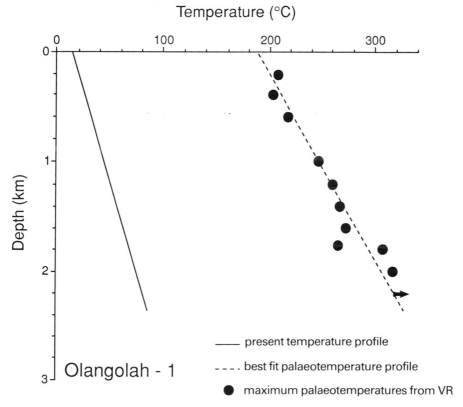

Fig. 12. Palaeotemperature estimates from VR in the Olangolah-1 well, plotted against depth. AFTA data in all samples from this well provide only minimum limits (not illustrated) for the maximum palaeotemperature, due to total annealing of all tracks formed prior to the onset of cooling.

not usually possible to reconstruct palaeogeothermal gradients and estimate amounts of section removed using formal methods. Nevertheless AFTA can reveal events which cannot be revealed by any other technique, and can provide at least an order of magnitude estimate of the amount of section removed.

Moore *et al.* (1986) showed that AFTA data in crystalline basement rocks from southeastern Australia revealed exposure to palaeotemperatures up to c. 110°C or above at around 100 Ma, while Dumitru *et al.* (1991) published a more recent compilation of data showing similar effects. As discussed in more detail by Duddy & Green (1992), these data reveal inversion of basement margins to the Otway, Bass and Gippsland Basins at c. 95 Ma (Fig. 14), synchronous with inversion in both the Otway and Gippsland Basins. As with the example of the Southern North Sea, Thermal History Reconstruction shows that in southeastern Australia, the effects of mid-Cretaceous inversion

were much more widespread than previously recognized.

Duddy & Green (1992) also commented that parts of the basins were inverted at the same time as the basin margins, particularly in the Otway Ranges (see previous section) and the western-most part of the Gippsland Basin onshore where Early Cretaceous sediments outcrop with VR levels around 0.7%, indicating maximum palaeotemperatures of over 100°C prior to mid-Cretaceous inversion. This early event represents a serious problem to prospects of preservation of hydrocarbons which might have been generated at the time, in basins which have been subjected to repeated episodes of tectonism.

The importance of Thermal History Reconstruction

While being by no means an exhaustive list of

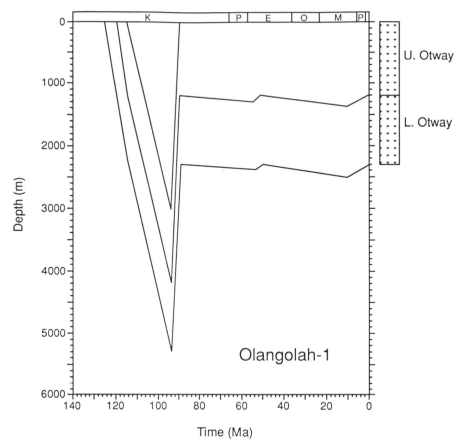

Fig. 13. Reconstructed burial and uplift history for the Olangolah-1 well, using the maximum likelihood values of removed section and palaeogeothermal gradient discussed in the text, derived from the palaeotemperature data shown in Fig. 12.

examples, the case histories discussed in preceding sections have been selected to demonstrate the importance of examining not only the stratigraphic and structural details of the preserved section but also the palaeothermal structure, and through that the section that has been removed by erosion, in reconstructing sedimentary basin histories. The major conclusions are as follows.

• Heating on a regional scale, together with subsequent cooling often synchronous with recognized basin inversion events, is a common occurrence and appears to reflect regional burial followed by uplift and erosion. Burial during such episodes is usually rapid, and although underlying processes are not understood, it is possible that the burial and subsequent uplift and erosion may be part of a common response to tectonic driving forces. As well as the examples discussed in this

paper, similar effects have been recognized in southern England and the offshore Netherlands region. Failure to allow for these regional effects can lead to a significant underestimation of regional maturity trends, while failure to take account of such events in tectonic studies of basin evolution ignores one of the major aspects of the subject.

• In basins where stratigraphic and structural relationships allow only broad constraints on the timing of inversion, Thermal History Reconstruction often shows that inversion took place at a different time to that which has been inferred on the basis of geological evidence from surrounding regions.

• In basins subjected to multiple episodes of inversion, Thermal History Reconstruction can readily define the event during which maximum hydrocarbon generation occurred.

• In areas where seismic data are either unavailable or allow only limited resolution, and

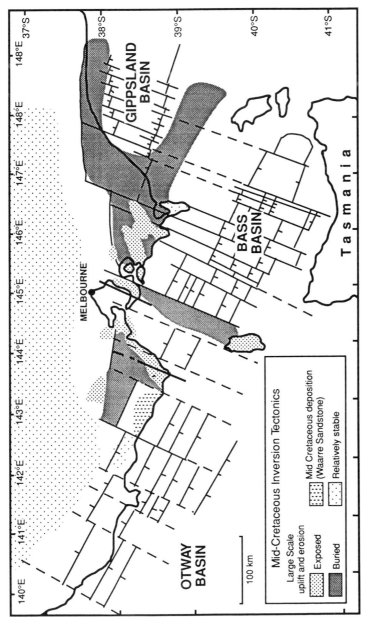

Fig. 14. Sketch map of the early Cretaceous rift basins of southeastern Australia, showing areas affected by mid-Cretaceous basin and basement margin inversion.

where inversion involves basement, Thermal History Reconstruction allows unique insights into basin development and hydrocarbon generation.

In summary, we suggest that reconstructions of basin histories based only on the stratigraphic and structural relationships within the preserved section provide only part of the story. The complete history of hydrocarbon generation can be reconstructed with greater confidence by integrating such information with data on the palaeo-thermal history of a section derived from direct measurements of palaeotemperature profiles.

We are grateful to Kerr-McGee Oil (UK) Ltd and their block 12/16 partners for permission to publish results and information from the 12/16-1 well. We are also grateful to Geoff Laslett, CSIRO Division of Mathematics and Statistics, Clayton, Victoria, for assistance with statistical analyses, to Kerry Hegarty, Geotrack International, for comments on an early draft of this paper and to other friends and colleagues for advice and assistance. AFTA is the registered trademark of Geotrack International.

References

BRAY, R. J., GREEN, P. F. & DUDDY, I. R. 1992. Thermal History Reconstruction using apatite fission track analysis and vitrinite reflectance: a case study from the UK East Midlands and the Southern North Sea. *In*: HARDMAN, R. F. P. (ed.) *Exploration Britain: Into the next decade*. Geological Society, London, Special Publication, **67**, 3–25.

BULAT, J. & STOKER, S. J. 1987. Uplift determination from interval velocity studies, UK southern North Sea. *In*: BROOKS, J. & GLENNIE, K. (eds) *Petroleum Geology of Northwest Europe*. Graham & Trotman, London, 293–305.

BUSHELL, T. P. 1986. Reservoir geology of the Morecambe field. *In*: BROOKS, J. & VAN HOORN, B. (eds) *Habitat of Palaeozoic Gas in N.W. Europe*. Geological Society, London, Special Publication, **23**, 189–208.

CHADWICK, R. A. 1986. Extension tectonics in the Wessex Basin, southern England. *Journal of the Geological Society, London*, **143**, 465–488.

COPE, J. C. W., INGHAM, J. K. & RAWSON, P. F. (eds) 1992. *Atlas of Palaeogeography and Lithofacies*. Geological Society, London, Memoir, **13**.

COPE, M. J. 1986. An interpretation of vitrinite reflectance data from the Southern North Sea basin. *In*: BROOKS, J., GOFF, J. C. & VAN HOORN, B. (eds) *Habitat of Palaeozoic Gas in N.W. Europe*. Geological Society, London, Special Publication, **23**, 85–98.

DUDDY, I. R. & GREEN, P. F. 1992. Tectonic development of the Gippsland Basin and environs: Identification of key episodes using Apatite Fission Track Analysis. *In: Proceedings of the Gippsland Basin Symposium, Melbourne, 1992,* 111–120.

DUMITRU, T. A., HILL, K. C., COYLE, D. A., DUDDY, I. R., FOSTER, D. A., GLEADOW, A. J. W., GREEN, P. F., KOHN, B. P., LASLETT, G. M. & O'SULLIVAN, A. J. 1991. Fission track thermochronology: application to continental rifting of southeastern Australia. *The APEA Journal,* 1991, 131–142.

ELLYARD, E. & BEATTIE, R. 1990. Inversion structures and hydrocarbon potential of the Southern Taranaki Basin. *In*: *1989 New Zealand Oil Exploration Conference Proceedings*. Petroleum and Geothermal Unit, Energy and Resources Division, Ministry of Commerce, 259–271.

FRASER, A. J. & GAWTHORPE, R. L. 1990. Tectono-stratigraphic development and hydrocarbon habitat of the Carboniferous in Northern England. *In*: HARDMAN, R. F. P. & BROOKS, J. (eds) *Tectonic Events Responsible for Britain's Oil and Gas Reserves*. Geological Society, London, Special Publication, **55**, 49–86.

——, NASH, D. F., STEELE, R. P. & EBDON, C. C. 1990. A regional assessment of the intra-Carboniferous play of Northern England. *In*: BROOKS, J. (ed.) *Classic Petroleum Provinces*. Geological Society, London, Special Publication, **50**, 417–440.

GLENNIE, K. W. & BOEGNER, P. L. E. 1981. Sole pit inversion tectonics. *In*: ILLING, L. V. & HOBSON, G. V. (eds) *Petroleum Geology of the Continental Shelf of North-west Europe*. The Institute of Petroleum, London, 110–20.

GREEN, P. F. 1986. On the thermo-tectonic evolution of Northern England: evidence from fission track analysis. *Geological Magazine*, **123**, 493–506.

—— 1989. Thermal and Tectonic history of the East Midlands shelf (onshore U.K.) and surrounding regions assessed by apatite fission track analysis. *Journal of the Geological Society, London*, **146**, 755–773.

——, DUDDY, I. R., BRAY, R. J. & LEWIS, C. L. E. 1993. Elevated palaeotemperatures prior to early Tertiary cooling throughout the UK region: implications for hydrocarbon generation. *In*: PARKER, J. R. (ed.) *Petroleum Geology of Northwest Europe: Proceedings of the 4th Conference*. Geological Society, London, 1067–1074.

HILLIS, R. 1991. Chalk porosity and Tertiary uplift, Western Approaches Trough, SW UK. *Journal of the Geological Society, London*, **148**, 669–679.

—— 1993. Tertiary erosion magnitudes in the East Midlands Shelf, onshore UK. *Journal of the Geological Society, London*, **150**, 1047–1050.

HOLLIDAY, D. W. 1993. Mesozoic cover over northern England: interpretation of apatite fission track data. *Journal of the Geological Society, London*, **150**, 657–660.

KAMP, P. J. J. & GREEN, P. F. 1990. Thermal and tectonic history of selected Taranaki Basin (New Zealand) wells assessed by apatite fission track analysis. *AAPG Bulletin*, **74**, 1401–1419.

KARNER, G. D., LAKE, S. D. & DEWEY, J. F. 1987. The thermal and mechanical development of the

Wessex Basin, southern England. *In*: Coward, M. P., Dewey, J. F. & Hancock, P. L. (eds) *Continental Extensional Tectonics*. Geological Society, London, Special Publication, **28**, 517–536.

Knox, G. J. 1982. Taranaki Basin, structural style and tectonic setting. *New Zealand Journal of Geology and Geophysics*, **25**, 125–140.

Lewis, C. L. E., Green, P. F., Carter, A. & Hurford, A. J. 1992. Elevated late Cretaceous to Early Tertiary paleotemperatures throughout North-west England: three kilometres of Tertiary erosion? *Earth and Planetary Science Letters*, **112**, 131–145.

Macchi, L., Curtis, C. D., Levison, A., Woodward, K. & Hughes, C. R. 1990. Chemistry, morphology and distribution of illites from Morecambe Gas Field, Irish Sea, Offshore United Kingdom. *AAPG Bulletin*, **74**, 296–308.

Marie, J. P. P. 1975. Rotliegendes stratigraphy and diagenesis. *In*: Woodland, A. W. (ed.) *Petroleum and the Continental Shelf of North-West Europe*. Applied Science Publishers, London, 205–11.

Moore, M. E., Gleadow, A. J. W. & Lovering, J. F. 1986. Thermal evolution of rifted continental margins: new evidence from fission tracks in basement apatites from southern Australia. *Earth and Planetary Science Letters*, **78**, 255–270.

Schmidt, D. S. & Robinson, P. H. 1990. The structural setting and depositional history of the Kupe South field, Taranaki Basin. *In*: *1989 New Zealand Oil Exploration Conference Proceedings*. Petroleum and Geothermal Unit, Energy and Resources Division, Ministry of Commerce, 151–172.

Stuart, I. A. & Cowan, G. 1991. The south Morecambe Field, blocks 110/2a, 110/3a, 110/8a, UK East Irish Sea. *In*: Abbots, I. L. (ed.) *United Kingdom Oil and Gas Fields, 25 Years Commemorative Volume*. Geological Society, London, Memoir, **14**, 527–541.

Van Hoorn, B. 1987. Structural evolution, timing and tectonic style of the Sole Pit inversion, *Tectonophysics*, **137**, 239–284.

—— 1989. Structural evolution, timing and tectonic style of the Sole Pit inversion (abstract). *In*: Cooper, M. A. & Williams, G. D. (eds) *Inversion Tectonics*. Geological Society, London. Special Publication, **44**, 356.

Whittaker, A., Holliday, D. W. & Penn, I. E. 1985. *Geophysical Logs in British Stratigraphy*. Geological Society, London, Special Report, **18**.

Woodward, K. & Curtis, C. D. 1987. Predictive modelling for the distribution of production-constraining illites – Morecambe Gas Field, Irish Sea, Offshore UK. *In*: Brooks, J. & Glennie, K. (eds) *Petroleum Geology of Northwest Europe*. Graham, & Trotman, London, 205–215.

Ziegler, P. A. 1990. *Geological Atlas of Western and Central Europe* (2nd edition). Shell International Petroleum Maatschappij BV, The Hague, Netherlands.

Regional Tertiary Exhumation in and around the United Kingdom

RICHARD R. HILLIS

Department of Geology and Geophysics, University of Adelaide, SA 5005, Australia

Abstract: Sonic velocities in the Upper and Middle Chalk, the Bunter Sandstone and the Bunter Shale were used independently to quantify apparent exhumation (amount of missing section) in the UK Southern North Sea (SNS). Apparent exhumation results derived from these units are statistically similar. The consistency of results from chalks, shales, and sandstones in the UK SNS, and in other areas (Inner Moray Firth and Celtic Sea/South-Western Approaches), suggests that, at a formation and regional scale, burial-depth is the primary control on velocity (and porosity) in these units, and validates the use of lithologies other than shale in maximum burial-depth determination. The consistency of apparent exhumation results from units of Early Triassic to Late Cretaceous age suggests that Tertiary exhumation was of sufficiently great magnitude to mask any earlier Mesozoic periods of exhumation, and that maximum Mesozoic–Cenozoic burial-depth was attained prior to Tertiary exhumation.

The analysis of sonic velocities, apatite fission tracks, vitrinite reflectance, normalized drilling rate and clay mineral diagenesis all suggest that there was a regional component of approximately 1 km of exhumation during Tertiary times, which affected structurally uninverted areas in and around the UK. The widely-recognized inversion-related exhumation is superimposed on regional Tertiary exhumation. Regional Tertiary exhumation need not have been contemporaneous with structural inversion. Further refinement beyond Tertiary age is not possible from the sonic velocity data. Regional exhumation, however, is most likely to be associated with regional Palaeocene or Oligocene/Miocene unconformities.

The evidence for regional exhumation implies that there was a pre-exhumational phase of burial (during which the eroded sedimentary rocks were deposited). Exhumation and prior burial associated with the regional Tertiary event need to be incorporated in aspects of basin analysis such as sediment decompaction and diagenetic and maturation modelling, and can, for example, account for the maturity of the Carboniferous of the onshore East Midlands Shelf, and the Lias exposed on the north coast of Somerset. The regional, Tertiary tectonic uplift associated with exhumation must have had a thick-skinned origin, and, in areas where there is no evidence for significant Tertiary igneous activity, a decoupled, two-layer model of lithospheric compression is invoked to account for the subsidence/uplift patterns of both inverted and uninverted areas.

Basin inversion is the process whereby basin-controlling extensional faults reverse their movement during compressional tectonics, and to varying degrees, basins are turned inside out to become positive features (Williams *et al.* 1989). Late Cretaceous–Tertiary basin inversion has long been recognized to have occurred over large areas of northwest Europe (e.g. Masson & Parson 1983; Ziegler 1987; Biddle & Rudolph 1988; Cooper & Williams 1989; Butler & Pullan 1990; Jensen *et al.* 1992; Chadwick 1993; this volume). Inversion axes can be recognized on geological maps of the southern United Kingdom and adjacent areas as major inliers of Jurassic–Lower Cretaceous (even Permo–Triassic in areas of extreme denudation) such as the Cleveland Hills, Sole Pit and Wealden axes. The Late Cretaceous and/or Tertiary inversion axes mapped for example by Ziegler (1990) are easily recognized both on reflection seismic

profiles and from stratigraphical data as areas where a thick Jurassic–Lower Cretaceous sequence has been structurally uplifted. The preferential Late Cretaceous–Tertiary uplift and exhumation of Mesozoic depocentres with respect to surrounding areas is elegantly accounted for by the compressional reactivation of crustal-scale faults, extension on which had previously formed the basins (Beach 1987; Kusznir *et al.* 1987; McClay 1989; Chadwick 1993).

Recently, several studies have argued that basin inversion and associated exhumation in and around the UK were only relatively localized processes, superimposed on Tertiary exhumation of a much more regional extent. Proponents of regional Tertiary exhumation have generally been trying to account for petrophysical properties which suggest that, even in areas unaffected by structural inversion,

From BUCHANAN, J. G. & BUCHANAN, P. G. (eds), 1995, *Basin Inversion*, Geological Society Special Publication No. 88, 167–190.

167

rocks are not at their maximum burial-depth. The following petrophysical properties have been cited as evidence for burial of rocks in excess of that currently observed, in areas which are not inverted:

- sonic velocities (e.g. Bulat & Stoker 1987; Hillis 1991; Hillis 1993a; Japsen 1993; Hillis et al. 1994; Hillis in press);
- apatite fission track analysis (e.g. Green 1989; Lewis et al. 1992; Keeley et al. 1993; Green et al. 1993);
- vitrinite reflectance (e.g. Cornford 1986; Bray et al. 1992; Thomson & Hillis in press);
- normalized drilling rate (e.g. Nyland et al. 1992; Jensen & Schmidt 1992); and
- clay mineral diagenesis (e.g. Hurst 1980; Nyland et al. 1992; Jensen & Schmidt 1992).

Despite evidence from the above petrophysical properties, there are many aspects of regional Tertiary exhumation (including whether it even occurred) that are enigmatic and currently under debate. Regional Tertiary exhumation is not associated with a regional, angular unconformity on reflection seismic profiles, hence its timing is unclear. Stratigraphic reconstruction of the sequence preserved in basins provides little evidence of the former presence of the approximately 1 km of sedimentary rocks which petrophysical properties require to have been eroded. Finally, unlike compressional structural reactivation of crustal-scale faults, which can account for basin inversion, there is no widely accepted tectonic mechanism for regional exhumation.

This paper addresses:

- the evidence for regional Tertiary exhumation in north–west Europe, concentrating on sonic velocity-based quantification of exhumation;
- the above enigmatic aspects of the case for regional Tertiary exhumation;
- the implications of regional Tertiary exhumation for hydrocarbon exploration; and
- possible tectonic (geodynamic) models for regional Tertiary exhumation.

A clear distinction should be made between regional exhumation and basin inversion. Quantification and mapping of exhumation alone, as presented in this paper, cannot demonstrate basin inversion. In order to demonstrate basin inversion it is necessary to show that maxima of Tertiary exhumation coincide with maxima of Jurassic–Lower Cretaceous burial. This is indeed the case: recognized zones of structural inversion such as the Sole Pit and Britanny and South-West Channel Basins are maxima of both

Jurassic–Lower Cretaceous burial and Tertiary exhumation. This paper deals not with that widely recognized phenomenon (e.g. Kent 1980; Glennie & Boegner 1981; Beach 1987; Badley et al. 1989; Hayward & Graham 1989; Butler & Pullan 1990; Chadwick 1993), but rather with the more contentious (e.g. discussions/replies of Brown 1991a/Green 1991; McCulloch 1994/Holliday 1994; Smith et al. 1994/Hillis 1994) evidence for regional Tertiary exhumation.

Regional Tertiary exhumation may be temporally, spatially and mechanistically unrelated with basin inversion. However, it is necessary to understand both phenomena if the Tertiary burial, thermal and tectonic history of the United Kingdom and surrounding areas is to be fully understood.

Estimating the magnitude of exhumation from sonic velocity data

Methodology with reference to the UK Southern North Sea

Since depth-controlled compaction is largely irreversible, units which are shallower than their maximum burial-depth will be overcompacted with respect to their present burial-depth (e.g. Magara 1976; Lang 1978; Bulat & Stoker 1987; Hillis 1991; Issler 1992; Hillis et al. 1994). It is assumed that all units follow a normal compaction trend with burial, and that compaction is not reversed by subsequent exhumation (Fig. 1). With these assumptions, the amount of elevation of exhumed sedimentary rocks above their maximum burial-depth (termed apparent exhumation) is given by the displacement, along the depth axis, of the observed compaction trend from the normal, undisturbed trend (Fig. 1). Although porosity directly describes compaction state, sonic velocity is widely used as an indicator of compaction because it is strongly dependent on porosity (e.g. Wyllie et al. 1956; Raiga–Clemenceau et al. 1988) and routinely logged in wells.

Stratigraphically-equivalent units that exhibit a vertically- and laterally-consistent relationship between depth and velocity are required for the determination of maximum burial-depth. Shales have been widely used because they are believed to show the most simple velocity/depth relationships (e.g. Jankowsky 1962; Marie 1975; Magara 1976; Lang 1978; Cope 1986; Wells 1990; Issler 1992; Japsen 1993). However, any unit which exhibits a vertically- and laterally-consistent velocity/depth relationship is suitable. Units with little or no lateral facies/lithological

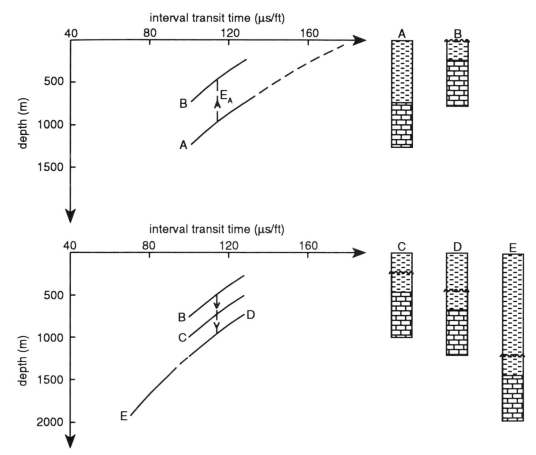

Fig. 1. Idealized evolution of sonic interval transit time (itt) during burial (A), subsequent exhumation (B), and post-exhumational burial (C, D and E). Apparent exhumation (E$_A$) is the amount of exhumation not reversed by subsequent burial. It can be measured by the displacement, along the depth axis, of the sonic interval transit time/depth relationship of the exhumed sequence (B or C) from that of a reference, or normally compacted sequence (A, D or E).

variation are used because changes in their compaction state across the region are unlikely to be due to facies/lithology variation. Ideally, reasonably thick units are analysed because thick units are less affected by any laterally variable intra-unit sedimentological or diagenetic phenomena (such as hardgrounds in chalk) that might cause deviation from the assumed normal velocity/depth relationship. Units should also not exhibit any overpressure or hydrocarbon-filled porosity as both these phenomena may inhibit normal compaction. Finally, units analysed should be geographically extensive because, the more extensive the unit, the greater the likelihood of finding wells which are at their maximum burial-depth to define a true normal velocity/depth relationship (i.e. one unaffected by exhumation).

The tops and bases of the stratigraphic units are consistently picked from vertically compressed (typically 1:4000–1:6000 depth scale) plots of the sonic and gamma ray logs. Such compressed plots facilitate the picking of tops and bases which, for consistency, often need to be modified from those on operators' composite logs. The mean sonic velocity of the resultant intervals is determined from digital sonic log data. Sonic log calibration is checked with reference to the continuous sonic velocity (check shot) data. Mean sonic velocity is then plotted against the burial-depth of the midpoint of the unit. Figure 2 shows sonic interval transit time, the reciprocal of velocity, plotted against burial depth for the Upper and Middle Chalk, the Bunter Sandstone and the Bunter Shale of the UK Southern North Sea (UK SNS). Since

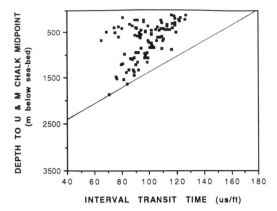

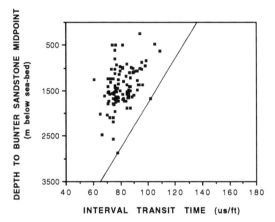

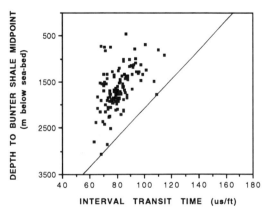

Fig. 2. Mean sonic itt/depth to unit midpoint plots for the three stratigraphic units analysed in the UK Southern North Sea. Interval transit time is in microseconds per foot (μs/ft). The normal compaction relationship for each unit (i.e. that unaffected by exhumation, determined as outlined in the text) is also shown.

apparent exhumation is given by the displacement, on the depth axis, of a given velocity/depth point from the normal velocity/depth trend relationship, the critical factor in determining the magnitude of apparent exhumation lies in the selection of the normal velocity/depth relationship. Following the standard exponential porosity/depth trend (e.g. Sclater & Christie 1980) and Wyllie *et al.*'s time average equation (1956), Bulat & Stoker (1987) showed that the theoretical form of velocity/depth relationship is close to linear for the burial-depths typically under consideration.

In an area subject to exhumation, the wells with the lowest velocity for their given burial-depth should be considered normally compacted, provided their relatively low velocity is not merely due to phenomena which may inhibit normal compaction (such as overpressure or hydrocarbon-filled porosity). For a linear increase of velocity with depth, the two wells (termed the reference wells) which can be linked by a straight line that has no points falling to its right (less compacted) side on Fig. 2 define the normal compaction relationship (Hillis 1993a; Menpes & Hillis this volume; Hillis in press). Alternatively, if there is a consistent intra-well rate of increase of velocity with depth in a given stratigraphic unit, then the mean intra-well velocity/depth gradient within the unit in the wells analysed can be combined with the single well that shows the lowest velocity for its burial-depth (allowing for the mean velocity/depth gradient) to determine the normal velocity/depth relationship (Hillis 1991; Hillis 1993a; Hillis *et al.* 1994).

The intra-well velocity/depth gradient within a unit should only be used in defining the normal velocity/depth relationship for units which exhibit no systematic vertical facies variation such as becoming sandier or shalier upwards. Units exhibiting vertical facies variation may be used in compaction-based studies of maximum burial-depth provided any vertical facies change is consistent throughout the area of study. Although the bulk or mean increase in velocity with depth in such units may still be primarily controlled by burial-depth, the compaction rate within the unit in an individual well may not represent the bulk or mean compaction rate of the unit.

The normal compaction relationship should be verified by comparison to published compaction relationships (e.g. Scholle 1977 for chalks; Baldwin & Butler 1985 for shales) and reasonable surface values (e.g. Pryor 1973 for sandstones). With transformation of velocity to porosity using Wyllie *et al.*'s (1956) time average

Table 1 *Correlation between apparent exhumation results from the Upper and Middle Chalk, Bunter Sandstone and Bunter Shale in the UK SNS*

	Upper and Middle Chalk (C)		Bunter Sandstone (BT)		
Bunter	C = 313 + 0.424BT				
Sandstone	BT = 539 + 0.701C				
(BT)	70	0.545	5.4		
Bunter	C = 292 + 0.507BH		BT = 207 + 1.029BH		
Shale (BH)	BH = 525 + 0.455C		BH = 198 + 0.661BT		
	80	0.481	4.8	109	0.825	15.1

Linear, best-fit, least-squares regression between apparent exhumation values derived from the Upper and Middle Chalk, Bunter Sandstone and Bunter Shale in the UK SNS. Since there is no dependent variable, apparent exhumation from the shallower unit was regressed on that from the deeper (first line of entry) and vice-versa (second line of entry). The third line of the entry for each pair of units indicates (from left to right) the number of wells in which both units occur, the co-efficient of correlation between apparent exhumation values derived from the two units in these wells, and the t-statistic for the co-efficient of correlation. In all three cases, comparison of the t-statistic with the one-tailed Student's t-distribution allows rejection of the null hypothesis (that the co-efficients of correlation come from a population the mean of which is zero) at the 99% confidence level.

equation, the normal compaction relationships and illustrated on Fig. 2 are consistent with the above.

Apparent exhumation (E_A) can be estimated graphically from the plots of interval transit time against depth to formation mid-point (Fig. 2). However, in practice, it is determined numerically using the simple equation:

$$E_A = 1/m(\Delta t_U - \Delta t_0) - d_U,$$

where m is the gradient of the normal compaction line, Δt_U is the mean interval transit time of the well under consideration, Δt_0 is the surface intercept of the normal compaction line, and d_U is the depth of the formation mid-point (here taken below sea-bed) of the well under consideration.

Regional studies of maximum burial-depth have been undertaken using the above methodology in the:

- UK SNS, based on velocities in the Chalk, the Bunter Sandstone and the Bunter Shale (as illustrated here and in Hillis (in press), and the onshore East Midlands Shelf (EMS), based on velocities in the Chalk, Kimmeridge Clay, Lias and Bunter Sandstone (Hillis 1993a);

- the Inner Moray Firth (IMF), based on velocities in the Hod and Tor Formations (both Chalk) and the Kimmeridge Clay (Hillis *et al.* 1994; Thomson & Hillis in press); and
- Celtic Sea/South-Western Approaches (CS/SWA), based on velocities in the Danian Chalk, the Upper Cretaceous Chalk, the Greensand/Gault Clay and the Mercia Mudstone (Hillis 1991; Menpes & Hillis this volume).

Comparing apparent exhumation results from different stratigraphic units/lithologies

If maximum burial-depth is the principal control on the velocity of the units analysed, and burial to and exhumation from maximum depth postdated the deposition of the youngest unit analysed, then all units analysed in the same well should yield similar apparent exhumation values. Apparent exhumation results derived from velocities in the different stratigraphic units in the same well are crossplotted as a test of the validity of using different stratigraphic units/lithologies in maximum burial-depth/exhumation studies. Figure 3 shows crossplots of apparent exhumation values from the Upper and Middle Chalk, the Bunter Sandstone and the Bunter Shale in the UK SNS. Regression analysis is used to determine least-squares, best-fit, linear relationships between the apparent exhumation values from the different units, and associated co-efficients of correlation (Table 1). The t-statistic of each co-efficient of correlation is calculated and tested against the one-tailed Student's t-distribution in order to test the null hypothesis that the co-efficients of correlation come from a population whose mean value is zero (e.g. Till 1974). This hypothesis can be rejected at the 99% confidence level in all three cases. Hence the results of apparent exhumation as determined from the Upper and Middle Chalk, the Bunter Sandstone and the Bunter Shale in the UK SNS are statistically similar.

Hillis *et al.* (1994) and Thomson & Hillis (in press) showed similarly that apparent exhumation values from velocities in the Hod, Tor (both within the Chalk) and Kimmeridge Clay Formations of the IMF are statistically similar. Menpes & Hillis (this volume) have shown that apparent exhumation values from the Danian Chalk, the Upper Cretaceous Chalk, the Greensand/Gault Clay and the Mercia Mudstone in the CS/SWA are also statistically similar.

It is unlikely that sedimentological and/or diagenetic mechanisms independent of burial-depth, such as those discussed by Bulat & Stoker

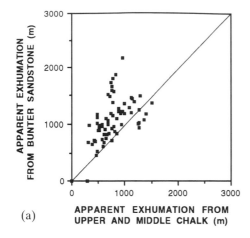

(a)

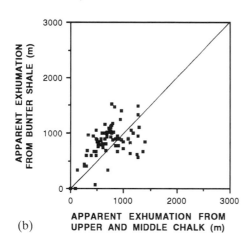

(b)

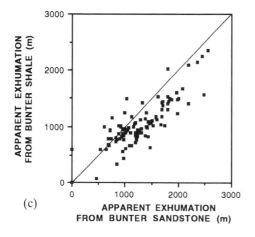

(c)

(1987) and Hillis (1991), could account for similar amounts of overcompaction/apparent exhumation in the variety of lithological units (chalk, shale, shaly sandstone and sandstone) analysed in the UK SNS/EMS, the IMF and the CS/SWA. The statistically significant correlation between the amount of overcompaction/apparent exhumation in these lithologically distinct units supports the hypothesis that burial at depth beyond that currently observed is responsible for the overcompaction, and that overcompaction may be used to determine exhumation magnitudes. A corollary is that lithologies other than shale may be used for estimating exhumation magnitude. The mean of the apparent exhumation values derived from such of the units as are present in each well is considered the most robust description of apparent exhumation because any laterally variable intra-unit sedimentological or diagenetic effects that might cause variation from the normal velocity/depth relationship in an individual unit will tend to be reduced. Figure 4a shows the principal structural elements of the UK SNS/EMS, CS/SWA and IMF and Fig. 4b shows mean apparent exhumation values for the UK SNS/EMS and the CS/SWA, and values based on velocities in the Kimmeridge Clay of the IMF (the Hod and Tor Formations are restricted to the eastern Inner Moray Firth).

The similarity of results from units of Triassic to Late Cretaceous/Danian age in the UK SNS/EMS, CS/SWA and IMF suggests that Tertiary exhumation in these areas was of sufficiently great magnitude to mask any earlier Mesozoic (e.g. 'Cimmerian') periods of exhumation, and that maximum Mesozoic–Cenozoic burial-depth was attained prior to Tertiary exhumation. These data, of course, reveal nothing about the magnitude of late Carboniferous (Variscan) exhumation prior to the deposition of the oldest units analysed. Maximum burial-depth of the pre-Variscan sequence may have been attained prior to

Fig. 3. Crossplots of apparent exhumation (in metres) derived from itt in the three stratigraphic units studies: **(a)** apparent exhumation from Bunter Sandstone itt against that from the Upper and Middle Chalk; **(b)** Bunter Shale against Upper and Middle Chalk; **(c)** Bunter Shale against Bunter Sandstone. The line illustrating the 1 : 1 relationship between apparent exhumation values from each of the pairs of units analysed is shown. Table 1 lists the linear, best-fit, least squares regression relationship between each of the pairs of apparent exhumation values and associated co-efficients of correlation.

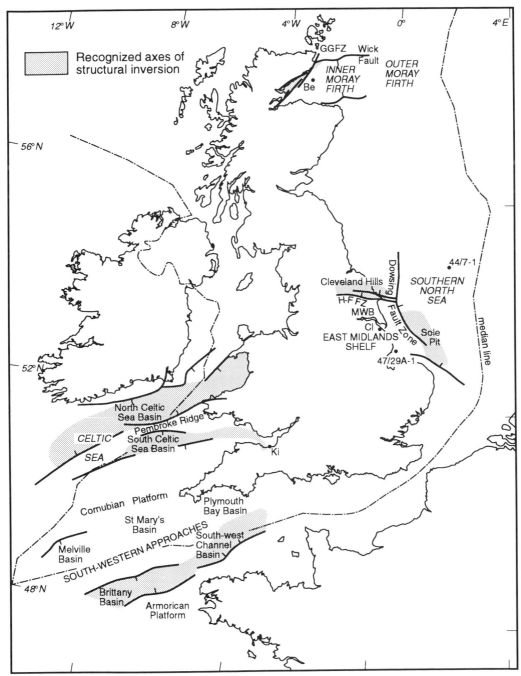

Fig. 4.(a) Map of the principal structural elements of the IMF, UK SNS/EMS and CS/SWA and locations and wells discussed in the text. Dotted areas are recognized axes of structural inversion. The recognized axes of structural inversion in the SNS/EMS are after Van Hoorn (1987*b*) and Ziegler (1990), those in the CS/SWA are after Ziegler (1990), and those in the IMF are from Thomson (pers. comm.). Ticks on faults indicate the Jurassic–Lower Cretaceous downthrown side. Be, Beatrice Oilfield; Cl, Cleethorpes-1 well; Ki, Kilve; MWB, Market Weighton Block; GGFZ, the Great Glen Fault Zone; H–F FZ, Howardian–Flamborough Fault Zone.

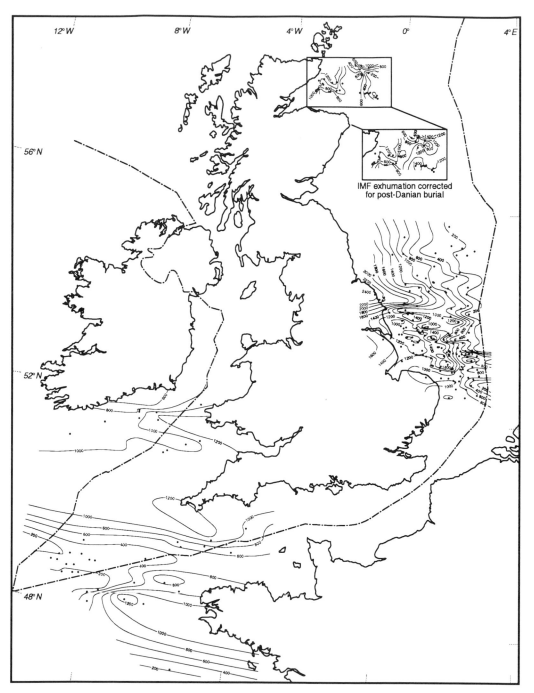

Fig. 4.(b) Compilation of apparent exhumation data from sonic velocities in and around the UK. The maps of apparent exhumation (i.e. elevation above maximum burial-depth) and apparent exhumation corrected for post-Danian burial (i.e. amount of exhumation that occurred during Palaeocene denudation) for the IMF are based on velocities in the Kimmeridge Clay (Thomson & Hillis in press). The map for the UK SNS is based on the mean of results from velocities in the Upper and Middle Chalk, Bunter Sandstone and Bunter Shale (this paper; Hillis in press). The map for the CS/SWA is based on velocities the Danian Chalk, the Upper Cretaceous Chalk, the Greensand/Gault Clay and the Mercia Mudstone (Menpes & Hillis this volume). If the normal velocity/depth trend for the Chalk of the SNS can be extrapolated to the CS/SWA (where there are considerably fewer wells), the reference wells in the CS/SWA are themselves approximately 350 m above maximum burial-depth, and the apparent exhumation values for the CS/SWA are approximately 300–400 m too low.

Variscan burial, although this does not usually seem to be the case (Cope 1986; Green 1989; Bray *et al.* 1992).

Apparent exhumation, maximum burial-depth and exhumation at the time of denudation

Apparent exhumation is exhumation not reversed by subsequent burial (Fig. 1). Hence it is not necessarily the same as the amount of exhumation that occurred at the time of denudation. If renewed burial follows exhumation, the magnitude of apparent exhumation is reduced by the amount of that subsequent burial (Fig. 1, well C). Once the unit again reaches its maximum burial-depth (Fig. 1, well D) it is normally compacted, and there is no evidence of the previous phase of exhumation:

total exhumation at the time of denudation (E_T) = apparent exhumation (E_A) + post-exhumational burial (B_E)

	E_T	=	E_A	×	B_E
Fig. 1, well B (m):	500	=	500	+	0
Fig. 1, well C (m):	500	=	250	+	250
Fig. 1, well D (m):	500	=	0	+	500

Maximum burial-depth (B_T) is given by the sum of present (B_P) burial-depth and apparent exhumation (E_A):

$$B_T = E_A + B_P.$$

The burial/exhumation histories for the base Cretaceous of the Cleethorpes-1 well on the EMS and well 44 /7-1 in the SNS are shown in Fig. 5 in order to illustrate these relationships. In these examples it is assumed that exhumation occurred during the Miocene, the time of the most pronounced Cenozoic stratigraphic break in well 44/7-1. At the Cleethorpes-1 well there is no post-Miocene burial, hence the amount of exhumation that occurred during the Miocene equals the 1.4 km of apparent exhumation (Fig. 5). At well 44/7-1 apparent exhumation is 0.6 km, and there was 0.4 km of post-exhumational, Pliocene–Quaternary burial, hence there was 1.0 km of exhumation during the Miocene (Fig. 5). The maximum burial-depth of the base Cretaceous in the Cleethorpes well is 1.7 km, the apparent exhumation plus the present burial-depth of 0.3 km. At well 44/7-1 the maximum burial-depth of the base Cretaceous is 2.0 km, the sum of apparent exhumation and the present burial-depth of 1.4 km.

It is important to remember these relationships when considering maps such as those of Fig. 4b. All regions of zero apparent exhumation

in Fig. 4b are characterized by the presence of a thick Tertiary sequence, and there may have been significant exhumation in these areas that has been masked by subsequent Tertiary burial.

Results of exhumation of magnitude studies based on sonic velocity data

Apparent exhumation

Apparent exhumation in the UK SNS shows a steady regional increase from zero in the centre of the SNS, where there is a thick Tertiary sequence, to approximately 1.5 km onshore eastern England (Fig. 4b). Superimposed on this regional trend are highs associated with the Cleveland Hills and Sole Pit inversion axes (Figs 4a, b). The southern/south-western margin of the Cleveland Hills/Sole Pit axes highs in apparent exhumation is marked by the basin-controlling Howardian–Flamborough and Dowsing Fault Zones of Kent (1980). There is a low in apparent exhumation between the Sole Pit and onshore eastern England, on the offshore EMS. However, it is significant that apparent exhumation on the offshore EMS, which is a relatively stable, platform area (Kent 1980), and shows no evidence of structural inversion, is 1.0–1.4 km (Figs 4a, b).

The growth of Zechstein salt structures in the UK SNS (e.g. Taylor 1990) has probably led to overlying units having been exhumed from, and remaining above maximum burial-depth. Hence apparent exhumation in the SNS (Fig. 4b) is probably influenced by salt growth structures (Brown *et al.* 1994). However, salt growth structures would only be witnessed by relatively localized noise superimposed on the more widespread trends of exhumation of the Cleveland Hills/Sole Pit inversion axes and regional, SNS-wide exhumation being considered here.

Apparent exhumation increases from zero around the Inner Moray Firth/Outer Moray Firth transition, at the eastern limit of the mapped area, where there is a thick Tertiary sequence, to about 1 km in the vicinity of the Beatrice Field in the south-west of the mapped area (Fig. 4b). Apparent exhumation exhibits a maxima near the junction of the Great Glen and Wick Faults in the north-west of the mapped area, and to a lesser extent along the Wick Fault at the northern boundary of the mapped area (Figs 4a, b). Inversion structures occur in the immediate vicinity of the Great Glen and Wick Faults (Thomson & Underhill 1993; Thomson & Hillis in press). However, there is no evidence of structural inversion away from these basin-controlling faults, for example, in the vicinity of

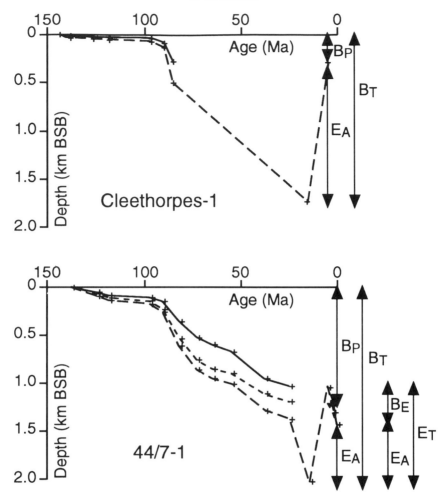

Fig. 5. Present observed (solid) and decompacted burial-histories for the base Cretaceous in the Cleethorpes-1 and 44/7-1 wells. The short dashed burial plot for well 44/7-1 was corrected for compactional effects without consideration of Tertiary exhumation, whereas the long dashed plots for both wells were corrected for compaction with allowance for Tertiary exhumation. The plot of decompacted burial-history without allowance for Tertiary exhumation for Cleethorpes-1 cannot be resolved from the present observed (solid) burial-history plot. Decompaction procedure was based on that of Sclater & Christie (1980), modified to account for exhumation as outlined in the text. Formation tops and ages were taken from the operators' composite logs and geochronologically calibrated after the time scale of Harland *et al.* (1989). In both wells, as is always the case, maximum burial-depth (B_T) equals present burial-depth (B_P) plus apparent exhumation (E_A), i.e. $B_T = E_A + B_P$. In Cleethorpes-1, where there was no subsequent burial, apparent exhumation (E_A) equals the exhumation that occurred at the time of denudation. However, in well 44/7-1 exhumation at the time of denudation (E_T) equals apparent exhumation (E_A) plus the post-exhumational burial (B_E), i.e. $E_T = E_A + B_E$ (again this is a general rule).

the Beatrice Field where apparent exhumation is approximately 1 km (Fig. 4a).

In the CS/SWA, the recognized inversion axis of the Brittany and South-West Channel Basins (Hayward & Graham 1989; Ziegler 1990) is witnessed by highs delimited by 800 and 1200 m

apparent exhumation contours (Figs 4a, b; Menpes & Hillis this volume). The North Celtic Sea Basin inversion axis (Tucker & Arter 1987; Ziegler 1990) is also witnessed by a high in apparent exhumation (Figs 4a, b). However, the less strongly inverted South Celtic Sea Basin

(Van Hoorn 1987a; Ziegler 1990) is not picked out on the apparent exhumation map. The Melville Basin, where the Tertiary sequence is thickest, is the major low in apparent exhumation in the CS/SWA area. Regional highs in apparent exhumation of approximately 1 km occur over the uninverted St. Mary's and Plymouth Bay Basins (Figs 4a, b), and on the uninverted margins of the Armorican and Cornubian Platforms and the Pembroke Ridge. Although contours over the platform areas are poorly constrained due to the paucity of data, highs in apparent exhumation are not restricted to the recognized inversion axes (Figs 4a, b), and in uninverted areas where there is little post-exhumational Tertiary burial, the regional component of exhumation is approximately 1 km.

If the normal velocity/depth trend for the Chalk of the SNS can be extrapolated to the CS/SWA (where there are considerably fewer wells), the reference wells in the CS/SWA are themselves approximately 350 m above maximum burial-depth, and the apparent exhumation values for this area may be 300–400 m too low. There are relatively few wells in the CS/SWA region and at such an early stage of exploration, wells tends to be concentrated on structural highs. However, especially in the northern part of the SWA which was not affected by structural inversion, the structures drilled pre-date the major Lower Cretaceous unconformity, and are not expressed in the Upper Cretaceous and Tertiary section (e.g. Bennet et al. 1985). Hence the inferred regional trends in exhumation are not considered to be significantly biased towards areas of Tertiary structuration.

Exhumation at the time of denudation: the post-exhumational burial correction

The regional nature of Tertiary exhumation is even more striking if apparent exhumation is corrected for post-exhumational burial. Regional exhumation in the IMF probably occurred in mid-late Danian times because this corresponds with the age of a major unconformity separating prograding deltaic clastic facies of the Montrose Group from the underlying Chalk in the eastern IMF and Outer Moray Firth (Milton et al. 1990).

Accepting this age of exhumation, the magnitude of exhumation during Danian denudation equals the apparent exhumation plus the thickness of the post-Danian sequence. The exhumation map corrected for post-Danian burial (Fig. 4b) suggests that the entire IMF underwent approximately 1 km of exhumation. One can only speculate on the magnitude of exhumation to the east of the mapped area, in the Outer Moray Firth, where apparent exhumation is zero. There may have been Danian exhumation of less than or equal to the amount of post-Danian burial in the Outer Moray Firth (Fig. 1). However, the sequence stratigraphic work of Milton et al. (1990) suggests that erosion did not extend across the entire Central North Sea Basin because there is a correlative conformity to the above Danian unconformity. Considering the work of Milton et al. (1990), it is unlikely that there was significant Danian erosion east of 0°E. Knox et al. (1981) and Mudge & Bliss (1983) suggested that a period of widespread uplift and erosion followed the deposition of the Danian chalk and marl, and that Danian sedimentary rocks were removed from all but the deepest parts of Central/Northern North Sea.

In the UK SNS, apparent exhumation decreases eastwards from the Sole Pit towards the median line between the UK and the Netherlands, complementing the increasing thickness of the Tertiary. It is difficult to correct the apparent exhumation map of the UK SNS for post-exhumational burial because the timing of the exhumation is unclear (as discussed below, Palaeocene and Miocene are the most likely ages), and because the available data on Tertiary stratigraphy is limited. Nonetheless, it is instructive to consider the five wells in Quadrant 38 (in the northeast of the mapped area, directly south of the 200 m contour label) which are in an area unaffected by structural inversion. If the mean apparent exhumation values for these wells are corrected for the thickness of the Plio-Pleistocene sequence (i.e. assuming Miocene exhumation), the magnitude of exhumation is 0.9 km ± 0.2. If exhumation occurred in the Palaeocene, then the corrected values of exhumation in Quadrant 38 are 1.5 km ± 0.2.

It appears that Tertiary exhumation throughout the UK SNS had a regional component of approximately 1.0 or 1.5 km (depending on the time of exhumation). These values are consistent with the mean apparent exhumation values of approximately 1.2 km off-structure, immediately southwest and northeast of the Sole Pit axis where there is little or no post-exhumational burial. Menpes & Hillis (this volume) similarly show that, after correction for post-exhumational burial, there was significant exhumation along the axis of the SWA where apparent exhumation reaches low values (Fig. 4b).

Further evidence for regional Tertiary exhumation in north-west Europe

The United Kingdom and its Continental Shelf

There have been a number of studies of the difference between present and maximum burial-depth (i.e. apparent exhumation) in the Southern North Sea and adjacent (onshore) eastern England. Marie (1975) and Glennie & Boegner (1981) determined apparent exhumation in the Sole Pit area from sonic velocities in the Bunter Shale. Whittaker et al. (1985) presented a map of 'post-Cretaceous uplift' in eastern England from sonic log data. In none of these studies are details of data and methodology presented. Cope (1986) used vitrinite reflectance data from the Carboniferous and interval velocities from the Bunter Shale to map exhumation in the SNS.

The maximum values of apparent exhumation (on the Sole Pit/Cleveland Hills inversion axes) derived from the above studies are similar to, or a few hundred metres less than, those presented here. However, the above studies show apparent exhumation decreasing to minimal values on the offshore East Midlands Shelf. Data presented here, in contrast, suggest approximately 1.2 km of apparent exhumation over the offshore East Midlands Shelf. Similarly, Marie (1975) and Cope (1986) suggest apparent exhumation decreases north–east of the Sole Pit inversion axis, reaching zero in the north–east corner of Quadrant 48, an area where data presented here again suggest that apparent exhumation is approximately 1.2 km.

Bulat & Stoker (1987) also estimated apparent exhumation in the UK SNS from velocity data. Like the study presented here, their analysis extended to the north–east of the UK SNS (Quadrants 38 and 44) where the reference wells defining their normal compaction relationship are at maximum burial-depth. Their maps show significant exhumation away from the Sole Pit/Cleveland Hills inversion axes, and the maps based on velocities in the Upper Cretaceous, Bunter Sandstone and Bunter Shale all suggest approximately 4000 ft (1200 m) of exhumation in the vicinity of the 47/29A-1 well. As suggested by Green (1989), Hillis (1991), Bray et al. (1992) and Hillis (1993a), many previous studies of apparent exhumation from sonic velocity in the UK SNS/EMS (Marie 1975; Glennie & Boegner 1981; Whittaker et al. 1985; Cope 1986), and elsewhere, may underestimate apparent exhumation because they use a normal velocity/depth relationship defined by wells which are not

at their maximum burial-depth. Comprehensive study of velocities in the UK SNS/EMS shows that only wells in the north–east of the UK SNS should be used as reference wells for defining the normal compaction relationship.

There is good agreement between the apparent exhumation results based on sonic velocities presented here for the UK SNS (Fig. 4) and the onshore EMS (Hillis 1993a), and those based on apatite fission track analysis and vitrinite reflectance. Green (1989) estimated 1.3–1.7 km of missing section at the Cleethorpes-1 well from apatite fission track analysis. Upper and Middle Chalk and Bunter Sandstone velocity in the Cleethorpes-1 well suggest approximately 1.5 and 1.4 km of section are missing respectively. Well 47/29A-1 is located on the offshore EMS (at the mouth of The Wash) where previous studies of velocity (Marie 1975; Glennie & Boegner 1981; Whittaker et al. 1985; Cope 1986), suggest that minimal section is missing. However, apatite fission track analysis suggests 1.31 km of missing section at well 47/29A-1, while vitrinite reflectance data suggest 1.51 km (Bray et al. 1992; Green et al. 1993). Sonic velocity in the Bunter Shale of well 47/29A-1 suggests approximately 1.2 km of section are missing.

It is interesting to note that well 47/29A-1 lies in an area where Cope (1986) suggested vitrinite reflectance data indicated anomalously large amounts of exhumation. Cope (1986) suggested that oxidation effects might be responsible for the anomalously high vitrinite reflectance values. The realization, from compaction, fission track and more recent vitrinite reflectance studies that the EMS has been subjected to a kilometre or more of regional Tertiary exhumation, may account for the reflectance data Cope (1986) considered anomalous.

Regional exhumation also clearly affected onshore northern England. Apatite fission track analysis suggests that some 3 km of section were removed from the Palaeozoic rocks forming the structural highs in northern England, such as the present-day upland areas of the English Lake District and Northern Pennines (Green 1986; Lewis et al. 1992). Arguments based on preserved sedimentary thicknesses suggest only 1200–1750 m of section was removed (Holliday 1993), and a number of re-analyses of the above apatite fission track data suggest somewhat less section was removed: 1 km (Brown 1991a), <1 km (McCulloch 1994) and 1500–2000 m (quoted in Holliday 1994). However, despite debate on its magnitude onshore northern England, the occurrence of regional exhumation appears to be accepted.

In the IMF, vitrinite reflectance and sonic velocity data in wells in the vicinity of the Beatrice Field both indicate approximately 1 km of exhumation (Thomson & Hillis in press). Although some other wells in the IMF show discrepancies in predicted exhumation values between the vitrinite and velocity data, the vitrinite data indicate significant exhumation away from the inverted basin-bounding faults (Thomson & Hillis in press). On the basis of vitrinite reflectance and clay mineralogy, Hurst (1980) also suggested that Middle Jurassic sedimentary rocks in the vicinity of the Beatrice Field had been raised by at least 1 km with respect to the sea-bed. However, Roberts et al. (1990) suggested a lower estimate of missing section, 300–750 m, based on raw seismic velocity spectra, and concluded that Chalk and Palaeocene sequences were never present in the western part of the basin. Roberts et al. (1990) do not present details of their methodology, but their lower exhumation values may reflect use of a normal velocity/depth trend which has itself been exhumed from maximum burial-depth.

Menpes & Hillis (this volume) discuss evidence supporting that from sonic velocities for the regional exhumation of areas of the CS/SWA unaffected by structural inversion. Vitrinite reflectance trends on the Cornubian Platform onshore southwest England (Cornford 1986), and the occurrence of flints (presumably derived from now eroded Chalk) suggest that the Cornubian Platform, which is not inverted, has undergone significant exhumation (Roberts 1989). Major regional exhumation of the Cornubian Platform is vividly illustrated by the presence of a retort house near Kilve on the north coast of Somerset (Fig. 4a), now disused but built to produce oil from the mature Liassic shale of the present-day coastal cliffs. To the north of the Celtic Sea, apatite fission track analysis suggests 1.3–2.5 km of post-Bajocian sedimentary rocks have been eroded from southern Ireland (Keeley et al. 1993).

In summary, there is a strong body of evidence, based largely on petrophysical properties, that supports the hypothesis that Tertiary exhumation, of around 1 km, occurred throughout the UK SNS/EMS, the IMF and the CS/SWA, including areas unaffected by structural inversion.

Biomarker maturities in the Kimmeridge Clay Formation following approximately the arc of its outcrop from the Wash to Wiltshire (Donington-on-Bain to Tisbury) consistently suggest approximately 1 km of exhumation (Scotchman 1994). Hopane maturity values at ten locations from this region unaffected by structural inversion suggest a mean exhumation of 1.07 km ± 0.11 (± one standard deviation), while maturity values from steranes in the same ten locations suggest a mean of 1.28 km ± 0.11. Green et al. (1993) review further evidence for regional kilometre-scale exhumation in and around the United Kingdom.

Brodie & White (this volume) note that fault-controlled (i.e. syn-rift) Permo-Triassic sedimentary rocks crop out close to the surface on mainland Britain, in the Irish Sea and north of Scotland. The lithospheric stretching model of basin formation (McKenzie 1978; Jarvis & McKenzie 1980) predicts that these syn-rift rocks should be overlain by 2–4 km of post-rift or thermal subsidence phase rocks. Hence Brodie & White (this volume) argue that there must have been widespread erosion of the rocks deposited during the post-rift phase of basin development.

Scandinavia and its Continental Shelf

Regional exhumation was not restricted to the UKCS. Sonic velocities in lowermost Jurassic claystones witness regional exhumation, interpreted to be Neogene, in northern Denmark, southern Norway, southern Sweden and the adjacent offshore Skagerrak area (Japsen 1993). Japsen (1993) argued that 1 km of Palaeogene sedimentary rocks must be missing at Stevns Klint, the type locality for the Danian, and where the Cretaceous/Tertiary boundary crops out just above sea-level.

The most complete published analysis of exhumation magnitudes in north–west Europe has been undertaken in the Norwegian area, using principally vitrinite reflectance, shale compaction and drilling parameters (Jensen et al. 1992; Skagen 1992). These data suggest major Neogene uplift and exhumation onshore, and offshore from the Skagerrak area, north along the Norwegian coast of the North and Norwegian Seas and into the Barents Sea (e.g. Jensen et al. 1992; Nyland et al. 1992). In the Skagerrak area, 'Neogene uplift was due to broad-scale warping without major faulting' (Jensen & Schmidt 1992). Cenozoic uplift in the Stord Basin, offshore west of Stavanger, southern Norway, was also accomplished by regional warping with no evidence of major fault activity (Ghazi 1992).

Enigmatic aspects of the case for regional Tertiary exhumation

When was the regional exhumation?

A more complete understanding of the regional exhumation described above requires an understanding of its timing. Since sonic velocities in the Chalk of the UK SNS/EMS, the IMF and the CS/SWA all yield similar apparent exhumation values to those based on the velocities of older formations, regional exhumation is believed to be of Tertiary age. The Danian Chalk of the SWA yields similar apparent exhumation values to the Upper Cretaceous Chalk, hence exhumation from maximum burial-depth post-dated the Danian Chalk in the SWA (Menpes & Hillis this volume).

In order to try to further constrain the timing of regional exhumation, the author and colleagues have attempted to analyse apparent exhumation in younger, Eocene–Miocene sequences. However no conclusive results have been obtained on whether these younger sequences have been exhumed from maximum burial-depth because they are preserved only where apparent exhumation is low.

Comparing Figs 4a and 4b, it appears that inversion-related exhumation is (spatially) superimposed on regional exhumation. Apparent exhumation values from the different stratigraphic units within the recognized inversion axes are statistically similar. Hence inversion-related exhumation is also inferred to have occurred during Tertiary times. However, regional exhumation and basin inversion are spatially unrelated, have entirely different structural/stratigraphic expression (the former quite obvious, the latter much less clear), and may be temporally unrelated. Measurements of apparent exhumation cannot distinguish whether regional exhumation was contemporaneous with inversion-related exhumation, or whether they occurred at different times in the Tertiary, as might be inferred, for example, from Green's (1989 fig. 9) and Green et al.'s (1993 fig. 4) burial/exhumation histories which show discrete Palaeocene–Eocene and Oligocene–Miocene exhumation events in the EMS.

The age of inversion-related exhumation can be dated from structural relationships on seismic records. However, since it may be unrelated to the age of regional exhumation, its dating is not discussed here, and for such discussion the reader is referred to Ziegler (1990); for UK SNS, Walker & Cooper (1987), Van Hoorn (1987b) and Badley et al. (1989); for IMF, Thomson & Underhill (1993) and Thomson & Hillis (in press); and for CS/SWA, Masson & Parson (1983), Tucker & Arter (1987), Van Hoorn (1987a) and Hillis (1991).

Regional exhumation did not necessarily occur at the same time in all the basins studied. However, the similarity of the magnitude of regional exhumation calculated from the techniques discussed above, which suggest 1 km of regional exhumation in widely separated areas in and around the UK, may be suggestive of a consistent timing.

Japsen (1993) argued that only the pre-Quaternary, Late Oligocene–Miocene unconformity in Denmark and its surrounding Continental Shelf has an areal extent similar to the regional exhumation pattern inferred from sonic velocities, and thus that regional exhumation occurred during Neogene times. Jensen & Schmidt (1992) also suggested that major regional exhumation in the Skagerrak area occurred during Neogene times because of the accelerated Neogene (post-Mid Miocene) sedimentation rate in the Central North Sea (Bjørslev Nielsen et al. 1986).

There is a well-developed Miocene unconformity in the UK SNS (Badley et al. 1989; Walker & Cooper 1987). However, the Miocene age of regional exhumation favoured, on the above structural/stratigraphic grounds, for the Scandinavian Continental Shelf is inconsistent with the Palaeocene age inferred from seismic stratigraphic grounds for the IMF (Knox et al. 1981; Mudge & Bliss 1983; Milton et al. 1990), if regional exhumation is contemporaneous in the two areas.

As suggested by Green (1989 fig. 9) and Green et al. (1993 fig. 4), Palaeocene/Eocene and Oligocene/Miocene are the two most likely ages of regional exhumation. However, the age of regional exhumation remains unclear. In order for the age of the regional Tertiary exhumation in and around the UK to be determined with more confidence, further published petrophysical studies, such as the compaction-based work presented here and apatite fission track analyses, will need to be combined with detailed bio- and sequence stratigraphic analyses of the Upper Cretaceous–Tertiary sequence, particularly in areas where the sequence is relatively complete.

What was eroded?

It is an important, indeed inescapable, but not always recognized, implication of the exhumation interpreted from petrophysical data that there must have been burial prior to the exhumation. Exhumation is inferred because petrophysical properties witness burial at greater depth than present. The anomalously fast velocities, anomalously high vitrinite reflectance, and annealed fission tracks, record not only exhumation from maximum burial-depth, but prior burial to that depth (Fig. 5). The evidence for regional Tertiary exhumation thus raises the question of what was deposited then eroded.

Holliday (1993) argued, on the basis of the greatest preserved sedimentary thicknesses in adjacent basins, that 1200–1750 m of Mesozoic sedimentary rocks formerly existed over the structural highs of northern England, and that estimates of Tertiary exhumation of 3 km based on apatite fission track analysis were probably an overestimate. If as petrophysical properties suggests, Tertiary exhumation was regional, then the technique of reconstructing maximum probable sediment thickness from preserved thicknesses is flawed. By analogy with the Market Weighton Block, Holliday (1993) suggests that the probable thickness of the Chalk over the structural highs in northern England is 300–600 m, and he makes no allowance for possible Palaeogene rocks. However, the Market Weighton Block is not itself at maximum burial-depth, and was subject to significant Tertiary exhumation (Whittaker et al. 1985). The results from the UK SNS/EMS suggests that approximately 1.5 km of section has been removed from the Cleethorpes-1 well, just south of the Market Weighton Block (Fig. 4; Hillis 1993a).

Despite the above, in areas where the exhumation is associated with a major unconformity, it is generally uncontentious to assign deposition of the eroded sequence to the time gap encompassed by the major unconformity. For example, Cornford (1986) ascribes the maturity of the outcropping Lias of south–west England to a now eroded Upper Jurassic–Lower Cretaceous sequence. However, in other areas, regional exhumation is not associated with such major unconformities. If regional exhumation occurred during Miocene times in well 44/7-1 in the UK SNS (the time gap encompassed by the most pronounced Tertiary unconformity in the well), then approximately 1 km of sedimentary rocks must have been deposited during the Early Miocene to be eroded during the Late Miocene (Fig. 5). Assuming similar burial and exhumation rates, this implies approximately 100 mm/my for both processes. An Upper Cretaceous Chalk sequence exists on the East Midlands Shelf: e.g. 233 m of Cenomanian–Coniacian/?Santonian Chalk is unconformably overlain by the Quaternary at the Cleethorpes-1 well. If the exhumation is again assumed to have occurred during Late Miocene times, then deposition of the eroded sequence occurred between approximately 87 (83 if the Santonian is present) and 15 Ma (Fig. 5). If exhumation occurred during Palaeocene times, then deposition of the eroded sequence is restricted to the (?late) Santonian–Palaeocene, between approximately 87/?83 and 60 Ma.

It is perhaps intuitively unsatisfactory to infer a rapid burial rate during a period of time from which there is no section preserved. However, if the petrophysical evidence of burial-depth in excess of that at the present is accepted, then this rapid burial phase is unavoidable. Even assuming Palaeocene exhumation in the Cleethorpes-1 well, which implies a more rapid prior burial rate than does a Miocene exhumation, the deposition of the oldest preserved Chalk in the Cleethorpes-1 well (Turonian–Coniacian/?Santonian) is associated with a burial rate greater than that during the inferred, now eroded, pre-exhumational burial phase (Fig. 5). Hillis (1991) and Hillis et al. (1994) similarly discuss the implied prior burial in the SWA and IMF respectively.

There is widespread evidence of increased burial/subsidence rates in Late Cretaceous–Eocene times in the North Sea (e.g. White & Latin 1993; Turner & Scrutton 1993 and references therein). Joy (1992) describes the enigmatic rapid increase in Palaeocene subsidence of the North Sea Basin, and of other basins bordering the North Atlantic in detail, Joy (1992) concludes that the rapid increase in subsidence is in the 'right place, wrong time', because it overlies extensional basins like a post-rift subsidence phase, but it post-dated the cessation of extension by a significant period (up to 80 my). Unlike White & Latin (1993), Joy (1992) argued that the rapid increase in Palaeocene subsidence cannot be explained by the mantle plume that existed during Palaeocene opening of the North Atlantic north–west of the UK, and concludes that 'Practically the only conclusion about these Palaeocene phenomena that can be reached with confidence is that, if they are related, their cause is a truly regional one.' The rapid burial phase prior to exhumation inferred from petrophysical properties may be closely linked to Joy's (1992) 'right place, wrong time' increase in subsidence rates. However, the controls on the relative timing and distribution of the regional exhumation and rapid burial phases is not sufficient to distinguish:

- whether the rapid Palaeocene burial phase was truly regional (not limited to deep graben areas unaffected by regional exhumation) and pre-dated exhumation i.e. is equivalent to the pre-exhumation burial phase inferred from petrophysical properties;
- or whether the rapid Palaeocene burial phase was restricted to the deep graben areas unaffected by regional exhumation and was

synchronous with regional exhumation i.e. represents the deposits of the regional phase of exhumation and erosion.

Where did the eroded sequence go?

If the case for regional Tertiary erosion, and the case for the regional deposition of the sequences that were eroded, is accepted, one may then reasonably ask where the eroded sediment was deposited. In the North Sea, the deepest parts of the Central and Northern North Sea Graben may have remained areas of deposition during regional Tertiary exhumation (Knox *et al.* 1981; Mudge & Bliss 1983; Milton *et al.* 1990). In the Celtic Sea/South-Western Approaches, eroded sedimentary rocks may have been carried down and indeed beyond the adjacent continental slope. Sub-aerial exhumation at this time may also have involved solution of calcium carbonate, particularly from the Upper Cretaceous and Danian Chalk.

Implications of regional Tertiary exhumation for hydrocarbon exploration

Sediment decompaction

Sediment decompaction is a widely performed technique which aims to restore the original depositional thickness of now buried and compacted stratigraphic units and thus recreate burial-history through time (e.g. Sclater & Christie 1980; Falvey & Deighton 1982). In order to restore the original thickness of a stratigraphic unit, its normal porosity/depth relationship must be determined. Restored thicknesses are calculated from the porosity (hence associated volume) increase indicated by the porosity/depth relationship as the rock unit is raised from its present burial-depth to the surface.

In areas such as the UK Southern North Sea, where sedimentary rocks are not at their maximum burial-depth, this procedure is more complex than generally considered. Firstly, a normal porosity/depth relationship is not simply an average one for the area. It should be determined only from wells where the rock unit concerned is at its maximum burial-depth. Secondly, the deposition and removal of the eroded rock unit must be added to the observed stratigraphy. Rock units deposited prior to exhumation will decompact more (increase in thickness more) than if allowance is not made for the eroded unit (Fig. 5). In the case of well 44/7-1, present-day stratigraphy suggests the

base Cretaceous was at 1.0 km depth prior to Miocene exhumation. Decompacting base Cretaceous burial-history with reference only to the preserved stratigraphy (i.e. without allowing for exhumation) suggests that the base Cretaceous was at 1.2 km depth at that time. The decompacted burial history with allowance for exhumation suggests the base Cretaceous was at 1.4 km depth prior to Miocene exhumation. Furthermore, if the sequence is not at its maximum burial-depth, then the units deposited after the exhumation, the Pliocene–Quaternary of well 44/7-1, will not cause further compaction of older units. Following exhumation, units that pre-date exhumation (pre-Miocene of well 44/7-1) will act as compaction basement (i.e. all follow parallel decompacted burial-histories) until their burial-depth exceeds the previous maximum.

Source rock maturity

Source rock maturity is a function of temperature and time (e.g. Sweeney & Burnham 1990; Suzuki *et al.* 1993). The source rock thermal history required to model observed maturity parameters is generally determined from burial history and palaeogeothermal gradient(s) or palaeoheatflow/thermal conductivity (e.g. Falvey & Deighton 1982; Bray *et al.* 1992). A parameter frequently used to express the maturity of source rocks is the quantitatively determined extent to which the kerogen maceral, vitrinite, reflects visible light (vitrinite reflectance). Vitrinite reflectance levels have been modelled in the Cleethorpes-1 well, with and without allowance for Tertiary exhumation, assuming a constant palaeogeothermal gradient of 30°C/km (Fig. 6). Clearly the assumed geothermal history is simplistic, and other parameters such as apatite fission track analysis and vitrinite reflectance should be utilized to determine palaeogeothermal gradients (e.g. Bray *et al.* 1992). However, the purpose is simply to illustrate the difference between incorporating and not incorporating regional Tertiary exhumation in modelling source rock maturity in an area not affected by structural inversion. Modelling was undertaken with the Platte River Associates BasinMod™ software which uses the kinetics of Sweeney & Burnham (1990) to calculate theoretical vitrinite reflectance.

Without allowance for burial/exhumation, the top Carboniferous of Cleethorpes-1 never reaches a vitrinite reflectance level of 0.5% R_0, equivalent to early maturity for oil generation (Fig. 6). However, with allowance for Tertiary exhumation, the top Carboniferous reaches a

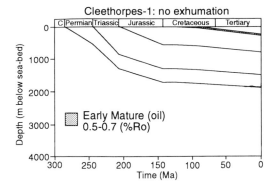

Cleethorpes-1: no exhumation

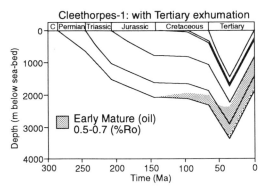

Cleethorpes-1: with Tertiary exhumation

Fig. 6. Burial and thermal geohistories for the Cleethorpes-1 well, with and without allowance for Tertiary exhumation. The burial histories were decompacted using the methodology of Sclater & Christie (1980) with allowance for the eroded unit as discussed in the text. Formation tops and ages were taken from the operators' composite logs and geochronologically calibrated after the time scale of Harland *et al.* (1989). A geothermal gradient of 30°C/km was assumed throughout basin history in all models. Burial and thermal geohistories were calculated with these parameters using the BasinMod™ software. The incorporation of Tertiary exhumation significantly increases the modelled level of thermal maturity of the same stratigraphic level which demonstrates that regional Tertiary exhumation, even in areas unaffected by structural inversion, has important consequences for hydrocarbon generation.

vitrinite reflectance level of 0.5% R_0 around the time of the Jurassic/Cretaceous boundary, and reaches almost 0.7% R_0 during Tertiary times. Deeper levels in the Carboniferous (the Cleethorpes-1 well terminated in the Carboniferous after penetrating approximately 200 m of Westphalian Coal Measures) would show higher levels of thermal maturity.

Kirby (1985) quoted observed reflectances in the Cleethorpes-1 well which include: 0.44%

$R_0 \pm 0.06$ at 700 m in the Lower Jurassic Coleby Mudstone (nine measurements, mean $\pm$ one standard deviation); 0.58% $R_0 \pm 0.10$ at 1866 m in the Permian Marl Slate Formation or Kupferschiefer (nine measurements); and 0.68% $R_0 \pm 0.10$ at 1990 m in the Westphalian Coal Measures (20 measurements). The maturation modelling which incorporates Tertiary exhumation is consistent with the observed reflectances in the Cleethorpes-1 well, while that which does not is inconsistent with the observed reflectances. More broadly, incorporating Tertiary exhumation in maturation modelling is consistent with the sourcing of the onshore oilfields of the East Midlands Shelf from the Carboniferous (Fraser & Gawthorpe 1990).

Combining any given palaeogeothermal gradient(s) with a burial-history plot for a potential hydrocarbon source rock that allows for exhumation will predict higher levels of organic maturity than the same palaeogeothermal gradient(s) combined with a burial-history plot that does not allow for exhumation. Thus estimates of exhumation, such as those presented, should be incorporated in maturation modelling of wells not at their maximum burial-depth. Due to the regional nature of Tertiary exhumation, not all wells above their maximum burial-depth are located on the widely recognized inverted highs, in which areas account has previously been taken of exhumation (e.g. Marie 1975; Glennie & Boegner 1981; Cope 1986). The effect of regional exhumation on the maturation of the Norwegian offshore basins has been discussed by Jensen & Schmidt (1992), Ghazi (1992) and Nyland *et al.* (1992).

Diagenesis

The Lower Permian Rotliegend sandstones are the principal reservoir of the UK SNS gas province. However, the Lower Triassic Bunter Sandstone is also a reservoir in the SNS (e.g. Esmond, Forbes and Gordon and Hewett fields). The statistically consistent apparent exhumation results derived from the Upper and Middle Chalk, Bunter Sandstone and Bunter Shale have important implications for the processes controlling the diagenesis of the Bunter Sandstone. While it is widely accepted that burial-depth is the primary control on shale compaction, burial-depth independent processes, such as early framework-generating cements, and secondary porosity generation are often considered important in sandstone diagenesis. If burial-depth is the primary control on Bunter Shale velocity, then the excellent correlation between apparent exhumation results derived from the Bunter Shale and Sandstone (Fig. 3) suggests that burial-depth is also the primary control on Bunter Sandstone velocity, and by inference porosity. It is suggested that, at the formation and regional scale, burial-depth is the primary control on Bunter Sandstone porosity.

Hillis (1993*b*) similarly demonstrated that apparent exhumation results from velocity in the Middle Jurassic Hutton Sandstone of the central Australian Eromanga Basin are very similar to those derived from the Lower Cretaceous Allaru/Oodnadatta mudstone in the same basin, and inferred that, at a formation and regional scale, burial-depth is the principal control on porosity in the Hutton Sandstone (an important hydrocarbon reservoir in the basin).

Tectonic models for basin inversion and regional Tertiary uplift

Since the scale of tectonic processes controlling uplift is at least that of crustal thickness, only uplifts that occur over a region of at least 10^3–10^4 km^2 are of potential tectonic significance (England & Molnar 1990). Clearly the scale over which exhumation occurred in the UK SNS/EMS, IMF and CS/SWA, and over a wider area as discussed above and in Green *et al.* (1993), is potentially significant *sensu* tectonic processes (England & Molnar 1990).

The tectonic component (U_T) of regional exhumation (E_T) can be calculated as outlined by Garfunkel (1988) and BNrown (1991b):

$$U_T = E_T(1 - \rho_s/\rho_m) - \Delta SL - \Delta H$$

where ρ_m and ρ_s are mantle (approximately 3.3 gcm^{-3}) and sedimentary rock (approximately 2.3 gcm^{-3}) densities, and ΔSL and ΔH are any change (pre-exhumational less post-exhumational) in sea-level and any change in surface elevation (with respect to contemporary sea-level) respectively.

Erosion induces isostatic crustal uplift to counterbalance the weight of the rocks removed. The first term of the above equation corrects the amount of section removed (derived from sonic velocities or other petrophysical data) for the amplification of the initial uplift caused by this isostatic rebound. Given the above densities, erosion of an initial surface uplift of 0.3 km would generate the observed regional exhumation of 1.0 km. The amount of tectonic uplift also needs to be corrected for any change in sedimentary base-level, here taken as sea-level, which might induce erosion and any change in surface elevation pre- and post-erosion. Assuming no change in elevation (with respect to sea-level before and after exhumation), a sea-level fall of 0.3 km is required alone to account for the observed regional exhumation (without an additional tectonic component). According to Hallam (1984) and the first order curve of Haq *et al.* (1987), The major eustatic sea-level falls of the Tertiary do not significantly exceed 100 m, and while these figures may be in error, Tertiary eustatic sea-level falls are unlikely to be as great as 0.3 km. Given that the timing of the regional exhumation is not constrained, and accepting maximum Tertiary sea-level changes of the order of 100 m, the tectonic component of regional Tertiary exhumation is 0.3 ± 0.1 km.

The key observations for which any tectonic model of Late Cretaceous–Tertiary evolution of the UK SNS/EMS, IMF and CS/SWA, and possibly a still wider area as discussed above and in Green *et al.* (1993), must account for are:

- tectonic uplift of the inverted areas which occurred in association with upper lithospheric compression/thickening (i.e. basin inversion);
- regional tectonic uplift (0.3 ± 0.1 km) without associated upper lithospheric compression/thickening;

Fig. 7. Decoupled, two-layer lithospheric compression and thickening. (**A**) Jurassic–Early Cretaceous depocentres such as the southern (French) sector of the Western Approaches Trough, the North and South Celtic Sea Basins, the Cleveland and Sole Pit axes formed above ramps in a major fault during extension (Beach, 1987; Kusznir *et al.* 1987). (**B**) Compression reactivates the major fault, inverting the Jurassic–Early Cretaceous depocentres. Mantle lithospheric thickening is decoupled and laterally displaced from that in the crust and causes initial subsidence: the subsidence phase prior to uplift. (**C**) The lithosphere thermally re-equilibrates to its pre-compressional level, uplifting the region of mantle lithospheric thickening: the regional uplift phase. (**D**) Balanced distribution of lithospheric thickening similar to that illustrated schematically in (**A**)–(**C**). The crustal thickening factor (f_c-solid line) is the ratio of thickness of the deformed crust to its initial thickness (Sandiford & Powell 1990). Similarly, f_{ml} is mantle lithospheric thickening (dashed) and f_l is the whole lithospheric thickening (dotted). (**E**) Surface elevation changes associated with the distribution of lithospheric thickening shown in (**D**), assuming local isostasy and no surface loading (Sandiford & Powell 1990). Syn-compressional elevation changes are shown by the solid lines, and post-compressional, thermal uplift is shown by the dashed line (from Hillis 1992).

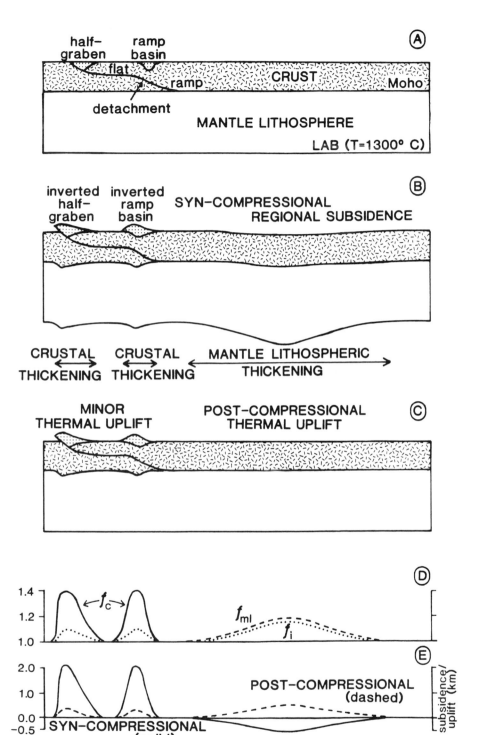

• burial prior to uplift, during which the rocks became overcompacted (Figs 5, 6).

In areas where there is evidence of crustal compression (basin inversion), hence crustal thickening, tectonic uplift is readily explained, indeed required, by the isostatic response to crustal thickening (Chadwick 1985; Murrell 1986). However, the origin of the tectonic force driving the regional component of 0.3 ± 0.1 km uplift where there is no evidence of crustal compression and thickening, is more enigmatic. The regional extent of the tectonic uplift points to a thick-skinned (lower lithospheric) source.

A decoupled, or two-layer model of lithospheric compression as proposed by Hillis (1992) can account for both the crustal thickening-driven uplift of the inversion axes, and the regional subsidence then uplift of areas where there is no evidence of crustal thickening (Fig. 7). In the proposed model, compression and thickening in the lower lithosphere (possibly equivalent to the mantle part of the lithosphere) is decoupled, and laterally displaced from that in the upper lithosphere (possibly equivalent to the crust). Such a two-layer model is directly analogous to the decoupled, two-layer models of lithospheric extension (e.g. Royden & Keen 1980; Hellinger & Sclater 1983; Kusznir et al. 1987; White & McKenzie 1988). Indeed, the model illustrated in Fig. 7 is based on Kusznir et al.'s (1987) model of lithospheric deformation. As proposed by Hillis (1992), reverse movement on the pre-existing, weak faults, extension on which had formed the basins, caused the recognized basin inversion (Beach 1987; Kusznir et al. 1987). Decoupled thickening of the mantle lithosphere (without thickening of the overlying crust) is invoked to account for the regional uplift (Fig. 7). Submersion of cold, dense mantle lithosphere into the surrounding asthenosphere would have caused an initial, isostatically-driven subsidence. Subsequent warming of the lithosphere would have caused uplift. Hence thickening of the mantle lithosphere without thickening of the overlying crust (which is concentrated in the inversion axes) can account for the initial subsidence then uplift of uninverted areas inferred from sonic velocity data (Fig. 7). The stresses responsible for two-layer lithospheric compression are considered to have resulted from the interaction of the stress fields set up by Alpine continent–continent collision, and the opening of the North Atlantic Ocean. The two-layer compressional model is discussed in more detail by Hillis (1992).

Heating associated with the mantle plume above which the North Atlantic opened northwest of the UK during early Tertiary times (White & McKenzie 1989) cannot account for the regional uplift. The United Kingdom is now removed from the present centre of the mantle plume beneath Iceland, and uplift due to early Tertiary heating would largely have thermally relaxed by Recent times. Hence no overcompaction would be observed. Furthermore, the UK SNS/EMS and CS/SWA were probably too far from the plume, even at the time of North Atlantic opening, to have been significantly uplifted (White & McKenzie 1989, fig. 8).

Although plume-related heating cannot account for the uplift, Brodie & White (this volume) have proposed that magmatic underplating, such as that associated with the plume, could have generated regional uplift. Magmatic underplating at the base of the crust by igneous material produced by adiabatic decompression of mantle material will generate uplift that, unlike dynamic thermal uplift, does not subsequently collapse (White & McKenzie 1989; Brodie & White this volume). Extensive magmatic underplating and basaltic volcanism are witnessed by refraction velocities and seaward dipping reflectors offshore northwest of Britain (White & McKenzie 1989 and references therein). The onshore British Tertiary Igneous Province of northwest Britain and Ireland also underwent extensive igneous activity (Mussett et al. 1988; Brodie & White this volume, fig. 3). Regional uplift and exhumation offshore northwest of Britain and within the British Tertiary Igneous Province is reasonably explained by magmatic underplating. However, there is no evidence of regional underplating, on a scale comparable to that of northwest Britain and Ireland and its continental margin, in the vicinity of the UK SNS/EMS or CS/SWA. Indeed, White & McKenzie (1989) quote the southwest UK margin as a classic non-volcanic case. Nor is there evidence of significant extrusive volcanism (with the exception of Lundy) in these areas. In the absence of evidence for magmatic underplating in these areas, regional uplift must have a different cause. Furthermore, underplating cannot account for the subsidence that precedes uplift in uninverted areas. While underplating is probably an important source of early Tertiary uplift over northwest Britain and Ireland and its continental margin, where it has demonstrably occurred, and may have played a role in the regional uplift of the Inner Moray Firth, it is not considered to be a significant source of uplift in the UK SNS/EMS or CS/SWA.

Successful quantitative tectonic modelling of Tertiary regional uplift in the UK SNS/EMS or CS/SWA will require a detailed knowledge of the magnitude and timing of tectonic uplift throughout northwest Europe. Unfortunately these data are not yet available. However, the decoupled lithospheric compressional model, and in areas of igneous activity, magmatic underplating, provide tectonic models which can be further tested against improved knowledge of the magnitude and timing of the regional and inversion-related components of uplift.

The author wishes to thank Rob Menpes and Ken Thomson who have worked extensively on determining exhumation in the CS/SWA and IMF respectively. Shell UK provided the majority of the funding for the SNS study, and are also thanked for providing the requisite digital well log data. Shell UK and Esso UK are both thanked for permission to publish the results from the SNS. The Royal Society and the Carnegie Trust for the Universities of Scotland provided additional financial support for the SNS study. The Regional Geophysics Group of the British Geological Survey (BGS), and those involved with BGS computing provided extensive support, and covered computing expenses for the SNS study. Andy Mitchell and Peter Tingate at the National Centre for Petroleum Geology and Geophysics, Adelaide University, provided access to, and help in using the Geolog and BasinMod™ software. Reviewers Hardman and Wayte are thanked for their constructive criticisms. The author also acknowledges the University of Adelaide for financial assistance in attending the Basin Inversion Conference in Oxford, UK, October 1993.

References

BADLEY, M. E., PRICE, J. D. & BACKSHALL, L. C. 1989. Inversion, reactivated faults and related structures: seismic examples from the southern North Sea. *In*: COOPER, M. A. & WILLIAMS, G. D. (eds) *Inversion Tectonics*. Geological Society, London, Special Publication, **44**, 201–219.

BALDWIN, B. & BUTLER, C. O. 1985. Compaction curves. *American Association of Petroleum Geologists Bulletin*, **69**, 622–626.

BEACH, A. 1987. A regional model for linked tectonics in north–west Europe. *In*: BROOKS, J. & GLENNIE, K. (eds) *Petroleum Geology of North West Europe*. Graham & Trotman, London, 43–48.

BENNET, G., COPESTAKE, P. & HOOKER, N. P. 1985. Stratigraphy of the Britoil 72/10-1A well, Western Approaches. *Proceedings of the Geologists' Association*, **96**, 255–261.

BIDDLE, K. T. & RUDOLPH, K. W. 1988. Early Tertiary structural inversion in the Stord Basin, Norwegian North Sea. *Journal of the Geological Society, London*, **145**, 603–611.

BJØRSLEV NIELSEN, O., SØRENSEEN, S., THIEDE, J. & SKARBØ, O. 1986. Cenozoic differential subsidence of North Sea. *American Association of Petroleum Geologists Bulletin*, **70**, 276–298.

BRAY, R. J., GREEN, P. F. & DUDDY, I. R. 1992. Thermal history reconstruction using apatite fission track analysis and vitrinite reflectance: a case study from the UK East Midlands and the Southern North Sea. *In*: HARDMAN, R. F. P. (ed.) *Exploration Britain: Geological Insights for the Next Decade*. Geological Society, London, Special Publication, **67**, 3–25.

BRODIE, J. & WHITE, N. 1995. The link between sedimentary basin inversion and igneous underplating. *This volume*.

BROWN, G., PLATT, N. H. & McGRANDLE, A. 1994. The geophysical expression of Tertiary dykes in the southern North sea. *First Break*, **12**, 137–146.

BROWN, R. W. 1991*a*. Discussion on thermal and tectonic history of the East Midlands shelf (onshore UK) and surrounding regions assessed by apatite fission track analysis. *Journal of the Geological Society*, London, **148**, 785–786.

—— 1991*b*. Backstacking apatite fission-track 'stratigraphy': a method for resolving the erosional and isostatic rebound components of tectonic uplift histories. *Geology*, **19**, 74–77.

BULAT, J. & STOKER, S. J. 1987. Uplift determination from interval velocity studies: *In*: BROOKS, J. & GLENNIE, K. (eds) *Petroleum Geology of North West Europe*. Graham & Trotman, London, 293–305.

BUTLER, M. & PULLAN, C. P. 1990. Tertiary structures and hydrocarbon entrapment in the Weald Basin of southern England. *In*: HARDMAN, R. F. P. & BROOKS, J. (eds) *Tectonic Events Responsible for Britain's Oil and Gas Reserves*. Geological Society, London, Special Publication, **55**, 371–391.

CHADWICK, R. A. 1985. Permian, Mesozoic and Cenozoic structural evolution of England and Wales in relation to the principles of extension and inversion tectonics. *In*: WHITTAKER, A. (ed.) *Atlas of Onshore Sedimentary Basins in England and Wales: Post-Carboniferous Tectonics and Stratigraphy*. Blackie, Glasgow, 9–25.

—— 1993. Aspects of basin inversion in southern Britain. *Journal of the Geological Society*, London, **150**, 311–322.

COOPER, M. A. & WILLIAMS, G. D. (eds) 1989. *Inversion Tectonics*. Geological Society, London, Special Publication, **44**.

COPE, M. J. 1986. An interpretation of vitrinite reflectance data from the Southern North Sea Basin. *In*: BROOKS, J., GOFF, J. C. & VAN HOORN, B. (eds) *Habitat of Palaeozoic Gas in N.W. Europe*. Geological Society, London, Special Publication, **23**, 85–98.

CORNFORD, C. 1986. The Bristol Channel Graben: organic geochemical limits on subsidence and speculation on the origin of inversion. *Proceedings of the Ussher Society*, **6**, 360–367.

ENGLAND, P. & MOLNAR, P. 1990. Surface uplift, uplift of rocks, and exhumation of rocks. *Geology*, **18**, 1173–1177.

FALVEY, D. A. & DEIGHTON, I. 1982. Recent advances in burial and thermal geohistory analysis. *Australian Petroleum Exploration Association Journal*, **22**, 65–81.

FRASER, A. J. & GAWTHORPE, R. L. 1990. Tectonostratigraphic development and hydrocarbon habitat of the Carboniferous in northern England. *In*:

HARDMAN, R. F. P. & BROOKS, J. (eds) *Tectonic Events Responsible for Britain's Oil and Gas Reserves*. Geological Society, London, Special Publication, **55**, 49–86.

GARFUNKEL, Z. 1988. Relation between continental rifting and uplifting: evidence from the Suez rift and northern Red Sea. *Tectonophysics*, **150**, 33–49.

GHAZI, S. A. 1992. Cenozoic uplift in the Stord Basin area and its consequences for exploration. *Norsk Geologisk Tidsskrift*, **72**, 285–90.

GLENNIE, K. W. & BOEGNER, P. L. E. 1981. Sole Pit Inversion Tectonics. *In*: ILLING, L. V. & HOBSON, G. D. (eds) *Petroleum Geology of the Continental Shelf of North–West Europe*. Heyden, London, 110–120.

GREEN, P. F. 1986. On the thermo-tectonic evolution of Northern England: evidence from fission track analysis. *Geological Magazine*, **123**, 493–506.

—— 1989. Thermal and tectonic history of the East Midlands Shelf (onshore UK) and surrounding regions assessed by apatite fission track analysis. *Journal of the Geological Society, London*, **146**, 755–773.

—— 1991. Reply to discussion on thermal and tectonic history of the East Midlands shelf (onshore UK) and surrounding regions assessed by apatite fission track analysis. *Journal of the Geological Society, London*, **148**, 786–787.

——, DUDDY, I. R., BRAY, R. J. & LEWIS, C. L. E. 1993. Elevated palaeotemperatures prior to Early Tertiary cooling throughout the UK region: implications for hydrocarbon generation. *In*: PARKER, J. R. (ed.) *Petroleum Geology of North West Europe, Proceedings of the 4th Conference*. Geological Society, London, 1067–1074.

HALLAM, A. 1984. Pre-Quaternary sea-level changes. Annual Review *Earth and Planetary Science*, **12**, 205–243.

HAQ, B. U., HARDENBOL, J. & VAIL, P. R. 1987. Chronology of fluctuating sea levels since the Triassic. *Science*, **235**, 1156–1167.

HARLAND, W. B., ARMSTRONG, R. L., COX, A. V., CRAIG, L. E., SMITH, A. G. & SMITH, D. G. 1989. *A Geologic Time Scale*. Cambridge University Press, Cambridge.

HAYWARD, A. B. & GRAHAM, R. H. 1989. Some geometrical characteristics of inversion. *In*: COOPER, M. A. & WILLIAMS, G. D. (eds) *Inversion Tectonics*. Geological Society, London, Special Publication, **44**, 17–39.

HELLINGER, S. J. & SCLATER, J. G. 1983. Some comments on two-layer extensional models for the evolution of sedimentary basins. *Journal of Geophysical Research*, **B88**, 8251–8269.

HILLIS, R. R. 1991. Chalk porosity and Tertiary uplift, Western Approaches Trough, SW UK and NW French continental shelves. *Journal of the Geological Society*, London, **148**, 669–679.

—— 1992. A two-layer lithospheric compressional model for the Tertiary uplift of the southern United Kingdom. *Geophysical Research Letters*, **19**, 573–576.

—— 1993a. Tertiary erosion magnitudes in the East

Midlands Shelf, onshore UK. *Journal of the Geological Society*, London, **150**, 1047–1050.

—— 1993b. Quantifying erosion in sedimentary basins from sonic velocities in shales and sandstones. *Exploration Geophysics*, **24**, 561–566.

—— 1994. Reply to discussion on the amount of Tertiary erosion in the UK estimated using sonic velocity analysis. *Journal of the Geological Society*, London, **151**, 1042–1044.

—— in press. Quantification of Tertiary exhumation in the UK Southern North Sea using sonic velocity data. *American Association of Petroleum Geologists Bulletin*.

——, THOMSON, K. & UNDERHILL, J. R. 1994. Quantification of Tertiary erosion in the Inner Moray Firth using sonic velocity data from the Chalk and the Kimmeridge Clay. *Marine and Petroleum Geology*, **11**, 283–293.

HOLLIDAY, D. W. 1993. Mesozoic cover over northern England: interpretation of apatite fission track data. *Journal of the Geological Society*, London, **150**, 657–660.

—— 1994. Reply to discussion on Mesozoic cover over northern England: interpretation of apatite fission track data. *Journal of the Geological Society*, London, **151**, 736.

HURST, A. 1980. *The Diagenesis of Jurassic Rocks of the Moray Firth, NE Scotland*. PhD thesis, University of Reading, UK.

ISSLER, D. R. 1992. A new approach to shale compaction and stratigraphic restoration, Beaufort-Mackenzie, Basin and Mackenzie Corridor, northern Canada. *American Association of Petroleum Geologists Bulletin*, **76**, 1170–1189.

JANKOWSKY, W. 1962. Diagenesis and oil accumulation as aids in the analysis of the structural history of the north-western German Basin. *Zeitschrift der Deutscher Geologischer Gesellschaft*, **114**, 452–460.

JAPSEN, P. 1993. Influence of lithology and Neogene uplift on seismic velocities in Denmark: implications for depth conversion of maps. *American Association of Petroleum Geologists Bulletin*, **77**, 194–211.

JARVIS, G. T. & MCKENZIE, D. P. 1980. Sedimentary basin formation with finite extension rates. *Earth and Planetary Science Letters*, **48**, 42–52.

JENSEN, L. N., RIIS, F. & BOYD, R. (eds) 1992. Post-Cretaceous uplift and sedimentation along the western Fennoscandian Shield. *Norsk Geologisk Tidsskrift*, **72**.

—— & SCHMIDT, B. J. 1992. Late Tertiary uplift and erosion in the Skagerrak area: magnitude and consequences. *Norsk Geologisk Tidsskrift*, **72**, 275–279.

JOY, A. M. 1992. Right place, wrong time: anomalous post-rift subsidence in sedimentary basins around the North Atlantic Ocean. *In*: STOREY, B. C., ALABASTER, T. & PANKHURST, R. J. (eds) *Magmatism and the Causes of Continental Break-up*. Geological Society, London, Special Publication, **68**, 387–393.

KEELEY, M. L., LEWIS, C. L. E., SEVASTOPULO, G. D., CLAYTON, G. & BLACKMORE, R. 1993. Apatite fission track data from southeast Ireland: implications for post-Variscan burial history. *Geological Magazine*, **130**, 171–176.

KENT, P. E. 1980. Subsidence and uplift in East Yorkshire and Lincolnshire: a double inversion. *Proceedings of the Yorkshire Geological Society*, **42**, 505–524.

KIRBY, G. A. 1985. *Cleethorpes No. 1 Geothermal Well: Geological Well Completion Report*. British Geological Survey, Keyworth.

KNOX, R. W. O. B., MORTON, A. C. & HARLAND, R. 1981. Stratigraphic relationships of Palaeocene sands in the U.K. sector of the central North Sea. *In*: ILLING, L. V. & HOBSON, G D. (eds) *Petroleum Geology in the Continental Shelf of North-West Europe*. Heyden, London, 267–281.

KUSZNIR, N. J., KARNER, G. D. & EGAN, S. 1987. Geometric, thermal and isostatic consequences of detachments in continental lithosphere extension and basin formation. *In*: BEAUMONT, C. & TANKARD, A. J. (eds) *Sedimentary Basins and Basin-Forming Mechanisms*. Canadian Society of Petroleum Geologists Memoir, **12**, 185–203.

LANG, JR., W. H. 1978. The determination of prior depth of burial (uplift and erosion) using interval transit time. *Society of Professional Well Log Analysts Nineteenth Annual Logging Symposium*, June 13–16, 1978, Paper B.

LEWIS, C. L. E., GREEN, P. F., CARTER, A. & HURFORD, A. J. 1992. Elevated K/T palaeotemperatures throughout northwest England: three kilometres of Tertiary erosion? *Earth and Planetary Science Letters*, **12**, 131–145.

McCLAY, K. R. 1989. Analogue models of inversion structures. *In*: COOPER, M. A. & WILLIAMS, G. D. (eds) *Inversion Tectonics*. Geological Society London, Special Publication, **44**, 41–59.

McCULLOCH, A. A. 1994. Discussion on Mesozoic cover over northern England: interpretation of apatite fission track data. *Journal of the Geological Society*, London, **151**, 735–736.

McKENZIE, D. 1978. Some remarks on the development of sedimentary basins. *Earth and Planetary Science Letters*, **40**, 25–32.

MAGARA, K. 1976. Thickness of removed sedimentary rocks, paleopore pressure, and paleotemperature, southwestern part of Western Canada Basin. *American Association of Petroleum Geologists Bulletin*, **60**, 554–565.

MARIE, J. P. P. 1975. Rotliegendes stratigraphy and diagenesis. *In*: WOODLAND, A. W. (ed.) *Petroleum and the Continental Shelf of North-West Europe, Volume 1, Geology*. Applied Science Publishers, London, 205–211.

MASSON, D. G. & PARSON, L. M. 1983. Eocene deformation on the continental margin SW of the British Isles. *Journal of the Geological Society*, London, **140**, 913–920.

MENPES, R. J. & HILLIS, R. R. 1995. Quantification of Tertiary exhumation from sonic velocity data, Celtic Sea/South-Western Approaches. *This volume*.

MILTON, N. J., BERTRAM, G. T. & VANN, I. R. 1990. Early Palaeogene tectonics and sedimentation in the Central North Sea. *In*: HARDMAN, R. F. P. & BROOKS, J. (eds) *Tectonic Events Responsible for Britain's Oil and Gas Reserves*. Geological Society, London, Special Publication, **55**, 339–351.

MUDGE, D. C. & BLISS, G. M. 1983. Stratigraphy and sedimentation of the Palaeocene sands in the Northern North Sea. *In*: BROOKS, J. (ed.) *Petroleum Geochemistry and Exploration of Europe*. Geological Society, London, Special Publication, **12**, 95–111.

MURRELL, S. A. F. 1986. Mechanics of tectogenesis in plate collision zones. *In*: COWARD, M. P. & RIES, A. C. (eds) *Collision Tectonics*. Geological Society, London, Special Publication, **19**, 95–111.

MUSSETT, A. E., DAGLEY, P. & SKELHORN, R. R. 1988. Time and duration of activity in the British Tertiary Igneous Province. *In*: MORTON, A. C. & PARSONS, L. M. (eds) *Early Tertiary Volcanism and the Opening of the NE Atlantic*. Geological Society, London, Special Publication, **39**, 337–348.

NYLAND, B., JENSEN, L. N., SKAGEN, J., SKARPNES, O. & VORREN, T. 1992. Tertiary uplift and erosion in the Barents Sea: magnitude, timing and consequences. *In*: LARSEN, R. M., BREKKE, H., LARSEN, B. T. & TALLERAAS, E. (eds) *Structural and Tectonic Modelling and its Application to Petroleum Geology*. Norwegian Petroleum Society Special Publication, **1**, 153–162.

PRYOR, W. A. 1973. Permeability–porosity patterns and variations in some Holocene sand bodies. *American Association of Petroleum Geologists Bulletin*, **57**, 162–189.

RAIGA-CLEMENCEAU, J., MARTIN, J. P. & NICOLETIS, S. 1988. The concept of acoustic formation factor for more accurate porosity determination from sonic transit time data. *The Log Analyst*, January–February, 54–59.

ROBERTS, A. M., BADLEY, M. E., PRICE, J. D. & HUCK, I. W. 1990. The structural evolution of a transtensional basin: Inner Moray Firth, NE Scotland. *Journal of the Geological Society*, London, **147**, 87–103.

ROBERTS, D. G. 1989. Basin inversion in and around the British Isles. *In*: COOPER, M. A. & WILLIAMS, G. D. (eds) *Inversion Tectonics*. Geological Society, London, Special Publication, **44**, 131–150.

ROYDEN, L. & KEEN, C. E. 1980. Rifting processes and thermal evolution of the continental margin of eastern Canada determined from subsidence curves. *Earth and Planetary Science Letters*, **51**, 343–361.

SANDIFORD, M. & POWELL, R. 1990. Some isostatic and thermal consequences of the vertical strain geometry in convergent orogens. *Earth and Planetary Science Letters*, **98**, 154–165.

SCHOLLE, P. A. 1977. Chalk diagenesis and its relation to petroleum exploration: oil from chalks, a modern miracle? *American Association of Petroleum Geologists Bulletin*, **61**, 982–1009.

SCLATER, J. G. & CHRISTIE, P. A. F. 1980. Continental stretching: an explanation of the post-mid-Cretaceous subsidence of the Central North Sea

Basin. *Journal of Geophysical Research*, **85**, 3711–3739.

SCOTCHMAN, I. C. 1994. Maturity and burial history of the Kimmeridge Clay Formation, onshore UK: a biomarker study. *First Break*, **12**, 193–202.

SKAGEN, J. I. 1992. Methodology applied to uplift and erosion. *Norsk Geologisk Tidsskrift*, **72**, 307–311.

SMITH, K., GATLIFF, R. W. & SMITH, N. J. P. 1994. Discussion on the amount of Tertiary erosion magnitudes in the UK estimated using sonic velocity data. *Journal of the Geological Society*, London, **151**, 1042–1044.

SUZUKI, N., MATSUBAYASHI, H. & WAPLES, D. W. 1993. A simpler kinetic model of vitrinite reflectance. *American Association of Petroleum Geologists Bulletin*, **77**, 1502–1508.

SWEENEY, J. J. & BURNHAM, A. K. 1990. Evaluation of a simple model of vitrinite reflectance based on chemical kinetics. *American Association of Petroleum Geologists Bulletin*, **74**, 1559–1570.

TAYLOR, J. C. M. 1990. Upper Permian–Zechstein. *In*: GLENNIE, K. W. (ed.) *Introduction to the Petroleum Geology of the North Sea*. Blackwell, Oxford, 153–90.

THOMSON, K. & HILLIS, R. R. in press. Tertiary structuration and erosion of the Inner Moray Firth. *In*: SCRUTTON, R. A., SHIMMIELD, G. B., STOKER, M. S. & TUDHOPE, A. W. (eds) *The Tectonics, Sedimentation and Palaeoceanography of the North Atlantic Region*. Geological Society, London, Special Publication, **90**.

—— & UNDERHILL, J. R. 1993. Controls on the development and evolution of structural styles in the Inner Moray Firth Basin. *In*: PARKER, J. R. (ed.) *Petroleum Geology of North West Europe, Proceedings of the 4th Conference*. Geological Society, London, 1167–1178.

TILL, R. 1974. *Statistical Methods for the Earth Scientist*. Macmillan, London.

TUCKER, R. M. & ARTER, G. 1987. The tectonic evolution of the North Celtic Sea and Cardigan Bay basins with special reference to basin inversion. *Tectonophysics*, **137**, 291–307.

TURNER, J. D. & SCRUTTON, R. A. 1993. Subsidence patterns in western margin basins: evidence from the Faeroe–Shetland Basin. *In*: PARKER, J. R. (ed.) *Petroleum Geology of North West Europe, Proceedings of the 4th Conference*. Geological Society, London, 975–983.

VAN HOORN, B. 1987a. The South Celtic Sea/Bristol Channel Basin: origin, deformation and inversion history. *Tectonophysics*, **137**, 309–334.

—— 1987b. Structural evolution, timing and tectonic style of the Sole Pit inversion. *Tectonophysics*, **137**, 239–284.

WALKER, I. M. & COOPER, W. G. 1987. The structural and stratigraphic evolution of the northeast margin of the Sole Pit Basin. *In*: BROOKS, J. & GLENNIE, K. (eds) *Petroleum Geology of North West Europe*. Graham & Trotman, London, 263–275.

WELLS, P. E. 1990. Porosities and seismic velocities of mudstones from Wairarapa and oil wells of North Island, New Zealand, and their use in determining burial history. *New Zealand Journal of Geology and Geophysics*, **33**, 29–39.

WHITE, N. & LATIN, D. 1993. Subsidence analyses from the North Sea 'triple junction'. *Journal of the Geological Society*, London, **150**, 473–488.

—— & McKENZIE, D. 1988. Formation of the 'steer's head' geometry of sedimentary basins by differential stretching of crust and mantle. *Geology*, **16**, 250–253.

WHITE, R. & McKENZIE, D. 1989. Magmatism at rift zones: the generation of volcanic continental margins and flood basalts. *Journal of Geophysical Research*, **B94**, 7685–7729.

WHITTAKER, A., HOLLIDAY, D. W. & PENN, I. E. 1985. *Geophysical logs in British stratigraphy*. Geological Society, London, Special Report, **18**.

WILLIAMS, G. D., POWELL, C. M. & COOPER, M. A. 1989. Geometry and kinematics of inversion tectonics. *In*: COOPER, M. A. & WILLIAMS, G. D. (eds) *Inversion Tectonics*. Geological Society, London, Special Publication, **44**, 3–15.

WYLLIE, M. R. J., GREGORY, A. R. & GARDNER, L. W. 1956. Theory of propagation of elastic waves in a fluid saturated porous solid. *Journal of the Acoustical Society of America*, **28**, 168–191.

ZIEGLER, P. A. (ed.) 1987. Compressional Intra-Plate Deformations in the Alpine Foreland. *Tectonophyics*, **137** (Special Issue).

—— 1990. *Geological Atlas of Western and Central Europe* (Second Edition). Shell Internationale Petroleum Maatschappij BV.

Quantification of Tertiary exhumation from sonic velocity data, Celtic Sea/South-Western Approaches

ROBERT J. MENPES & RICHARD R. HILLIS

Department of Geology and Geophysics, The University of Adelaide, SA 5005, Australia

Abstract: Sonic velocities from the Danian Chalk, the Upper Cretaceous Chalk, the Lower Cretaceous Greensand/Gault Clay, and the Triassic Mercia Mudstone were used to quantify apparent exhumation (i.e. amount of missing section) in the Celtic Sea/South-Western Approaches (southern Irish, south-western United Kingdom, and north-western French continental shelves). Analysis of all four of these stratigraphic units yield similar results. Since it is unlikely that sedimentological and/or diagenetic processes are responsible for similar amounts of overcompaction in the chalks, shaly sandstones and shales analysed, burial at depth greater than that currently observed is the most likely cause of overcompaction (anomalously high velocities). The use of lithologies other than shales in maximum burial-depth/exhumation studies is thus validated. The consistency of results from units of Triassic to Danian age suggests that Tertiary exhumation was of sufficiently great magnitude to mask any earlier Mesozoic periods of exhumation, and that the maximum Mesozoic–Cenozoic burial-depth in the Celtic Sea/South-Western Approaches was attained prior to Tertiary exhumation.

The recognized inversion axes of the Brittany Basin, South-West Channel Basin and North Celtic Sea Basin are highs in apparent exhumation, however highs in apparent exhumation of approximately 1 km also occur in the uninverted St. Mary's and Plymouth Bay Basins, and on the margins of the Armorican and Cornubian Platforms and the Pembroke Ridge. Tertiary exhumation was regional and not restricted to inversion axes, and thus had a thick-skinned origin. Heterogeneous lithospheric compression is the geodynamic model preferred to account for the regional exhumation of the Celtic Sea/South-Western Approaches. The identification of regional exhumation of approximately 1 km has important implications for modelling thermal maturation in the Celtic Sea/South-Western Approaches, with source rocks previously believed to be immature shown to be potentially mature for oil generation.

When corrected for exhumation, the post-rift sequence in the Celtic Sea/South-Western Approaches area shows relatively little localization over the syn-rift Jurassic–Lower Cretaceous sequence, suggesting that during basin formation, the lower lithosphere, thermal re-equilibration of which controls post-rift subsidence, did not experience localized thinning on the scale of the syn-rift basins.

Most discussions of the Late Cretaceous–Tertiary development of the basins of the southern United Kingdom and surrounding areas have focused on the widely recognized inversion structures/axes such as the classic Purbeck monocline and Wealden anticline. In and around the Celtic Sea/South-Western Approaches area (Fig. 1a), inversion structures have been described in the Wessex Basin (Lake & Karner 1987), the Brittany and South-West Channel Basins (Ziegler 1987; Hayward & Graham 1989), the South Celtic Sea–Bristol Channel Basin (Van Hoorn 1987; Roberts 1989), the North Celtic Sea–Cardigan Bay Basin (Tucker & Arter 1987), the Fastnet Basin (Robinson *et al.* 1981), and the adjacent continental slope (Masson & Parson 1983). However, there is clear evidence of regional Tertiary exhumation in areas removed from the recognized inversion axes. The widespread occurrence of flints across the Cornubian Platform of southwest England suggests regional erosion of the Upper Cretaceous Chalk (Roberts 1989). Cornford (1986) postulated an eroded thickness of approximately 1 km of Upper Jurassic–Lower Cretaceous sediment on the Dorset and Somerset coasts on the basis of vitrinite reflectance levels in the now exposed Lias.

The aims of this study are to:

- determine the magnitude of Tertiary exhumation in the Celtic Sea/South-Western Approaches area, using sonic velocity data from 43 released (i.e public-domain) wells (Fig. 1b);
- assess whether lithologies other than shales

From BUCHANAN, J. G. & BUCHANAN, P. G. (eds), 1995, *Basin Inversion*, Geological Society Special Publication No. 88, 191–207.

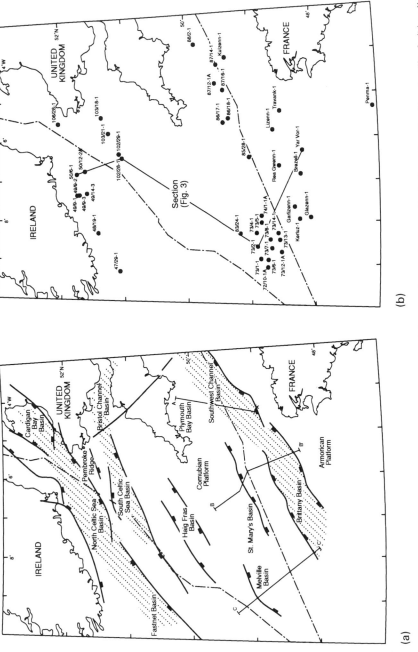

(b)

(a)

Fig. 1. Location maps for the Celtic Sea/South-Western Approaches: (**a**) structural elements, inverted regions (dotted), and section lines for Fig. 2; (**b**) wells used in the study and section line for Fig. 3.

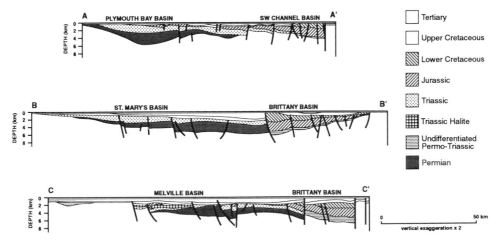

Fig. 2. Cross-sections interpreted from seismic lines in the South-Western Approaches, illustrating the geometry of the inversion axes of the Brittany Basin and SW Channel Basin.

may be used to estimate exhumation magnitude (e.g. Bulat & Stoker 1987; Hillis 1991; Hillis *et al.* (1994));

• discuss the implications of the apparent exhumation results with respect to thermal maturation of potential source rocks and tectonic models of basin formation and inversion.

The term exhumation (as opposed to erosion or uplift) is used here in the sense of England & Molnar (1990), to describe displacement of rocks with respect to the surface.

Regional stratigraphy

Development of the sedimentary basins in the Celtic Sea/South-Western Approaches area began following the Variscan orogeny with deposition of a thick Permo–Triassic redbed sequence on the eroded Variscan land surface (Kamerling 1979; Ziegler 1987; Brooks *et al.* 1988; Petrie *et al.* 1989; Hillis & Chapman 1992) (Figs 2 & 3). A Jurassic–Lower Cretaceous sequence of up to 4 km thickness was then deposited in the Brittany and South-West Channel Basins, whilst in the Melville, St. Mary's and Plymouth Bay Basins this sequence is relatively thin or absent, and a major, angular Mesozoic ('Cimmerian') unconformity is developed (Figs 2 & 3). Approximately 500 m of Upper Cretaceous Chalk was deposited throughout the South-Western Approaches, then the Jurassic–Lower Cretaceous depocentres of the Brittany and South-West Channel Basins were inverted.

The regional stratigraphy in the Celtic Sea across the axis of the Cornubian Platform is similar to that of the South-Western Approaches, with only relatively minor Late Cretaceous–Tertiary inversion of the South Celtic Sea Basin where the Jurassic–Lower Cretaceous is relatively thin, and stronger inversion of a thicker Jurassic–Lower Cretaceous sequence in the North Celtic Sea Basin (Fig. 3 and Van Hoorn's 1987, fig. 2). A similar regional pattern is also observed in the Wessex Basin, with the Jurassic–Lower Cretaceous sequence thick in the inverted areas, and relatively thin in the uninverted areas (Simpson *et al.*'s 1989, figs 4–7). Regionally, maximum Late Cretaceous–Tertiary inversion occurred in the regions of maximum Jurassic to Early Cretaceous subsidence (Fig. 2).

Methodology

Figure 4 illustrates the evolution of interval transit time with burial-depth that is assumed in order to calculate maximum burial-depth/exhumation magnitude from interval transit time (the reciprocal of velocity). Interval transit time decreases during burial according to a normal compaction relationship which must be determined for each unit under analysis. Since depth-controlled compaction of sediments is largely irreversible, formations which are shallower than their maximum burial-depth have an anomalously low interval transit time with respect to their present burial-depth (i.e. they are overcompacted (e.g. Magara 1976; Lang 1978; Bulat & Stoker 1987). For a given unit the

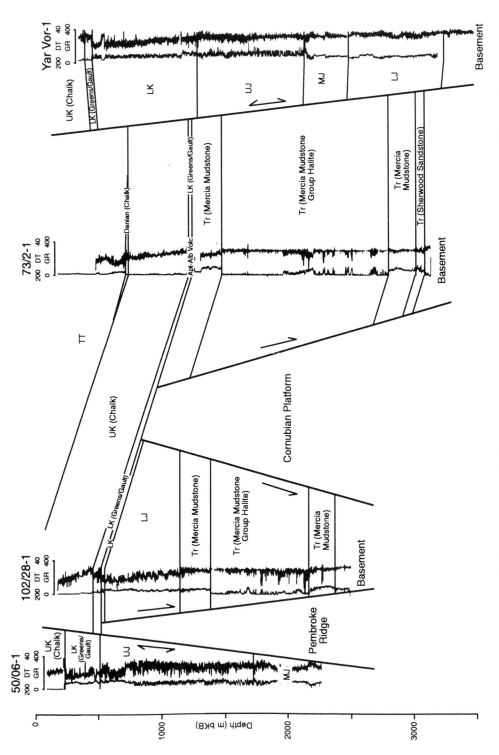

Fig. 3. Representative sonic transit time (DT) and gamma ray (GR) log correlation for the Celtic Sea/South-Western Approaches. GR (API units) is plotted on the left, and DT (μs/ft) on the right. Reference datum is Kelly Bushing (0 metres). See Fig. 1 for section line. (TT, Tertiary; UK, Upper Cretaceous; LK, Lower Cretaceous; Greens/Gault; Greensand/Gault Clay; Apt–Alb Volc, Aptian–Albian Volcanics; UJ, Upper Jurassic; MJ, Middle Jurassic; LJ, Lower Jurassic; Tr, Triassic)

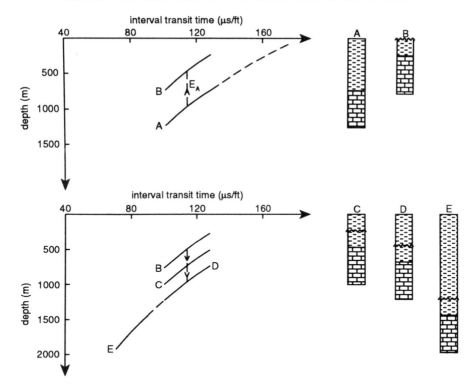

Fig. 4. Interval transit time evolution during burial (A), subsequent uplift and exhumation (B), and post-exhumation burial (C, D and E). The apparent exhumation (E_A) is the amount of exhumation not reversed by subsequent burial (i.e. height above maximum burial-depth).

amount of apparent exhumation (E_A), or exhumation above maximum burial-depth, is given by the vertical displacement of its interval transit time from the normal interval transit time/depth relationship (Fig. 4).

There are simple relationships between apparent exhumation (E_A), present burial-depth (B_P), maximum burial-depth (B_T), post-exhumation burial (B_E) and exhumation at the time of denudation (E_T):

$$B_T = E_A + B_P$$

i.e. maximum burial-depth = apparent exhumation + present burial-depth, and

$$E_T = E_A + B_E$$

i.e exhumation at the time of denudation = apparent exhumation + post-exhumation burial. These relationships are illustrated in Fig. 5.

Stratigraphic units analysed

Regionally extensive stratigraphic units are the most useful for this type of study because the more widely occurring the unit, the greater the

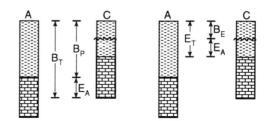

Fig. 5. Illustrated relationship between apparent exhumation (E_A), present burial-depth (B_P), maximum burial-depth (B_T), post-exhumation burial (B_E) and exhumation at the time of denudation, for the chalk mid-point. Well A is prior to exhumation, well C post-dates exhumation and post-exhumation burial.

likelihood of the reference wells being at or near their maximum burial-depth (e.g. Jurassic–Lower Cretaceous units that are restricted to the inversion axes are never found at their maximum burial-depth, and hence reference wells defined by the above methodology would also be shallower than their maximum burial-depth). In addition, units with little to no lateral facies

Table 1. *Formation mid-point depth (metres below sea-bed), mean interval transit time, and apparent exhumation data for the Celtic Sea/South-Western Approaches*

Well Number	Danian Chalk			Upper Cretaceous Chalk			Lower Cretaceous Greensand/Gault Clay			Triassic Mercia Mudstone			mean app. exhumation (km)
	mid-point depth (m)	mean DT (μs/ft)	app. exhumation (km)	mid-point depth (m)	mean DT (μs/ft)	app. exhumation (km)	mid-point depth (m)	mean DT (μs/ft)	app. exhumation (km)	mid-point depth (m)	mean DT (μs/ft)	app. exhumation (km)	
47/29-1				501.97	85.14	0.76	683.97	95.71	1.06				0.91
48/19-1				384.81	84.23	0.89	768.25	93.69	1.04				0.96
49/9-1				422.90	101.69	0.59	622.78	107.54	0.77				0.68
49/9-2				285.45	106.67	0.65	594.21	112.88	0.64				0.64
49/9-3				319.73	106.04	0.62	589.63	99.76	1.04				0.83
49/14-3				379.17	99.61	0.66	763.84	95.54	0.99				0.82
50/6-1				107.45	109.03	0.79	337.57	123.07	0.59				0.69
50/12-2A				131.65	84.15	1.14	374.60	101.50	1.20				1.17
72/10-1A				1030.83	87.17	0.20				1562.42	82.92	0.53	0.36
73/01-1	891.40	102.46	0.00	1136.61	93.16	0.00				2084.38	82.99	0.00	0.00
73/02-1	550.47	107.82	0.28	799.19	92.68	0.34				1177.44	90.38	0.56	0.39
73/04-1				557.86	102.43	0.44	842.11	113.10	0.38	968.11	106.63	0.00	0.27
73/05-1				949.76	87.86	0.27	1221.33	108.87	0.13	1472.18	86.83	0.43	0.28
73/06-1				843.53	97.04	0.23	1028.54	97.10	0.68	1222.55	92.90	0.39	0.39
73/07-1	650.75	100.03	0.27	1078.23	87.36	0.15	1372.82	92.04	0.49	1803.05	79.62	0.44	0.36
73/08-1				878.44	92.92	0.26				1390.09	87.08	0.50	0.38
73/12-1A				1249.68	85.68	0.00				1559.21	85.75	0.39	0.20
73/13-1	1054.31	88.41	0.00	1282.91	83.44	0.00				2207.06	72.55	0.37	0.12
73/14-1	928.43	91.19	0.09	1171.51	84.44	0.10				1693.78	80.09	0.53	0.24
74/01-1A				773.58	92.64	0.37	978.56	121.26	0.00	1145.33	92.02	0.51	0.29
83/24-1				473.50	104.46	0.49	667.50	113.25	0.55				0.52
85/28-1				655.93	96.66	0.43	951.59	119.53	0.08	1037.54	102.68	0.12	0.21
86/17-1				367.90	95.78	0.73	547.73	114.00	0.65	747.53	99.25	0.57	0.65
86/18-1				576.99	87.76	0.64	815.80	105.17	0.65				0.65
87/12-1A				151.70	76.47	1.24	261.70	111.38	1.01	458.70	89.13	1.34	1.20
87/14-1				368.00	89.15	0.83				755.95	96.00	0.71	0.83
87/16-1				395.70	93.02	0.75	689.45	110.99	0.60	706.50	94.52	0.83	0.69
88/02-1				76.50	100.13	0.96				1184.76	76.08	1.23	0.90
102/28-1				207.95	108.58	0.70	328.17	95.71	1.37				1.10
102/29-1				346.72	90.83	0.83	431.15	102.09	1.13				0.98
103/18-1										1280.47	74.83	1.19	1.19
103/21-1				272.50	98.85	0.78	320.97	112.16	0.93	1410.17	76.85	0.96	0.89
106/28-1										1382.73	79.99	0.84	0.84
Brezell-1				248.50	64.07	1.33							1.33
Gartizenn-1	912.50	100.41	0.00	1224.00	85.64	0.03	1521.00	103.26	0.00				0.01
Glazenn-1	614.00	105.52	0.24	884.00	97.97	0.18	1398.50	73.50	0.69				0.37
Kerluz-1				746.10	101.56	0.26							0.26
Kulzemn-1	238.50	97.24	0.71	457.50	95.07	0.65				818.01	93.84	0.75	0.70
Lizemn-1				433.55	78.05	0.93							0.93
Penma-1	883.50	89.85	0.15	1096.50	84.05	0.18	1358.50	94.36	0.43				0.25
Re-Gwenn-1				1236.00	79.61	0.10	1542.00	77.76	0.72	2205.00	73.61	0.32	0.38
Travank-1							228.85	122.32	0.72				0.72
Yar Vor-1				237.00	78.26	1.12	340.00	104.12	1.16				1.14

Table 2. *Data defining normal compaction relations (ITT, interval transit time; D, mid-point depth (in km); mbsb, metres below sea-bed)*

Stratigraphic Unit	Reference Wells	Mean interval transit time (μs/ft)	Mid-point depth (mbsb)	Equation of Normal Compaction Relationship
Danian Chalk	73/01-1	102.5	891.4	ITT = 179–86D
	73/13-1	88.4	1054.3	
Upper Cretaceous	73/01-1	93.2	1136.6	ITT = 169–67D
Chalk	73/13-1	83.4	1282.9	
Lower Cretaceous	74/01-1A	121.3	978.6	ITT = 154–33D
Greensand/Gault Clay	Garlizenn-1	103.3	1521.0	
Triassic Mercia	73/01-1	83.0	2084.4	ITT = 127–21D
Mudstone	73/04-1	106.6	968.1	

variation should be sought as they retain a regionally consistent (sonic) log character, and changes in mean interval transit time for the unit across the region are unlikely to be facies-related. Finally, given that log data are recorded every 15.2 cm (6 inches), thicker units are more desirable for the analysis because they tend to average out the effects of localised anomalous (sonic) log values (e.g. chalk hardground beds), when the mean interval transit time for the entire unit is calculated.

This study analyses the Danian Chalk and Upper Cretaceous Chalk, the mixed sandy/shaly Lower Cretaceous Greensand/Gault Clay, and the shales of the Upper Triassic Mercia Mudstone. Since salt does not compact with depth, the analysis of the Mercia Mudstone has been limited to the shales above the salt.

As previously discussed (Figs 2 & 3), thick Jurassic–Lower Cretaceous Wealden sequences in the Celtic Sea/South-Western Approaches are localized in the regions of maximum inversion, and hence are nowhere at maximum burial-depth. Any compaction relationship determined for the Jurassic–Lower Cretaceous Wealden sequences would reflect exhumation from maximum burial-depth and would yield low estimates of apparent exhumation. Hence the Jurassic–Lower Cretaceous Wealden sequences have not been considered in this study. With the exception of the Danian, the Tertiary has also been omitted from this study, as it tends to be of laterally variable facies and is often not logged. The Danian Chalk has been included, even though it is restricted to a few wells in the South-Western Approaches, in an attempt to constrain the timing of the exhumation (i.e. pre- or post-Danian).

Interval transit time/depth relations and the determination of apparent exhumation

The tops and bases of the Danian Chalk, the Upper Cretaceous Chalk, the Lower Cretaceous Greensand/Gault Clay, and the Triassic Mercia Mudstone (above salt) were picked from sonic/gamma ray log plots at 1 : 4000 scale (e.g. Fig. 3). The formation picks are basically consistent with those on operator's composite logs, however some were adjusted to ensure a consistent sonic log signature was compared between wells. Digital sonic log data were averaged over the picked interval to determine the mean interval transit time. It should be noted that the mean interval transit time refers to the quantity measured by the sonic log, the mean time required for a sound wave to traverse a foot of formation, and not to the average velocity of a sonic wave in the formation under consideration, as determined from one or two-way travel times. Table 1 lists the mean interval transit time and depth to formation mid-point (below sea-bed) for each of the formations in the wells used in this study.

The normal compaction relationship for each unit was defined by the straight line linking the two wells (termed the reference wells) with the highest interval transit time for their burial-depth. The surface intercept of the normal compaction trend was checked to ensure it was less than 189 μs/ft, the interval transit time for salt water. The reference well data and the relationships describing normal compaction for each of the units analysed are given in Table 2. Although the velocity/depth relationship is exponential, over the depth range being considered, a linear relationship between interval

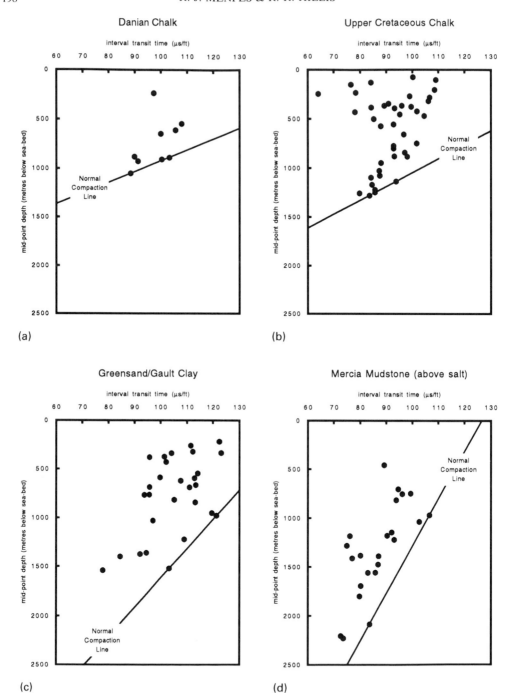

Fig. 6. Interval transit time/mid-point depth plots for (**a**) the Danian Chalk, (**b**) the Upper Cretaceous Chalk, (**c**) the Lower Cretaceous Greensand/Gault Clay, (**d**) and the Triassic Mercia Mudstone (above salt).

transit time and burial-depth yields equivalent results (e.g. Bulat & Stoker 1987).

An empirical estimate of the accuracy of the

normal compaction relationship was determined by examining the spread of the sonic logs of the reference wells around the normal compaction

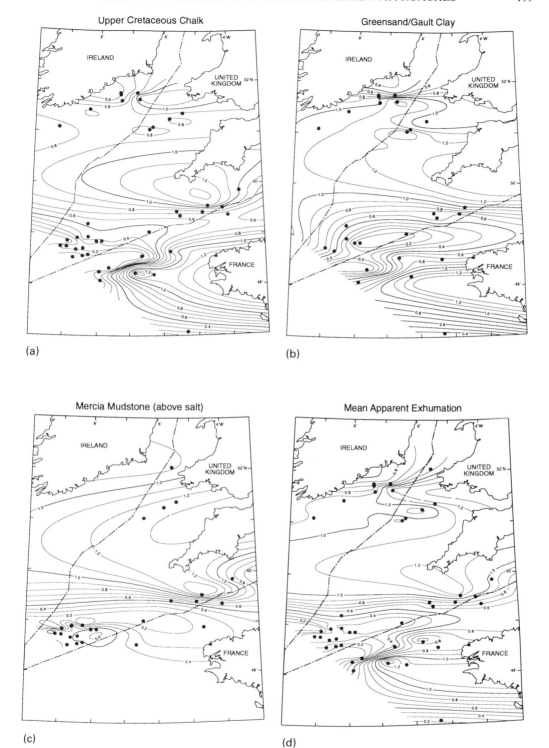

Fig. 7. Apparent exhumation maps for (**a**) the Upper Cretaceous Chalk, (**b**) the Lower Cretaceous Greensand/Gault Clay, (**c**) the Triassic Mercia Mudstone (above salt), and (**d**) the mean of the four formations. Contour values are in kilometres.

trend. The error estimates are such that 95% of
the sonic log data lie within the error margins,
hence only large data spikes are outside the error
margins. For the Danian the error estimate is
±150 m, for the Upper Cretaceous Chalk the
error is ±250 m, for the Lower Cretaceous
Greensand/Gault ±450 m, and for the Mercia
Mudstone ±400 m. The latter two error estimates are probably excessively high, as the
spread around the normal compaction trend is
predominantly due to lithological variations
within the formations assessed, and hence
should be considered to be the maximum
potential error in the normal compaction relationship. If a particularly fast or slow portion
of formation were removed during exhumation,
the mean interval transit time calculated would
correspondingly decrease or increase, however
the error induced by section removal would not
exceed the estimated errors above.

To address whether there is a systematic error
in the placement of the normal compaction
relationship (i.e. whether the reference wells are
indeed at maximum burial-depth), this study
should be compared with the results of similar
studies. For the Upper Cretaceous Chalk, the
results presented here are very similar to those
of Hillis (1991), where the normal compaction
trend was determined by averaging the Chalk
porosity/depth trends within individual wells.
This approach requires a relatively homogeneous lithology, and hence could not be
applied in this study. In particular it is unsuitable
for the lithologically variable Greensand/Gault
Clay. The authors are unaware of any other
detailed published studies of this region using
apatite fission track analysis, vitrinite reflectance, or other quantitative measures of apparent exhumation with which the results
presented herein might be compared. However,
other studies using this methodology such as
Hillis (1993) on the East Midlands Shelf have
given good correlation with apatite fission track
analyses (see also Hillis this volume).

The normal compaction relationship used here
for the Upper Cretaceous Chalk is at most 100 m
above the wet chalk sonic velocity/depth curve of
Scholle (1977). Recent studies involving the
Upper Cretaceous Chalk in the Southern North
Sea (Hillis this volume) suggest that the Chalk
normal compaction relationship for the Celtic
Sea/South-Western Approaches may underestimate apparent exhumation by about 350 m.

Apparent exhumation (E_A) can be estimated
graphically from the plots of interval transit time
against depth to formation mid-point (Fig. 6).
However, in practice, it was determined numerically using the equation:

$$E_A = 1/m(\Delta t_U - \Delta t_O) - d_U$$

where m is the gradient of the normal compaction relation given in Table 2, Δt_U is the mean
interval transit time of the well under consideration, Δt_O is the surface intercept of the normal
compaction relation (Table 2), and d_U is the
depth of the formation mid-point (below sea-
bed) of the well under consideration. The
resultant apparent exhumation values have been
calculated (Table 1) and plotted and contoured
(Fig. 7).

Discussion

Comparison of apparent exhumation values from different units

Apparent exhumation values from the four
formations were plotted against each other in
order to check their consistency (Fig. 8).
Least-squares, best-fit, linear relations between
the apparent exhumation values and associated
coefficients of correlation were determined by
regression of apparent exhumation values from
the deeper formation on the shallower formation (Table 3). The t-statistic of the co-
efficients of correlation were calculated and
tested against the one-tailed Student's t-distribution in order to determine whether the
coefficients of correlation were significant (e.g.
Till 1974). The correlation coefficient of the
linear relationship between apparent exhumation values from any two lithologies is
significant at the 95% confidence level in all
cases, and in all but one case it is significant at the
99% confidence interval.

Having shown that a linear relationship exists
between apparent exhumation values for each of
the four formations, these regression lines were
then tested for a statistically significant deviation
from the one-to-one lines shown in Fig. 8. The
methodology for this test is discussed in the
Appendix and the results are given in Table 4.
As shown in Table 4, the Danian Chalk, Upper
Cretaceous Chalk and Triassic Mercia Mudstone all yield equivalent results, and the Lower
Cretaceous Greensand/Gault Clay also gives
similar results to the Mercia Mudstone. The
Greensand/Gault Clay does not however have a
one-to-one correlation with the two chalk units,
due to the larger errors (±450 m) in the apparent
exhumation estimates from the Greensand/
Gault Clay which result from lithological variations.

It is unlikely that sedimentological or diagenetic processes could be responsible for the
similar amounts of overcompaction in the

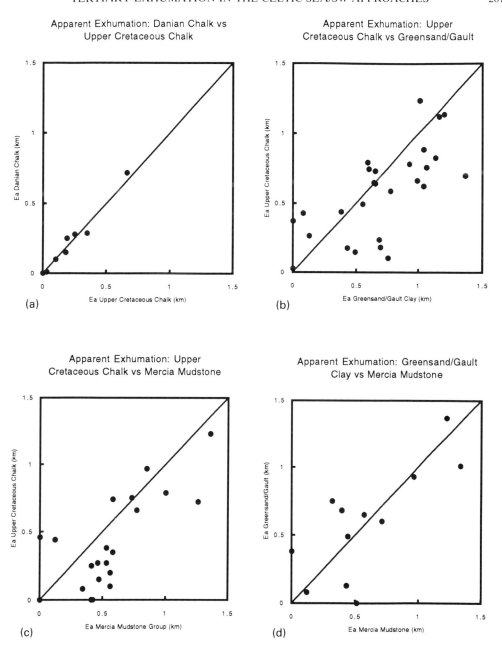

Fig. 8. Representative crossplots of apparent exhumation (E_A, in kilometres) derived from the four stratigraphic units studied: (**a**) Danian Chalk versus Upper Cretaceous Chalk, (**b**) Upper Cretaceous Chalk versus Lower Cretaceous Greensand/Gault Clay, (**c**) Upper Cretaceous Chalk versus Triassic Mercia Mudstone (above salt), and (**d**) Lower Cretaceous Greensand/Gault Clay versus Triassic Mercia Mudstone (above salt). Table 3 lists the data for all six combinations of the stratigraphic units analysed.

carbonate, shaly and sandy units analysed. Hence it is suggested that overcompaction (i.e anomalously fast sonic velocity) is primarily controlled by exhumation from previously greater burial-depth. A corollary of this is that lithologies other than shales may be used for estimating apparent exhumation magnitude. There is no evidence from any of the units in the wells studied that abnormal pore pressure effects have resulted in a decrease in sonic

Table 3. *Comparison between apparent exhumation results from different stratigraphic units. Y* The correlation coefficient of the linear relationship between the E_A values from the two formations is significant at the 95% confidence level. Y† The correlation coefficient of the linear relationship between the E_A values from the two formations is significant at the 99% confidence level. N indicates that the correlation coefficient is not significant at the 99% confidence level*

Units	Danian Chalk		Upper Cretaceous Chalk		Lower Cretaceous Greensand/Gault Clay	
Upper Cretaceous Chalk	9 14.92	0.985 Y:Y				
Lower Cretaceous Greensand/Gault Clay	4 9.14	0.994 Y:Y	27 4.69	0.684 Y:Y		
Triassic Mercia Mudstone	6 2.38	0.766 Y:N	21 4.91	0.748 Y:Y	12 3.59	0.751 Y:Y

number of wells	correlation coefficient
t-statistic	Y* : Y†

Table 4. *Comparison between regression relationships from different stratigraphic units; column headers are regressed against row headers. Y* A one-to-one relationship is statistically feasible at the 97.5% confidence level. N indicates that the regression line between column header and row header is statistically different to a one-to-one relationship at the 97.5% confidence level*

Units	Danian Chalk		Upper Cretaceous Chalk		Lower Cretaceous Greensand/Gault Clay		Triassic Mercia Mudstone	
Danian Chalk			9 6.54	0.531 Y	4 39.00	881.9 N	6 10.65	3.621 Y
Upper Cretaceous Chalk	9 6.54	1.255 Y			27 4.32	3.335 Y	21 5.93	4.792 Y
Lower Cretaceous Greensand/Gault Clay	4 39.00	120.6 N	27 4.32	8.629 N			12 5.10	0.653 Y
Triassic Mercia Mudstone	6 10.65	4.347 Y	21 5.93	4.846 Y	12 5.10	0.777 Y		

number of wells	F-statistic
Critical F-value	Y†

velocity, or that hydrocarbon entry has inhibited depth-controlled mechanical and chemical compaction.

The similarity of results from units of Triassic or Danian age suggests that Tertiary exhumation was of sufficiently great magnitude to mask any earlier Mesozoic periods of exhumation, and that the maximum Mesozoic-Cenozoic burial-depth in the Celtic Sea/South-Western Approaches was attained prior to Tertiary exhumation. Where Danian Chalk is present in the South-Western Approaches it shows a similar amount of overcompaction to the older units (Table 4). Hence it is inferred that exhumation in the South-Western Approaches post-dated the Danian. However, no Danian Chalk was analysed in the Celtic Sea and exhumation there only demonstrably post-dates the deposition of the Upper Cretaceous Chalk, and may have occurred earlier than in the South-Western Approaches, around the Cretaceous–Tertiary boundary as suggested by Tucker & Arter (1987).

Exhumation in the Celtic Sea/South-Western Approaches

There are several features to note in the apparent exhumation maps of Fig. 7, in particular the mean apparent exhumation map, which is the most robust estimate of apparent exhumation, being the average of estimates from all formations present in any given well. Firstly, regional highs in apparent exhumation of around 1 km occur on the uninverted areas of the Celtic Sea/South-Western Approaches, and on the margins of the Armorican and Cornubian Platforms and the Pembroke Ridge. The inversion of the Brittany and South-West Channel Basins is witnessed by the relatively localized apparent exhumation highs at Brezell-1 and Yar Vor-1, and the inversion axis of the North Celtic Sea Basin is also an apparent exhumation high. The Melville Basin is the major low in apparent exhumation in the Celtic Sea/South-Western Approaches area and the South Celtic Sea Basin is a low in apparent exhumation with respect to the surrounding platform margins. The paucity of data over the platform areas themselves has resulted in poorly constrained contours over the platforms, however the important feature to note is that highs in apparent exhumation are not restricted to inversion axes.

It should be remembered that apparent exhumation is the amount of exhumation that is not reversed by subsequent burial. There is no Tertiary sequence preserved, and therefore no post-exhumational burial on the uninverted platform margin areas which exhibit the regional highs in apparent exhumation, hence their apparent exhumation values reflect the total exhumation at the time of denudation. However in the Melville Basin, assuming Eocene–Oligocene erosion, there has been typically around 500 m of burial since exhumation. Hence exhumation at the time of denudation in the Melville Basin was around 500 m greater than the calculated apparent exhumation. In most of the wells in the Celtic Sea area there was relatively little burial subsequent to exhumation, except at 49/9-1 (340 m) and 106/28-1 (910 m), so again in these wells exhumation at the time of denudation was significantly greater than the calculated apparent exhumation.

While there was significant exhumation of the recognized inversion axes, more regional highs in exhumation coincide with the platform margins. Regional exhumation and erosion of the platforms is supported by Roberts (1989), who argued that the base of the Upper Cretaceous Chalk on the Cornubian Peninsula has been elevated by approximately 1 km. In addition, Keeley et al. (1993) suggested 1.3–2.5 km of (now eroded) post-Bajocian burial in southern Ireland, based on AFTA results. Cornford (1986) postulated an eroded thickness of approximately 1 km of Upper Jurassic–Lower Cretaceous sediment on the Dorset and Somerset coasts on the basis of vitrinite reflectance levels in the now exposed Lias. It is clear that Late Cretaceous–Tertiary exhumation was more widespread than are inversion structures in the Celtic Sea/South-Western Approaches area.

Maturation modelling in the Celtic Sea/South-Western Approaches

It is crucial to take account of exhumation magnitudes in modelling the thermal maturity of potential source rocks in the Celtic Sea/South-Western Approaches area. Wells 87/12-1A and 106/28-1 are modelled in Fig. 9, showing the thermal maturities predicted with and without incorporation of exhumation into the well geohistory. In all models, surface temperature was assumed to be 15°C, and geothermal gradients 30°C km^{-1} throughout basin development. While the paleogeothermal history is simplistic, the purpose of the modelling is to illustrate the effects of incorporating exhumation into maturation modelling. Decompaction was performed using the methodology of Sclater & Christie (1980), with allowance for the effects of Tertiary exhumation as described in Hillis et al. (1994).

It can clearly be seen from Fig. 9a that the top of the Carboniferous reaches early maturity for oil generation in well 87/12-1A if exhumation is incorporated into the well geohistory, but would be predicted not to have generated any hydrocarbons if exhumation is not incorporated. Similarly, the base of the Triassic (TD) in well 106/28-1 reaches mid-maturity levels for oil generation allowing for exhumation, but reaches only early maturity if exhumation is not incorporated into the well geohistory (Figure 9b).

Maximum burial-depth in the Celtic Sea/South-Western Approaches: implications for models of basin development

The evidence that pre-Tertiary units in the Celtic Sea/South-Western Approaches area are not at their maximum burial-depth has a profound effect on the inferred pre-exhumation burial history of the region. As can be seen in Fig. 9, the occurrence of Tertiary exhumation requires a period of burial prior to exhumation not witnessed by the preserved stratigraphy.

The maximum burial-depth of any pre-exhumation horizon in the Celtic Sea/Western Approaches area may be calculated by adding apparent exhumation to present burial-depth. Present and maximum burial-depth maps of the base of the Upper Cretaceous Chalk, based only on data from well localities, are shown in Fig. 10. The Chalk was selected as the onset of chalk deposition corresponds approximately with the

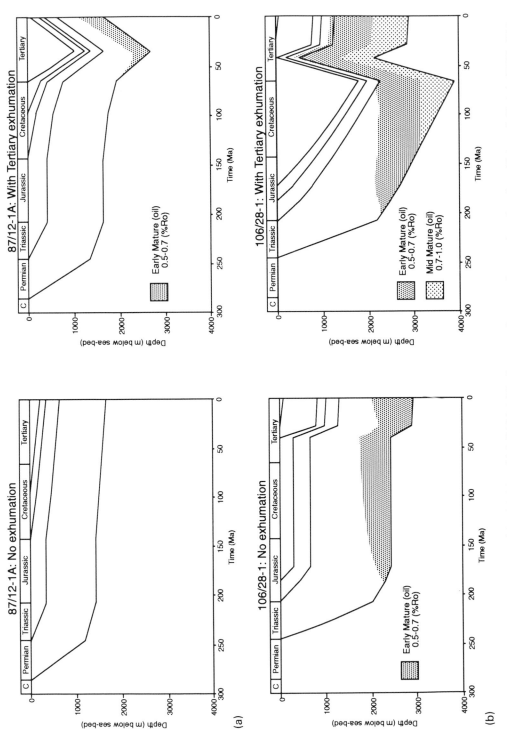

Fig. 9. Decompacted burial histories for wells (**a**) 87/12-1A and (**b**) 106/28-1. The decompaction procedure was based on that of Sclater & Christie (1980). With allowance for Tertiary exhumation, the Carboniferous of 87/12-1A in the Plymouth Bay Basin reaches early oil maturity, whilst the Triassic in well 106/28-1 in the Cardigan Bay Basin is elevated from early to mid-oil maturity. Modelling was undertaken using the BasinMod™ software.

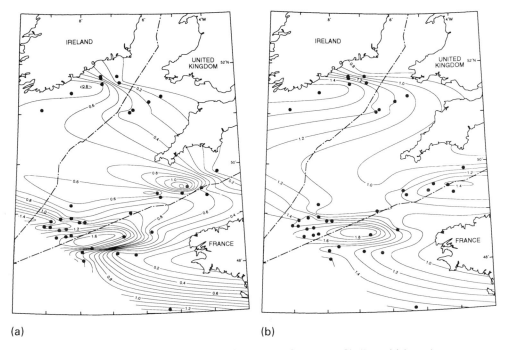

(a)   (b)

Fig. 10. (**a**) Present burial-depth map for the base of the Upper Cretaceous Chalk, and (**b**) maximum burial-depth (B) map for the base of the Upper Cretaceous Chalk based on the addition of the mean apparent exhumation (E_A) estimates to present burial-depth (B_P).

end of extensional, basin-forming rifting in the area. From Fig. 10, it is clear that the maximum burial-depth of the Chalk shows a more even, regional pattern of post-rift subsidence than its present burial-depth uncorrected for exhumation. When corrected for exhumation, the post-rift sequence in the Celtic Sea/South-Western Approaches area shows relatively little localization over the syn-rift Jurassic–Lower Cretaceous sequence.

The 'rift-drift' Avalon Unconformity of the conjugate, but largely uninverted Canadian margin is horizontal. There is no localization of the post-rift sequence over the fault-controlled syn-rift basins of the Canadian margin (Keen *et al.* 1987). As Keen *et al.* (1987) proposed for the Canadian margin, corrected post-rift burial thicknesses in the Celtic Sea/South-Western Approaches area suggest that during basin formation the lower lithosphere, thermal re-equilibration of which controls post-rift sub-sidence, did not experience localized thinning on the scale of the syn-rift basins. Hence extension was decoupled between the upper and lower lithosphere. Alternatively, post-rift lithospheric flexural rigidities may have been sufficiently high

that post-rift subsidence was regionally distrib-uted.

The distribution of Late Cretaceous–Tertiary exhumation: implications for models of basin inversion

Only regional uplift of the earth's surface (as opposed to exhumation or local uplift of rocks) demonstrates work done against gravity (England & Molnar 1990). Hence it is necessary to prove regional surface uplift if isostatic forces requiring uplift are to be invoked (England & Molnar 1990). Marine shelf sediments occur above and below the Late Cretaceous–Tertiary unconformities during which major exhumation in the Celtic Sea/South-Western Approaches occurred. Hence, ignoring eustatic variation, exhumation indicates uplift of the rock sequence with respect to mean sea-level. Since the eroded material was not redeposited in the Celtic Sea/South-Western Approaches area, regional uplift of the surface is inferred. Ignoring eustatic sea-level variation, regional exhumation of around 1 km requires approximately 0.3 km of tectonic uplift.

The key observations for which any tectonic model of Late Cretaceous–Tertiary evolution of the Celtic Sea/South-Western Approaches area must account for are:

- exhumation (uplift) of the inverted areas of the Brittany and South-West Channel Basins, the South Celtic Sea–Bristol Channel Basin, the North Celtic Sea–Cardigan Bay Basin, and the Fastnet Basin, which occurred in association with upper lithospheric compression/thickening (i.e. basin inversion);
- regional exhumation (and associated tectonic uplift of approximately 0.3 km) of the uninverted St. Mary's and Melville Basins and the Armorican Platform, Cornubian Platform, and Pembroke Ridge margins, without associated upper lithospheric compression/thickening;
- burial prior to exhumation, during which the rocks became overcompacted (Fig. 9).

Eustatic sea-level falls and in plane stresses (Cloetingh et al. 1990) cannot account for the amount of exhumation/uplift (1 km of exhumation corresponds to approximately 300 m of tectonic uplift). However, hotspot-related lithospheric underplating (see Brodie & White this volume) such as that associated with the Iceland hotspot could generate uplift during the Paleocene. The hotspot model does not however account for subsidence to maximum burial-depth prior to uplift, nor do the authors believe it can account for regional exhumation in the Celtic Sea/South-Western Approaches, distant from the hotspot, and adjacent to the classic non-volcanic continental margin of the Western Approaches (White & McKenzie 1989).

Heterogeneous lithospheric compression, as proposed by Hillis (1992), in which compressional deformation in the upper lithosphere is decoupled from compression in the lower lithosphere, can account for the observed subsidence and uplift patterns of the inverted and uninverted areas. (?Upper) Crustal thickening is responsible for uplift of the inverted basins, and decoupled lower lithospheric compression can generate both the initial subsidence and subsequent uplift of the uninverted regions such as the St. Mary's and Melville Basins, and the Armorican and Cornubian Platform margins. This model is discussed further by Hillis (this volume).

The authors wish to thank Andy Mitchell and Peter Tingate at the National Centre for Petroleum Geology and Geophysics, Adelaide University, for access to and help in using the Geolog and BasinMod software. Sherry Proferes is thanked for her help with drafting. Chevron UK are thanked for open-file well log data for the Irish sector of the area. French log data were purchased with the aid of an Australian Research Council Grant to RRH. RJM is in receipt of a University of Adelaide postgraduate research scholarship. RRH and RJM acknowledge the University of Adelaide for financial assistance in attending the Basin Inversion Conference.

Appendix

To test a one-on-one linear hypothesis against a fitted regression line (Draper & Smith 1981):
- fit the regression of y on x (i.e. the E_A estimates from one formation against the other formation), calculating the residual sum of squares:

$$SS_1 = \sum_1^n (y_i - \hat{a} - \hat{b} x_i)^2$$

where y_i is the 'i'th E_A estimate from one formation, x_i is the 'i'th E_A estimate from the other formation, $\hat{a}$ and $\hat{b}$ are the coefficients of the regression line.

Under the null hypothesis that a one-to-one relationship exists between the E_A estimates from different formations:
- calculate the sum of squares of the differences between individual E_A data pairs:

$$SS_2 = \sum_1^n (y_i - x_i)^2$$

At all times $SS_2 \geqslant SS_1$.
- calculate the test statistic

$$F = \frac{(SS_2 - SS_1)/2}{SS_1/(n-2)}$$ which is distributed as $F(2, n-2)$

where n is the number of E_A data pairs.

Reject the hypothesis that a one-to-one relationship exists between the E_A estimates from different formations if $F > F(2, n-2)$ at the chosen confidence level.
- repeat the above three steps, but regress x on y to examine consistency.

References

BROOKS, M., TRAYNER, P. M. & TRIMBLE, T. J. 1988. Mesozoic reactivation of Variscan thrusting in the Bristol Channel area, UK. *Journal of the Geological Society, London*, **145**, 439–444.

BULAT, J. & STOKER, S. J. 1987. Uplift determination from interval velocity studies. *In*: BROOKS, J. & GLENNIE, K. (eds) *Petroleum Geology of North West Europe*. Graham & Trotman, London, 293–305.

CLOETINGH, S., GRADSTEIN, F. M., KOOI, H., GRANT, A. C. & KAMINSKI, M. 1990. Plate reorganisation: a cause of rapid late Neogene subsidence and sedimentation around the North Atlantic? *Journal of the Geological Society, London*, **147**, 495–506.

CORNFORD, C. 1986. The British Channel Graben: organic geochemical limits on subsidence and speculation on the origin of inversion. *Proceedings of the Ussher Society*, **6**, 360–367.

DRAPER, N. R. & SMITH, H. 1981. *Applied regression analysis*, 2nd edition, John Wiley & Sons, Toronto, 709p.

ENGLAND, P. & MOLNAR, P. 1990. Surface uplift, uplift of rocks, and exhumation of rocks. *Geology*, **18**, 1173–1177.

HAYWARD, A. B. & GRAHAM, R. H. 1989. Some geometrical characteristics of inversion. *In*: COOPER, M. A. & WILLIAMS, G. D. (eds) *Inversion Tectonics*. Geological Society, London, Special Publication, **44**, 17–39.

HILLIS, R. R. 1991. Chalk porosity and Tertiary uplift, Western Approaches Trough, SW UK and NW French continental shelves. *Journal of the Geological Society, London*, **148**, 669–679.

—— 1992. A two-layer lithospheric compressional model for the Tertiary uplift of the southern United Kingdom. *Geophysical Research Letters*, **19**, 573–576.

—— & CHAPMAN, T. J. 1992. Variscan structure and its influence on post-Carboniferous basin development, Western Approaches Basin, SW UK Continental Shelf. *Journal of the Geological Society, London*, **149**, 413–417.

—— 1993. Tertiary erosion magnitudes in the East Midlands Shelf, onshore UK. *Journal of the Geological Society, London*, **150**, 1047–1050.

——, THOMSON, K. & UNDERHILL, J. R. 1994. Quantification of Tertiary erosion in the Inner Moray Firth using sonic velocity data from the Chalk and Kimmeridge Clay. *Marine and Petroleum Geology*, **11**, 283–293.

KAMERLING, P. 1979. The geology and hydrocarbon habitat of the Bristol Channel Basin. *Journal of Petroleum Geology*, **2**, 75–93.

KEELEY, M. L., LEWIS, C. L. E., SEVASTOPULO, G. D., CLAYTON, G. & BLACKMORE, R. 1993. Apatite fission track data from southeast Ireland: implications for post-Variscan burial history. *Geological Magazine*, **130**, 171–176.

KEEN, C. E., BOUTILIER, R., DE VOOGD, B., MUDFORD, B. & ENACHESCU, M. E. 1987. Crustal geometry and extensional models for the Grand Banks, eastern Canada: constraints from deep seismic reflection data. *In*: BEAUMONT, C. & TANKARD, A. J. (eds) *Sedimentary Basins and Basin-forming Mechanisms*, Canadian Society of Petroleum Geologists, Memoir **12**, 101–115.

LAKE, S. D. & KARNER, G. D. 1987. The structure and evolution of the Wessex Basin, southern England: an example of inversion tectonics. *Tectonophysics*, **137**, 347–378.

LANG, JR., W. H. 1978. The determination of prior depth of burial (uplift and erosion) using interval transit time. *Society of Professional Well Log Analysts Nineteenth Annual Logging Symposium*, June 13–16, 1978, Paper B.

MAGARA, K. 1976. Thickness of removed sedimentary rocks, paleopore pressure, and paleotemperature, southwestern part of Western Canada Basin. *American Association of Petroleum Geologists Bulletin*, **60**, 554–565.

MASSON, D. G. & PARSON, L. M. 1983. Eocene deformation on the continental margin SW of the British Isles. *Journal of the Geological Society, London*, **140**, 913–920.

PETRIE, S. H., BROWN, J. R., GRANGER, P. J. & LOVELL, J. P. B. 1989. Mesozoic history of the Celtic Sea Basins. *In*: TANKARD, A. J. & BALKWILL, H. R. (eds) *Extensional Tectonics and Stratigraphy of the North Atlantic Margins*. American Association of Petroleum Geologists Memoir, **46**, 433–444.

ROBERTS, D. G. 1989. Basin inversion in and around the British Isles. *In*: COOPER, M. A. & WILLIAMS, G. D. (eds) *Inversion Tectonics*. Geological Society London, Special Publication, **44**, 131–150.

ROBINSON, K. W., SHANNON, P. M. & YOUNG, D. G. G. 1981. The Fastnet Basin: an integrated analysis. *In*: ILLING, L. V. & HOBSON, G. D. (eds) *Petroleum Geology of the Continental Shelf of North-West Europe*. Heyden, London, 444–454.

SCLATER, J. G. & CHRISTIE, P. A. F. 1980. Continental stretching: an explanation of the post-mid-Cretaceous subsidence of the Central North Sea Basin. *Journal of Geophysical Research*, **85**, 3711–3739.

SIMPSON, I. R., GRAVESTOCK, M., HAM, D., LEACH, H. & THOMPSON, S. D. 1989. Notes and cross-sections illustrating inversion tectonics in the Wessex Basin. *In*: COOPER, M. A. & WILLIAMS, G. D. (eds) *Inversion Tectonics*. Geological Society London, Special Publication, **44**, 123–129.

SCHOLLE, P.A. 1977. Chalk diagenesis and its relation to petroleum exploration: oil from chalks, a modern miracle? *American Association of Petroleum Geologists Bulletin*, **61**, 982–1009.

TILL, R. 1974. *Statistical Methods for the Earth Scientist*. London, Macmillan, 154p.

TUCKER, R. M. & ARTER, G. 1987. The tectonic evolution of the North Celtic Sea and Cardigan Bay basins with special reference to basin inversion. *Tectonophysics*, **137**, 291–307.

VAN HOORN, B. 1987. The South Celtic Sea/Bristol Channel Basin: origin, deformation and inversion history. *Tectonophysics*, **137**, 309–334.

WHITE, R. & MCKENZIE, D. 1989. Magmatism at rift zones: the generation of volcanic continental margins and flood basalts. *Journal of Geophysical Research*, **B94**, 7685–7729.

ZIEGLER, P. A. 1987. Evolution of the Western Approaches Trough. *Tectonophysics*, **137**, 341–346.

Case studies: Americas

Inversion structures and hydrocarbon occurrence in Argentina

MIGUEL A. ULIANA[1], MARCELO E. ARTEAGA[2], LEONARDO LEGARRETA[1], JORGE J. CERDÁN[1] & GUSTAVO O. PERONI[1]

[1]ASTRA C.A.P.S.A. Exploration and Production, Tucumán 744 (Piso 7), 1049 Buenos Aires, Argentina

[2]Present address: Pluspetrol S.A., La Rioja 301, 1214 Buenos Aires, Argentina

Abstract: Positive inversion features are a common structural ingredient in broad regions of western and central Argentina. They show up in several geological provinces distributed from Salta at 24°S down to Tierra del Fuego below 54°S. Considerable variety in inversion style and intensity developed due to regional positioning within the Andean orogenic belt, near the mountain edge, or well within the South American foreland. Weakly inverted grabens, doubly plunging buckles and compressional anticlines may site as far as 600 km away from the present day oceanic trench and 300 km east of the Cordilleran front. Most of the studied examples are genetically linked to high-angle or listric normal fault families that developed across southern South America, induced by pervasive Triassic to Cretaceous extensional stress conditions. Old extensional structures involved in the inversion process include: segments of Triassic successor basins linked to extensional collapse of the Late Paleozoic orogenic tract, Jurassic faults belonging to a multicomponent rift system developed after incipient Basin-and-Range style deformation of Patagonia, segments of a network of linear Jurassic–Neocomian troughs and half-troughs related to pre-breakup fragmentation of the South American slab, and roughly linear Cretaceous–Palaeogene fault-bounded troughs of the arc-crest type induced after supracrustal extension driven by easterly subduction along the Pacific edge of the plate.

As in other deformed areas around the world the inversion process did not advance as a wave from the orogenic front ('bulldozer' mode), but apparently occurred as a more or less synchronous reactivation of faults distributed across the inverted tract ('accordion' mode). In the Neuquén basin middle Jurassic to Cretaceous reversal from block subsidence to local uplift occurred along an intra-plate megashear aligned with the inboard projection of the South Atlantic oceanic transforms. Protracted strike-slip motion and related syndepositional block-inversion favoured pre-migration structuring and stratal anomalies prone to provide sedimentary and combination traps. Over most of the oil-producing Argentine basins inversion took place late, during the Cenozoic. Thus timing, with respect to the local maturation histories and onset of regional hydrocarbon charge, was a key control on field size and oil richness in each region. Late tectogenesis and late regional migration combined in the non-marine Cuyo basin to provide fields in the 250 MMBO range, out of inverted Triassic half-grabens. Less favourable convergence of inversion and migration is recorded in the western San Jorge basin, where regional contraction occurred 20–40 Ma after regional oil emplacement, and late inversion resulted in dispersive hydrocarbon reaccommodation. Finally, migration in the southern Magallanes basin happened at the time when burial beneath a thick foreland wedge had reduced reservoir properties and thermally degraded the source sections into the gas generation stage.

During the last decade, work associated with exploration for hydrocarbons has demonstrated that a number of oil and gas fields across western Argentina (Noroeste, Cuyo, Neuquén and San Jorge basins, Fig. 1) are related to traps controlled by inversion structures. Most of these occurrences are linked to Late Cenozoic (Andean) compressional and transpressional inversion of families of normal faults developed when the Gondwana supercontinent collapsed and split apart, and South America drifted away from Africa.

A majority of the oil bearing traps developed in the course of the last 20 My or so in a geographical setting that involved the western fringe of the South American hinterland and parts of the adjacent Central and Southern Andes. The productive tracts were structured in

From BUCHANAN, J. G. & BUCHANAN, P. G. (eds), 1995, *Basin Inversion*, Geological Society Special Publication No. 88, 211–233.

211

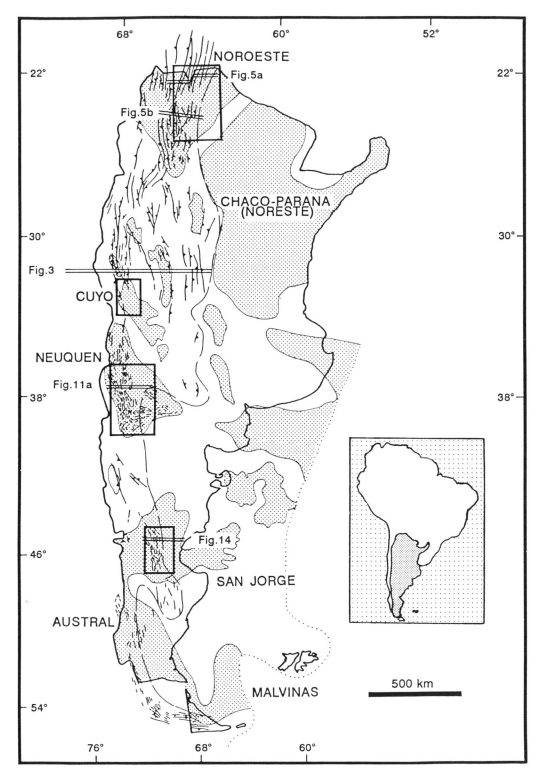

Fig. 1. Simplified map of Argentina illustrating distribution of Mesozoic sedimentary basins (shaded), selected structural elements over the western side of the country, and location of the oil-bearing areas and detailed cross-sections discussed in the text.

a tectonic context that involved the back side of the relatively simple and non-collisional Nazca–South America convergent plate junction, at a time when the system was subject to an advancing and nearly ortogonal subduction regime (Isacks 1988; Ramos 1989). Inverted features developed both at the lower (foreland) and upper (Cordilleran) slabs of the A subduction zone that frames the eastern side of the Andean ranges. Oil pools hosted by inversion structures include some 50 fields scattered from NW Argentina at 24°S, down to Patagonia at 54°S. They range from very small to large size accumulations related to several Mesozoic petroleum systems. Volumetric estimates suggest recoverable reserves in the range of 2–3 BBO and 15 TCFG. The present contribution reviews the structural setting and trapping styles that control these inversion-related hydrocarbon occurrences.

Geological framework

Pre-Mesozoic basement underpinning

The southern South American basement complex is formed by a piece of the relatively stable cratonic mass that was once at the core of Gondwana, and a series of smaller crustal blocks presumably accreted to its periphery during the late Precambrian to middle Palaeozoic period (Ramos et al. 1986; Ramos 1988; Forsythe et al. 1993). Over broad areas the old crystalline rocks are overlain by Lower to Upper Palaeozoic sedimentary deposits. Along western Argentina and Chile these accumulations are particularly thick and locally show evidence of polyphase deformation and metamorphic overprint.

A protracted history of convergence against the Pacific edge of Gondwana resulted in several episodes of contractional deformation and substantial margin-parallel displacement of crustal slivers (Hervé 1988; Hervé & Mpodozis 1990).

The final shaping of the Late Palaeozoic Pangea developed in a local context of shortening across the palaeo-Pacific margin of Gondwana ('San Rafael' orogeny, Llambías & Sato 1990; 'Gondwanide' orogeny, Storey & Alabaster 1991). Early Permian contractional deviatoric stress drove crustal thickening and set up conditions for widespread extensional deformation of the South American slab. By Late Permian, when the Gondwana supercontinent was almost stationary (Rapalini 1990), the orogenic tract experienced regional elevation. Shortly after, the overthickened Late Palaeozoic welt became gravitationally unstable and began to collapse (Llambías & Sato 1990; Storey & Alabaster 1991).

Mesozoic extensional deformation

The oldest extensional structures are related to Middle to Late Triassic sedimentary wedges, which appear confined to NNW–SSE linear trends (Fig. 2a; Uliana et al. 1989). They contrast with coeval depocentres such as the Chaco–Paraná and Karoo basins, located at the Gondwana interior (e.g. Uliana & Biddle 1988; Cole 1992) on top of the older and more stable Gondwana–core basement province. We interpret them as a family of successor fault-bound troughs, induced by gravitational collapse of the Late Palaeozoic orogenic tract (Mpodozis & Kay 1990; de Wit & Ransome 1992) that fringed the Pacific margin of Gondwana.

By Early and Middle Jurassic crustal stretching was preferentially centred on Patagonia (Gust et al. 1985; Uliana et al. 1989). Diffuse extension was associated with a magmatically active region where multiple and relatively short rifts developed. Structural and magmatic style might be described as an incipient Basin-and-Range extensional system (Fig. 2b).

Close to the end of the Jurassic and by earliest Cretaceous, when the South America–Africa decoupling began to develop from south to north (Simpson 1977; Lawver & Scotese 1987), tensional tectonic instability and magmatic outpourings shifted toward the Gondwana interior (Uliana et al. 1989; Conceicao et al. 1988). The basement was broken by a series of multidirectional narrow and linear troughs (e.g. Bianucci & Homovc 1982; Gómez Omil et al. 1989) akin to the modern East-African rift types (Fig. 2c).

During the Middle and Late Cretaceous, when global spreading reached a peak and South America drifted apart from Africa, most of local intra-plate faulting faded out. Extension and related magmatism persisted only near the Pacific margin, where plate interaction under a retreating subduction regime promoted active fault-driven subsidence, since at least the Middle Jurassic (Fig. 2d).

Cenozoic contractional deformation and tectonic inversion

Late Cenozoic Pacific plate-reorganization near 25 Ma (Handschumacher 1976; Pilger 1983) brought a shift into a dominantly advancing (Royden 1993) subduction mode (Mpodozis 1984). The new regime was associated with widespread contractional deformation of the Andean belt and portions of the South American foreland (e.g. Isacks 1988; Ramos 1989),

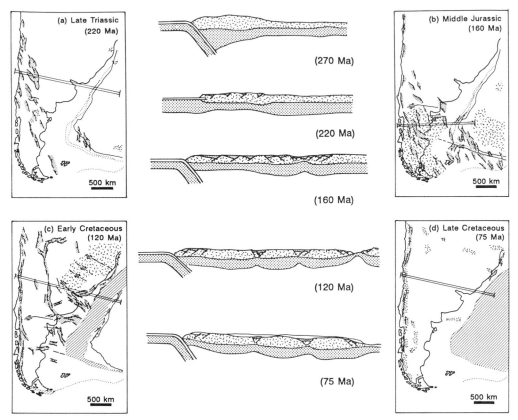

Fig. 2. Palaeotectonic sketches showing Mesozoic evolution of extensional structures in southern South America. Maps slightly modified after Uliana *et al.* (1989).

and induced widespread appearance of inversion structures (e.g. Mancini & Serna 1989; Yrigoyen *et al.* 1989; Grier *et al.* 1991). Most authors envision that, as the orogenic episode developed during the Late Cenozoic, the focus of contractional deformation shifted eastward. East–West shortening was associated or quickly followed by uplift and mountain building (e.g. Vicente 1972*a*).

Structural restorations across central Argentina (Fig. 3) point out amounts of total contraction in the 50–70 km range (Isacks 1988), and detailed studies demonstrate substantial along-strike variation in the local magnitude and style of shortening. While the Subandean fold-thrust belt near Bolivia (Fig. 4) was contracted by 22% (Allmendinger *et al.* 1983), the broad thick-skinned Pampean Ranges at the centre of Argentina became contracted by only 2% (Jordan & Allmendinger 1986; Fig. 3). Definite along-strike tectonic segmentation and changes in the style and location of the shortening-related fabrics are thought to be the response to

several factors (Allmendinger *et al.* 1983; Jordan *et al.* 1983): among others, dip of the subducting Nazca plate, irregular Palaeozoic through Cenozoic sedimentary development, and pre-existing structural discontinuities.

Noroeste (Oran) basin

Regional setting

Cretaceous oil accumulations of NW Argentina occur close to the outer fringe of the central Andean Cordillera near 24°S. This area shows the juxtaposition of quite diverse structural styles (Fig. 4). North of 24°S the edge of the Cordilleran belt known as the Subandean Ranges (Mingramm *et al.* 1979), consists of a series of easterly-vergent thrusted slices and anticlines locally stacked along listric faults that sole-out into an Early Palaeozoic master detachment zone (Fig. 5a). These detached structures, which form an uninterrupted trend that can be followed to the north into central–eastern

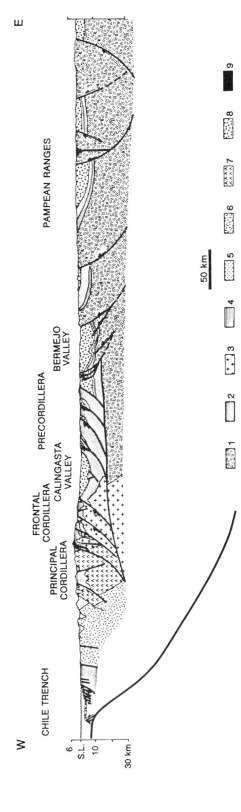

Fig. 3. Structural cross-section through west–central Argentina and Chile, near 31°S. Compiled after data in Gordillo & Lencinas (1979), Ortiz & Zambrano (1981), Moscoso & Mpodozis (1988), and Zapata & Allmendinger (1993). Notice style variability across the Andean belt, and thick-skinned involvement at the Pampean foreland.

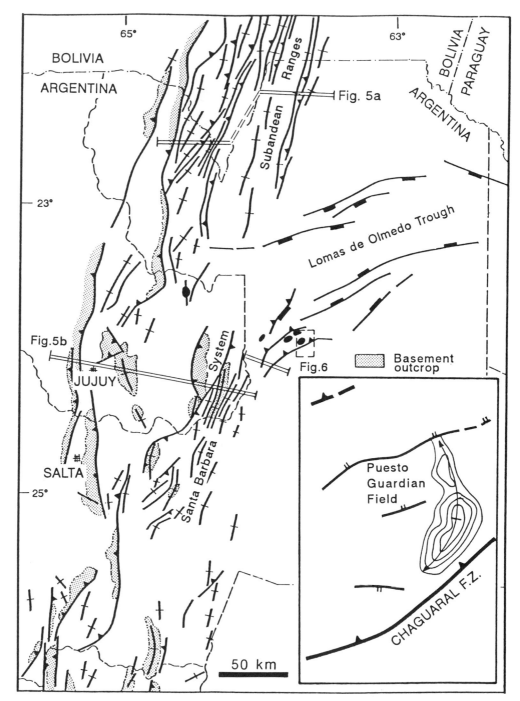

Fig. 4. Structural sketch map of north–western Argentina. Inset map outlines shape and structural elements bounding Puesto Guardian Field. (Data after Di Persia *et al.* 1991.)

Bolivia, are underlain by an unbroken basement ramp. East from the mountains, the west-dipping basement ramp and the Palaeozoic sedimentary rocks are covered by an undisturbed foreland wedge that lies beneath the extensive Chaco plains.

W   E

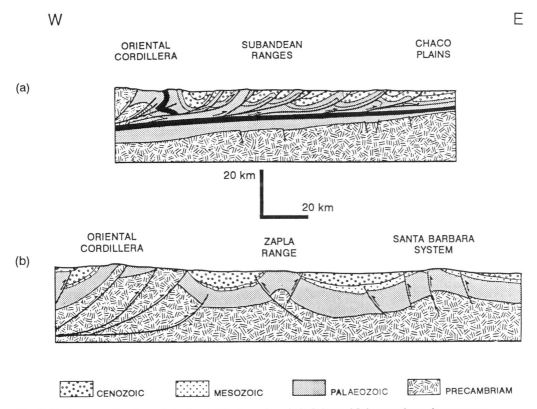

Fig. 5. Regional profiles through the frontal Andean domain in Salta and Jujuy provinces demonstrate contrasting deformation styles. (**a**) Slightly modified after Mingramm *et al.* (1979), and (**b**) after Cahill *et al.* (1992).

South of 24°S the 600 km long Subandean trend is replaced by a tectonic pattern that comprises substantial basement involvement, along high-angle reverse faults, and the presence of westerly-vergent structural components (Fig. 5b). The Pampean Ranges, a series of irregularly spaced and diversely orientated compressional wedges and tilted blocks, dominate the Argentine foreland down to 33°S and beyond (Caminos 1979; Jordan & Allmendinger 1986).

The regional change from the thin-skinned Subandean style to the thick-skinned Pampean architecture is interpreted as a reflection of a shallowing of the Benioff zone (Isacks 1988). On a closer view, the style variation shown at the frontal Andean tract between 23 and 25°S is understood as an interference effect linked to the presence of the obliquely-trending, aborted Cretaceous rift structures (Fig. 6; Lowell 1985; Grier *et al.* 1991).

Trapping style

Oil occurrences are mainly centred near the vicinity of the NE–SW to ENE–WSW master

faults that outline the southern flank of the Lomas de Olmedo rift system (Fig. 4; Gómez Omil *et al.* 1989; Chiarenza & Ponzoni 1989; Di Persia *et al.* 1991). Closures at the level of the otherwise little deformed Maastrichtian–Palaeocene reservoirs are interpreted as a result of transpressional inversion induced by dextral slip along the graben-bounding faults (Lowell 1985; Letouzey 1990). For example, the Puesto Guardián field is a low relief anticlinal closure (Di Persia *et al.* 1991) developed on the hangingwall of a NW-dipping rift-fault (Fig. 4 inset), subsequently reactivated as a reverse fault during the Cenozoic. Magnitude of the transpression-induced inversion effect and involvement of the younger cover rocks diminishes along the trend, away from the Cordilleran front.

Cuyo (Cacheuta) basin

Regional setting

Located in west–central Argentina, the Triassic through Cretaceous Cuyo basin (Figs 1, 7) is

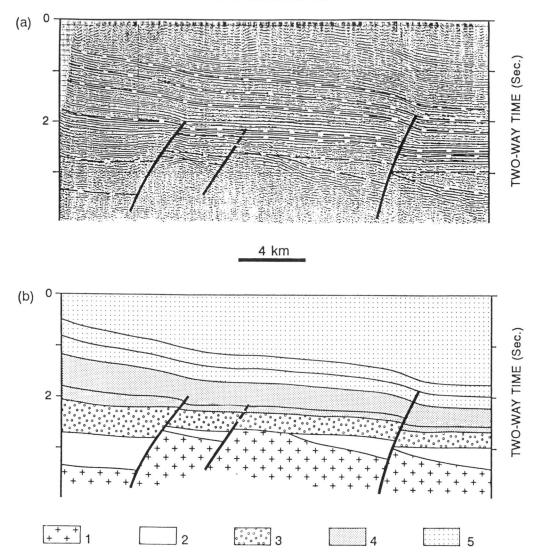

4 km

Fig. 6. (a) Seismic section near Puesto Guardián Field illustrates Late Cenozoic inversion of Chaguaral and associated Mesozoic extensional faults (see Fig. 3 for location). (b) Line drawing depicts interpreted position of: (1) basement; (2) Palaeozoic pre-rift strata; (3) Jurassic–Cretaceous rift-fill; (4) Cretaceous–Palaeogene sag deposits; and (5) Neogene foreland basin strata.

largely a subsurface feature (Rolleri & Criado Roque 1969; Rolleri & Garrasino 1979). It lies beneath a 2000–3000 m thick Cenozoic clastic cover, contained in a regionally low-standing and relatively little deformed piece of the foreland. The western side of the basin is flanked by the Precordillera mountain range, a 3500 m high decoupled thrust-stack (Ortiz & Zambrano 1981), disclosed on at the outer margin of the Andean orogenic complex. To the east and southeast, the Cuyo basin connects with the Los

Tordillos and Alvear basins (Fig. 7b). Farther to the east and to the north, these Triassic basins are surrounded by the Pampean Ranges (Caminos 1979; Gordillo & Lencinas 1979), a series of thick-skinned crystalline uplifts that disrupt the central Argentine foreland (Jordan & Allmendinger 1986).

The Cuyo and coeval Triassic half-graben sedimentary wedges are arranged following NNW–SSE trends (Fig. 7b). This linear development is thought to be reflecting the

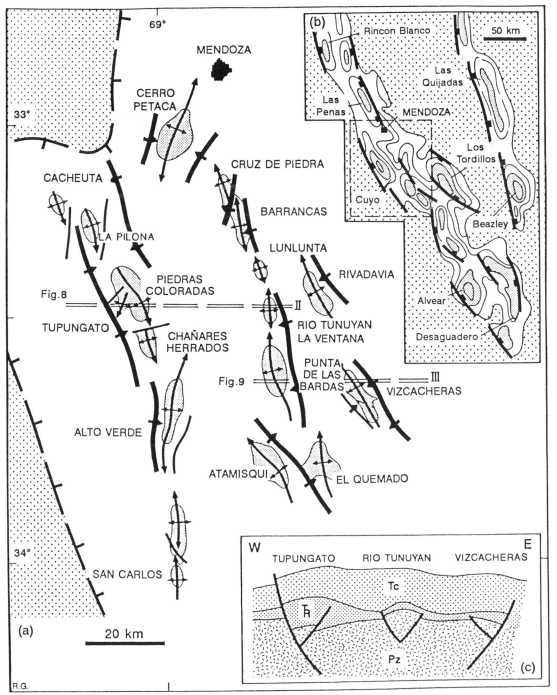

Fig. 7. (**a**) Triassic Cuyo basin, principal tectonic elements and oil-bearing structures (see Fig. 1 for location). Inset (**b**) shows position of the Cuyo with respect to other Triassic depocentres (shaded) in the region. Inset (**c**) is a schematic cross-section illustrating inverted basin architecture.

regional extension nucleated on two major crustal discontinuities: the Chilenia–Precordillera and the Precordilleran–Pampean terrane boundaries (Ramos & Kay 1991). Transtensional conditions (Criado Roqué *et al.* 1981) or, more likely, extensional collapse of the Late

W   E

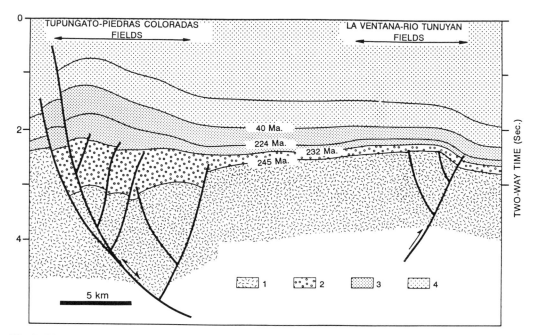

Fig. 8. Line drawing of seimic data across the inverted western half of the Cuyo depocentre (see Fig. 7 for location). (1) Palaeozoic and older 'Basement'; (2) Triassic syn-rift fill; (3) Triassic through Early Cenozoic late to post-rift fill; (4) Cenozoic foreland basin strata.

Palaeozoic orogenic belt (Vicente 1972*b*; Legarreta *et al.* 1992) induced failure and local fault-driven subsidence on the upper plate of the collisional ensemble (Ramos 1992).

Because the old suture zones and the superimposed Mesozoic structures are obliquely orientated with respect to the NS-trending modern Andes, the Triassic successor basins became differentially involved in the process of Cenozoic deformation. The Rincón Blanco and Las Peñas depocentres, more northerly and westerly located, became severely contracted (Figueroa & Ferraris 1989), ramped-up and exposed to denudation. The centrally located Cuyo basin was only marginally uplifted and subject to widespread but moderately intense structural inversion. At the southern end of the basin complex and farther removed from the Cordilleran front, the Alvear and Desaguadero depocentres were much less disturbed and only display the development of low-relief transpressional structures (Vicente 1975).

The large scale architecture that controlled the deposition of the non-marine Triassic wedges near the city of Mendoza, has been described in terms of east- and west-dipping, arcuate, half-graben compartments (Legarreta

et al. 1992). On the northern part of the basin, or Cacheuta depocentre (Fig. 7b), Triassic subsidence was dictated by a zone of east-dipping and convex to the west master faulting. At the latitude of Mendoza master faulting was transferred to a compound half-graben system featuring westerly polarity, known as the Las Peñas depocentre.

Trapping styles

Exploration during the last 50 years has disclosed the presence of some 15 oilfields in the northern Cuyo basin. Pools are hosted by doubly plunging anticlines, hemi-anticlines and structural noses. They are crudely arranged following an en echelon pattern that follows three NNW–SSE striking trends (Fig. 7a). Most oil-bearing traps show asymmetric cross-section. Steeper flanks may be at eastern or western side of the structures, and they are commonly linked to more or less obvious reverse faults that may disappear at depth. Seismic control demonstrates that shallow folds, at the level of the Tertiary or Cretaceous strata, become replaced at depth by a series of differentially displaced and rotated fault blocks.

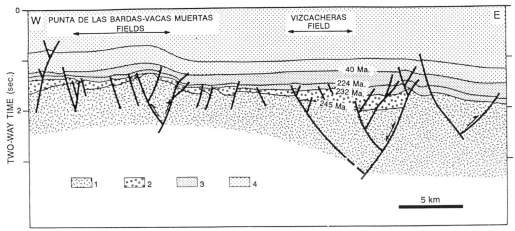

Fig. 9. Line drawing of seismic data across the mildly inverted western half of the Cuyo depocentre (see Fig. 7 for location). Key as in Fig.8.

Typically the roots of the structural highs involve locally expanded Triassic deposits, originally related to master faults with normal separations up to 3500 m (Fig. 7c). The late Cenozoic Andean compression induced half-graben arching and reverse displacements along many Mesozoic normal faults. A number of the reactivated master faults are linked to zones of antithetic faulting, resolved as converging back-thrusts that locally play a substantial trapping role.

Structures showing the most advanced deformation stage, to the west and north within the basin (Fig. 8), closely resemble the 'Sunda' folds described by Eubank & Makki (1981) and Letouzey (1990). Fold cores began to be penetrated by faulting and some pop-up wedges developed, presumably reflecting inversion along second order antithetic and synthetic accommodation faulting. Vizcacheras field, located at a position comparatively removed from the mountain front, was less affected by contractional deformation and only shows an incipient amount of structural inversion (Fig. 9). The structural culmination is controlled by the proximity to mildly inverted antithetic faults, and dips away into a structural low coincident with the hangingwall side of the original half-trough. The present trapping configuration might be considered as a gently inverted rollover feature.

Neuquen basin

Regional setting

The Neuquén basin consists of a 2000–6000 m thick Mesozoic synrift and postrift wedge,

locally overlain by moderately thick to thin Tertiary foreland deposits. The structural pattern is punctuated by a series of NS-orientated basement uplifts (Fig. 10) that break through both the relatively simply structured foreland and the Cordilleran tract, where these larger scale features interfere with thin-skinned detachments. As shown in Fig. 11a, while the sedimentary cover appears detached and shortened by folding, the pre-Triassic 'basement' contracted by reverse slip along a number of higher-angle faults (see also Kozlowski 1991; Manceda & Figueroa 1993). Because the Cenozoic basement shortening was distributed along several faults and spread areally, the Andean telescoping was unable to generate enough local load to flex the foreland severely. This distributed or 'accordion' shortening mode departs widely from the frontal or 'bulldozer' mode recorded in other segments along the Andean mountain front (Fig. 5a). Unlike those segments, development of a Tertiary foreland wedge is surprisingly limited across the central and southern Neuquén basin.

As in other compressional provinces around the world (Hayward & Graham 1989), indirect evidence across the fold belt suggests the influence of older extensional fault systems on the Late Cenozoic deformation. Location of the basement uplifts relates to Mesozoic palaeogeographical peculiarities (Gerth 1931; Legarreta & Kozlowski 1984), induced by Triassic–Jurassic collapse structures. In addition, basement uplifts show arcuate shapes and appear bounded by high to medium dipping thrusts that may verge toward or away from the foreland (Groeber 1929; Lambert 1947).

Recent studies (Kozlowski 1991; Manceda &

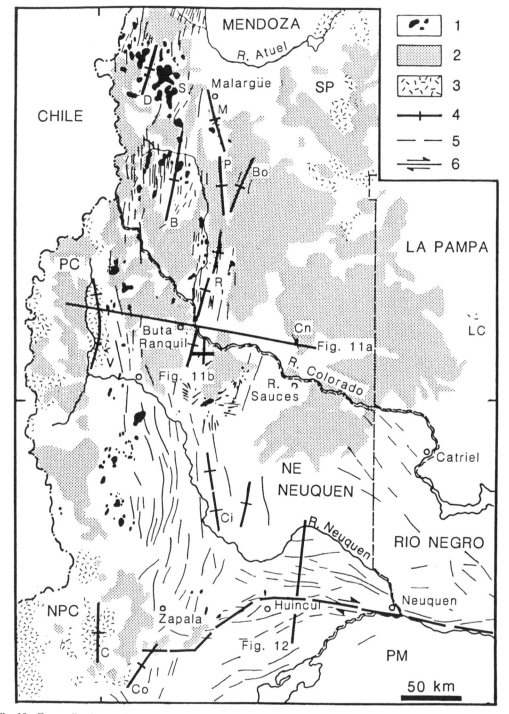

Fig. 10. Generalized tectonic elements map of the Neuquén basin, modified after Uliana & Legarreta (1993) (see Fig. 1 for location). PC, Principal Cordillera; NPC, North-Patagonian Cordillera; NE, Neuquén Embayment; SP, Sierra Pintada; PM, Patagonian Massif; (1) Cenozoic intrusives; (2) Cenozoic volcanics; (3) Pre-Triassic 'basement'; (4) first-order compressional (inverted) uplifts; (5) second-order positive features; (6) shear zones.

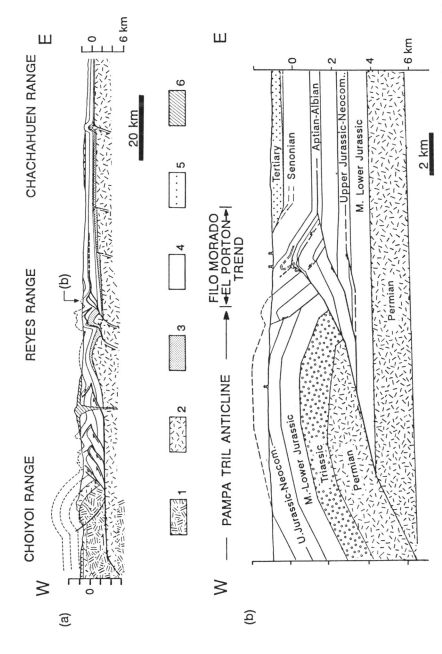

Fig. 11. (a) Structural cross-section central Neuquén basin depicting pre-Mesozoic involvement along inverted Triassic extensional faults (see Figs 1, 10 for location). (1) pre-Permian; (2) Permian volcanics; (3) Triassic–E. Jurassic syn-rift fill; (4) Jurassic–Cretaceous postrift succession; (5) Cenozoic sediments and volcanics; (6) Cenozoic intrusives. **(b)** Detail cross-section shows first-order Pampa Tril anticline and possible linkage between inverted Mesozoic faults and oil-bearing closure in the adjacent El Portón Field. Location on trend, some 10 km south of regional cross-section in **(a)**.

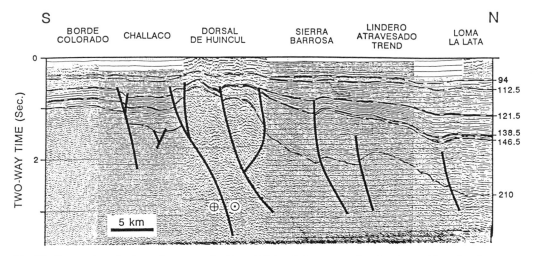

Fig. 12. Composite seismic section across the Dorsal Huincul shear zone, showing the position of early Mesozoic faults, and related asymmetrical folds that control hydrocarbon occurrences after transpressional reactivation (see Fig. 10 for location).

Figueroa 1993) show the northern part of the basin as an inverted tectonic province, where Cenozoic compressional deformation operated preferentially by reactivation of master faults bounding Early Mesozoic half-graben complexes. According to Manceda & Figueroa (1993) the strain system was ultimately rooted into an intra-crustal detachment that operated as a sled-runner domain that connected the reverse faults, lifting-up and locally extruding the half-graben wedges.

The 'Dorsal de Huincul' oil-bearing district, over the south–central part of the Neuquén foreland tract, encompasses structural features that evolved within a pre-Andean (Jurassic–Cretaceous) transpressional setting. A series of extensionally rotated Triassic blocks became inverted (Fig. 12) after being caught in the area influenced by the convex-to-the-north restraining bend of an EW-trending intraplate megashear (Ploszkiewicz *et al.* 1984). Tectonic drive and right lateral slip are understood as far-field effects derived from alignment with the inboard projection of the South Atlantic transforms (Francheteau & Le Pichon 1972; Unternehr *et al.* 1988; Uliana *et al.* 1989).

Trapping styles

Commercial oil and gas production linked to inverted structures at the Neuquén basin is known to occur in at least three distinct geological situations:

(1) Broad anticlinoria inverted by reverse motion along master faults bounding the original

half-graben complexes, like the Palauco–Ranquilcó and the Chihuidos trends. Multiple closures are arranged as en echelon culminations, that presumably reflect syn-inversional block jostling and differential accommodation of the internal compartments segmenting the extensional structures. Individual culminations like the Palauco field (Andrada 1993) show asymmetric sections and eventual development of shortcut faults (Manceda & Figueroa 1993), and an outer carapace transected by extensional faults. In spite of a high relief and favourable trapping geometry, most of these first-order inverted anticlinoria (e.g. Pampa Tril, Reyes, Cara Cura, Bardas Blancas and Malargüe anticlines) are barren of hydrocarbons. They have been forcefully uplifted, perhaps near the point of total inversion (Cooper *et al.* 1989), and thus became prone to erosional breaching. The most distal intraforeland compressional inversions consist of small isolated anticlines, that involve relatively sharp folding of the post-rift succession (e.g. Rincón Amarillo, Cerro Boleadero and Cerro Fortunoso anticlines; Fig. 11a).

(2) Away from the edges of some of the first-order antiform uplifts that have been activated along basement-penetrating reverse faults or thrusts, strain was laterally transferred to the post-rift series. Structural style became dominated by thin-skinned thrust ramps and stacks that locally contain prolific oil pools (Ploszkiewicz 1987; Viñes 1990). For example, Filo Morado and El Portón fields (Fig. 11b) involve complex antiformal duplexes developed

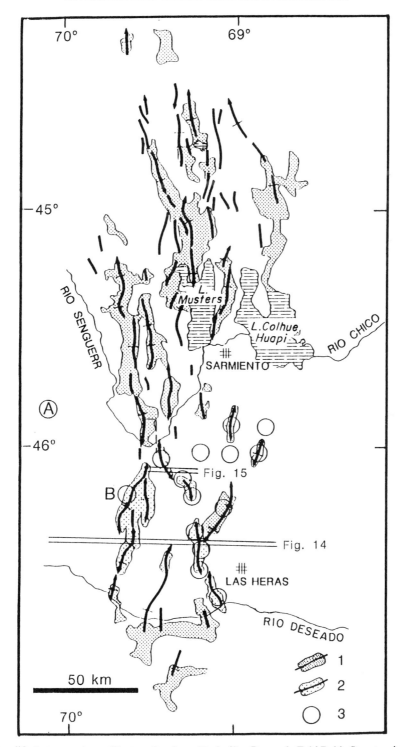

Fig. 13. Simplified structural map Western San Jorge Basin (San Bernardo Fold Belt). Structural culminations form a segmented pattern linked to deep half-graben trends (see Fig. 1 for location). (1) Exposed features; (2) Buried features; (3) Hydrocarbon occurrences.

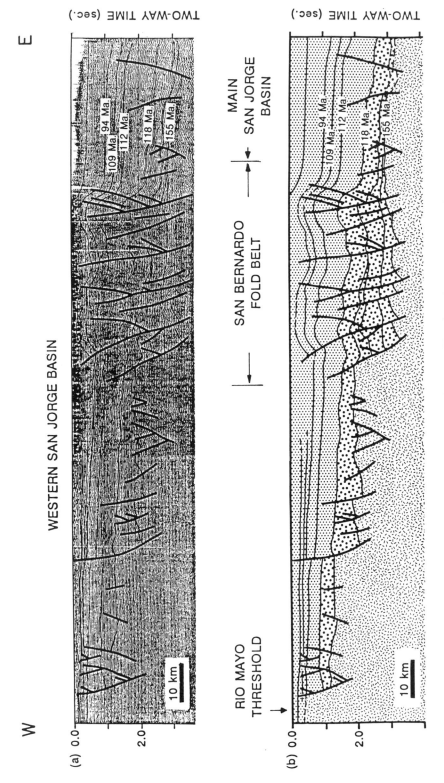

Fig. 14. Seismic profile (**a**) and line drawing (**b**) crossing the southern San Bernardo inverted belt (see Figs 1, 13 for location). Interval between 155 Ma and 118 Ma sequence boundaries represents syn-rift fill.

where the inverted normal faults propagated into a sedimentary cover that includes evaporitic members providing a roof detachment zone and a master top seal.

(3) A number of fields near the right-lateral Huincul shear zone (Figs 10, 12) are linked to rotational hangingwall anticlines (e.g. Sierra Barrosa–Aguada Toledo and Cerro Bandera fields). These transpressionally-induced inversion features (Ploszkiewicz *et al*. 1984) show a classic 'Sunda'-style asymmetric section with the crest located above the syn-rift depocentre and a gentle flank locally broken by subsidiary accommodation faults. Frequent internal stratigraphic discontinuities reveal a protracted history of syndepositional growth, punctuated by erosional events during the Middle Jurassic through Senonian period. The stratal peculiarities involve sub-unconformity reservoir truncations near crestal positions and up-flank onlap terminations that introduce reservoir discontinuities and add stratigraphic trapping components. Trap development clearly predated hydrocarbon migration in the area (Uliana & Legarreta 1993).

Western San Jorge Basin

Regional setting

The San Bernardo Belt is a 600 km long and 100 km wide polyphase deformed trend, that transects the central Patagonian foreland from NNW to SSE, distinctly segregated from the Cordilleran system to the west (Ugarte 1966; Fitzgerald *et al*. 1990; Fig. 13). Most obvious surface structures are located in south-central Chubut and consist of narrow and box shaped folds (Feruglio 1929; Sciutto 1981), locally segmented by tears (Barcat *et al*. 1984) and linked to forward- and backward-verging reverse faults.

The San Bernardo Belt shares, with other Patagonian provinces, the presence of grabens and troughs which developed under tensional-transtensional stress regimes during the Jurassic Period. These Early Mesozoic structures represent the existence of an incipient Basin-and-Range style multicomponent rift system, referred to as the Chon Aike–Lonco Trapial or 'Tobífera' tectonomagmatic event (Gust *et al*. 1985; Uliana *et al*. 1989). Fígari & Courtade (1993, figs 6, 7) have tied Jurassic fault-induced subsidence to crustal attenuation through local shearing of the entire crustal lithosphere, perhaps guided by inhomogeneities related to an older structural grain (Ugarte 1966; Frutos & Tobar 1975).

During the Cretaceous and Early Cenozoic the synextensional wedges were buried under a 2000–4000 m thick succession filling the epicratonic San Jorge Basin (Lesta 1968; Fitzgerald *et al*. 1990). Late in the Cenozoic a number of faults, which bound troughs close to a northerly orientation, were reactivated under compression (Homovc *et al*. 1993), coeval with EW-orientated Cordilleran shortening that took place 200 km to the west. As a rule the inverted faults retain a normal separation at depth and only show reverse offsets at intermediate levels (Fig. 14), implying a modest level of total shortening (Peroni *et al*. in press).

Trapping style

Structural style can best be reconstructed by combining surface observations and subsurface data derived from seismic sections tied to well control. The area prospective for hydrocarbon accumulation shows broad synclines, long and arcuate anticlines and a diverse array of associated faults. Folding is discontinuous and irregularly spaced (Fig. 13), locally following relay or en echelon patterns (Feruglio 1929; Sciutto 1981). Deformation at the level of the younger Cretaceous strata featuring reservoir-quality members is dominated by narrow and high-relief buckles. Major anticlinal trends are box shaped and may be transected by tear faults that offset fold axes at several spots. East and west fault plane dipping orientations alternate along and across the folded tract, and some structures show along strike changes in facing direction (Homovc *et al*. 1993; Peroni *et al*. in press). Steep to slightly overturned fold limbs evolve locally into longitudinal thrusts, and fault plane dipping flips associate to relay transfers between thrusts cutting the flanks. Thus, seismic sections across the fold culminations may show the core zones as popped-up bounded by oppositely dipping faults thrust slices (Peroni *et al*. in press; Fig. 15).

Hydrocarbon systems and petroleum occurrence

Argentine established ultimate recoverable reserves are in the order of 6.3 BBO and 34 TCFG (Yrigoyen 1993). These volumes were found in over 250 fields scattered in six oil and gas provinces, as shown in Fig. 16. The majority of the occurrences relate to Mesozoic source rocks and reservoirs and to relatively young structural traps formed after Cenozoic tectogenesis (Table 1).

E W

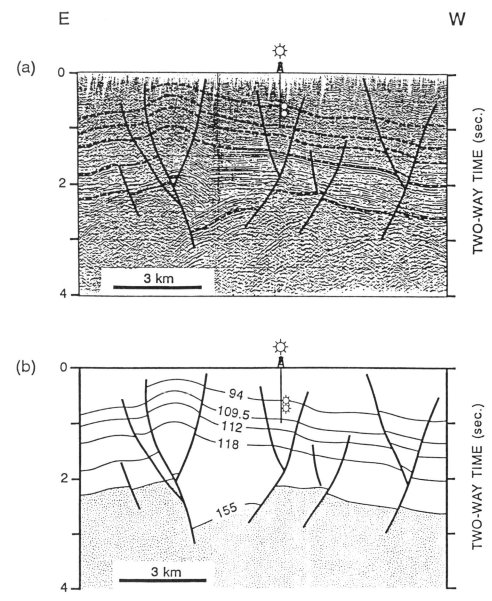

Fig. 15. Seismic profile (**a**) and line drawing (**b**) across the Barranca Yancowsky anticline showing a typical Sunda-style asymmetric configuration (see Fig. 13 for location). Interval between 155 and 118 Ma sequence boundaries represents syn-rift fill.

Hydrocarbons, other than those produced in the Palaeozoic Tarija basin near Bolivia (Turic *et al.* 1987), were derived from Mesozoic source rocks. In the Neuquén and Austral (Magallanes) basins kerogenous members developed during episodes of marine flooding, when Pacific waters drowned wide segments of the South American slab. In other provinces like the Cuyo and San Jorge basins, however, organic-rich accumulations were formed after fresh-water or saline-

alkaline lake systems (Rosso *et al.* 1987; Van Nieuwenhuise & Ormiston 1989).

Most source rocks were deposited after the main stage of fault-related mechanical subsidence. More specifically during the rift-to-sag transition in the Austral, San Jorge and Cuyo basins, and well within the thermal subsidence stage in the Neuquén and Orán basins. Tectonic segmentation along the Pacific margin induced variable subsidence patterns and presumably

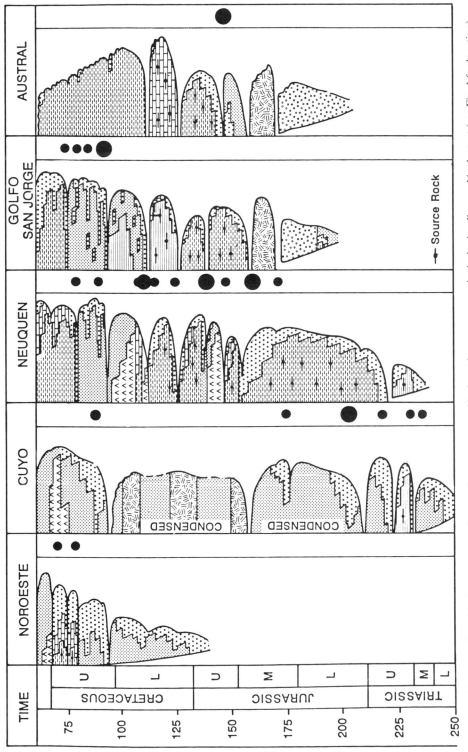

Fig. 16. Chronostratigraphic chart compares the petroleum systems and hydrocarbon occurrences in the basins discussed in the text (see Fig. 1 for location).

Table 1. *Major characteristics of the Argentine basins (as discussed in the text)*

Features	Noroeste Basin	Cuyo Basin	Neuquen Basin	W. San Jorge Basin
Tectonic environment	on the foreland adjacent to the mountain front	structural depression between the Cordillera and the thick-skinned Pampean ranges	outer edge of the fold and thrust belt on locally broken foreland	isolated intra-foreland inversion belt
structural pattern	low relief block edge closures after block flexing and minor drag folding	en echelon doubly plunging anticlines and noses after opposing and partially overlapping half-graben complexes	truncated hangingwall anticlines after synsedimentary inversion compressional rotated blocks and associated antiformal duplexes	opposingly facing short wavelength and medium amplitude buckles cut by reverse fault and segmented by tears
hydrocarbon system	undercharged laterally drained low impedance	supercharged lat. & vert. drained moderate impedance	normally charged lat. & vert. rained variable impedance	normally charged vertically drained high impedance
inversion	Miocene–Pliocene	Miocene–Pleistocene	Jurassic–Cretaceous Senonian–Palaeogene	Late Cenozoic
migration	Miocene	Miocene–Pliocene	Miocene–Pliocene Late Cenozoic	Palaeogene
hydrocarbon occurrences	sparse 5–15 mmbo	multiple (dominant style) 20–250 mmbo	areally restricted 20–350 mmbo multiple, irreg. spaced 10–50 mmbo	multiple, irregular spaced 5–20 mmbo

heat-flow histories, and thus each basin features a distinct maturation–migration record.

As in several prolific Subandean hydrocarbon provinces, critical P–T ranges were achieved during late Cenozoic rapid burial related to downbending and asymmetric subsidence of the South American foreland. Chiefly in the Austral, Cuyo and Orán basins, these conditions resulted from regional shortening and local supracrustal loading, coupled with sediment supply derived from the newly rising Andean ranges. In contrast, basins like the Neuquén and San Jorge record a protracted and uninterrupted subsidence process that promoted early entrance and long-term residence in the hydrocarbon window.

Overview and conclusion

Argentine commercial oilfields controlled by inversion features occur at different positions that range from the Andean fold and thrust belt (Neuquén basin), to the outer foreland edge near the mountain front (Noroeste, Neuquén basin), to isolated inverted trends that transect non-contracted portions of the foreland (San Jorge, Neuquén), and to little deformed foreland slabs that lie between the mountain front and foreland segments broken by thick-skinned compressional blocks (Cuyo). Inverted structures across the foreland are at least partially

coeval with the main period of Andean tectogenesis. In consequence most authors found it reasonable to assume that the horizontal stress field generated by plate interaction along the west edge of South America, propagated into the plate interior using the lithosphere as a stress guide. In this context inversion might be linked to a regional mid crustal detachment (e.g. 'orogenic float', Oldow *et al.* 1989), or perhaps to an even deeper effect involving crustal thickening along pre-existing faults cutting the entire crust (e.g. Ziegler 1989). Late Cenozoic traps linked to compressional reactivation of earlier normal faults comprise an array of structures varying from low-relief embrionic features (Noroeste, Cuyo and Neuquén basins), to moderately inverted forced folds and buckles (Cuyo and San Jorge basins), to large basement-involved anticlinoria at the scale of the western US Laramide uplifts (Neuquén). Regional distribution over the Argentine backarc and foreland provinces is irregular and discontinuous, reflecting a heterogeneous basement architecture, with an orientation oblique to the prevalent Andean trend.

From the previous discussion it is apparent that oil bearing inversion structures in Argentina relate primarily to Mesozoic rift-sag pairs, locally overlain by Cenozoic foreland wedges. In most districts master inversion was achieved late during the Cenozoic. Beyond this shared basic architecture, discontinuities in the location and timing of subsidence and superimposed palaeogeographical variability resulted in unequal petroleum systems. Thus, hydrocarbon potential of geometrically similar structures located in comparable tectonic settings record large disparity. Most of the effective traps consist of Sunda-style hangingwall anticlines located near the outer fringe of the foreland province, where regional basement downwarp and burial beneath foredeep deposits have provided adequate levels of organic maturity and prevented erosional breaching.

The authors wish to thank ASTRA C.A.P.S.A. Exploration and Production for encouragement and permission to publish this paper. Our appreciation is extended to Dr. James G. Buchanan and Dr. Kevin T. Biddle for critical reading, to Raul Genovesi for drafting preparation and to Edith Fernández Rocha for text processing.

References

ALLMENDINGER, R. W., RAMOS, V. A., JORDAN, T. E., PALMA, M. & ISACKS, B. L. 1983. Paleogeography and Andean structural geometry, Northwest Argentina. *Tectonics*, **2**, 1–16.

ANDRADA, L. 1993. Yacimiento Pampa Palauco. *In:* RAMOS, V. A. (ed.) *Geología y Recursos Naturales de Mendoza.* Doceavo Congreso Geológico Argentino, Relatorio, Buenos Aires, 459–462.

BARCAT, C., CORTINÁS, J. S., NEVISTIC, V. A., STACH, N. H. & ZUCCHI, H. E. 1984. Geología de la región comprendida entre los lagos Musters–Colhue Huapi y la Sierra Cuadrada, Departamentos Sarmiento y Paso de Indios, Provincia del Chubut. *Noveno Congreso Geológico Argentino, Actas,* **2**, 263–282.

BIANUCCI, H. A. & HOMOVC, J. F. 1982. Tectogénesis de un sector de la cuenca del Subgrupo Pirgua. *Quinto Congreso Latinoamericano de Geología, Actas,* **I**, Buenos Aires, 539–546.

CAHILL, T., ISACKS, B.L., WHITMAN, D., CHATELAIN, J-L., PEREZ, A. & CHIU, J. M. 1992. Seismicity and Tectonics in Jujuy Province, Northwestern Argentina. *Tectonics*, **11**, 944–959.

CAMINOS, R. 1979. Sierras Pampeanas noroccidentales, Salta, Tucumán, Catamarca, La Rioja y San Juan. *In:* TURNER, J. C. M. (ed.) *Geología Regional Argentina.* Academia Nacional de Ciencias, Córdoba, **I**, 225–292.

CHIARENZA, D. G. & PONZONI, E. 1989. Contribución al conocimiento geológico de la cuenca cretácica en el ámbito oriental de la subcuenca de Olmedo, Pcia de Salta, República Argentina. *Primer Congreso Nacional de Exploración de Hidrocarburos, Actas, Mar del Plata,* **I**, 209–228.

COLE, D. I. 1992. Evolution and development of the Karoo basin. *In:* DE WIT, M. J. & RANSOME, G. D. (eds) *Inversion tectonics of the Cape Fold Belt, Karoo and Cretaceous basins of Southern Africa. Karoo and Related basins*, Balkema, 87–100.

CONCEICAO, J. C., ZALAN, P. V. & WOLFF, S. 1988. Mecanismo, Evolucao e cronologia do rift Sul-Atlantico. *Boletim Geociencias Petrobrás,* **2**, 255–265.

COOPER, M. A., WILLIAMS, G. D., DE GRACIANSKY, P. C., MURPHY, R. W., NEEHAM, T., DE PAOR, D., STONELEY, R., TODD, S. P., TURNER, J. P. & ZIEGLER, P. A. 1989. *Inversion Tectonics – A discussion. In:* COOPER, M. A. & WILLIAMS, G. D. (eds) *Inversion Tectonics.* Geological Society, London, Special Publication, **44**, 335–347.

CRIADO ROQUE, P., MOMBRU, C. A. & RAMOS, V. A. 1981. Estructura e Interpretación Tectónica. *Octavo Congreso Geológico Argentino, Relatorio*, Buenos Aires, 155–192.

DE WIT, M. J. & RANSOME, G. D. 1992. Regional inversion tectonics along the southern margin of Gondwana. *In:* DE WIT, M. J. & RANSOME, G. D. (eds) *Inversion tectonics of the Cape Fold Belt, Karoo and Cretaceous basins of Southern Africa. Regional overviews.* Balkema, 15–22.

DI PERSIA, O. E., CARLE, R. J. & BELOTTI, H. 1991. Geología petrolera en la subcuenca Lomas de Olmedo. *Boletín de Informaciones Petroleras,* **25** 14–29.

EUBANK, R. T. & MAKKI, A. C. 1981. Structural geology of the central Sumatra back-arc basin. *Tenth Indonesian Petroleum Convention, Proceedings.* Jakarta.

FERUGLIO, E. 1929. Apuntes sobre la constitución geológica de la región del Golfo San Jorge. *Boletín de Informaciones Petroleras,* **63**, 925–1025.

FIGARI, E. G. & COURTADE, S. F. 1993. Evolución tectosedimentaria de la cuenca de Cañadón Asfalto, Chubut, Argentina. *Doceavo Congreso Geológico Argentino, Actas,* Buenos Aires, **I**, 66–77.

FIGUEROA, D. E. & FERRARIS, O. R. 1989. Estructura del margen oriental de la Precordillera Mendocino–Sanjuanina. *Primer Congreso Nacional de Exploración de Hidrocarburos, Actas,* Buenos Aires **I**, 515–529.

FITZGERALD, M. G., MITCHUM, R. M., ULIANA, M. A. & BIDDLE, K. T. 1990. Evolution of the San Jorge Basin Argentina. *AAPG Bulletin,* **72**, 49–72.

FORSYTHE, R. D., DAVIDSON, J., MPODOZIS, C. & JESINKEY, C. 1993. Lower Paleozoic relative motion of the Arequipa block and Gondwana; Paleomagnetic evidence from Sierra de Almeida of Northern Chile. *Tectonics,* **12**, 219–236.

FRANCHETEAU, J. & LE PICHON, X. 1972. Marginal fracture zones as structural framework of continental margins in South Atlantic Ocean. *AAPG Bulletin,* **56**, 991–1007.

FRUTOS, J. & TOBAR, A. 1975. Evolution of the southwestern continental margin of South America. *Third Gondwana Symposium, Gondwana Basins and Continental Margins,* **39**, 655–578.

GERTH, E. 1931. La estructura geológica de la Cordillera Argentina entre el río Grande y el río Diamante en el sur de la Provincia de Mendoza. *Academia Nacional de Ciencias, Actas,* **X**, 123–172.

GOMEZ OMIL, R. J., BOLL, A. & HERNANDEZ, R. M. 1989. Cuenca Cretácico–Terciária del Noroeste argentino (Grupo Salta). In: SPALLETTI, L. & CHEBLI, G. A. (eds) Cuencas Sedimentarias Argentinas, Tucumán, 43–64.

GORDILLO, E. & LENCINAS, A. N. 1979. Sierras Pampeanas de Córdoba y San Luis. In: TURNER, J. C. M. (ed) Geología Regional Argentina. Academia Nacional de Ciencias, Córdoba I, 577–650.

GRIER, M. E., SALFITY, J. A. & ALLMENDINGER, R. W. 1991. Andean reactivation of the Cretaceous Salta rift, Northwestern Argentina. Journal of South American Earth Sciences, 4, 351–372.

GROEBER, P. 1929. Líneas fundamentales de la geología del Neuquén, Sur de Mendoza y regiones adyacentes. Dirección Minería, Geología e Hidrología, Buenos Aires, Publicación 58, 1–109.

GUST, D. A., BIDDLE, K. T., PHELPS, D. W. & ULIANA, M. A. 1985. Associated Middle to Late Jurassic volcanism and extension in Southern South America. Tectonophysics, 116, 223–253.

HANDSCHUMACHER, D. W. 1976. Post-Eocene Plate Tectonics of the Eastern Pacific. American Geophysical Union Monograph, 19, 177–202.

HAYWARD, A. B. & GRAHAM, R. 1989. Some geometrical characteristics of inversion. In: COOPER, M. A. & WILLIAMS, G. D. (eds) Inversion Tectonics. Geological Society, London, Special Publication, 44, 17–40.

HERVE, F. 1988. Late Paleozoic subduction and accretion in Southern Chile. Episodes, 11, 183–188.

—— & MPODOZIS, C. 1990. Terrenos tectonoestratigráficos en la evolución geológica de los Andes chilenos: una revisión. Décimo Congreso Geológico Argentino, Actas, San Juan 2, 319–323.

HOMOVC, J. F., CONFORTO, G. A. & LAFOURCADE, P. A. 1993. Faja plegada en el subsuelo de la cuenca del Golfo San Jorge, Ejemplo de Inversión Tectónica. Doceavo Congreso Geológico Argentino, Actas, Buenos Aires, III, 233–240.

ISACKS, B. L. 1988. Uplift of the central Andean plateau and bending of the Bolivian Orocline. Journal of Geophysical Research, 93, 3211–3231.

JORDAN, T. E. & ALLMENDINGER, R. W. 1986. The Sierras Pampeanas of Argentina: A modern analogue of Rocky Mountain foreland deformation. American Journal of Science, 286, 737–764.

——, ISACKS, B. L., RAMOS, V. A. & ALLMENDINGER, R. W. 1983. Mountain building in the Central Andes. Episodes, 3, 20–26.

KOZLOWSKI, E. 1991. Structural Geology of the NW Neuquina Basin, Argentina. Cuarto Simposio Bolivariano, Bogotá, I, 1–10.

LAMBERT, L. R. 1947. Contribución al conocimiento de la sierra de Chacay-Co. Asociación Geológica Argentina, Revista, I, 231-256.

LAWVER, L. A. & SCOTESE, C. R. 1987. A revised reconstruction of Gondwanaland. In: McKENZIE, G. D. (ed.) Gondwana Six: Structure, Tectonics and Geophysics. A.G.U. Geophysical Monographs, 40, 17–23.

LEGARRETA, L. & KOZLOWSKI, E. 1984. Secciones condensadas del Jurásico-Cretácico de los Andes del Sur de Mendoza: Estratigrafía y significado tectosedimentario. Noveno Congreso Geológico Argentino, Actas, Buenos Aires, I, 286–297.

——, KOKOGIAN, D. A. & DELLAPE, D. A. 1992. Estructuración terciaria de la Cuenca Cuyana: Cuánto de inversión tectónica? Asociación Geológica Argentina, Revista, 47, 83–86.

LESTA, P. J. 1968. Estratigrafía de la cuenca del Golfo San Jorge. Terceras Jornadas Geológicas Argentinas, Actas, III, 251–289.

LETOUZEY, J. 1990. Fault reactivation, inversion and fold-thrust belt. In: LETOUZEY, J. (ed.) Petroleum and tectonics in mobile belts. Editions Technip, Paris, 101–128.

LOWELL, J. D. 1985. Structural styles in petroleum exploration. Oil and Gas Consultants International Inc., Tulsa.

LLAMBIAS, E. J. & SATO, A. M. 1990. El batolito de Colanguil (29-31°S); Cordillera Frontal Argentina: estructura y marco tectónico. Revista Geológica de Chile, 17, 89–108.

MANCEDA, R. & FIGUEROA, D. 1993. La inversión del rift mesozoico en la faja fallada y plegada de Malargue. Provincia de Mendoza. XII Congreso Geológico Argentino, Actas, III, 219–232.

MANCINI, C. D. & SERNA, M. 1989. Evaluación petrolera de la cuenca de Ñirihuao. Primer congreso Nacional de Exploración de Hidrocarburos, Buenos Aires, 2, 739–762.

MINGRAMM, A., RUSSO, A., POZZO, A. & CAZAU, L. 1979. Sierras Subandinas. In: TURNER, J. C. M. (ed.) Geología Regional Argentina. Academian Nacional de Ciencias, Córdoba, I, 95–138.

MOSCOSO, R. & MPODOZIS, C. 1988. Estilos estructurales en el Norte Chico de Chile (28–31°S), Regiones de Atacama y Coquimbo. Revista Geológica de Chile, 15, 151–166.

MPODOZIS, C. 1984. Dinámica de los márgenes continentales activos. Seminario Actualización de la Geología de Chile. Servicio Nacional de Geología y Minería, Miscelánea No 4, Santiago, A1–A22.

—— & KAY, S. 1990. Provincias magmáticas ácidas y evolución tectónica de Gondwana: Andes Chilenos (28-31°S). Revista Geológica de Chile, Santiago, 7, 153–180.

OLDOW, J. S., BALLY, A. W., AVE LALLEMANT, H. G. & LEEMAN, W. P. 1989. Phanerozoic evolution of the North American Cordillera; United States and Canada. In: BALLY, A. W. & PALMER, A. R. (eds) The Geology of North America: An Overview. Geological Society of America, 139–232.

ORTIZ, A. & ZAMBRANO, J. J. 1981. La Provincia Geológica Precordillera Oriental. Octavo Congreso Geológico Argentino, Actas, Buenos Aires, III, 59–74.

PERONI, G. O., HEGEDUS, A. G., LAFFITE, G., CERDAN, J., LEGARRETA, L. & ULIANA, M. A. in press. Hydrocarbon accumulations in an inverted segment of the Andean foreland.

PILGER, R. H. 1983. Kinematics of the South American subduction zone from global plate reconstructions. *In*: CABRÉ, R. (ed.) *Geodynamics of the Eastern Pacific Region, Caribbean and Scotia Arcs*. A.G.U., Washington DC, Geodynamic series, **9**, 113–125.

PLOSZKIEWICZ, J. V. 1987. Las zonas triangulares de la faja fallada y plegada de la Cuenca Neuquina, Argentina. *Décimo Congreso Geológico Argentino, Actas*, **I**, Tucumán.

—, ORCHUELA, I. A., VAILLARD, J. C. & VIÑES, R. F. 1984. Compresión y desplazamiento lateral en al zona de falla Huincul, estructuras asociadas. Provincia del Neuquén. *Noveno Congreso Geológico Argentino, Actas*, Buenos Aires, **I**, 163–169.

RAMOS, V. A. 1988. Tectonics of the Late Proterozoic–Early Paleozoic: a collisional history of Southern South America. *Episodes*, **11**, 168–174.

—— 1989. The birth of Southern South America. *American Scientist*, **77**, 444–450.

—— 1992. Control geotectónico de las cuencas triásicas de Cuyo. *Boletín de Informacionas Petroleras*, **Septiembre 1992**, 2–9.

—, JORDAN, T. E., ALLMENDINGER, R. W., MPODOZIS, C., KAY, S. M., CORTES, J. M. & PALMA, M. A. 1986. Paleozoic terranes of the central Argentine-Chilean Andes. *Tectonics*, **5**, 855–880.

—— & KAY, S. M. 1991. Triassic rifting and associated basalts in the Cuyo basin, Central Argentina. *In*: HARMON, R. S. & RAPELA, C. W. (eds) *Andean magmatism and its tectonic setting*. Geological Society of America, Special Paper, **265**, 79–91.

RAPALINI, A. E. 1990. Variaciones paleolatitudinales de America del sur en el Carbonífero-Pérmico. *Decimoprimer Congreso Geológico Argentino, Actas*, San Juan, **2**, 259–262.

ROLLERI, E. O. & CRIADO ROQUE, P. 1969. La cuenca triásica del norte de Mendoza. *Terceras Jornadas Geológicas Argentinas, Actas*, Buenos Aires, **I**, 1–76.

—— & GARRASINO, C. A. F. 1979. Comarca septentrional de Mendoza. *In*: TURNER, J. C. M. (ed.) *Geología Regional Argentina*. Academia Nacional de Ciencias, Córdoba, **I**, 771–810.

ROSSO, M., LABAYEN, I., LAFFITTE, G. & ARGUIJO, M. 1987. La generación de hidrocarburos en la cuenca Cuyana. *Décimo Congreso Geológico Argentino, Actas*, **II**, 267–270.

ROYDEN, L. H. 1993. The tectonic expression of slab pull at continental convergent boundaries. *Tectonics*, **12**, 303–325.

SCIUTTO, J. C. 1981. Geología del Codo del Río Senguerr, Chubut, Argentina. *Octavo Congreso Geológico Argentino, Actas*, Buenos Aires, **III**, 203–219.

SIMPSON, E. S. W. 1977. Evolution of the South Atlantic. *Geological Society of Southafrica, Transactions*, **80**, 1–15.

STOREY, B. C. & ALABASTER, T. 1991. Tectonomagmatic control on Gondwana break-up models: Evidence from the proto-Atlantic margin of Antarctica. *Tectonics*, **10**, 1274–1288.

TURIC, M., ARAMAYO FLORES, F., GOMEZ OMIL, R., POMBO, R., PERONI, G., SCIUTTO, J., ROBLES, D. & CACERES, A. 1987. Geología de las cuencas petroleras de la Argentina. *In*: FELDER, B. A. (ed.) *Evaluación de Formaciones en La Argentina*. YPF-SCHLUMBERGER, 1–44.

UGARTE, F. R. 1966. La cuenca compuesta carbonífera-jurásica de la Patagonia meridional. *Universidad de la Patagonia San Juán Bosco, Ciencias Geológicas*, **2**, 37–68.

ULIANA, M. A. & BIDDLE, K. T. 1988. Mesozoic–Cenozoic Paleogeographic and Geodynamic Evolution of Southern South America. *Revista Brasileira de Geociencias*, **18**, 172–190.

—, —— & CERDAN, J. J. 1989. Mesozoic extension and the formation of Argentine sedimentary basins. *In*: TANKARD, A. J. & BALKWILL, H. R. (eds) *Extensional tectonics and stratigraphy of the North Atlantic margins*. AAPG Memoir, **46**, 599–614.

—— & LEGARRETA, L. 1993. Hydrocarbon habitat in a Triassic–Cretaceous Sub-Andean setting: Neuquén basin, Argentina. *Journal of Petroleum Geology*, **16**, 393–420.

UNTERNEHR, P., CURIE, D., OLIVET, J. L. & BEUZART, P. 1988. South Atlantic fits and intraplate boundaries in Africa and South America. *Tectonophysics*, **155**, 169–179.

VAN NIEUWENHUISE, D. S. & ORMISTON, A. R. 1989. A model for the origin of source-rich lacustrine facies. San Jorge basin, Argentina. *Primer Congreso Nacional de Exploración de Hidrocarburos*, Buenos Aires, **2**, 853–883.

VICENTE, J. C. 1972a. Reflexiones sobre la porción meridional del Sistema Peripacífico Oriental. *Conferencia sobre problemas de la Tierra sólida*. Proyecto Internacional del Manto Superior, **37**, 158–184.

—— 1972b. Essai d'organisation paléogéographique et structurale du Paleozoique des Andes Méridionales. *Geologische Rundschau*, **64**, 343–394.

VICENTE, O. M. 1975. Caracteres estructurales del área sur de General Alvear. *Segundo Congreso Iberoamericano de Geología Económica, Actas*, Buenos Aires, **I**, 197–214.

VIÑES, R. F. 1990. Productive duplex imbrication at the Neuquina Basin Thrust Belt Front, Argentina. *In*: LETOUZEY, J. (ed.) *Petroleum and Tectonics in Mobile Belts*. Edition Technip, 68–80.

YRIGOYEN, M. R. 1993. The history of hydrocarbons exploration and production in Argentina. *Journal of Petroleum Geology*, **16**, 371–382.

—, ORTIZ, A. & MANONI, R. 1989. Cuencas sedimentarias de San Luís. *In*: CHEBLI, G. & SPALLETTI, L. (eds) *Cuencas Sedimentarias Argentinas*, Universidad Nacional de Tucumán, 203–219.

ZAPATA, T. & ALLMENDINGER, R. 1993. Central and Eastern Precordillera in the Jachal Area: The interaction of two systems of opposing vergence. *Doceavo Congerso Geológico Argentino, Actas*, Buenos Aires **III**, 149–159.

ZIEGLER, P. A. 1989. A geodynamic model for Alpine intra-plate compressional deformation in Western and Central Europe. *In*: COOPER, M. A. & WILLIAM, G. D. (eds) *Inversion Tectonics*. Geological Society, London, Special Publication, **44**, 63–86.

Fold belt in the San Jorge Basin, Argentina: an example of tectonic inversion

J. F. HOMOVC, G. A. CONFORTO, P. A. LAFOURCADE & L. A. CHELOTTI

YPF S.A. Exploration, Comodoro Rivadavia, Chubut, Argentina

Abstract: The San Jorge basin, located in southern Argentina, extends over the central part of the Patagonian region. The sedimentary fill of the basin is related to different rift and sag tectonic phases, from Triassic to Cretaceous. During Late Cretaceous–Early Tertiary, a marine transgression from the Atlantic developed; Tertiary sediments completed the basin fill. Late Tertiary compression produced a narrow structural belt that crops out in the San Bernardo Range and is present in the subsurface of the Santa Cruz and Chubut Provinces. The fold belt is 150 km long and 50 km wide. In its northern zone, it is formed by a series of anticlinal structures related to high angle reverse faults; in the southern part, these structures have similar characteristics, being associated with reverse faults which are the result of the reactivation of pre-existing normal faults. This report describes the structural features of a wide subsurface area of the fold belt, with high hydrocarbon potential. The development of the Neocomian sequence is used to establish the relationship between tectonism and sedimentation.

The San Jorge basin is predominantly an extensional basin, located in the inner part of the South American plate, between the Nordpatagonic Massif and the Deseado Massif (Fig. 1).

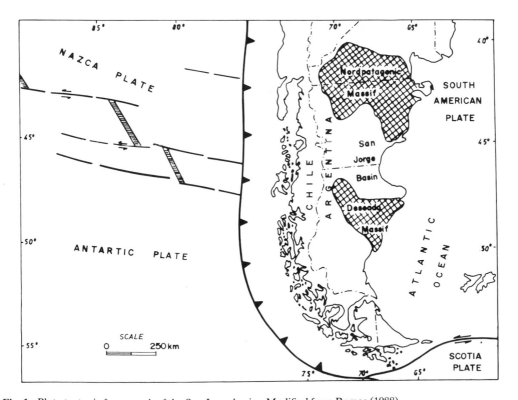

Fig. 1. Plate tectonic framework of the San Jorge basin. Modified from Ramos (1988).

From BUCHANAN, J. G. & BUCHANAN, P. G. (eds), 1995, *Basin Inversion*, Geological Society Special Publication No. 88, 235–248.

235

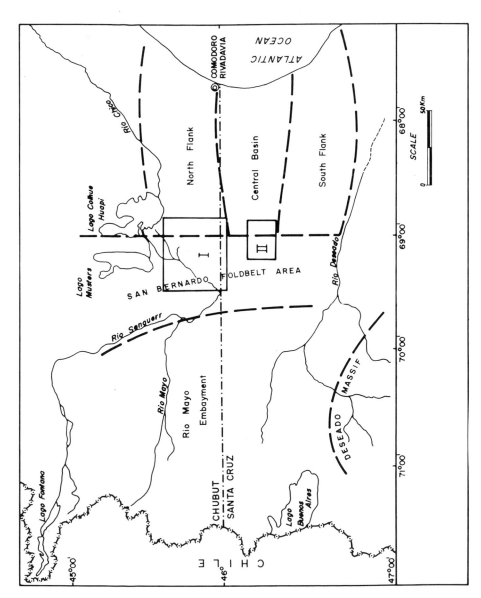

Fig. 2. Geographical divisions of the San Jorge basin related to the structural styles. (Detail of areas I and II in Figs 5 and 10.)

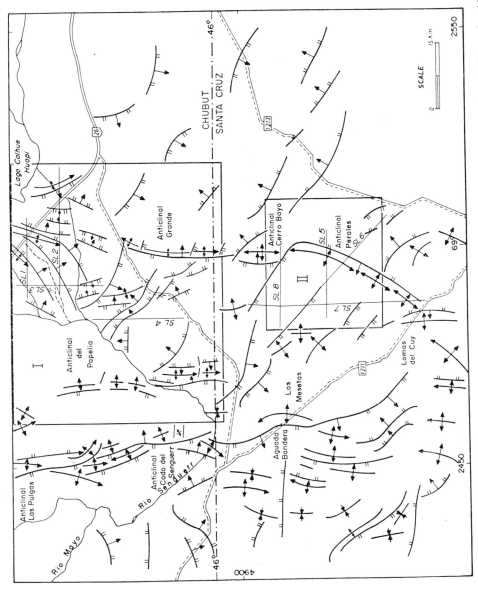

Fig. 3. Distribution of main structures of the studied area. Surface and subsurface data. Seismic data base map. SL: Seismic line (Detail of areas I and II in Figs 5 and 10.)

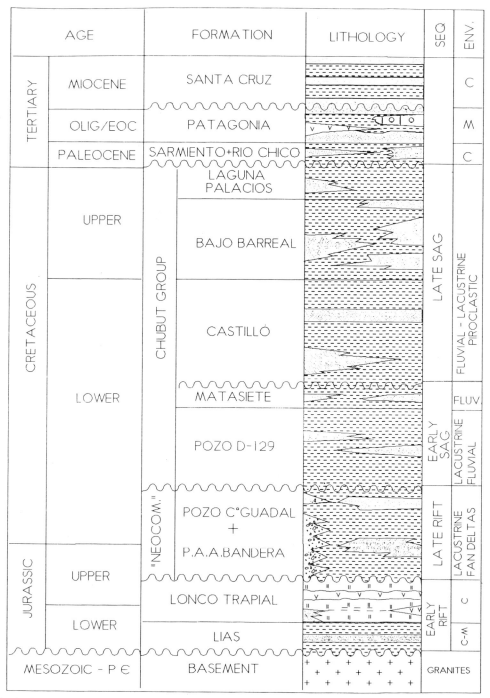

Fig. 4. Generalized stratigraphic column of the San Jorge basin. Abbreviations: seq, sequence; env, environment; c, continental; m, marine; fluv, fluvial.

Several types of extensional structures can be found in the eastern and western segments of the basin; compressional structures extended over the central area along the narrow north trending San Bernardo structural belt (Fig. 2).

The San Bernardo Range southern and

eastern continuations in the Santa Cruz and Chubut Provinces consist of structural elements such as the Perales, Grande, Cerro Guadal and Funes anticlines, among others, situated in the San Jorge basin fold belt (Fig. 3). Because of their high hydrocarbon storing potential, these structures have attracted interest in the oil industry and have been the subject of several seismic surveys and wildcat drilling.

Variations in the faulting style and characteristic geometric assemblages mark the different tectonic stages of the San Jorge basin. The Late Jurassic extensional event formed a series of narrow halfgraben filled with lacustrine sediments. This Neocomian tectonosedimentary episode with a NNW–SSE orientation was followed by a regional sag, during the Late Cretaceous, with widely distributed deposition of fluvio lacustrine sediments of the Chubut Group.

During the Tertiary, the basin received both marine ingressions and a continental sediment supply. A compressive event, mainly in the Late Tertiary, generated the San Bernardo fold belt with a N–S orientation trend.

This paper describes the structural characteristics along the southern subsurface extension of the San Bernardo Range, documenting the relationship between tectonics and sedimentation in a wide area of the San Jorge fold belt. Identification of the early extensional stage sequences is the main argument used to establish the relationship between tectonic inversion and original normal faults (Cooper & Williams 1989).

Database and methodology

Seismic reflection data acquired by Yacimientos Petrolíferos Fiscales during the 1980's were used for this report. The available coverage was mostly good quality 6 fold CDP. Electric logs of exploration wells were also available for the project. This allowed an accurate correlation of the seismic sequences, particularly those in the Upper Cretaceous. Eight seismic sections (Fig. 3) are presented to show the main structural characteristics.

The major sequence boundaries were defined by the correlation of log characteristics. Reflection terminations were also used to define the most important sequence boundaries and were correlated throughout the area of seismic control.

Three units were used in order to obtain a better definition of the tectonosedimentary history. These units are: Bajo Barreal, Castillo and Pozo D-129 Formations (Fig. 4). The term

'Neocomian' was used to distinguish sequences originated during the Late Jurassic–Early Cretaceous extensional event. The term Jurassic was employed to name Aalenian–Bathonian volcanic rocks of the Lonco Trapial Group. The boundary between 'Jurassic' and 'Neocomian' was seismically defined on the criteria of erosional truncation. Only one well drilled the Jurassic interval. Deep reflections shown in some sections are defined as basement although these rocks have not been drilled in this part of the basin.

Results

Structural features of the area are described below.

Major tectonic features

Tectonics had a great influence in the development of the sedimentary column and in the architecture of the principal deformational segments. These major features comprise extensional faults, bounding Jurassic blocks with three principal orientations: NE–SW, N–S and NW–SE (Figs 5 and 10). These blocks formed depocentres in the Late Jurassic–Early Cretaceous, during the late rift stage (Fitzgerald *et al.* 1990); locally they host more than 3000 m of sediments, mostly of lacustrine origin and Neocomian in age.

From tectonic analysis, it can be seen that in places where basement faults have had a significant throw, they have a characteristic growth halfgraben profile (Figs 6, 7, 8 and 9). Where those depocentres have a suitable orientation with respect to the Tertiary E–W compressive stress, they are inverted by the reactivation of the fault in a thrust sense.

In the study area some conditions must occur in order to initiate tectonic inversion: firstly, orientation of the basement block relative to the E–W stress propagation direction, the existence of a major throw in the original normal fault, development of thick 'Neocomian' sediments and the presence of large faults in the strike direction. Where these tectonic elements coexist, the inversion process is important.

Generally, the faults that are less than 3 km long and poorly orientated to the compressive stress, are not inverted, regardless of their original throw. On the other hand, faults with a great N–S trending elongation and a minor original throw (200 m), do produce inversion. There are faults with an orientation of no more than 45° from the North and 10 km with an original throw of 1000 m which are inverted.

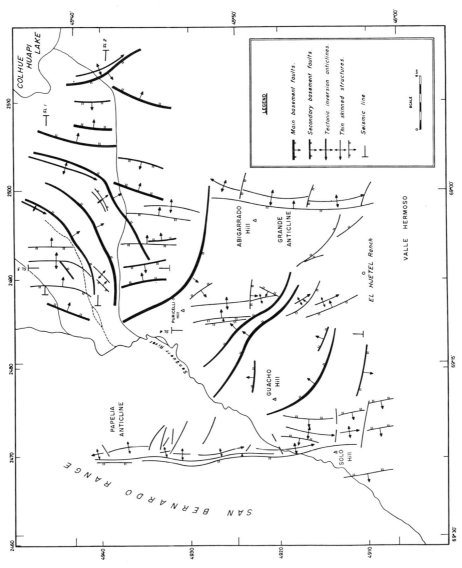

Fig. 5. Map of principal structural elements of the northern part of the study area. (Corresponds to Area I located in Figs 2 and 3.)

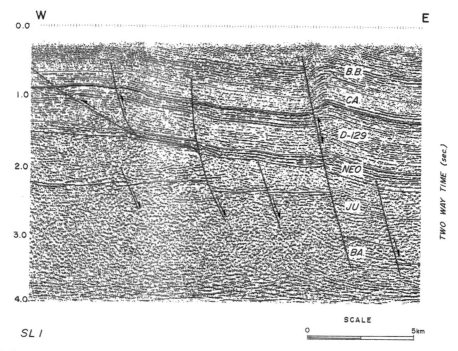

W   E

TWO WAY TIME (sec.)

B.B.

CA.

D-129

NEO

JU

BA

SCALE
0   5km

SL I

Fig. 6. Interpreted E–W seismic line, showing the tectonic inversion of a normal fault and fault propagation anticline with detachment zone in Neocomian strata. B.B, Bajo Barreal Formation; C.A, Castillo Formation. D-129, Pozo D-129 Formation; NEO, Neocomian; JU, Jurassic; BA, Basement. (Location of line shown in Figs 3 and 5.)

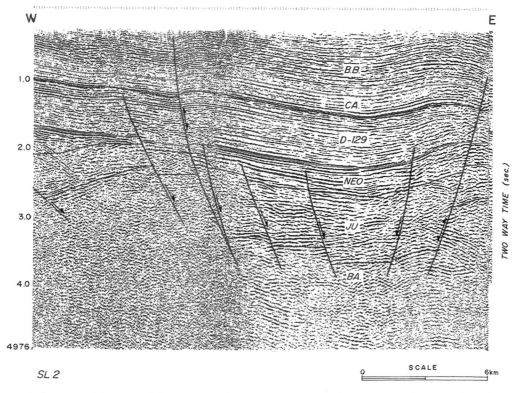

W   E

TWO WAY TIME (sec.)

B.B.

CA.

D-129

NEO

JU

BA.

4976

SL 2

SCALE
0   6km

Fig. 7. Interpreted E–W seismic line, showing the tectonic inversion of the margins of a Jurassic graben. (See Fig. 6 for stratigraphic references. Location of line shown in Figs 3 and 5.)

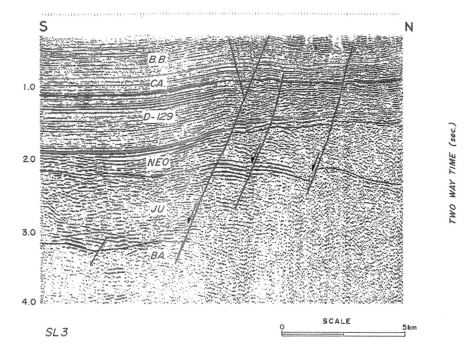

Fig. 8. Interpreted N–S seismic line, showing the original extensional features with no inversion. (See Fig. 6 for stratigraphic references. Location of line shown in Figs 3 and 5.)

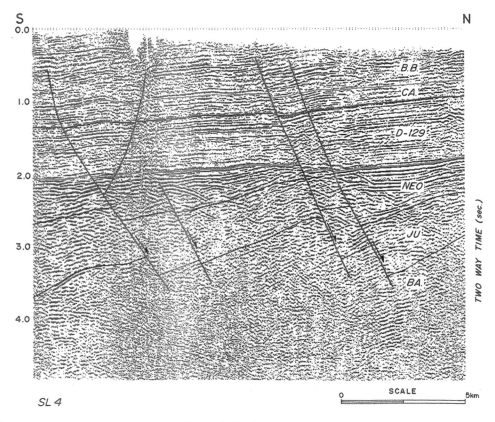

Fig. 9. Interpreted N–S seismic line showing the halfgraben profile of the Neocomian stratigraphic units with no inversion. (See Fig. 6 for stratigraphic references. Location of line shown in Figs 3 and 5.)

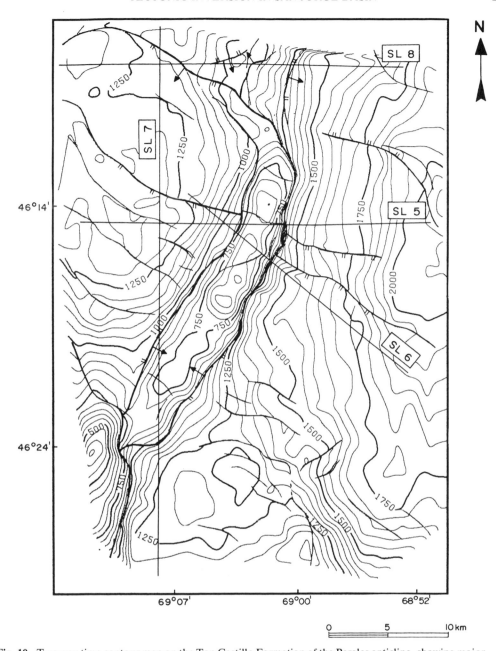

Fig. 10. Two way time contour map on the Top Castillo Formation of the Perales anticline, showing major deformation along a zone with a NE–SW trend, and minor deformation along a zone with a NW–SE trend. (Corresponds to Area II located in Figs 2 and 3.)

Figures 6, 7 and 12 show the relationship between normal faults of the graben and their influence in the asymmetric anticlinal generation in the units that post-date the graben fill.

Structures generated by this process of tectonic inversion partially lose their expression in places where the basement faults are orientated obliquely to the direction of stress propagation, and they 'disappear' where the fault becomes parallel (E–W) (Figs 7 and 8). The Perales anticline is related to two main deformational features with NNE–SSW and NW–SE trends.

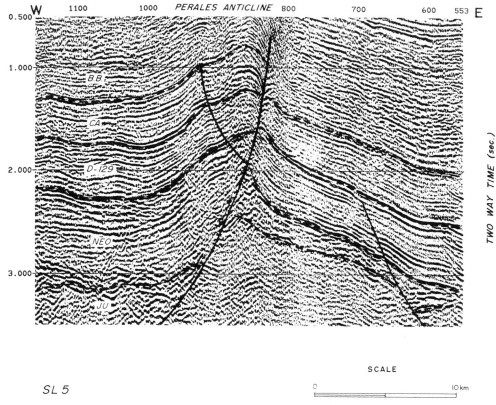

Fig. 11. Interpreted E–W seismic line showing the most deformed part of Perales anticline with high angle reverse fault and an asymmetric fold. (See Fig. 6 for stratigraphic references. Location of line shown in Figs 3 and 10.)

Both of them represent old scarps of the pre-Jurassic blocks and are major depocentres. They are most strongly deformed in the central zone (Figs 11 and 12), with high angle reverse faulting and an associated asymmetric fault propagation fold. The greater shortening in this part of the area is due to the favourable orientation of the original fault to the regional compressive stress. The fault retains its extensional character at depth, reflected by thickness variations of the Neocomian sequence which can be identified on both sides of the fault. The disposition of the inversion structure is north–south, coincident with the direction of the regional compressive stress. An antithetic fault in the hangingwall is well developed.

Figure 13 shows the heterogeneous distribution of the Neocomian sequence. The thickest sequence is located in the central part of the structure and adjacent to the main fault in the northern area. The principal structure has

an east–west strike, still showing the original extensional configuration without signs of inversion. The positive feature, located at the northern part of the structure, is interpreted as a rollover of the post-rift strata. The fault on the opposite side of the graben has a predominant NE–SW orientation, which is more oblique to the stress. As a consequence, compressive features are better defined because of an incipient inversion process.

There are two main deformational segments in the Perales area (Fig. 10): firstly, NNE–SSW trending alignments with strong compressive characteristics which correspond with Neocomian depocentres, and now form major anticlinal axes. Secondly, NW–SE trending alignments which may be related to the original graben margins or low relief normal faults. These structures do not show any major folding, but show growth faulting and a low relief anticlines.

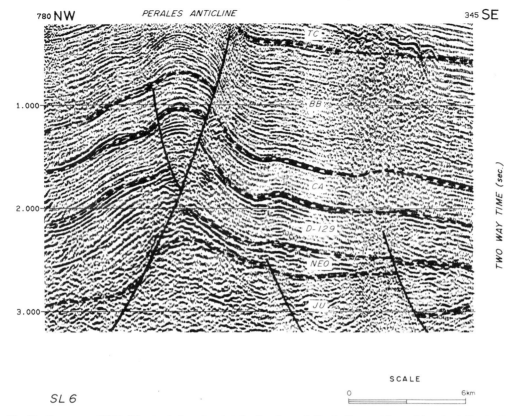

PERALES ANTICLINE

SL 6

SCALE

0   6km

Fig. 12. Interpreted NW–SE seismic line showing the Perales anticline in the most favorable orientation to the regional compressive stress. TC: Tertiary. (See Fig. 6 for other stratigraphic references. Location of line in Figs 3 and 10.)

Intermediate tectonic features

Structures that record an indirect genetic relationship with basement blocks are included in this category. They comprise extensive symmetrical folds associated with reverse faults (Figs 6 and 14). They have a detachment zone in the Neocomian or Upper Cretaceous which roots into basement faults. These faults represent synthetic secondary faults generated at a favourable orientation to the subhorizontal compressive stress (Cooper & Williams 1989). In Fig. 14, the basement fault from which the thrust originates, presents an important regional development being one of the principal structures of sedimentary control in the basin.

Minor tectonic features

These features comprise basement faults with minor regional development and stratigraphic importance. They mainly form synthetic faults that do not deform in the post-rift units

regardless of their orientation. East–west trending strike slip faults, which displace the second order anticline axes, are also included in this category.

Discussion

One topic that remains controversial is the origin of the stress that forms the San Bernardo foldbelt.

A regional palaeogeographic analysis indicates that the origin of the San Jorge basin corresponds with the break-up of Gondwana and the creation of the Atlantic Ocean, with a series of basins located on its Atlantic margins, during the Triassic period. Volcanic activity increased during the Middle to Late Jurassic, with maximum expression between 165–150 Ma (Uliana *et al.* 1985). In the Neocomian, the late extensional event (Fitzgerald *et al.* 1990) resulted in the reactivation of pre-existing faults, with the fragmentation of Triassic basins. The architecture and distribution of Neocomian basins are

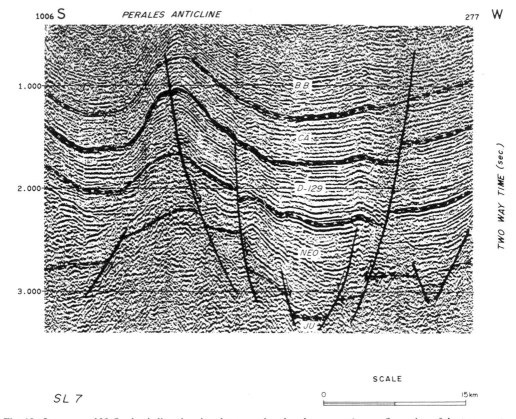

PERALES ANTICLINE   277 W

SL 7

Fig. 13. Interpreted N–S seismic line showing the extensional and compressive configuration of the two most important deformational elements. (See Fig. 6 for stratigraphic references. Location of line in Figs 3 and 10.)

related to a change in the orientation of the regional stresses, due to a penetration of transcurrent regional faults into the inner part of the continent. Francheteau & Le Pichon (1972), and Rapela (1990) discussed the San Jorge example.

The San Bernardo foldbelt is the result of a Tertiary tectonic reactivation of pre-existing normal faults. Ramos (1988) related its origin to tectonics of the Andean Cordillera at these latitudes, and showed that the intensity of the compressive tectonics between 47° and 49° were not severe and were restricted to the lateral slip along pre-existing normal faults. These stresses were transmitted east by a series of lineaments which originated by extensional faulting during Neocomian.

Barcat *et al.* (1984) linked the origin of the foldbelt to a mid-crustal sled-runner domain in a left lateral transpressive model.

Figari & Courtade (1993) proposed for the northern area of the San Jorge basin an origin related to local crustal thickening along old shear zones, penetrating the entire lithosphere and with selective reactivation of some branches.

Conclusions

The major compressional structures of the San Jorge foldbelt are the product of the tectonic reverse reactivation of pre-existing basement normal faults, which formed the steepest margins of asymmetric grabens. These Jurassic–Cretaceous depocentres are mostly filled with Neocomian sediments, deposited during the post-rift stage of the basin.

The origin of the basement normal faults can be found in the propagation of extensional stresses during the Late Jurassic to the intra-cratonic region along major transcurrent faults and controlled by the regional fracture arrangement.

The Tertiary compressional stresses produced inversion in the original normal faults, generating uplift of part of the San Jorge basin and creating the foldbelt.

W   E

0.000

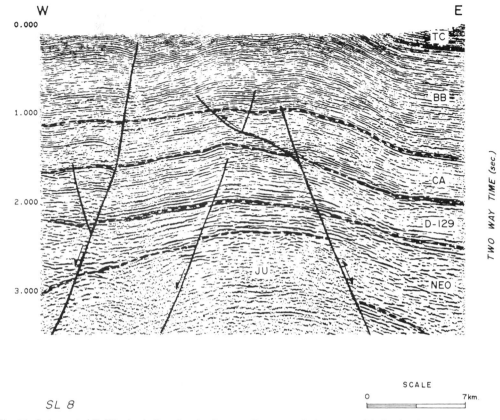

1.000

2.000

3.000

SL 8

SCALE

0   7 km.

Fig. 14. Interpreted E–W seismic line showing intermediate tectonic features, with the development of a secondary synthetic fault. TC: Tertiary. (See Fig. 6 for other stratigraphic references. Location of line in Figs 3 and 10.)

The structures analysed show the best developed compressive features in places where their orientation is north–south. The extensional characteristics remain well developed in the east–west uninverted segments.

The intermediate tectonic features constitute fault propagated anticlines with detachment zones in Neocomian or Upper Cretaceous strata which root into the pre-existing normal faults.

The foldbelt has formed favourable structures for hydrocarbon accumulation in traps which are related mainly to stratigraphic features. Nevertheless it has been found that some structures with huge oil accumulations are related to a minor deformation, or a mild tectonic inversion in which the reservoir units have remained at depth, thus avoiding processes of degradation and erosion.

We thank YPF S.A. Exploration authorities for giving permission to publish this work. Sincere thanks to Mrs. Halina Mischtschenko for drafting the figures.

Early versions of this manuscript benefited from critical comments from Dr Jennifer Urquhart (British Gas) and Dr Miguel A. Uliana (Astra).

References

BARCAT, C., CORTIÑAS, J. S., NEVISTIC, V. A., STACH, N. H. & ZUCCHI, H. E. 1984. Geología de la región comprendida entre los Lagos Musters-Colhué Huapi y la Sierra Cuadrada, Departamentos Sarmiento y Paso de Indios, Provincia del Chubut. *IX Congreso Geológico Argentino*, Actas, **2**, 263–282.

COOPER, M. A. & WILLIAMS, G. D. 1989. (eds) *Inversion tectonics*. Geological Society, London, Special Publication, **44**, 375.

FIGARI, E. G. & COURTADE, S. 1993. Evolución tectosedimentaria de la Cuenca de Cañadón Asfalto, Chubut, Argentina. *XII Congreso Geológico Argentino, y II Congreso de Exploración de Hidrocarburos*. Actas, Tomo **I**, 66–77.

FITZGERALD, M. G., MITCHUM, R. M. JR., ULIANA, M. A. & BIDDLE, K. T. 1990. Evolution of the San Jorge Basin, Argentina. *American Association of Petroleum Geologists Bulletin*, **74**, 879–920.

FRANCHETEAU, J. & LE PICHON, X. 1972. Marginal fracture zones as structural framework of continental margins in South Atlantic Ocean: *American Association of Petroleum Geologists Bulletin,* **56**, 991–1007.

RAMOS, V. A. 1988. La estructura de la Cordillera Patagónica (47°–49°lat. sur) de Argentina y Chile. *V Congreso Geológico Chileno*, Tomo **I**, 101–114.

RAPELA, C. W. 1990. El magmatismo gondwánico y la megafractura de Gastre. *XI Congreso Geológico Argentino, San Juan, 1990*, Actas **I**, 113–116.

ULIANA, M. A., BIDDLE, K. T., PHELPS, D. W. & GUST, D. A. 1985. Significado del vulcanismo y extensión mesojurásicos en el extremo meridional de Sud-América. *Asociación Geológica Argentina*. Revista XL **3–4**, 231–253.

Transpressional inversion due to episodic rotation of extensional stresses in Jeanne d'Arc Basin, offshore Newfoundland

IAIN K. SINCLAIR

University of Aberdeen and Canada-Newfoundland Offshore Petroleum Board, 140 Water Street, St John's, Newfoundland, Canada, A1C 6H6

Abstract: Variations in structural architecture indicate that a ninety degree rotation of extensional stresses occurred between multiple Mesozoic rift phases, resulting in local structural inversion in the Jeanne d'Arc Basin on the Grand Banks of Newfoundland, offshore eastern Canada.

NW–SE-oriented extension caused dip-slip movement on NE-SW-trending en echelon faults during the Late Triassic to earliest Jurassic rift episode. These en echelon faults appear to have been separated by tilted basement relay ramps in accommodation zones rather than by cross-strike transfer faults.

NE–SW-orientated extension during the mid-Aptian to late Albian rift episode resulted in oblique-slip reactivation and linkage of the earlier normal faults. Transpressional stresses were generated by oblique-slip motion at restraining bends in Jeanne d'Arc Basin faults. Structural responses include: 'pop-up' blocks between upward-diverging faults; reverse faults separating overlapping 'shingles' of strata; forced folds in strata overlying deep-seated faults; and wrench-related folds trending oblique to through-going transfer fault zones. Salt and shale layers are seen to act as horizons of fault detachment, modifying the transmission of tectonic stress and strain from the basement into the sedimentary cover. The transpressional mechanism and direction of strike-slip motion are confirmed through recognition of associated transtensional structures located along fault strike at nearby releasing bends.

Transpressional structures in the Jeanne d'Arc Basin are recognizable over a wide range of scales from relatively large-scale reverse faults visible on seismic data to small-scale reverse faults seen in core.

The Jeanne d'Arc Basin is a fault-bounded basin located on the broad continental shelf area of eastern North America known as the Grand Banks of Newfoundland. The Grand Banks area of North America was situated between and connected to the continental plates of Africa, Iberia and northwest Europe at the start of the Mesozoic (Fig. 1). Mesozoic basins of this area experienced polyphase histories of rifting and passive subsidence related to opening of the North Atlantic Ocean (e.g. Ziegler 1988; Hiscott *et al.* 1990). Fifteen hydrocarbon discoveries have been made in the Jeanne d'Arc Basin since drilling began in 1971 in this portion of the Grand Banks region with the Murre G-67 well (Fig. 2; CNOPB 1990; Chipman 1992).

Previous studies have recognized that the orientations of both extensional stresses and normal fault strains were different during successive episodes of rifting in the Jeanne d'Arc Basin area (e.g. Hubbard *et al.* 1985; Enachescu 1987; Sinclair 1988). Though extensional faulting is predominant, indications of strike-slip movements in grabens of the Grand Banks have also been reported as early as 1973 (Amoco and Imperial). Mesozoic strike-slip deformation has been identified in areas bordering the Jeanne d'Arc Basin and this deformation has been attributed to changing extensional stress regimes through discrete episodes of rifting (Enachescu 1987; Foster & Robinson 1993). However, there has been a lack of direct evidence for the pattern of faults active during the initial Mesozoic rift phase of Late Triassic to Early Jurassic age (i.e. Early Cimmerian of Ziegler 1988) or for the involvement of basement blocks in subsequent strike-slip deformation. This paucity of definitive data has been due, in great measure, to the limitations of seismic resolution in areas of thick sediment accumulation. Specifically, two-way time (TWT) to basement is greater than 7.0 seconds beneath much of the Jeanne d'Arc Basin, whereas seismic resolution on most industry surveys is poor below 4.0 to 4.5 seconds. Consequently, additional evidence for the presence, age and associated stress regime of strike-slip basement deformation is required in the Jeanne d'Arc Basin area.

From BUCHANAN, J. G. & BUCHANAN, P. G. (eds), 1995, *Basin Inversion*, Geological Society Special Publication No. 88, 249–271.

I. K. SINCLAIR

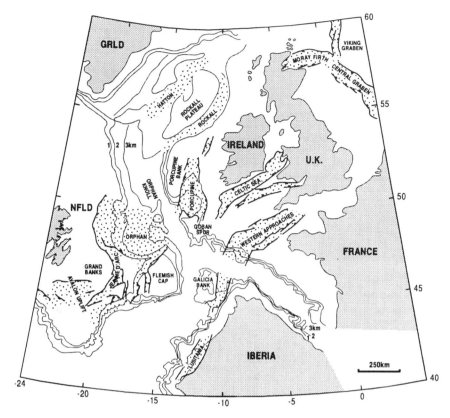

Fig. 1. Mesozoic basin locations on a Late Jurassic plate reconstruction of the North Atlantic margins (modified from Masson & Miles 1986).

This study was undertaken in order to establish the structural architecture of the Triassic–Jurassic sediments (Fig. 3) in an area of the Jeanne d'Arc Basin with relatively thin sediments preserved over basement at 2 to 4 seconds TWT. The determination of structural styles through reflection seismic mapping of lower Mesozoic horizons provides new data for identification of basement-involved inversion structures and for interpretation of the polyphase tectonic history of the broad Grand Banks area. Data from areas of almost continuous sedimentation in the Jeanne d'Arc Basin during the Cretaceous provide additional control on the timing of oblique-slip induced transpression. Strike-slip deformation is also suggested to have controlled both small and large-scale structuring in portions of at least two oil fields (i.e. Hibernia and Terra Nova).

Structural architecture of lower Mesozoic strata

The portion of the Jeanne d'Arc Basin located south of the Egret fault (Fig. 2; Enachescu 1987)

lies on the northeastern flank of a broad basement arch known as the Avalon Uplift (Fig. 1; Sherwin 1973; Jansa & Wade 1975). The Avalon Uplift was formed during two distinct phases of uplift, erosion and non-deposition during Tithonian to early Valanginian and Barremian to Albian (Sinclair 1988). Consequently, Cenomanian and younger sediments overlie tilted and truncated Upper Jurassic to Upper Triassic sediments within the southern portion of the Jeanne d'Arc Basin (Fig. 7).

The lithostratigraphy for basal Mesozoic sediments of Grand Banks basins (Fig. 3) was introduced by Jansa & Wade (1975) and slightly modified by McAlpine (1990), while the character and extent of lower Mesozoic igneous units have been described by Pe-Piper et al. (1992). Three reflective horizons have been mapped at a scale of 1:100 000 using more than 6000 line kilometres of industry seismic data. These horizons correspond to near top of the Whale Member of the Downing Formation, the top Iroquois Formation, and the top of a basal

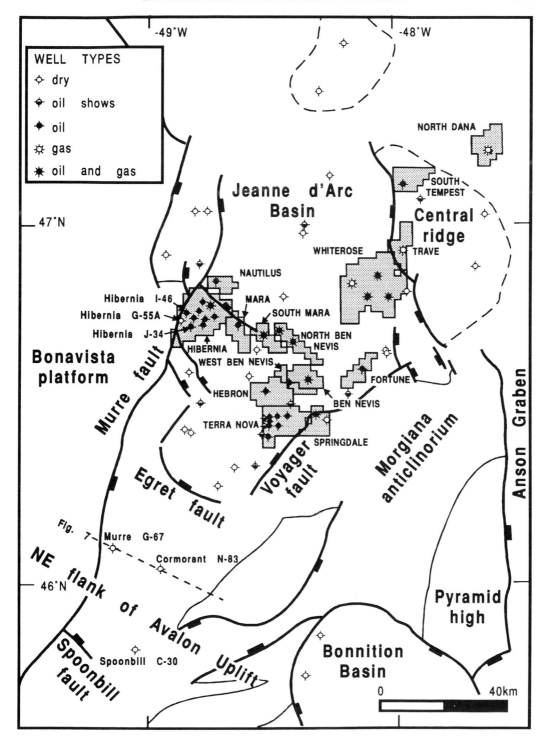

Fig. 2. Main structural features of the Jeanne d'Arc Basin (modified from CNOPB 1989) with nomenclature taken mainly from Enachescu (1987). Shaded areas indicate 15 hydrocarbon discoveries (CNOPB 1993), including the Hibernia field (Hurley *et al*. 1992) which has been projected to see first production in 1997.

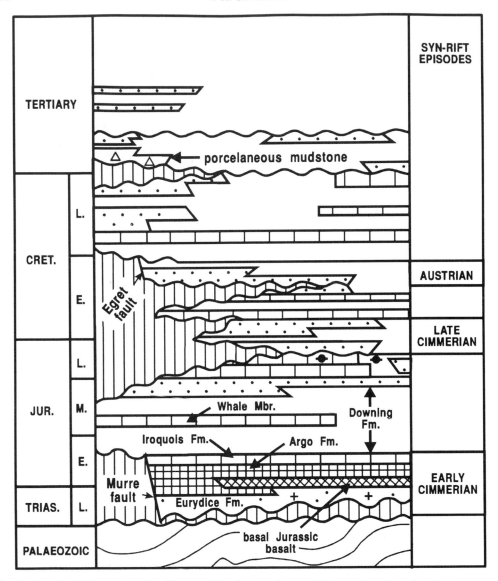

Fig. 3. Simplified lithostratigraphy of Jeanne d'Arc Basin sediments. Lithostratigraphic units which respectively correspond to structural maps of Figs 4, 5 and 6 are: near Whale member limestone; Iroquois Formation limestone and dolomite; basal Jurassic basalt. More detailed lithostratigraphic charts are provided in McAlpine (1990) and Sinclair (1993). Rift phase terminology is from Ziegler (e.g. 1975, 1988) while data on timing and sedimentological responses to Late Cimmerian and Austrian rift tectonism in the Jeanne d'Arc Basin are respectively provided in Sinclair & Riley (1994) and Sinclair (1993). The chart left and right represent west and east except for the uppermost Jurassic through Lower Cretaceous portion where left and right represent south and north, respectively.

Jurassic basalt (Figs 4, 5 & 6). The unconformable contact at the base of Mesozoic sediments does not yield a clear, consistent seismic reflection. The basal Jurassic basalt (i.e. Hettangian, Pe-Piper *et al.* 1992) which was penetrated in the Cormorant N-83 well from 2943 to 2972 m is the oldest Mesozoic unit that generates a reflector which can be mapped seismically with confidence over a significant area of the Jeanne d'Arc Basin. The structural map of the top of this basalt is therefore used here as an approximation of the structure on the closely underlying

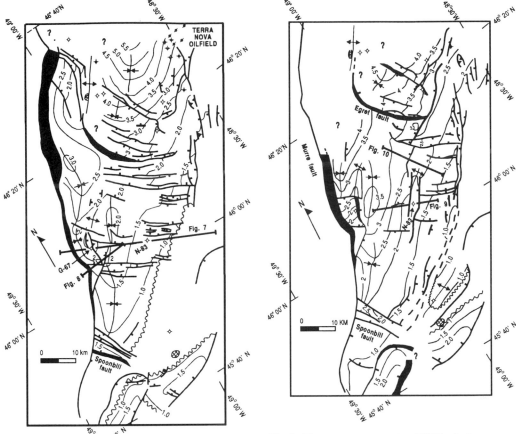

Fig. 4. Time structure map (sec TWT) of a Middle Jurassic reflector (i.e. near top of Whale Member).

Fig. 5. Time structure map (sec TWT) of a Lower Jurassic reflector (i.e. top Iroquois Formation).

but weakly to non-reflective top-of-basement metasediments. A TWT map of an interpreted top of pre-Mesozoic basement is provided, however, in Edwards (1989).

Deep crustal reflection seismic data demonstrate the presence of a simple half-graben bounded by a major listric fault (i.e. Murre fault; Figs 2 & 7) which is thought to have formed initially in response to orthogonal NW–SE-directed extension of the crust (Keen et al. 1987; Enachescu 1987). However, complex local structures indicate that the original half-graben and sediment fill were subsequently deformed. Figure 8, for example, shows Middle Jurassic shale and limestone beds folded into a narrow anticline between two upward-diverging reverse fault strands. This 'pop-up' or 'positive flower' structure (Harding 1985) is situated on the western side of the Jeanne d'Arc Basin half-graben, adjacent to the narrow basement block

drilled by the Murre G-67 well (Fig. 4). Bedded salt of the Upper Triassic-Lower Jurassic Argo Formation was not mobilized at this location, indicating that halokinesis was not a driving mechanism in formation of this feature. The apparent throw on the western fault strand of the 'pop-up' is reversed in Middle Jurassic strata (e.g. near top of Whale Member), opposite to the normal sense of throw at the level of the Lower Jurassic top Iroquois Formation (Figs 4 & 5). Similar structural features are commonly ascribed to a two-stage tectonic history of extension followed by compression (e.g. Williams et al. 1989). This structural style suggests partial inversion of an extensional structure through reverse sense reactivation of a normal fault. Analysis of regional data is required, however, to determine whether this reactivated structure is due to thrust (i.e. compressional) or oblique-slip (i.e. transpressional) fault movement (Harding 1990).

Regional mapping at the top basal Jurassic

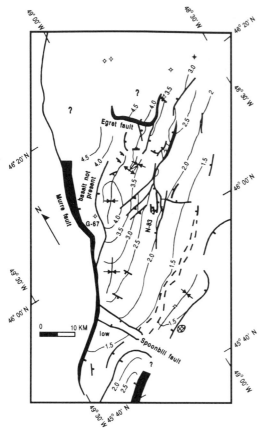

Fig. 6. Time structure map (sec TWT) of a basal Jurassic reflector (i.e. top basalt layer).

basalt has defined a sinuous and branching NE-SW-trending normal fault which formed antithetic to the Murre fault near the Cormorant N-83 well location (Figs 6 & 7). Two examples of the changing structural architecture along the length of the antithetic-to-Murre fault zone are described in the following section. These structures provide the data needed to determine the mechanisms of deformation.

Cormorant anticline

Industry seismic data offer an expanded view of the anticlinal structure drilled at Cormorant N-83 (Fig. 9) compared to that seen on the deep crustal seismic section 85-4A (Fig. 7). Figure 9 illustrates the following structural features of the Cormorant anticline.

(i)   Middle Jurassic strata are gently folded over a faulted basement structure, while Upper Cretaceous and Tertiary sediments only show subtle drape across the underlying structure.

(ii) Lower Jurassic shales and marlstones display increasing thicknesses toward NE-SW-trending en echelon faults which detach within the Argo Formation bedded salt interval (Figs 5 & 9).

(iii) The Argo Formation salt layer shows two opposing thickness trends: bedded salt is generally thicker on the downthrown hanging wall block to the northwest (300 msec TWT versus 150 on the footwall); but the salt layer is thin (less than 200 msec TWT) on the hangingwall block immediately adjacent to the interpreted intra-basement fault.

(iv) The structure on the top Iroquois Formation defines the original asymmetrical anticlinal target of the Cormorant N-83 well (Figs 5 & 9).

(v)   The basal Jurassic basalt layer is deformed into an asymmetrical fold that is segmented by a set of reverse faults trending obliquely toward the main through-basement normal fault trend (Figs 6 & 9).

(vi) The amount of normal throw on the fault offsetting the basal Jurassic basalt is small (about 100 msec TWT) relative to the throw on the overlying antithetic-to-Murre detached fault at the top Iroquois Formation (about 250 msec TWT).

Large basement 'step' northeast of Cormorant N-83

While a number of structural features of the type described above continue along the length of the sinuous antithetic-to-Murre fault zone, distinctly different features have also been developed to the northeast. A seismic profile located about 11 km northeast of Fig. 9 (Figs 5 & 10) illustrates the following common and distinct features.

(i)   The fold in Middle Jurassic strata seen on 83-1300 has been replaced by subtle drape above the antithetic-to-Murre faults on seismic section 83-1306 (Figs 9 & 10 respectively).

(ii) Lower Jurassic shales and marlstones thicken into a normal fault which detaches downward within the salt layer at the northeastern location, analogous to the stratigraphic thickening seen on the southwest. However, the lower portion of the detached fault is oversteepened in the area of Fig. 10, unlike the planar to listric nature of the fault plane seen in Fig. 9.

(iii) Large-scale variations in the Argo Formation salt layer are similar to those

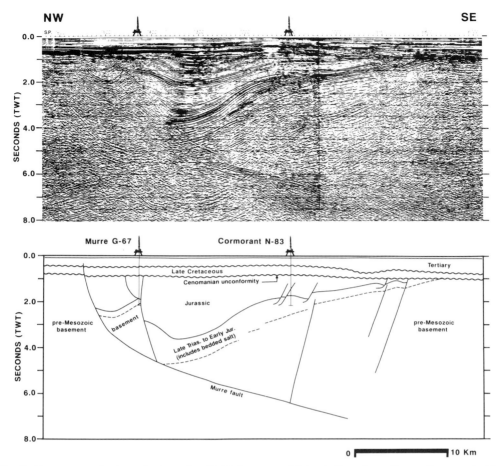

Fig. 7. Deep crust reflection seismic section 85-4A (courtesy of Atlantic Geoscience Centre, Dartmouth, Nova Scotia, Canada) previously published in Keen *et al.* (1987; location shown on Figs 2 & 4).

patterns seen to the southwest, with greater thicknesses present on the basement hangingwall block relative to the footwall block of the antithetic-to-Murre fault. In marked contrast to section 83-1300, however, 83-1306 shows an anomalously thick salt section in the downthrown area immediately adjacent to the antithetic-to-Murre intra-basement fault (650 vs <200 msec TWT).

(iv) The anticline seen at the top Iroquois Formation, which was drilled by the Cormorant N-83 well, does not extend into the area of Fig. 10.

(v) The reverse faults which offset the top basal Jurassic basalt of the Cormorant anticline area (Fig. 6) also do not appear to extend to the area of the large basement 'step'.

(vi) The degree of normal throw is much greater on the antithetic-to-Murre basement fault

in the northeast than that observed to the southwest (about 550 versus about 100 msec TWT).

NW–SE-oriented normal faults

One important structural feature that spans the southern portion of the Jeanne d'Arc Basin is not clearly represented on NW–SE-oriented seismic sections such as 80-1300 and 80-1306. This feature is numerous NW–SE-oriented normal faults which offset Middle Jurassic strata and generally detach downward within Lower Jurassic shales or the Lower Jurassic/Upper Triassic Argo Formation salt. Figures 5 & 6 can be compared to identify the numerous faults of this trend which detach below the top near Whale Member and above the top Iroquois Formation. Jurassic reflectors do not diverge toward NW–SE-trending faults in contrast to the stratal

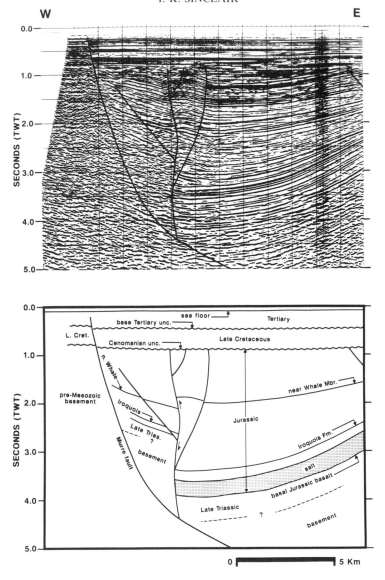

Fig. 8. Portion of seismic section 83-4912 (SOQUIP 1983) showing a 'pop-up' anticline between two upward-diverging fault strands (location on Fig. 4).

thickening toward NE–SW-trending intra-salt detached faults observed on Figs 9 & 10. The former pattern indicates that the NW–SE-trending faults were only active after deposition of Jurassic strata. Upper Cretaceous and Tertiary sediments are almost completely unaffected by these faults, indicting that termination of fault growth pre-dates the Late Cretaceous. Therefore, growth of NW–SE-orientated normal faults through Jurassic sediments occurred synchron-

ous with the deep erosion of strata in the southern Jeanne d'Arc Basin during some period of Early Cretaceous time. Timing of this faulting is crucial to the tectonic history of the area and is discussed further below.

The similarities and differences in structural architecture across the southern Jeanne d'Arc Basin described above were used to deduce the following three stages of the area's tectonic history (Figs 11 & 12).

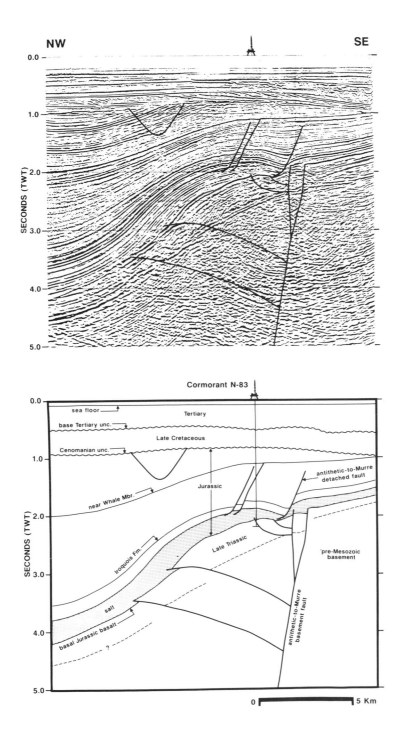

Fig. 9. Portion of seismic section 80-1300 (Petro-Canada 1980) intersecting the Cormorant anticline and the Cormorant N–83 well location (location on Fig. 5).

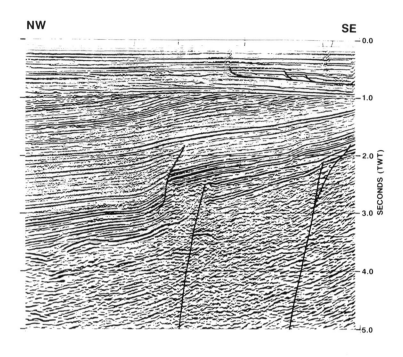

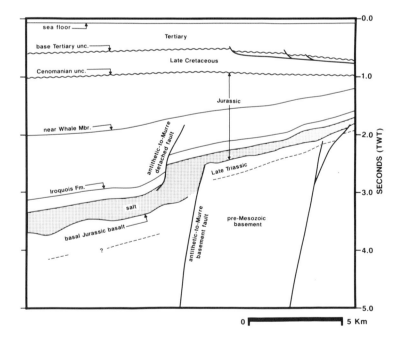

Fig. 10. Portion of seismic section 80-1306 (Petro-Canada 1980) located along structural strike intersecting a large step in the top of basement to the northeast of Fig. 9 (location on Fig. 5).

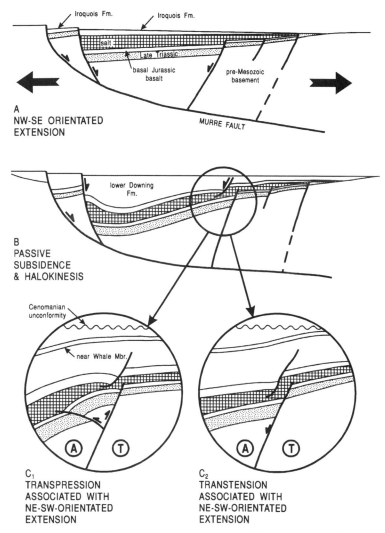

Fig. 11. Schematic cross-sections illustrating the three-stage tectonic history of the southern Jeanne d'Arc Basin (see text for discussion). The A and T represent motion respectively away and toward the viewer, perpendicular to the plane of the diagram.

Interpretation of a polyphase tectonic history

Stage 1, NW–SE-orientated extension

The first stage was a Late Triassic to Early Jurassic period of extension which corresponds to the Early Cimmerian rifting phase of Ziegler (1988; Fig. 11a). Differential rates of rift-induced subsidence on adjacent fault blocks resulted in broad variations in the regional thicknesses of syn-rift strata such as the Argo Formation salt layer above the basal Jurassic basalt (Figs 9 & 10). NW–SE-orientated extension

(Jansa & Wade 1975; Hubbard *et al.* 1985; Enachescu 1987) generated the large half-graben which is bounded by the Murre fault and which characterizes the southern Jeanne d'Arc Basin (Fig. 7, Keen *et al.* 1987). The Murre fault, however, may not have been a single continuous fault during the initial rift phase (Edwards 1989). Numerous bends in strike of the Murre fault and the sinuous and branching nature of the anti-thetic-to-Murre fault mapped at the top basal Jurassic basalt (Fig. 6) suggest each originated as a series of en echelon normal faults which only became linked during a later stage of deformation. Current structural mapping provides no

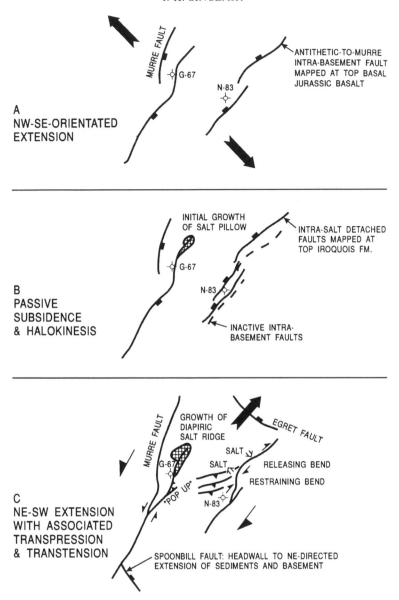

Fig. 12. Schematic plan-view illustrating the three-stage tectonic history of the southern Jeanne d'Arc Basin.

support for the presence of contemporaneous cross-strike transfer faults (*sensu* Gibbs 1984) offsetting the normal faults of this initial syn-rift phase, though such a pattern was suggested in Sinclair (1988). The en echelon pattern of faulting is even more clearly recognizable at the top Iroquois Formation level (Fig. 5) as these overlying detached faults of the second tectonic phase did not become linked during subsequent rift episodes. The left-stepping faults which offset the Iroquois Formation are separated by

tilted relay ramps of the type described by Larsen (1988) in East Greenland and termed overlapping synthetic transfer zones by Morley *et al.* (1990) in their study of East African rifts.

Stage 2, passive subsidence and halokinesis

The second tectonic stage (Figs 11b & 12b) was characterized by passive subsidence during decay of the thermal anomaly associated with rifting ('thermal' subsidence, McKenzie 1978).

Basement-involved faults appear to have been mainly inactive during the late Early to early Late Jurassic sedimentation. Sediment loading, however, caused continued growth on a few major intra-basement faults during this phase of 'thermal' subsidence (Figs 7 & 8). The combination of density inversion, strata tilting, and load-stress differentials on salt layers buried to various depths across the half-graben caused local salt mobilization and diapirism (Jenyon 1985, 1988) along pre-existing Jeanne d'Arc Basin margin structures (Figs 7 & 12b). This halokinesis caused deformation of the overlying post-rift sedimentary sequence. For example, migration of salt toward and up the eastern bounding fault of the Murre G-67 basement block caused uplift of the overlying sediments and additional subsidence on the eastern flank of the salt 'pillow' (Figs 7 & 12b). The growth of this rim syncline appears to have caused extensional forces restricted to the sedimentary column. Thus, Early to Middle Jurassic halokinesis along the basin's northwestern margin appears to have initiated and/or facilitated growth of gravity-driven listric detached faults on the opposite basin margin in the area above the antithetic-to-Murre intra-basement faults near the Cormorant N-83 well (Figs 9, 10 & 12b).

Salt migration and associated intra-sedimentary faulting were temporarily halted during the latter half of the Jurassic Stage 2 as indicated by the unbroken character of the near top Whale Member reflector across NE–SW-trending faults (Figs 9 & 10). Parker & McDowell (1955) have modelled a process which may account for such a termination of salt diapirism where the thickness of overburden exceeds a certain value or 'critical overburden thickness'.

Stage 3, NE–SW-orientated extension with associated transpression and transtension

The third tectonic stage (Fig. 11c & 12c) caused a variety of structural distortions along the main basement fault trends created during the initial Late Triassic to Early Jurassic rift stage. NE–SW-orientated extension (i.e. rifting) caused growth of the NW–SE-orientated normal faults which offset the previously deposited Middle Jurassic strata (Fig. 4). While many NW–SE-orientated faults detach within lower Jurassic shales and salt, a few major faults of this trend are continuous into basement (see Sinclair 1993 for a published seismic profile across the Egret fault). One example is the Spoonbill fault (Figs 2, 4–6) which also acted as a headwall limit to basement extension (sensu Gibbs 1990) during this third tectonic stage. The combined pattern

of numerous normal fault detachments in the Mesozoic sedimentary column with a few major faults being continuous through Jurassic sediments, salt and pre-Mesozoic basement is consistent with physical modelling studies of crustal extension by Vendeville et al. (1987). Their model of two sand layers separated by a layer of mobile silicone showed that, though extensional faulting in upper and lower brittle layers had been partially isolated by the intervening mobile detachment layer, a few faults penetrated the entire three-layer model (ibid.).

NE–SW-trending en echelon faults which offset basement during Late Triassic to Early Jurassic rifting (stage 1) were orientated preferentially for strike-slip reactivation under the new NE–SW-orientated extensional stress regime of stage 3. However, the en echelon strands of basement faults had to become linked to accommodate strike-slip motion (Fig. 12c). Linkage of the left-stepping faults created a series of restraining and releasing bend pairs (sensu Woodcock & Fischer 1986).

Seismic section 80-1300 (Fig. 8) intersected a restraining bend where northeast-directed extension of the crust resulted in oblique-slip motion and local transpression, a term introduced by Harland (1971). As a consequence, the hangingwall basement block moved obliquely upward and to the northeast (Fig. 11c). This movement decreased the pre-existing normal throw on the antithetic-to-Murre basement fault both at the top of basement and at the top basal Jurassic basalt.

Growth of reverse faults, trending oblique to the main fault trend (Figs 6, 9 & 12c), created overlapping 'shingles' of the basal Jurassic basalt. Examination of cores taken from the Lower Cretaceous strata along the western bounding fault of the Hibernia structure (i.e. wells Hibernia I-46, J-34 and G-55A; Figs 2 & 13) has identified the presence of small-scale reverse faults. Though the direction of block motion (up or down) on large-scale faults is practically impossible to identify in core, these microfaults provide physical evidence confirming the existence of reverse faults which resulted in repetition of sedimentary beds at restraining bends on the margins of the Jeanne d'Arc Basin. Tankard & Welsink (1987) interpreted 1.5 km of strike-slip movement to have occurred along the western fault boundary of the Hibernia structure equivalent to the amount of extension that they calculated to have occurred southwest of Hibernia.

Another characteristic of transpression seen at the sinstral restraining bend of Fig. 9 is the migration of salt out of the high compressive

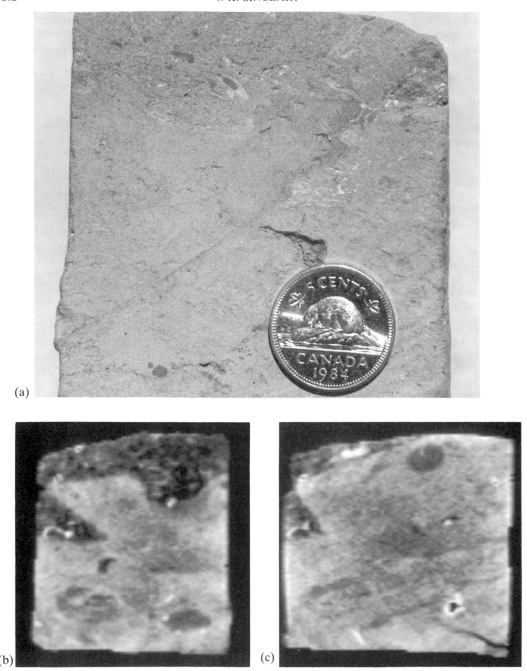

(a)

(b)   (c)

Fig. 13. Photo (**a**) and x-ray sections (**b** & **c**) of a small-scale reverse fault from the Hibernia I-46 well (Fig. 2), core 19 at about 2530 m drill depth. The photo shows the scoured contact between a bioclastic-rich, storm-generated porous sandstone and an underlying fairweather, bioturbated, argillaceous sandstone deposited in the offshore-to-shoreface transition (coin diameter is 2 cm). The x-ray CAT scans were generated by staff of the Aberdeen Rock Imaging Centre and each images a 5 mm 'slice' of core. CAT scan (**b**) was taken parallel to the slabbed surface of core which is oblique to the plane of the reverse fault. CAT scan (**c**) was taken perpendicular to the plane of the reverse fault to show its true dip (about 40° from the vertical wall of the core). The CAT scans give a reverse image to the photo and porous sandstone appears dark (low density), while the 'tight' argillaceous sandstone appears light grey (high density).

stress zone formed alongside the fault segment which is interpreted to link two en echelon, NE-trending intra-basement antithetic-to-Murre faults. The mobile salt layer also acted as a structural detachment horizon between the upper sedimentary strata and the lowermost strata and basement. However, a forced fold (Stearns 1978; Withjack *et al.* 1989, 1990) was created in the Middle and Upper Jurassic strata by the oblique-upward reactivation of the basement hangingwall block. Similarly, partial inversion of an extensional structure by oblique-upward slip at a restraining bend would account for the 'pop-up' block seen on Fig. 8.

In contrast, section 80-1306 (Fig. 10) intersected a releasing bend where northeast-directed extension of the crust resulted in local transtension (Fig. 12c). As a consequence, the hangingwall block moved obliquely downward to the northeast (Fig. 11c). This movement added to the pre-existing normal throw of the antithetic-to-Murre basement fault at the levels of both top basement and top basal Jurassic basalt. This additional normal throw did not cause fault propagation through the Middle Jurassic strata due to the following synchronous halokinetic response.

The local transtensional setting also caused salt to migrate into the low compressive stress zone adjacent to the dextral releasing bend. This lateral migration created an anomalously thick section of salt unrelated to vertical diapirisim. As a consequence of salt 'influx' compensating for increased fault throw, Middle Jurassic strata only drape across the underlying complex structure. Elsewhere, renewed growth of salt diapirs was accentuated and modified by oblique-slip deformation as occurred alongside the Murre G-67 block (Figs 7 & 12c).

Oblique subsidence of the basement hangingwall block at the releasing bend during tectonic stage 3 also resulted in deformation of the NE–SW-trending, intra-salt detached fault which offsets lower Jurassic strata (Figs 10 & 11c). Renewed subsidence of the hangingwall block caused bending of the fault plane so that the lower portion of the fault became oversteepened.

The age of the tectonic stage 3 (i.e. time of local inversion) is not tightly constrained by data from the southern Jeanne d'Arc Basin. Since the Jurassic strata are deformed and Upper Cretaceous strata are dominantly underformed, all that can be directly determined is that transpressional and transtensional deformation occurred during some portion of Early Cretaceous time represented by the lacuna at the base Late Cretaceous (i.e. Cenomanian) unconformity

south of the Egret fault. The following section presents data which allow for precise identification of the time of inversion.

Age of oblique slip deformation and inversion

Mapping of Lower Cretaceous seismic reflectors over the central Jeanne d'Arc Basin defines a dominant set of NW–SE-trending normal faults in an area of hydrocarbon entrapment and more nearly continuous Cretaceous sedimentation (e.g. Fig. 14). These NW–SE-orientated faults have been referred to as trans-basin faults by Grant *et al.* (1986) and one of the largest individual faults of this set is the Egret fault (e.g. Figs 2 & 14). The Egret fault has been identified by Enachescu (1987) as a basin-bounding fault which limits the areal extent of a reactivated Lower Cretaceous portion of a larger Jurassic basin. Rift-induced growth of the Egret and related trans-basin faults was initiated in the mid-Aptian and terminated in the late Albian as evidenced by thickness variations in syn-rift sediments (see Sinclair 1988 & 1993 for well and seismic evidence of this period of extension). This period of rifting corresponds with the Austrian tectonic phase of Ziegler (1988).

The NW–SE-trend of the trans-basin normal faults mapped in Lower Cretaceous sediments to the north of the Egret fault and the parallel trend of the dominant normal fault set mapped in Middle Jurassic sediments in the southern Jeanne d'Arc Basin (Figs 14 & 4 respectively) suggest a shared response to extension in mid-Aptian to late Albian times. The dramatic differences in sediment preservation and thickness between the northern and southern areas, however, can also be explained by large-scale patterns of crustal deformation during Austrian phase tectonism. Mid-Aptian to late Albian extension of the southern Jeanne d'Arc area occurred on the northeastern flank of the crustal arch known as the Avalon Uplift. The Avalon Uplift arch was raised first during the Tithonian to Early Valanginian and again during the Barremian to Late Albian (Sinclair 1988) based on megasequence architecture and variations in sediment input. Consequently, the southern Jeanne d'Arc Basin acted as a clastic provenance area with mid-Aptian to late Albian extensional faulting accompanied by erosion. The combination of coeval uplift, extension and erosion resulted in differential preservation of Jurassic sediments below a peneplain. Synchronous extension of basement and sediments north of the Egret fault caused rapid subsidence of fault

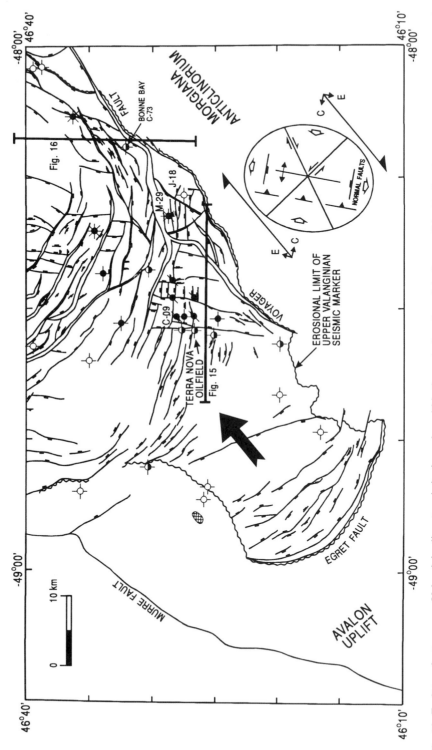

Fig. 14. Fault trends at an upper Valanginian limestone/seismic marker within the central portion of the Jeanne d'Arc Basin. The strain ellipse is from Harding (1974).

blocks to below sea-level and accumulation of a thick sequence of transgressive coastal to marine syn-rift sediments (Sinclair 1993). Both basins and basin margins became flooded at the beginning of the post-rift Late Cretaceous (Hubbard *et al.* 1985).

Identification of the mid-Aptian to late Albian as the time of NE–SW-directed extension provides considerable control for the most likely time of basement-involved oblique-slip motion seen along northeast-trending faults in the southern Jeanne d'Arc Basin. However, direct evidence for the presence and age of transpressional and transtensional structures in an area of synchronous sedimentation in the central Jeanne d'Arc Basin is needed to confirm the time of oblique-slip reactivation of basement blocks and local inversion. Such evidence supporting a mid-Aptian to late Albian age of oblique-slip is found in the following examination of structural variations along the Voyager fault.

The Voyager fault (Fig. 14; modified from use by Enachescu 1987) is a major lineament along the southeastern margin of the Jeanne d'Arc Basin and which continues to the northeast through the Flemish Pass Basin (Sinclair 1988; Foster & Robinson 1993). The NE-trending strike of the Voyager fault is parallel to the Late Triassic/Early Jurassic intra-basement faults of the southern Jeanne d'Arc Basin, suggesting these faults were all initiated by a common stress regime during the Early Cimmerian rift phase. However, the presence of thick Jurassic and Lower Cretaceous sediments in the area of the Voyager fault results in the oldest Mesozoic sediments being situated below the range of clear seismic resolution, making direct confirmation of a common Early Cimmerian origin impossible.

Late Jurassic and Early Cretaceous structural deformation along the Voyager fault are, however, resolvable using seismic and well data. Also, the structural architecture of Jurassic and Lower Cretaceous sediments is recognized to vary dramatically along strike of the Voyager fault. For example, seismic sections 83-4992 and 83-2656 illustrate structural variations on either side of a major bend in the Voyager fault (Figs 14–16). Line 83-4992 crosses the Terra Nova arch, while line 83-2656 crosses the Bonne Bay graben. These two structures are described and contrasted in the next two sections, followed by an interpretation of their tectonic origin.

Terra Nova arch

A prominent northerly-trending arch with a graben located nearly along its anticline crest

(Fig. 15) provides a major structural component of the hydrocarbon-trap at the Terra Nova oil field. The Terra Nova arch and a smaller adjacent anticline are located to the west of a sinistral bend in the Voyager fault formed around the fault block drilled at the Springdale M-29 and Voyager J-18 well locations (Fig. 14).

Increases in the seismic travel time interval of the Austrian depositional sequence to the east and west of the Terra Nova arch demonstrate that folding of strata alongside the Voyager fault was restricted to mid-Aptian to late Albian times (Figs 3 & 15). The term 'depositional sequence' is used for packages of genetically-related sediments bound by unconformities and their correlative conformities as defined by Mitchum *et al.* (1977). Tectonism, however, provided the dominant genetic relationship for syn-rift depositional sequences or megasequences (Hubbard *et al.* 1985; Sinclair 1993; Sinclair & Riley 1994).

Travel time interval variations in the Late Cimmerian depositional sequence indicate that the N–S-trending fault defining the western boundary of the Terra Nova crestal graben may have been initiated during the Tithonian to early Valanginian rift period (i.e. Late Cimmerian of Ziegler 1988). However, additional growth on the graben-bounding faults during the mid-Aptian to late Albian period of folding (Austrian) is also recognizable (Fig. 15).

The NW–SE strike which is characteristic of trans-basin faults over most of the basin swings to an E–W trend in the area of the Terra Nova arch. N–S-orientated seismic profiles show that the Austrian syn-rift depositional sequence thickens toward the downthrown side of the E–W-trending faults, demonstrating that development of these normal sense faults also occurred during the mid-Aptian to late Albian. The use of sediment thickness variations as indicators of basin tectonism is valid here as the Jeanne d'Arc Basin was receiving an abundant supply of clastic material into mainly coastal to shallow marine settings during the mid-Aptian to late Albian. As a consequence, sedimentation was nearly able to keep pace with subsidence and the caveats of Bertram & Milton (1989) concerning determination of basin evolution in deep sediment-starved rift basins are not strongly applicable in this instance.

Bonne Bay graben

The Bonne Bay graben is located to the north of a dextral bend in the Voyager fault (Fig. 14). This complex graben comprises a set of NE–SW-trending, downthrown fault blocks (Fig. 16).

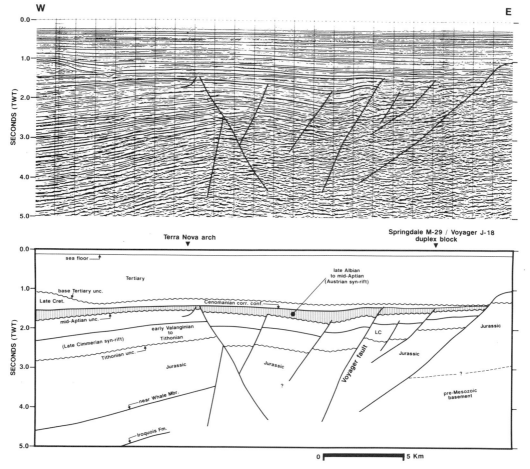

Fig. 15. Portion of seismic section 83-4992 (SOQUIP 1983) showing the north-trending Terra Nova arch formed during mid-Aptian to late Albian extension of the Austrian tectonic phase (location on Fig. 14).

Isochron trends of pre-Aptian depositional sequences continue without change across faults of the Bonne Bay graben, indicating that this graben was initiated after formation of the mid-Aptian unconformity. A gradual decrease in fault throw at successive levels demonstrates that growth of the graben continued through the late Aptian and into the late Albian but ceased prior to deposition of Upper Cretaceous and Tertiary strata.

Both well and seismic indicate that there were large differences in the rates of subsidence over the Terra Nova arch relative to within the Bonne Bay graben. For example, the Bonne Bay C-73 well penetrated 772 m of mid-Aptian to upper Albian sediments on a block which flanks the central axis and deepest portion of the graben complex (Fig. 14). In contrast, equivalent-age sediments are only 175 m thick over the crest of the Terra Nova arch at the C-09 well location.

Coupled with the above structural patterns, thickening of the Late Cimmerian (Tithonian to early Valanginian) depositional sequence towards the Voyager fault (Fig. 16) indicates that this lineament experienced a polyphase history. In addition to its postulated origin in Late Triassic/Early Jurassic times, the Voyager fault was active as a simple, normal fault during the Tithonian to early Valanginian, prior to additional reactivation and development of the complex Bonne Bay graben and the Terra Nova arch.

Tectonic interpretation of Voyager fault structures

The areal extent of trans-basin faults, as mapped in Cretaceous sediments, indicates that the Jeanne d'Arc Basin was affected by NE-directed

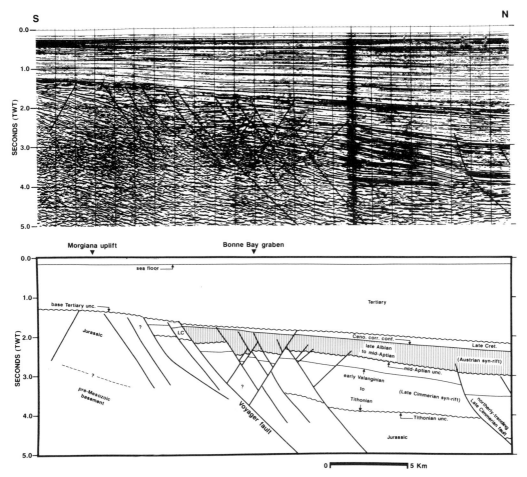

Fig. 16. Portion of seismic section 83-2656 (SOQUIP 1983) showing the northeast-trending Bonne Bay graben formed during mid-Aptian to late Albian extension of the Austrian tectonic phase (location on Fig. 14).

extension but that the adjacent Morgiana anticlinorium (Enachescu 1987) area was not (Fig. 14). The pre-existing Voyager fault was preferentially orientated to accommodate dextral oblique-slip motion during the northeast-directed extension of mid-Aptian to late Albian times. Consequently, the Voyager fault appears to have formed the boundary between the extending basin and its stable margin during the Austrian rift phase.

The left-stepping bend in the Voyager fault where it is deflected around the Voyager J-18/Springdale M-29 block east of the Terra Nova oilfield is interpreted to have acted as a restraining bend during Austrian phase rifting. The N–S orientation of the Terra Nova arch is consistent with growth of a transpressional fold trending oblique to the through-basement Voyager fault zone during NE-directed extension

(see strain ellipse on Fig. 14). A similar mega-sequence architecture across the smaller anticline located east of the Terra Nova arch (Fig. 15) suggests that it also formed due to mid-Aptian to late Albian transpressional oblique slip reactivation of a pre-existing normal fault.

Kinematic models indicate that similar convergent wrench settings can generate large extensional forces parallel to fold axes (Jamison 1991; see also strain ellipse of Fig. 14). Therefore, the anomalous east–west trend of trans-basin faults in the area of the Terra Nova arch (Fig. 13) was likely to have been caused by the rotation of mid-Aptian to late Albian extensional stress from the NE–SW direction characteristic of most of the Jeanne d'Arc Basin to north–south in this area of transpressional folding (i.e. local convergent wrench setting).

The right-stepping fault bend which forms the northern margin of the Voyager J-18/Springdale M-29 block is interpreted to have acted as a releasing bend so that dextral oblique-slip motion led to the local generation of trans-tensional stresses synchronous with transpress-ional folding to the southwest. The resultant high subsidence rates adjacent to a basin margin provence area (i.e. Morgiana anticlinorium) led to the accumulation of great thicknesses of mid-Aptian to late Albian syn-rift sediment in the Bonne Bay graben.

The lenticular Springdale M-29/Voyager J-18 block is bounded by the Voyager fault and a second subparallel fault located to the southwest (Fig. 14). This structural form, with associated transpressional and transtensional structures, is consistent with the development of this fault block as a transfer duplex in the style of Gibbs (1990) during the mid-Aptian to late Albian rift phase.

The above data and interpretation suggest that the Voyager fault acted as a transfer fault (*sensu* Gibbs 1984) accommodating differential rates of shear in Jurassic and Lower Cretaceous sediments on adjacent blocks in the Jeanne d'Arc Basin as was also documented in the Flemish Pass Basin (Foster & Robinson 1993). Deformation along the Voyager fault shows that oblique-slip motion occurred in the upper sedimentary column in a similar fashion to the oblique-slip motion that has been recognized along basement blocks of the southern Jeanne d'Arc Basin.

Many often conflicting interpretations of relationships of various rift episodes (i.e. ages and stress orientations) and halts in crustal extension to the opening history of the North Atlantic have been widely published. Interpre-tations generally consistent with the polyphase history proposed here, have been published by Hubbard et al. (1985), Enachescu (1987) and Sinclair (1988).

Conclusions and speculations on wider application

Thin sediment preservation on the northeastern flank of the Avalon Uplift allows for direct identification of the Late Triassic into Early Jurassic as the rift period during which a large NE-trending half graben formed in the area of the southern Jeanne d'Arc Basin. Coeval en echelon, normal faults formed antithetic to the half graben-bounding Murre fault.

Mapping of seismic reflections from the oldest Mesozoic strata in the southern Jeanne d'Arc Basin demonstrates the presence of transpress-ional and transtensional structures involving basement. These structures were respectively developed at restraining and releasing bends respectively, while the bends were formed by oblique-slip reactivation and linkage of Late Triassic/Early Jurassic NE–SW-trending en echelon normal faults.

Mapping to the north, in an area of more nearly continuous sedimentation, provides con-clusive proof that oblique-slip fault reactivation of basement blocks and local structural inversion occurred during the mid-Aptian to late Albian under conditions of NE–SW-directed extension. Therefore, local structural inversion resulted from the 90-degree rotation of extensional stresses between two separate rift phases (i.e. Early Cimmerian and Austrian). Structural responses to transpression include: 'pop-up' blocks between upward-diverging faults; reverse faults separating overlapping 'shingles' of strata; forced folds in overlying strata; and wrench-related folds trending oblique to through-going (Harding 1990) fault zones.

The Late Triassic to Early Jurassic normal faults were separated by tilted ramps while a number of mid-Aptian to late Albian normal faults terminated at strike-slip fault zones. It is conjectured that accommodation zones of the style of Morley et al. (1990) may be the preferred mode of fault linkage in basins that experienced a single rift phase, while transfer faults of the style of Gibbs (1984, 1990) may be more prevalent in basins that experienced a realign-ment of extensional stresses during successive rift phases such that faults were reactivated with a different sense of motion.

The identification of both small-scale reverse faults in core from the Hibernia I-46 well and large-scale reverse faults on seismic data over the southern Jeanne d'Arc Basin supports the concept that large-scale transpressional struc-tures may have developed locally along the basin margins within the arca of certain oil fields. The possible presence of a few large-scale reverse faults in reservoir horizons at the Hibernia field and the involvement of both wrench folds and the inferred deflection of regional extensional stresses in development of the Terra Nova structure, indicate that identification of restrain-ing bends and ages of deformation may be beneficial to development studies of these two oil fields.

The age and type of oblique-slip deformation identified in the Jeanne d'Arc Basin may also have wider application into distant but tec-tonically linked rift basins of the North Atlantic margins. The clear recognition of inversion structures in a region which only experienced rift deformation (i.e. Mesozoic Grand Banks area) provides an alternative and older mechanism to Tertiary compressive events for development of some structural highs in the North Sea. While the misalignment of inherited structure and extensional stress during Late Cimmerian rifting has already been identified as a means of transpressional uplift (Bartholomew *et al.* 1993), I believe that inversion and formation of hydrocarbon traps due to rotation of extensional forces through successive Mesozoic rift episodes may be a less widely appreciated, but economically important, phenomenon. One candidate for possible application may be the Tern field, East Shetland Basin of the UK sector North Sea, whose published polyphase tectonic history and complex fault style (Van Panhuys-Sigler *et al.* 1991) suggest that the northwestern-bounding fault of this structural trap may have been oblique-slip reactivated during the mid-Aptian to late Albian. Secondly, the structure of the Gullfaks field within the Norwegian sector of the East Shetland Basin has been partially ascribed to compression in response to dextral strike-slip movement along northeast–southwest fault system (Fossen 1989; Petterson *et al.* 1992). Comparison of the ages and styles of deformation in these two North Sea fields with that of the Jeanne d'Arc Basin could prove instructive.

The author gratefully acknowledges the support of BP Exploration's Les Adamson Memorial grant. Thanks to Iain Mackenzie and Jeff Wilson of the Aberdeen Rock Imaging Centre for the generous provision of their equipment and expertise in the x-ray imaging of reverse faults in core. The manuscript was greatly improved through review by Judith McIntyre (CNOPB), Aidan Joy (Kerr McGee Oil PLC) and Duncan MacGregor (BP Exploration Operating Company). David Oliver and Winston Howell are thanked for drafting the bulk of the diagrams, and Mary Glynn for typing of the manuscript.

References

AMOCO CANADA PETROLEUM COMPANY LTD. & IMPERIAL OIL LIMITED 1973. Regional geology of the Grand Banks. *Bulletin of Canadian Petroleum Geology*, **21**, 479–503.

BARTHOLOMEW, I. D., PETERS, J. M. & POWELL, C. M. 1993. Regional structural evolution of the North Sea: oblique slip reactivation of basement lineaments. *In*: PARKER, J. R. (ed.) *Petroleum Geology of Northwest Europe: Proceedings of the 4th Conference*, Geological Society, London, 1109–1122.

BERTRAM G. T. & MILTON, N. J. 1989. Reconstructing basin evolution from sedimentary thickness; the

importance of palaeobathymetric control, with reference to the North Sea. *Basin Research*, **1**, 247–257.

CHIPMAN, W. 1992. Oil and gas fields offshore Newfoundland and Labrador: development potential and scenarios. *In*: *Proceedings of the Newfoundland Offshore Industries Association (NOIA), June 23–25, 1992*, St. John's, Newfoundland, 207–232.

CNOPB (CANADA-NEWFOUNDLAND OFFSHORE PETROLEUM BOARD) 1989. *Regional Geophysical Mapping, North Grand Banks*. CNOPB Report GP-CNOPB-89-2.

—— 1990. *Schedule of Wells, Newfoundland Offshore Area*, St. John's.

—— 1993. *Annual Report 1992–1993*, St. John's.

EDWARDS, A. 1989. *Seismic studies of the Jeanne d'Arc Basin*. Geological Survey of Canada, Open File 2098.

ENACHESCU, M. E. 1987. The tectonic and structural framework of the northeast Newfoundland margin. *In*: BEAUMONT, C. & TANKARD, A. J. (eds) *Sedimentary Basins and Basin-forming Mechanisms*, Canadian Society of Petroleum Geologists, Memoir, **12**, 117–146.

FOSSEN, H. 1989. Indication of transpressional tectonics in the Gullfaks oilfield, northern North Sea. *Marine and Petroleum Geology*, **1**, 22–30.

FOSTER, D. G. & ROBINSON, A. G. 1993. Geological history of the Flemish Pass Basin, offshore Newfoundland. *American Association of Petroleum Geologists Bulletin*, **77**, 588–609.

GIBBS, A. D. 1984. Structural evolution of extensional basin margins. *Journal of the Geological Society*, London, **141**, 609–620.

—— 1990. Linked fault families in basin formation. *Journal of Structural Geology*, **12**, 795–803.

GRANT, A. C., McALPINE, K. D. & WADE, J.A. 1986. The continental margin of eastern Canada: geological framework and petroleum potential. *In*: HALBOUTY, M. T. (ed.) *Future Petroleum Provinces of the World*, American Association of Petroleum Geologists, Memoir, **40**, 177–205.

HARDING, T. P. 1974. Petroleum traps associated with wrench faults. *American Association of Petroleum Geologists Bulletin*, **58**, 1290–1304.

—— 1985. Seismic characteristics and identification of negative flower structures, positive flower structures, and positive structural inversion. *American Association of Petroleum Geologists Bulletin*, **69**, 582–600.

—— 1990. Identification of wrench faults using subsurface structural data: criteria and pitfalls. *American Association of Petroleum Geologists Bulletin*, **74**, 1590–1609.

HARLAND, W. B. 1971. Tectonic transpression in Caledonian Spitsbergen. *Geological Magazine*, **108**, 27–42.

HISCOTT, R. N., WILSON, R. C. L., GRADSTEIN, F. M., PUJALTE, V., GARCÍA-MONDÉJAR, J., BOUDREAU, R. R. & WISHART, H. A. 1990. Comparative stratigraphy and subsidence history of Mesozoic rift basins of North Atlantic. *American Association of Petroleum Geologists Bulletin*, **74**, 60–76.

HUBBARD, R. J., PAPE, J. & ROBERTS, D. G. 1985. Depositional sequence mapping to illustrate the evolution of a passive continental margin. *In*: BERG, O. R. & WOOLVERTON, D. (eds) *Seismic Stratigraphy II*. American Association of Petroleum Geologists, Memoir, **39**, 93–115.

HURLEY, T. J., KREISA, R. D., TAYLOR, G. G. & YATES, W. R. L. 1992. The reservoir geology and geophysics of the Hibernia Field, offshore Newfoundland. *In*: HALBOUTY, M.T. (ed.) *Giant Oil and Gas Fields of the Decade 1977–1988*, American Association of Petroleum Geologists, Memoir, **54**, 35–54.

JAMISON, W. R. 1991. Kinematics of compressional fold development in convergent wrench terranes. *Tectonophysics*, **190**, 209–232.

JANSA, L. F. & WADE, J. A. 1975. Geology of the continental margin off Nova Scotia and Newfoundland. *In*: VAN DER LINDEN, W. J. M. & WADE, J. A. (eds) *Offshore Geology of Eastern Canada*, Geological Survey of Canada, Paper 74–30, **2**, 51–105.

JENYON, M. K. 1985. Basin-edge diapirism and updip salt flow in Zechstein of southern North Sea. *American Association of Petroleum Geologists Bulletin*, **69**, 53–64.

—— 1988. Overburden deformation related to pre-piercement development of salt structures in the North Sea. *Journal of the Geological Society, London*, **145**, 445–454.

KEEN, C. E., BOUTILIER, R., DE VOOGD, B., MUDFORD, B. & ENACHESCU, M. E. 1987. Crustal geometry and extensional models for the Grand Banks, eastern Canada: constraints from deep seismic reflection data. *In*: BEAUMONT, C. & TANKARD, A. J. (eds) *Sedimentary Basins and Basin-Forming Mechanisms*, Canadian Society of Petroleum Geologists, Memoir, **12**, 101–115.

LARSEN, P.-H. 1988. Relay structures in a Lower Permian basement-involved extension system, East Greenland. *Journal of Structural Geology*, **10**, 3–8.

McALPINE, K. D. 1990. *Mesozoic Stratigraphy, Sedimentary Evolution, and Petroleum Potential of the Jeanne d'Arc Basin, Grand Banks of Newfoundland*. Geological Survey of Canada, Paper 89–17, 50 pp.

McKENZIE, D. 1978. Some remarks on the development of sedimentary basins. *Earth and Planetary Science Letters*, **40**, 25–32.

MASSON, D. G. & MILES, P. R. 1986. Development and hydrocarbon potential of Mesozoic sedimentary basins around margins of North Atlantic. *American Association of Petroleum Geologists Bulletin*, **70**, 721–729.

MITCHUM, R. M. JR., VAIL, P. R. & THOMPSON, S. III. 1977. Part 2: The depositional sequence as a basic unit for stratigraphic analysis. *In*: PAYTON, C. E. (ed.) *Seismic stratigraphy – applications to hydrocarbon exploration*, American Association of Petroleum Geologists, Memoir, **26**, 53–62.

MORLEY, C. K. NELSON, R. A., PATTON, T. L. & MUNN, S. G. 1990. Transfer zones in the East Africa Rift system and their relevance to hydrocarbon exploration in rifts. *American Association*

of Petroleum Geologists Bulletin, **74**, 1234–1253.

PARKER, T. J. & McDOWELL, A. N. 1955. Model studies of salt-dome tectonics. *American Association of Petroleum Geologists Bulletin*, **39**, 2384–2470.

PE-PIPER, G., JANSA, L. F. & LAMBERT, R. ST. J. 1992. Early Mesozoic magmatism on the eastern Canadian margin: petrogenetic and tectonic significance. *In*: PUFFER, J. H. & RAGLAND, P. C. (eds) *Eastern North America Mesozoic Magmatism*, Geological Society of America, Special Paper, **268**, 13–36.

PETRO-CANADA 1980. Marine gravity and magnetic survey South Hibernia, 1980. C-NOPB Project No. 8624-P28-4E.

PETTERSON, O., STORLI, A., LJOSLAND, E., NYGAARD, O., MASSIE, I. & CARLSEN, H. 1992. The Gullfaks Field. *In*: HALBOUTY, M. T. (ed.) *Giant Oil and Gas Fields of the Decade 1977–1988*, American Association of Petroleum Geologists, Memoir, **54**, 429–446.

SHERWIN, D. F. 1973. Scotian Shelf and Grand Banks. *In*: McCROSSAN, R. G. (ed.) *Future Petroleum Provinces of Canada*. Canadian Society of Petroleum Geologists, Memoir, **1**, 519–559.

SINCLAIR, I. K. 1988. Evolution of Mesozoic-Cenozoic sedimentary basins in the Grand Banks area of Newfoundland and comparison with Falvey's (1974) rift model. *Bulletin of Canadian Petroleum Geology*, **36**, 255–273.

—— 1993. Tectonism: the dominant factor in mid-Cretaceous sequence stratigraphy in the Jeanne d'Arc Basin, Grand Banks. *Marine and Petroleum Geology*, **10**, 530–549.

—— & RILEY, L. A. 1994. Separation of Late Cimmerian rift and post-rift megasequences: a comparison of the Jeanne d'Arc Basin, Grand Banks and the Outer Moray Firth, North Sea. *In*: STEEL, R. J., FELT, V., JOHANNESSEN, E. & MATHIEU, C. (eds) *Sequence Stratigraphy: advances and applications for exploration and production in North West Europe*. Norwegian Petroleum Society, Special Publication, **5**, 347–363. Elsevier, Amsterdam.

SOQUIP 1993. Marine reflection seismic survey over the Jeanne d'Arc-Flemish Pass Area, 1983. C-NOPB Project No. 8620-S14-8E.

STEARNS, D. W. 1978. Faulting and forced folding in the Rocky Mountains foreland. *In*: MATTHEWS, V. (ed.) *Larimide Folding Associated With Basement Block Faulting in the Western United States*, Geological Society of America, Memoir, **151**, 1–37.

TANKARD, A. J. & WELSINK, H. J. 1987. Extensional tectonics and stratigraphy of Hibernia Oil Field, Grand Banks, Newfoundland. *American Association of Petroleum Geologists Bulletin*, **71**, 1210–1232.

VAN PANHUYS-SIGLER, M., BAUMANN, A. & HOLLAND, T. C. 1991. The Tern Field, Block 210/25a, UK North Sea. *In*: ABBOTTS, I. L. (ed.) *United Kingdom Oil and Gas Fields, 25 Years Commemorative Volume*, Geological Society, London, Memoir, **14**, 191–197.

VENDEVILLE, B., COBBOLD, P. R., DAVY, P., BRUN, J. P. & CHOUKROUNE, P. 1987. Physical models in extensional tectonics at various scales. *In*: COWARD, M. P., DEWEY, J. F. & HANCOCK, P. L. (eds) *Continental Extensional Tectonics*, Geological Society, London, Special Publication, **28**, 95–107.

WILLIAMS, G. D., POWELL, C. M. & COOPER, M. A. 1989. Geometry and kinematics of inversion tectonics. *In*: COOPER, M. A. & WILLIAMS, G. D. (eds) *Inversion Tectonics*, Geological Society, London, Special Publication, **44**, 3–15.

WITHJACK, M. O., MEISLING, K. E. & RUSSELL, L. R. 1989. Forced folding and basement-detached normal faulting in the Haltenbanken area, offshore Norway. *In*: TANKARD, A. J. & BALKWILL, H. R. (eds) *Extensional tectonics and stratigraphy of the North Atlantic Margins*, American Association of Petroleum Geologists, Memoir, **46**, 567–575.

——, OLSON, J. & PETERSON, E. 1990. Experimental models of extensional forced fold. *American Association of Petroleum Geologists Bulletin*, **74**, 1038–1054.

WOODCOCK, N. H. & FISCHER, M. 1986. Strike-slip duplexes. *Journal of Structural Geology*, **8**, 725–735.

ZIEGLER, P. A. 1975. Geologic evolution of North Sea and its tectonic framework. *American Association of Petroleum Geologists Bulletin*, **59**, 1073–1097.

—— 1988. *Evolution of the Arctic–North Atlantic and the Western Tethys*, American Association of Petroleum Geologists, Memoir, **43**, 198 pp.

Case studies: Europe

Late Jurassic–Early Cretaceous inversion of the northern East Shetland Basin, northern North Sea

D. W. THOMAS & M. P. COWARD

Department of Geology, Imperial College, London, SW7 2BP, UK

Abstract: Current research into the structural evolution of the East Shetland Basin, northern North Sea, indicates that a significant phase of structural inversion occurred during the Latest Jurassic–Early Cretaceous.

Structural effects of this tectonic phase include the pronounced uplift of pre- and syn-rift sequences along the western margin of the basin and the partial inversion of Mesozoic half-graben along intra-basinal NE–SW controlling faults. In addition, the localized occurrence of flower geometries and pop-ups along NE–SW trending faults suggests a component of strike-slip during the reactivation of these pre-existing structures.

Work presented stresses the importance in the role of pre-existing lineaments and the effects of superimposing subsequent non-coaxial extensional events upon them. Lineaments may be prominent reactivated basement shear zones or fault systems from earlier extensional phases.

Inversion will be analysed in detail with respect to two areas within the northern East Shetland Basin, the Tern sub-basin and the northern Penguin ridge. A regional tectonic model of northern North Sea rift evolution is presented and its implications to observed Latest Jurassic–Early Cretaceous structure discussed.

During the past twenty-five years of North Sea oil exploration the structural evolution of the East Shetland Basin has received a great amount of attention and the resultant rifted fault block model is well documented. Both Late Jurassic and Triassic rift events have been recognized (Badley *et al.* 1984, 1988; Roberts *et al.* 1991; Yielding *et al.* 1992; Lee & Hwang 1993). Late Jurassic structuration is, for the most part well imaged on seismic data. In contrast, Triassic structures are often poorly imaged and as a result the kinematics of this earlier event and its timing are poorly understood. In addition, the pre-Triassic history of the basin is relatively unknown (Coward 1990).

Detailed structural analysis over the past few years (aided by the enhanced acquisition and processing of available 2D data and the advent of high quality 3D data) has noted the occurrence of anomalous Late Jurassic–Early Cretaceous structural features within the basin (Gabrielsen & Robinson 1984; Fossen 1989; Booth *et al.* 1992; Bartholomew *et al.* 1993). These structural anomalies include partial reactivation and back-steepening of normal faults, localised 'pop-up' structures and flower geometries. In addition, there has been widespread uplift and erosion of pre-rift sediments along the western margin of the basin in Late Jurassic–Early Cretaceous times and more so in Late Cretaceous–Early Tertiary times (Brodie

& White 1993). Various mechanisms have been proposed to account for such structures, including strike-slip tectonics and wholescale basin inversion (Frost 1987; Speksnijder 1987; Demyttenaere *et al.* 1993).

Location and geological setting

The East Shetland Basin is situated approximately 150 km to the northeast of the Shetland Islands, in the U.K. sector of the northern North Sea. It occupies a lozenge-shaped, faulted terrace, bounded to the West by the Palaeozoic East Shetland Platform, to the East and South by the deeper Viking Graben and to the North by the Magnus Basin (see Fig. 1). Economically it is one of the most prolific oil provinces in Western Europe and has received much attention over the past twenty-five years of U.K. North Sea oil exploration and production.

The general stratigraphy of the region is illustrated in Fig 2. Basin sediments, ranging in age from Devonian to Recent, were deposited upon an underlying basement of Caledonian metamorphic rocks. Oil reservoirs are essentially restricted to limited sand systems within the Triassic, Middle and Upper Jurassic, and the Palaeocene-Eocene. To date, some fifty oil fields have been discovered in the East Shetland Basin, with an estimated fifty billion barrels of recoverable oil in the Middle Jurassic sequences

From BUCHANAN, J. G. & BUCHANAN, P. G. (eds), 1995, *Basin Inversion*,
Geological Society Special Publication No. 88, 275–306.

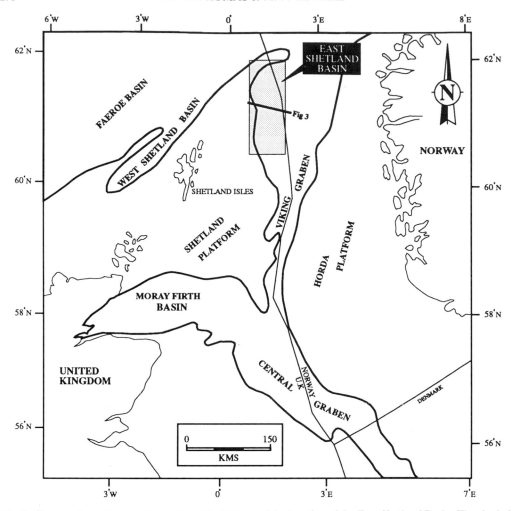

Fig. 1. Structural provinces in the northern North Sea and the location of the East Shetland Basin. The shaded box denotes the area of the regional map illustrated in Fig. 4.

alone (Lee & Hwang 1993). Although sometimes modified by erosion, the majority of traps are structural, thus an understanding of the structural evolution of the basin is of prime importance, both from an academic and an industrial point of view and can be considered in two stages:

i) development of the Palaeozoic framework of the basin, encompassing Caledonian orogenesis, subsequent orogenic collapse and Devonian/Carboniferous basin formation;

ii) inherent control of this early structure on the later Mesozoic basin evolution.

The main aim of this paper is to describe the Late Jurassic–Early Cretaceous evolution of the basin and to present a model which helps explain the occurrence of anomalous structural features of this age, observed on seismic sections. The model will illustrate how the orientation, attitude and distribution of pre-existing planes of weakness ultimately control the geometry and form of subsequent rift structures. Such a model may be of significance to hydrocarbon exploration and production.

Database

Approximately 5000 km of seismic data have been interpreted within the basin. These include Nopec ESBT 89 lines from Oryx UK, which provide the regional framework, together with more detailed data packages provided by Nopec (UK) and Ranger Oil (UK). The seismic data

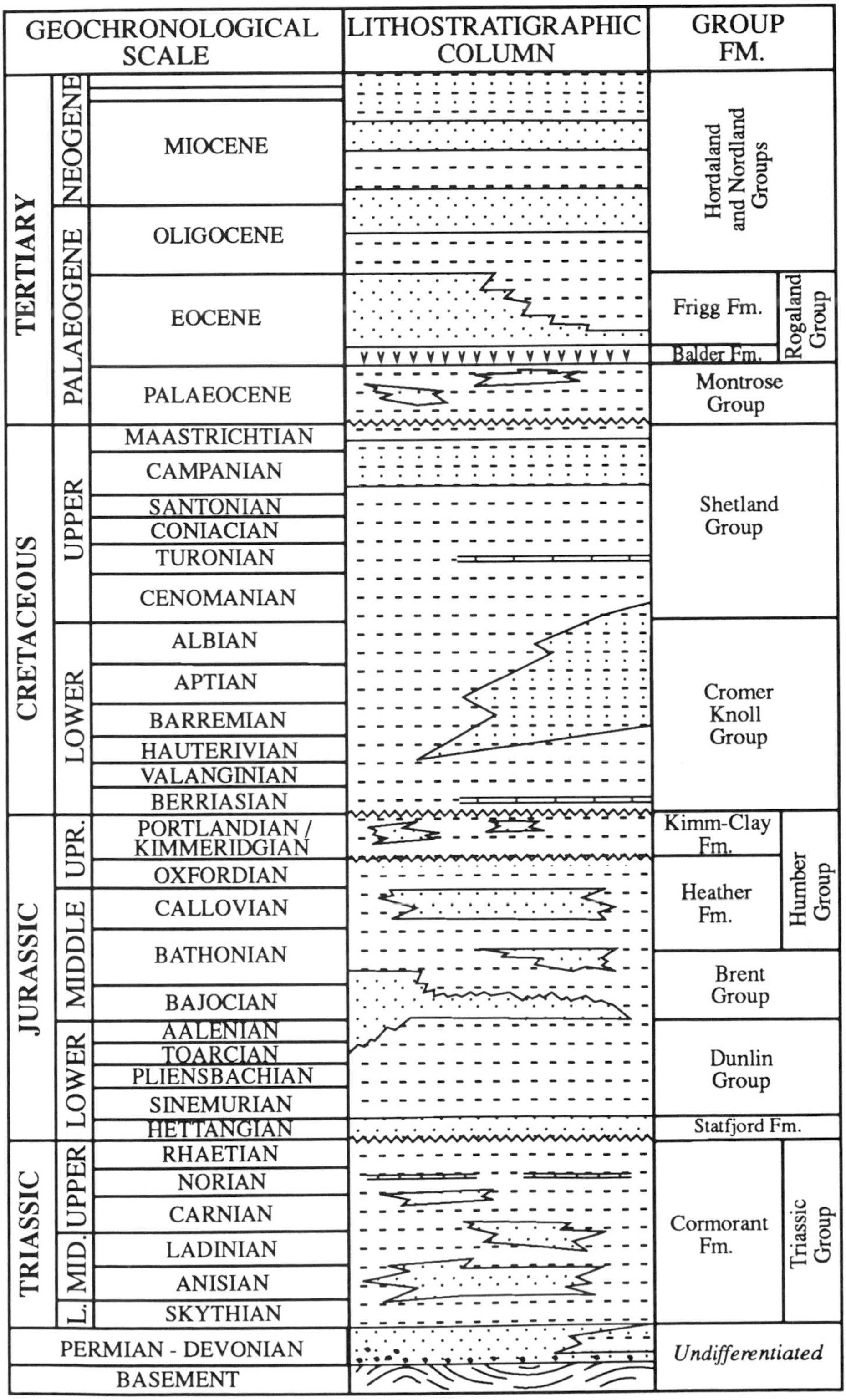

GEOCHRONOLOGICAL SCALE				LITHOSTRATIGRAPHIC COLUMN	GROUP FM.	
TERTIARY	PALAEOGENE	NEOGENE	MIOCENE		Hordaland and Nordland Groups	
			OLIGOCENE			
			EOCENE		Frigg Fm.	Rogaland Group
					Balder Fm.	
			PALAEOCENE		Montrose Group	
CRETACEOUS	UPPER		MAASTRICHTIAN		Shetland Group	
			CAMPANIAN			
			SANTONIAN			
			CONIACIAN			
			TURONIAN			
			CENOMANIAN			
	LOWER		ALBIAN		Cromer Knoll Group	
			APTIAN			
			BARREMIAN			
			HAUTERIVIAN			
			VALANGINIAN			
			BERRIASIAN			
JURASSIC	UPR.		PORTLANDIAN / KIMMERIDGIAN		Kimm-Clay Fm.	Humber Group
	MIDDLE		OXFORDIAN		Heather Fm.	
			CALLOVIAN			
			BATHONIAN		Brent Group	
			BAJOCIAN			
	LOWER		AALENIAN		Dunlin Group	
			TOARCIAN			
			PLIENSBACHIAN			
			SINEMURIAN			
			HETTANGIAN		Statfjord Fm.	
TRIASSIC	UPPER		RHAETIAN		Cormorant Fm.	Triassic Group
			NORIAN			
	MID.		CARNIAN			
			LADINIAN			
	L.		ANISIAN			
			SKYTHIAN			
PERMIAN - DEVONIAN					Undifferentiated	
BASEMENT						

Fig. 2. Generalized stratigraphic column for the East Shetland Basin (modified from Lee & Hwang 1993).

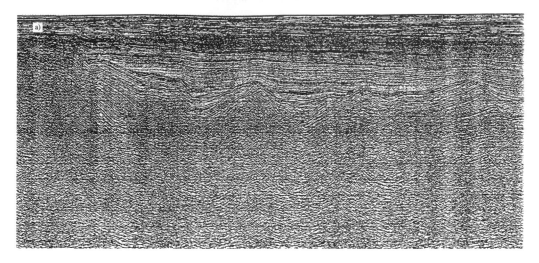

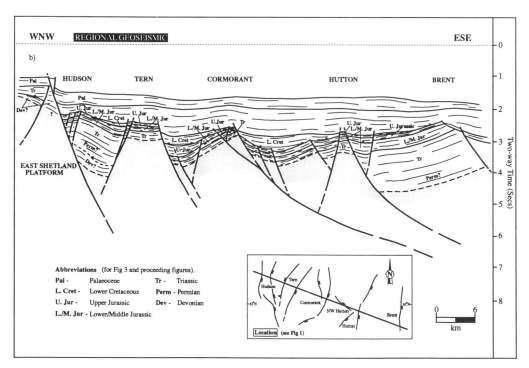

Fig. 3. (a) Uninterpreted, and (b) geoseismic interpretation of a WNW–ESE regional seismic line across the central region of the East Shetland Basin. The locations of major oilfields are indicated. Basement rocks shaded. Inset map shows the section location (see Fig. 1).

have been correlated to some 100 wells distributed throughout the basin.

Tectonic history

Structurally the East Shetland Basin is located on the western flank of the Viking Graben system and grossly consists of west-tilted fault blocks, bounded by down to the east normal faults. Two phases of rifting have been recognized, in the Early Triassic and the Late Jurassic (Badley *et al.* 1984, 1988; Ziegler 1988; Yielding *et al.* 1992), each followed by an associated phase of thermal subsidence, in the Early–

Middle Jurassic and the Cretaceous–Tertiary respectively. The amount of crustal extension within the Viking Graben has been estimated from deep seismic profiling (Klemperer 1988) and gravity modelling (Zervos 1987). Analyses of BIRPS deep seismic reflection surveys across the region indicate that crustal thinning beneath the Viking Graben is in the order of a factor approaching 2 (Klemperer 1988). Such crustal attenuation has been explained by various workers in terms of pure, simple and hetero-geneous shear mechanisms of basin formation (Mackenzie 1978; Barton & Wood 1984; Beach 1985; Coward 1986).

An interpreted WNW–ESE seismic section across the central region of the basin is illus-trated in Fig. 3. Geologically the section can be divided into two: a lower rifted succession characterised by west-tilted fault blocks, base-ment horsts and large east-dipping normal faults, overlain by an onlapping succession of thermal subsidence-related sediments. Obser-vations made from the analysis of the section include the following.

i) Thickening of Triassic sequences ba-sinwards towards the east, cut by major east-dipping normal faults. There is very little evidence of divergent geometries within these sequences, although there is a possible Late Triassic growth sequence within the Tern sub-basin, related to move-ment across the Northern Tern Fault (see later discussion).

ii) An increase in the thickness of Lower/ Middle Jurassic sequences towards the east. For the most part this thickening is gradual, though abrupt changes occur across the Hutton Fault Block (Yielding *et al.* 1992). These rapid thickness changes are too great to result from differential compaction and can be interpreted as representing local normal fault movement as a precursor to the Oxfordian–Kimmeridgian rift phase. It should also be noted that throw across the Hutton Fault is important in the Triassic but only minor within the Late Jurassic.

iii) Thickening of Upper Jurassic sequences towards the west within isolated, fault-bounded half-grabens. Divergent ge-ometries are poorly developed within the Oxfordian–Kimmeridgian succession and slight aggradation occurs against certain bounding faults. The thickness of syn-rift packages is often overestimated (Prosser 1993) and it is probable that the majority of Upper Jurassic sequences have parallel geometries, which would reflect passive infill of sediment starved depressions.

iv) The high amplitude, continuous nature of the Base Cretaceous unconformity reflector and its breaching by faulting across the section. Such relationships may be a result of differential compaction across fault boundaries, or alternatively may represent continued fault activity within the Lower Cretaceous. Indeed the basin margin faults appear to have been active up until Late Palaeocene/Early Eocene time.

v) The apparently anomalous structural ge-ometry of the Hudson Field; the impli-cations of which will be discussed within a later section.

Stretching estimates calculated from fault heaves for the Triassic and Late Jurassic se-quences across this section, give B factors of 1.3 and 1.1 respectively. Such values attest to a more intense Triassic stretching event, but do not explain the amount of crustal thinning across the basin shown by deep seismic data. Assuming deep seismic B factors are correct, additional extension may be accomplished via sub-resolution faults (Marrett & Allmendinger 1992), or by incorporating earlier Devonian/ Carboniferous extensional phases into the evol-ution of the basin (Coward 1990, 1993).

Late Jurassic/Early Cretaceous basin structure

The structure of the Base Cretaceous uncon-formity is illustrated in Fig. 4. Faults that intersect and cut the unconformity surface have been mapped, to give the likely basin structure in Late Jurassic/Early Cretaceous times. Con-tour patterns are complex, but depict a general deepening of the basin towards the east, into the Viking Graben proper. Three trends of regional structure are seen to exist:

i) NE–SW striking faults in the northern regions of the basin;

ii) NW–SE striking faults in the southern regions of the basin;

iii) N–S striking faults in the central and eastern regions of the basin.

Each group expresses itself in a particular structural style on the seismic data. NE–SW faults are often associated with structural highs, (Tern ridge, Penguin ridge). These ridges are bounded by steep, locally overturned faults, which often display convergent and divergent structural geometries. Well developed Ox-fordian–Kimmeridgian growth sequences lo-cally abut these faults. NW–SE faults are geometrically, less complex than the NE–SW faults, and generally there is a lack of Late

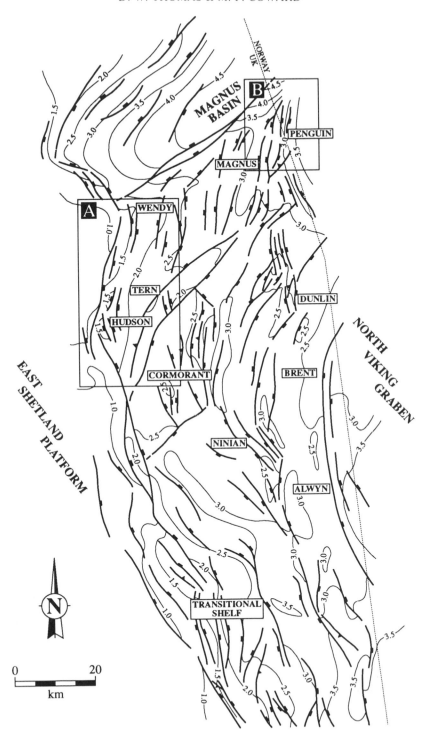

Fig. 4. Base Cretaceous two-way time structure map of the East Shetland Basin. Note the dominance of NE–SW faults in the North and NW–SE faults in the South. The location of the Tern sub-basin (Fig. 7) and the northern Penguin ridge (Fig. 13) are indicated. The kinematic significance of localized, reverse faulting will be discussed in the text.

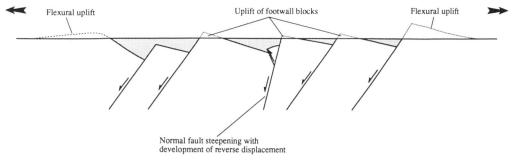

Fig. 5. Models of inversion applicable to rift basins (after Gillcrist *et al.* 1987; Kusznir *et al.* 1991; Roberts *et al.* 1991; Yielding *et al.* 1992; Coward 1994).

Jurassic half-graben development across these structures (Transitional Shelf). N–S faults form the majority of the major east-dipping faults imaged on seismic data across the eastern fields of the basin (Brent, Alwyn). The throw on these faults is generally large and it is these structures that have been used by previous workers to deduce an E–W extension direction across the basin in Late Jurassic times (Hay 1978; Badley *et al.* 1988; Roberts *et al.* 1991; Yielding *et al.* 1992). Interpretation of 3D data also shows the presence of minor E–W striking faults in crestal regions of the eastern fault blocks. These E–W structures have small throws and a range of attitudes and are an important element of most oil field traps. Unfortunately such structures are extremely difficult to pick out on regional data, however, their structural importance should not be underestimated.

It is important to quantify the strike extent of individual faults, a problem inherent to all scales of seismic interpretation and mapping. Recent work on the Cormorant Field by Demyttenacre *et al.* (1993) has noted that the eastern block-bounding fault (originally interpreted as a N–S fault), consists of a complex array of NE-SW and NW–SE fault segments. This interpretation has implications on the validity of the presumed E–W extension in Late Jurassic times.

Late Jurassic/Early Cretaceous kinematics

The interpretation of Late Jurassic–Early Cretaceous kinematics along the structures discussed above is dependent upon what type of fault they are deemed to represent (normal or strike-slip) and the predicted extension direction. Bartholomew *et al.* (1993) assume that the range of fault orientations is a direct consequence of underlying basement anisotropy. They propose that superimposed Late Jurassic E–W extension has resulted in the reactivation

of underlying basement shear zones, which are expressed in the cover by a range of complex oblique-slip fault geometries. Alternatively, the contrasting Late Jurassic half-graben development across NE–SW and NW–SE faults outlined above, suggests selective reactivation of basement structures, possibly attributed to temporal changes in the regional stress field.

The Base Cretaceous structure map (Fig. 4) shows the localized occurrence of NE–SW reverse faults within the basin, bounding basin highs (Tern, Penguin), and also in field crests (Dunlin, Cormorant, Penguin). The structural significance of such anomalous features within an extensional basin requires a thorough understanding of the Late Jurassic–Early Cretaceous kinematic development of the basin.

Basin inversion

The main origins of basin inversion, as we understand the phenomenon at the present time, can be summarized as follows (after Coward 1994).

i)   Related to isostatic rebound – inversion is caused by removal of a load (glacial overburden) or by the erosion of a mountain belt.

ii) Related to the diapiric action of salt – inversion is caused by a) movement of salt up or down the dip of a tilted fault block, b) slip of the post-salt carapace into the basin.

iii) Related to lithospheric heating – inversion occurs where plates migrate across hot spots or where magmatic underplating occurs (Brodie & White 1993).

iv) Related to extensional faulting – inversion includes footwall uplift of rotated fault blocks and rift flank uplift at the margins of large grabens (Marsden *et al.* 1990; Kusznir *et al.* 1991).

v) Related to horizontal plate movements (tectonic inversion) – these include:
   a) inversion due to collisional processes, especially oblique plate collision;
   b) inversion due to strike-slip tectonics, particularly at restraining bends and offsets along lateral fault zones;
   c) inversion due to the rotation of earlier fault blocks, shortening can occur across rotated blocks between reactivated bounding faults.

Models of inversion applicable to rift basins are illustrated in the schematic diagram of Fig. 5. The quantification of uplift forms an integral part of basin modelling and its determination is relying increasingly on new technologies, including fission track analysis (Bidstrup et al. 1993) and sonic velocity studies (Hillis 1993). However, from seismic and field data, inversion structures can be recognized by (i) the development of reverse fault displacement in response to excess steepening of pre-existing normal faults and associated folding of sediments; and (ii) the associated uplift of sediment horizons above their regional level for the particular part of the basin.

Inversion structures within the northern North Sea

Two main phases of regional scale tectonic inversion have been documented from the northern North Sea. They have been attributed to Variscan-related and Alpine-related orogenesis (Lee & Hwang 1993).

i) Variscan inversion is centred on the possible existence of folded and thrusted Devonian sequences flooring isolated Mesozoic half-graben upon the East Shetland Platform (Holloway et al. 1991). Thrusted successions of similar age are encountered sporadically along the Transitional Shelf (see Fig. 4), in the southern regions of the East Shetland Basin. Such observations imply that Early Mesozoic rifting acted upon a thickened and highly irregular crust. Unfortunately the basinal extent of these Palaeozoic successions is unknown.

ii) Alpine inversion structures have been described from the Norwegian sector of the northern North Sea. These include transpressional reactivation of Jurassic fault blocks (Pegrum & Ljones 1984; Caselli 1987) and reutilization of early normal faults as reverse faults (Biddle & Rudolph 1988).

We propose a third phase of tectonic inversion in Late Jurassic–Early Cretaceous times. Resultant structures are far more localized but are nevertheless numerous, and are believed to represent an integral part of the structural framework of the basin. Structural effects will be analysed in detail with respect to two areas within the northern East Shetland Basin, the Tern sub-basin and the northern Penguin ridge (see Fig. 4).

Area A (Tern sub-basin)

Area A is located in the vicinity of the western margin of the East Shetland Basin (see Fig. 4), at the confluence of the NE–SW trending Tern/Eider ridge with the NNE–SSW/NW–SE trending western margin fault complex. The condensed nature of conformable Mesozoic successions upon the Tern/Eider ridge and the shallow nature of basement (2.6–3.0 secs TWT), attest to the fact that it acted as a structural high during both Triassic and Jurassic rift events (Yielding et al. 1992). The Tern ridge separates NW-facing half-grabens to the north, from SE-facing half-grabens to the south, and has been described as representing an accommodation zone between offset extensional faults (Lee & Hwang 1993) and possibly the expression of the Iapetus suture (Frost 1987).

The area has been chosen to illustrate the relationship between the Mesozoic half-graben to the north of the ridge (Tern sub-basin) and its continuation to the southwest, into the area of interaction with the western margin faults of the East Shetland Basin. The structural form of this half-graben, was controlled during both the Triassic and Jurassic rift phases by movement on the NE–SW trending northern bounding-fault of the Tern/Eider ridge (Northern Tern Fault). The precise age of initiation of this fault is unknown, however, extremely thick Permo-Triassic and possibly older sequences are recognized on seismic data at depth within the Tern sub-basin. Overprint by later Jurassic movement along the fault is minimal and, therefore, the Early Mesozoic structure of the half-graben can be analysed in detail.

A generalized map of Triassic structure in the northern North Sea is illustrated in Fig 6, with the location of the Tern sub-basin highlighted. The fault pattern is dominated by NE–SW/NNE–SSW trending structures. Sequences in the northern region thicken essentially to the east–southeast within asymmetric graben controlled by major westward dipping normal faults, producing clastic-rich basins (West Shetland, Fair Isle, Unst, Tern, North Viking)

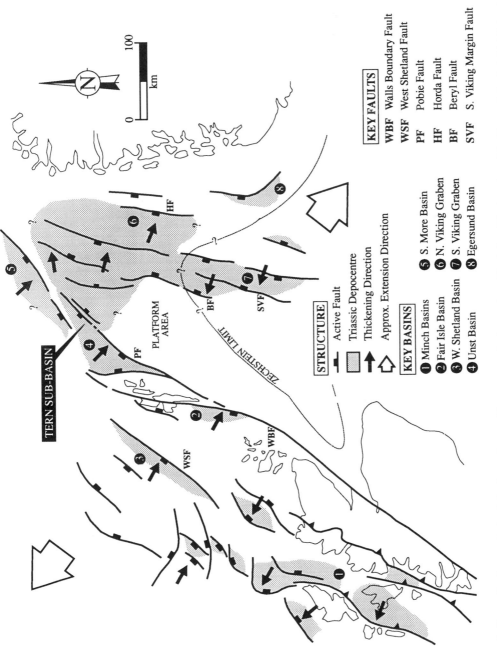

Fig. 6. Generalized map of Triassic basins and controlling structure in the northern North Sea. The location of the Tern sub-basin is highlighted.

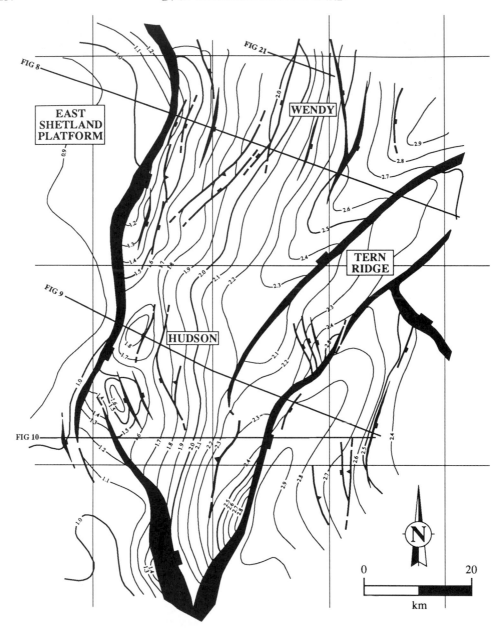

Fig. 7. Base Cretaceous two-way time structure map of the Tern sub-basin, showing the location of seismic lines illustrated in Figs 8, 9, 10 & 21.

separated by areas in which more distal sedimentation predominates. Basin controlling faults include the West Shetland Fault, the Walls Boundary Fault, the Pobie Fault, the Northern Tern Fault and the Horda Fault system. The overall eastward thickening of Triassic sequences, evident from analysis of regional seismic data across the East Shetland Basin and

North Viking Graben, defines an asymmetric half-graben controlled by the major westward dipping faults of the Horda Platform (Badley *et al*. 1984, 1988). The accumulation of thick clastic sequences (Skagerrak) within the Egersund and North Danish Basins is probably attributed to the footwall uplift and erosion of the Norwegian Platform during the Triassic extension of this

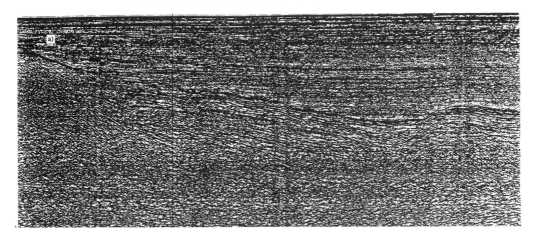

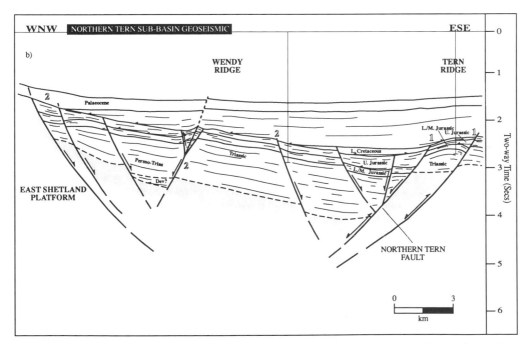

Fig. 8. (**a**) Uninterpreted, and (**b**) geoseimic interpretation of a seismic line across the northern region of the Tern sub-basin. Basement rocks shaded. See text for discussion.

during the Triassic extension of this Horda Fault system. The NE–SW strike of the basins is noteworthy. Development of Devonian basins off the NW coast of Scotland, clearly involves the reutilization of low-angle Caledonian thrust faults (Coward & Enfield 1987). It is expected that similar zones of weakness were important during Permo-Triassic basin evolution.

Triassic extension within the northern North

Sea appears to be primarily concentrated along westward dipping normal faults. The extension direction and its probable variation is difficult to constrain. Regional mapping and analysis of residual gravity data within the West Shetland Basin, Faeroe Basin and Unst Basin have noted the occurrence of approximately NW–SE compartmentalizing transfer faults to the Triassic structure (Rumph *et al.* 1993). This suggests a more or less orthogonal NW–SE extension across the northern North Sea region.

A Base Cretaceous structure map of Area A is illustrated in Fig. 7, along with the orientation and position of selected seismic lines used in the discussion. These sections are presented in time format. The NE–SW trending high which marks the Tern/Eider ridge can easily be picked out on the map. Within the Tern sub-basin, structural contours depict a deepening to the northeast and intra-basin faults define ridges which strike N–S to NNE–SSW, i.e. Wendy ridge. The western margin fault complex is composed of a right-stepping series of alternating NW–SE and NNE–SSW fault segments, and complex structures occur within the vicinity of overlap zones, the Hudson field being a prime example. The occurrence of NE–SW to NNE–SSW striking reverse faults will be discussed below.

Seismic observations

Figure 8 illustrates a seismic line across the northern region of the Tern sub-basin (see Fig. 7) and its geoseismic interpretation. The Northern Tern Fault dips approximately 50° to the NW and has been the controlling influence in the development of the Tern sub-basin during Triassic and Oxfordian/Kimmeridgian extensional phases.

Thick Permo-Triassic sequences, dipping approximately 10° to the SE, are well imaged within the sub-basin. Their depth extent, however, is uncertain as the interpreted top of acoustic basement cannot be calibrated with well control. Divergence of sequences into the Northern Tern Fault is not seen and hence Permo-Triassic growth cannot be inferred from analysis of this section alone. Divergence of sequences at depth into the NW-dipping bounding fault of the Wendy ridge, attest to an earlier, significant growth phase possibly within the Devonian–Carboniferous.

Lower and Middle Jurassic sequences show little thickness change, both across the sub-basin and the Tern ridge, which indicates tectonic quiescence within the region during these times. They do thicken, however, very slightly to the southeast into the East Shetland Basin. Such relationships can be attributed to a broad, thermal subsidence-related deposition in Early and Middle Jurassic times. Wedge shaped seismic geometries are better developed within the Oxfordian/Kimmeridgian sequences of the sub-basin. Reflectors exhibit well defined divergence within early infill deposits, with associated northwest onlap onto the top Middle Jurassic horizon. Overlying Volgian sequences display

parallel infill geometries. Seismically, the lower wedge shaped sequences are characterized by low amplitude discontinuous events, whereas the later parallel sequences exhibit higher amplitude, more continuous reflection events.

Shallow basement (2.9–3.0 secs TWT) is interpreted beneath the Tern ridge. Triassic sequences are highly condensed on this structure, and attest to its presence during the Triassic development of the area. In contrast, Upper Jurassic sequences appear to be only slightly condensed across the structure. The throw estimates in two-way time across the Northern Tern Fault (Base Triassic: 0.85 secs (c. 1200 m); Late Jurassic: 0.36 secs (c. 400 m)), indicate the dominance of the earlier Triassic phase of faulting and the lesser importance of the subsequent Jurassic rift overprint. Western margin faults which bound the distal margin of the sub-basin, dip towards the southeast and show a lack of growth in both the Triassic and Late Jurassic. They are difficult to assess kinematically, due to large amounts of footwall erosion at Base Cretaceous level, however, appear to post-date the development of the Tern sub-basin oulined above.

Figure 9 illustrates a seismic line across the central region of the Tern sub-basin (see Fig. 7) and its geoseismic interpretation. The Northern Tern Fault dips 64° to the northwest and has exerted a similar control on the structural development of the Tern sub-basin as was outlined in the interpretation of Fig. 8. A thick sequence of Permo-Triassic deposits dipping approximately 16° to the southeast are well imaged within the sub-basin. Triassic reflectors exhibit strong seismic character and there is evidence for slight divergence of a high amplitude series of reflectors into the Northern Tern Fault. A package of discontinuous high amplitude reflectors at depth possibly represent Devonian sequences. The unconformable nature of overlying Permo-Triassic basin fill, suggests an earlier, totally unrelated phase of Palaeozoic basin evolution.

No observable Upper Jurassic growth is encountered on this section and it appears that either growth along NW-dipping faults in this area was minor, or that substantial erosion of Upper Jurassic syn-rift sequences has occurred. Sequences display a low amplitude character and thus stratigraphic relationships are difficult to assess.

The Tern ridge is underlain by shallower basement (2.6–2.7 secs TWT) than in Fig. 8 and across it Triassic sequences are severely condensed. Lower and Middle Jurassic sequences show little thickness change across the section

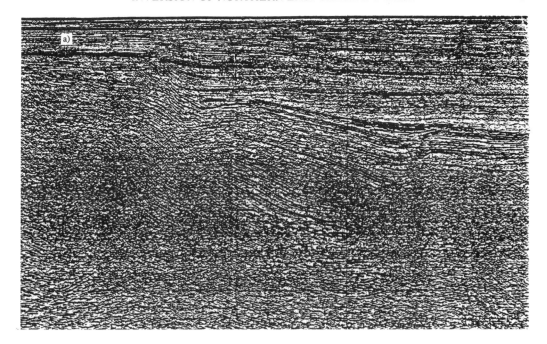

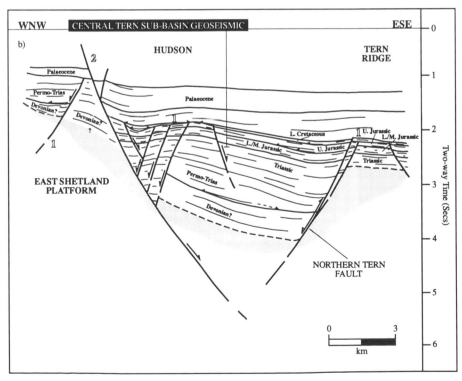

Fig. 9. (a) Uninterpreted, and (b) geoseimic interpretation of a seismic line across the central region of the Tern sub-basin. Note the structural high of the Hudson field. Basement rocks shaded. See text for discussion.

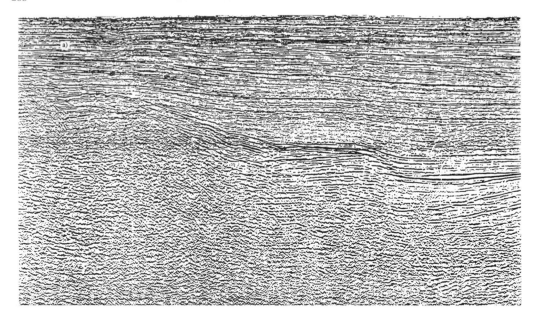

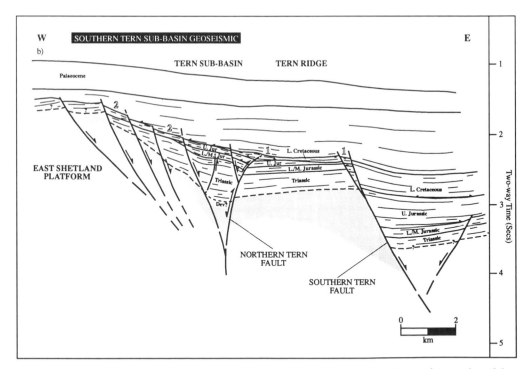

Fig. 10. (a) Uninterpreted, and (b) geoseimic interpretation of a seismic line across the southern region of the Tern sub-basin. Note the inversion of the Tern sub-basin along the Northern Tern Fault. Basement rocks shaded. See text for discussion.

and attest to a tectonically stable period during these times. Renewed tectonic activity in the Late Jurassic is evident from differential growth across the ridge.

Throw estimates in two-way time across the Northern Tern Fault (Base Triassic: 1.28 secs (*c.* 1500 m), Late Jurassic: 0.07 secs (*c.* 80 m)), indicate a larger component of Triassic extension than in the northern section, but a significantly smaller Late Jurassic component. This further supports the concept that Triassic rifting exerted the major control upon the structural development of the basin and that Oxfordian/Kimmeridgian re-utilization of controlling northwest dipping faults was minimal.

A SE-dipping, western margin fault is well imaged within this section. The fault dips at approximately 65° and attains a throw of 0.8 secs two-way time (*c.* 1200 m) at Base Cretaceous level. No Triassic or Jurassic growth is observed towards the fault, however, displacement exists on the fault high up into the Tertiary section. Whether this displacement represents tectonic faulting or differential compaction remains a matter of debate. Upper Jurassic–Lower Cretaceous sequences within the Tern sub-basin show a pronounced uplift towards the western margin fault in the form of an apparent hangingwall fold, which forms the structural high of the Hudson field.

Figure 10 shows a seismic line across the southern region of the Tern sub-basin (see Fig. 7) and its geoseismic interpretation. The section shows that here the Northern Tern Fault has a dip of 76° and is concave in profile. The relationship between this fault and its control on Triassic and Jurassic sub-basin structure is difficult to assess, due to the narrowing of the basin and the poor resolution of the seismic data. The distal margin of the Tern sub-basin is bounded by a terrace of E-dipping normal faults which comprise a synthetic relay zone along the western margin fault complex (see Fig. 7). No Triassic or Jurassic thickness changes are interpreted across these faults, however, observable displacement is evident up into the early Late Cretaceous section.

Permo-Triassic megasequences dip approximately 20° to the east and consist of high amplitude, continuous reflectors. Divergent packages seem to be absent from the Triassic section, the lateral and vertical extent of which is interpretable within the sub-basin. The gradual SW-thickening of Triassic section evident from seismic interpretation throughout the region, suggests substantially thick sequences are envisaged to exist within this area of the sub-basin. Shallow basement (2.9 secs TWT) is interpreted beneath the Tern ridge, across which Triassic sequences thin.

Jurassic sequences are also difficult to interpret with absolute confidence across the section. As interpreted, Lower and Middle Jurassic sequences show little thickness change across the section and are unaffected by the Tern ridge. A slight thinning towards the southern margin of the ridge has resulted from footwall uplift, attributed to active faulting along the Southern Tern Fault during Oxfordian/Kimmeridgian times. This is supported by divergence in Oxfordian/Kimmeridgian sequences in the adjacent half-graben to the east. There is no evidence for divergent geometries within the Upper Jurassic of the Tern sub-basin and only minor thickness changes occur across the Northern Tern Fault.

A pronounced Late Jurassic–Early Cretaceous uplift of the distal margin of the Tern sub-basin occurs towards the west, with resultant erosional truncation of Jurassic sequences. In addition, minor folding of Lower Cretaceous sequences in the hangingwall of the Northern Tern Fault, is attributed to a coeval phase of normal fault inversion.

Kinematic conclusions

NW-dipping, NE–SW faults are interpreted as having controlled the Permo-Triassic development of the Tern sub-basin. They were co-axially reactivated to a minor extent in Oxfordian/Kimmeridgian times. These faults are labelled 1 in Figs 8, 9 and 10. SW-dipping, NE–SW and NW–SE fault segments which form the western margin of the East Shetland Basin appear to postdate Late Jurassic growth in the Tern sub-basin, and in places along the margin have remained active until the Eocene. These faults are labelled 2 in Figs 8, 9 and 10. It is difficult to interpret with certainty the zone of interaction between the two sets of faults due to poor resolution of the seismic at depth.

Of prime interest is the occurrence of the anomalous Late Jurassic–Early Cretaceous structures, outlined above, within the Tern sub-basin. These include: (i) apparent inversion along the Northern Tern Fault and (ii) uplift along the distal margin of the sub-basin. The three-dimensional relationship of these structures to one another is illustrated in the schematic block diagram of the Tern sub-basin shown in Fig. 11.

(i) Inversion along NE–SW faults A progressive increase to the SW in Late Jurassic–Early Cretaceous inversion of the Northern Tern Fault is apparent from analysis of seismic sections across the Tern sub-basin (see Figs 8–10). This is associated with a gradual SW-steepening of the Northern Tern Fault, coupled with the development of a flower structure well imaged on seismic data. In the southern-most seismic section (Fig. 10) and from map analysis (Fig. 7), the Tern sub-basin appears to have

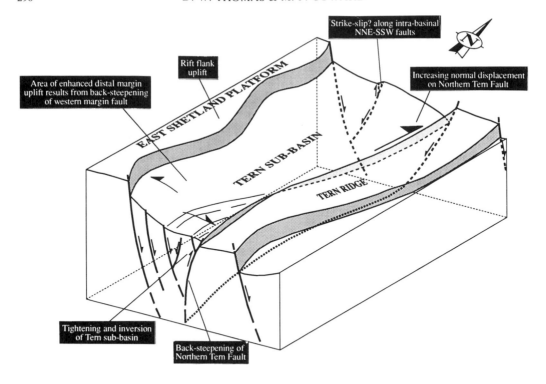

Fig. 11. Schematic block diagram of the Tern sub-basin illustrating the steepening and SW-inversion of the Northern Tern Fault and the uplift of the distal margin of the sub-basin.

been tightened and wedged out. Figure 12 shows a two-way time separation diagram constructed for throw of the Top Triassic reflector horizon along the Northern Tern Fault. This clearly illustrates the increase in normalized fault displacement towards the NE and its gradual inversion to the SW. Using an average interval velocity of 12500 ft/sec, the reverse offset at Top Triassic level is calculated to be approximately 180 m.

Additional effects include the apparent back-steepening of NW-dipping NNE–SSW striking intra-basin normal faults, i.e. Wendy ridge, with associated footwall folding and the development of hangingwall thrusts (see Fig. 8).

(ii) Uplift along the western margin of the Tern sub-basin Assuming constant interval velocities within the post-rift succession, the observed SW-increase in dip of the sub-basin fill towards the Northern Tern Fault (10–20°) is attributed to a SW-increase in distal margin uplift of the Tern sub-basin in Latest Jurassic–Early Cretaceous times. This uplift is believed to result from back-steepening of the developing western margin fault, with associated clockwise rotation of

its hangingwall succession, and appears to be controlled by the constraining effects of the Tern ridge to western margin extension (see Fig. 11). Modification of this uplift by basin subsidence is expressed on seismic data by the progressive marginward onlap of overlying Cretaceous sequences onto the tilted surface of the resultant unconformity.

The Hudson structure (Fig. 9) is significant, since it is interpreted as representing a pre-existing NW-dipping fault structure, which has subsequently been uplifted and tilted during Latest Jurassic–Early Cretaceous times.

Area B (northern Penguin ridge)

This area is located in the northern reaches of the East Shetland Basin (see Fig. 4), in the vicinity of the confluence of the NNE–SSW trending Penguin ridge with the WSW–ENE trending Magnus Fault.

The Penguin ridge is underlain by shallow basement (3.2–3.4 secs TWT) and like the Tern ridge described earlier, is believed to have acted as a structural high during both the Triassic and Jurassic rift events. Both the Penguin and

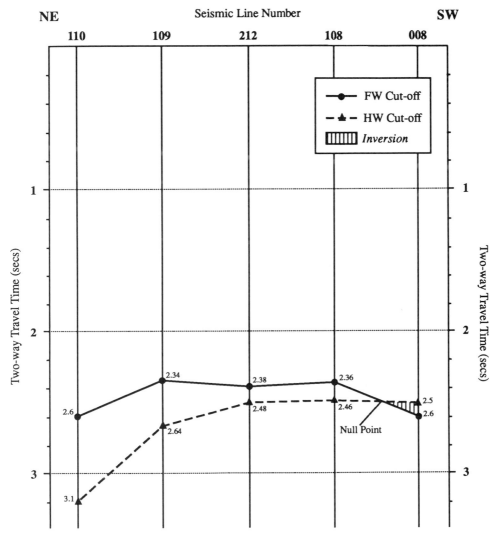

Fig. 12. Two-way time separation diagram for the Northern Tern Fault at Top Triassic level. Note the increase of normal displacement to the NE and its inversion to the SW.

Tern/Eider ridges lie on a NE–SW trend of structural highs, which can be traced north into the Haltenbanken and Nordland terraces of offshore Norway.

A Base Cretaceous structure map of the area is illustrated in Fig. 13 along with the location and orientation of two seismic profiles used in the discussion. The ridge itself is delineated by the prominent NNE–SSW trending West Penguin Fault and is structurally complex, with the development of both normal and reverse faults. In contrast, the Penguin half-graben is structurally less expressive, however, mapping clearly shows differential amounts of half-

graben rotation within northern and southern sub-basins, controlled by an interpreted NW–SE transfer fault which cross-cuts the area. Both the Penguin ridge and Penguin half-graben are seen to narrow towards the northeast.

The area illustrates the interaction of the Magnus Fault with the Penguin ridge. Booth *et al.*(1992) argue for a Late Jurassic–Early Cretaceous strike-slip origin for the Penguin ridge. Evidence of Late Jurassic–Early Cretaceous transpression has been proposed for the Halten-banken and Nordland terraces of offshore Norway (Gabrielsen & Robinson 1984; Larsen 1987; Caselli 1987; Fossen 1989), and for the

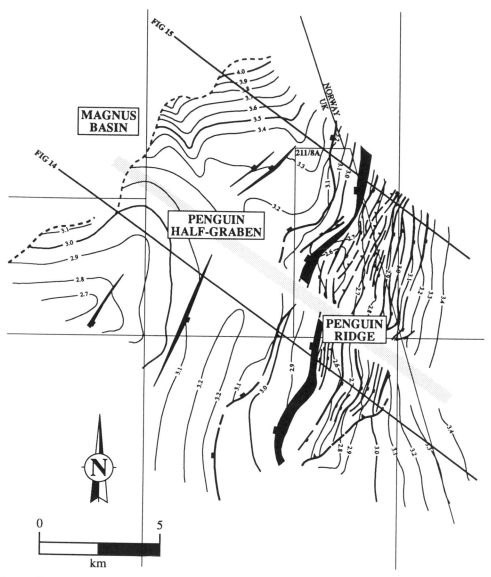

Fig. 13. Base Cretaceous two-way time structure map of the northern Penguin ridge, showing the location of seismic lines illustrated in Figs 14 & 15. The orientation of an interpreted NW–SE transfer zone is indicated.

More–Trondelag Fault Zone of onshore Central Norway (Gronlie & Roberts 1989).

The ridge is bounded on its western side by the Penguin half-graben. Thick Oxfordian–Volgian infills within the graben attest to substantial Late Jurassic movement upon the NW-dipping, West Penguin Fault. The Triassic and Jurassic development of the ridge is difficult to assess, due to large amounts of erosion along the crest, coupled with poor resolution of the flanks at depth. Analysis of NE–SW striking, SE-dipping faults within the Penguin half-graben indicate

greater Triassic throw than Late Jurassic throw. Structures are predominantly dip-slip, and an orthogonal NW–SE extension direction has been assumed. This is further supported by the NW–SE trend of the interpreted transfer fault outlined earlier.

Seismic and well analysis of the Magnus Basin (see Fig. 13) indicates that it is a very large half-graben, with a basin fill approximately 6 km thick. It is characterized by a thick infill of sub-parallel Lower Cretaceous sequences, which progressively onlap tilted, high amplitude

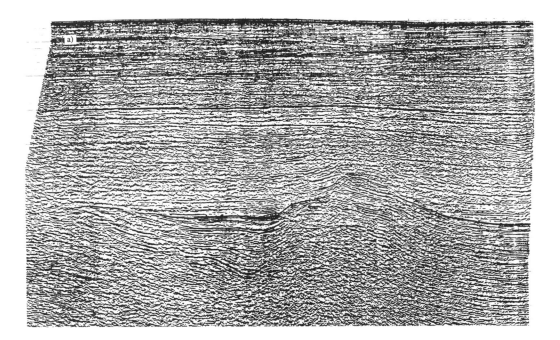

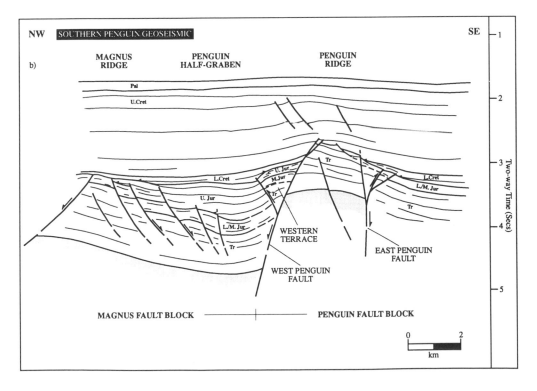

Fig. 14. (a) Uninterpreted, and (b) geoseimic interpretation of a seismic line across the southern region of the northern Penguin ridge. Basement rocks shaded. See text for discussion.

Triassic and older sequences to the northwest. Seismic sections analysed by the authors, across the basin, are severely undermigrated at depth and it is difficult to interpret stratigraphic relationships confidently. However, what appear to be aggradation geometries are evident within the vicinity of the Magnus Fault (Prosser 1993). Unfortunately it is not known if these aggradation features are Late Jurassic or Early Cretaceous in age. The great thickness of the Lower Cretaceous succession within the basin and the sub-parallel nature of its reflectors suggests passive infill of a rapidly developed depocentre.

Seismic observations

Figure 14 shows a NW–SE trending seismic line and geoseimic interpretation across the southern region of the Penguin ridge. The ridge forms a pronounced, highly eroded, structural high, composed of alternating high and low amplitude Triassic sequences, which dip gently towards the southeast. It is not known whether Permian or older sequences occur at depth. The underlying acoustic basement is interpreted to occur at a shallow level (3.3–3.5 secs TWT) and defines a broad, faulted antiform.

The adjacent Penguin half-graben is approximately 6.5 km in width, within which the geometries of Upper Jurassic reflectors clearly indicate infill. Seismically the infill can be divided into a lower package of low amplitude, discontinuous reflectors and an upper package of higher amplitude, more continuous reflectors. The sub-parallel nature of the reflectors suggests infill of a rapidly developed depocentre controlled by normal faulting along the West Penguin Fault during Oxfordian/Kimmeridgian times. The Western Penguin Fault dips at approximately 70° within its upper section, but reflector cut-offs at depth suggest a flattening towards the northwest. Of particular interest is the anomalous hangingwall upwarp of the Base Cretaceous Unconformity reflector, which defines the Western Terrace of the ridge.

Underlying Middle and Lower Jurassic sequences within the Penguin half-graben dip towards the southeast and are onlapped towards the northwest by Upper Jurassic sequences. Footwall uplift in the Late Jurassic–Early Cretaceous along the Magnus Fault is apparent from the erosional truncation of Lower and Middle Jurassic sequences in its crest. This is clearly evident from NW-onlap of overlying Lower Cretaceous sequences onto the tilted Magnus fault block.

The eastern margin of the Penguin ridge is seismically chaotic and difficult to interpret with absolute confidence. The East Penguin Fault is interpreted to possess a concave geometry. Well information in the hangingwall of this fault, a few kilometres south, confirms that Middle Jurassic sequences sub-crop the Base Cretaceous unconformity. Lower Jurassic sequences maintain a similar thickness to those in the Penguin half-graben.

Figure 15 shows a NW–SE trending seismic line and geoseismic interpretation across the northern region of the Penguin ridge. As in the southern section, the ridge is expressed on seismic data as a structural high, underlain by shallow (3.3–3.4 secs TWT) acoustic basement, fault bounded to the northwest and southeast. The ridge crest, Western Terrace and adjacent Penguin half-graben are much narrower in this section and denote a tapering of the structure towards the north.

Stratigraphic relationships within the Penguin half-graben are similar to those in the southern section, but are geometrically contrasting. Late Jurassic infill is minor and reflectors appear to be more chaotic, particularly within the Volgian section. Lower and Middle Jurassic sequences have substantially thinned, and along with the underlying Triassic sequences of the half-graben and the Penguin ridge itself, possess an increased southeast dip, compared to Fig. 14. Half-graben sequences are erosionally truncated towards the northwest into the Magnus ridge. The Base Cretaceous unconformity dips into the Magnus Basin and is broadly convex in form. Lower Cretaceous sequences onlap this surface to the southeast.

The West Penguin Fault has a similar angle of dip to that in the southern section within its upper regions, but flattens considerably at depth to approximately 36°. The interpreted low-angle nature of the fault is consistent with the higher degree of rotation evident within the Penguin half-graben. In addition, reflector cut-offs in the intra-Triassic and deeper section of the half-graben and high amplitude NW-dipping reflections at 5.0–5.5 secs TWT, are the expression of synthetic low-angle normal faults.

In contrast, the eastern flank of the Penguin ridge is seismically chaotic and faults define a SE-dipping terrace. Further interpretation is difficult, due to the lack of proximal well data, however, it is assumed that stratigraphic relationships are similar to those in the south.

Of note in this section is the development of intense faulting, concentrated within Upper Cretaceous and Palaeocene sequences. Faults have a range of dips and orientations and are both normal and reverse in form. In map view they trend approximately NNE–SSW, i.e., sub-parallel to the ridge crest. Faulting appears to be associated with enhanced flexure clearly visible at Upper Cretaceous and Palaeocene levels, centred over an axis slightly west of the ridge crest.

Kinematic conclusions

(i) Localized anomalous Late Jurassic–Early Cretaceous structural geometries The Western Terrace of the Penguin ridge in the southern area (see Fig. 14) defines a localized, marked upwarp of the Base Cretaceous reflector with subsequent Early Cretaceous onlap. The West Penguin Fault bounds the terrace to the east and displays Late Jurassic–Early Cretaceous collapse within its footwall. In the crest of the terrace, chaotic seismic reflectors are interpreted as representing reworked Jurassic or possibly Triassic sequences resulting from this footwall collapse of the ridge flank. Similar chaotic reflectors within the Upper Jurassic sequences of the Penguin half-graben infill represent the basinward propagation of these deposits.

Of interest in this area is the anomalous folding of Penguin half-graben sequences up into the Western Terrace (see Fig. 14). The reverse offset of the Base Cretaceous Unconformity reflector is attributed to the development of a 'pop-up' zone along the West Penguin Fault. Such a geometry can be interpreted as resulting from oblique/strike-slip kinematics along the western margin of the Penguin ridge in Late Jurassic–Early Cretaceous times.

The East Penguin Fault is locally interpreted as possessing a downward steepening geometry, with associated W-dipping, WSW–ENE-striking, antithetic reverse fault splays (see Figs 13, 14). Such a geometry is characteristic of oblique-slip kinematics (Nelson 1993) and is in agreement with the interpretations of Booth *et al.* (1992). A well drilled on the eastern margin of the ridge a few kilometres south of this section, penetrates a Lower Jurassic succession, which is double the thickness of any other similar age succession in the region. The well passes directly through the inferred position of a reverse splay fault, and provides further evidence to the presence of these features.

(ii) Early Cretaceous onlaps and ridge development Interpretation of Figs 14 and 15 clearly show how a change in polarity of Early Cretaceous onlap occurs along the strike of the Magnus ridge. In the south, NW-Early Cretaceous onlap (see Fig. 14) indicates preferential

uplift of the Magnus ridge in Late Jurassic–Early Cretaceous times. To the north, SE-Lower Cretaceous onlap (see Fig. 15) indicates subsidence of the Magnus ridge coupled with uplift of the Penguin ridge during this time.

Such observations suggest a marked change in Late Jurassic–Early Cretaceous structural processes along the strike of the Magnus and Penguin ridges, possibly compartmentalized by the NW–SE transfer zone illustrated in Fig. 13. This is indicated by the observed change in structural attitude and geometry of the West Penguin Fault and the Magnus fault block. Low-angle faulting in the northern region is the possible expression of slumping of the Magnus block to the northwest, resulting from pronounced uplift and tilting along the western margin of the Penguin ridge. In contrast, localized pop-up zones along the West Penguin Fault and reverse fault splays along the East Penguin Fault, suggests reactivation of horst bounding faults in the southern region.

The origin of the Latest Jurassic uplift of the Penguin ridge is an important issue. The widespread occurrence of reverse faults within the ridge crest (see Fig. 13), coupled with localized reactivation of margin faults, indicates inversion of a pre-existing horst block.

Discussion

Inversion mechanisms

(i) Late Jurassic–Early Cretaceous fault reactivation The Late Jurassic–Early Cretaceous reactivation structures of the Tern sub-basin and the northern Penguin ridge are dissimilar to the total half-graben inversions in the Tertiary of the southern North Sea (Badley *et al.* 1989), where classic 'harpoon' and 'keystone block' structures are developed. In contrast, effects are mild and are expressed at different scales in a range of forms and styles.

Large-scale effects encountered within the Tern sub-basin include the partial inversion of NE–SW trending Mesozoic half grabens with associated backsteepening of pre-existing graben controlling faults (Figs 8, 9, 10). At the opposite end of the spectrum, detailed 2D interpretation has noted the presence of localized pop-ups along the western margin of the Penguin ridge (Fig. 14), folding, reverse faults and minor thrusts within the crestal regions of fault blocks (Fig. 13).

These structures are an important aspect of the basin evolution and have been attributed to: basement block rotations during rifting (Hay 1978); basement shear zone reactivation and

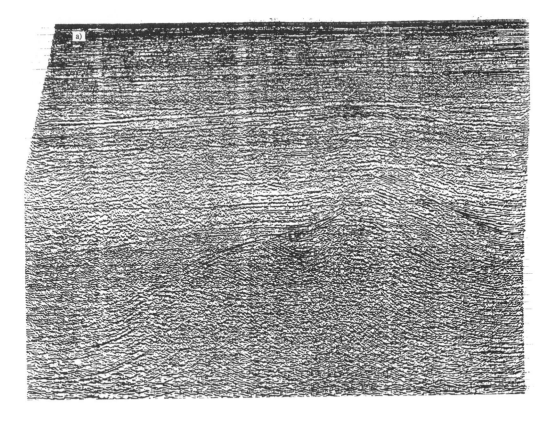

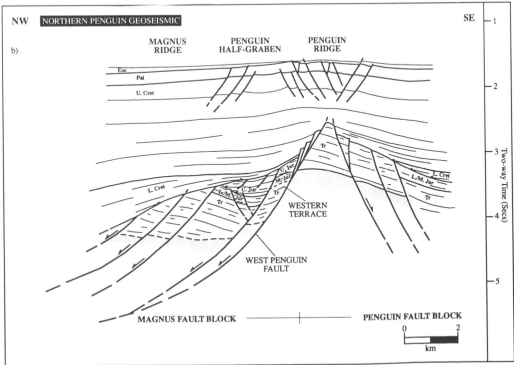

Fig. 15. (a) Uninterpreted, and (b) geoseimic interpretation of a seismic line across the northern region of the northern Penguin ridge. Basement rocks shaded. See text for discussion.

oblique-slip during rifting (Bartholomew *et al.* 1993); strike-slip transpression along reactivated NE–SW structures (Gabrielsen & Robinson 1984; Speksnijder 1987; Fossen 1989); segmentation of differential rates of extension and associated compression (Pegrum *et al.* 1984; Booth *et al.* 1992); oblique-shearing along fault block edges during extension (Lee & Hwang 1993); and post-rift shortening following failed rifting (Knott *et al.* 1993). In order to arrive at a realistic mechanism for the kinematic development of these Late Jurassic/Early Cretaceous structures, it is essential to understand the tectonic evolution of the northern North Sea from the regional point of view.

(ii) A tectonic model A proposed model for the tectonic evolution of the northern North Sea in Latest Jurassic–Early Cretaceous time is illustrated in Fig. 16. The diagram shows the spatial relationships of interpreted structures attributed to Triassic and Jurassic extensional phases within the region along with the postulated extension direction for each case. The model is based on extensive seismic analysis and compliments the observations made by Thomson & Underhill (1993) and by Hibbert & Mackertich (1993) in their work on the structural evolution of the Inner and Outer Moray Firth Basins respectively. Both studies show that an initial NW–SE extension was followed by a NE–SW extension. The same kinematics appear to apply in the East Shetland Basin and Viking Graben.

NE–SW structures were active in the Triassic and their predominantly dip-slip nature attests to a NW–SE extension direction, consistent with the separation of the Greenland plate from the Scandinavian plate during this time. This localization and the trend of the structures possibly indicate the reactivation of underlying Caledonide/Devonian crustal features. Faults in the west of the basin formed part of the West Shetland trend of Permo/Triassic basins. The more NNE–SSW trend of structures in the east of the basin may represent a broader range in strike of the controlling basement structures. Seismic analysis in the Tern sub-basin has shown that coaxial overprint of these Triassic structures occurred in Late Middle Jurassic/Early Late Jurassic times, under similar regional stress conditions. Extension on individual faults was less during this latter event, but well developed Oxfordian/Kimmeridgian growth sequences occur across NE–SW faults throughout the basin.

During the Latest Jurassic–Early Cretaceous, a switch in extension direction to the NE–SW is proposed, with dominant rifting concentrated to the south within the Central Graben and the Outer Moray Firth. This switch in extension direction is attributed to the complex interaction between extension directions of the Artic and Atlantic Oceans, plus possibly a NE–SW rift trend emanating from the Neo-Tethyan system in Central and Eastern Europe. Regional structure mapping by the authors has shown that resultant NW–SE trending dip-slip structures are abundant throughout the East Shetland Basin and are particularly evident along the western margins of the basin and the Transitional Shelf in the south (see Fig. 4). Implications of this model are the distribution of non-coaxial extensional stress across the pre-existing NE–SW structure of the East Shetland Basin and Viking Graben.

Constraining the timing of this interaction is obviously important. Regional interpretations in the East Shetland Basin have noted the occurrence of Oxfordian/Kimmeridgian growth sequences and half-graben geometries associated with NE–SW faults, in contrast with the overall lack of Upper Jurassic half-graben development across NW–SE faults. In addition, we believe that NW–SE faults remain active into the Early Cretaceous. These observations suggest differing times of development for the two fault trends. Further evidence for the time selective development of these fault sets has come from 3D work in the Cormorant Field (Demyttenaere *et al.* 1993), in which azimuth time-slice maps clearly show a switching in dominance from NE–SW trending faults at Top Middle Jurassic level to NW–SE trending faults at Base Cretaceous level. A similar picture emerges from studies of the Snorre Field, within the Norwegian Sector of the northern North Sea (Dahl & Solli 1993). A switch from NE–SW to NW–SE trending faults occurs at the Base Cretaceous horizon, which remains unchanged up into the Early Tertiary.

Additional extension is expressed to the north of the East Shetland Basin by large-scale ENE–WSW trending faults. These extend from the southern reaches of the Magnus Basin, west into the Faeroe Basin. Fault throws are larger than any others seen within the northern North Sea and from seismic analysis appear to have been fairly rapid. Structures are for the most part dip-slip in geometry and transfer faults trend NW–SE. An approximate NW–SE extension direction is assumed. The precise age for the initiation of this northern faulting remains a matter of debate. The trend of the faults suggest the re-utilization of Triassic structures, which possibly had an earlier origin. A coaxial Latest Jurassic event is evident from seismic stratigraphical analysis in the Magnus Basin. Thus,

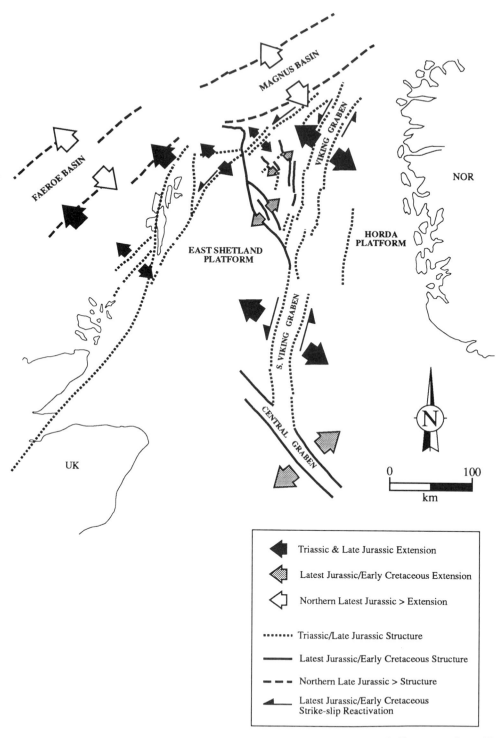

Fig. 16. Tectonic model for northern North Sea evolution from Permo-Trias to Early Cretaceous times. The following phases of Mesozoic extension are proposed: (i) Triassic NW–SE extension; (ii) Oxfordian–Early Kimmeridgian NW–SE extension; (iii) Volgian–Ryazanian NE–SW extension; (iv) NW–SE extension from the Latest Jurassic onwards in the Magnus and Faeroe Basins. The switch in Late Jurassic extension direction with resultant strike-slip reactivation of pre-existing fault systems, evident within the East Shetland Basin, Moray Firth Basin and Central Graben, is believed to account for localized, anomalous inversion structures encountered throughout the region.

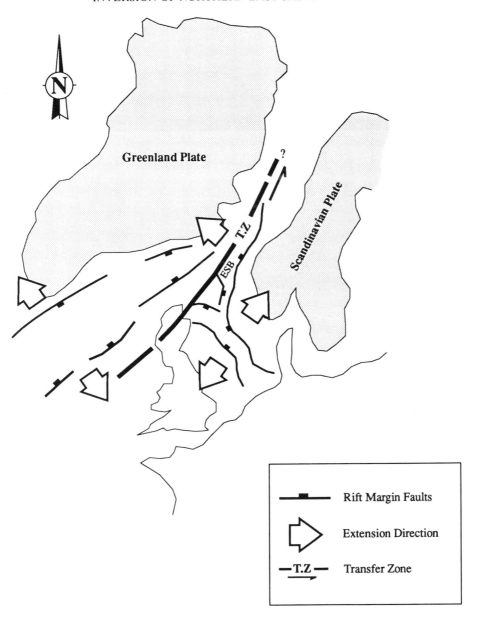

Fig. 17. Proposed tectonic picture in Latest Jurassic/Early Cretaceous times. Note the requirement of a NE–SW transfer fault separating NW–SE extension in the N from NE–SW extension in the South. The line of basement highs stretching from the Tern ridge in the southwest, to the Haltenbanken terrace in the northeast, marks the possible expression of this accommodation structure.

there must be the requirement for a transfer element in Latest Jurassic–Early Cretaceous times, separating NW–SE extension in the Magnus Basin to the north, from NE–SW extension in the East Shetland Basin to the south. The line of NE–SW trending structural highs which comprise the Tern and Penguin ridges of the East Shetland Basin, and the Nordland and Haltenbanken terraces of offshore Norway, may be the surface expression of part of this crustal transfer zone (see Fig. 17), though the Great Glen/Walls Boundary Fault system is a major crustal lineament which could have been reactivated at this time. On a regional

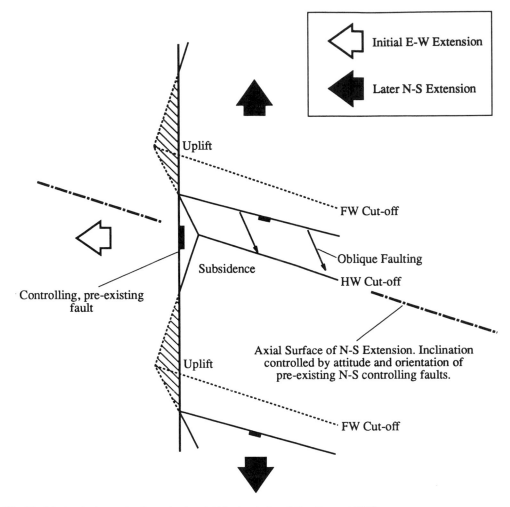

Fig. 18. Mechanical aspects of non-horizontal block rotation (after Coward 1990).

scale the transfer is likely to form part of the suite of NE–SW crustal lineaments which cross-cut the region, including the Great Glen Fault of Northwest Scotland and the Bodo and More–Trondelag Fault complexes of off/onshore Norway. The lateral extent of these lineaments suggests the existence of a fundamental crustal structure, which is likely to have been important throughout the evolution of NW Europe (Coward 1990, 1993).

(iii) Non-coaxial superimposed extension The overall picture is that the NE–SW basin structure used by Triassic and Oxfordian/Kimmeridgian rift phases (see Figs 8–10), is believed to have exerted a control upon subsequent non-coaxial NE–SW extension in Latest Jurassic–Early Cretaceous times. Triassic–Kimmeridgian

structures possessed a range of structural orientations and attitudes, both within map and profile view, which when reactivated, led to complexities within the fault block rotation of the ongoing NE–SW extension. The axis of fault block rotation may be controlled by the orientation and attitude of the pre-existing structure. Figure 18 (after Coward 1990) illustrates a theoretical example where the dip of a pre-existing N–S fault, has controlled the axis of subsequent N–S block rotation. Such non-horizontal block rotation has resulted in an obliquity of the later faults to their extension direction and, in addition, has produced localized areas of extension and compression within the fault block structure. The pattern of variable block rotation is illustrated by two field examples, Ninian and Brent, presented in Fig. 19,

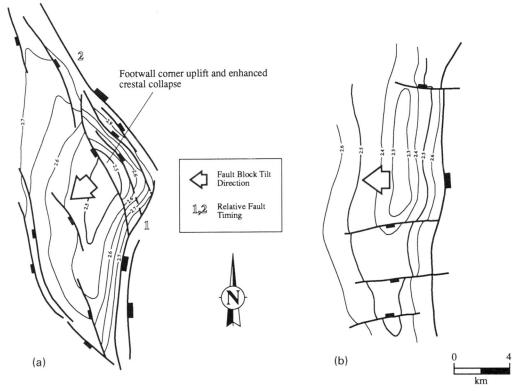

Footwall corner uplift and enhanced
crestal collapse

Fault Block Tilt
Direction

1,2 Relative Fault
Timing

(a)   (b)   0   4

km

Fig. 19. Field examples chosen to represent realistic end-members of complex fault block rotation
mechanisms: (**a**) Ninian; (**b**) Brent. See text for discussion.

believed to represent end members of the
model. Base Cretaceous structure maps (from
Speksnijder 1987) display contrasting contour
patterns. Structural contours in the Ninian field
are oblique to the main bounding faults and are
indicative of complex tilting during block rotation. This is expressed by well developed
footwall corner uplift and associated gravitational collapse of the northeast crest of the
field. Brent field contours on the other hand
show simple west-tilting away from the block
bounding fault. The eastern bounding fault of
Ninian is composed of a northern NW–SE
segment and a southern NE–SW segment,
whereas the eastern bounding fault of the Brent
field appears from regional interpretation to be a
simple N–S feature. The fault pattern at Ninian
is proposed to indicate time selective activation
of orthogonal fault segments during the development of the structure. Conversely, the
single bounding fault of the Brent field, suggests
a less complex extensional phase of formation,
portrayed by the simple field structure. If
boundary conditions were added suppressing
NE–SW movement, the initial NE–SW fault

may backsteepen and tighten/invert within areas
of highest confinement. An example of this
inversion occurs within the southern reaches of
the Tern sub-basin (see Fig. 11).

As well as rotational block movement, sinistral oblique/strike-slip reactivation of the pre-
existing NE–SW structure will have occurred,
the amount of strike-slip movement depending
on the angle of interaction between the pre-
existing structure and the later NE–SW extension. Large-scale effects of this reactivation
include widespread, oblique inversion along the
western margin of the South Viking Graben,
with associated out of graben propagation
thrusting (Thomas & Coward, unpublished
data). Effects progressively diminish northwards away from the locus of extension within
the Central Graben, however, in the East
Shetland Basin appear to be tramlined by
deep-seated, Palaeozoic structure.

*(iv) Significance of the northern East Shetland
Basin structure* A kinematic interpretation of
Base Cretaceous structure encountered within
the northern East Shetland Basin is illustrated in

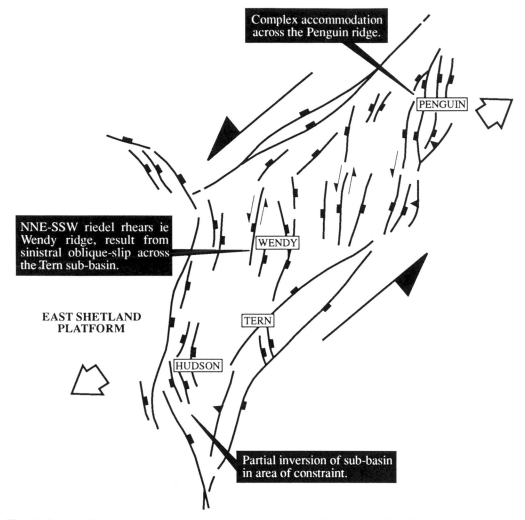

Fig. 20. Proposed Latest Jurassic–Early Cretaceous development of the northern East Shetland Basin as a sinistral shear system, tramlined by pre-existing basement-involved Palaeozoic fault systems.

Fig. 20. NNE–SSW intra-basin faults are interpreted as representing riedel shears, the orientation of which are indicative of a sinistral shear component across the region, bounded by the Northern Tern/West Penguin and Magnus Faults. The development of this strike-slip system is attributed to Latest Jurassic–Early Cretaceous, non-coaxial, NE–SW extension, centred across the western margin of the East Shetland Basin. The partial inversion of the Tern sub-basin within the Hudson vicinity is attributed to a combination of the constraining effect of the Tern ridge to western margin extension, and buttressing of shear by the developing East Shetland Platform. Complex structuration and uplift along the northern

Penguin ridge indicate strike-slip accommodation along the margins of this shear system.

A geoseismic interpretation of a WNW–ESE line across the western margin of the Wendy ridge is illustrated in Fig. 21, the location of which is shown in Fig. 7. The overall ESE-thickening of Triassic and Jurassic sequences across the section represents the gradual deepening of the Tern sub-basin into its controlling Northern Tern Fault (see Fig. 7). Interpreted riedel shears downthrow to the WNW, and the lack of any growth within the Triassic and Jurassic section, attests to rapid fault development in the Latest Jurassic–Early Cretaceous. Variable throw along the easternmost fault attests to a strike-slip origin, as does the

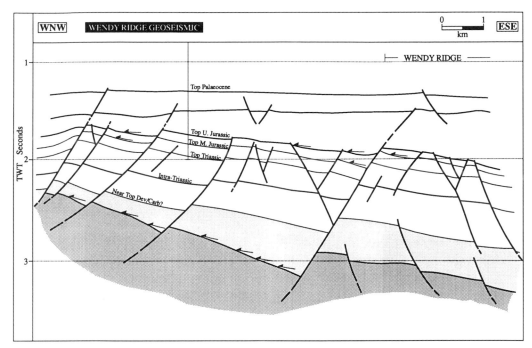

Fig. 21. Geoseismic interpretation of a WNW–ESE seismic line across the Wendy ridge (see Fig. 7 for location). Note the lack of Triassic and Jurassic growth across the faults, in contrast with interpreted Devonian–Carboniferous half-graben development at depth.

development of associated folds and thrusts in the southern regions of the ridge, outlined earlier (see Fig. 8). However, the occurrence of well developed growth packages within interpreted Devonian–Carboniferous half-grabens at depth (also apparent in Fig. 8), suggests the reactivation and controlling effect of pre-existing Palaeozoic fault systems during riedel shear development. The probable sub-parallel trend of these Palaeozoic systems to the NNE–SSW riedel shear system, is consistent with the sinistral pull-apart model of proto-East Shetland Basin development in the Upper Palaeozoic, proposed by Coward (1993).

Conclusions

(1) The tectonic evolution of the northern North Sea can be considered in two stages: (i) the initial development of the Palaeozoic framework of the basin; (ii) the inherent control of this early structure on the later stages of Mesozoic basin formation.

(2) The Palaeozoic stratigraphic/structural setting of the northern North Sea is poorly understood, however, folded Devonian strata upon the East Shetland Platform, Transitional Shelf and within the Tern sub-basin, are possible remnants of an early phase of basin formation, controlled by NNE–SSW faults. Initial extensional mechanisms may have involved the collapse of Caledonide thrustal structure with subsequent basin formation during Devonian/ Carboniferous times.

(3) Two phases of Mesozoic rifting can be recognized from regional seismic analysis: (i) Triassic; (ii) Late Jurassic; each followed by an associated phase of thermal subsidence. The age(s) of Triassic rifting is difficult to constrain, however, NW–SE dip-slip extension along NE–SW structures is proposed for the northern regions of the East Shetland Basin. East-thickening of Lower and Middle Jurassic sequences is attributed to thermal subsidence following this rift event and supports the idea of a basin depocentre over the Horda Platform during this time. Jurassic rifting is better constrained and is observed in the Late Middle Jurassic/Early Late Jurassic and in the Latest Jurassic–Early Cretaceous. A NW–SE extension direction is proposed for the initial stage of Jurassic stretching and involves the extensional reactivation of existing NE–SW Triassic structures. A switch in extension direction occurs for the later stage of Jurassic stretching, with the development of NW–SE structures. This situation is further

complicated to the north of the area by the coeval onset of NW–SE spreading within the Magnus Basin. The resultant cross-hatch pattern of faulting seen at Base Cretaceous level is believed to be the product of these separate NW–SE and NE–SW extensional phases. Faulting is the expression of favourably orientated basement structure, selectively reactivated during each extensional episode.

(4) B factors calculated from fault heaves for Triassic and Late Jurassic extension across the basin, give figures of 1.3 and 1.1 respectively. Such values are in complete contrast with deep seismic studies across the basin which show a crustal thinning factor approaching 2. If it is assumed that the estimates from deep seismic profiling are accurate, then the missing extension may be attributed to (i) an early phase of basin formation in Devonian/Carboniferous times; (ii) incorrect calculation of extensional factors, particularly for the Late Jurassic rift event; (iii) the integral role of sub-seismic resolution faulting.

(5) Latest Jurassic–Early Cretaceous strike-slip reactivation of NE–SW structures in the northern East Shetland Basin, is attributed to the change in extension direction during the Latest Jurassic in the northern North Sea. Effects include (i) the partial inversion of the southwest regions of the Tern sub-basin, in the vicinity of the Hudson field, with associated backsteepening of graben controlling faults; (ii) localized flower geometries, folding, reverse faulting, and differential block uplift, along the strike of the reactivated structure. NNE–SSW riedel shears, i.e. Wendy ridge, attest to an expected sinistral component of shear across the basin. The rooting of these riedel shears into interpreted Devonian–Carboniferous growth faults at depth, indicates the coaxial overprint and reactivation of NNE–SSW Palaeozoic fault systems.

The authors would like to thank Oryx UK Energy Company, Nopec UK Ltd and Ranger Oil UK Ltd for providing seismic data and also giving permission to publish the included sections. D.T. would personally like to thank John Robbins and Jim Stockley of Oryx, Mike Pepper of Ranger and Andy McGrandle of Ark Geophysics for discussion and helpful advice. J. Booth (Amerada Hess) is thanked for a thorough and extremely helpful review of the original manuscript. D. T. is funded by an Oryx studentship.

References

BADLEY, M. E., EGEBERG, T. & NIPEN, O. 1984. Development of rift basins, illustrated by the structural evolution of the Oseberg feature, Block 30/6, offshore Norway. *Journal of the Geological Society*, London, **141**, 639–49

——, PRICE, J. D., RAMBECH DAHL, C. & AGDESTEIN, T. 1988. The structural evolution of the northern Viking Graben and its bearing upon extensional modes of basin formation. *Journal of the Geological Society, London*, **145**, 455–472

——, —— & BLACKSHALL, L. C. 1989. Inversion, reactivated faults and related structures – seismic examples from the southern North Sea. *In*: COOPER, M. A. & WILLIAMS, G. D. (eds) *Inversion Tectonics*. Geological Society of London Special Publication, **44**, 201–219

BARTHOLOMEW, I. D., PETERS, J. M. & POWELL, C. M. 1993. Regional structural evolution of the North Sea: oblique-slip and the reactivation of basement lineaments. *In*: PARKER, J. R. (ed.) *Petroleum Geology of Northwest Europe: Proceedings of the 4th Conference*. Geological Society, 1109–1122

BARTON, P. & WOOD, R. 1984. Tectonic evolution of the North Sea Basin: crustal stretching and subsidence. *Geophysical Journal of the Royal Astronomical Society*, **79**, 987–1022.

BEACH, A. 1985. Some comments on sedimentary basin development in the northern North Sea. *Scottish Journal of Geology*, **21**, 493–512.

BIDDLE, K. T. & RUDOLPH, K. W. 1988. Early Tertiary structural inversion in the Stord Basin, Norwegian North Sea. *Journal of the Geological Society* London, **145**, 603–612.

BIDSTRUP, T., CHRISTIANSEN, F. G. & MATHIESEN, A. 1993. Jameson Land Basin, East Greenland: Modelling of uplift history from Maturity and Apatite Fission Track data. *In*: *Basin Inversion Conference Abstracts*, Oxford, October 1993.

BOOTH, A., STOCKLEY, F. J. & ROBBINS, J. A. 1992. *Late Jurassic structural inversion in the North Viking Graben and East Shetland Basin, U.K. North Sea*. ORYX UK Energy Company, Internal Publication.

BRODIE, J. & WHITE, N. 1993. Sedimentary basin inversion around the British Isles due to igneous underplating. *In*: *Basin Inversion Conference Abstracts*, Oxford, October 1993.

CASELLI, F. 1987. Oblique-slip tectonics, Mid-Norway Shelf. *In*: BROOKS, J. & GLENNIE, K. W. (eds) *Proceedings of the 3rd Conference on the Petroleum Geology of NW Europe*. Geological Society, London, 1049–1063.

COWARD, M. P. 1986. Heterogeneous stretching, simple shear and basin development. *Earth and Planetary Science Letters*, **80**, 325–336.

—— 1993. The effect of Late Caledonian and Variscan continental escape tectonics on basement structure, Palaeozoic basin kinematics and subsequent Mesozoic basin development in NW Europe. *In*: PARKER, J. R. (ed.) *Petroleum Geology of Northwest Europe: Proceedings of the 4th Conference*. Geological Society, London, 1095–1108.

—— 1994. Inversion Tectonics. *In*: HANCOCK, P. *(ed.) Continental Deformation*. Pergamon Press, 289–304.

—— & ENFIELD, M. A. 1987. The structure of the West Orkney and adjacent Basins. *In*: BROOKS, J.

& GLENNIE, K. W. (eds) *Proceedings of the 3rd Conference on the Petroleum Geology of NW Europe*. Geological Society, London, 687–696.

—— 1990. The Precambrian, Caledonian and Variscan framework to NW Europe. *In*: HARD-MAN, R. F. P. & BROOKS, J. (eds) *Tectonic Events Responsible for Britain's Oil and Gas Reserves*. Geological Society, London, Special Publication, **55**, 1–34.

DAHL, N. & SOLLI, T. 1993. The structural evolution of the Snorre Field and surrounding areas. *In*: PARKER, J. R. (ed.) *Petroleum Geology of Northwest Europe: Proceedings of the 4th Conference*. Geological Society, London, 1159–1166.

DEMYTTENAERE, R. R. A., SLUIJK, A. H. & BENTLEY, M. R. 1993. A fundamental reappraisal of the structure of the Cormorant Field and its impact on field development strategy. *In*: PARKER, J. R. (ed.) *Petroleum Geology of Northwest Europe: Proceedings of the 4th Conference*. Geological Society, London, 1151–1157.

FOSSEN, H. 1989. Indication of transpressional tectonics in the Gullfaks oil-field of the northern North Sea. *Marine and Petroleum Geology*, **6**, 22–30.

FROST, R. E. 1987. The evolution of the Viking Graben tilted fault block structures: a compressional origin. *In*: BROOKS, J. & GLENNIE, K. W. (eds) *Proceedings of the 3rd Conference on the Petroleum Geology of NW Europe*. Geological Society, London, 1009–1024.

GABRIELSEN, R. H. & ROBINSON, C. 1984. Tectonic inhomogeneities of the Kristiansund-Bodo Fault Complex, offshore Mid-Norway. *Petroleum Geology of the North European Margin, Norwegian Petroleum Society*, 397–406.

GILLCRIST, R., COWARD, M. P. & MUGNIER, J. L. 1987. Structural inversion, examples from the Alpine Foreland and the French Alps. *Geodinimica Acta*, **1**, 5–34.

GRONLIE, A. & ROBERTS, D. 1989. Resurgent strike-slip duplex development along the Hitra-Snasa and Verran Faults, More-Trondelag Fault Zone, Central Norway. *Journal of Structural Geology*, **11**, 295–305.

HAY, J. T. C. 1978. Structural development in the northern North Sea. *Journal of Petroleum Geology*, **1**, 65–77.

HIBBERT, M. J. & MACKERTICH, D. S. 1993. The structural evolution of the eastern end of the Halibut Horst, Block 15/21, Outer Moray Firth, UK North Sea. *In*: PARKER, J. R. (ed.) *Petroleum Geology of Northwest Europe: Proceedings of the 4th Conference*. Geological Society, London, 1179–1188.

HILLIS, R. R. 1993. Quantification of Tertiary erosion in the UK southern North Sea by sonic velocity data. *In*: *Basin Inversion Conference Abstracts*, Oxford, October 1993.

HOLLOWAY, S., REAY, D. M., DONATO, J. A. & BEDDOE-STEPHENS, B. 1991. Distribution of granite and possible Devonian sediments in part of the East Shetland Platform, North Sea. *Journal of the Geological Society*, London, **148**, 635–638.

KLEMPERER, S. 1988. Crustal thinning and nature of extension in the northern North Sea from deep seismic reflection profiling. *Tectonics*, **7**, 803–821.

KNOTT, S., BEACH, A., WELBON, A., BROCKBANK, P., REKSNES, P. & OLSEN, T. 1993. Basin inversion in the Gulf of Suez – implications for exploration and development in failed rifts. *In*: *Basin Inversion Conference Abstracts*, Oxford, October 1993.

KUSZNIR, N. J., MARSDEN, G. & EGAN, S. S. 1991. A flexural cantilever simple shear/pure shear model of continental extension: application to the Jeanne d' Arc Basin, Grand banks and Viking Graben, North Sea. *In*: ROBERTS, A. M., YIELD-ING, G. & FREEMAN, B. (eds) *The Geometry of Normal Faults*. Geological Society, London, Special Publication, **56**, 41–60.

LARSEN, V. B. 1987. A synthesis of tectonically-related stratigraphy in the North Atlantic-Artic region from Aalenian to Cenomanian time. *Norsk Geologisk Tidsskrift*, **67**, 281–293.

LEE, M. J. & HWANG, Y. J. 1993. Tectonic evolution and structural styles of the East Shetland Basin. *In*: PARKER, J. R. (ed.) *Petroleum Geology of Northwest Europe: Proceedings of the 4th Conference*. Geological Society, London, 1137–1149.

MCKENZIE, D. P. 1978. Some remarks on the development of sedimentary basins. *Earth and Planetary Science Letters*, **40**, 25–32.

MARRETT, R. & ALLMENDINGER, R. W. 1992. Amount of extension on 'small faults': An example from the Viking Graben. *Geology*, **20**, 47–50.

MARSDEN, G., YIELDING, G., ROBERTS, A. M. & KUSZNIR, N. J. 1990. Application of a flexural cantilever simple shear/pure shear model of continental lithosphere to the formation of the northern North Sea. *In*: BLUNDELL, D. J. & GIBBS, A. D. (eds) *Tectonic Evolution of the North Sea Rifts*. Oxford Science Publication, 236–257.

NELSON, A. 1993. Wrench and inversion structures in the Timor Sea region. *Petroleum Exploration Society of Australia Journal*, **21**, 3–30.

PEGRUM, R. M. 1984. The extension of the Tornquist Zone in the Norwegian North Sea. *Norsk Geologisk Tidsskrift*, **64**, 39–68.

—— & LJONES, T. E. 1984. 15/9 Gamma Gas Field, offshore Norway, new trap type for the North Sea Basin with regional structural implications. *American Association of Petroleum Geologists Bulletin*, **68**, 874–902.

PROSSER, S. 1993. Rift-related linked depositional systems and their seismic expression. *In*: WILLIAMS, G. D. & DOBB, A. (eds) *Tectonics and Seismic Sequence Stratigraphy*. Geological Society, London, Special Publication, **71**, 35–66.

ROBERTS, A. M. & YIELDING, G. 1991. Deformation around basin margin faults in the North Sea/mid-Norway rift. *In*: ROBERTS, A. M., YIELDING, G. & FREEMAN, B. (eds) *The Geometry of Normal Faults*. Geological Society, London, Special Publication, **56**, 61–78.

——, —— & BADLEY, M. E. 1991. A kinematic model for the orthogonal opening of the Late Jurassic

North Sea rift system, Denmark–Mid Norway. *In*: BLUNDELL, D. J. & GIBBS, A. D. (eds) *Tectonic Evolution of the North Sea Rifts*. Oxford Science Publication, 180–199.

RUMPH, B., REAVES, C. M., ORANGE, V. G. & ROBINSON, D. L. 1993. Structuring and transfer zones in the Faeroe Basin in a regional tectonic context. *In*: PARKER, J. R. (ed.) *Petroleum Geology of Northwest Europe: Proceedings of the 4th Conference*. Geological Society, London, 999–1011.

SPEKSNIJDER, A. 1987. The structural configuration of Cormorant Block IV in context of the northern Viking Graben structural framework. *Geologie en Mijnbouw*, **65**, 357–379.

THOMSON, K. & UNDERHILL, J. R. 1993. Controls on the development and evolution of structural styles in the Inner Moray Firth Basin. *In*: PARKER, J. R. (ed.) *Petroleum Geology of Northwest Europe: Proceedings of the 4th Conference*. Geological Society, London, 1167–1178.

YIELDING, G., BADLEY, M. E. & ROBERTS, A. M. 1992. The structural evolution of the Brent Province. *In*: MORTON, A. C., HASZELDINE, R. S., GILES, M. R. & BROWN, S. (eds) *Geology of the Brent Group*. Geological Society, London, Special Publication, **61**, 27–43.

ZERVOS, F. 1987. A compilation and regional interpretation of the northern North Sea gravity map. *In*: COWARD, M. P., DEWEY, J. F. & HANCOCK, P. L. (eds) *Continental Extensional Tectonics*. Geological Society, London, Special Publication, **28**, 477–493.

ZIEGLER, P. A. 1988. *Evolution of the Artic–North Atlantic and the Western Tethys*. American Association of Petroleum Geologists, Memoir, **43**.

The inversion history of the northeastern margin of the Broad Fourteens Basin

ROBERT J. HOOPER[1], LENG SIANG GOH[2] & FIONA DEWEY[2]

[1]Conoco Inc, PO Box 2197, Houston, TX 77252-2197, USA

[2]Conoco Netherlands Oil Company, PO Box 1122, 2260 BC Leidschendam, The Netherlands

Abstract: The Broad Fourteens Basin experienced a long and complicated developmental history, involving extension, salt tectonics, inversion and subsidence, that began in the Triassic and continued into the middle Tertiary. Inversion in our study area on the northeast margin of the basin was accommodated not only by backward expulsion of the basin-fill (creating a series of northeast-vergent thrusts and related lobate folds) but also reactivation of basement faults. Restorations indicate that the basin-fill did not behave independently of the basement during inversion. At least 2200 m of inversion-related basement uplift must have occurred at the basin margin in addition to backward expulsion of the basin-fill. The most important factors controlling the final 'inversion style' appear to be the pre-inversion structural configuration of the margin and the resolved inversion direction.

The Broad Fourteens Basin, located in the Dutch sector of the southern North Sea, is one of a series of Mesozoic extensional basins in the Alpine foreland that have been affected by Late Cretaceous and younger inversion. The basin, which trends broadly southeast–northwest, is bound to the southwest by the Winterton–Ijmulden High and to the northeast by the Texel–Ijsselmeer High (Fig. 1). The Broad Fourteens Basin experienced a long and complicated developmental history, involving extension, salt tectonics, inversion, and subsidence, that began in the Triassic and continued into the middle Tertiary. Recent publications that contain discussions on the tectonostratigraphy of the basin include those of Hayward & Graham (1989), Huyghe (1992), Roelofsen & deBoer (1989), and Van Wijhe (1987a,b).

Previously published basin-scale restorations to illustrate the pre-inversion configuration reveal that the Broad Fourteens Basin formed a large asymmetrical half-graben, approximately 60 km wide, throughout most of the Jurassic and Cretaceous (Hayward & Graham, 1989). The geometry of this half-graben is inferred to have been controlled by a major listric extensional fault along the southwestern margin of the basin. The northeastern margin of the basin is defined by a series of broadly en echelon Jurassic to Cretaceous normal-separation faults that dip towards the basin axis. These faults may have formed on the flexural margin of the half-graben and be part of a crestal-collapse-graben system. The Triassic to Cretaceous section, which can be as much as 5000 m in the basin axis, thins dramatically onto the adjacent platform, the Texel–Ijsselmeer High, where the equivalent section is rarely over 500 m thick.

Inversion of the Broad Fourteens Basin is interpreted to be diachronous. In the centre of the basin inversion may have begun as early as the Albian, as evidenced by the onlap of Chalk onto an eroded section of Holland Formation; inversion on the margins of the basin began somewhat later, most likely early in the Maastrichtian. The main phase of inversion throughout the basin was over by the Base Tertiary unconformity, as evidenced by the dramatic truncation of units in the centre of the basin and the erosion of inversion-related fold-crests on the basin margin (Fig. 2a). Minor movements related to continued contraction and/or post-inversion salt tectonics distorted the Base Tertiary and Oligocene/Miocene unconformities but these effects are small and localized. Cumulative shortening of the Broad Fourteens Basin as a result of inversion was approximately 10%.

The principal focus of this paper is the inversion history of the northeastern margin of the Broad Fourteens Basin. Inversion of that margin was accommodated by basement uplift and backward expulsion of the basin fill. The precise nature of the deformation, however, shows considerable variation. We present data from a recently acquired 3D seismic survey across the northeastern margin of the basin (Fig. 2a) that have a bearing on the character of the

From BUCHANAN, J. G. & BUCHANAN, P. G. (eds), 1995, *Basin Inversion*, Geological Society Special Publication No. 88, 307–317.

307

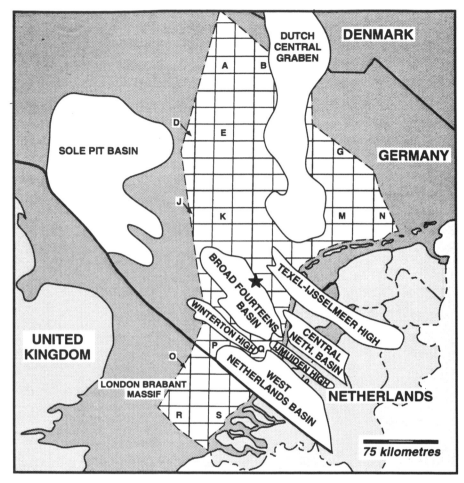

Fig. 1. Schematic map of the southern North Sea showing the location of the Broad Fourteens Basin in relation to the main structural elements. The study area, indicated by a star (*) is located on the northeastern margin of the Broad Fourteens Basin in the Dutch Sector.

inversion-related deformation and discuss the role of basement during inversion.

Local setting – NE-margin Broad Fourteens Basin

The northeastern margin of the Broad Fourteens Basin can be separated into two broad regions, a platformal area (NE) and a basinal area (SW) (Fig. 2b). The principal tectonostratigraphic sequences are presented in Fig. 3; a stratigraphic correlation chart is presented in Fig. 4.

Basin

The basinal area was affected by multiple phases of deformation that generated a complex fault array. Basement (hereafter referring to all strata below the Zechstein Group) under the basin steps down to the southwest via a series of southwest-dipping normal-separation faults (Fig. 2b). The cover sequence (hereafter referring to all strata above the Zechstein Group) was deformed by three prominent southwest-dipping faults that were active during both extension and inversion (Fig. 2b). These faults divide the basinal section into three blocks. Normal and thrust-separation faults related to extension and inversion can be identified in the section.

The pre-rift basin stratigraphy above the Carboniferous basement comprises a Permo–Triassic section (Rotliegend–Zechstein–Lower Bunter Shale–Triassic) of fairly uniform thickness (Figs 2b, 3). The syn-rift section, the base of

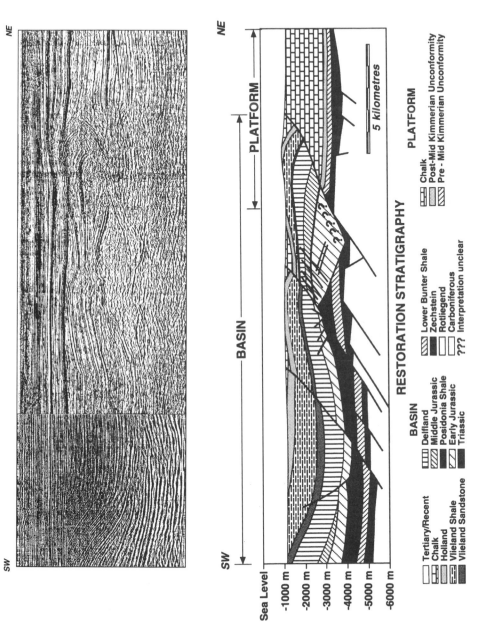

Fig. 2. (a) Uninterpreted composite seismic section across the northeastern margin of the Broad Fourteens basin. The Base Tertiary unconformity causes the dramatic truncation of units at the left end of line, and the erosion of inversion-related fold-crests on the basin margin. (b) Depth-converted geoseismic section (corresponding to (a). The section can be divided into two broad regions – a platformal area (NE) and a basinal area (SW). The basinal section is cut by three major normal-separation faults that dip southwesterly towards the basin axis. The section is punctuated by several major unconformities. The Mid Kimmerian unconformity cuts out Middle Jurassic Fm., Posidonia shale and into the Early Jurassic in the middle of the line. The Base Tertiary unconformity truncates the crests of inversion-related folds at the basin margin (NE) and cuts down to the Jurassic in a spectacular angular unconformity towards the basin axis at the SW-end of the line.

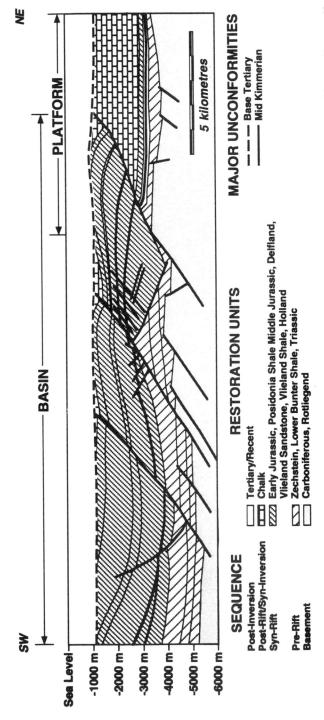

Fig. 3. Major tectonostratigraphic sequences in the present-day geoseismic section. Five major sequences can be identified: basement, pre-rift, syn-rift, post-rift – syn-inversion and post-inversion. The syn-rift section displays an 'harpoon' shape typical of inverted basins.

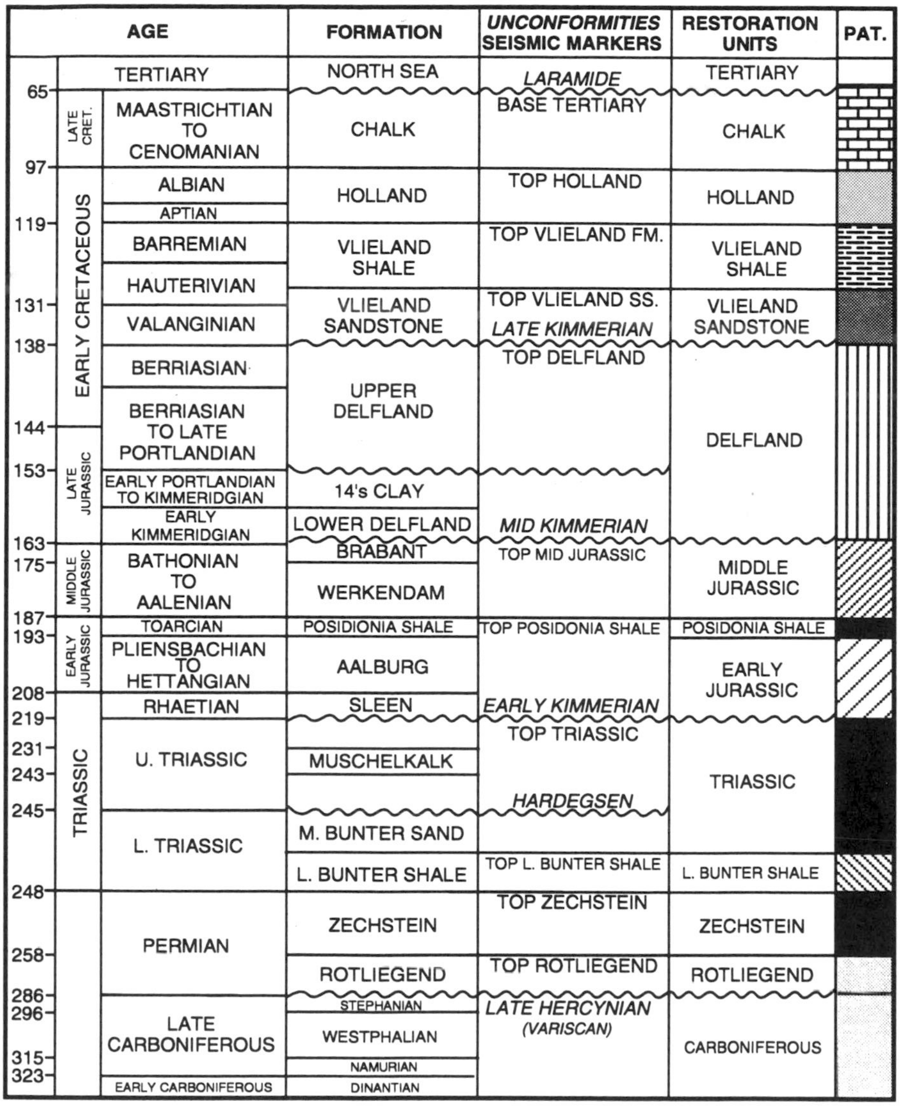

AGE			FORMATION	UNCONFORMITIES SEISMIC MARKERS	RESTORATION UNITS	PAT.
65		TERTIARY	NORTH SEA	LARAMIDE	TERTIARY	
	LATE CRET.	MAASTRICHTIAN TO CENOMANIAN	CHALK	BASE TERTIARY	CHALK	
97						
	EARLY CRETACEOUS	ALBIAN	HOLLAND	TOP HOLLAND	HOLLAND	
119		APTIAN				
		BARREMIAN	VLIELAND SHALE	TOP VLIELAND FM.	VLIELAND SHALE	
		HAUTERIVIAN				
131		VALANGINIAN	VLIELAND SANDSTONE	TOP VLIELAND SS. LATE KIMMERIAN	VLIELAND SANDSTONE	
138						
		BERRIASIAN		TOP DELFLAND		
144		BERRIASIAN TO LATE PORTLANDIAN	UPPER DELFLAND		DELFLAND	
153	LATE JURASSIC	EARLY PORTLANDIAN TO KIMMERIDGIAN	14's CLAY			
		EARLY KIMMERIDGIAN	LOWER DELFLAND	MID KIMMERIAN		
163			BRABANT	TOP MID JURASSIC		
175	MIDDLE JURASSIC	BATHONIAN TO AALENIAN	WERKENDAM		MIDDLE JURASSIC	
187						
193	EARLY JURASSIC	TOARCIAN	POSIDONIA SHALE	TOP POSIDONIA SHALE	POSIDONIA SHALE	
		PLIENSBACHIAN TO HETTANGIAN	AALBURG		EARLY JURASSIC	
208						
219		RHAETIAN	SLEEN	EARLY KIMMERIAN TOP TRIASSIC		
231	TRIASSIC	U. TRIASSIC	MUSCHELKALK		TRIASSIC	
243				HARDEGSEN		
245		L. TRIASSIC	M. BUNTER SAND			
248			L. BUNTER SHALE	TOP L. BUNTER SHALE	L. BUNTER SHALE	
			ZECHSTEIN	TOP ZECHSTEIN	ZECHSTEIN	
258	PERMIAN					
286			ROTLIEGEND	TOP ROTLIEGEND	ROTLIEGEND	
296			STEPHANIAN	LATE HERCYNIAN (VARISCAN)		
	LATE CARBONIFEROUS		WESTPHALIAN		CARBONIFEROUS	
315			NAMURIAN			
323	EARLY CARBONIFEROUS		DINANTIAN			

Fig. 4. Stratigraphic correlation chart for the northeastern margin of the Broad Fourteens Basin.

which is difficult to define, comprises the upper part of the Triassic and a Jurassic to early Cretaceous section (Jurassic–Posidonia Shale– Middle Jurassic–Delfland–Vlieland Sandstone– Vlieland Shale–Holland) that thickens basinward (i.e. to the southwest) (Fig. 3). The section is punctuated by several unconformities (Fig. 4). The most prominent unconformity, visible in the centre of the section, is the Mid Kimmerian, which cuts down through the Middle Jurassic rocks, Posidonia shale and into Early Jurassic rocks (Figs 2b, 3). The pre- to syn-inversion section in the basin is represented by small

remnants of Chalk preserved in faulted synforms between inversion-related antiforms (Figs 2a, 3). A post-inversion Tertiary to Recent section overlies the whole basin.

Platform

The platform area comprises a relatively uniform basement surface overlain by thin parallel units separated by regional unconformities (Fig. 2b). The syn-rift section present in the basin is largely missing or thinned on the platform (Fig. 2b). The section from Base Lower

Bunter Shale to Top–Holland in the basin is in excess of 5000 m; this compares with only 500 m of equivalent section preserved on the platform (Fig. 2b). The lower, syn-rift platform section is overlain by an 1800 m thick pre-to syn-inversion Chalk sequence; the equivalent section in the basin is present only in small erosional remnants (Fig. 2b). The basinward edge of the platform sequence forms the footwall block for the frontal inversion fault. The platform edge has been mildly affected by inversion in the basin resulting in folding and possible overturning of the footwall (Fig. 2b) and the development of a complex footwall fault-array.

Deformational history: extension

The principal problem in defining the extensional history of the northeastern margin of the basin using a balanced cross-section approach is the nature and extent of erosion at major unconformities in the section, principally the Hardegsen, Lower Kimmerian, Mid Kimmerian and Base Tertiary (Figs 2b, 3). There are no continuous seismic markers in the Middle Jurassic, Posidonia and part of the Lower Jurassic section because of erosion by the Mid Kimmerian unconformity (Figs 2b, 3); likewise, there are no continuous seismic markers between the Top–Vlieland Sandstone and the Base Tertiary unconformity (Figs 2b, 3). It is thus not possible to make restorations for the extensional history of the basin without making some assumptions as to the original sediment distributions.

Extension can be considered to have occurred in two phases. An early intracratonic phase of extension began in the Upper Triassic with slip on the basin margin fault system and development of a pronounced hangingwall rollover (Fig. 5a). Deformation of the cover sequence was largely decoupled from deformation in the basement by the Zechstein Group evaporites. Rotation of blocks in the hangingwall was accentuated by salt tectonics and resulted in the pattern of truncation of units (either by non-deposition or erosion) at the Hardegsen, Lower Kimmerian and Mid Kimmerian unconformities (Fig. 5a). The later phase of extension was accommodated primarily over the basin margin fault and the two major outboard faults (Fig. 5b). Block rotation was not as pronounced as in the early phase of extension and independent effects of salt tectonics are difficult to resolve. Units above the Mid Kimmerian unconformity thicken basinward across each of the major faults (Fig. 5b). The pattern of growth on the major faults indicates that extension was over by mid-Holland (Fig. 5b).

Deformational history: inversion

Previously published interpretations of the structural style across the northeast margin of the basin suggest that the character of the inversion is quite variable. Hayward & Graham (1989, figure 9) show a series of imbricate slices detached in the Zechstein Group evaporites, that end in a triangle zone. They note that 'In areas where there are no Zechstein (Group) evaporites to form a suitable detachment, the additional shortening required in the cover sequence is accommodated by the development of a series of small pop-up structures.' Roelofsen & deBoer (1989), describing the Q/1 oil fields, suggest that inversion took place above basal detachment faults within the Zechstein Group. The structure of the Helder field is described as a 'relatively simple detached anticline'. The structure of the Hoorn field is far more complicated. Roelofsen & deBoer (1989) interpret the structure of the Hoorn field to be an antiformal stack involving Triassic to Holland Formation rocks. The Hoorn structure is shown in their interpretation to comprise part of a complex duplex floored in the Zechstein Group with a roof thrust in the upper part of the Holland Formation. The deformed-state section in Roelofsen & deBoer (1989) shows, in conceptual form, that the upper part of the section is detached along the Zechstein Group and that basement is not involved in the inversion.

Detachment above the Zechstein Group evaporites is, however, not the only mode of shortening. Inversion in our study area on the northeastern margin of the Broad Fourteens Basin was accommodated by both folding and thrusting of the upper basinal section over the platform, and uplift of the basement. Two kinds of inversion-related faults can be identified in the seismic section (Fig. 2a): normal-separation faults active during Mesozoic extension that were subsequently reactivated during inversion, and thrust-separation faults with relatively small displacements related solely to inversion. Inspection of the geoseismic section (Fig. 2b) and the restoration to Top–Holland (Fig. 5b) reveals that the three major Mesozoic normal-separation faults were reactivated as thrusts. Their null points separate regions with different slip senses (Fig. 6). Several other faults in the section, however, show only thrust separation and were thus only active during inversion (Fig. 2a). The majority of contraction during inversion was accommodated on the frontal fault where the null point has moved down into the Jurassic/Triassic part of the section (Fig. 6). Reactivation was somewhat less on the outer

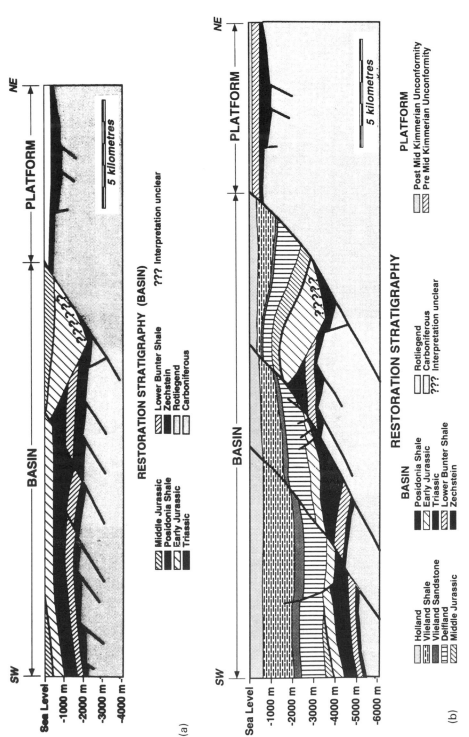

Fig. 5. (a) Schematic restoration to Mid Kimmerian unconformity. An early intracratonic phase of extension began in the Upper Triassic with slip of the basin margin fault system and development of a pronounced hangingwall rollover. Rotation of blocks in the hangingwall was accentuated by salt tectonics and resulted in the pattern of truncation of units (either by non-deposition of erosion) at the Mid Kimmerian unconformity. (b) Schematic restoration to Top-Holland. Extension, post Mid Kimmerian unconformity, was accommodated primarily over the basin margin fault and the two major outboard faults. Block rotation was not as pronounced as in the early phase of extension. Units above the Mid Kimmerian unconformity (Delfland and younger) thicken basinward across each of the major faults. The pattern of growth on the major faults indicates that extension was over by mid-Holland.

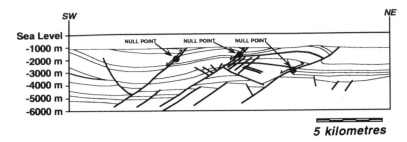

Fig. 6. Depth-converted geoseismic section (corresponding to Fig. 2a) indicating the position of the null points on the inversion-related faults. The majority of contraction during inversion was accommodated on the frontal fault where the null point has moved down into the Jurassic/Triassic part of the section. Reactivation was somewhat less on the outer two faults where the null point has only moved down into the Cretaceous Vlieland Shale.

two faults where the null point has only moved down into the Cretaceous Vlieland Shale (Fig. 6).

Role of basement during inversion

Basement in the present-day deformed-state section steps down fairly gently from the platform towards the southwest (i.e. towards the depositional centre of the basin) via a series of faults that still have normal-separation at the basement level (Fig. 2b). Several units in the inverted blocks are, however, above their regional level on the platform as indicated by the positions of their null points (Fig. 6). The depth to Top–Holland in the syncline on the backlimb of the frontal fold is *c.* 1500 m higher than the Top–Holland on the adjacent platform; the depth to Top–Holland on the crest of the frontal fold is over 2500 m above the Top–Holland on the platform (Fig. 2b). If it is assumed that the platform has remained relatively stable during the inversion phase, then the excess material above regional level (i.e. above the platform level) must be restored to that level. In the present day section, the difference in elevation between basement under the platform and basement under the basin margin is only 500 m. There is thus insufficient accommodation space to restore the cover sequence in the basin to its regional level with respect to the platform without involving the basement. The basement must have been active, inversion taking place along faults that were linked into the basement as has been assumed in the restoration to Top–Holland (Fig. 5b). The cover sequence did not behave entirely independently of the basement during inversion. The inversion process thus has two components: uplift along faults linked into the basement, and overthrusting of the cover sequence (Zechstein and above) over the platform.

Amount of basement uplift during inversion

The amount of basement uplift can be determined from comparison of the pre- and post-inversion sections (Fig. 7). At points H_1 and H_2 in the pre-inversion section, basement under the basin is *c.* 2700 m and *c.* 4600 m below the basement crest on the platform respectively. In the post-inversion section, points H_1 and H_2 are only *c.* 500 m and *c.* 2100 m respectively, below the basement crest on the platform. Points H_1 and H_2 have thus been uplifted. The amount of basement uplift at the edge of the basin (point H_1) is *c.* 2700–*c.* 500 = *c.* 2200 m. The amount of inversion-related uplift progressively increases to the southwest towards the basin axis. At point H_2 the uplift increases to *c.* 4600–*c.*2100 = 2500 m. Inversion-related uplift in the centre of the basin is as much as 3500 m (Van Wijhe 1987*b*).

The amount of basement uplift should be compatible with the amount of uplift of the overlying cover sequence. The Top–Holland Formation at the outboard end of the frontal inversion-related fold is *c.* 1600 m above the Top–Holland Formation on the adjacent platform. This represents the minimum amount of inversion-related uplift of the cover at the margin of the basin. There is, however, a discrepancy of *c.* 600 m between the amount of uplift determined from the basement (*c.* 2200 m), and that determined from the Top–Holland (*c.* 1600 m). Several factors could contribute to this apparent discrepancy in uplift. The basinal section is considerably thickened relative to the equivalent section on the platform

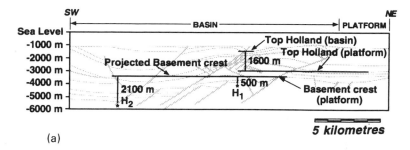

(a)

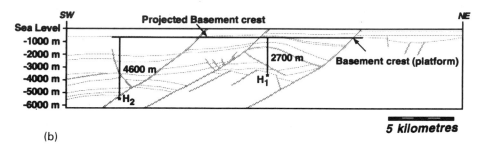

(b)

Fig. 7. Annotated cross-sections to illustrate the estimation of the magnitude of inversion-related uplift in the basin (after Figs 2b, 5b). Basement: At points H_1 and H_2 in the restoration to Top–Holland (**b**), basement under the basin is 2700 m and 4600 m below the basement crest on the platform respectively. In the present-day section (**a**), points H_1 and H_2 have been uplifted and are now only 500 m and 2100 m below the basement crest on the platform. The amount of basement uplift at the edge of the basin (point H_1) is thus 2700–500 = 2200 m. The amount of inversion-related uplift progressively increases to the southwest towards the basin axis. At point H_2 the uplift is 4600–2100 = 2500 m. Inversion-related uplift in the centre of the basin is as much as 3500 m (Van Wijhe, 1987*b*). Holland formation: The Top–Holland Formation at the outboard end of the frontal inversion related fold (**a**) is 1600 m above the Top–Holland Formation on the adjacent platform. This represents the minimum amount of inversion related uplift at the margin of the basin. The difference between the amount of uplift determined from the basement, and that from the Top–Holland (2200–1600 = 600 m) can be accounted for by the increased compaction of the basinal section relative to the platform, and the migration of salt during inversion.

(namely *c*. 500 m on the platform vs. *c*. 3000 m at the basin margin and *c*. 5000 m in the basin centre). The basinal section would have experienced increased compaction relative to the platformal section. This would be manifest as a relative increase in uplift of the less compactable basement section relative to the more compactable cover sequence. Migration of salt away from the frontal area during inversion could also allow increased uplift of the basement relative to the cover and further contribute to the apparent discrepancy between uplift-estimates. The value of uplift as determined from the Top–Holland thus underestimates the amount of uplift of the basin because it fails to account for compaction and salt migration. The best estimate of uplift of the basin is determined from movement of the basement. We estimate *c*. 2200 m of basement uplift at the basin margin; that value increases to as much as 3500 m in the basin centre (Van Wijhe 1987*b*).

Character of the accommodation to inversion

Several previous interpretations for the structural configuration indicate that the basin-fill was thrust over the platform along low-angle detachments without basement involvement (e.g. Hayward & Graham 1989; Roelofsen & deBoer 1989). Our restorations, however indicate that the basin-fill (Zechstein and above) did not behave independently of the basement during inversion; *c*. 2200 m of inversion-related basement uplift must have occurred at the basin margin. The style of inversion may thus vary along the margin. A general explanation for the range in inversion style is that it is related to stratigraphy, specifically the presence or absence of Zechstein Group evaporites. Where the Zechstein Group is present, detachment faulting occurs; where absent or thin, pop-ups occur. Our studies, however, indicate that substantial basement uplift does occur in areas where relatively thick Zechstein Group is present. While the presence or absence of a weak lower detachment helps decouple the cover from the basement and consequently influences the character of inversion, it cannot be the only controlling factor. We attribute this deformation style to two factors: the pre-inversion configuration of the margin and the resolved inversion direction.

Footwall shortcutting has been identified as an important deformation process during inversion (Huyghe & Mugnier 1992). Footwall shortcut faults have been recognized in seismic and outcrop data (e.g. Dart & McClay 1993), and reproduced in analogue model experiments (e.g. Buchanan 1991; Buchanan & McClay 1991). Such faults, however, have not been interpreted in our data from the northeastern margin of the Broad Fourteens Basin. Our data indicate that the principal responses to inversion were uplift of the basement along pre-existing extensional faults and backwards expulsion of the cover over the platform.

The main bounding fault controlling the Mesozoic extensional history of the Broad Fourteens Basin, and thus defining the half-graben character, is interpreted to lie along the southwestern margin of the basin. The fault array along the northeastern margin is thus developed on the flexural margin of the basin and may define a crestal-collapse-graben system. Analogue model experiments (e.g. Buchanan 1991; Buchanan & McClay 1991) reveal that deformation of a crestal-collapse-graben at low percentage inversions (i.e. below 100%) is dominated by vertical uplift along the pre-existing normal faults defining the graben. New syn-inversion faults nucleate at the tips of pre-existing normal faults, and while these typically propagate at a lower angle than the original normal-fault, footwall shortcutting does not appear to be an important deformation process. The pre-existing configuration of the northeastern margin of the Broad Fourteens Basin (i.e. on the flexural margin of a half-graben) would thus favour reactivation of the basement faults without the development of footwall shortcut faults.

The most important factor contributing to the inversion style, however, may be the resolved inversion direction. High angle extensional faults are generally not favourably orientated for orthogonal reactivation as contractional faults. Inversion of the Broad Fourteens Basin, however, has been interpreted to result from sinistral transpression with a Late Cretaceous contraction direction of 125° (+/−15°) (Huyghe & Mugnier 1993). The trend of faults on the northeastern margin of the basin is 140° (+/−15°). The reactivation angle (0°–30°) will thus be considerably lower than the fault dips of *c.* 60°. The strong oblique component of the deformation suggested by Huyghe & Mugnier (1993) could thus easily allow reactivation of pre-existing normal faults without the development of new, lower-angle contractional faults.

Curiously however, the basin-fill appears to have been expelled broadly orthogonal to the trend of the associated basement faults (i.e. NE) and does not follow the inferred contraction direction (NW–SE). Interpretation of our 3D data volume reveals that the frontal inversion-related folds have strongly curved axial traces and appear as 'tongues' thrust over the platform. The inversion-related deformation may thus be partitioned into two components. Basement blocks could have a sinistral strike-slip motion component to their oblique motion that ultimately resulted in uplift amounts of at least 2200 m at the basin margin. The basement must have been decoupled from the cover sequence by the Zechstein Group evaporites. Deformation in the cover was a simple response to the orthogonal contraction component of the deformation as the sub-structure of the basin became shorter.

Summary

Basin-wide inversion of the Broad Fourteens Basin during the Late Cretaceous and early Tertiary resulted in as much as 3500 m of vertical uplift along the inversion axis. Inversion was diachronous beginning in the centre of the basin as early as Albian but somewhat later (Maastrichtian) on the northeastern margin. Inversion on the northeastern margin was accommodated not only by backward expulsion of the basin-fill (creating a series of northeast-vergent thrusts and related lobate folds) but also reactivation of the basement faults. While inversion was accommodated primarily by the reuse of pre-existing normal-separation faults, some new fore- and back-thrusts developed particularly in the frontal area. Several previous interpretations indicate that the basin-fill (Zechstein and above) was thrust over the platform along low-angle detachments without basement involvement. Our restorations, however, indicate that the basin-fill did not behave independently of the basement during inversion; at least 2200 m of inversion-related basement uplift must have occurred. Substantial basement uplift does therefore occur in areas where the Zechstein Group is present. While the presence or absence of a weak lower detachment undoubtedly influences the character of inversion, the most important factors controlling final 'style' appear to be the pre-inversion structural configuration of the margin and the resolved inversion direction.

We would like to thank the management of Conoco Netherlands Oil Company, Conoco Inc. and our partners in the study area for their support and approval to publish this paper. The interpretations and conclusions, however, are those of the authors and do not necessarily reflect the views of Conoco Netherlands Oil Company, Conoco Inc. or our partners. The comments of an anonymous reviewer resulted in significant improvements to the clarity of the manuscript and are gratefully acknowledged.

References

BUCHANAN, P. G. 1991. *Geometries and kinematics analysis of inversion tectonics from analogue model studies*. PhD thesis, University of London.
—— & McCLAY, K. R. 1991. Sandbox experiments of inverted listric and planar fault systems. *In*: COBBOLD, P. R. (ed.) *Experimental and numerical modelling of inverted listric and planar fault systems. Tectonophysics*, **188**, 97–115.
DART, C. & McCLAY, K. R. 1993. 3D analysis of inverted extensional fault systems, southern Bristol Channel Basin, U.K. (abstract). *Abstracts with Program, 1993 Basin Inversion Conference*, Oxford, p. 31.

HAYWARD, A. B. & GRAHAM, R. H. 1989. Some geometrical characteristics of inversion. *In*: COOPER, M. A. & WILLIAMS, G. D. (eds) *Inversion Tectonics*. Geological Society, London, Special Publication, **44**, 17–39.
HUYGHE, P. 1992. *Enregistrement sédimentaire des déformations intraplaques: l'example de l'inversion structurale d'un bassin de la Mer du Nord*. PhD thesis, Université Joseph Fourier, Grenoble.
—— & MUGNIER, J. L. 1992. Short-cut geometry during structural inversions: competition between faulting and reactivation. *Bulletin de la Société Géologique de France*, **163**, 691–700.
—— & —— 1993. Inversion tectonics: a view from intra-plate basins and structure incorporated in the Alpine collision belt (abstract). *Abstracts with Program, 1993 Basin Inversion Conference*, Oxford, p. 70.
ROELOFSEN, J. W. & DEBOER, W. D. 1989. Geology of the Lower Cretaceous Q/1 oil fields, Broad Fourteens Basin, The Netherlands. *Proceedings of the EAPG Conference, Berlin, 1989*.
VAN WIJHE, D. H. 1987a. Structural evolution of inverted basins in the Dutch offshore. *Tectonophysics*, **137**, 171–219.
—— 1987b. The structural evolution of the Broad Fourteens Basin. *In*: BROOKS, J. & GLENNIE, K. (eds) *Petroleum Geology of North West Europe*. Graham and Trotman, London, 315–323.

Basin inversion in a strike-slip regime: the Tornquist Zone, Southern Baltic Sea

NIGEL R. DEEKS[1] & STEFAN A. THOMAS[2]

[1]*Department of Geology, Royal Holloway, University of London, Egham, Surrey, TW20 0EX, UK*

[2]*Institut für Geophysik, Universität Kiel, Olshausenstr. 40, D-24098 Kiel, Germany*

Abstract: The southern section of the deep seismic reflection profile BABEL crosses the Tornquist Zone between the mainland of Southern Sweden (Scania) and the island of Bornholm. This marks the boundary between the Sorgenfrei Tornquist Zone to the northwest and the Teisseyre Tornquist Zone to the southeast. The region represents a large-scale releasing bend in the dextral strike-slip system of Tornquist and has led to the development of pull-apart basins such as the Rønne and Arnager grabens, from the Carboniferous to the Late Jurassic.

The area around BABEL is covered by proprietary surveys (courtesy of JEBCO Seismic Ltd, BGR Germany and the Swedish Geological survey) which allow the three-dimensional interpretation of sediment/structure interactions within these basins. Within the area various classical (positive) inversion structures are seen on seismic sections, including harpoon structures and inversion monoclines. These features are found in association with both basin-bounding and intra-basinal faults. There is also evidence for inversion of basement blocks within the upper crust. Inversion took place between the late Cretaceous and early Tertiary, probably as a response to Carpathian/Alpine movements to the south. Strike-slip structures indicate that movements were dominantly dextral with rigid blocks such as Bornholm channelling deformation around them. At the same time, progradational sequences of sediment are seen to build off these blocks.

The BABEL section images a number of features believed to have resulted from the post-Carboniferous strike-slip movements of the Tornquist fault system. Structures imaged within the upper part of the section are consistent with those interpreted in three dimensions from the Jebco survey, as a brittle response within the upper crust. Below these, between 7 and 12 seconds TWT, a zone of high reflectivity is observed. This is similar in character to regions of lower crust imaged on either side to the northeast and southwest, but it is shallower below the Tornquist Zone. This is interpreted as a region of lower crust in which deformation is transferred in a more ductile fashion. Below this again there are coherent reflectors that may represent a more brittle transfer of the deformation into the upper mantle. The extent of upper crustal deformation imaged on this profile is around 40 km in width whilst in the lower crust it is at least 90 km.

A number of marine seismic reflection surveys have been carried out in the Southern Baltic Sea during the 1970s and 1980s as part of the search for hydrocarbon resources in the area. Results of such a commercial survey (courtesy of JEBCO Seismic Ltd.) have been combined with detailed sections from the marine deep seismic reflection profile BABEL (Baltic and Bothnian Echoes from the Lithosphere). In conjunction with selected lines from the PETROBALTIC survey (courtesy of Bundesanstalt für Geowissenschaften und Rohstoffe) this gives good coverage of much of the Southern Baltic region (Fig. 1). Although the three seismic datasets differ from each other in a number of ways, it was still possible to combine them in a useful and accurate way. Supplementary well data were used to constrain the geological interpretation. As a result, a regional geological interpretation has emerged. This has given a new understanding of the interplay between structural movement on a complex strike-slip fault system and the sedimentation history of the associated grabens. Overprinted on the earlier basin development are the effects of late Cretaceous to Tertiary inversion in the area.

It is not possible to date accurately the onset of inversion in the study area as the wells studied were drilled on relative structural highs and do not show complete sections. However similar inversion phases are recorded during the Late Cretaceous into the Tertiary from other areas of NW Europe. This phase referred to as the Laramide phase, was responsible for the

From BUCHANAN, J. G. & BUCHANAN, P. G. (eds), 1995, *Basin Inversion*, Geological Society Special Publication No. 88, 319–338.

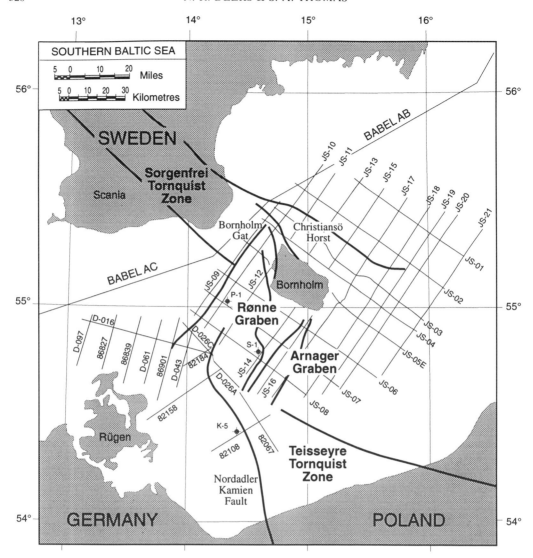

Fig. 1. Regional structure and seismic grid. Lines marked JS represent the JEBCO Seismic Ltd, Offshore Bornholm survey: the location of the BABEL profile is shown with the remaining lines representing a sample of the GO PETROBALTIC dataset. P-1, Pernille-1; S-1, Stina-1; K-5, K5-1/88.

development of up to 3000 m of structural relief in the Polish Trough (Pozaryski & Brochwicz-Lewinski 1978) and over thrusting in the Lower Saxony basin (Baldschuhn *et al*. 1985). In many cases the onset of inversion over most of NW Europe was diachronous. For example, to the northwest of Scania from the Kattegat to the Egersund basin, Mogensen & Jensen (1994) note a north–westward progradation of inversion during the Late Cretaceous. They suggest that inversion probably started in the Turonian.

This inversion phase commonly affected basinal areas that had developed along long lived fault trends. Other well documented examples include the Danish Central graben (Cartwright 1989), the Broad Fourteens basin in the Dutch sector of the North Sea (Dronkers & Mrozek 1991) and within the UK sector around the South Hewitt fault zone (Bradley *et al*. 1989) and into the Sole Pit (Van Hoorn 1987). For these areas as well as for the Rønne graben Ziegler (1983) reports inversion between the Senonian and the Palaeocene. Strike-slip movements are commonly inferred but they are generally not as clearly evident as in the area around Bornholm.

The geotectonic setting of the SW Baltic Sea area is characterized and dominated by the Tornquist Zone. This is a long lived structural lineament that has been active at least since the Palaeozoic. Brochwicz-Lewinski *et al.* (1984) and Pegrum (1984) infer from earlier work that there were most likely several episodes of sinistral strike-slip movements during the Palaeozoic. They also note that this was the result of transverse motions during the Appalachian–Caledonian orogeny and that this motion probably did not cease until the Early Devonian (Siegenian).

From the Permo-Carboniferous the Tornquist Zone developed as part of a dextral transform system within Northern Europe, linking the Uralides with the Appalachians (Arthaud & Matte 1977, Ziegler 1981). The region around Bornholm represents a large-scale releasing overstep in the dextral strike-slip system between the Sorgenfrei Tornquist Zone to the northwest and the Teisseyre Tornquist Zone to the southeast (Fig. 1). Mogensen & Jensen (1994) also suggest that there may have been minor amounts of dextral movement along the Tornquist Zone throughout the Mesozoic.

Seismic and well data

The location map in Fig. 1 shows the area studied around Bornholm and the seismic profiling performed by JEBCO, BABEL and PETROBALTIC. It also shows major geological features and the location of three of the wells used in this study.

Offshore Bornholm survey

The JEBCO survey includes approximately 1830 km of seismic lines spaced between 7 and 20 km (Fig. 1). The survey was shot in 1988 by Digital Exploration Ltd using airgun sources of 59.65 litre capacity and a shot interval of 25 m. The 6 s record (processed to 5 s) was sampled at a 2 ms rate using a 3 km streamer incorporating 240 groups giving a 60 fold CMP stack. Relatively standard processing was carried out using Digicon's DISCO software. This included resampling to 4 ms, pre-stack F–K filtering, NMO and deconvolution. After creation of the 60 fold CMP gathers and stacking, post-stack deconvolution and F–K filtering were applied. The data then underwent wave equation migration using 95% stacking velocities prior to time variant bandpass filtering and display.

Although the line spacing of this survey is generally much greater than that of the PETRO-BALTIC survey, it has still been possible to map both basin bounding and intrabasinal faults with some accuracy. The second big advantage that this dataset has over the other two is that it alone has been migrated. This has made it possible to distinguish clearly the complex structural features associated with the basin development and more importantly inversion and strike-slip derived structures in the area. Further work on this dataset aims to clarify the true geometries of the fault systems by depth migration of the data. Ziegler (1983) shows a true scale interpretation of an example from an earlier survey of the area in which most of the faults have maintained their listric profiles.

GO PETROBALTIC survey

This survey was shot between 1975 and 1987 as a joint venture between the former German Democratic Republic, Soviet Union and Poland as 'Gemeinsame Organisation PETROBALTIC Gdansk' (for details see Rempel 1992). It comprises a very dense seismic network (over 11 000 km), selected profiles from which are shown in Fig. 1. The data were kindly made available by the Bundesanstalt für Geowissenschaften und Rohstoffe. The survey was shot with an airgun source and a 48 channel streamer to give a record length of between 3 and 5 s (dependant on local geology), with a line spacing of between 0.5 and 2 km. This gives very tight control on the structure. The data underwent standard processing using the COMECON processing system although unfortunately they were not migrated. For this study selected lines from the northeast of the survey have been used to tie in with the Bornholm survey to the north and as such gives a more complete structural picture of the area.

BABEL profile

The BABEL deep seismic reflection profile was carried out by Prakla Seismos AG on contract for BIRPS on behalf of the BABEL Working Group in 1989. Sections AB and AC cross the Tornquist Zone orthogonally through the Bornholm Gat (Fig. 1). The data were acquired using a tuned airgun array of 48 guns of 120.6 litres total volume, towed at 7.5 m depth, with a 3000 m, 60-channel streamer, towed at 15 m. The record length in the study area was 18 s, allowing a 30-fold stack per 25 m CMP trace (see also BABEL Working Group 1993). The sections used in this interpretation are based on

Table 1 *Additional processing carried out on final stack sections*

Process	Parameters	Objective
Band pass filtering (zero phase)	12–28 Hz with cut-off frequencies of 8 and 32 Hz	Attenuating those frequency bands of the airgun bubble in the shallowest part of the section corresponding to two-way times of less than 1 s as well as the high frequency noise of intra-crustal and upper mantle levels between 9 and 18 s.
Gain function	Time variant scaling, 5000 ms steps	The scaling function is derived from the data and brings up weak reflection zones on a trace-by-trace basis.
2D Filter	3.0 to $-3.0\,\text{ms}^{-1}$ per trace, 20 adjacent traces Window 200 ms^{-1}	Enhanced lateral amplitude relationships within a specified range of defined dips.
Display	Variable area/wiggle trace	Wavelet is more easily interpreted. There is no vertical exaggeration at an average crustal velocity of about $6500\,\text{ms}^{-1}$.

the unmigrated final stack sections as produced by BIRPS and GECO-PRAKLA. In order to help interpretation some additional processing was carried out on this final stack data. This allowed better imaging of dipping reflectors throughout the 18 second trace. Brief details of each additional step are given in Table 1. Although shot as a single line, 3D control within the top 5 s can be given by combining the BABEL section with the Bornholm survey.

Well data

Within the offshore Bornholm survey area two wildcat wells were drilled and completion logs were kindly made available by Norsk Hydro (Pernille-1) and Amoco Denmark (Stina-1). Both wells were drilled within the Rønne graben; Stina-1 on the southeast flank, adjacent to the bounding fault of the graben, Pernille-1 towards the west of the graben above a complex basement high and into the hangingwall of an inverted normal fault. The wells are approximately 34 km apart (Fig. 1). Neither well drilled to basement, with both TDs being around 400 m into the Graptolitic Mudstone group of the Silurian. Formation tops and sonic log traces were used to correlate with seismic phases from the survey. Although both wells are within the Rønne graben it was possible to use them to tie seismic stratigraphic units over the majority of the southern area of the survey. To the north within the Hanö Bay basin and the Christiansø half-graben, interpretation is based solely on seismic character and seismic stratigraphic principles. However, well data from northeast of the survey area in Swedish waters have recently become available, enabling control on the interpretation.

Within the area covered by the PETROBAL-TIC survey a series of wells has also been drilled and details made available to accurately tie this area, for example the stratigraphy on and around line 82108 (Fig. 3) has been tied to the nearby well K5-1/88 (Fig. 1).

Geological succession pre-inversion

The geological evolution of the area since the beginning of the Palaeozoic is linked to the development of a series of basins with predominantly similar fills, which later undergo inversion. The progressive development of the area can be viewed as being a sequential system from deposition on a uniform basement of pre-, syn- and post-rift sequences and then a syn- to post-inversion sequence associated with the most recent phase of structuring.

Basement

The basement rocks of the area are predominantly Precambrian in age and form part of the Gothian Terrain (1.77–1.6 Ga) (Johansson & Larsen 1989) and outcrop both in southern Sweden and on Bornholm (Berthelsen 1992). In addition to the Gothian rocks, younger granites (c. 1.4 Ga) are likely to be present similar to those seen onshore Bornholm, but there is little seismic evidence for this.

In most cases the acoustic impedance contrast between these mainly igneous and metamorphic rocks and the overlying sedimentary sequences is such that a clear reflection is produced. This is particularly true in areas where relatively thin Mesozoic sediments rest directly on the basement. In areas where significant thicknesses of

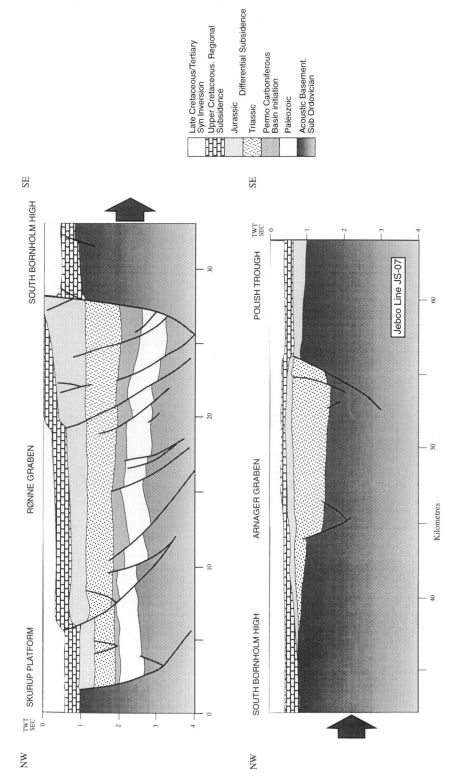

Fig. 2. Geoseismic cross-section along Jebco line JS-07. Representative basin fills and inversion-derived structures within the Rønne and Arnager Grabens. Note both preferential inversion of intra-basinal fault and well developed harpoon structure in the Rønne Graben, also the 'pop-up' of the Arnagar Graben.

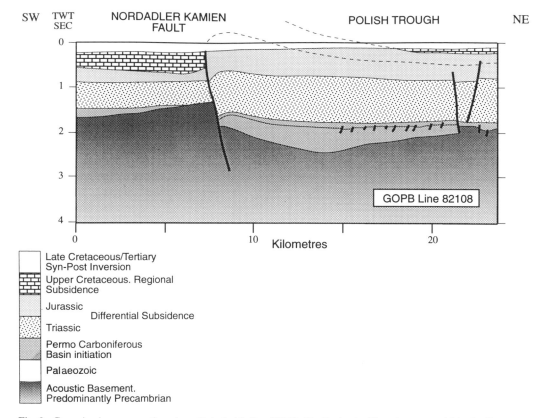

Fig. 3. Geoseismic cross-section along Petrobaltic line 82108. Similar basin fill to that seen within the Rønne and Arnager Grabens. Here the Nordadler Kamien fault marks the inverted margin of the Polish trough. Also note lack of preserved Palaeozoic deposits.

Palaeozoic deposits are present the boundary is more difficult to define, especially where basal Cambrian quartzites are preserved, due to the weak acoustic impedance contrast between them. This boundary is largely inferred in the area as no wells penetrate to basement. However this basal unconformity is seen onshore throughout Scania (Bergström *et al.* 1982).

Pre-rift

The pre-rift sequence in the area is represented by a series of Palaeozoic deposits. These range in age from Cambrian to Silurian as a sequence of marine deposits marking deposition on the stable continental shelf of the Baltic plate. In common with mainland Sweden no deposits of Devonian age are found in the area (Bergström *et al.* 1982). This is believed to be the result of non-deposition on what was then the Old Red Sandstone continent (Ziegler 1982).

Palaeozoic rocks are locally preserved within down-faulted zones as in the Rønne graben

(Fig. 2). Other Palaeozoic fault bounded remnants occur on the Skurup Platform and within the Colonus Shale Trough across the Bornholm Gat into Scania (Fig. 4). To the north of the area within the Hanö Bay basin, Kumpas (1980) notes thin Lower Palaeozoic deposits preserved in depressions in the basement. Wells Pernille-1 and Stina-1 both encountered Silurian shales. The preserved thickness of Lower Palaeozoic rocks within the area is very variable with a general thickening towards the south, particularly within the southern part of the Rønne graben where the Lower Palaeozoic sequence reaches a thickness of greater than 4 km (1.7 s TWT). However further to the southeast within the Polish trough, no preserved Lower Palaeozoic deposits are seen.

Syn-rift

The beginning of the syn-rift sequence is represented by Permo-Carboniferous deposits which rest with marked unconformity on the underlying Lower Palaeozoics. Largely as a result of their limited thickness, the Carboniferous, Rotliegende and Zechstein are treated as a single seismic unit. Although there is none of the rapid thickening evident in the Triassic and Jurassic succession, there are indications of syn-sedimentary fault movements. This is taken as an indication of the initiation of rifting at this time (see also Ziegler 1983). This would be consistent with other areas of Europe where a number of localized wrench basins with varying amounts of fill were developing at this time (Ziegler 1981). Many of these newly formed basins were associated with extrusion of varying amounts of volcanics, as in the case of the Oslo graben. However there is no indication of volcanic activity within the Rønne graben from either of the available well logs. A beta factor for the Rønne graben has been calculated to be at least 1.32 after taking into account the effects of the later inversion phase. Although only preserved around Bornholm within the Rønne graben, there are similar deposits interpreted to the southeast within the Polish trough (Fig. 3). This period is believed to mark the onset of dextral movements on the Tornquist, within the Late Variscan wrench setting (Ziegler 1981). Arthaud & Matte (1977) also infer dextral movements in this area during the Permo-Carboniferous, attributing them to the development of a right-lateral shear zone between the Appalachians and the Urals. This was coincident with the emplacement of a set of NW–SE orientated basic dykes paralleling the newly developing fault zone (Berthelsen 1992, Liboriussen et al. 1987).

The start of the Triassic saw the development of a releasing overstep at the junction of the Teisseyre and Sorgenfrei Tornquist Zones. Basin development resulted from dextral NW–SE strike-slip movement on the system. The Rønne and Arnager grabens developed rapidly as characteristic pull-apart basins, which extended between opposing normal faults oriented NE–SW that are well imaged on the survey. Seismic packages are observed to thicken into the active basin-bounding faults (Fig. 2). The seismic character of these units is generally slightly chaotic, it being difficult to follow coherent individual phases over great distances (Figs 6, 7). This is particularly noticeable when compared with the seismic character of, for example, the marine deposits of the Upper Cretaceous (Figs 6, 7). This is a characteristic response of terrestrial deposits and is consistent with evidence from both wells and onshore surface geology (Gravesen et al. 1982). In addition to downthrow of the basinal areas

there is some evidence of footwall uplift directly adjacent to the graben. This is inferred from combining the JEBCO data with the adjacent part of the BABEL line A and selected lines from a nearby Swedish survey. From these it is possible to observe two NW–SE trending, down-faulted remnant Palaeozoic units, preserved within the Skurup platform (Fig. 4). These units thin dramatically to pinch out towards the Rønne graben along the length of the bounding fault. This, together with the very texturally and mineralogically immature sediments of this age found from well data, suggest a very proximal source, likely to be the eroding Skurup platform itself. Both wells show coarse clastic fills deposited during the early Triassic. These are believed to represent deposition from alluvial fans building directly off the rapidly developing graben bounding faults. In both grabens the bounding faults converge towards the north giving a triangular shape to these basins (Fig. 4). Within the northern part of the Rønne graben the basin edge is formed by a more complex set of terraces. These comprise a series of broadly en echelon and commonly curved faults accommodating extension within the graben floor (Fig. 4). This complex structure is important during the later inversion phase.

Subsidence histories from the wells Stina-1 and Pernille-1 (both within the Rønne Graben (Fig. 1)) indicate that subsidence was most rapid during the lowest most Triassic with rapid subsidence into the Lower Jurassic. Unfortunately due to a pre-Cretaceous erosional event no Jurassic deposits less than 200 Ma are preserved. However the Lower Cenomanian, marls of the Arnager greensand equivalent in Pernille-1 have a much less steep subsidence history (these rest unconformably on the underlying Lower Jurassic deposits), which may be taken as an indication that subsidence rates in the area at this time were much reduced. By the onset of chalk deposition in the Coniacian, deposition rates had increased greatly once more.

At the same time as the Rønne and Arnager grabens were filling with sediment, the Polish trough was developing further to the southeast. A similar syn-tectonic thickening is seen across the Nordadler Kamien fault, the southern bounding fault of the Tornquist Zone, which here forms the southwestern margin of the Polish trough (Fig. 3). In the north of the area the Hanö Bay basin was also developing in the same strike-slip regime but with a different response. No Triassic deposits are recorded from this area although there was differential subsidence during the Jurassic.

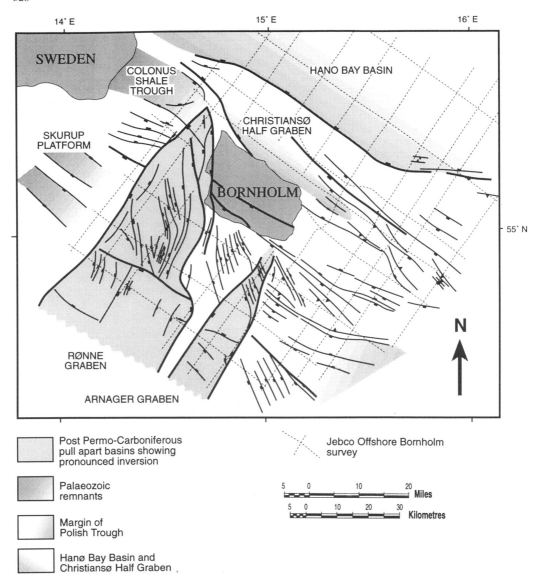

Fig. 4. Basement structure map for the Bornholm survey area, showing major fault trends and location of various basins. Note predominantly NW–SE orientated faulting outside the major grabens. This trend paralleling the edges of the Tornquist zone. Within the floors of the two main grabens the faults are orientated N–S to NNW–SSE. The location of down faulted Palaeozoic sediments within the Colonus Shale Trough and within the floor of the Skurup platform are also indicated. Onshore structure after Gry (1960–64).

Sedimentation continued into the Jurassic with further thickening of units and some further syn-sedimentary fault movement. In the central Rønne Graben listric faults developed that sole in the Permo-Carboniferous (see McClay this volume). These faults probably detach in lacustrine shales of the Carboniferous, as interpreted from Pernille-1. Within the Arnager graben the transition is clearly seen from brittle rifting

rifting during initiation of the pull-apart basin to a later sag phase involving a more regional subsidence in the Late Triassic and into the Jurassic. Here the sediments progressively onlap on to the South Bornholm high (Fig. 2). Outside the grabens there was general subsidence eastwards towards the Polish trough. It is unclear whether areas such as the Skurup platform to the west of the Rønne graben were areas of

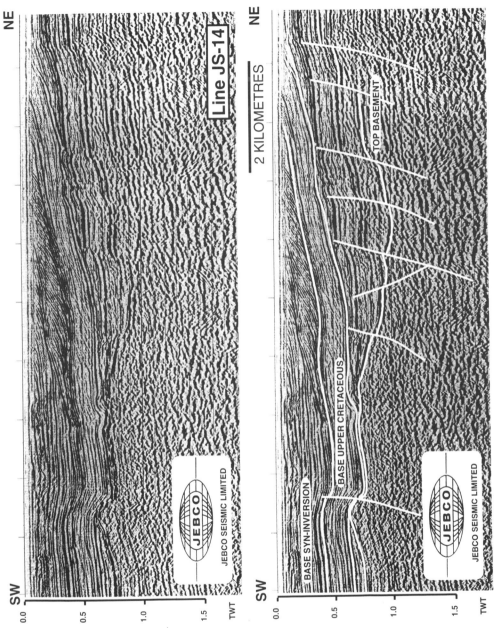

Fig. 5. Syn-inversion offlap with progradational sets building off the uplifting and eroding Bornholm block to the northeast. Detail of line JS-14.

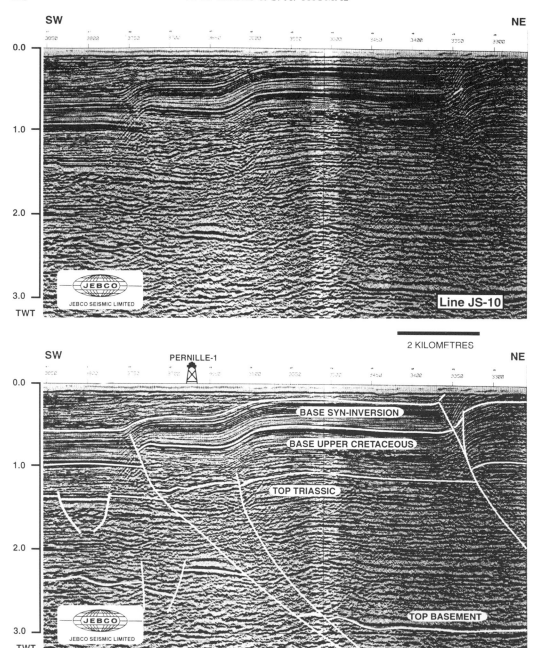

Fig. 6. Detail of line JS-10 from within the central Rønne Graben. To the southwest a pair of inverted listric faults have developed fault propagation folds giving rise to 'inversion monoclines'. To the northeast a third inverted fault has developed a minor footwall shortcut. Note faults at the basement level remain in net extension or have not undergone inversion.

deposition at this time as the overlying Upper Cretaceous unconformity lies directly on basement (Figs 2, 7).

Sedimentation continued in the area up into the Lower Cretaceous at least in the Rønne graben and presumably within the Polish Trough. The Pernille-1 well indicates that these Early Cretaceous deposits (dated as being equivalent to the Arnager Greensand) may represent the early stages of a return to marine conditions in the area.

During the Early Cretaceous, sedimentation was interrupted with the development of an angular unconformity prior to the deposition of the Upper Cretaceous. This is best seen in dip sections that cross the axis of the Rønne (Fig. 7) and Arnager grabens. Over the relative basement highs of the Skurup platform and the South Bornholm high the Upper Cretaceous deposits rest directly on crystalline basement (Figs 2, 7). This unconformity may represent an earlier phase of transpressional movement but this is unlikely as there appears to have been no structuring associated with it. It is more likely that the hiatus and unconformity are the result of more regional uplift and associated changes in relative sea level.

Post-rift

The re-establishment of sedimentation during the Late Cretaceous marks a short period of quiescence and regional subsidence in the region. This sequence can be viewed as a post-rift or even pre-inversion unit separating the extensional phase from the compressional phase to come. Although now eroded in some areas (principally in the northern part of the Rønne graben), deposition appeared to have been fairly uniform over the whole area with no indications of the fault controlled thickening seen throughout the Triassic and Jurassic. Upper Cretaceous reflectors are mostly laterally persistent and coherent, characteristic of a relatively low energy marine depositional system of the chalk as seen in Pernille-1. This differs markedly from the much less coherent reflectors of terrestrial and deltaic deposits underlying them (Figs 6, 7). This is in good agreement with the observations of Thomas et al. (1993).

To the north of Bornholm the base Upper Cretaceous reflector is also seen but is not as pronounced and in general the unconformity surface does not show as much angularity as to the south. The base Upper Cretaceous reflector, here marked over most of the area by a strong doublet, is a widely correlatable horizon. Maps of this horizon give a good indication of the structural effects of the later inversion, as in most cases this reflector is now above the null point of most of the faults in the inverted system.

Late Cretaceous inversion

During the Late Cretaceous and into the Tertiary, this area of NW Europe is believed to have been affected by both the opening of the North Atlantic and the onset of compression from the Alps to the south (Biddle & Rudolph 1988). This would have resulted in right lateral transpression along the Tornquist Zone (Mogensen 1992) and as such gives a method of inverting the basins in this study area while maintaining net dextral movement on the Tornquist Zone.

Syn-inversion sedimentation

Sedimentation was also continuing whilst this inversion was taking place. Over the Skurup platform at this time there appears to have been broadly uniform subsidence. Within the basins some syn-inversion sequences are recognized. To the south of Bornholm these take the form of progradational sets building off the uplifting and presumably eroding Bornholm block (Fig. 5). In addition to the sedimentary build out from the uplifted basement blocks growth folds are seen to develop onto rising inversion monoclines. As a monocline continues to develop sediment thins onto its flanks and earlier deposited units become progressively more tilted.

Inversion structures

Within this strike-slip environment it appears that individual crustal blocks acted in different ways. The sediment-filled basins acted more plastically and underwent shortening, giving rise to the characteristic inversion structures observed (Figs 2, 3, 6, 7). These structures can be directly compared with those described by McClay (1989) and others from analogue model studies. These include very well imaged harpoon structures (asymmetric anticlines) (Figs 2, 3), inversion monoclines (Fig. 6) developed over the tip lines of blind faults as fault propagation folds, and more detailed faulting including footwall shortcut faults (Fig. 5). Harpoon structures are seen on various scales associated with both the intra-basinal faults and at the larger scale associated with the basin bounding faults of the grabens and also on the Nordadler Kamien fault (Fig. 3) to the southeast. In most cases it is possible to pick null points for inverted faults, below which the faults are in extension. This means that at the basement level the majority of the faults remain in net extension regardless of whether they have undergone inversion (Fig. 4). In some areas in addition to fault reactivation the inversion has resulted in varying degrees of buttressing. This is seen on the southwest side of the Rønne graben against the Precambrian basement of the South Bornholm High, as

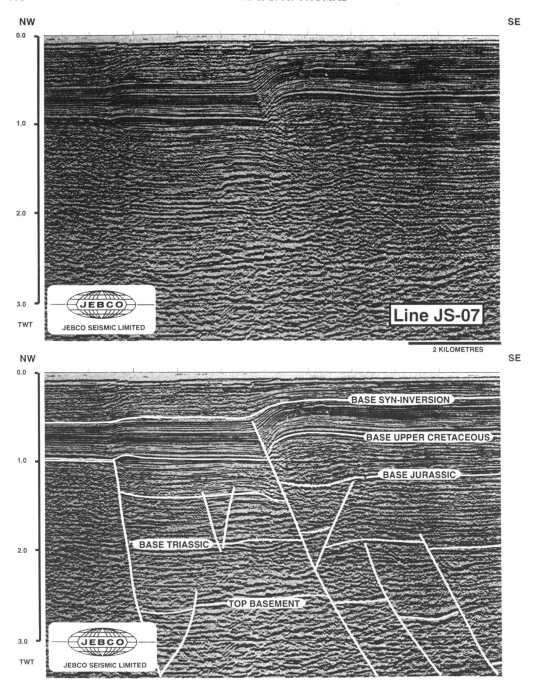

Fig. 7. Preferential inversion of NNW–SSE orientated intrabasinal fault over NE–SW orientated basin-bounding fault. Detail of line JS-07 from the NW side of the Rønne graben.

shown by McClay (this volume). In addition to true inversion, developing faults have in some cases propagated laterally as thrusts which are not utilizing pre-existing normal faults. Within the northern portion of the Rønne graben where the basin floor comprises a series of en echelon faults (Fig. 4), inversion of a number of these structures has led to a basin wide uplift.

There is evidence that the area as a whole continued with the same net dextral movement as during the original opening of the graben. An Alpine–Carpathian derived force from the South that is then partitioned into two orthogonal components has been suggested as a mechanism. This would result in continued dextral movement and compression from the southwest (Ruelland 1992).

There is apparent preferential inversion of some faults over others, for example as seen on the western side of the Rønne graben (Figs 2, 7). Here an intrabasinal fault orientated NNW–SSE over the NE–SW orientated basin-bounding fault. The basin-bounding fault does show very minor inversion, with a slight kinking of the Upper Cretaceous sediments, indicating that it was not fully locked. Even so, the majority of shortening is taken up on the adjacent more shallowly dipping fault. In other parts of the study area buttressing of the sedimentary sequences of the Rønne Graben against the basement highs is seen. This is best seen on the east side of the graben. The most pronounced buttressing is seen against the bend in the bounding fault around the cross over of lines JS-08 and JS-14 (Figs 1, 4; McClay this volume, fig. 13c). This may point to the inversion being oblique to the direction of original extension. With this in mind it seems likely that the compressional forces resulting in the inversion were from further to the south rather than a simple reversal of the NW–SE strike-slip forces that resulted in the opening of the grabens. However other factors may also be affecting whether or not one fault is preferentially inverted compared to another. These include the steepness of the fault and whether or not it has become locked.

Seismic sections along lines perpendicular to the axis of both the Rønne and Arnager grabens clearly show the wider effects of inversion. In the case of the Arnager graben this has resulted in the entire sedimentary basin being popped up to form a broad, low amplitude anticline with its axis trending along the axis of the graben (Fig. 2).

The Nordadler Kamien fault (Figs 1, 3) has a long history similar to those imaged to the northwest and also exhibits the characteristic large-scale harpoon structure (asymmetric anticline) derived from the Late Cretaceous to Early Tertiary inversion, whilst minor crestal collapse structures adjacent to the fault, within the Polish Trough, are still in net extension. Unlike the inverted faults within the pull-apart basins around Bornholm, this fault has a predominantly NW–SE orientation. The inversion seen here is also believed largely to be the result of transpressional movements, with a component of the compression acting from the southwest.

Strike-slip and block movements

In contrast some areas acted as rigid blocks channelling the deformation around them as appears to have been the case with Bornholm (Fig. 8). In this case it appears that the uplifted Bornholm block has remained relatively rigid, buttressing against the mainland of Scania via the basement ridge of the Northern Bornholm Gat.

A series of broadly NW–SE orientated faults are seen to the southeast of Bornholm. Many of these have undergone reactivation during the Alpine–Carpathian inversion phase. In this area there are indications that much of the deformation was derived from strike-slip rather than pure dip-slip movement. Some of the NW–SE trending structures observed outside the grabens can be interpreted as flower structures, some of which can be traced laterally for many kilometres (Figs 8a, b). Although broadly orientated NW–SE, these structures commonly have curved traces and as a result it is possible to recognize features resulting from deformation at a restraining bend. Within the restraining bend the complexity of the flower structure increases. This gives the direction of strike-slip movement. Although these structures are laterally persistent they do eventually die out along strike, in doing so becoming broader with less structural relief. This broadening takes the form of a fault splay in which, unlike the true flower structure, the individual faults need not be traced to a single fault surface at depth. It is likely that this fault was in the form of a flower structure during the earlier Mesozoic as the grabens were opening and that it was modified by the later transpressional movements. This may help to explain why, unlike more classical flower structures, the primary basement cutting fault is not near vertical.

To the northeast the movements, although still regionally dextral, resulted in sinistral movements relative to the Bornholm basement block. This resulted in the development of a minor sinistral, positive flower structure to the north of the Bornholm block (Fig. 8a). As such the rigid basement block of Bornholm acted as an indentor against the dextral movements within the plate, channelling movements to the north and south of it.

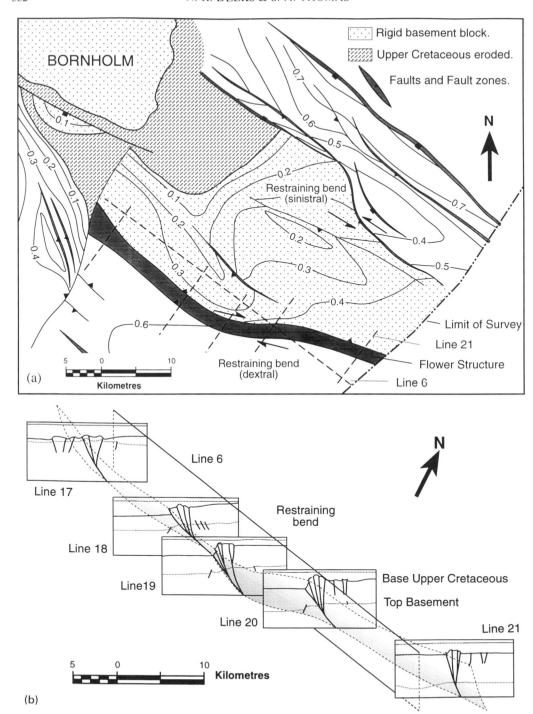

Fig. 8. Detail of strike-slip derived structures to the southeast of Bornholm. (**a**) Structure map of base Upper Cretaceous reflector showing subsurface extension of the uplifted Bornholm block. Note the sinistral sense of movement on a restraining bend (poorly developed flower structure) to the north of the structure and well developed dextral feature of to the south. This dextral feature has also developed on a complex flower structure, detailed in (b). Contours are in seconds TWT. (**b**) 3D representation of lateral variation along dextral flower structure on southern side of uplifted block in (a). Lines 6 and 17–21 are as located in (a).

Deep structure

The improved processing of this section has enabled most of the faulting seen to be linked into a coherent system within this part of the Tornquist Zone (see also Blundell and the BABEL Working Group 1993). As such this is an improvement on previous interpretations based on line drawings of the section (e.g. Thybo *et al.* 1993). The structures imaged on the BABEL profile where it crosses the Tornquist Zone cannot simply be viewed as products of the recent phase of inversion in the area. These structures are more likely to have developed throughout the history of the Tornquist Zone, subsequently overprinted or modified by the later phases of movement on the system. Structures imaged within the upper part of the section are consistent with those interpreted in three dimensions from the JEBCO survey.

A number of near-surface faults are imaged on the section shown in Fig. 9. Most are steep to near vertical and some are planar to slightly curved. They bound down-faulted blocks of both Palaeozoic and Mesozoic sediments as well as uplifted blocks of basement. Particular structures can be identified over the width of the Bornholm Gat. These include the Cretaceous Vomb Trough, the down faulted Palaeozoic sediments preserved in the Colonus shale trough and uplifted basement highs such as the Christiansø horst (Figs 4, 9). Flanking the uplifted blocks of the Bornholm Gat are the Skurup platform to the southwest and the Hanö Bay basin to the northeast. Both remained as areas of sediment accumulation during the Late Cretaceous to Tertiary inversion.

Between 4.0 and 7.0 s TWT a low angle NE-dipping reflector passes through the relatively transparent middle crust to the top of a zone of high lower-crustal reflectivity (Fig. 9). This reflective lower crust is not only related to the Tornquist Zone itself but extends with slightly varying patterns for more than 70 km to both sides along the BABEL profile. This band of sub-horizontal reflections dominates, with varying thickness, the depth range between 7.0 and 12.0 s TWT and within the Tornquist Zone is a domal uplifted region, 1.0–1.5 s above regional. Immediately below the Tornquist Zone the thickness of lower crustal reflectivity appears to be reduced, down to a minimum of 2.0 s, compared with a thickness of between 2.5 and 3.5 s on either side.

There are various pieces of evidence suggesting that ductile deformation in the lower crust acts to enhance lower crustal reflectivity. As recently summarized by Mooney & Meissner (1992) most suitable causes of crustal reflectivity are igneous intrusions, lithologic and metamorphic layering, and reflectivity from mylonite zones. Additional evidence shows that underplating of the lower crust with mantle-derived mafic rocks and compositional layering by common rock types also contribute to reflectivity. In addition lower crustal ductile flow enhances reflectivity by aligning and stretching lower crustal bodies into a sub-horizontal geometry. Ductile deformation may be ancient as in Precambrian suture zones, or young, as in recently extended crust. Zoback *et al.* (1989) noted that there is stress-strain evidence for major displacements in the lower crust. Large-scale stress systems in the upper crust are related to plate boundary forces that must also affect the lower crust, and can apparently be transferred over horizontal distances of more than 1000 km. As shown by Strehlau and Meissner (1987) and Nur *et al.* (1986) tectonic rotation of upper crustal blocks can be modelled by stress concentrations in the rigid upper crust above a weak, low viscosity, lower crust, i.e. brittle upper crustal deformation is decoupled from a more uniform deformation in the middle and lower crust.

The general observation of lower crustal reflectivity (see Matthews 1986) leads to the suggestion that a common process acts in its formation. We favour a model whereby regional and global plate stresses act to enhance reflectivity by inducing lower crustal ductile flow, producing sub-horizontal lamination. This ordering process would enhance reflectivity that has its primary cause in igneous intrusions, or compositional, metamorphic, or mineralogical layering. This process also requires elevated temperatures to promote ductile flow.

Thus, the observation of lower crustal reflectivity and NE-dipping intracrustal reflectors leads to the suggestion that this intracrustal reflector may represent a brittle to semi-ductile shear zone. Movements being transferred through the crust via faulting in the brittle upper crust and more ductile flow on top or within the reflective lower crust. The slightly uplifted lower crust immediately below the Bornholm Gat may be a direct result of transpressive movements along Tornquist, the compressional forces being translated into vertical movements on a crustal scale (Fig. 9). In the study area the Tornquist Zone is relatively narrow with a surface breadth of only 45 km, whereas at depth, the crust has been affected over a width of around 90 km. In addition this interpretation precludes the need for through-going steep faults cutting vertically

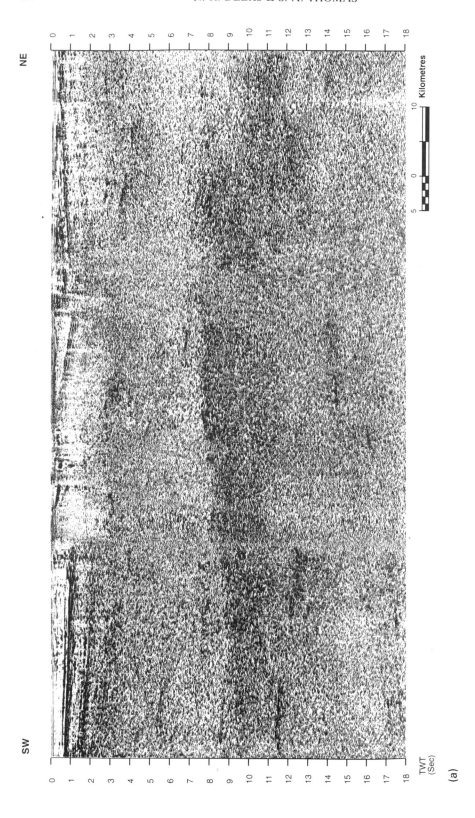

(a)

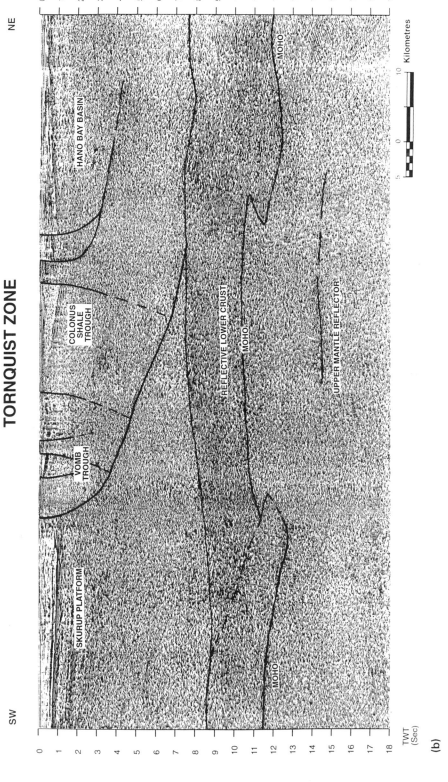

Fig. 9. Portion of deep seismic line BABEL AB/AC where it crosses the Tornquist zone3 orthogonally at the Bornholm Gat. True scale profile for $V = 6500 \, \text{ms}^{-1}$. (**a**) uninterpreted section; (**b**) interpreted section.

through the entire crust and forming distinct northern and southern boundaries to the Tornquist Zone.

This section of the BABEL line is orientated perpendicular to the movement directions on the Tornquist Zone so that, over time, various movements into and out of the plane of section must have taken place. This may go some way to explain the juxtaposition of different units. In a similar way within the strike-slip system the near-surface faults that make up the geometry seen today need not all have been active at the same time. This would allow for the Vomb trough to be active during the Cretaceous whereas the Colonus Shale trough now only preserves Palaeozoic deposits.

Within the less reflective zone towards the base of the BABEL section, there are further coherent dipping reflectors. In a similar way to those imaged within the upper crust these dip at low angle to the northeast and may represent a more brittle transfer of the deformation into the upper mantle. The Moho is interpreted as the boundary between the highly reflective region of lower crust and the broadly less reflective region directly below. Only in a few parts of the section is the Moho marked by distinctive coherent reflectors.

Conclusions

Faults and pull-apart basins associated with the Tornquist Zone, such as the Rønne and Arnager Grabens, were initiated during Permo-Carboniferous activity and can be regarded as the local expression of late Variscan wrenching in Europe (Vejbœk 1985, Liboriussen et al. 1987, Thybo & Schönharting 1991). Uplift was the result of intraplate tectonics in response to the Carpathian/Alpine orogens building to the south. This mountain building resulted in a shift from a former transtensional regime to one of transpression. However the movements along the Tornquist Zone remained dextral. This is in agreement with the work of Mogensen (1992) and Mogensen & Jensen (1994) who pointed out that the interaction of Alpine compressional tectonics and the opening of the North Atlantic would give rise to right lateral transpression along the Tornquist Zone (see also Blundell & BABEL Working Group 1993). In addition to folding and faulting within the sediment fill of the pull-apart basins, individual basement blocks were locally uplifted, with others remaining as sites of sedimentation.

Various classical (positive) inversion structures are seen on seismic sections, including harpoon structures (asymmetric anticlines),

flower structures and footwall shortcut faults. Inversion monoclines developed as fault propagation folds above inverting normal faults. Syn-inversion sedimentation gives rise to the development of growth folds and progradational systems building away from uplifted basement blocks. These features are found at various scales in association with both basin-bounding and intra-basinal faults.

Fault reactivation is also seen in areas where deformation is predominantly strike-slip rather than transpressional. In these cases the faults are orientated almost parallel to the direction of strike-slip movement, rather than oblique to it as in the grabens. Here the features that develop take the form of complex flower structures that can be followed along strike for many kilometres. Structure along these features is often most complex where the fault traces are slightly arcuate leading to the development of restraining and releasing bends. These features are valuable in determining the nature of the strike-slip movement. There is also evidence for inversion of basement blocks within the upper crust. Strike-slip structures indicate that movements were dominantly dextral with rigid blocks such as Bornholm apparently channelling deformation. This channelling means that in some cases both dextral and sinistral movements can be seen on opposing sides of the uplifted block. This is demonstrated to the southeast of Bornholm where localized poorly developed sinistral structures are seen on the northeast flank of the block with dextral structures to the south (Fig. 8).

The uplift that results in the inversion structures seen within the upper crust can be imaged throughout the crust. Through-going dipping reflectors within the upper crust, interpreted as brittle to semi-brittle shear zones, are seen to link with zones of high reflectivity in the lower crust and then tentatively on to dipping structures within the upper mantle. In common with other regions, these high reflectivity zones within the lower crust are interpreted as ductile shear zones, transferring deformation between the upper crust and mantle.

The combination of the broad 3-dimensional control provided by the proprietary surveys and the deep structural control given by the BABEL profile make possible a better understanding of the inversion tectonics within this part of the Tornquist Zone.

This research is being undertaken within the R and D programme JOULE II 'Deep Reservoir Geology' (JOU2-CT92-0082) of the Commission of the European Communities. The authors gratefully acknowledge both Jebco Seismic Ltd and Bundesanstalt für Geowissenschaften und Rohstoffe for making the proprietary seismic data available and allowing its publication. In addition we would like to thank Norsk Hydro and Amoco for making well data available and the members of the BABEL II working group, especially Professor Blundell and Professor Meissner, for their support. J. Cartwright and N. White are thanked for their very useful suggestions. Thanks are also due to P. Ruelland for his early work on the area and to Professor K. R. McClay and other members of the Fault Dynamics Project at Royal Holloway for their useful suggestions and use of their computing facilities.

References

ARTHAUD, F. & MATTE, P. 1977. Late Palaeozoic strike-slip faulting in Southern Europe and North Africa: results of a right-lateral shear zone between the Appalachians and the Urals. *Geological Society of America Bulletin*, **88**, 1305–1320.

BABEL WORKING GROUP 1991. Deep seismic survey images crustal structure of Tornquist Zone beneath southern Baltic Sea. *Geophysical Research Letters*, **18**, 1091–1094.

— 1993. Deep seismic reflection/refraction interpretation of BABEL A and B in the southern Baltic sea. *Geophysical Journal International*, **112**, 325–343.

BADLEY, M. E., PRICE, J. D. & BACKSHALL, L. C. 1989. Inversion, reactivated faults and related structures: seismic examples from the southern North Sea. *In*: COOPER, M. A. & WILLIAMS, G. D. (eds) *Inversion Tectonics*. Geological Society, London, Special Publication, **44**, 201–219.

BALDSCHUHN, R., FRISCH, U. & KOCKEL, F. 1985. Inversionsstrukturen in NW-Deutschland und ihre Genese. *Zeitschrift der Deutschen Geologischen Gesellschaft*, **136**, 139–139.

BERGSTRÖM, J., HOLLAND, B., LARSSON, K., NORLING, E. & SIVHED, U. 1982. Guide to excursions in Scania. *Sveriges Geologiska Undersökning*, Ser. Ca., **54**.

BERTHELSEN, A. 1992. From Precambrian to Variscan Europe. *In*: BLUNDELL, D. J., FREEMAN, R. & MUELLER, S. (eds) *A Continent Revealed: The European Geotraverse*, Cambridge University Press, 153–163.

BIDDLE, K. T. & RUDOLPH, K. W. 1988. Early structural inversion of the Stord Basin, Norwegian North Sea. *Journal of the Geological Society*, London, **145**, 603–611.

BLUNDELL, D. J. & BABEL WORKING GROUP 1993. Seismic reflectivity of the crust transition from basin to platform regions in Europe. *Bulletin de la Société Géologique de France*, **164**, 343–351.

BROCHWICZ-LEWINSKI, W., POZARYSKI, W. & TOMCZYK, H. 1984. Sinistral Strike-Slip movements in Europe in the Paleozoic. *Publ. Inst. Geophys. Pol. Acad. Sc.*, **A13**, 3–13.

CARTWRIGHT, J. A. 1989. The kinematics of inversion in the Danish Central Graben. *In*: COOPER, M. A. & WILLIAMS, G. D. (eds) *Inversion Tectonics*, see Badley *et al.* 1988 in full, 153–176.

DRONKERS, A. J. & MROZEK, F. J. 1991. Inverted basins of The Netherlands. *First Break*, **9**, 409–425.

GRAVESEN, P., ROLLE, F. & SURLYK, F. 1982. Lithostratigraphy and sedimentary evolution of the Triassic, Jurassic and Lower Cretaceous of Bornholm, Denmark. *Geological Survey of Denmark, Series B*, **No. 7**, 51.

GRY, H. 1960. Geology of Bornholm. Guide to excursion Nos. A45 and C40. *International Geological Congress*, **21**, 16.

JOHANSSON, Å. S. & LARSEN, O. 1989. Radiometric age determination and Precambrian geochronology of Blekinge, southern Sweden. Geologiska Föreningens I Stockholm Förhandlingar, **111**, 35–90.

KUMPAS, M. G. 1980. Seismic stratigraphy and tectonics in Hanö Bay, Southern Baltic. *Stockholm Contributions in Geology*, **34**, 36–168.

LIBORIUSSEN, J., ASHTON, P. & TYGESEN, T. 1987. The tectonic evolution of the Fennoscandian Border Zone in Denmark. *Tectonophysics*, **137**, 21–29.

McCLAY, K. R. 1989. Analogue models of inversion tectonics. *In*: COOPER, M. A. & WILLIAMS, G. D. (eds) *Inversion Tectonics*. Geological Society London, Special Publication, **44**, 41–59.

—— 1995. The geometries and kinematics of inverted fault systems: a review of analogue model studies. *This volume*.

MATTHEWS, D. H. 1986. Seismic reflections from the lower crust around Britain. *In*: DAWSON, J. B., CARSWELL, D. A., HALL, J. & WEDEPOHL, K. H. (eds) *Nature of the Lower Continental Crust*. Geological Society, London, Special Publication, **24**, 11–21.

MOGENSEN, T. E. 1992. Præ Kænozoisk strukturel analyse af Kattegat området. *Dansk Geologisk Forening Årsskrift for 1990–91*, 129–134.

—— & JENSEN, L. N. 1994. Cretaceous subsidence and inversion along the Tornquist Zone from Kattegat to the Egersund Basin. *First Break*, **12**, 211–222.

MOONEY, W. D. & MEISSNER, R. 1992. Multi-genetic origin of crustal reflectivity: a review of seismic reflection profiling of the continental lower crust and Moho. *In*: *Continental Lower Crust*. FOUNTAIN D. M., ARCULUS, R. & KAY, R. W. (eds) Elsevier, Amsterdam, 45–79.

NUR, A., RON, H. & SCOTTI, O. 1986. Fault mechanics and the kinematics of block rotations. *Geology*, **14**, 746–749.

PEGRUM, R. M. 1984. The extension of the Tornquist zone into the Norwegian North Sea. *Norsk Geologisk Tidsskrift*, **64**, 39–68.

POZARYSKI, W. & BROCHWICZ-LEWINSKI, W. 1978. On the Polish Trough: *In*: VAN LOON, A. J. (ed.) *Key-Notes of the MEGS II Geologie en Mijnbouw*, **57**, 545–557.

REMPEL, H. 1992. Erdölgeologische Bewertung der Arbeiten der GO 'Petrobaltic' im deutschen Schelfbereich. *Geologisches Jahrbuch*, **D99**, 3–32.

RUELLAND, P. J. 1992. *The Tornquist Zone around Bornholm: Inversion structures in the context of lateral extrusion.* MSc thesis, University of London.

STREHLAU, J. & MEISSNER, R. 1987. Estimation of crustal viscosities and their role in geodynamic processes. *In*: FUCHS, K. & FROIDEVAUX, C. (eds) The composition, structure and dynamics of the Lithosphere-Asthenosphere system, AGU. *Geodynamique*, **16**, 69–87.

THOMAS, S. A., SIVHED, U., ERLSTRÖM, M. & SEIFERT, M. 1993. Seismostratigraphy and structural framework of the SW Baltic Sea. *Terra Nova*, **5**, 364–374.

THYBO, H. & SCHÖNHARTING, G. 1991. Geophysical evidence for Early Permian igneous activity in a transtensional environment, Denmark. *Tectonophysics*, **189**, 193–208.

——, FLÜH, E. & BABEL WORKING GROUP. 1993. Project Babel – Integrated interpretation of Seismic reflection and refraction data across the narrow Tornquist fan in the Baltic Sea. *In*: GEE, D. G. & BECKHOLMEN, M. (eds) *Europrobe Symposium, Jablonna 1991.* Publication of the Institute of Geophysics, Polish Academy of Sciences, A-20, (255): 147–152.

VAN HOORN, B. 1987. Structural evolution, timing and tectonic style of the sole pit inversion. *Tectonophysics*, **137**, 239–284.

VEJBÆK, O. V. 1985. Seismic stratigraphy and tectonics of sedimentary basins around Bornholm. *Danmarks Geologiske Undersogelse Series A*, **8**, 30.

ZIEGLER, P. A. 1981. Evolution of Sedimentary Basins in North–West Europe. *In*: ILLING, L. V. & HOBSON, G. D. (eds) *Petroleum Geology of the Continental shelf of North–West Europe.* The Institute of Petroleum, London, 3–39.

—— 1982. *Geological atlas of western and central Europe.* Shell Internationale Petroleum Mij. B. V. Elsevier, Amsterdam.

—— 1983. Inverted Basins in the Alpine Foreland. *In*: BALLY, A. W. (ed.) *A picture work atlas, Seismic Expressions of Structural Styles.* Studies in Geology series, American Association of Petroleum Geologists, **15**, 3.3-3–12.

ZOBACK, M. L. & 28 others. 1989. Global patterns of tectonic stress. *Nature*, **341**, 291–298.

A comparison of inverted basins of the Southern North Sea and inverted structures of the external Alps

PASCALE HUYGHE[1] & JEAN-LOUIS MUGNIER[2]

[1]*Département de Géologie et d'Océanographie, URA CNRS 197, Avenue des Facultés,
33405 Talence cedex, France*
[2]*Laboratoire de Géodynamique des Chaînes Alpines, URA CNRS 69, rue M. Gignoux,
38031 Grenoble cedex, France*

Abstract: For several years, the relationships between extensional and contractional features have been studied in detail by numerous authors, and structural rules suggested regardless of the size and the geodynamic context of the inverted structures. Examples from the external part of the Western Alps and its foreland follow these rules (oblique-slip reactivation, buttress effect of normal faults, short-cuts, preserved half-grabens beneath decollement, forced folds, etc.), but are not controlled by the same geodynamic conditions as intra-plate inverted basins. The Jura platform, which is not controlled by extensional tectonics, shows fault reactivation, buttress effects and preserved structures beneath the decollement. Inverted structures are nicely illustrated in the Dauphinois domain of the stretched western Tethyan margin, but the effect of lithospheric weakening inherited from the stretching was reduced at the time of the Neogene shortening, and the maximum burial is induced by nappe loading. Therefore the relative timing of inversion tectonics, maximum burial and maximum temperature are very different in the Bourg-d'Oisans half-graben of the Dauphinois domain than for intra-plate inverted basins of the Southern North Sea, e.g. the Broad Fourteens Basin. Nonetheless, fold structures overlapped by major unconformities developed during early deformation events in restricted areas of the Western Alps. The example of structures in the Devoluy is linked to the inversion of the Vocontian basin, and is followed by a renewal of subsidence prior to the later Alpine shortening. Therefore, the Alps provide geometrical field examples of the interaction between stretching and shortening events, and may give some insight into the geometry of structures on the seismic scale. Intra-plate inverted basins provide examples of folded and faulted structures overlapped by sediments and provide a means of explaining the early shortening events that occurred in the Western Alps which were not related to the alpine evolution.

A succession of stretching and shortening events leads to typical geometries in tectonic structures, regardless of their size and geodynamic context. Many outcrops showing these geometries have been described in numerous studies that detail the relationships between extensional and contractional features (Davies 1982; Gillcrist *et al.* 1987; Tricart & Lemoine 1986; Baby *et al.* 1988; Hayward & Graham 1989; Butler 1989; de Graciansky *et al.* 1989; Guellec *et al.* 1990; and many others) and structural rules have been suggested.

For the past few years, the term 'inversion' has often been used to describe within-plate inverted basins and also major orogens resulting from convergence of lithospheric plates (Cooper & Williams 1989).

The aim of this paper is to outline the similarities and differences of the structures in these different plate settings. For this purpose,

the examples of the external zone of the Western French Alps, and the Broad Fourteens Basin in the Southern North Sea located within the European plate, were chosen.

Examples from the external Western Alps

The external thrust belt of the Western Alps is Neogene in age. The main thrust fronts divide the belt into distinct tectonostratigraphic units that: (1) are characterized by their specific Mesozoic and Cenozoic cover sequences; (2) reflect different ages of deformation; and (3) were sequentially accreted onto the allochton. From west to east, in a progressively more allochthonous position, the following units can thus be distinguished (Fig. 1). (1) The Jura unit, which comprises not only the present physiographic province of the Jura mountains proper,

From BUCHANAN, J. G. & BUCHANAN, P. G. (eds), 1995, *Basin Inversion*,
Geological Society Special Publication No. 88, 339–353.

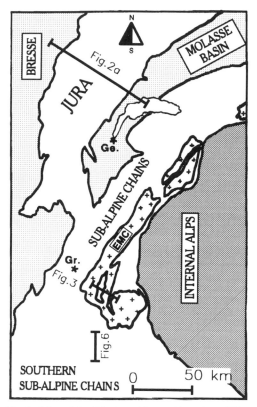

Fig. 1. Map of the Western Alps and location of Fig. 2a, Fig. 3, Fig. 6. EMC, External Crystalline Massifs; Ge, Genève; Gr, Grenoble.

but also the displaced sedimentary cover underlying the Molasse basin; (2) the sub-alpine units, a series corresponding to the Dauphinois domain of sedimentation; (3) the Ultra-Helvetic and Penninic units, which are largely eroded. The basement of the external Alps is also involved in the thrust tectonics (Goguel 1944; Ménard 1979; Butler 1984) and an imbrication of upper crustal rocks results in the basement culmination of the external crystalline massifs of the Western Alps. The Alps result from closure of the Ligurian Tethys, and the external Western Alps from the shortening of its western passive margin (de Graciansky et al. 1979). Consequently, the Mesozoic palaeogeographic domains of the Western Alps underwent successive tectonic and thermal subsidence (Rudkiewicz 1988; Loup 1992) followed by tectonic uplift of their substratum during Alpine shortening. There, the Alpine belt provides an excellent example, at a crustal scale, of a strongly inverted passive margin (Tricart & Lemoine 1986; Gillcrist et al. 1987; de Graciansky et al. 1989).

Balanced cross-sections drawn up along ECORS-CROP deep seismic profiles suggest that the basement of the external Alps has been subjected to 35–50 km of shortening, with approximatively 10 km resulting from the inversion of earlier extensional structures (Mugnier et al. 1990). Major extensional events occurred during the Upper Triassic and Jurassic (de Graciansky et al. 1979), but other pulses were also locally recorded during the Lower Cretaceous and Oligocene (Loup 1992). Therefore, the cover of the external Alps was subjected to succession of extensional and contractional events that led to reverse or transpressive movements along previously extensional faults and the folding of their hanging-walls. These examples will be described in detail and then compared with inverted structures from an intra-plate setting.

Cover structures above a decollement: the 'pinched' structures of the Jura platform

Most of the Mesozoic cover of the Jura moved over the underlying basement during Neogene shortening, along decollement levels located in Triassic evaporites (Fig. 2a). The flat external part of the Jura contains a number of peculiar structures comprising a set of parallel, tight steep faults extending more than 100 km along the strike direction, that are termed 'faisceau' structures (Glangeaud 1949). In these structures, Upper Jurassic or Cretaceous sediments have been preserved from erosion by Oligocene normal faulting. These 'pinched' structures are the most common effect of normal-faulting during Oligocene extension within the Jura (Glangeaud 1949). A classic interpretation suggests that the basement was involved during stretching but not during shortening (Chauve et al. 1980). An alternative interpretation (Guellec 1987; Guellec et al. 1990) suggests that an extensional decollement already existed during the Oligocene (Fig. 2b). Such an Oligocene decollement is inferred from seismic data in the adjacent Bresse graben (Bergerat et al. 1990), and the extensional structures in the cover are not located over the extensional structures in the basement. Few Oligocene faults are located in the basement beneath the Jura platform itself (Guellec et al. 1990; Wildi & Huggenberger 1993), and most of the basement stretching occurred beneath the Bresse basin. In this area, the amount of basement stretching may have been greater than that in the cover, resulting in a decoupling movement of the cover towards the Bresse basin from its eastern shoulder (Laubscher 1982; Bergerat et al. 1990).

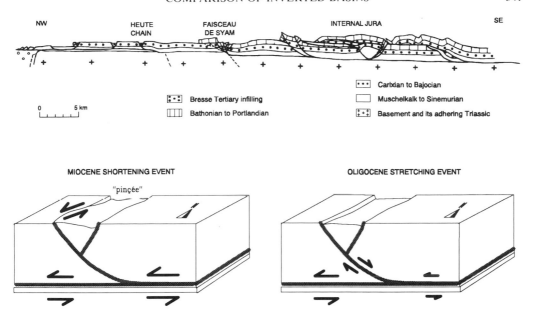

Fig. 2. (a) Cross-section through the Jura; (b) sketch of the inversion of a 'pincée structure' of the external Jura.

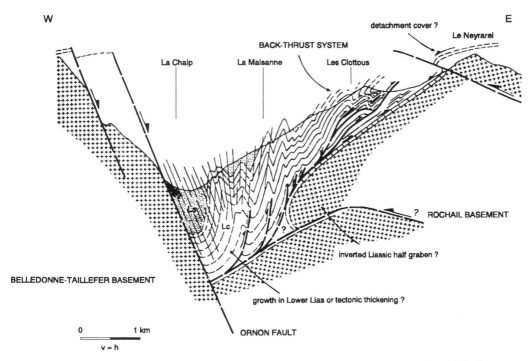

Fig. 3. Cross-section across the central part of the Bourg-d'Oisans half-graben (from Gillcrist 1988). Crosses, crystalline basement; brickwork, Triassic dolomites and spillites; Lc, Lias calcaire; Ls, Lias schisteux; olistoliths adjacent to the Ornon fault also shown as is the main cleavage.

Fig. 4. Folded unconformity in the Bourg-d'Oisans area (East to the left). The unconformity is Domerian in age.

Fig. 5. View from the Devoluy discordance (West to the left).

During the Miocene, the Alpine shortening event led to a weak structural inversion of the earlier extensional features within the cover of the external Jura. The faults above the de-collement show mainly strike-slip reactivation (Chauve *et al.* 1980) and the cover moved in a W–NW direction (Laubscher 1965) (Fig. 3b). Therefore, inversion of the stress field from extension to compression does not imply any reversal of the sense of motion along the basal decollement of the Jura. Nonetheless, the geometry of the basal decollement in detail is complex and at least two decollement levels (one at the base of the Lower Keuper, the other at the base of the Middle Muschelkalk) were active in the Triassic leading to ramps and duplex structures (Mugnier & Vialon 1986) beneath the 'Heute chaine' (Fig. 2) (Mascle 1964; Mathis 1973). Basement fault scarps inherited from the Oligocene extension locally cause buttressing (Mugnier & Vialon 1986) and back-thrusts in the cover like those of the 'faisceau' de Syam (Guillaume 1961), or the geometry of half-grabens is preserved beneath the decollement. This is the case of the structure reached by the Poisoux borehole along the ECORS transect (Guellec *et al.* 1990). Furthermore, earthquake focus data show that thrust and strike-slip motions occurred in the basement (Pavoni 1961), suggesting small reactivations of pre-existing faults (Mugnier & Vialon 1986). Seismic data suggest that Carboniferous basins located beneath the decollement are slightly inverted (Philippe 1994) and induced gentle bending of the decollement during late Alpine events (Guellec *et al.* 1990; Roure *et al.* 1990).

Cover structures in the external basement massifs of the Western Alps: the Bourg-d'Oisans example

The external basement massifs have been considered as thrust sheet culminations (Goguel 1944; Ménard 1979; Butler 1984) and as crustal scale tilted fault blocks of extensional origin that controlled Mesozoic sedimentation patterns during the evolution of the European Tethyan margin (de Graciansky *et al.* 1979; Lemoine *et al.* 1981; Tricart 1984; Tricart & Lemoine 1986). These interpretations have compared the tilted fault blocks of the external Alps to structures found at the Atlantic margin (Montadert *et al.* 1979).

The Bourg-d'Oisans structure comprises pre-rift to early syn-rift Upper Triassic sediments and syn-rift Jurassic sediments deposited in a half-graben (Fig. 3 after Gillcrist 1988; Gillcrist

et al. 1987). These sediments have been considerably shortened by folding, faulting and penetrative internal deformation above a basement/cover detachment that tipped against a major basement scarp ('Faille d'Ornon') inherited from the Jurassic extension (Vialon 1986). Numerous structures demonstrate the influence of the syn-extensional depositional geometry on later contractional structures (Gillcrist *et al.* 1987; Gillcrist 1988). Figure 4 illustrates the fold of 'Pré gentil' (Barfety & Gidon 1983) affecting the asymmetrical Domerian, Toarcian and Aalenian sedimentary bodies. These bodies onlap upon Carixian levels and the geometry of the unconformity suggest a sedimentary infilling of a syncline structure. This syncline could have formed above a normal basement fault dipping to the northwest ('lac du vallon' fault from Barfety & Gidon 1983), and the continuity of the Carixian strata suggests that the 'lac du vallon' fault induced a step for a Jurassic extensional basement/cover detachment. During Alpine shortening, the pre-existing step acted as a forced ramp along the detachment.

The penetrative deformation of the Jurassic cover has been studied by Gratier (1979) and Gratier & Vialon (1980). These studies demonstrate the internal deformation to be very strong with a stretching ratio in places exceeding 200% with the principal stretching axis close to the vertical adjacent to the basement fault scarp ('Ornon' fault). A large part of the shortening is accommodated by diffusional mass transfer processes and a pressure-solution cleavage is well developed in the shales. The pressure-temperature conditions prevalent during deformation have been estimated from fluid inclusions (Bernard *et al.* 1977; Jenatton 1981), dating the clay minerals found in veins using the Potassium/Argon method (Nziengui 1993), from mineral phase assemblages within pebbles embedded in Triassic sediments (Bohorquez 1993), and remagnetization of pyrhotite (Lamarche *et al.* 1988; Crouzet 1993).

These different studies support the existence of high pressures (200–250 MPa) during deformation (Vialon 1986), implying an 8–10 km overburden was present over the Bourg-d'Oisans structure. As the maximum measured thickness of the Mesozoic sequence (less than 5 km, from Debelmas 1974) and Cenozoic sequence in the Western Alps (less than 2–4 km, from Debrand-Passard 1984) are insufficient to explain this overload, it is postulated that the Bourg-d'Oisans structure was located in the footwall of overlying thrust sheets. Furthermore, a maximum recorded temperature of

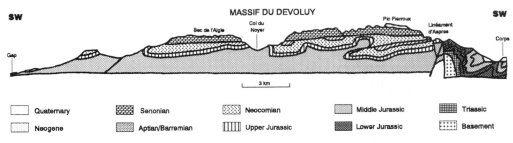

Fig. 6. Cross-section through the Devoluy structures (modified from Gidon 1983).

335–400° C post-dates the deformation (Crouzet 1993), a situation commonly found in the footwalls of major thrusts (Shi & Wang 1987). These inferred overlying thrust sheets could correspond to the frontal climb continuation of thrusts that affected the Pelvoux crystalline massif (Vialon & Pecher 1974; Butler 1985), and the northward Ultra-Dauphinois sedimentary cover (Beach 1981). Alternatively, these nappes could also be the prolongation of the internal Penninic units (Davies 1982). Thrust klippen are however nowhere documented in the area of Bourg-d'Oisans, and the nappes responsible are therefore completely eroded. The deformation age is not well constrained, but is probably older than 24 Ma, and therefore Oligocene or Miocene in age (Lamarche et al. 1988). It has been suggested that this area also underwent some pre-Eocene deformation (Gillcrist 1988), observed southeastward along the southern and eastern margin of the Pelvoux and to the North of Pelvoux massif (Gidon 1979). It cannot be precluded that this pre-Eocene deformation is the same as the pre-Senonian deformation observed in the Devoluy. Nonetheless, the southwest margin of the external crystalline massifs is located above a major N140° trending lithospheric boundary in the Alps (Guyoton 1991) that acts as a lateral ramp for the whole Alpine thrust belt (Ménard 1979; Butler 1984) and for the Mesozoic extensional fault system (Lemoine et al. 1981; Gillcrist 1988). Therefore, the major tectonic events in the southern sub-Alpine chains are presumably slightly different from those of the external crystalline basement massifs.

The upper Cretaceous structure of the Devoluy area

The Devoluy massif is a part of the southern sub-Alpine zone of the Western Alps (Fig. 6). It is mainly formed of Jurassic and Cretaceous sediments deposited in the Vocontian basin which was a part of the southeastern basin of France (Baudrimont & Dubois 1977). The Vocontian basin is situated south of the Vercors Massif (Arnaud 1979) and west of the Dauphinois–Durance High, as defined by Baudrimont & Dubois (1977). Geometrical relationships between the Vocontian basin and its margins are poorly known, however the influence of NE–SW and NW–SE trending Late Hercynian lineaments is inferred (Arthaud & Matte 1975; Boudon et al. 1978). These fault systems may delineate the northwest and northeast boundaries of the Vocontian basin and they controlled sedimentation from the Triassic to Upper Cretaceous (Baudrimont & Dubois 1977). The effects of tectonics on subsidence in the Vocontian basin is difficult to pinpoint accurately, however, a thick sequence of marine carbonates deposited in the basin during the Lower Cretaceous, and a N–S Lower Cretaceous extensional event observed and analysed to the south in the Baronies and in the French Southern Alps (Dardeau 1987; Gerlier & Mascle 1987), suggest that the tectonics responsible for basin formation in southern sub-Alpine Chains ended within the Albian.

Locally strongly folded strata are eroded and a thick Late Cretaceous marine sequence was deposited over the erosion surface (Lory 1860) (Figs 5, 6). The trend of the folds ranges from E–W to NE–SW and the angular unconformity locally exceeds 90°. The unconformity surface is composite, and the age of the sediments above the unconformities varies from Lower Turonian to Lower Campanian (Porthault 1974), suggesting several pulses of folding, often referred to as pre-Senonian (Lemoine 1986).

These N60–N90° folds are interpreted to have formed during a regional N/S shortening by buttressing against a corner of the basin delineated by Late Hercynian faults (Baudrimont & Dubois 1977). Therefore, these fold structures probably formed during the inversion of the Vocontian basin, although the deep structure of the basin is still poorly understood. These Upper Cretaceous structures situated in the external

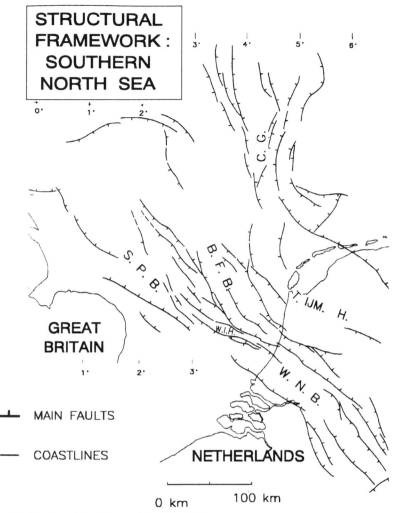

Fig. 7. Map of the Southern North Sea and location of Fig. 8. B.F.B., Broad Fourteens Basin; C.G., Central Graben; L.B.M., London–Brabant-Massif; S.P.B., Sole Pit Basin; T.IJH.H., Texek Ijsselmeer High; W.I.H., Winterton High; W.N.B., West Netherland Basin. Modified from Van Wijhe (1987) and Van Hoorn (1987).

part of the Alpine chain may be a direct expression of stress within the European craton induced by the anticlockwise rotation of the Iberian and African plates (de Graciansky *et al.* 1989; Hibsch *et al.* 1992).

Structural inversion in the Southern North Sea

The Southern North Sea area (Fig. 7) has been subjected to Ceno-Mesozoic multiphase tectonics superimposed on older Variscan and Caledonian structures (Ziegler 1987). Several basins in the area have been inverted. One of

them, the Broad Fourteens Basin, shows moderate inversion and provides a good illustration of the complex structural patterns that form in the cover of an inverted basin.

The Broad Fourteens example

The Broad Fourteens Basin is a NE/SW present-day trending structure located off the Netherlands coast (BFB on Fig. 7) that developed from the Mid Jurassic to Lower Cretaceous (Van Wijhe 1987). During Cretaceous times, the basin subsidence changed into uplift. In order to document the timing of main tectonic events, the Cretaceous sedimentary record has been

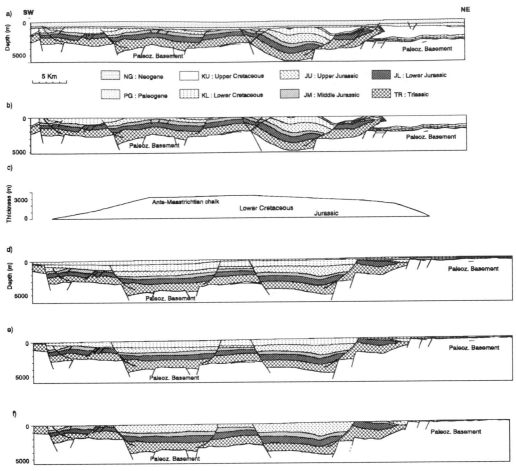

Fig. 8. Balanced cross-section through the Broad Fourteens basin. (**a**) present state; (**b**) Palaeocene state; (**c**) reconstitution of sedimentary thickness eroded during Upper Cretaceous structural inversion (estimated from compaction anomalies); (**d**) Cretaceous pre-inversion geometry; (**e**) geometry of the basin after Lower Cretaceous tectonics; (**f**) geometry of the basin before Lower Cretaceous tectonics.

analysed in detail using approximately 3000 km of seismic reflection profiles and fifty well logs (Huyghe 1992). Sixteen sedimentary sequences are distinguished in the Cretaceous and special attention has been given to their geometrical relationships. This study shows that the extensional tectonics ended during the Early Cretaceous. Sag subsidence was interrupted during the Upper Cretaceous by the structural inversion of the Broad Fourteens Basin when its main faults underwent reverse movement and basin infill was strongly eroded. Balanced cross-sections constructed across the basin (Fig. 8) indicate around 10–12% of horizontal shortening (Hayward & Graham 1989; Huyghe 1992), but neglect the strike-slip component of displacement during inversion. From the analysis of sonic velocity anomalies in overcompacted

and homogeneous pre-rift series, it has been estimated that more than 3000 m of erosion took place in the centre of the basin, decreasing towards the basin margins (Trichon *et al.* 1990) (Fig. 8c). Stratigraphic studies of Upper Cretaceous sediments performed in the northern part of the SW margin of the Broad Fourteens Basin where they were locally preserved from erosion show the existence of several unconformities (Huyghe 1992; Huyghe & Mugnier 1994). They indicate that the onset of structural inversion began as early as the end of the Turonian. In the northern part of the Broad Fourteens Basin, presence of Zechstein evaporite levels in the basin resulted in a 'decollement' of the cover giving rise to major thrust structures (Hayward & Graham 1989) during the Coniacian to Lower Santonian (Fig. 8b). The centre of the basin seems to have suffered erosion prior to the margins where the existence of channels in the chalk suggests sub-aerial erosions occurred only during the Maastrichtian. The strong unconformity between the dip of Maastrichtian chalks and underlying deposits indicates that the whole of the Broad Fourteens Basin underwent its strongest uplift at the end of the Upper Cretaceous (Oele *et al.* 1981; Trichon pers. comm.). This uplift is followed by subsidence (Fig. 8a) and small tectonic events during the Tertiary.

Other basins of the Southern North Sea that formed during Jurassic/Lower Cretaceous times have also been inverted since the Upper Cretaceous but the intensity and timing of their inversions vary. The Dutch Central Graben (CG on Fig. 7) was inverted during the Upper Cretaceous (Kooi *et al.* 1989), but the total amount of uplift is nearly 0.5 km and the amount of shortening very small (1–2%) (Cartwright 1989). The Sole Pit Basin, which is approximately NW–SE trending (SPB on Fig. 7) was subjected to a Ceno-Mesozoic tectonic evolution rather similar to the Broad Fourteens Basin situated 70 km to its east. However, the Sole Pit Basin underwent its maximum inversion during the Oligocene (Van Hoorn 1987) and the total amount of uplift is close to 1.5 km for a shortening value of about 3% (Badley *et al.* 1989). It appears that the Broad Fourteens Basin is one of the most strongly inverted structures of the northwest intra-plate Mesozoic basins. Structural controls interpreted to be responsible for these different structural responses to a same regional compression (Huyghe & Mugnier 1994) are discussed below.

Discussion

Structural controls on basin inversion

Of the several uplift mechanisms inferred for the uplift of sedimentary basins (Chadwick 1993), the term 'inversion' is mainly ascribed to uplift induced by the compressive reactivation of pre-existing faults. The inversion is mainly controlled (Gillcrist *et al.* 1987) by (1) the strength of the deep crustal levels; (2) the geometry of pre-existing faults in the upper crust and their ability to reactivate.

The distribution and style of deformation within the upper crust is mainly controlled by the brittle behaviour of its rocks. In the case of structural inversion, the distribution of the deformation is mainly related to the ability of pre-existing faults to reactivate. Therefore, the inversion geometry

in the upper crust is partly controlled by the branch pattern of original faults. For example, it has been suggested that inverted structures developed above listric faults are different from those developed above a domino-type fault-system (Gillcrist *et al.* 1987; Williams *et al.* 1989). In addition to the geometrical characteristics (strike and dip) of pre-existing faults, changes in mechanical properties of the sediment within the basin control the structural style of the inversion (Chadwick 1993). It has also been found that strength variations with depth influence the ability of faults to reactivate (Huyghe & Mugnier 1992a); in most cases, the domain where faults are the most difficult to reactivate is usually located above weakness zones (Huyghe & Mugnier 1992b). Therefore, short-cut geometries (Beach 1981) controlled by the competition between reactivation and new faulting are branched at upper tips of weak layers or decollement levels. These short-cut geometries develop at the scale of both sedimentary cover and crust. In the latter case, short-cut faults could cross the whole brittle crust (Butler 1989).

The strength of the deep crustal levels is not well known, though it is generally suggested that it is dramatically temperature-dependent (Kusznir & Park 1987; Banda & Cloetingh 1992). Numerous models of basin development are suggested, depending on the inferred rheologies and kinematics of the stretching. Kinematic models of stretching coupled with cooling of the lithosphere give a simple model of basin evolution (McKenzie 1978; Coward 1986; White & McKenzie 1988) that is not completely accurate in describing the strength evolution of the lithosphere. The subsidence phase that follows basin extension is induced by the thermal contraction of the lithosphere (McKenzie 1978). The homogeneous stretching model of the whole lithosphere (Le Pichon & Sibuet 1981) and the displacement model along a lithospheric-scale shear zone (Wernicke 1985) are two end members of a range of possible extensional models. The model of asymmetric extension proposed by Coward (1986) can be used to explore numerous intermediate solutions, and gives subsidence curves for different stretching models. These curves certainly do not provide exact fits for any real basin evolution and neglect the extra-subsidence due to the sediment supply, but all of them reflect the attenuation of subsidence with time. At the beginning of cooling, the subsidence rate is high, and the deep lithosphere is still hotter and weaker in the zone affected by extensional tectonics. But with time, the lithosphere cools back to its original geo-thermal gradient, and the thermal subsidence tends to zero. Lithospheric strength is also composition-dependent, and stretching could induce variations in lithospheric petrographical layering. Therefore the area where lithospheric mantle rocks have risen up becomes stronger than before the stretching event when thermal subsidence ends (Kusznir & Park 1987). If regional compression occurs, deformation local-izes preferentially in the previously stretched zone only if the delay between the stretching and shortening event is short (Gillcrist *et al.* 1987). The weakening of the whole lithosphere is more important in the case of homogeneous extension of the crust and mantle than in the case of heterogeneous extension where mantle weaken-ing is shifted from basin area (Coward 1986; Gillcrist *et al.* 1987). Therefore, basins con-trolled by homogeneous stretching and having a fairly symmetrical geometry are more easily inverted than basins developed by heterogene-ous stretching (Gillcrist *et al.* 1987). Pull-apart basins controlled by wrench faults affecting the whole lithosphere could in the same way be easily reactivated. To summarize therefore, at a large scale, the location of shortened areas could be restricted to previously stretched areas if the delay between extension and shortening events preserves weakening of the deep lithosphere.

The differences between intra-plate inverted basins and collision belts affected by pre-existing extensional structures

When changes of stress regime affect previously stretched basins, the so-called inversion struc-tures can be classified by estimating the time delay between the last stretching and shortening events and by studying the pre-, syn-, and post-inversion subsidence history (positive or negative). This classification is based not only on geometrical similitudes, but also on the burial history of the basin and thermal relaxation of the lithosphere (Fig. 9). Vitrinite, maturity or compaction data may provide additional con-straints on the burial history and thermal model.

When the delay between lithospheric stretching and shortening is short (less than 30–40 Ma), the strength of stretched lithosphere at the time of the shortening event could still be less than that of unstretched lithosphere and make it easier for the basin to be inverted and its infill strongly uplifted. Consequently, the sea floor uplift leads to erosion that affects the inverted area. If the inversion is moderate, thermal contraction

continues at depth and subsidence resumes immediately after tectonic uplift ceases. This gives birth to a major unconformity above the shortened structures and both physical proper-ties of ante-inversion sediments and subsidence pattern of syn-inversion and post-inversion sediments do record the uplift (Huyghe 1992). If the rate of thermal subsidence is greater than the rate of tectonic uplift, a slow subsidence con-tinues during the inversion with attenuation of deposits above weak inverted structures. These conditions lead to a succession of stretching and shortening events that (1) are restricted to areas that subside more than the surrounding domain when long periods are considered, and (2) affect the whole lithosphere beneath the inverted area though the mechanism and partitioning of deformation at depth is unknown. These charac-teristics are typically inferred in intraplate inverted basins and are exemplified by the Broad Fourteens Basin in the Southern North Sea.

If inversion is stronger and crustal roots are created beneath the inverted area, i.e. the amount of contraction exceeds the value of early extension, then thermal contraction will cease, or its subsidence effects will be hidden, and no sedimentation will overlap the inverted struc-tures. Quantitative geodynamic models (Stock-mal *et al.* 1986) indicate that these characteristics may be inferred for strong inversion of a passive margin and that 'hot' lithospheres characterized by a thin elastic thickness are consistent with modest topography overlying thick overthrusts (Stockmal *et al.* 1986). Strong inversion of the Tethyan margin in the inner zones of the Alps during Eocene–Paleocene time (Tricart & Le-moine 1986) exemplified collision belt showing high pressure metamorphism in the axial zone (Gillet *et al.* 1986) and bounded by narrow starved foreland basins (Deville *et al.* 1992) located above a thin elastic lithosphere.

When the delay between lithospheric stretching and shortening event is longer, or when the extension is limited in the upper crust, the strength of the lithosphere is high, and shorten-ing deformation propagates by gradual accretion of thrust slabs (Dewey *et al.* 1986). Maximum burying of sediments results from additional thickness of tectonic nappes, and no residual thermal subsidence controls sedimentation. Tectonic loading of the lithosphere allows development of a large flexural basin (Karner & Watts 1983) in which thin-skinned thrusts may locally control piggy-back basins and erosion surfaces at the top of ramp anticlines (Ori & Friend 1984). Internal deformation is intensive and penetrative within the sediments beneath a

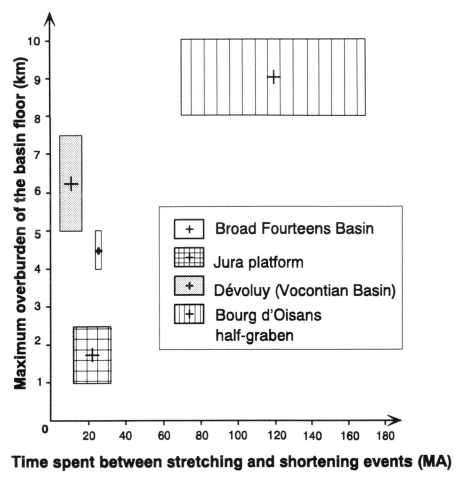

Fig. 9. Plot of the delay between extension and shortening versus the maximum thickness of sediments at the beginning of the inversion for the studied examples. Size of the rectangles is proportional to the error bar.

pile of thrust sheets, even when preserved in half-graben structures. The pressure–temperature path at the footwall of large nappes shows that the maximum burial is reached during shortening events while maximum temperature post-dates the deformation events. These conditions are typically recognized in the collision belts where previous passive margins were either controlled by lithospheric shear zones, or shortened after a long delay. This is the case of the external part of the Alps that are supported by a strong lithosphere showing a Hercynian type (Meissner & DEKORP Research Group 1991) lower crust (Mugnier *et al.* 1990), and the Bourg d'Oisans area illustrates such an evolution.

When the stretched cover is supported by an unstretched lithosphere, the thermal pattern and strength of the lithosphere are not affected by the delay between stretching and shortening event. If this zone is located on the margin of a stretched lithosphere, it could be affected by an uplift shoulder effect (Chery *et al.* 1990) during an extensional event and an uplift due to thickening during the shortening event. These conditions are recognized for the external Jura structures.

Conclusion

In summary, all the areas discussed in this paper display some geometric elements of inversion in different geodynamic setting. Comparisons of the style and deformation conditions of the areas mentioned above underline that the Alps show a very complicated deformation history which led to a wide range of structures and many relationships of the interaction between thrust tectonics and inversion tectonics. The external Alps show numerous field examples of interactions between extensional and shortening structures and suggest useful rules for the analysis of inverted structures, whereas intra-plate inverted basins provide a means of interpreting early folding events in the Alps that are difficult to integrate in any other way in the Alpine collision story.

The study of maximum burial versus delay between last stretching and shortening events (Fig. 9) suggests some remarks concerning the thermo-mechanical conditions during the different 'inversions' exemplified in the above sections.

(1) The Bourg d'Oisans area shows numerous examples of relationships between extensional and compressional features (buttress effect, short-cut, forced anticline, etc.), and its high internal deformation and high burial is due to an allochthonous overburden (alpine nappes).

(2) The 'pinched' structures of the Jura are induced by an inversion of the stress field and show reversal of sense of movement along reactivated faults within the displaced cover.

(3) The Devoluy massif is an intraplate basin (Vocontian Basin) inverted during the Late Mesozoic. The thermo-mechanical conditions for the inverted Vocontian Basin are the same as for the inverted Broad Fourteens Basin. Their inversion may be controlled by intra-plate stresses generated by the dynamic events all around the European plate. Therefore, the inversion of the Vocontian Basin is not strictly related to the Alpine orogen before its Neogene incorporation into the Alpine thrust belt.

The study of the Jura has been supported by the 'Action Accompagnement ECORS' of INSU-CNRS, the study of the Bourg-d'Oisans area by a grant of the British Council paid to J.-L. M. during a stay at Imperial College and the study of the Broad Fourteens Basin by a research agreement between Elf-Petroland B.V. and Grenoble University.

References

ARNAUD, H. 1979. Surfaces d'ablation sous-marines et sédiments barrémo–bédouliens remaniés par gravité du Barrémien au Cénomanien entre le Vercors et le Dévoluy. *Géologie alpine*, **55**, 5–21.

ARTHAUD, F. & MATTIE, P. H. 1975. Les décrochements tardy–hercyniens du sud-ouest de l'Europe, et essai de reconstitution des conditions de la déformation. *Tectonophysics*, **25**, 139–171.

BABY, P., CROUZET, C., SPECHT, M., DERAMOND, J., BILOTE, M. & DEBROAS, J. 1988. Rôle des paléostructures albo-cénomaniennes dans la géométrie des chevauchements frontaux nord-pyrénéens. *Compte-Rendus Académie Sciences, Paris*, **309**, 1717–1722.

BADLEY, M. E., PRICE, J. D. & BACKSHALL, L. C. 1989. Inversion, reactivated faults and related structures: seismic examples from the southern North Sea, *In*: COOPER, M. A. & WILLIAMS, G. D. (eds) *Inversion Tectonics*, Geological Society, London, Special Publication, **44**, 201–217.

BANDA, E. & CLOETINGH, S. 1992. Europe's lithosphere – Physical properties. *In*: BLUNDELL D., FREEMAN R. & MUELLER S. (eds) *A continent revealed: the European Geotraverse*, 71–91.

BARFETY, J. C. & GIDON, M. 1983. La stratigraphie et la structure de la couverture dauphinoise au sud de Bourg-d'Oisans. Leurs relations avec les déformations synsédimentaires jurassiques. *Géologie Alpine*, **59**, 5–32.

BAUDRIMONT, A. & DUBOIS, P. 1977. Un bassin Mésogéen du domaine péri-alpin: le Sud-Est de la France, *Bulletin du Centre Recherche Exploration Production Elf-Aquitaine*, **1**, 261–308.

BEACH, A. 1981. Thrust structures in the eastern Dauphinois zone (French Alps), north of the Pelvoux massif. *Journal of Structural Geology*, **3**, 299–308.

BERGERAT, F., MUGNIER, J. L., GUELLEC, S., TRUFFERT, C., CAZES, M., DAMOTTE, B. & ROURE, F. 1990. Extension tectonics and subsidence of the Bress basin: a view from Ecors data. *Mémoires Société géologique France*, **156**, 145–157.

BERNARD, D., GRATIER, J. P. & PECHER, A. 1977. Application de la micro-thermométrie des inclusions fluides des cristaux syncinématiques à un problème tectonique. *Comptes Rendus Sommaires Séances Société Géologique France*, **5**, 284–288.

BOHORQUEZ, F. 1993. Les matériaux du contact socle-couverture au Sud-Est de Bourg d'Oisans (Alpes occidentales): géométrie, structure et pétrographie. *Mémoire D.E.A.*, Université de Savoie.

BOUDON, J. & 7 others. 1978. L'arc alpin occidental: Réorientation de structures primitivement E–W par glissement et étirement dans un système de compression global N-S. *Eclogae Geologicae Helvetiae*, **69**, 509–519.

BUTLER, R. W. H. 1984. Balanced cross-section and their implications for the deep structure of the northwest Alps: discussion. *Journal of Structural Geology*, **6**, 603–606.

—— 1985. Thrust tectonics, deep structure and crustal subduction in the Alps and Himalayas. *Journal of the Geological Society*, London, **85**, 1437–1473.

—— 1989. The influence of pre-existing structures on thrust system evolution in the Western Alps. *In*: COOPER, M. A. & WILLIAMS, G. D. (eds) *Inversion Tectonics*, Geological Society, London, Special Publication, **44**, 105–122.

CARTWRIGHT, J. A. 1989. The kinematics of inversion in the Danish Central Graben. *In*: COOPER, M. A. & WILLIAMS, G. D. (eds) *Inversion Tectonics*, Geological Society, London, Special Publication, **44**, 153–175.

CHADWICK, R. A. 1993. Aspects of basin inversion in southern Britain. *Journal of the Geological Society*, London, **150**, 311–322.

CHAUVE, P., ENAY, R., FLUCK, P. & SITTLER, C. 1980. Vosges, Fossé Rhénan, Bresse, Jura. *In: Géologie des pays Européens*. Dunod Paris, 357–431.

CHERY, J., LUCAZEAU, F., DAIGNIERES M. & VILOTTE, J. P. 1990. The deformation of the continental crust in extensional zones: a numerical approach. *In*: PINET, B. & BOIS (eds) *The Potential of Deep Seismic Profiling for Hydrocarbon Exploration*. Technip, Paris, 35–44.

COOPER, M. A. & WILLIAMS, G. D. 1989. *Inversion Tectonics*, Geological Society, London, Special Publication, **44**.

COWARD, M. P. 1986. Heterogeneous stretching, simple shear and basin development. *Earth and Planetary Science Letters*, **80**, 325–336.

CROUZET, C. 1993. Etude magnétique de l'évolution tectonique et thermique tardive dans la zone Delphino–Helvétique. (Alpes Occidentales, France). *Mémoire D.E.A.*, Université de Savoie.

DARDEAU, G. 1987. Inversion du style structural et permanence des unités structurales dans l'histoire mésozoïque et alpine du bassin des Alpes maritimes, partie de l'ancienne marge passive de la Tethys. *Comptes-rendus Académie des sciences, Paris*, **305**, 483–486.

DAVIES, V. M. 1982. Interaction of thrust and basement faults in the French external Alps. *Tectonophysics*, **88**, 325–331.

DEBELMAS, J. 1974. *In: Géologie de la France*. Doin, Paris.

DEBRAND-PASSARD, S. 1984. Synthèse géologique du Sud-Est de la France. *Mémoire BRGM France*, **125**.

DEVILLE, E., FUDRAL, S., LAGABRIELLE, Y., MARTHALER, M. & SARTORI, M. 1992. From oceanic closure to continental collision: A synthesis of the "Schistes lustrés" metamorphic complex of the Western Alps. *Geological Society of America Bulletin*, **104**, 127–139.

DEWEY, J. F., HEMPTON, M. R., KIDD, W. S. F., SAROGLU, F. & SENGOR, A. M. C. 1986. Shortening of continental lithosphere: the neotectonics of Eastern Anatolia. A young collision zone. *In*: COWARD, M. P. & RIES, A. C. (eds) *Collision Tectonics*, Geological Society, London, Special Publication, **19**, 3–36.

GERLIER, A. & MASCLE, G. 1987. *Distensions d'âge Crétacé inférieur dans les Baronnies, 11ème Réunion des Sciences de la Terre, Clermont-Ferrand*, 77.

GIDON, M. 1979. Le rôle des étapes successives de la déformation dans la tectonique alpine du massif du Pelvoux. *Compte-rendus Académie des Sciences Paris*, **288**, 803–806.

—— 1983. *In*: DEBELMAS, J. *Guides géologiques régionaux, Alpes du Dauphiné*, 96. Masson, Paris.

GILLET, P., CHOUKROUNE, P., BALLEVRE, M. & DAVY, P. 1986. Thickening history of the Western Alps. *Earth and Planetary Science Letters*, **78**, 4–52.

GILLCRIST, R. 1988. *Mesozoic basin development and structural inversion in the external French Alps*. PhD Thesis, Imperial College, London.

——, COWARD, M. P. & MUGNIER, J. L. 1987. Structural inversion and its controls: examples from the Alpine foreland and the French Alps. *Geodynamica Acta, Paris (1)*, **1**, 5–34.

GLANGEAUD, L. 1949. Les caractères structuraux du Jura. *Bulletin Société Géologique France*, **19**, 669–688.

GOGUEL, J. 1944. La tectonique de fond dans la zone externe des Alpes, *Bulletin Société Géologique France*, **XIV**, 201–218.

GRACIANSKY DE, P.-C., BOURBON, M., CHARPAL, O., CHENET, P. Y. & LEMOINE, M. 1979. Genèse et évolution comparée de 2 marges continentales passives: marge ibérique de l'océan atlantique et marge européenne de la tethys. *Bulletin Société Géologique France*, **21**, 663–674.

——, DARDEAU, G., LEMOINE, M. & TRICART, P. 1989. The inverted margin of the French Alps and foreland basin inversion. *In*: COOPER, M. A. & WILLIAMS, G. D. (eds) *Inversion Tectonics*, Geological Society, London, Special Publication, **44**, 87–104.

GRATIER, J. P. 1979. Mise en évidence des relations entre changement de composition chimique des roches et intensité de leur déformation. *Bulletin Société Géologique France*, **7**, **XXI**, 95–104.

—— & VIALON, P. 1980. Deformation pattern in a heterogeneous material: folded and cleaved sedimentary cover immediately overlying a crystalline basement (Oisans, French Alps). *Tectonophysics*, **65**, 151–180.

GUELLEC, S. 1987. Polyphasage tectonique dans l'avant-pays alpin. *Mémoire de D.E.A.* Université de Paris VI.

——, MUGNIER, J.-L., ROURE, F. & TARDY, M. 1990. Neogene evolution of the alpine foreland: a view from Ecors data and balanced cross-section. *Mémoires Société Géologique France*, **(8)**, **156**, 165–184.

GUILLAUME, A. 1961. La partie méridionale du faisceau de Syam, *Bulletin Service Carte Géologique France*, **264**, 49–68.

GUYOTON, F. 1991. *Sismicité et structure lithosphérique des Alpes occidentales*. Thèse, Université de Grenoble.

HAYWARD, A. B. & GRAHAM, R. 1989. Some geometrical characteristics of inversion. *In*: COOPER, M. A. & WILLIAMS, G. D. (eds) *Inversion Tectonics*. Geological Society, London, Special Publication, **44**, 17–40.

HIBSCH, C., KANDEL, D., MONTENAT, C. & ESTEVOU, P. 1992. Evénéments tectoniques crétacés dans la partie méridionale du bassin subalpin. Implications géodynamiques. *Bulletin de la Société Géologique de France*, **163**, 147–158.

HUYGHE, P. 1992. *Enregistrement sédimentaire des déformations intraplaques: l'exemple de l'inversion structurale d'un bassin de la Mer du Nord*. Thèse, Université de Grenoble.

—— & MUGNIER, J. L. 1992a. The influence of depth on reactivation in normal faulting. *Journal of Structural Geology*, **14**, 991–998.

—— & —— 1992b. Short-cut geometry during structural inversions: competition between faulting and reactivation. *Bulletin Société Géologique France*, **6**, 691–700.

—— & —— 1994. Intra-plate stresses and basin inversion: a case from the Southern North Sea. *In*: ROURE, F. (ed.) *Peri-Tethyan Platform*. Eds Technips, Paris, 211–226.

JENATTON, L. 1981. *Microthermométrie des inclusions fluides des cristaux associés à l'ouverture de fentes alpines.* Thèse 3eme cycle, Université Grenoble.

KARNER, G. D. & WATTS, A. B. 1983. Gravity anomalies and flexure of the lithosphere. *Journal of Geophysical Research*, **88**, 445–477.

KOOI, H., CLOETINGH, S. & REMMELTS, G. 1989. Intraplate stresses and the stratigraphic evolution of the North Sea Central Graben. *Geologie en Mijnbow*, **68**, 49–72.

KUSZNIR, N. J. & PARK, R. G. 1987. The extensional strength of the continental lithosphere: its dependence on geothermal gradient, crustal composition and crustal thickness. *In*: COWARD, M. P., DEWEY, J. F. & HANCOCK, P. (eds) *Continental extensional tectonics*. Geological Society, London, Special Publication, **28**, 35–52.

LAMARCHE, G., MENARD, G. & ROCHETTE, P. 1988. Données paléomagnétiques sur le basculement tardif de la zone dauphinoise interne (Alpes occidentales). *Compte-Rendus Académie Sciences, Paris*, **306**, 711–716.

LAUBSCHER, H. P. 1965. Ein kinematisches Modell der Jurafaltung. *Eclogae Geologicae Helvetiae*, **58**, 232–318.

—— 1982. Die Südostecke des Rheingrabens– ein kinematisches und dynamisches Problem. *Eclogae Geologicae Helvetiae*, **75**, 101–116.

LEMOINE, M. 1986. Rythme et modalités des plissements superposés dans les chaînes subalpines méridionales des Alpes occidentales françaises. *Geologische Rundschau*, **61**, 975–1010.

——, GIDON, M. & BARFETY, J. C. 1981. Les massifs cristallins externes des Alpes occidentales: d'anciens blocs basculés nés au Lias lors du rifting téthysien. *Comptes Rendus Académie Sciences, Paris*, **292**, 917–920.

LE PICHON, X. & SIBUET, J. C. 1981. Passive margins: a model of formation. *Journal of Geophysical Research*, **86**, 3708–3720.

LORY, P. 1860. Mouvement du sol et sédimentation en Dévoluy durant le Crétacé supérieur. *Bulletin de la Société Géologique de France*, **38**, 780–782.

LOUP, B. 1992. Mesozoic subsidence and stretching models of the lithosphere in Switzerland (Jura, Swiss plateau and Helvetic realm). *Eclogae Geologicae Helvetiae*, **85**, 541–572.

McKENZIE, D. P. 1978. Some remarks on the development of sedimentary basins. *Earth and Planetary Science Letters*, **40**, 25–32.

MASCLE, G. 1964. Un exemple de pincée de la bordure externe du Jura. *Annales scientifiques de l'université de Besançon, 2ème série, Géologie*, **18**, 24–38.

MATHIS, M. 1973. *La chaîne de l'Heute (Jura).* Thèse 3ème cycle, Besançon university.

MEISSNER, R. & DEKORP RESEARCH GROUP. 1991. The DEKORP Surveys: Major Achievements for tectonical and Reflective styles. *In: Continental Lithosphere: Deep Seismic Reflections.* American Geophysical Union, Geodynamics series, **22**, 69–78.

MENARD, G. 1979. *Relation entre structures profondes et structures superficielles dans le Sud-Est de la France. Essai d'utilisation de données géophysiques.* Thèse, Université de Grenoble.

MONTADERT, L., ROBERTS, D. G. & DE CHARPAL, O. 1979. *Rifting and subsidence of the northern continental margin of the bay of Biscay.* Initial report of the deep sea drilling project n°48 US government printing office, Washington, 1025–1060.

MUGNIER, J. L. & VIALON, P. 1986. Deformation and displacement of the Jura cover on its basement. *Journal of Structural Geology*, **8**, 373–387.

——, GUELLEC, S., MENARD, G., ROURE, F., TARDY, M. & VIALON, P. 1990. Crustal balanced cross-sections through the external Alps deduced from ECORS profile. *Mémoires Société géologique de France*, **8**, **156**, 203–216.

NZIENGUI, J. J. 1993. *Excès d'argon radiogénique dans les quartz des fissures tectoniques: implications pour la datation des séries métamorphiques. L'exemple de la coupe de la Romanche, Alpes occidentales françaises.* Thèse Université de Grenoble.

OELE, J. A., HOL, A. & TIEMANS, J. 1981. Some roetliegend gas fields of the K. and L. Blocks, Netherlands offshore (1968–1978) – A case history. *In*: ILLING, L. V. & HOBSON, G.D. (eds) *The petroleum Geology of the continental shelf of Northwest Europe*. Institute of Petroleum, London & Heydons Ltd, 289–300.

ORI, G. G. & FRIEND, P. F. 1984. Sedimentary basin formed and carried piggyback on active thrust sheets. *Geology*, **12**, 475–478.

PAVONI, N. 1961. Faltung durch horizontal Verschiebung. *Eclogae Geologicae Helvetiae*, **54**, 515–534.

PHILIPPE, Y. in press. Transfer zone in the Southern Jura thrust belt (Eastern France): geometry, development and comparison with analogue modelling experiments, *In*: MASCLE, A. (ed.) *Exploration and Petroleum Geology of France* EAPG Memoir 4.

PORTHAULT, B. 1974. *Le Crétacé supérieur de la fosse vocontienne et des régions limitrophes.* Thèse, Lyon University.

ROURE, F., HOWELL, D. G., GUELLEC, S. & CASERO, P. 1990. Shallow structures induced by Deep-Seated Thrusting. *In*: LETOUZEY, J. (ed.) *Petroleum and Tectonics in Mobile Belts*. Editions Technip, Paris, 15–30.

RUDKIEWICZ, J. L. 1988. Quantitative subsidence and thermal structure of the European margin of the Tethys during early and middle Jurassic times in the western Alps. *Bulletin Société Géologique France*, **4**, 623–632.

STOCKMAL, G. S., BEAUMONT, C. & BOUTILIER, R. 1986. Geodynamic Models of convergent margin Tectonics: transition from rifted margin to overthrust belt and consequences for foreland-basin development. *The American Association of Petroleum Geologists Bulletin*, **70**, 181–190.

SHI, Y. & WANG, C. Y. 1987. Two dimensional modeling of the PT paths of regional metamorphism in simple overthrust terrains. *Geology*, **15**, 1048–1051.

TRICART, P. 1984. From passive margin to continental collision: a tectonic scenario for the western Alps. *American Journal of Science*, **284**, 97–120.

—— & LEMOINE, M. 1986. From faulted blocks to megamullions and megaboudins: Tethyan heritage in the structure of the Alps, *Tectonics*, **5**, 95–118.

TRICHON, H., HUYGHE, P., MUGNIER, J. L. & ARNAUD, H. 1990. *Détermination de l'épaisseur des séries érodées par l'étude des anomalies de surcompaction: application à l'Offshore sud hollandais*. 13è Réunion des Sciences de la Terre, abstrat, Grenoble.

VAN HOORN, B. 1987. Structural evolution, timing and tectonic style of Sole Pit inversion. *Tectonophysics*, **137**, 239–284.

VAN WIJHE, D. H. 1987. Structural evolution of inverted basins in the Dutch offshore. *Tectonophysics*, **137**, 171–219.

VIALON, P. 1986. Les déformations alpines de la couverture sédimentaire de blocs du socle cristallin basculés de Belledonne, Grandes Rousses et Pelvoux dans la région de Bourg d'Oisans. Réunion Extraordinaire de la Société Géologique de France: de la marge océanique à la chaine de collision dans les Alpes du Dauphiné. *Bulletin Société Géologique France*, **II, 2**, 202–204.

—— & PECHER, A. 1974. Métamorphismes, clivages syn-schisteux et dérivées dans le cadre pétrostructural des massifs cristallins du Haut-Dauphiné et de leur couverture. *Bulletin Société Géologique France*, **7, XVI**, 266–268.

WERNICKE, B. 1985. Uniform-sense normal simple shear of the continental lithosphere. *Canadian Journal of Earth Sciences*, **22**, 108–125.

WHITE, N. & McKENZIE, D. 1988. Formation of the "steer's head" geometry basin by differential stretching of the crust and mantle. *Geology*, **16**, 250–253.

WILDI, W. & HUGGENBERGER, P. 1993. Reconstitution de la plate-forme européenne anté-orogénique de la Bresse aux chaînes subalpines; éléments de cinématique alpine (France et Suisse occidentale). *Eclogae Geologicae Helvetiae*, **86**, 47–64.

WILLIAMS, G. D., POWELL, C. M. & COOPER, M. A. 1989. Geometry and kinematics of inversion tectonics, *In*: COOPER, M. A. & WILLIAMS, G. D. (eds) *Inversion Tectonics*. Geological Society, London, Special Publication, **44**, 3–16.

ZIEGLER, P. A. 1987. Compressional intra-plate deformations in the Alpine foreland. *Tectonophysics*, **137**, 420.

Structural analysis of the inverted Bristol Channel Basin: implications for the geometry and timing of fracture porosity

MICHAL NEMČOK, ROD GAYER & MARIOS MILIORIZOS

Laboratory for Strain Analysis, Department of Earth Sciences, University of Wales, Cardiff, CF1 3YE, UK

Abstract: The Bristol Channel basin was developed as an early Mesozoic half-graben, with the Bristol Channel fault zone (BCFZ) representing the principal down-to-the-south basin controlling structure. The basin was inverted during the Tertiary. This paper illustrates the application of structural analysis to the northern and southern margins of the Bristol Channel basin, where on shore medium- to small-scale structures provide analogues for the BCFZ. Throughout the outcrop four distinct structural mechanisms and associated fracture systems can be recognised. i) The oldest are normal faults, in many cases resulting from the reactivation of underlying WNW–ESE to NE–SW Late Palaeozoic Variscan thrusts. These are linked by N–S transfer faults and associated with WNW–ESE to ENE–WSW extensional veins. Evidence for synsedimentary development of both normal faults and extensional veins has been found in the Triassic and Lower Liassic sequences. Palaeostress analysis suggests an approximately NE–SW oriented σ_3. ii) During inversion the normal faults acted as buttresses, forming WNW–ESE and younger NW–SE trending folds. Associated thrusts and oblique- to strike-slip faults reactivated the earlier transfer faults. The orientation of σ_1 was NE–SW during this deformation. iii) A regionally consistent system of NE–SW striking extensional veins formed at the close of the strike-slip inversion event. iv) Following primary inversion, lateral escape structures formed against the buttresses, with the development of oblique- and strike-slip faults in a variable and locally controlled stress field. Fracture porosity determined at sites throughout the outcrop show highest levels, up to 20%, formed in association with normal faults during the rifting event, but also significant amounts developed during the inversion event. Veins associated with strike-slip faulting of the latter give average porosities of 6.5%, whilst the later inversion-related extensional veins give average porosities of 0.8%. All these porosities show a high degree of directional permeability. It is argued that the oil generation window was reached during burial in the later stages of the rifting event, and that rapid polyphase fluid discharge from over-pressured fracture-bounded compartments allowed hydrocarbon migration into normal fault-related fracture porosity to form traps. The development of fracture porosity during inversion produced a long-lasting directed permeability and allowed many of the traps to drain. Only those traps associated with normal faults not directly affected by the inversion will have survived. A strategy to discover such a play requires a well targeted and detailed structural study.

The Bristol Channel basin, filled with up to 3.4 km of Mesozoic and Tertiary sediments (Kamerling 1979), is a relatively small WNW–ESE trending structure, 155 km E–W and 30 km N–S (Fig. 1). To the north and south it is bordered by Variscan structures at outcrop, and lies above a major WNW–ESE striking Variscan thrust, the Bristol Channel Thrust (Brooks *et al.* 1988). Extensional reactivation of the Variscan structures controlled the location and orientation of the basin-forming faults during the Mesozoic, such that the main basin synform occurs in the hangingwall of the Bristol Channel Fault Zone (Mechie & Brooks 1984), representing a half-graben formed by reactivation of the Bristol Channel Thrust (Brooks *et al.* 1988,

1993). In addition, two major faults (the Sticklepath-Lustleigh Fault in the west and the Cothelstone Fault in the east), and several minor NW–SE striking faults transect the basin. These were originally Variscan dextral strike-slip faults (Dearman 1963), which were reactivated during the Mesozoic and Tertiary tectonic history of the basin, and had a major influence on the structures developed in the basin-fill. The basin merges ESE, onshore across the Mendip Hills, into the northern part of the Wessex Basin, and westwards into the South Celtic Sea Basin (Van Hoorn 1987).

Previous investigations of the Bristol Channel Basin have involved interpretation of regional gravity and seismic refraction surveys (Brooks

From BUCHANAN, J. G. & BUCHANAN, P. G. (eds), 1995, *Basin Inversion*, Geological Society Special Publication No. 88, 355–392.

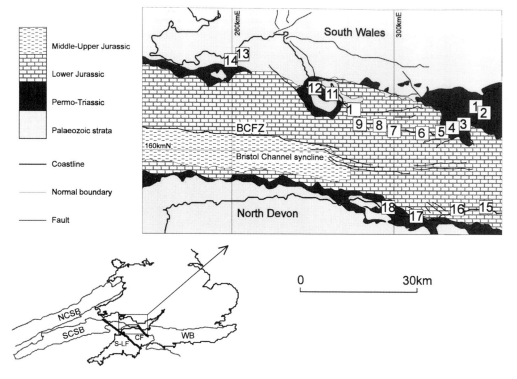

Fig. 1. Sketch map of the area investigated along the northern and southern margins of the Bristol Channel, showing the principal geological units and structure. The location of data collection sites, referred to in the text, is indicated by numbers: 1 – Penarth Marina; 2 – Penarth Head; 3 – St Mary's Well Bay; 4 – Bendrick Rock; 5 – Barry Old Harbour; 6 – Rhoose Point; 7 – St Donats east; 8 – St Donats west; 9 – Nash Point; 10 – Trwyn-y-Witch; 11 – Newton-Black Rocks; 12 – Sker Point; 13 – Caswell Bay; 14 – Pwll Du; 15 – Kilve; 16 – St Audries Bay; 17 – Watchet; 18 – Blue Anchor Bay. The inset shows the location of the study area. BCFZ – Bristol Channel Fault Zone; CF – Cothelstone Fault; NCSB – North Celtic Sea Basin; SCSB – South Celtic Sea Basin; S-LF – Sticklepath-Lustleigh Fault; WB – Wessex Basin.

et al. 1977; Bayerly & Brooks 1980; Llewllyn 1981; Mechie & Brooks 1984), seismic reflection surveys (Kamerling 1979, Van Hoorn 1987; Brooks *et al.* 1988; Miliorizos 1992). These have demonstrated a rift-related Triassic to Aptian sequence, interrupted by a late Cimmerian unconformity and succeeded by a transgressive Upper Cretaceous section before main basin inversion during the Tertiary. This represents a similar history to the onshore Wessex basin (Chadwick 1986; Karner *et al.* 1987; Lake & Karner 1987), where rifting occurred from Triassic through Aptian, with inversion from Late Cretaceous through Oligocene.

This paper describes the results of a structural investigation along the northern and southern margins of the Bristol Channel Basin, where excellent coastal exposures allow an analysis of the sequence of tectonic events, from rifting through inversion, to be determined. The fracture geometry and mechanisms involved

with each event will be discussed and models developed for the timing of hydrocarbon migration and trap formation in relation to the inversion sequence.

Methods

Fourteen sub-areas mainly along the northern shore of the Bristol Channel have been studied in detail. These were chosen to maximize the range of lithostratigraphic/lithotectonic units and palaeo-depth levels (Fig. 1). The following information was obtained at sampling stations in each sub-area: 1) orientations of faults, slickenside lineations, folds, extensional veins, and stylolites; 2) mineralization on fault surfaces and extensional veins; 3) sense of motion on faults, using the criteria of Petit (1987); Means (1987) and Hancock (1985); 4) fold hinges and axial surfaces (Ramsay 1967); 5) fold geometries; 6) parameters associated with syntectonic veins, such as fibre orientation and phases of vein development (Ramsay 1980; Ramsay & Huber 1983; Thomas & Pollard 1993; Kirschner *et al.* 1993); 7) cross-cutting relationships of the structures referred to above. The orientation of structures was plotted stereographically to identify local and regional orientation patterns and constrain palaeostress configurations.

About 1000 fault and slickenside lineation orientations and relative displacement senses were used to determine palaeostress fields. The inversion stress computation methods of Hardcastle & Hills (1991) were used to characterize principal stress orientations, the ratio of their magnitudes ($\phi = \sigma_2 - \sigma_3/\sigma_1 - \sigma_3$), and to separate fault-striae data related to the different tectonic events. The method performs a grid search, identifying all palaeostress tensors that are consistent with all, or a subset of the data. This is achieved by testing more than 10^4 tensors against each fault datum. For each σ_1 orientation σ_2 and σ_3 and the stress ratio are varied incrementally. After each stress test a calculation based on the Coulomb–Mohr failure criterion is performed. We used a zero value for cohesion as most of the faults studied were reactivated and a value of 0.4 for the friction as the majority of faults were mineralized with calcite. Data separation is then made as follows. The largest fault data set related to a particular stress tensor derived from the grid search is taken as a nucleus. This stress tensor is then applied to the whole fault population and all faults producing a divergence greater than a chosen limit (15° in this study) between the measured and predicted shear stress vectors are stored separately. These are then used in a second round of the grid search, producing a second subset of stress tensors, and the process is repeated until a subset contains less than four faults (the minimum number required for the method).

About 800 syntectonic veins and associated mineral fibre orientations were used to determine σ_3 orientations. A few tens of fold hinge and axial surface orientations were used to give an approximate orientation of $\sigma_1\sigma_2\sigma_3$, as a check of the other determinations.

Sampling localities affected by a single structural event were rare (localities 1 and 11). At most localities multiple structural overprints were observed. In some cases polyphase reactivation occurred during the same tectonic event, such that faults show multiple slip events with a similar sense of movement. This was observed in the case of normal faults at localities 5, 6 and 10, and of strike-slip faults at localities 3, 6, 7, 8, 9, 10 and 12. The majority of sites showed at least

two (localities 7, 12 and 14), three (localities 4, 8 and 9) or four (localities 3, 5, 6 and 10) faulting phases related to distinct tectonic events. Superposition of striations on fault planes and intersecting fault patterns allowed the determination of relative movement (stress configuration) chronology at each outcrop. Structural sequences could be correlated throughout the region, and, therefore, a synthesis of these results provided a regional relative structural chronology. More rigorous age constraints were established using the stratigraphic ages of the deformed rocks and various published geological constraints, discussed later in the paper.

Values of fracture porosity have been estimated from open fractures and mineral-filled veins. Veins showing synkinematic, stretched crystal fills, which would have developed minimal porosity during their formation, were not included. The estimates were made, from field tracings of planar surfaces, either parallel or perpendicular to bedding, in which the percentage of the plane occupied by the fracture or vein was determined using the digitized tracings and an image analysis computer package. The area percentages approximate to volume percentages as the fractures and veins continue in the third dimension. It is unlikely that all the veins would have been open simultaneously so that the estimates represent a maximum porosity. Fracture and vein development are very heterogeneous, being most intensely developed close to faults and in the more competent limestone lithologies. The range in values determined reflects the inhomogeneity of the fracture systems, even over small areas. The porosity estimates were made in locations related to specific structures so as to determine the effect of the various tectonic events on fracture porosity.

Data and results

Rifting-related event

Structures formed during the rifting event were recorded at all the localities in the study area. Orientation data for meso- and macro-scale normal fault populations and related extensional veins are shown in Figs 2 & 3. Table 1 shows the computed palaeostress configurations for the rifting event, and Figure 4 shows the extension vectors for the same event. Some of the faults, such as the normal faults at Nash Point (locality 9), formed above reactivated Variscan structures. Other faults, such as those cutting the Mercia Mudstones at Penarth Head (locality 2) and at Watchet (locality 17), are syn-sedimentary faults with evidence of growth. The σ_1/σ_3

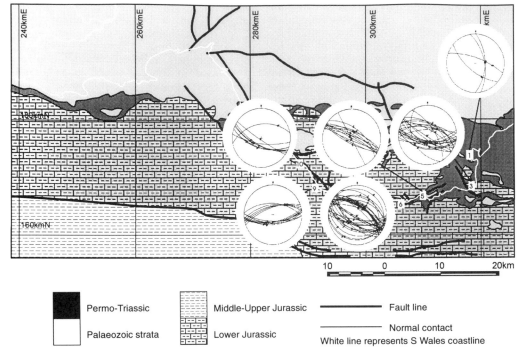

Fig. 2. Map of the studied northern margin of the Bristol Channel with rifting-related normal faults shown as great circles on lower hemisphere equal area stereographic projections at the numbered localities.

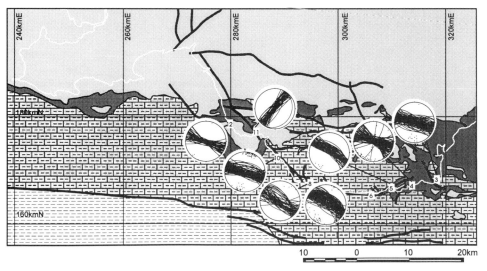

Fig. 3. Map of the studied northern margin of the Bristol Channel with rifting-related extensional veins shown as great circles on lower hemisphere equal area stereographic projections at the numbered localities. Poles indicate orientation of mineral fibres in the veins. See Fig. 2 for key.

planes determined for normal faults at localities 1, 3, 5, 6, 9 and 10 show a consistent regional pattern. At Nash Point (locality 9, Fig. 2), this stress configuration generated neoformed (*sensu* Angelier 1994) oblique-slip normal faults with an ENE–WSW strike instead of the typical WNW–ESE strike. At Black Rock, Newton (locality 11) neptunian dykes, with Norian fills

Table 1 *Computed palaeostress configurations for the rifting event in the north Bristol Channel study area. The method used is that of Hardcastle & Hills (1991)*

| Site | Age of rock | No of faults | Stress tensor | | | | Method |
			sigma 1	sigma 2	sigma 3	stress ratio	
Sully, 3	Triassic	18	095/80	285/10	195/02	0.2	H&H
Marina, 1	Triassic	04	101/70	318/16	225/11	0.2	H&H
Barry, 5	Liassic	07	000/80	116/04	207/09	0.1	H&H
Nash, 9	Liassic	13	045/80	136/00	226/10	0.5	H&H
Trwyn, 10	Liassic	07	029/70	137/06	229/19	0.7	H&H
Rhoose, 6	Liassic	23	051/85	297/02	207/05	0.6	H&H
		08	210/70	301/00	031/20	0.9	H&H

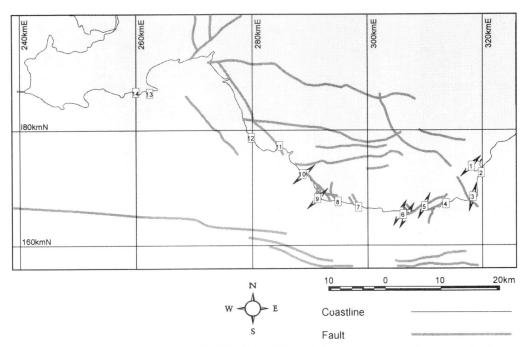

Fig. 4. Map of the studied northern margin of the Bristol Channel showing the computed σ_3 vectors for the rifting-related extension at the numbered localities.

(Ivimey-Cook 1974) and major extensional veins, filled with idiomorphic calcite, indicate an earlier more NW–SE extension, almost perpendicular to the dominant trend. At Rhoose Point (locality 6) two normal faults, dipping towards NNE, are preserved in their embryonic stages. They occur in a region of more dense normal faulting and record only the initial Riedel shears and associated extensional veins. In these zones the extensional veins strike parallel to the regional trend of normal faults (NW–SE), with calcite fibres developed normal to the vein walls. The Riedel shears occur in en echelon arrays, each with minimal displacement, and in these

cases no associated through-going displacement surface. Both the extensional veins and associated Riedel shear arrays have been rotated about a horizontal axis through approximately 30° towards the south, in the southerly limb of a major roll-over antiform. An array of bed-parallel decollements within shale horizons has developed within the roll-over, recording layer-parallel displacements. Calcite growth within the decollement horizons took place by a crack-seal mechanism, and shows evidence for multiple reactivation (Fig. 5a). Similar features are also present at locality 5.

Extensional veins are ubiquitously developed

(a)

Fig. 5. (**a**) Calcite filled bed-parallel decollement zone in Liassic Shale at Rhoose Point (locality 6, located in Fig. 1). The decollement was developed by a crack-seal mechanism during the rifting event. (**b**) The inverted Blue Ben Fault between Kilve and St Audrie's Bay, Watchet, showing preserved rift-related extensional structures in the footwall of the fault in the foreground, and normal drag in the hangingwall of the fault in cliff section. The footwall synform was developed during the inversion of the fault.

in association with the rift related normal faults. The most common mineral filling the veins is calcite. Some of the veins developed in Triassic and Liassic sediments are partly filled by barite, e.g. Barry Old Harbour (locality 5), St Audrie's Bay, Lavernock Point (to the east of locality 3) (Waters *et al.* 1987) and Ogmore (to the west of locality 10) (Lee 1991). At St Audrie's Bay, on the southern margin of the Bristol Channel, the barite shows growth during early stages of extensional fracturing. Extensional veins in Triassic mudstones commonly contain gypsum in the form of satin spar, that in some cases show vertical injection structures, e.g. at Penarth Head (locality 2).

Evidence of several deformation mechanisms during the rifting event are present. For example, at locality 6, arrays of pressure-solution seams and stylolites curve from lying perpendicular to σ_1, distant from the shear plane, to sub-parallel slickolitic structures along the shear plane. Many veins are filled by syn-deformational calcite fibres. Evidence for their syn-deformational relationship with normal faulting is present at Rhoose and Trwyn-y-Witch

(localities 6 & 10), where the veins were formed as pull-aparts, perpendicular to the sub-horizontal layering, and linked by normal faults with slickolites, the likely source of the calcite. A typical feature of all the localities with normal faults (localities 1, 3, 5, 6, 9 & 10) is that the fault shear planes change orientation as they pass through differing lithologies. Normal faults occur in shales, linked by extensional veins in limestones, indicating contrasting strength properties for the two lithologies. These veins are often filled by calcite lacking fibrous form. Frequently voids are still present in the veins and are lined by idiomorphic calcite and rarer barite crystals. At St Audrie's Bay, in the downthrown hangingwall of a major north-dipping normal fault, an extensional vein-related boudinage structure has V-shaped clefts that cut through the earlier calcite-filled veins. The re-opened centres of these veins are filled by barite and calcite, in which the latter is clearly separated from the earlier calcite-fill event. Some of the clefts are filled by an untextured limestone/shale mixture derived from the underlying beds. The normally bed-parallel fabric of the underlying

(b)

shale is tightly folded into the base of the clefts, suggesting an origin as a soft sediment injection structure, discussed below. The boudinage occurred along discrete minor normal faults. Displacements in the shales were so high that these high-angle faults lose their discrete character at the limestone/shale boundary. The shale layers have accommodated this stretching by plastic deformation, with no discrete shear planes being visible. Differential movement of adjacent shale horizons and movement along brittle fractures in the limestone layers have produced a significant rotation of the bedding that has sometimes detached lozenge-shaped blocks.

Inversion event

Thrusting Meso-scale thrust populations formed during the inversion event are shown in Fig.6. They are characteristically related to strike-slip fault zones, although some have reactivated pre-existing low-angle normal faults or normal fault-related decollements. One such decollement at Barry Old Harbour (locality 5) cuts and displaces a high angle planar normal fault. Reactivation of normal fault-related decollements has also been recorded at Rhoose (locality 6). Thrusts within the Carboniferous Limestone at Caswell Bay (locality 13) show two sets of thrust striae; one set NW-vergent and a younger set NE-vergent. Their significance is interpreted below.

Approximately dip-slip thrusts, formed at low dip angles have been recorded at Caswell Bay and Sker Point (localities 13, 12). Thrust populations with dip maxima around 30° occur at Rhoose and Trwyn-y-Witch (localities 6, 10). At none of the localities studied along the

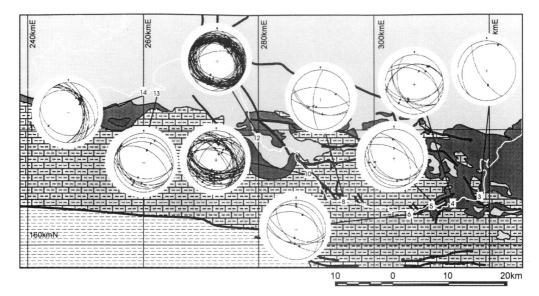

10   0   10   20km

Fig. 6. Map of the studied northern margin of the Bristol Channel, showing thrusts developed during early stages of inversion as great circles on lower hemisphere equal area stereographic projections at the numbered localities. See Fig. 2 for key.

northern margin of the Bristol Channel have we observed any instances of direct reactivation of pre-existing, high-angle normal faults. However, along the southern margin of the Bristol Channel at Kilve, Watchet, and St Audrie's Bay, typically inverted normal faults occur, cut by lower-angle neoformed thrusts. Localities 3, 5, 6 and 10 show a buttressing effect of the pre-existing normal faults. Figures 7a & 8 show such buttressing at Trwyn-y-Witch (locality 10) along the profile indicated in Fig. 9. The rock mass has been shortened against pre-existing normal faults, forming a set of folds, each of which has formed against a normal fault buttress. Detailed analysis indicates that the two southern buttresses are affected by shortening, but that the northernmost buttress has preserved unmodified rift-related features. Fold axes are slightly oblique to the pre-existing normal faults, and poles of calcite-filled extensional vein arrays also show the slight sinistral transpressional character of these folds. These relations are discussed below. At Barry Old Harbour (locality 5) evidence for a stronger sinistral transpression is provided by the patterns of fold-related extensional calcite veins at two locations (Fig. 10). These veins have developed by tangential longitudinal strains in the outer arc of the associated buckle fold. Synkinematic fibres within the veins in the overturned fold limb indicate a sinistral deflection of the σ_1 azimuth of about 15° as a result of transpression. The computed regional palaeostress orientation of σ_1 at this locality is 035° (Fig. 11). The hinge zone of the north vergent fold has been passively rotated by about 29°, as the mass in the hinge region was able to adjust more freely than the mass in the overturned limb. However, at Rhoose (locality 6) extensional veins, of the same origin, are parallel to the fold axis and show no evidence for transpression (Fig. 10c).

Strike-slip faults at all localities containing thrusts are abundant, and show pre-, syn- and post-thrusting age relationships. A minor number reactivate pre-existing rifting-related transfer faults that link normal faults (e.g. locality 6).

The computed stress configurations (Table 2) indicate that σ_1 azimuths range from 015° to 045°. At Trwyn-y-Witch (locality 10), and at a few other localities with insufficient data to allow palaeostress computations, some thrust planes preserve at least 3 generations of the thrust striae. Fairly constant stress ratios between 0.1 and 0.5 indicate a stress state between strong compression to plane stress, with localities recording strong compression prevalent. Deformation varies between flattening and plane strain.

All thrust-related folds show a parallel geometry and have been produced by bed-parallel slip. Some thrusts (e.g. at locality 10) have associated subhorizontal extensional calcite veins.

Fig. 7. (a) Folds formed in Liassic shales and limestones during the early stages of inversion at Trwyn-y-Witch (locality 10, located in Fig.1). The fold has developed by progressive transpression against a buttress formed by a rift-related normal fault, seen in the right (north) of the photograph. (b) Folds with steeply north-dipping axial surfaces, formed in Upper Triassic, Blue Anchor Beds at Warren Bay, Watchet. The folds have developed during the inversion event, by buttressing against the south-dipping normal fault, seen in the left (north) of the photograph, with Upper Triassic Mercia Mudstones, containing bed-parallel gypsum veins, in the footwall.

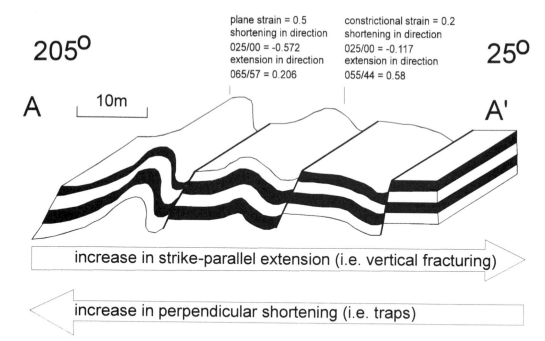

205⁰

plane strain = 0.5
shortening in direction
025/00 = -0.572
extension in direction
065/57 = 0.206

constrictional strain = 0.2
shortening in direction
025/00 = -0.117
extension in direction
055/44 = 0.58

25⁰

A   10m   A'

increase in strike-parallel extension (i.e. vertical fracturing)

increase in perpendicular shortening (i.e. traps)

Fig. 8. Block diagram to illustrate the inversion-related buttressing effect in Liassic shales and limestones against rift-related normal faults, along section A-A' to the south of Trwyn-y-Witch (locality 10, located in Fig. 9). The diagram records N–S lateral variations in: i) syn-rift extension, recorded as trend/plunge of extension direction and values of extensional strain for the indicated faults, measured by summing the widths of exensional veins; and ii) inversion-related shortening, recorded as trend/plunge of shortening direction and values of shortening strain for the indicated buttress structures, determined from bed-length measurements.

Strike-slip faulting Meso- and macro-scale strike-slip faults that formed during inversion are shown in Fig. 12. Some of them reactivated large, NW–SE striking, Variscan strike-slip faults; such as the Cothelstone–Watchet fault zone (Watchet), and the Dinas Powys fault zone at St Mary's Well Bay (locality 3), or unnamed regional faults at Barry (locality 5), Newton-Black Rocks (locality 11) and Sker Point (locality 12).

The strike-slip faults thus show a wide range of orientations and geometries. The neoformed faults usually occur in conjugate N–S dextral and NE–SW sinistral sets, that clearly offset pre-existing rifting related structures (Fig. 13a).

The reactivated strike-slip faults frequently show other than pure strike-slip geometries. At Bendrick Rock and Rhoose (localities 4 and 6) NNE–SSW dextral strike-slip faults experienced a significant amount of extension, as indicated by the fact that calcite is not formed in pressure shadows on the fault surface, as would be expected for pure strike-slip faults (Fig. 13b), but shows fibres grown at a low angle to the fault surface. At Watchet (the Cothelstone–Watchet

fault), St Mary's Well Bay (the Dinas Powys fault, locality 3), and some of the structures at localities 8, 9, 10 and 11, NNW–SSE to NW–SE strike-slip faults show evidence of transpression with a significant amount of shortening. Each strike-slip fault developed its own pattern of fracturing; most frequently combinations of Riedel shears and extensional veins. NW–SE to N–S dextral strike-slip faults frequently develop associated contractional duplexes, e.g. at Watchet, St Audrie's Bay, localities 7, 8 and 10. In the cases of larger duplexes (localities 7, 8), faults within the duplexes are commonly rotated, and some show overprinting evidence of several generations of strike-slip movement. A small-scale strike-slip duplex, at locality 10, shows a series of fault-bound horses stacked at a fault bend, and subsequent strong contractional deformation.

Strike-slip faulting is common and dominates the structural character at most localities. The computed palaeostress configurations for the strike-slip fault populations are shown in Table 3, and the σ_1 vectors are shown in Fig. 14. Azimuths for σ_1 range from 345° to 065°, and the

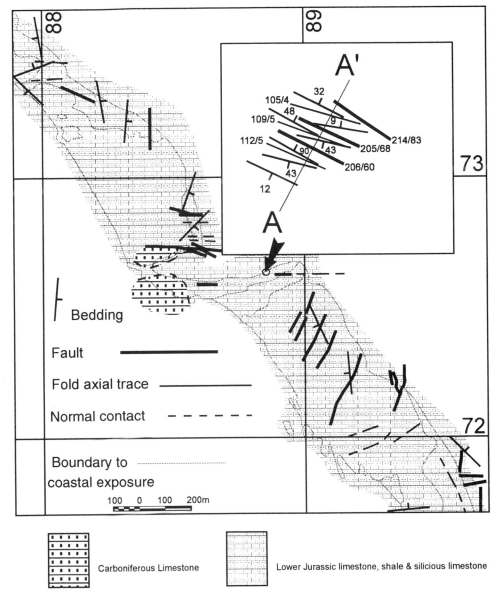

Fig. 9. Geological map of the Trwyn-y-Witch peninsula (locality 10, located in Fig. 1), showing the major structure. The northern boundary of the Carboniferous Limestone inlier is a Variscan thrust that has been reactivated as a Mesozoic extensional fault during rifting, and later inverted by oblique sinistral transpression. The inset shows the detailed structure along line A-A', illustrated as a block diagram in Fig. 8.

stress ratio ($\phi = \sigma_2 - \sigma_3/\sigma_1 - \sigma_3$) indicates a rather broad range of stress systems, from those with σ_1 dominant (ratios 0.1–0.3), through those close to plane stress (ratios 0.4–0.6), to configurations with less pronounced σ_1 (ratios 0.7, 0.8). Fracture porosity associated with the strike-slip faulting determined at St Donats and

Nash Point (localities 8 & 9) averages 6.5% (Fig. 20).

There is a regionally consistent system of extensional veins developed at localities 3, 5, 6, 8, 9 and 12 (Fig. 15), which, based on overprinting relationships, is contemporaneous with, or younger than, the last strike-slip fault

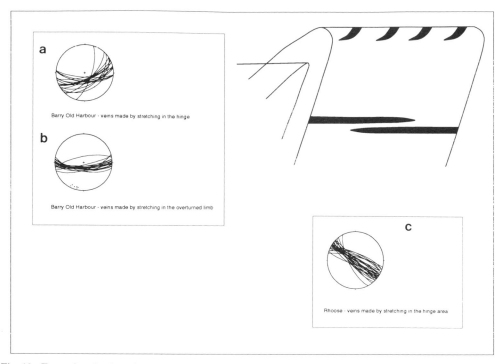

Fig. 10. Extensional veins, shown as great circles on lower hemisphere equal area stereographic projections, developed in the outer arc of folds formed during the early stages of inversion, (**a**) and (**b**) show transpressional inversion at Barry Old Harbour (locality 5), discussed in the text; (**c**) shows pure compression at Rhoose Point (locality 6).

Table 2 *Computed palaeostress configurations for the inversion-related thrusting event in the north Bristol Channel study area. The method used is that of Hardcastle & Hills (1991)*

Site	Age of rock	No of faults	Stress tensor				
			sigma 1	sigma 2	sigma 3	stress ratio	Method
Hayes, 4	Triassic	10	030/05	298/21	132/68	0.1	H&H
Barry, 5	Liassic	05	035/10	132/36	292/52	0.5	H&H
Trwyn, 10	Liassic	38	015/00	285/14	105/76	0.3	H&H
		05	036/35	127/01	218/55	0.5	H&H
Donats, 8	Liassic	05	221/25	315/08	061/64	0.2	H&H
Sker, 12	Carbon	80	030/00	300/14	120/76	0.2	H&H
Caswel, 13	Carbon	09	025/05	115/01	219/85	0.1	H&H
Pw.Du, 14	Carbon	11	045/05	136/16	299/73	0.2	H&H
Rhoose, 6	Liassic	06	200/30	351/57	102/13	0.1	H&H

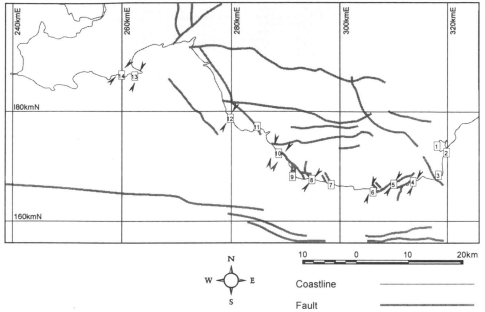

Fig. 11. Map of the studied northern margin of the Bristol Channel showing the computed σ_1 vectors for the inversion-related thrusting event at the numbered localities.

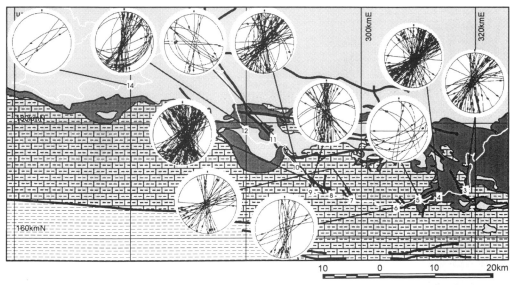

Fig. 12. Map of the studied northern margin of the Bristol Channel, showing strike-slip faults developed during inversion as great circles on lower hemisphere equal area stereographic projections at the numbered localities.

(a)

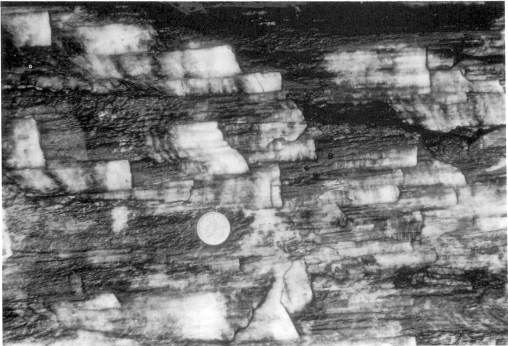

(b)

Fig. 13. (**a**) Sinistral strike-slip faulting, with minor calcite veining, associated with the strike-slip inversion event, offsetting calcite veins developed during the rifting event. Liassic limestones at Trwyn-y-Witch (locality 10, located in Fig. 1). (**b**) Detail of sinistral strike-slip fault surface, with calcite mineralization showing features of crack-seal mechanism of formation. Trwyn-y-Witch (locality 10).

Table 3 *Computed palaeostress configurations for the inversion-related strike-slip event in the north Bristol Channel study area. The method used is that of Hardcastle & Hills (1991)*

| Site | Age of rock | No of faults | Stress tensor | | | | |
			sigma 1	sigma 2	sigma 3	stress ratio	Method
Sully, 3	Triassic	30	200/15	293/11	058/71	0.1	H&H
		09	182/25	329/61	085/14	0.8	H&H
		05	015/15	275/61	112/24	0.1	H&H
Hayes, 4	Triassic	75	020/05	227/84	110/03	0.1	H&H
		11	245/10	127/70	338/17	0.4	H&H
Barry, 5	Liassic	08	046/20	211/69	314/05	0.1	H&H
Nash, 9	Liassic	24	180/15	333/73	088/07	0.2	H&H
		07	345/25	154/65	253/05	0.4	H&H
Trwyn, 10	Liassic	22	011/20	248/56	111/26	0.1	H&H
		14	206/30	346/53	104/20	0.2	H&H
		08	016/20	243/62	113/19	0.2	H&H
		06	226/20	100/58	325/24	0.2	H&H
Donats, 8	Liassic	89	010/10	163/79	279/05	0.4	H&H
		24	180/10	280/44	080/45	0.2	H&H
		09	026/20	253/62	123/19	0.6	H&H
Sker, 12	Carbon	31	030/05	120/01	224/85	0.1	H&H
		06	246/40	092/47	347/14	0.5	H&H
Stout, 7	Liassic	22	210/20	016/69	119/05	0.7	H&H
		04	218/50	031/40	124/03	0.3	H&H
Rhoose, 6	Liassic	14	215/25	041/65	307/02	0.5	H&H

movements. The veins are calcite-filled, consistently straight, usually only 1–2 mm wide (Fig.16a) and have very gentle curvature in the areas of overlapping joint traces. Fracture porosity associated with these veins give values ranging from 0.4–2.6% with a mean of 0.8% (Fig. 21).

Escape Meso-scale strike-slip, oblique-slip and rare normal fault populations in the study area, that formed at the end of the inversion event, are shown in Fig. 17. They either reactivated pre-existing normal faults (localities 3, 5, 6, 9 and 10; Fig. 16b) or, in localities lacking normal faults, they offset pre-existing strike-slip patterns at acute angles as neoformed faults (localities 4, 7 and 8).

The computed stress configurations at localities 3, 5, 6, 9 and 10 (Table 4, Fig. 18) show considerable variation, with different orientations of principal stress axes and different stress ratios ($\phi = \sigma_2 - \sigma_3/\sigma_1 - \sigma_3$). Three main stress configurations have been recorded: 1) a range of subhorizontal σ_1 orientations and low ratios indicating a dominant σ_1; 2) a range of subhorizontal σ_1 orientations, but with high ratios indicating a decreasing magnitude of σ_1; and 3) vertical σ_1, indicating an extensional stress field. Localities 4, 7 and 8 lack normal faulting, and localities 7 and 8 also lack rifting-related extensional veins. The typical

relationships at locality 4 is that neoformed ENE–WSW sinistral strike-slip faults cut and offset the pre-existing conjugate system of N–S and NE–SW oriented dextral and sinistral strike-slip faults. The situation at localities 7 and 8 is more complex, with the development of strike-slip related transpressional duplexes, usually connected with dextral strike-slip faults, and shortened by folding. The internal strike-slip patterns form several cross-cutting generations that developed during different stages of this shortening. The computed stress configurations for this type of escape indicate two main conditions similar to those of 1) and 2) above for reactivated normal faults (Table 4, Fig. 18).

Interpretation

Rifting event

The structures associated with the rifting event (Figs 2, 3) were formed in sediments that ranged from uncompacted through partly to completely lithified. Many normal faults, such as those at Nash Point (locality 9), were formed in the Mesozoic sequence by reactivation of pre-existing Variscan structures in the basement. They thus developed an unusual geometry, with oblique-slip displacements (Fig. 2).

Extensional veins in Triassic and Liassic sediments related to basin rifting are sometimes

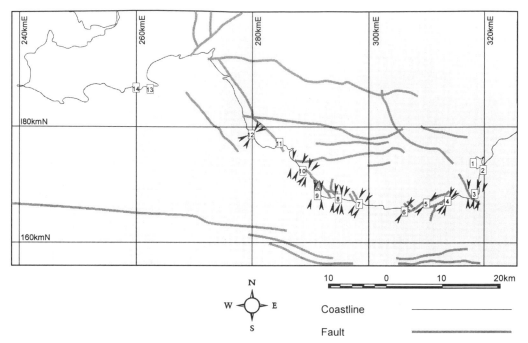

Fig. 14. Map of the studied northern margin of the Bristol Channel showing the computed σ_1 vectors for the inversion-related strike-slip event at the numbered localities. Compare with Fig. 11.

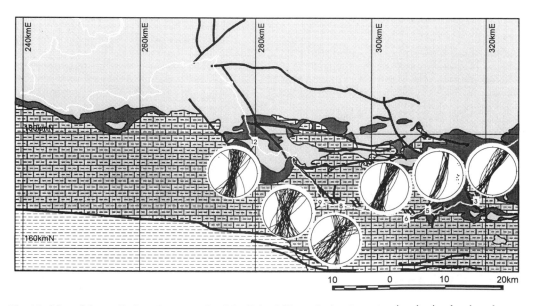

Fig. 15. Map of the studied northern margin of the Bristol Channel, showing extensional veins developed as a distinct extensional event at the end of the inversion-related strike-slip faulting. The veins are shown as great circles on lower hemisphere equal area stereographic projections at the numbered localities. Poles indicate orientation of mineral fibres in the veins.

(a)

(b)

Fig. 16. (a) Regularly spaced calcite veins formed during the extensional event at the end of the inversion-related strike slip faulting. Upper Triassic marginal facies, St Mary's Well Bay (locality 4). (b) Extensional vein, with idiomorphic calcite, formed during the rifting event, reactivated by strike-slip faulting during inversion. Liassic limestone, Rhoose Point (locality 6).

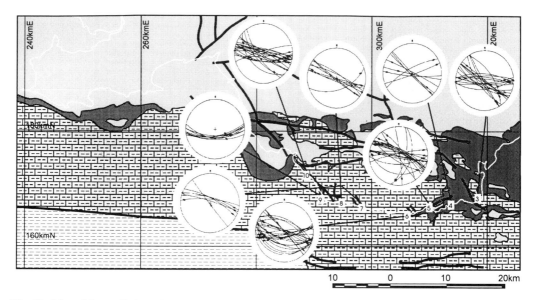

Fig. 17. Map of the studied northern margin of the Bristol Channel, showing strike-slip faults developed during inversion-related escape event. The faults are shown as great circles on lower hemisphere equal area stereographic projections at the numbered localities.

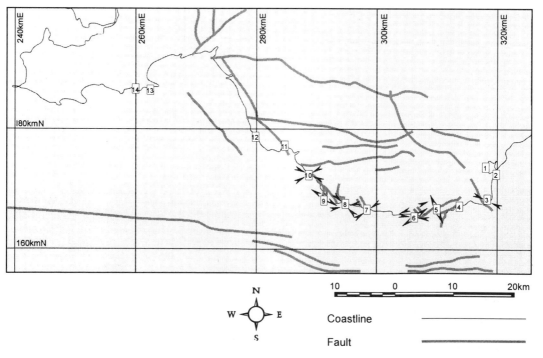

Fig. 18. Map of the studied northern margin of the Bristol Channel showing the computed σ_1/σ_3 vectors for the escape structures at the end of the inversion event at the numbered localities.

filled by barite. The sulphate ions (SO_4^{-2}) are probably of evaporitic origin, derived from the Triassic sequence. The barium (Ba^{+2})may have been derived from an earlier mineralization in the Carboniferous Limestones, or more probably from the South Wales Variscan clastic sequences (Lee 1991). In order to precipitate barite, either significant volume exchange through the pore space has taken place, suggesting that the sediments had a rather high permeability, or fluid flow has taken place through the fracture system. The presence of barite in extensional veins within the Liassic sequence thus suggests significant fluid flow transport.

Calcite in extensional veins has many sources. The earliest available is a calcitic cement formed during the initial stages of diagenesis, before the Liassic sediments had gained a stable grain structure. Bjorkum & Walderhaug (1990) have shown from offshore Norway that the derivation of calcite cement was not assisted by ground water flow, sea water or organically derived CO_2. This diagenetic calcite was probably transported by diffusion, and precipitated in pore spaces at depths of about 1–2 km and at temperatures of about 50–70° C. The isotopic composition of calcite cement from Liassic Bridport Sands along the Dorset coast indicates that it was probably sourced by carbonate fossils (Bjorkum & Walderhaug 1990). The critical degree of supersaturation necessary for nucleation of calcite cement was first achieved within carbonate fossil-rich layers. The local transport of this calcite into extensional veins indicates a short range diffusional process.

Some of the extensional fractures were initially incompletely filled by mineralized water. Stress concentration at the tips of these elastically opened fractures can induce pressure solution (Pollard & Avdin 1988), the products of which will be precipitated in the central parts of the fracture. Whether an open fracture is completely filled depends on the type of fluids available. Such partly filled open fractures suggest that two immiscible liquids, one of which could have been hydrocarbons, were present during their propagation.

Many fully filled fractures are also developed. Some have been filled with stretched crystals by nonsystematic crack-seal vein growth (Durney & Ramsay 1973; Kirschner et al. 1993). Other veins have developed synchronously with syn-deformational growth of calcite fibres, in some cases with evidence that the vein-fill was derived by diffusional mass transfer from related slicko-litic normal faults.

Continuously increasing burial depths during the pre-Aptian rifting event are indicated by vitrinite reflectance in Liassic source rocks (Cornford 1986), suggesting pre-Aptian temperatures of about 80° C, at depths of 2.25–3.2 km.

An increasing overburden load is recorded by various features. The most mobile material, and the most sensitive to loading, is gypsum which records increasing load by hydraulic injection (Waters et al. 1987). Evidence for such vertical injection is present at Penarth Head. The increased load was also recorded by pressure solution features, which are enhanced by rising temperature. At the same depth level, these are more likely to develop in fine grained parts of the sequence, as a result of diffusional mass-transfer processes.

Failure in the Liassic sequence was often influenced by contrasting material properties. Normal faults and extensional veins developed selectively in different materials, as illustrated by faults in shales which had already reached their shear strength limit, and extensional veins in limestones that had only reached their tensional strength limit (see also Fig. 7a, Hancock 1985).

Some structures, such as the boudinaged limestone layer between shale horizons at St Audrie's Bay, suggest tectonic activity within shales and limestones that were not fully lithified. Soft sediment injection structures at the base of the limestone layer were formed by independent particulate flow. Complete disaggregation during the deformation is indicated by the loss of primary structures (e.g. lamination) during flow. This feature shows that high fluid pressures were developed at the base of the limestone, which was lithified at that time. It acted as a hydraulic seal above the overpressured shale horizon. These features were formed by a cyclic increase of fluid pressure in the shale horizon, sealed laterally by calcite encrusted normal faults, due to the increased deposition focused in the fault hangingwall following each displacement event. Injection structures are confined to the shale horizon, suggesting that it was less lithified than the limestone at the time of deformation, due to the frequent over-pressuring cycles which slowed diagenesis. The discrete fractures in the limestone show no evidence of grain breakage characteristic of lithified sediments. Dilation, as shown by fibrous calcite, is a common feature of such fracturing which would allow liquid hydrocarbons still present in the Lower Liassic shales to penetrate between grains of the previously well compacted sediment. Such fracturing in incompletely lithified limestone has been attributed to grain-boundary sliding.

The effect of overpressuring during the normal faulting is indicated also by the stress ratios from the computed stress configurations at localities 1, 3 and 5 (Table 1). The ratio ($\phi = \sigma_2 - \sigma_3/\sigma_1 - \sigma_3$) between 0.1 and 0.2 would imply multidirectional extension in a subhorizontal plane, i.e. coaxial flattening driven by nearly uniaxial vertical stress. The same effect is also suggested by scattered orientations of rifting-related minor extensional veins at Trwyn-y-Witch (locality 10), and also by the layer-parallel extension observed at Barry and Rhoose (localities 5, 6).

A progressively developing overpressure along the base of the sequence affected by high-angle normal faults is indicated at locality 6. The presence of regularly spaced normal faults at the onset of faulting in this sequence suggests that their spacing was controlled by the mechanical properties of the sequence, in particular by basal friction. Two of the normal faults remained in their embryonic stage and experienced no further development, probably as a result of decreasing basal friction following an increase in overpressure.

The calculated stress configurations show consistent σ_3 orientations. There is no evidence for a change of the regional tectonic stress field across the area studied. The two palaeostresses calculated at locality 6 are due to the rotation of an earlier set of normal faults by an amount sufficient to cause data separation.

Stress computations from localities 6, 9 and 10 show higher stress ratios than those from localities 1, 3 and 5 (Table 1), and higher values than normally expected from rift areas. The values of stress ratios from New Hampshire are about 0.1–0.2 (Hardcastle 1989), and similar values have been recorded from the Red Sea, and Basin and Range Province (Angelier 1985, 1989). The stress ratio fluctuations recorded at localities 1, 3 and 5 (Table 1) can be explained by the fact that the stresses associated with Triassic to Aptian extension at localities 6, 9 and 10 (Fig. 4) were determined from Liassic rocks. Thus, the stress ratios could be affected by the same regional extension at a whole range of depths, from the time of their grain consolidation (with about 40% porosity) until the Aptian. Such a great difference in the depth of overburden during various stages of the same rifting event could explain the changes in the stress ratio.

All the features described above indicate that the Triassic to Aptian rifting-related event (Figs 2, 3), driven by NNE–SSW regional extension (Fig. 4, Table 1), was locally the result of an interaction between: 1) regional tectonic stress, 2) changing frictional properties of the affected

rocks, and 3) local sedimentation rates and fluid pressure cycles.

Observations of the Permo–Triassic rifting (Fig. 3) indicate that this event was driven by a differently orientated extension (σ_3 NW–SE) from that controlling the rifting during the Triassic to Aptian events (σ_3 NNE–SSW).

Inversion event

Thrusting Meso-scale thrusts at all the localities studied (Fig. 6) are subordinate to the other inversion structures and formed either in association with strike-slip faults or by reactivation of pre-existing low-angle faults. All the observed structures suggest that they formed in lithified sediments.

No cases of direct reactivation of pre-existing, high-angle normal faults have been found at localities along the northern margin of the Bristol Channel, and this suggests that pre-existing normal faults with calcite along the fracture zones remained sealed during the thrusting event. However, inverted normal faults have been observed at Watchet, Kilve and St Audrie's Bay (Fig. 5b), along the southern margin of the Bristol Channel. This implies a smaller thickness of rifting-related cover in the south, as compared with that in the north of the Bristol Channel at the time of inversion. However, localities 3, 5, 6, 10 and 17 showed the effect of buttressing against pre-existing normal faults (Figs 7a, 8 & 9). The example from locality 10, described above, showed from a study of the folds against the three buttressed pre-existing normal faults that only the largest buttress at the northern end of profile remained rigid. The observed features indicate that the inversion was insufficiently strong to affect the largest buttress. The slight obliquity of the thrusting to the pre-existing normal faults caused the weak sinistral transpression, described above. Stronger sinistral transpression was recorded at locality 5 (Fig. 10, and described above). The relationship between strike-slip and reverse faults at various localities suggests synchronous development of thrusting and strike-slip faulting.

The computed stress configurations (Table 2) indicate that the thrust event was regionally very consistent, with σ_1 azimuths ranging from 015° to 045° (Fig. 11). Inhomogeneous thrusting was observed at Trwyn-y-Witch (locality 10) and also at a few other localities where insufficient data were available to perform palaeostress computations. Within a stable regional stress, local thrust slices developed slightly different displacement paths due to the movement and frictional constraints induced by surrounding structures. The stress ratios also indicate relatively stable conditions from 0.1 to 0.5 (Table 2), indicating stress configurations between prolate to plane stress types, with prolate prevalent. Fluid pressure oscillations were not indicated during this event. Deformation varied between flattening and plane strain.

Comparison of the σ_1 azimuth distribution (Fig. 11) and the morphology produced by the pre-existing, predominantly E–W trending Variscan fold structure suggests that the regional stress field, controlling thrusting, was only slightly modified by the Variscan structure. For example, at localities 4, 5, 6 and 8 where the northward-vergent thrusts should be deflected slightly sinistrally, the average azimuth of σ_1 is 031.5°. A similar effect would be expected at localities 13 and 14, where the average σ_1 is 035°. At localities 10 and 12, where the thrusts should be deflected dextrally, the average σ_1 was 027°. However, the differences are too slight to derive any firm conclusions. At locality 13 the Carboniferous limestone is affected by NW-vergent thrusts with dip-slip striations (Gayer & Nemcok 1994) that have been overprinted by younger NE-vergent thrust striations. Analysis of this younger event gave a palaeostress that was indistinguishable from palaeostresses determined from the inversion thrust event in the Mesozoic rocks throughout the study area, and we infer that the older Variscan structures have been reactivated.

All the thrust-related folds were produced by flexural slip mechanisms, with no bed thickness variations and thus no associated flattening. Thus, they mostly experienced volume increase. Some thrusts (e.g. at locality 10) were accompanied by the formation of subhorizontal extensional calcite veins.

Strike-slip faulting The inversion-related strike-slip faults in the area (Fig. 12) were formed in lithified sediments. They either reactivated pre-existing Variscan strike-slip faults or developed as neoformed structures, and thus show various orientations and geometries. As with the normal faults (see above) the strike-slip faults were sealed by calcite after each fracture event. Their orientation within the regional stress field controls their kinematic development, ranging from transtensional through pure strike-slip (Fig. 13b), to transpressional.

In the case of the larger contractional strike-slip duplexes (localities 7, 8), faults within the duplexes frequently show evidence of rotation. Some of them developed 'forced' internal strike-slip patterns which were unrelated to the regional stress field, but were controlled by an intra-duplex stress generated by the development of the duplex. This mechanism sometimes produced several generations of cross-cutting conjugate sets of strike-slip faults. Some of these duplexes travelled along the strike-slip faults, e.g. at locality 10. In this case they have suffered either volume decrease or increase depending on the fault geometries they met during their transport.

Inversion-related strike-slip faulting was the dominant mechanism acting along the northern margin of the Bristol Channel (Fig. 12). At the localities affected by strike-slip faulting, thrusting was subordinate and formed as accommodation structures related to the strike-slip faulting. Thrusting, however, was prevalent along the southern Bristol Channel margin. We attribute this difference to a thicker sequence in the north, where the overburden was sufficient to exchange the σ_2 and σ_3 axes, favouring strike-slip faulting. Table 3 shows that the consistency of the stress field during the strike-slip event was slightly less than shown by the contemporaneous thrusting, with σ_1 azimuths ranging from 345° to 065°. The stress ratios indicate a rather large variation of stress configurations, from those with σ_1 dominant (ratios 0.1–0.3), through configurations close to plane stress (ratios 0.4–0.6), to configurations with less pronounced σ_1 (ratios 0.7, 0.8).

The computed stress configurations from the strike-slip faulting show a clearer effect of pre-existing Variscan morphology on the NE-directed mass transport. Localities where a slight sinistral deflection of the NE directed mass transport is expected, have an average σ_1 azimuth = 029.7°, while at localities 8, 9, 10 and 12, which should have dextrally deflected the mass transport, the average σ_1 azimuth = 017.0°. This effect is clearly seen in Fig. 14 by comparing the σ_1 azimuths from localities 4–6 and 7–9.

A regionally consistent system of extensional veins, either contemporaneous with, or younger than the last strike-slip faulting, developed at localities 3, 5, 6, 8, 9 and 12 (Fig. 16a). Their traces are very gently curved in the overlap zones indicating that they were formed by a large, remote differential stress (Thomas & Pollard 1993), in agreement with the stress ratio computations that suggest prevalent stress configurations with σ_1 and σ_3 subhorizontal and σ_1 dominant.

The fracture porosities developed during both the strike-slip faulting and the distinct extensional event formed long lasting and very effective directed permeability zones (Figs 20 & 21). Fracture porosity associated with the

Table 4 Computed palaeostress configurations for the escape event in the north Bristol Channel study area. The method used is that of Hardcastle & Hills (1991)

Site	Age of rock	No of faults	Stress tensor				
			sigma 1	sigma 2	sigma 3	stress ratio	Method
Sully, 3	Triassic	13	115/15	276/74	024/05	0.1	H&H
Barry, 5	Liassic	11	089/70	245/18	337/08	0.9	H&H
Nash, 9	Liassic	05	040/60	211/30	303/04	0.4	H&H
Trwyn, 10	Liassic	07	295/15	069/69	201/15	0.2	H&H
		05	060/15	167/47	318/39	0.5	H&H
Donats, 8	Liassic	05	077/25	212/57	337/21	0.3	H&H
		04	135/20	301/69	044/05	0.6	H&H
Stout, 7	Liassic	05	055/20	243/70	146/02	0.8	H&H
Rhoose, 6	Liassic	16	325/60	145/30	055/00	0.7	H&H
		07	275/15	130/72	008/10	0.1	H&H
		04	023/65	152/16	248/18	0.7	H&H

strike-slip faulting determined at St Donats and Nash Point (localities 8 & 9) averages 6.5% (Fig. 20). Fracture porosity associated with these veins give values ranging from 0.4–2.6% with a mean of 0.8% (Fig. 21).

Escape Structures formed during the escape event are either re-activated pre-existing normal faults (localities 3, 5, 6, 9 and 10), or neoformed faults post-dating the strike-slip fault event (localities 4, 7 and 8) (Fig. 17). The nature of faults indicates that they were formed in fully lithified sediments.

Where pre-existing normal faults occur, the local escape was caused by the build-up of high energy conditions resulting from the north-eastward transport of the rock mass being buttressed by the normal fault or by the underlying Variscan morphology. The escape of the squeezed rock mass against the obstacle followed a path of minimum physical work towards a lower energy condition. The computed stress configurations at localities 3, 5, 6, 9 and 10 indicate different modes of this process (Table 4 & Fig. 18). Thus, the computed escape stress configurations suggest the following distinct situations: at localities 3, 6 and 10, the generation of strong compression and the development of atypical local rotation; at locality 10, unusual rotation of σ_1, and a progressive change towards strong extension; and at localities 5, 6 and 9, extensionally driven escape in atypical directions (Table 4, Fig. 18).

In the case of localities 7 and 8, the local escape was effected by release of the rock mass trapped in strike-slip fault constrictional bends and oversteps zones. The higher stress ratios (0.7–0.8) show that the stress configuration had developed fairly strong extension, whilst the

stress ratio of 0.3 indicates build-up of strong compression following rotation of the σ_1 axis (Table 4, Fig. 18). Locality 4 contained insufficient escape-related data to allow computation of the palaeostress configuration. However, the younger, escape-related pattern shows a sinistral rotation of the pre-existing stress field, with σ_1 orientated subhorizontally, probably caused by the sub-Mesozoic Variscan morphology.

Porosities resulting from the escape-related fracture patterns produced relatively well directed zones of permeability.

Fracturing and porosity development

Figure 19 shows values of fracture porosity associated with the rifting event, varying from 8.45% at St Donats (locality 7) to 20.3% at Kilve (locality 15). Fracture porosity associated with the inversion event has been subdivided into i) that associated with strike-slip faults, ii) that formed by the distinct set of extensional veins towards the end of the strike-slip event, and iii) that formed by the escape structures at the close of inersion. Figure 20 shows that the former, determined at St Donats and Nash Point (localities 8 & 9) averages 6.5%, whilst fracture porosity associated with the distinct extensional veins give values ranging from 0.4–2.6% with a mean of 0.8% (Fig. 21) Porosities resulting from the escape-related fracture patterns range from 1.68–4.83% (Fig. 20g–j), and produced relatively well directed zones of permeability. The variation in these values appears to depend on: i) lithology (Fig. 22); ii) bed thickness (Fig. 23); and iii) the association with a major fracture zone (Fig. 23). These relationships are critical in determining the effects on hydrocarbon migration, trap formation and trap drainage (see

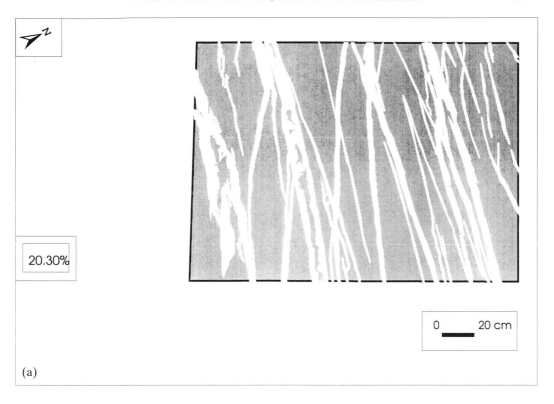

20.30%

0 _____ 20 cm

(a)

Fig. 19. Plans of extensional vein systems used to calculate fracture porosity developed during the rifting event; (**a**) calcite veins (unshaded) in Lower Liassic limestone, showing 20.3% porosity, Kilve, E of Watchet; (**b**) calcite veins in Lower Liassic limestone, showing 8.45% porosity, St Donats (locality 8).

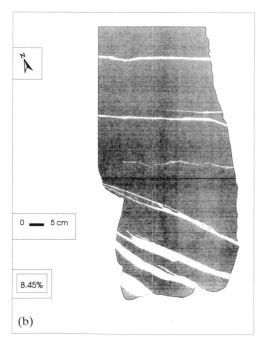

0 ▬ 5 cm

8.45%

(b)

below). The fracture porosities developed during both the strike-slip faulting and the distinct extensional event formed long lasting and very effective directed permeability zones (Figs 20 & 21).

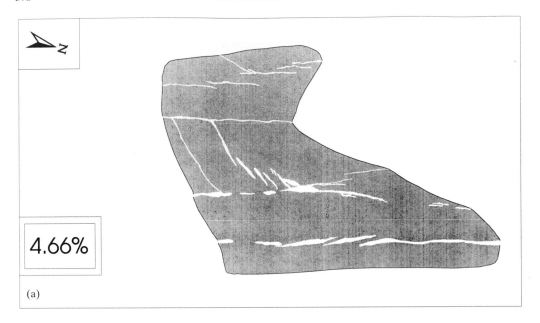

4.66%

(a)

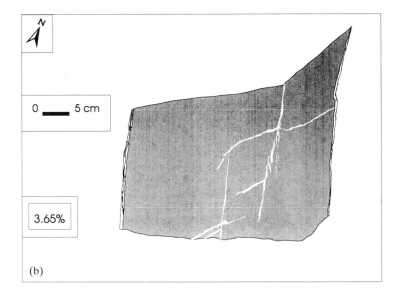

0 ▬▬ 5 cm

3.65%

(b)

Fig. 20. Plans of extensional vein systems used to calculate fracture porosity developed during the inversion-related strike-slip faulting event in Lower Liassic limestone; (a) & (b) Nash Point (locality 9), (c–j) St Donats (locality 8). Figures in boxes represent calculated percentage fracture porosity. (g–j) also show, with oblique ornament in both plans and boxes, porosity developed during the escape event at the end of inversion.

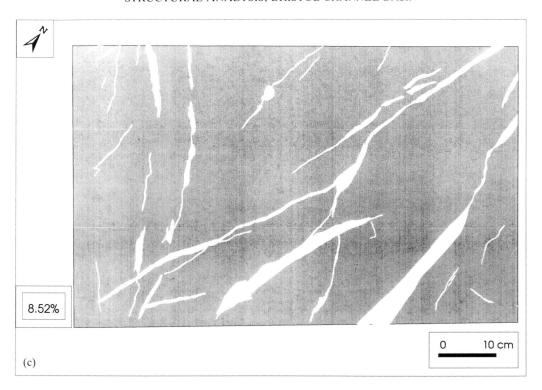

8.52%

0   10 cm

(c)

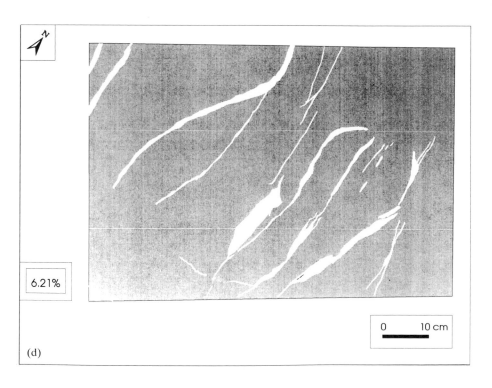

6.21%

0   10 cm

(d)

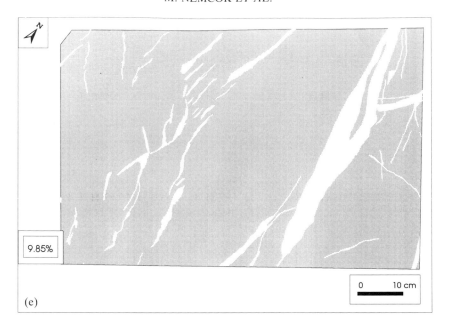

9.85%

0 10 cm

(e)

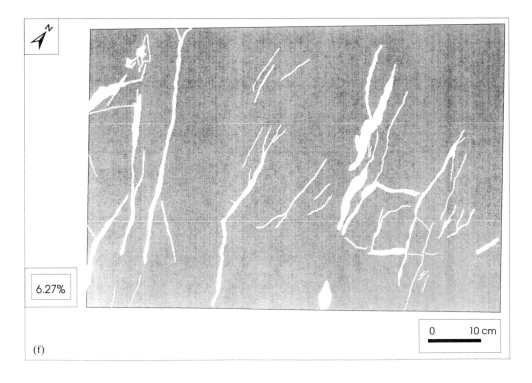

6.27%

0 10 cm

(f)

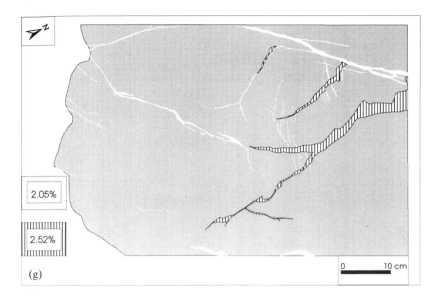

(g)

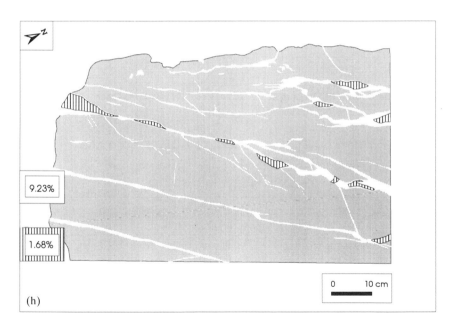

(h)

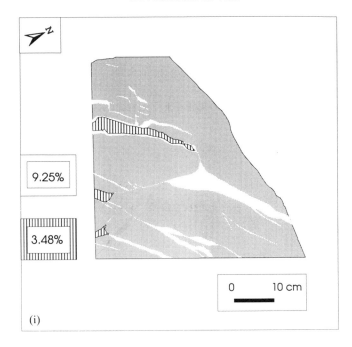

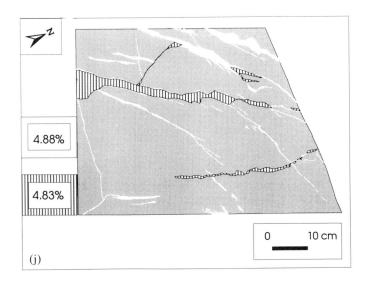

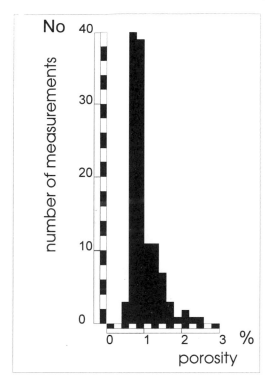

Fig. 21. Histograms showing the frequency distribution of fracture porosity developed during the extensional event at the end of inversion in the north Bristol Channel study area.

Discussion

Timing and global framework of events

Phase 1 rifting – Permo–Triassic The timing of the Permo–Triassic NW–SE rifting is based on our measurements of neptunian dykes at Black Rocks, Newton (locality 11). Similar data are known from the elongate N–S trending Permo–Triassic Worcester Basin (Chadwick & Smith 1988), where a N–S trending strike-slip fault zone was inverted by Permo–Triassic transtension. Kinematic analysis of the bounding normal faults indicates NW–SE extension. A broadly similar, NNW–SSE direction of extension is also recorded in the South Celtic Sea (Van Hoorn 1987), where pull-apart structures formed along the sinistrally reactivated Stickle-path strike-slip fault.

This early episode of rifting that disrupted Variscan structures caused basin formation at discrete depocentres; most of them of pull-apart origin (Karner *et al.* 1987; Lake & Karner 1987). This first post-Variscan phase took place after the Stephanian–Early Permian as a

wrench faulting event driven by dextral translations between Europe and Africa (Ziegler 1989).

Phase 2 rifting – Triassic–Aptian The youngest sediments exposed onshore in the study area are of Liassic age. The extent of the Triassic–Aptian rifting phase cannot therefore be wholly based on our data, but also by analogy with neighbouring basins. Two cases of syn-sedimentary growth normal faulting were observed in the Triassic Mercia Mudstone sequence at Watchet (locality 17), and syn-sedimentary Rhaetian–Liassic normal faulting was recorded at Penarth Head (locality 2). The syn-sedimentary nature of the latter is shown by two unconformities in the hangingwall of the normal fault that separate sequences in which bedding dips progressively less steeply upwards. The dip azimuth and dip values in the respective blocks are: 019/36, 034/18 and 298/8. The broad time span of the Triassic–Aptian rifting phase is indicated by the presence of Toarcian and Bajocian neptunian dykes in the Wessex Basin, and by Hettangian, Sinemurian, Plienbachian and Bajocian neptunian dykes in the Mendip region (Jenkins & Senior 1991). Synsedimentary normal faults in the Wessex Basin indicate active rifting during the Hettangian-Bajocian and Oxfordian-Kimmeridge time periods (Jenkins & Senior 1991). Stoneley & Selley (1986) described rifting in the Wessex Basin taking place from at least the Early Jurassic until the Neocomian, based on tilted fault-bounded sequences sealed by younger sediments. Other evidence suggesting Early Jurassic rifting, based on syn-depositional faulting, has been cited from the Mendip area by Cornford (1986). A roughly similar rifting history has been documented from the Celtic Sea region by Roberts (1989). Here, pre-Triassic initial rifting was followed by a Late Triassic rifting phase and ended in Late Jurassic–Early Cretaceous rifting (Roberts 1989). Another example has been described from the Northern Porcupine Seabight, W of Ireland in which Early Cretaceous strata are affected by rifting-related tilting and sealed by Late Cretaceous sediments (Roberts 1989). A similar situation occurs in the Wessex Basin (Stoneley & Selley, 1986). Normal faults, such as those at Kimmeridge Bay, indicate Jurassic–Early Cretaceous growth (Stoneley & Selley 1986). The direction of regional extension (σ_3), determined in this study as NNE–SSW, is similar to that of Ziegler (1988) affecting the Tethys during the same period.

This rifting phase was ultimately controlled by the NW–SE (337°–157°) sinistral relative motion of Africa and Europe during the Stephanian to

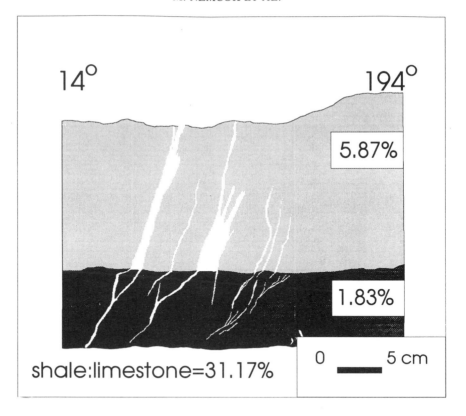

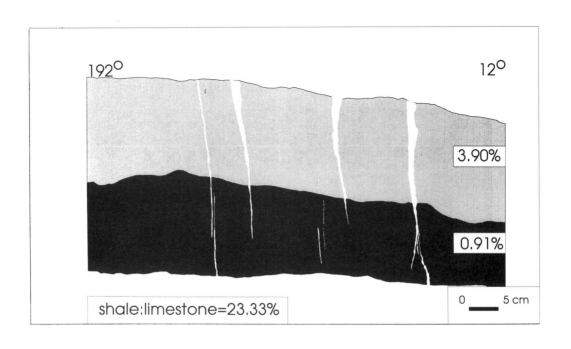

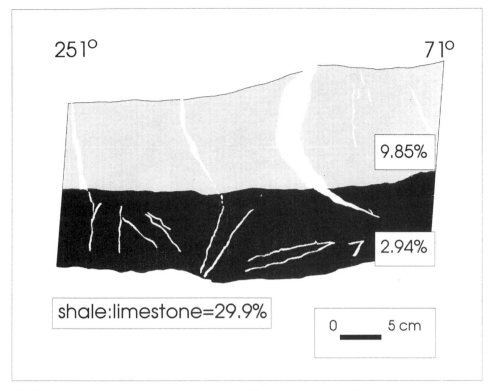

Fig. 22. Three examples of fracture porosity determined from extensional calcite veins formed during the inversion-related strike-faulting event developed across limestone shale contacts within the Lower Liassic sequence of the north Bristol Channel study area. The percentage porosity is shown in boxes against the relevant lithology and illustrates the effect of contrasting mechanical properties of shale and limestone. The percentage drop in porosity from limestone to shale is given beneath each diagram and averages 28%. Calcite veins – no ornament, limestone – pale grey, shale – dark grey.

Aptian, and Bathonian to Aptian (Lake & Karner 1987; Livermore & Smith 1985). Polyphase extension utilized available Variscan crustal detachments (Karner *et al.* 1987; Brooks *et al.* 1988; Roberts 1989; Lake and Karner 1987) which thus controlled depocentre distribution by their configuration (Karner *et al.* 1987; Stoneley & Selley 1986).

Late Cretaceous quiescence No faulting has been described from the Late Cretaceous time interval in the UK area and thus the Bristol Channel Basin is likely to have been quiescent at this time.

The Late Cretaceous is characterized by major tectono-eustatic sea level rise resulting from a global acceleration of sea-floor spreading and commensurate reduction of the ocean basin volume (Ziegler 1988). This caused a broad onlap of deposition across Early Cretaceous basin margins (Ziegler 1988). Much of the Early Cretaceous land areas became inundated, and

clastic influx into basins became drastically reduced, resulting in essentially clear water conditions and the deposition of the Chalk (Ziegler 1982). Based on regional depositional patterns during this time interval (e.g. Kamerling 1979; Cope 1984; Cornford 1986; Roberts 1989), it is likely that up to 200 m of Chalk were deposited in the Bristol Channel study area.

The ultimate control of the tectonics was the E–W (076°–256°) sinistral motion of Africa relative to Europe during the Aptian to Cenomanian (Lake & Karner 1987; Livermore & Smith 1985). This resulted from the Late Cretaceous opening of the South Atlantic–Indian Ocean (Dewey *et al.* 1973; Dercourt *et al.* 1986; Le Pichon *et al.* 1988) that induced a change in the drift pattern of Africa, which began to converge gradually with Laurasia (Ziegler 1988).

Inversion – Tertiary Finally, an extended period of inversion affected the northern Bristol

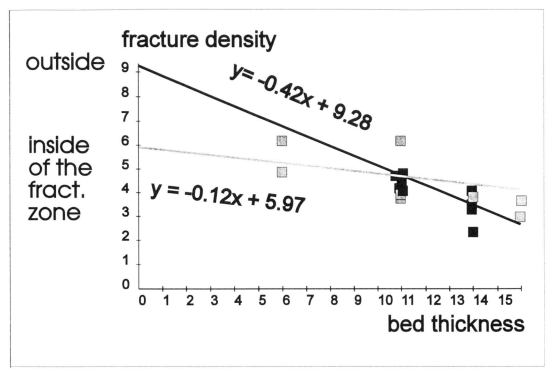

Fig. 23. Graph demonstrating an inverse relationship between fracture density and bed thickness for extensional veins developed during the extensional event at the end of the inversion-related strike-slip faulting. Separate relationships are shown for fractures developed inside and outside fracture zones. The equations represent linear regressions for the two data sets.

Channel region, with the development of a distinctive sequence of structural events. A pattern of dominant strike-slip faulting and subordinate thrusting is present throughout the studied localities, accompanied by a late, distinct extensional vein forming event, and followed by a locally constrained, escape-related system of structures. The age of these events is not well constrained as no rocks younger than Liassic are exposed onshore in the basin. However the Eocene–Oligocene sediments in the Stanley Bank Basin, a pull apart basin associated with sinistral slip along the major NW–SE Sticklepath–Lustleigh Fault imply N–S extension was operating during the early Tertiary (Holloway & Chadwick 1986; Arthur 1989). These authors suggested that dextral slip along the fault during inversion of the basin did not occur until Miocene times.

Simpson *et al.* (1989) related uplift along E–W striking, southward dipping, normal faults in the Wessex Basin to inversion during the Maastrichtian to Late Palaeocene. Stronger inversion in the centre of the Weald Basin was equated with a change of strike along the northern

controlling southward dipping fault from W–E to NW–SE, although no explanation was given. A possible cause, based on the NNE–SSW to NE–SW compression direction, calculated by us for the Bristol Channel Basin, is that the increase in vertical uplift results from the transition from an area of sinistral transpression along the fault, to an area of frontal compression. Other evidence for the timing of inversion is based on the absence of Danian sediments in the Celtic Sea indicating that inversion started in the Palaeocene, with a major inversion event during the Oligocene, comparable with that in the English Channel (Roberts 1989). Maastrichtian and Lower Palaeocene strata are missing in the Wessex Basin which is attributed to the initiation of inversion (Stoneley & Selley 1986). Structures in the Wessex Basin indicate northward folding against south-dipping, pre-existing normal faults (Stoneley & Selley 1986); a situation seen by us in the Bristol Channel Basin at localities 5, 6, 10, St Audrie's Bay, Kilve and Watchet. Some minor post-Cretaceous structures indicate N–S compression, i.e. sinistral NE–SW and normal N–S trending faults (Stoneley & Selley 1986). The Northern Porcupine Seabight, West of Ireland, has the Late Cretaceous and Early Tertiary strata folded and sealed by Miocene strata, with the youngest folded reflector being of Eocene age (Roberts 1989).

Lake & Karner (1987) gave evidence for a Late Cretaceous to Early Tertiary age of inversion in the Channel, Weald and Vale of Pewsey Basins with a culmination in the Oligocene–Miocene, when NW–SE dextral strike-slip faults developed local pull-apart basins (Lake & Karner 1987). WNW–ESE faults, such as the Pay de Bray fault, formed transpressional antiforms, and W–E faults, such as the south-dipping pre-existing Vale of Pewsey, Mere and Purbeck–Isle of Wight normal faults, were inverted.

Our calculated stress configurations can be directly compared with those determined by Bergerat (1987) from the European platform regions in front of the Alps, who showed that the stresses are transmitted over large distances without significant modification. Our results are roughly in accordance with Bergerat's late Eocene stress fields, characterized by approximately N–S compression, and Lower Miocene stress fields, characterized by approximately NE–SW compression. The lack of any preserved Tertiary lithostratigraphic units in the northern Bristol Channel area may have resulted in a grouping of events that elsewhere are separated. It may be that some of the escape structures in localities lacking either normal faults, causing a buttressing effect, or transpressional duplexes, causing a forced fault structure, may in reality belong to the different Tertiary events determined by Bergerat (1987). The slight misfit of orientations of 10–20° may be due to stress transfer over such a distance and so we have not attempted to distinguish the various Tertiary tectonic episodes that were recognised by Bergerat (1987) who had far greater control of dating events in the platform sequences of the alpine foreland.

Traps versus maturity timing

The only part of the lithostratigraphy investigated that is likely to be an oil source is the lower part of the Liassic argillaceous calcilutite/carbonaceous mudstone sequence, where the ratio of black organic-rich shale to limestone is very high (localities 2, 3, 5, 6, 7, 8, 9 and 10). Liassic shales are known to form hydrocarbon source rocks in surrounding basins, such as the Paris Basin (Matray et al. 1993), and the Wessex Basin (Stoneley & Selley 1986). The available data from the Wessex Basin indicate that the total organic carbon content (TOC) of these shales is up to 7.36% (Stoneley & Selley 1986). Type II kerogen indicates an autochthonous marine phytoplankton and zooplanktonic origin (Stoneley & Selley 1986).

Recent experiments on the maturation and expulsion of hydrocarbons from Liassic organic-rich shales (Leythaeuser 1993; Kroos & Hanebeck 1993; Hanebeck et al. 1993; Kroos et al. 1993) indicate that kerogen is load-bearing. Compaction of source rocks occurs, not only due to the loss of original porosity, but also by conversion of load-bearing kerogen into bitumen, which is extruded into the water-filled pore space. The product is oil-in-water emulsion which separates rapidly (Kroos & Hanebeck 1993). Short range transport (up to several metres) through the effectively impermeable (<1–2nd) kerogen network is very efficient, and achieved by diffusion (Hanebeck et al. 1993). Kerogen conversion thus caused either a fluid phase flow in the contemporaneously compacting Liassic sequence during the time when the system was open, or overpressuring when the associated compartment was hydraulically sealed.

The structural results combined with other published data (e.g. Kamerling 1979; Cornford 1986; Stoneley & Selley 1986) suggest that the Liassic sequence of the Bristol Channel Basin experienced maximum burial depth during Aptian time. Vitrinite reflectance data (Cornford 1986) from Lavernock Point, close to locality 2, give average values of $R_m = 0.51\%$ from both limestone and shale. Based on available models (e.g. Moore 1989), these rocks will have entered the oil generation window. However, as mentioned earlier, the period of Triassic to Aptian rifting was followed by uplift during the Aptian to Albian (e.g. Ziegler 1982, 1988); and Late Cretaceous deposition probably did not add any significant overburden in this area (e.g. Kamerling 1979; Cope 1984; Cornford 1986). Later inversion of the normal faults that initially controlled deposition, caused further uplift. It has been argued for the neighbouring basins that 0.5 to 1.5 km of uplift occurred in the various sub basins of the Wessex Basin (Simpson et al. 1989), and 0.35–2 km of uplift in different parts of the Celtic sea (Van Hoorn 1987; Roberts 1989).

England (1993) argued that the rate of secondary migration is rapid on a geological timescale and is usually limited by the rate of supply from the source rock. Petroleum expulsion efficiencies vary widely and evolve with progressive maturity (Leythaeuser 1993). Thus, the only traps likely to contain hydrocarbons in the Bristol Channel are those traps that were available at the end of the Triassic to Aptian rifting phase. This conclusion, apart from maturity considerations (e.g. Kamerling 1979; Cornford 1986; Stoneley & Selley 1986), is supported by the normal fault-related reservoir of the Wytch Farm oilfield in the Wessex Basin (Stoneley & Selley 1986).

The pre-Albian mechanisms related to maturity/migration/trap timing in the Bristol Channel Basin can be described as follows. Evidence for the primary origin of the varying proportions of shale-limestone alternations in the Liassic sequence, such as the burrowing relationships of the Dorset section, is largely lacking (Waters *et al.* 1987). Irregular bedding planes bounding limestone beds, nodular limestone horizons, anastomosing mudstone beds and pressure solution partings rather indicate a diagenetic redistribution. The calcilutites commonly contain bioclastic debris, suggesting, together with their other characteristics, a diagenetic history of freshly deposited Liassic shelf mudstones during the pre-Albian rifting phase. Recent mudstones have a porosity of about 70% (e.g. Chilingarian 1983; Robertson 1967). Based on knowledge of recent depositional environments (e.g. Moore 1989; Kukal 1990), Liassic mudstones will have undergone strong initial compaction and dewatering during their progressive burial. It is likely that a few hundred metres was sufficient for them to reach a stable grain framework, i.e. to gain elastic properties. This fact would mean that, during continuing rifting, these sediments were able to record and preserve the synsedimentary faulting recorded from most localities in this study. The diagenetic features of the Liassic shale/limestone interbeds suggest that lithification, progressively accelerated by increasing temperature, was initiated soon after deposition of the limestone layers. Bjorkum & Walderhaug (1990) have shown that calcite supersaturation, required for cementation, is first reached in fossil-rich layers, such as the Liassic calcilutites containing bioclastic debris. The early lithification of limestone results in calcite cementation of the pore spaces and the initiation of a hydraulic seal. The upper parts of the Liassic sequence, with very high limestone/shale ratios, acted as a very efficient hydraulic seal. Further burial induced overpressuring in the still highly porous shales, as indicated at St Audrie's Bay by the hydraulic injection of shale into the brittle fractured limestones. These overpressured shale compartments were laterally sealed by high-angle normal faults filled with precipitated calcite. Each episode of rifting caused a new

episode of fracturing along these normal fault zones, allowing fluids to migrate from the overpressured shale compartment upwards along the open fracture zones. Values of normal faulting-related porosities, as described earlier, were frequently very high (Fig. 19a,b) and indicate that effective directed permeability conduits occurred along the normal fault zones after each displacement event. Each episode of fluid flow caused a new event of mineral precipitation. As mentioned above, some of the extensional veins show evidence for multiple reopening and calcite precipitation. Kukal (1990) suggested that sealing by calcite precipitation takes place very rapidly. Thus, pore-fluid pressures in the shales will have increased from hydrostatic to lithostatic pressure between each episode of sealing, generating a new fracture event along the normal faults. If the regional stress configuration is considered to have been essentially stable, affected only by continuing deposition, each faulting increment will have been triggered by a progressive decrease in the normal stress acting on the fault plane, brought about by the increase in fluid pressure, until its further reactivation.

We presume that this stage of cyclic reactivation was fairly long-lived, because water will have been supplied not only from the pore space of Liassic sediments, but also from underlying gypsum horizons, and later also from hydrocarbon expulsion itself. Kroos *et al.* (1993) showed experimentally that large amounts of water, together with additional pore space, are generated during the early stages of oil production from type II kerogen. Another reason for thinking that this stage would have been protracted is that the overpressured zones will have isolated the pore fluids from surrounding diagenetic waters, preventing fluid and ion transfer, and thus slowing diagenesis (Moore 1989).

As burial progressed, the rock material properties of the shale/limestone sequence changed. Limestones, lithified early in the diagenesis and surrounded by incompletely compacted shales, became fractured to produce the extensional vein systems observed at most of the localities in this study. Fluid, expelled from the shales adjacent to such fractured limestone bodies, will have caused an abrupt decrease of porosity in these zones, because fractured limestones provide a better drainage of fluids to the fractured normal fault zones after each displacement event (Chilingarian 1983). Such a process of fluid expulsion can favour the possibility of hydrocarbon accumulation in these fractured zones (Chilingarian 1983), as also shown by the presence of only partially filled calcite veins. Thus, shales progressively started to gain their seal properties. It is likely that they had established their sealing characteristics by the end of the pre-Albian rifting phase. Published values of hydraulic parameters for this type of sediment (Thomas 1962) suggest a porosity of 0% and a permeability of less than 0.1 md.

Inversion Subsequent tectonic events acted to destroy the structures generating the pre-existing traps. They also provided effective mechanisms to drain previous hydrocarbon accumulations.

Subordinate thrusting and dominant strike-slip faulting gave rise to the following very efficient drainage mechanisms. Thrust patterns in recently active accretionary wedges (Westbrook *et al*. 1982; Moore *et al*. 1987; Lallemant *et al*. 1990), and experimental work on thrusting in mudstone (Brown *et al*. 1993), suggest that thrusts develop very highly permeable zones capable of controlling significant fluid flow. The open space is created by increased fluid pressure jacking-up the wall rocks of subhorizontal tension gashes during thrusting (e.g. Cosgrove 1993). Fracture patterns associated with strike-slip faulting, can have porosity greater than 6%, producing an effective drainage system. This porosity is both highly permeable, as in some commercial reservoirs (Moore 1989), and long lasting. The latter point is emphasized by the length of the inversion event, accommodated in the northern Bristol Channel Basin by strike-slip faulting, that lasted from the Cretaceous to Miocene. Despite the fact that the faults were quickly sealed by calcite precipitation after each displacement episode, the large number of movement increments observed at outcrop will have made this mechanism very effective.

However, this is not the only fluid-draining mechanism available. Faults are able actively to suck or expel fluids (e.g. Sibson 1989, 1993; Cosgrove 1993; Byerlee 1993). The valve action of a fault occurs when a rupture cuts a sealed overpressured compartment and destroys the hydraulic seal, leading to fluid discharge along the permeable fracture zone and local re-establishment of the normal hydrostatic fluid pressure (Byerlee 1993; Sibson 1993). Such a valve action is characteristic of thrusts and transpressional strike-slip fault zones (Sibson 1993), i.e. subordinate thrusts and frequent NW–SE striking dextral strike-slip faults, such as those described here from the northern Bristol Channel Basin inversion event.

While the valve action of the faults could be effective during the early stages of inversion through the operation of fluid pressure cycles, a more dominant and longer lasting effect would be produced by static stresses controlling an induced directional permeability. In the Bristol Channel Basin this directional permeability was produced by a distinct extensional event that followed the strike-slip episode, producing an escape-related fracture system.

Thus, in the case of the inverted Bristol Channel Basin, the probability of discovering a preserved hydrocarbon source is quite low, taking into consideration both that the source rocks have not passed through the complete oil generation process, and that reservoirs are only likely to be preserved as remnants of normal fault-related plays unaffected by either strike-slip faulting or its related thrusting. A strategy to discover such a play would require a well targeted and detailed structural study.

Conclusions

The Bristol Channel Basin experienced a prolonged rifting event from the Triassic through to the Aptian, although direct evidence from onshore exposures is restricted to the earliest stages of the sequence. NNE–SSW rifting generated dominantly normal faults and extensional veins in both partially and fully lithified sediments, with evidence for multiple reactivation. Stress configurations show the effect of increasing burial on the regional stress field. Cyclically fluctuating fluid pressures affected local stresses.

The basin was inverted during the Tertiary by a NE–SW oriented σ_1. Evidence from the neighbouring Wessex basin suggests inversion from the end Cretaceous through to Late Oligocene and Early Miocene. Inversion occurred in fully lithified rocks. The earliest stages resulted in pronounced strike-slip faulting with subordinate thrusting and buttressing against normal faults. Stress configurations are prolate. During later stages of inversion extensional veining became significant, due to high differential stress. Buttressed structures induced lateral escape into less restrained regions under locally variable stress conditions.

Fracture porosity determined for each of the tectonic events has a high directional permeability. Porosity is dependent on lithology, bed thickness and relationship to fault zones. The highest values of fracture porosity, up to 20%, were developed during the rifting event, and the fractures allowed rapid polyphase fluid discharge from over-pressured compartments. Fracture porosity developed during inversion averaged 6.5% and formed a long-lasting fluid drainage system.

The oil generation window was reached at the end of the rifting event, and traps associated with fracture porosity that resulted from the interaction of normal faulting and differential lithification were fully formed by this time. Oil generation was stopped by the end Cretaceous due to regional uplift. The inversion event allowed the traps to drain.

The only traps likely to have survived the inversion drainage event would be those associated with steeply dipping normal faults that had been bypassed by inversion structures.

This research was undertaken whilst MN was a Royal Society Research Fellow at University of Wales Cardiff. MM was supported by a Research Studentship from Shell Expro who also supplied seismic reflection profiles of the Bristol Channel and adjoining areas of the Vale of Glamorgan. Funds for fieldwork were made available by the University of Wales Cardiff. RAG wishes to thank John Underhill for help in earlier fieldwork and valuable discussion. MN is grateful to Don Secor for showing him the world of hydrofracturing. Richard Lisle, Sara Vandycke and Mike Brooks contributed helpful advice and discussion. The paper was much improved by constructive reviews by Chris Dart and Paul Hancock.

References

ANGELIER, J. 1985. Extension and rifting: Zeit region, Gulf of Suez. *Journal of Structural Geology*, **7**, 605–612.

—— 1989. From orientation to magnitudes in paleostress determinations using fault slip data. *Journal of Structural Geology*, **11**, 37–50.

—— 1994. Fault Slip Analysis and Palaeostress Reconstruction. *In*: HANCOCK, P. L. (ed.) *Continental Deformation*. Pergamon Press, Oxford, 53–100.

ARTHUR, M. J. 1989. The Cenozoic evolution of the Lundy Pull-Apart Basin into the Lundy Rhomb Horst. *Geological Magazine*, **126**, 187–198.

BAYERLY, M. & BROOKS, M. 1980. A seismic study of deep structure in South Wales using quarry blasts. *Geophysical Journal of the Royal Astronomical Society*, **60**, 1–19.

BERGERAT, F. 1987. Stress fields in the European platform at the time of Africa–Eurasia collision. *Tectonics*, **6**, 99–132.

BJORKUM, P. A. & WALDERHAUG, O. 1990. Geometrical arrangement of calcite cementation within shallow marine sandstones. *Earth Science Reviews*, **29**, 145–161.

BROOKS, M., BAYERLY, M. & LLEWELLYN, D. J. 1977. A new geological model to explain the gravity gradient across Exmoor, north Devon. *Journal of the Geological Society, London*, **133**, 385–393.

——, HILLIER, B. V. & MILIORIZOS, M. 1993. New seismic evidence for a major geological boundary at shallow depth under north Devon. *Journal of the Geological Society, London*, **150**, 131–135.

——, TRAYNER, P. M. & TRIMBLE, T. J. 1988. Mesozoic reactivation of Variscan thrusting in the Bristol Channel area, U.K. *Journal of the Geological Society, London*, **145**, 439–444.

BROWN, K. M., DEWHURST, D. N., WESTBROOK, G. K. & CLENNELL, M. B. 1993. Experimental determination of anisotropic hydraulic conductivity in artificially reconstituted mudstone: comparison of shear zones and wall rocks. *In*: PARNELL, J., RUFFELL, A. H. & MOLES, N. R. (eds) *Proceedings of the Geofluids '93 conference*, Torquay, 169–171.

BYERLEE, J. 1993. Model for episodic flow of high-pressure water in fault zones before earthquakes. *Geology*, **21**, 303–306.

CHADWICK, R. A. 1986. Extension tectonics in the Wessex Basin, southern England. *Journal of the Geological Society, London*, **143**, 465–488.

—— & SMITH, N. J. P. 1988. Evidence of negative structural inversion beneath central England from new seismic reflection data. *Journal of the Geological Society, London*, **145**, 519–522.

CHILINGARIAN, G. V. 1983. Compactional Diagenesis. *In*: PARKER, A. & SELLWOOD, B. W. (eds) *Sediment Diagenesis*. Reidel Publishing Company, 57–168.

COPE, J. C. W. 1984. The Mesozoic history of Wales. *Proceedings of the Geologists Association*, **95**, 373–385.

CORNFORD, C. 1986. The Bristol Channel Graben: organic geochemical limits on subsidense and speculation on the origin of inversion. *Proceedings of the Ussher Society*, **6**, 360–367.

COSGROVE, J. W. 1993. The interplay between fluids, folds and thrusts during the deformation of a sedimentary succession. *Journal of Structural Geology*, **15**, 491–500.

DEARMAN, W. R. 1963. Wrench-faulting in Cornwall and South Devon. *Proceedings of the Geologists' Association, London*, **74**, 265–287.

DERCOURT, J. *et al.* 1986. Geological evolution of the Tethys belt from Atlantic to Pamir since Liassic. *Tectonophysics*, **123**, 241–315.

DEWEY, J. F. PITMAN III, W. C., RYAN, W. B. F. & BONNIN, J. 1973. Plate tectonics and the evolution of the Alpine System. *Geological Society of America Bulletin*, **84**, 3137–3180.

DURNEY, D. W. & RAMSAY, J. G. 1973. Incremental strains measured by syntectonic crystal growth. *In*: DE JONG, K. & SCHOLTEN, R. (eds) *Gravity and Tectonics*. John Wiley, New York, 67–96.

ENGLAND, W. A. 1993. Petroleum Migration. *In*: PARNELL, J., RUFFELL, A. H. & MOLES, N. R. (eds) *Proceedings of the Geofluids '93 conference*, Torquay, 54–55.

GAYER, R. A. & NEMCOK, M. 1994. Transpressionally driven rotation in the external Variscides of south-west Britain. *Proceedings of the Ussher Society*, **8**, 224–227.

GLENNIE, K. W. 1990. Outline of North Sea History

and Structural Framework. *In*: GLENNIE, K. W. (ed.) *Introduction to the Petroleum Geology of the North Sea*. Third Edition, Blackwell Scientific Publications, Oxford, 34–77.

HANCOCK, P. L. 1985. Brittle microtectonics: principles and practice. *Journal of Structural Geology*, **7**, 437–457.

HANEBECK, D., KROOS, B. M. & LEYTHAEUSER, D. 1993. Experimental investigation of petroleum generation and migration under elevated pressure and temperature conditions. *In*: PARNELL, J., RUFFELL, A. H. & MOLES, N. R. (eds) *Proceedings of the Geofluids '93 conference*, Torquay, 50–53.

HARDCASTLE, K. C. 1989. Possible paleostress tensor configurations derived from fault-slip data in Eastern Vermont and Western New Hampshire. *Tectonics*, **8**, 265–284.

—— & HILLS, L. S. 1991. BRUTE3 and SELECT: Quickbasic 4 programs for determination of stress tensor configurations and separation of heterogeneous populations of fault-slip data. *Computing Geoscience*, **17**, 23–43.

HOLLOWAY, S. & CHADWICK, R. A. 1986. The Sticklepath–Lustleigh fault zone: Tertiary sinistral reactivation of a Variscan dextral strike-slip fault. *Journal of the Geological Society, London*, **143**, 447–452.

IVIMEY-COOK, H. C. 1974. The Permian and Triassic deposits of Wales. *In*: OWEN, T. R. (ed.) *The Upper Palaeozoic and post-Palaeozoic rocks of Wales*. University of Wales Press, Cardiff, 295–321.

JENKINS, H. C. & SENIOR, J. R. 1991. Geological evidence for Intra-Jurassic faulting in the Wessex Basin and its margins. *Journal of the Geological Society, London*, **148**, 245–260.

KAMERLING, P. 1979. The geology and hydrocarbon habitat of the Bristol Channel Basin. *Journal of Petroleum Geology*, **2**, 75–93.

KARNER, G. D., LAKE, S. D. & DEWEY, J. F. 1987. The thermal and mechanical development of the Wessex Basin, southern England. *In*: COWARD, M. P., DEWEY, J. F. & HANCOCK, P. L. (eds) *Continental Extensional Tectonics*. Geological Society, London, Special Publications, **28**, 517–536.

KIRSCHNER, D. L., SHARP, Z. D. & TEYSSIER, C. 1993. Vein growth mechanisms and fluid sources revealed by oxygen isotope laser microprobe. *Geology*, **21**, 85–88.

KROOS, B. M. & HANEBECK, D. 1993. Investigation of compositional variations during petroleum generation and expulsion by open system hydrous pyrolysis. *In*: PARNELL, J., RUFFELL, A. H. & MOLES, N. R. (eds) *Proceedings of the Geofluids '93 conference*, Torquay, 46–49.

——, LEYTHAEUSER, D. & HANEBECK, D. 1993. Volume balance of a maturity sequence of the Lower Jurassic Toarcian (Lias E) source rock reflects progress of petroleum generation/expulsion. *In*: PARNELL, J., RUFFELL, A. H. & MOLES, N. R. (eds) *Proceedings of the Geofluids '93 conference*, Torquay, 56–59.

KUKAL, Z. 1990. The rate of geological processes. *Earth-Science Reviews*, **28**, 1–284.

LAKE, S. D. & KARNER, G. D. 1987. The structure and evolution of the Wessex Basin, southern England: an example of inversion tectonics. *Tectonophysics*, **137**, 347–378.

LALLEMANT, S. J. C., HENRY, P., LE PICHON, X. & FOUCHER, J. P. 1990. Detailed structure and possible fluid paths at the toe of the Barbados accretionary wedge (ODP Leg 110 area). *Geology*, **18**, 854–857.

LE PICHON, X., BERGERAT, F. & ROULET, M. J. 1988. Plate kinematics and tectonics leading to the Alpine belt formation; a new analysis. *Geological Society of America, Special Paper*, **218**, 111–131.

LEE, C. W. 1991. Baryte and calcite cements in the 'breccias' of Ogmore-by-Sea, South Wales. *Geology Today*, **7**, 133–136.

LLEWELLYN, D. J. 1981. *Geophysical investigations of the deep structure of the Bristol Channel and South Wales*. PhD thesis, University of Wales.

LEYTHAEUSER, D. 1993. Petroleum generation and expulsion. *In*: PARNELL, J., RUFFELL, A. H. & MOLES, N. R. (eds) *Proceedings of the Geofluids '93 conference*, Torquay, 42–45.

LIVERMORE, R. A. & SMITH, A. G. 1985. Some boundary conditions for the evolution of the Mediterranean region. *In*: STANLEY, D. J. & WEZEL, F. C. (eds) *Geological Evolution of the Mediterranean Basin*. Springer Verlag, Berlin, 83–93.

MATRAY, J. M., FOUILLAC, C. & WORDEN, R. H. 1993. Thermodynamic control on the chemical composition of fluids from the Keuper aquifer of the Paris Basin. *In*: PARNELL, J., RUFFELL, A. H. & MOLES, N. R. (eds) *Proceedings of the Geofluids '93 conference*, Torquay, 12–16.

MEANS, W. D. 1987. A newly recognized type of slickenside striation. *Journal of Structural Geology*, **9**, 585–590.

MECHIE, J. & BROOKS, M. 1984. A seismic study of deep geological structure in the Bristol Channel area. *Geophysical Journal of the Royal Astronomical Society*, **87**, 661–689.

MILIORIZOS, M. 1992. *The tectonic evolution of the Bristol Channel borderlands*. PhD thesis, University of Wales.

MOORE, C. H. 1989. Carbonate Diagenesis and Porosity. *Developments in Sedimentology*, **46**, 1–338.

MOORE, J. C. & ODP Leg 110 Scientific party (22 others), 1987. Expulsion of fluids from depth along a subduction-zone decollement horizon. *Nature*, **326**, 785–788.

PETIT, J-P. 1987. Criteria for the sense of movement on fault surfaces in brittle rocks. *Journal of Structural Geology*, **9**, 597–608.

POLLARD, D. D. & AYDIN, A. 1988. Progress in understanding jointing over the last century. *Geological Society of America Bulletin*, **100**, 1181–1204.

RAMSAY, J. G. 1967. *Folding and fracturing of rocks*. McGraw-Hill, New York, 1–568.

—— 1980. The crack-seal mechanism of rock deformation. *Nature*, **284**, 135–139.

—— & HUBER, M. I. 1983. *The Techniques of Modern*

Structural Geology. Vol. 1: Strain Analysis. Academic Press, London, 1–307.

ROBERTS, D. G. 1989. Basin inversion in and around the British Isles. *In*: COOPER, M. A. & WILLIAMS, G. D. (eds) *Inversion Tectonics.* Geological Society, London, Special Publication, **44**, 123–129.

ROBERTSON, E. C. 1967. Laboratory consolidation of carbonate sediment. *In*: RICHARDS, A. F. (ed.) *Marine Geotechnique.* University of Illinois Press, Urbana, 118–127.

SIBSON, R. H. 1989. Earthquake faulting as a structural process. *Journal of Structural Geology,* **11**, 1–14.

—— 1993. Crustal stress, faulting, and fluid flow. *In*: PARNELL, J., RUFFELL, A. H. & MOLES, N. R. (eds) *Proceedings of the Geofluids '93 conference,* Torquay, 137–140.

SIMPSON, I. R., GRAVESTOCK, M., HAM, D., LEACH, H. & THOMPSON, S. D. 1989. Notes and cross-sections illustrating inversion tectonics in the Wessex Basin. *In*: COOPER, M. A. & WILLIAMS, G. D. (eds) *Inversion Tectonics.* Geological Society, London, Special Publication, **44**, 123–129.

STONELEY, R. & SELLEY, R. C. 1986. *A field guide to the petroleum geology of the Wessex Basin.* Imperial College of Science and Technology, London, 1–44.

THOMAS, A. L. & POLLARD, D. D. 1993. The geometry of en echelon fractures in rock: implications from laboratory and numerical experiments. *Journal of Structural Geology,* **15**, 323–334.

THOMAS, G. E. 1962. Grouping of carbonate rocks into textural and porosity units for mapping purposes. *In*: HAM, W. E. (ed.) *Classification of Carbonate Rocks.* American Association of Petroleum Geologists, Memoirs, **1**, 193–223.

VAN HOORN, B. 1987. The South Celtic Sea/Bristol Channel Basin: origin, deformation and inversion history. *Tectonophysics,* **137**, 309–334.

WATERS, R. A., LAWRENCE, D. J. D., IVIMEY-COOK, H. C., MITCHELL, M., WARRINGTON, G., WHITE, D. E. & LEWIS, M. A. 1987. *Geology of the South Wales Coalfield, Part II, the country around Cardiff.* Memoirs of the Geological Survey, Great Britain, HMSO, London, 1–114.

WESTBROOK, G. K., SMITH, M. J., PEACOCK, J. H. & POULTER, M. J. 1982. Extensive underthrusting of undeformed sediment beneath the accretionary complex of the Lesser Antilles subduction zone. *Nature,* **300**, 625–628.

ZIEGLER, P. A. 1982. *Geological atlas of Western and Central Europe.* Shell International Petroleum Mij BV, The Hague, Elsevier, Amsterdam, 1–133.

—— 1988. *Evolution of the Arctic-North Atlantic and the Western Tethys.* American Association of Petroleum Geologists, Memoirs, **43**, 1–198.

—— 1989. Geodynamic model for Alpine intra-plate compressional deformation in Western and Central Europe. *In*: COOPER, M. A. & WILLIAMS, G. D. (eds) *Inversion tectonics.* Geological Society, London, Special Publication, **44**, 63–85.

3D analysis of inverted extensional fault systems, southern Bristol Channel basin, UK

CHRIS J. DART[1], KEN McCLAY & PETER N. HOLLINGS

Fault Dynamics Project, Geology Department, Royal Holloway, University of London, Egham, Surrey, TW20 0EX, UK

[1]*Present address: Z & S Geologi a/s, Sverdrupsgate 23, N-4007, Stavanger, Norway*

Abstract: East–West trending inverted extensional fault systems offset Triassic to Lower Jurassic strata close to the southern margin of the Bristol Channel basin along the north Somerset coast. Field mapping, using exceptionally detailed aerial photographs, has revealed a three phase tectonic evolution. (i) North South orientated stretching, resulting in a well developed extensional fault system. The faults are segmented, linked by relay ramps and horsetail toward their tips. (ii) North–South oriented compression, resulting in partial inversion of the extensional fault system, with the development of hangingwall buttress anticlines and zones of intense folding. (iii) North–South orientated compression, resulting in NW–SE trending dextral and NE–SW trending sinistral strike-slip faults. Comparison of hangingwall buttress anticlines exposed in North Somerset with similar larger scale structures observed on seismic profiles from the Bristol Channel shows that both are directly analogous. Correlation with regional data on the tectonic evolution southwest England, and the Bristol Channel, indicates that extension occurred during the Jurassic–Lower Cretaceous, and contractional inversion and strike-slip deformation during the Tertiary.

Exceptional coastal exposures of inverted extensional fault systems crop out along the southern margin of the E–W trending Mesozoic to Tertiary Bristol Channel basin, UK (Lloyd *et al.* 1973; Kamerling 1979; Van Hoorn 1987) (Fig. 1). Cliffs and wave cut platforms within the Triassic/Lower Jurassic dominate several kilometres of the North Somerset coast, from east of Kilve to west of Watchet (Fig. 1). Cliffs are up to 50 m high and the unusually high tidal range in the area (up to 12 m) provides a wave cut platform width of up to 400 m at low tide and spectacular exposures of inverted extensional fault systems (Fig. 2). In this analysis, field mapping was undertaken using a detailed set of aerial photographs to construct maps and cross sections, in order to determine accurately the 3D geometry of inverted extensional fault systems along 10 km of foreshore on the north Somerset coast at Kilve and Watchet (Fig. 1).

Exposed faults have net displacements that range from the microscopic to up to 200 m, and those with the largest displacements would be clearly resolved on seismic sections. However, most of the faults have net displacements of <20 m and would not be resolved using conventional seismic reflection techniques, but do provide appropriate analogues for faults and fault related deformation at the scale of an individual hydrocarbon field. Many fault system geometries are similar regardless of scale over

several orders of magnitude. In this study outcrop scale structures are compared with larger scale structural geometries observed on seismic sections from the offshore Bristol Channel basin.

The Bristol Channel basin sits on a basement of Carboniferous limestones and Devonian sandstones and slates that were deformed by N–S oriented compression during the Variscan orogeny. The surface trace of the Variscan Frontal Thrust lies to the north of the basin (Chadwick 1986; Lake & Karner 1987) (Fig. 1), and this and other related thrusts give the basement a distinct E–W trending structural grain (Brooks *et al.* 1977; Chadwick *et al.* 1983; Chadwick 1986; Donato 1988; Brooks *et al.* 1988). Important phases of extension occurred during the Permian (Anderton *et al.* 1979), Lower Jurassic and Upper Jurassic to Lower Cretaceous (Kamerling 1979; Chadwick 1986; Lake & Karner 1987; Karner *et al.* 1987). Uplift and erosion followed in the Lower Cretaceous with the development of the 'late Cimmerian Unconformity' (Chadwick 1986; Ruffell 1992). Inversion and erosion occurred during the Tertiary (Stoneley 1982).

Along strike, the Bristol Channel basin is divided by major NW–SE trending strike-slip faults, such as the Sticklepath Fault (Holloway & Chadwick 1986) (Fig. 1). A similar structure, the NW–SE trending Watchet–Cothelstone

From BUCHANAN, J. G. & BUCHANAN, P. G. (eds), 1995, *Basin Inversion*, Geological Society Special Publication No. 88, 393–413.

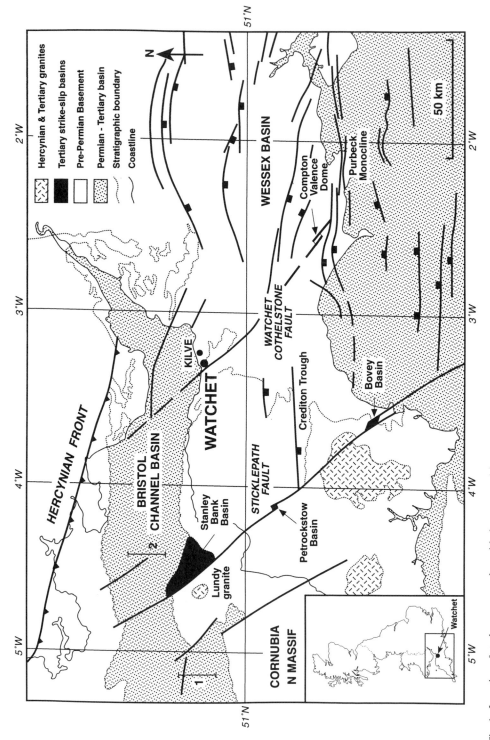

Fig. 1. Location of study areas together with the structural framework of southwest England. Compiled from Van Hoorn (1987) and Lake & Karner (1987). 1 & 2, location of seismic sections illustrated in Fig. 10.

Fig. 2. Photograph looking WSW along the foreshore at Kilve at low tide. The well bedded Lower Lias limestones and shales are cut by three east–west striking down to the north extensional faults. Photograph taken from the cliff top on cross-section line D – D' on Fig. 4.

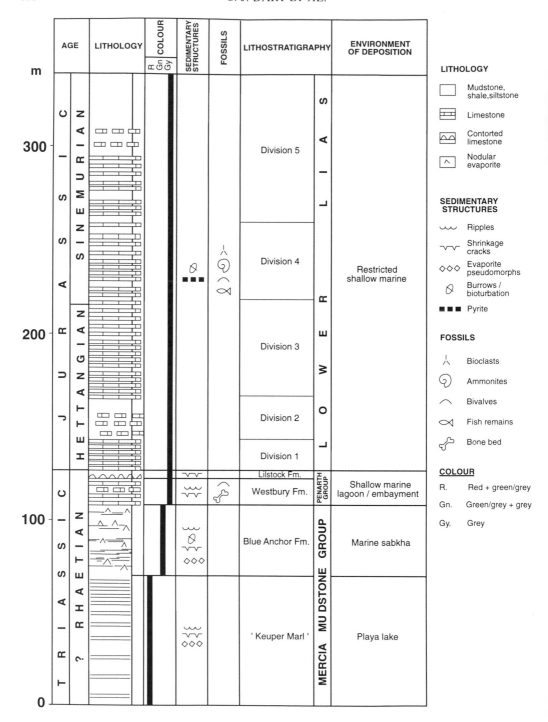

Fig. 3. Foreshore stratigraphy exposed in the Watchet and Kilve study areas. Compiled from Anderton *et al.* (1979); Whittaker & Green (1983). Thicknesses of individual limestone beds in the Westbury Formation through to the Lower Lias are schematic, the lithology column only illustrates the relative proportions of limestones and shale/mudstones.

Fault (Whittaker 1972) cuts the Watchet study area. These strike-slip faults were active during the Tertiary, when pull-apart basins were developed along the Sticklepath fault, and may have a history that extends back to the Variscan orogeny.

Triassic/Lower Jurassic stratigraphy of the North Somerset Coast

The coastal exposures of inverted extensional fault systems of North Somerset deform the Triassic to Lower Jurassic stratigraphy summarized in Figure 3. The following summary is based on Whittaker & Green (1983), Anderton et al. (1979) and the authors' own observations.

The oldest strata exposed in foreshore outcrop are red mudstones and siltstones of the 'Keuper Marl' (up to 85 m exposed). These contain nodular beds and vein networks of gypsum, and 1–2 m thick levels of intra-formational conglomerate and gypsum cemented sandstones. Toward the top the red lithologies become inter-bedded with reduced grey/green beds. A playa lake environment of deposition for these units has been suggested by Tucker (1977).

The top of the 'Keuper' is defined by the topmost red coloured bed. This is overlain by alternating green/grey and dark grey mud/siltstones of the Blue Anchor Formation (Tea Green and Grey Marls) (26–31 m), which are also rich in nodular and vein network gypsum. Marine fossils, desiccation cracks and burrows occur, indicating deposition of these units in a marine sabkha.

The Blue Anchor Formation is conformably overlain by inter-bedded light grey limestones, sandstones and dark grey shales of the Westbury Formation (6–7 m), the boundary being marked by the first limestone bed. Current ripples and shrinkage cracks are common, as are mono-specific bivalve assemblages and concentrations of bioclastic debris. The Westbury Formation also contains the Rhaetian bone beds (Whittaker & Green 1983). Further up section are inter-bedded grey mudstones, siltstones and limestones of the Lilstock Formation (Cotham and Langport Members) (4 m). U-shaped burrows occur. Deposition of these units is interpreted to have occurred in a shallow marine lagoon or embayment and the bone beds may represent littoral deposits.

The Triassic/Jurassic boundary occurs within the first few metres of the overlying inter-bedded grey mudstones, dark grey/black bituminous shales and light grey limestones of the Lower Lias (up to 203 m exposed). Following earlier work by Whittaker & Green (1983), five lithostratigraphic divisions, with varying mudstone/shale to limestone ratios, have been mapped in this study. There is no evidence of shallow water sedimentary structures and the Lower Lias is bioturbated, contains ammonites, thin shelled bivalves, fish remains and bioclastic debris. The bituminous shales are potential hydrocarbon source rocks (TOC up to 18%, average 2%; Cornford 1990). The general depositional environment for these units is interpreted as a restricted marine sea, subject to climatically controlled cyclical clastic input. Black shale deposition can be attributed to planktonic blooms, leading to the creation of anoxic bottom waters.

Geometries of inverted extensional fault systems

Maps and cross sections presented in Figs 4 and 5 are extremely well constrained by foreshore exposure which commonly approaches 100% (Figs 2 & 6). The kinematic evolution of individual faults is determined by combining information such as stratigraphic offset, drag folding, growth of fibrous vein fills, the orientation of slickenfibres and meso-scale shear bands. Contoured stereographic displays of poles to fault sets are illustrated in Fig. 7a, b & c.

Kilve

The faults systems in the Kilve area are predominately exposed in Lower Lias strata. They lie within the hangingwall of a major E–W trending, north-dipping fault which crosses the foreshore to the west of Blue Ben (Fig. 4a), and preserves greater than 200 m of net extension. The system is composed of an array of E–W-trending north-dipping synthetic, and south-dipping antithetic faults with dips that generally range between 40° and 70° (Fig. 7a). In detail most faults with significant displacements are formed by a zone of closely spaced offsets which are usually less than a metre wide. Along strike, faults are segmented. Individual segments are up to a few hundred metres in length. Displacement on each segment is greatest at its centre and dies towards its tip line both horizontally and vertically, where the fault trace may split into divergent strands or horsetails. Individual segments interact along strike and may link, either directly, or via relay ramps (see Peacock & Sanderson 1991), to form fault traces that are up to several kilometres in length. Away from

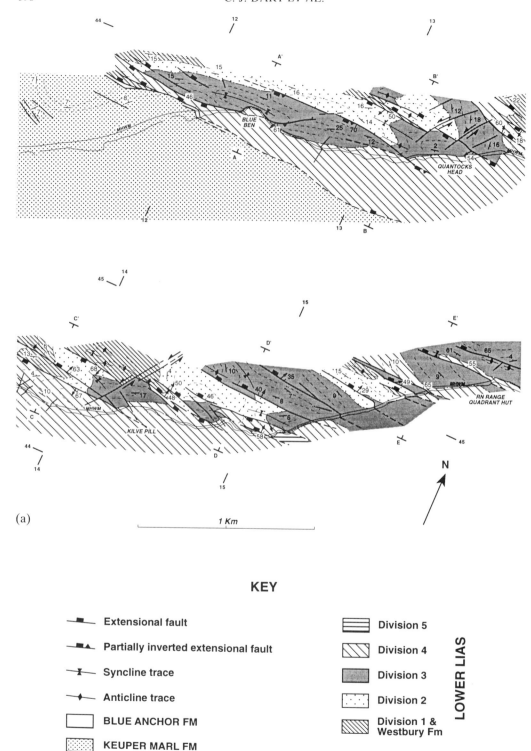

KEY

▬▬ Extensional fault	▤ Division 5	
▬▲ Partially inverted extensional fault	▨ Division 4	
✕ Syncline trace	▦ Division 3	**LOWER LIAS**
✦ Anticline trace	⫶ Division 2	
☐ BLUE ANCHOR FM	▧ Division 1 & Westbury Fm	
⫶ KEUPER MARL FM		

Fig. 4. (**a**) Maps of foreshore at Kilve illustrating partially inverted extensional, strike-slip fault systems, (**b**) cross-sections. The maps are constructed from aerial photographs and have not been undistorted.

Key

Partially inverted
extensional fault

Extensional fault

200 m

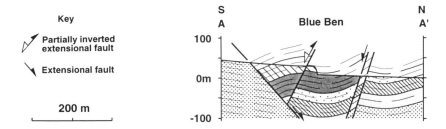

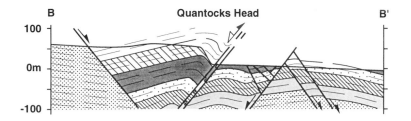

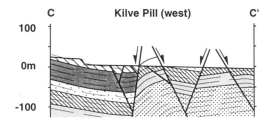

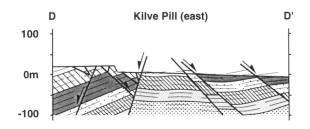

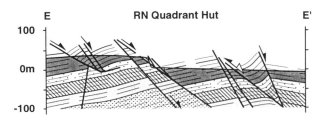

(b)

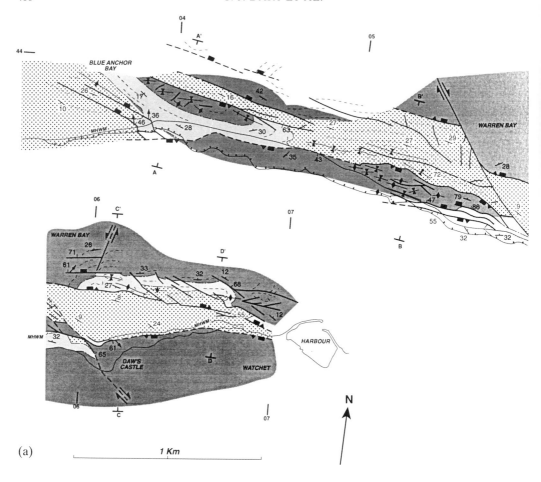

KEY

WESTBURY FM - LOWER LIAS

BLUE ANCHOR FM

KEUPER MARL FM

 Extensional fault

Partially inverted extensional fault

Syncline trace

Anticline trace

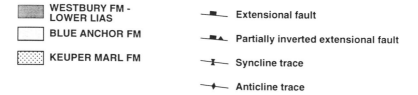

Fig. 5. (**a**) Maps of foreshore at Watchet illustrating partially inverted extensional, strike-slip fault systems. (**b**) cross-sections. The maps are constructed from aerial photographs and have not been undistorted.

areas of linkage, major faults have smooth plan and profile views at the scale of the illustrated maps and sections. Some of the major faults exhibit listric profiles and associated roll-over anticlines in cliff section (RN Range Quadrant Hut section, Fig. 4b) and shallow into shale-rich levels where early steep faults are offset by bedding parallel slip.

In the Kilve area almost all the faults preserve net extensional displacements, but some exhibit

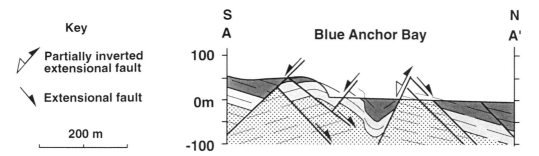

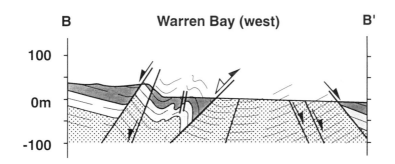

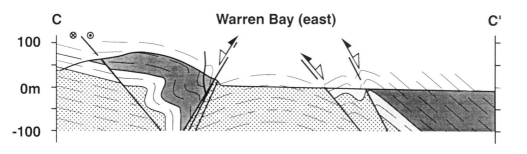

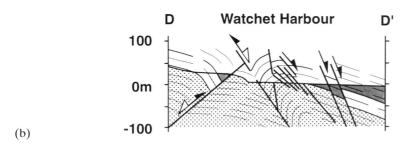

(b)

Fig. 6. Photographs of inversion related structures. (**A**) Reverse fault, Kilve; (**B**) hangingwall buttress anticline, Kilve. Cliff exposure is approximately 30 m high; (**C**) intense hangingwall buttress folding, Watchet.

evidence of contractional inversion (Fig. 6a). The fault that cuts through the headland known as Blue Ben in the Kilve area (Fig. 4a) has a region of net contractional displacement (up to 15 m), but this passes along strike to net extension (Fig. 4b). The fault just to the west of Quantocks Head (Fig. 4a) has a hangingwall buttress anticline that verges toward the main fault (Fig. 6b) and is associated with hangingwall vergent back thrusts. Within some fault zones there are strands with contractional offset and hangingwall and footwall contractional drag folds adjacent to faults are relatively common. Locally intense hangingwall buttress folding creates steeply dipping beds which extend up to a few metres from the fault. Fold axes are parallel to the main fault, but steep plunges give the folds complex patterns in plan. On a broader scale the entire foreshore exposure is deformed into gentle elongate periclines and synclinal basins with shallow dipping limbs and plunges of less than 5°. Bedding dips to the south are generally more common and steeper, possibly reflecting an initial rotation toward the major Lower Lias bounding fault during extension.

A NE–SW trending set of steeply–dipping sinistral strike-slip faults cross cut and offset steeply-dipping inverted extensional faults by up to a few metres (Figs 4a & 7b). Calcite slickenfibres preserved on their surfaces exhibit strike-slip/oblique-slip orientations.

Watchet

The fault systems exposed at Watchet crop out at a lower stratigraphic level than those at Kilve, and offset the Keuper Marl through to the basal portion of the Lower Lias (Fig. 5a). In places the highly fractured and jointed Keuper Marl and Blue Anchor Formation are shot through with veins of remobilized gypsum. The zones of gypsum veining are abruptly bounded by certain stratigraphic horizons and faults. These probably acted as permeability barriers, partitioning regions of elevated pore fluid pressures during burial in which hydro-fracturing and gypsum precipitation took place.

In the Watchet study area an E–W trending horst, predominantly composed of Keuper Marl and bounded by faults with net extensional displacement to the north and south, runs along the length of the foreshore (Fig. 5a). To the west, the horst widens and the structure changes from an anticline to a syncline, which is cut by additional faults. The character of the faults is broadly similar to those found in the Kilve area; they are also zones, segmented along their length and linked by relay structures.

Structures associated with contractional inversion of the extensional faults are better displayed at Watchet than at Kilve. As at Kilve, all of the major E–W trending faults (Fig. 7a) preserve net extensional offsets and exhibit dip-slip/oblique-slip oriented slickenfibres interpreted as the product of extension and contractional inversion. The hangingwalls of the faults that bound the northern and southern margin of the central horst display well developed buttress anticlines which verge toward the horst and locally have overturned forelimbs (Fig. 5b). Buttress fold axial surfaces may dip as low as 20°. The intensity of buttressing on the southern margin of the horst is greatest in the centre of the study area and decreases to the east and the west while on the northern margin it is greatest in the east. The southern buttress folding is the most complex with parasitic anticlines and synclines developed around the crest and a zone of intense folding immediately adjacent to the main fault (Fig. 5b, Warren Bay section & Fig. 6c). The buttress zones are also cut by numerous minor faults, some at high angles to the main fault. The minor faults may have formed either prior to, together with, or after the folding.

The NW–SE trending Watchet–Cothelstone dextral strike-slip fault, and an attended set of faults with similar orientations (Fig. 7c), cross the foreshore in the Watchet study area (Fig. 5a). It has a dip of 65° to the southwest and offsets steeply dipping inverted extensional faults, and their attendant buttress fold zones, laterally by up to 300 m. A drag fold zone, visible in map view, extends several metres from the strike-slip fault trace. Associated with this fault is a NNE–SSW trending sinistral strike slip fault, also with an attendant drag zone, located on the outer foreshore to the east of Warren Bay (Fig. 5a). This structure has an antithetic relationship to the Watchet–Cothelstone fault and is probably related to similar trending sinistral strike-slip faults exposed in the Kilve study area.

Extension/inversion direction

The extension/inversion direction for the Kilve and Watchet study areas were determined using fault slip data (fault plane orientation, striae lineation orientation and sense of slip). On most faults along the north Somerset coast it is not possible to distinguish between calcite slickenfibres generated during extension and those generated by compression. Other kinematic indicators (i.e. drag folds) indicate fault motion on individual faults under both stress regimes. Therefore, it is only possible to calculate a combined principle extension/inversion axes,

(a) **Poles to E-W trending partially inverted extensional faults**

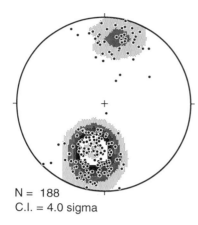

N = 188
C.I. = 4.0 sigma

(b) **Poles to NE-SW trending sinistral strike-slip faults**

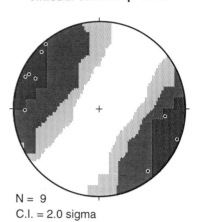

N = 9
C.I. = 2.0 sigma

(c) **Poles to NW-SE trending dextral strike-slip faults**

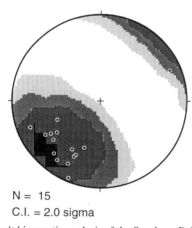

N = 15
C.I. = 2.0 sigma

(d) **Extension/inversion axes, all measured faults**

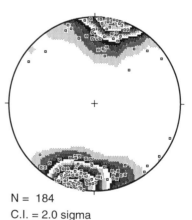

N = 184
C.I. = 2.0 sigma

Fig. 7. Fault kinematic analysis of the Southern Bristol Channel basin. N, number of measurements used in each plot. Both scatter and contour plots are shown on each projection. Combined extension/inversion axes were calculated using the method of Marrett & Allmendinger (1990).

which is interpreted to be the product of both the extensional and subsequent compressional phase of deformation. The true principle extension and principle inversion axes will lie within the range of the combined axes.

Combined principle extension/inversion axes were calculated for 184 faults using the method of Marrett & Allmendinger (1990) and the results were contoured to provide an average axis, which is oriented N–S (004/184°; Fig. 7d).

There is minimal scatter in the combined axes within or between the study areas, or between the dip-slip/oblique-slip E–W trending and strike-slip/oblique-slip NW–SE and NE–SW trending faults. This indicates that, in general, all the faults measured essentially operated as a coherent system during extension and compression. However, a few of the E–W trending faults do exhibit up to three separate sets of dip-slip/oblique through to strike-slip/oblique

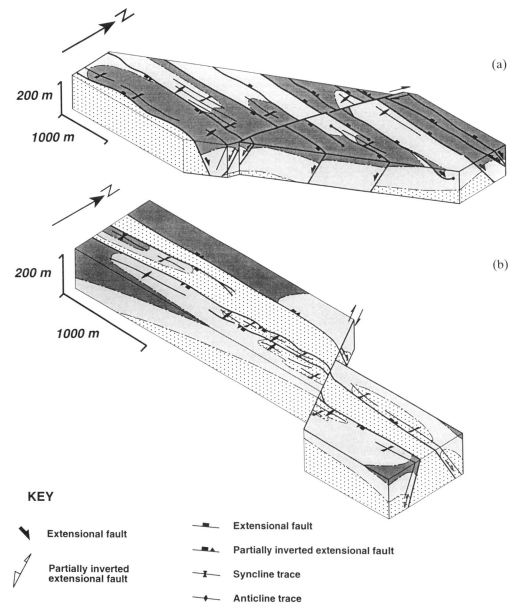

(a)

200 m

1000 m

(b)

200 m

1000 m

KEY

Extensional fault

Partially inverted
extensional fault

Extensional fault

Partially inverted extensional fault

Syncline trace

Anticline trace

Fig. 8. Schematic block diagram illustrating the geometries of fault systems from the North Somerset Coast,
(**a**) Watchet, (**b**) Kilve.

orientations. This indicates that at least some of
the faults within the system have been reutilized
by different stress fields throughout their history.

Tectonic evolution of the North Somerset coast

The three dimensional geometry of the inverted
extensional fault systems exposed in the
Watchet and Kilve study areas are summarized
in Fig. 8. Both study areas have a three phase
evolution (Fig. 9). An E–W trending extensional
fault system developed first, in response to
broadly N–S orientated stretching. This fault
system then suffered contractional inversion in
response to broadly N–S orientated compression. This was followed by a phase of
NW–SE trending dextral, and NE–SW trending

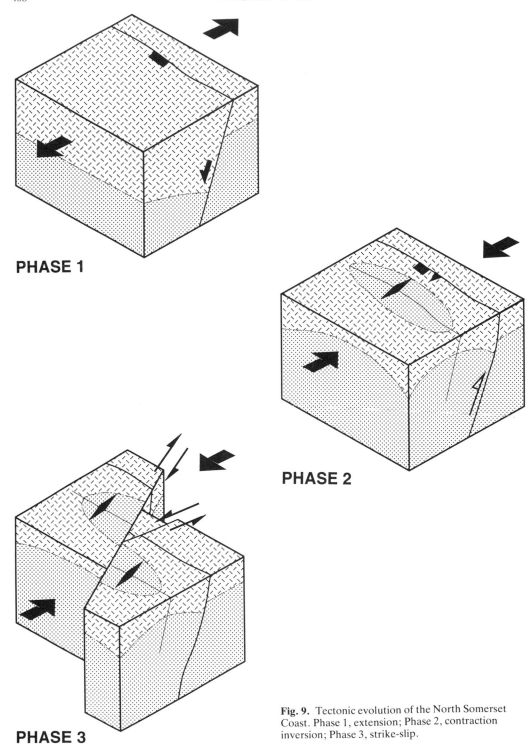

PHASE 1

PHASE 2

PHASE 3

Fig. 9. Tectonic evolution of the North Somerset Coast. Phase 1, extension; Phase 2, contraction inversion; Phase 3, strike-slip.

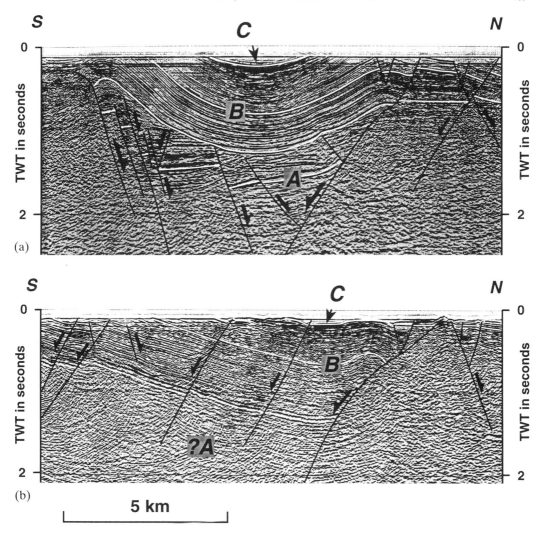

Fig. 10. Seismic sections from the offshore Bristol Channel: (**a**) western profile, (**b**) eastern profile. Seismic data courtesy of Geco Prakla Exploration Services. Sections located on Fig. 1. These inversion structures are at a scale an order of magnitude greater than those exposed on the foreshore, but have a very similar geometry. Sequence A, ?pre mid-Triassic; sequence B, ?mid-Triassic to lowermost Cretaceous; sequence C, Lower Cretaceous (Aptian) to Upper Cretaceous.

sinistral, strike-slip faulting, also the result of N–S compression.

Bristol Channel seismic profiles

Cross sections through the inverted extensional fault systems exposed in the Watchet and Kilve study areas (Figs 4b & 5b) are compared with those observed on two offshore seismic profiles from the Bristol Channel (Fig. 10). The profiles lie to the north and west of the onshore study areas (Fig. 1) and are oriented N–S.

On the western profile (Fig. 10a) three sequences can be identified. The deepest resolvable sequence (A) is characterized by relatively continuous moderate amplitude reflectors arranged in a series of southward-tilted fault blocks, which are offset by steep, northward-dipping faults with net extensional displacements. These reflectors are overlain by a sequence (B) characterized by very continuous reflectors of moderate to high amplitude. The basal reflector of this sequence is a prominent angular unconformity which truncates the crests of the tilted fault blocks beneath. Sequence B lies within a broad parallel syncline that dominates the profile segment illustrated (Fig. 10a). Both the southern and the northern margins of this syncline are bounded by inward-dipping, partially inverted extensional faults, which exhibit buttress anticlines in their hangingwalls. The southern buttress anticline is asymmetric and verges toward the main fault with a steeply dipping forelimb. The northern anticline is more symmetrical. Both of these structures show remarkable geometrical similarity with buttress anticlines exposed in the onshore study areas at Watchet (Watchet Harbour & Warren Bay sections, Fig. 4b) and Kilve (Quantocks Head section, Fig. 5b). However, the buttress zones in the seismic profiles are up to 1 km in width, and are thus approximately an order of magnitude larger than those exposed at Watchet and Kilve, which are generally up to 100 m across. This emphasizes the geometrical scale independence of inverted extensional fault systems. To the north of the northern partially inverted extensional fault, sequence B is cut by southward- and northward-dipping faults with net extensional displacements and no evidence for contractional inversion.

Only a limited region of the third sequence (C) is preserved, which lies within the core of the parallel syncline (Fig. 10a). Elsewhere in the survey a greater thickness of this sequence is preserved. It is characterized by continuous moderate amplitude reflectors which onlap truncated reflectors of the underlying sequence.

On the eastern seismic profile (Fig. 10b), sequence A lies below the depth of seismic resolution. The deepest resolvable sequence (B) is characterized by continuous to discontinuous, moderate to high amplitude reflectors that are cut into a series of northward tilted fault blocks, divided by southward dipping faults with net extensional displacements. A buttress anticline is developed in the hangingwall of a major partly inverted extensional fault close to the northern margin of the illustrated profile segment. This has similar dimension to those of the western profile. Similarly, the third sequence (C) is only preserved as thin poorly resolved remnants. It is not possible to date these seismic sequences accurately as the sections are located far from any well control.

Discussion

The field data from the North Somerset coast provide information on the relative timing of extension, contraction and strike-slip defor-

mation in the Bristol Channel basin, but do not constrain the geological age of these events, other than that they all post-date deposition of the youngest rocks exposed (Lower Jurassic–Lower Lias). The ages of these events can, however, be deduced by considering areas where younger strata are preserved in southwest England (e.g. Chadwick 1986, 1993; Lake & Karner 1987), and the Bristol Channel (Lloyd *et al.* 1973; Kamerling 1979; Van Hoorn 1987; and the seismic data presented here). A regional tectonic evolutionary sequence is summarized in Fig. 11.

Permian–Lower Cretaceous extension

Wedge thickening stratal geometries indicate several major phases of extension to the east of the Bristol Channel basin, in the Wessex basin (Stoneley 1982; Chadwick 1986; Karner *et al.* 1987; Jenkyns & Senior 1991). These were punctuated by periods of thermally related basin expansion and onlap of the surrounding Palaeozoic basement. In the Wessex basin the total amount of extension achieved is up to 10–15% (Chadwick 1993). The location of some major extensional faults appears to coincide with important Variscan thrusts and it is now thought that these thrusts were reutilized and inverted during extensional basin formation (Chadwick 1986; Lake & Karner 1987) (Fig. 11a, b). The first phase of extension occurred during the Permian with the development of small intermontane basins (Anderton *et al.* 1979) (Fig. 11b) which are now exposed along the eastern margin of the Cornubian Massif (Fig. 1). This phase of extension may have contributed significantly to the formation of depocentres in which the Triassic strata exposed along the North Somerset coast were later deposited. Permian deposition commenced with a brief period of volcanism (Exeter Volcanics), followed by alluvial fan sandstones and conglomerates, and playa lake mudstones and siltstones (Edmonds *et al.* 1975). Onshore exposures show that similar deposits laid down during the Triassic spread beyond the basin margins (Chadwick 1986). This phase of basin margin overlap can be correlated with the boundary between sequences A and B recognized in the segment of seismic profile from the Bristol Channel illustrated in Fig. 10a.

The passage from the Triassic to Jurassic is marked by a regional transgression which can be recognized throughout Europe (Anderton *et al.* 1979) and is not associated with extensional basin formation. Triassic playa lake sedimen-

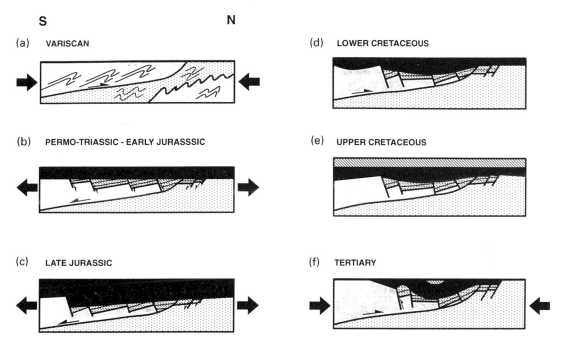

Fig. 11. Tectonic evolution of the Bristol Channel basin, modified after Roberts (1989).

tation gave way to Jurassic deposition in re-stricted marine basins (Fig. 3). Wedge thickening stratal geometries provide evidence for a second phase of extension during the Lower Jurassic (Hettangian–Sinemurian) in both the Bristol Channel and western Wessex Basins (Kamerling 1979; Chadwick 1986). However, convincing evidence for this phase of extension is not present in the Watchet and Kilve study areas.

Wedge thickening stratal geometries in the Bristol Channel and Wessex basins also provide evidence for a third, Upper Jurassic phase of extension (Kamerling 1979; Chadwick 1986) (Fig. 11c). In the western Wessex Basin an important period of extension occurred during the deposition of the Upper Jurassic Kimmeridge Clay. In the Watchet and Kilve study areas the initial phase of extensional faulting may be correlated with this regional phase of extension. This phase of extension is also interpreted to be responsible for the faults with net extensional displacement that cross cut sequence B in the seismic profiles from the Bristol Channel (Figs 10a, b). The Lower Lias black shales of the North Somerset coast have geochemical maturity indices that correspond with a vitrinite reflectance of 0.4%, which may be interpreted as

the result of a maximum of 2.4 km of burial (Cornford 1986). This figure provides an estimate of the depth of the North Somerset exposures during extensional deformation, prior to subsequent uplift and exhumation associated with the later phase(s) of contractional inversion.

Lower Cretaceous/Tertiary inversion

In the western Wessex and Bristol Channel basins the base of the Aptian Lower Greensand Formation is marked by a prominent angular unconformity (Kamerling 1979; Chadwick 1986; Van Hoorn 1987; Lake & Karner 1987). This was created by pre-Lower Greensand broad scale folding that may have been accompanied by regional erosion (Fig. 11d). This unconformity is the boundary between sequences B and C on the seismic profiles from the Bristol Channel (Figs 10a, b). Most of the Lower Cretaceous is absent across the Bristol Channel basin, and was either never deposited or has been subsequently eroded. Many of the faults that cut pre-Lower Greensand strata are truncated at the Aptian–Albian unconformity. Those that do transect the unconformity do not generate significant offsets in younger strata. The angular unconformity may be interpreted as a response to thermal effects associated with the Bay of Biscay and North Atlantic opening during the Lower Cretaceous (Lake & Karner 1987).

In the western Wessex and Bristol Channel Basins the pre-Lower Greensand unconformity is overlain by a significant thickness of Lower to Upper Cretaceous strata, including the Chalk (Kamerling 1979; Chadwick 1986; Van Hoorn 1987; Lake & Karner 1987) (Fig. 11e). Renewed deposition after the Lower Cretaceous phase of folding and uplift may be related to thermal relaxation of the continental margin at the onset of ocean floor spreading in the Bay of Biscay (Pegrum & Mounteney 1978; Van Hoorn 1987), combined with a high global sea-level stand.

A Tertiary phase of inversion (Fig. 11f) has been recorded in southwest England and is responsible for net contractional offsets and monoclinal folding at Upper Cretaceous levels in the Wessex basin (e.g. the Purbeck monocline; Stoneley 1982; Chadwick 1993). Tertiary basins were developed on what were Mesozoic footwall highs (Lake & Karner 1987). The total amount of shortening across the Wessex basin is up to 2%, considerably less than the magnitude of extension (Chadwick 1993). Lake & Karner (1987) suggest that inversion culminated in the Oligo-Miocene concurrent with the Helvetic phase of Alpine deformation. Widespread remanié deposits suggest that the Chalk may have been deposited over the entire region (Cornford 1986). Its subsequent removal during the Tertiary is probably related to a combination of falling global sea level and inversion associated uplift.

Tertiary strike-slip deformation

Major NW–SE trending fault lines divide the E–W-trending post-Variscan extensional basins in and around southwest England (e.g. Sticklepath and Watchet–Cothelstone faults) (Kamerling 1979; Chadwick 1986; Holloway & Chadwick 1986) (Fig. 1). These faults are deep-seated structures, which may have initiated during the Variscan orogeny (Holloway & Chadwick 1986), and have experienced strike-slip/oblique slip motions of both sinistral and dextral sense throughout their evolution. The N–S orientated combined principle extension/compression axis calculated from the Kilve and Watchet study areas is consistent with NW–SE-trending faults operating with transtensional sinistral motions during extension and transpressional dextral motions during compression.

Sinistral motions on the Sticklepath fault are indicated by Oligocene pull-apart basins which occur at left stepping offsets (Bovey Tracey & Petrockstow basins; IGS 1979; Holloway & Chadwick 1986). The Eocene Lundy island granite which lies along the fault trace (50–55 Ma; Miller & Fitch 1962; Dodson & Long 1962) (Fig. 1) may also be related to this phase of deformation. Compressional deformation of the pull-apart basin sediments (Bristow & Hughes 1971) indicates dextral reactivation of the Sticklepath fault after the Oligocene and the margins of an E–W trending Permian intermontane Crediton Trough (Fig. 1) are also offset with a dextral sense.

The Watchet–Cothelstone fault does not exhibit direct evidence for sinistral transpression. It displays dextral offset where it crosses the Watchet study area. Dextral motion is also indicated by an important uplift (Compton Valence dome; Chadwick 1993) between left stepping fault segments. The NW–SE-trending strike-slip fault set exposed in the Watchet study area exhibits dextral offset, and their NE–SW trending antithetic counterparts, exposed at Watchet and Kilve, exhibit sinistral offsets.

Slickenfibre orientations from the faults along the north Somerset coast imply sinistral transtension on NW–SE trending structures during the Upper Jurassic extensional phase. Pull-aparts along the Sticklepath fault also indicate sinistral transtension on NW–SE trending faults during the Oligocene. A phase of dextral transpression on these faults is the final response to N–S directed contraction along the north Somerset coast, and may be concurrent with post-Oligocene compressional structures in pull-apart basin sediments along the Sticklepath fault.

Conclusions

Triassic to Lower Jurassic coastal exposures along the north Somerset coast are cut by partially inverted E–W trending extensional faults. Deformation occurred in three phases. (i) An extensional fault system developed during the Upper Jurassic–Lower Cretaceous in response to N–S orientated extension; (ii) During the Tertiary some of the faults within this system were partially inverted in response to N–S orientated compression. This resulted in the formation of hangingwall buttress anticlines and zones of intense folding. (iii) Post-Oligocene, the partially inverted extensional faults were offset by NW–SE trending dextral, and NE–SW trending sinistral, strike-slip faults, also in response to N–S orientated compression.

Geometrically similar inverted extensional fault systems are observed on seismic profiles from the Bristol Channel, although these buttress anticlines may be up to an order of magnitude wider than those exposed onshore. This comparison emphasizes the geometrical scale independence of inverted extensional fault systems.

This research forms part of the Fault Dynamics Project funded by ARCO British, BRASOIL U.K. Ltd., BP Exploration, Conoco (U.K.) Limited, Mobil North Sea Limited and Sun Oil Britain. We thank Ian Davison, João Keller, Nigel Platt and the participants of numerous Fault Dynamics courses for helpful discussion. Kevin de Souza is especially thanked for carrying out essential photographic work at short notice. (Fault Dynamics Publication No. 38.) Seismic data by courtesy of Geco Prakla Exploration Services.

References

ANDERTON, R., BRIDGES, P. H., LEEDER, M. R. & SELLWOOD, B. W. 1979. *A dynamic stratigraphy of the British Isles*. Allen & Unwin.

BRISTOW, C. M. & HUGHES, D. E. 1971. A Tertiary thrust fault on the southern margin of the Bovey basin. *Geological Magazine*, **108**, 61–67.

BROOKS, M., BAYERLEY, M. & LLEWELLYN, D. J. 1972. A new geological model to explain the gravity gradient across Exmoor, North Devon. *Journal of the Geological Society*, London, **133**, 385–393.

——, TRAYNER, P. M. & TRIMBLE, T. J. 1988. Mesozoic reactivation of Variscan thrusting in the Bristol Channel area, U. K. *Journal of the Geological Society*, London, **145**, 439–444.

CHADWICK, R. A. 1986. Extensional tectonics in the Wessex Basin, southern England. *Journal of the Geological Society*, London, **143**, 465–488.

—— 1993. Aspects of basin inversion in southern Britain. *Journal of the Geological Society*, London, **150**, 311–322.

——, KENOLTY, N. & WHITTAKER, D. J. 1983. Crustal structure beneath southern England from deep seismic reflection profiles. *Journal of the Geological Society*, London, **144**, 893–911.

CORNFORD, C. 1986. The Bristol Channel Graben: Organic geochemical limits on subsidence and speculation on the origin of inversion. *Proceedings of the Ussher Society*, **6**, 360–367.

—— 1990. Source rocks and hydrocarbons of the North Sea. *In*: GLENNIE, K. W. (ed.) *Introduction to the Petroleum Geology of the North Sea*. Blackwell, London.

DODSON, M. H. & LONG, L. E. 1962. Age of the Lundy granite, Bristol Channel. *Nature*, **195**, 975–976.

DONATO, J. A. 1988. Possible Variscan thrusting beneath the Somerton anticline, Somerset. *Journal of the Geological Society*, London, **145**, 431–438.

EDMONDS, E. A., McKEOWN, M. C. & WILLIAMS, M. 1975. *South-west England, British regional geology*, 4th Edition. Institute of Geological Science.

HOLLOWAY, S. & CHADWICK, R. A. 1986. The

Sticklepath–Lustleigh fault zone: Tertiary sinistral reactivation of a Variscan dextral strike-slip fault. *Journal of the Geological Society*, London, **143**, 447–452.

IGS. 1979. Geological survey ten mile map, South Sheet. 1 : 625,000.

JENKYNS, H. C. & SENIOR, J. R. 1991. Geological evidence for intra-Jurassic faulting in the Wessex basin and on its margins. *Journal of the Geological Society*, London, **148**, 245–260.

KAMERLING, P. 1979. The geology and hydrocarbon habitat of the Bristol Channel Basin. *Journal of Petroleum Geology*, **2**, 75–93.

KARNER, G. D., LAKE, S. D. & DEWEY, J. F. 1987. The thermo-mechanical development of the Wessex basin, southern England. *In*: HANCOCK, P. L., DEWEY, J. F. & COWARD, M. P. (eds) *Continental extensional tectonics*. Geological Society, London, Special Publication, **28**, 517–536.

LAKE, S. D. & KARNER, G. D. 1987. The structure and evolution of the Wessex basin, southern England: An example of inversion tectonics. *Tectonophysics*, **137**, 347–378.

LLOYD, A. J., SAVAGE, R. J. G., STRIDE, A. H. & DONOVAN, D. T. 1973. The geology of the Bristol Channel floor. *Philosophical Transactions of the Royal Society*, **A274**, 595–626.

MARRETT, R. & ALLMENDINGER, R. W. 1990. Kinematic analysis of fault slip data. *Journal of Structural Geology*, **12**, 973–986.

MILLER, J. A. & FITCH, F. J. 1962. Age of the Lundy granites. *Nature*, **195**, 975–976.

PEACOCK, D. C. P. & SANDERSON, D. J. 1991. Displacements, segment linkage and relay ramps in normal fault zones. *Journal of Structural Geology*, **13**, 721–733.

PEGRUM, R. M. & MOUNTENEY, N. 1978. Rift basins flanking the North Atlantic ocean and their relation to the North Sea. *American Association of Petroleum Geologists Bulletin*, **62**, 419–441.

ROBERTS, D. G. 1989. Basin inversion in and around the British Isles. *In*: COOPER, M. A. & WILLIAMS, G. D. (eds) *Inversion tectonics*. Geological Society, London, Special Publication, **44**, 131–150.

RUFFELL, A. H. 1992. Early to mid-Cretaceous tectonics and unconformities of the Wessex Basin (Southern England). *Journal of the Geological Society*, London, **149**, 443–454.

STONELEY, R. 1982. The structural development of the Wessex basin. *Journal of the Geological Society*, London, **139**, 543–554.

TUCKER, M. E. 1977. The marginal Triassic deposits of South Wales: Continental facies and palaeogeography. *Geological Journal*, **12**, 169–188.

VAN HOORN, B. 1987. The south Celtic Sea/Bristol Channel basin: Origin, deformation and inversion history. *Tectonophysics*, **137**, 309–334.

WHITTAKER, A. 1972. The Watchet Fault – a post Liassic transcurrent reverse fault. *Geological Survey of Great Britain Bulletin*, **41**, 75–80.

—— & GREEN, G. W. 1983. *Geology of the country around Weston-super-Mare*. IGS Memoir for 1 : 50,000 geological sheet 279. HMSO.

Inversion of a Lower Cretaceous extensional basin, south central Pyrenees, Spain

R. M. G. BOND & K. R. McCLAY

Fault Dynamics Project, Department of Geology, Royal Holloway, University of London, Egham, Surrey, TW20 0EX, UK

Abstract: The structure of the Boixols thrust sheet is dominated by kilometric-scale E–W trending folds, the most important of which is the Sant Corneli–Boixols–Nargo fault-propagation, growth anticline at the leading edge of the thrust sheet. Five megasequences have been identified within the Boixols thrust sheet and these may be related to phases of extension, tectonic quiescence and phases of inversion. The Boixols thrust sheet incorporates the Lower Cretaceous Organyá basin which is bounded by a north-dipping extensional fault that underwent inversion in the Late Campanian–Maastrichtian and was incorporated into the Pyrenean south-vergent thrust system during the Eocene–Oligocene. The inversion history is recorded by the syn-inversion Arén growth sequence that is only preserved in the footwall syncline of the main fault-propagation growth fold. During inversion a short-cut thrust fault developed in the footwall to the Lower Cretaceous extensional fault. This now forms the Boixols thrust at surface. An inversion model for the Boixols thrust sheet and the Organyá basin is proposed and compared to analogue and numerical models of half-graben inversion. This model for inversion in the Boixols thrust sheet may be applied to other thrust structures in the Pyrenees and to other foreland fold and thrust belts.

The Pyrenees orogenic belt is a doubly vergent, east–west trending fold and thrust belt located between the Iberian and European plates. It was formed by Late Cretaceous to Miocene contractional (thrust) tectonics that deformed a Lower Cretaceous extended carbonate platform and extensional basins. The Pyrenees are flanked by the Aquitaine foreland basin to the north and by the Ebro foreland basin to the south (Fig. 1). From the Triassic to the end of the Early Cretaceous, separation of Iberia and Europe opened the Bay of Biscay (Choukroune *et al.* 1973; Masson & Miles 1984) and the Ligurian ocean, and formed a series of extensional and transtensional basins along the northern margin of Iberia and the southern margin of Europe (Peybernés & Souquet 1984; Puigdefabregas & Souquet 1986). One of these basins, the Organyá basin (Berastegui *et al.* 1990), is incorporated in the Boixols thrust sheet of the south central Pyrenees ('Unité sud Pyrénée Centrale', Séguret 1972).

The Pyrenean orogeny resulted from convergence of the Iberian and European plates. This started in the Late Santonian and progressed through to the Miocene (Roest & Srivastava 1991). Early convergence had a strong rotational component, whereas the later stages of the Pyrenean orogeny were the result of almost orthogonal motion (Roest & Srivastava 1991).

In the southern Pyrenees imbrication and stacking of south-vergent thrust sheets resulted in the development of the Late Cretaceous–Tertiary Ebro foreland basin (Fig. 1).

This paper discusses the tectonostratigraphic evolution of the Boixols thrust sheet together with an analysis of the inversion of the Lower Cretaceous Organyá extensional basin that forms the thrust sheet. The research is based upon detailed field mapping, structural, stratigraphic and sedimentological analyses, and numerical modelling of inversion geometries. Syn-inversion sedimentary sequences are developed in the basin in front of the Boixols thrust and in part onlap, and are folded by, the thrust-related folds. An inversion model is proposed for the evolution of the Boixols thrust sheet and is compared with analogue and numerical models of inversion structures.

Regional geology

The Boixols thrust sheet is one of the cover thrust sheets of the South Central Pyrenean unit (Fig. 1; Muñoz 1992). It is bordered by the Montsec thrust sheet to the south and by the Morreres backthrust to the north (Figs 1, 2). The structure of the Boixols thrust sheet is dominated by large, east–west trending kilometric scale folds, typically 4–8 km in wavelength and

From BUCHANAN, J. G. & BUCHANAN, P. G. (eds), 1995, *Basin Inversion*, Geological Society Special Publication No. 88, 415–431.

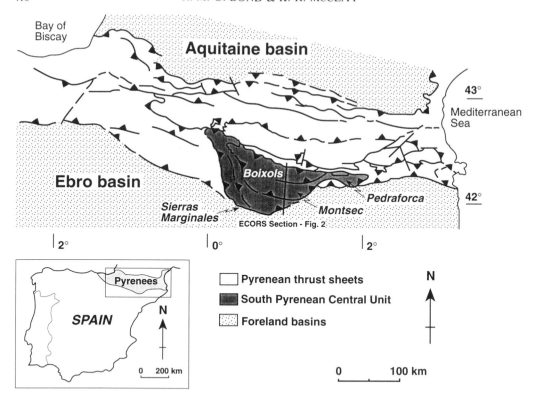

Fig. 1. Location of the Boixols thrust sheet, South Pyrenean Central Unit, Spain. The line of section (Fig. 2) of the ECORS seismic line is shown.

with amplitudes of 1–3 km (Figs 2, 3). The dominant folds, the Sant Corneli–Boixols–Nargo anticline, the Santa Fe syncline and the Boumort anticline are shown on Fig. 3. In the footwall to the Boixols thrust sheet are the broad Isona anticline and the tight, asymmetric Tremp–Sallent syncline (Fig. 3). The Boixols thrust is largely blind (Fig. 2) and is only emergent in the region of Boixols and along strike to the east towards the village of Coll de Nargo (Fig. 3).

Total stratigraphic thickness in the Boixols thrust sheet exceeds 5 km (Berastegui *et al.* 1990). Strata range from Triassic through to Oligocene in age. Five megasequences bounded by unconformities (see Hubbard *et al.* 1985) have been identified and mapped (Berastegui *et al.* 1990). The Jurassic to Upper Cretaceous units consist of platformal carbonates and their basinal time-equivalent marls and mudstones (Fig. 4). The Late Cretaceous–Palaeocene syn-inversion strata of the Arén sequence (Fig. 4) are marine siliciclastics which are overlain by continental red beds of the Tremp–Graus basin. Unconformably overlying most of these older units are the Upper Eocene to Oligocene

Collegats alluvial and fluvial conglomerates (Fig. 4) (Mellere 1993).

Stratigraphy of the Bioxols thrust sheet and Organyá basin

A summary stratigraphic sequence diagram of the Jurassic to the Oligocene of the Boixols thrust sheet/Organyá basin is shown in Fig. 4. Detailed descriptions of the five unconformity-bound megasequences are given in Puigdefabregas & Souquet (1986), Puigdefabregas *et al.* (1992), Simo (1985, 1986, 1989) and Berastegui *et al.* (1990) and only a brief review will be given here.

Megasequence 1

Triassic to Jurassic Triassic strata form the detachment level of the Boixols thrust (Fig. 2) and crop out below the Morreres back-thrust. They are continental facies: red micaceous sandstones and multicoloured mudstones with local gypsum and evaporite beds. In the Boixols thrust sheet, Jurassic strata are exposed above the Morreres backthrust and locally in the core

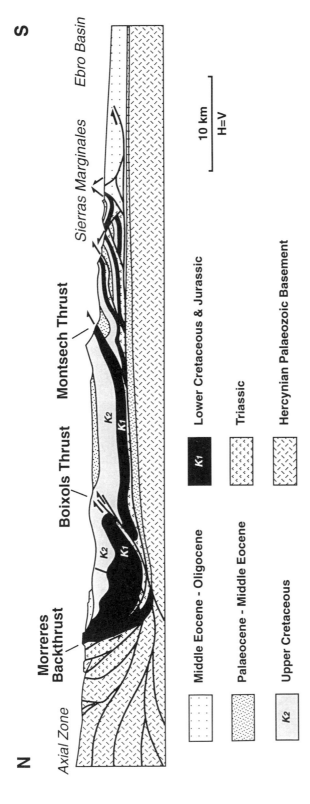

Fig. 2. Geo-seismic section along the ECORS line across the southern Pyrenees (after Losantos *et al.* 1988).

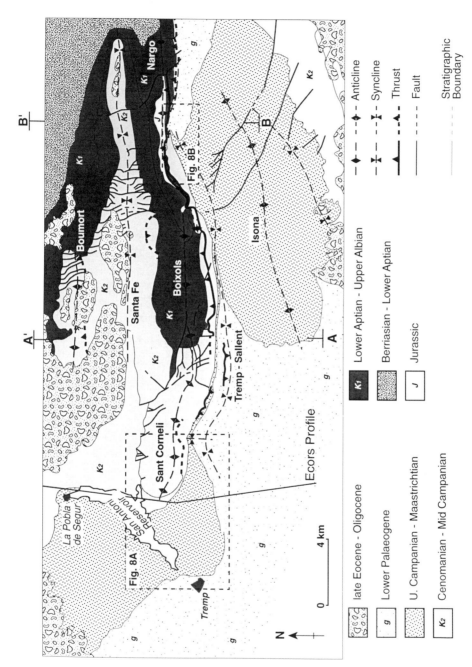

Fig. 3. Simplified geological map of the Boixols thrust sheet showing the major tectono-stratigraphic sequences.

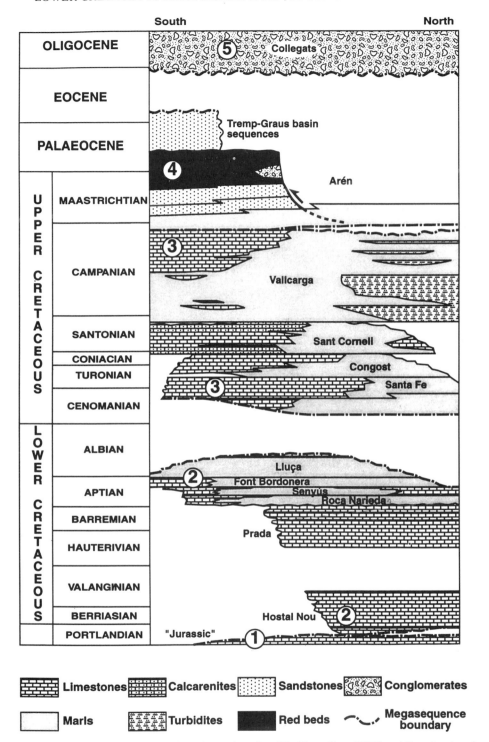

Fig. 4. Simplified stratigraphy of the Boixols thrust sheet (modified from Simo (1989) and Berastegui *et al.* (1990)). Numbers in circles refer to the megasequences defined in the text.

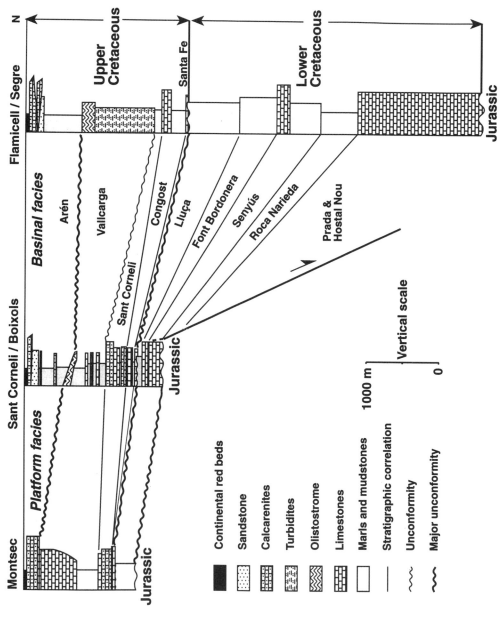

Fig. 5. South to north correlation of Cretaceous sections across the study area (after Simo (1985, 1986) and from the authors' work).

of the Boixols–Nargo anticline (Fig. 3). The Jurassic lithologies are shallow marine Liassic marls and fossiliferous limestones together with dark weathering, fine to coarse-grained dolomites of Dogger–Malm age (Fig. 4).

Megasequence 2

Lower Cretaceous The Lower Cretaceous of megasequence 2 is bounded by the Upper Portlandian unconformity at the base and the Cenomanian unconformity at the top (Fig. 4). These strata consist of platformal carbonates accumulated on the basin flanks and their time equivalent basinal marls deposited in the deeper water of the Organyá Basin (Berastegui *et al.* 1990). The Lower Cretaceous succession can be divided into five sequences as shown in Fig. 5.

The lowermost Lower Cretaceous sequences, the Prada and Hostal Nou limestones (*c.* 1700 m thick, Berastegui *et al.* 1990), and the Cabó marls (Roca Narieda sequence, Berastegui *et al.* 1990; 850–1000 m thick) are of Berriasian to Lower Aptian age and unconformably overlie the Jurassic dolomites to the north of the area in the Flamicell and Segre sections (Fig. 5), in the hangingwall of the Morreres backthrust. In contrast, at the leading edge of the Boixols thrust sheet the Prada and Hostal Nou limestones, and Cabó marls are absent (Sant Corneli/ Boixols section, Fig. 5) and the Jurassic is unconformably overlain by the Aptian Senyús limestones. The Senyús limestones south of the Santa Fe syncline axis have a thickness of approximately 350 m, thinning to a virtual pinch-out southwards, whereas the equivalent marls to the north of the syncline axis have a thickness of around 850 m. The overlying Aptian–Albian Font Bordonera and Lluça sequences thin and become more platformal in character southwards from the Boixols section to the Montsec section (Fig. 5).

Megasequence 3

Upper Cretaceous: Cenomanian to Campanian The Santa Fe, Congost and Sant Corneli sequences (Simo 1985, 1986, 1989) have a total thickness of 200–500 m in the study area and unconformably overlie the strata of megasequence 2 (Berastegui *et al.* 1990). These units consist of carbonate platform limestones, and slope and basinal facies marls. The units of megasequence 3 display a gradual thinning to the south and thickening to the north of the Boixols area (Fig. 5). The depocentre during this period was located further northwest, and the platform further north, with respect to the Lower Cretaceous depocentre. A locally developed intraformational angular unconformity at the top of the Sant Corneli sequence marks the start of a rapid deepening of the basin and deposition of the 800–1500 m thick Vallcarga sequence (Simo 1985, 1986, 1989). The Vallcarga turbiditic marls and clays are exposed within the Boixols thrust sheet and isolated time equivalent reefal limestones are found on the forelimb of the Boixols anticline. Palaeocurrents in the Vallcarga turbidites indicate the presence of a palaeohigh in the vicinity of the present day location of the Sant Corneli anticline (Van Hoorn 1970; Nagtegaal 1972).

Megasequence 4

Upper Cretaceous: Upper Campanian–Maastrichtian A regional angular unconformity separates the Vallcarga strata from the overlying Upper Campanian to Maastrichtian Arén sequence (Simo 1985, 1986, 1989). Near La Pobla de Segur (Fig. 3), immediately overlying the unconformity surface in the Sant Corneli area, is the Puymanons olistostrome which is interpreted to indicate significant tectonic instability within the basin. The Arén sequence consists of basinal and slope marls and turbidites, and shallow-marine estuarine siliciclastics. Local fan-delta deposits are found within the Arén sandstones. The depocentre of the Arén sequence was located to the south of the present day expression of the Boixols thrust-sheet, a major change with respect to the more northerly positions of the earlier Cretaceous depocentres. The Arén Sequence thins northwards to pinch-out in the footwall Tremp–Sallent syncline (Fig. 3) of the Boixols thrust system, but is over one kilometre thick in the Isona anticline. In detail, the Arén Sequence can be divided into three distinct subsequences (P. Arbués pers. comm. 1991), separated by locally developed angular unconformities and corresponding regional disconformities and erosional surfaces.

Upper Cretaceous–Tertiary: Maastrichtian–Palaeocene At the top of the Arén sequence there is a gradational change from slope marls and marginal marine sandstones to the distinctive continental red bed deposits of the 'Garumnian' facies (Kraus 1990). These are typically red and varicoloured fluvial sandstones and overbank mudstones with local alluvial fan systems, for example at Sallent (Fig. 3) in the footwall to the Boixols thrust.

Megasequence 5

Oligocene to Miocene In the region of the Boixols thrust and its immediate footwall (Fig.

Fig. 6. (a) Photograph of the Boixols anticline viewed from the south. The form of the anticline is depicted by the resistant Upper Cretaceous limestones of the Congost and Sant Corneli units (Fig. 4). (b) Photograph of the Boixols tip thrust at the village of Abella de Conca (viewed looking west) which is built on the footwall to the thrust. Upper Cretaceous limestones are displaced about 100 m to the left (south) by the Boixols thrust which runs just above the village.

3) the Palaeocene and older strata are unconformably overlain by the massive, thick-bedded
Oligo-Miocene conglomerates of the Collegats
formation (Figs 3, 4) (Mey *et al.* 1968; Mellere
1993). The Collegats conglomerates are fluvial
deposits that infill a very rugged palaeotopogra-
phy and display both local and more distal
provenances (e.g. from the uplifted hinterland
of the Pyrenees). They have a cumulative thickness of over 3500 m (Mellere 1993).

Structure of the Boixols thrust sheet

The internal structure of the Boixols thrust
sheet is dominated by a number of large east–
west trending, generally shallowly west-
plunging, fault-related folds. From north to
south these are the Boumort anticline, the
Santa Fe syncline and, at the leading edge of
the Boixols thrust sheet, the most prominent
fold, the Sant Corneli–Boixols–Nargo anticline
(Fig. 3). This anticline is spectacularly exposed
in the central part of the area (Fig. 6) and is
south-vergent and strongly asymmetric, with a
steep to overturned southern frontal limb, and
a more gently dipping (*c.* 40°) northern back
limb (Figs 6, 7). The corresponding footwall
syncline, the Tremp–Sallent syncline (Fig. 3)
contains a syn-folding wedge of Arén strata and
is also strongly asymmetric (Figs 6, 7). The
Boixols thrust itself is a small displacement,
tip-line structure that is emergent in the region
of Boixols to Coll de Nargo (Fig. 3) and cuts
through the steep frontal limb of the Boixols
fold (Fig. 6b). Cross-sections through two parts
of the leading edge of the Boixols thrust sheet
are shown in Fig. 7.

Fault-related folding

The contractional folds within the Boixols thrust
sheet at both outcrop scale (Fig. 6b) and in
cross-section (Fig. 7) are clearly thrust fault-
related. In particular the strongly asymmetric
Sant Corneli–Boixols–Nargo anticline with a
steep to overturned front limb with thrusting
through the forelimb has the geometric characteristics of a fault-propagation (Suppe & Medwedeff 1984; 1990) or possibly 'breakthrust' fold
(Fischer *et al.* 1992) developed above the tip-line
of the Boixols thrust (Fig. 7). The other large
folds in the thrust sheet may be fault-bend,
fault-propagation or breakthrust folds (e.g. Fig.
2). The difference in competencies between the
overlying Upper Cretaceous limestones and the
underlying incompetent Lower Cretaceous
marls may account for geometric deviations
from idealised fault-related fold models.

Stratal geometries and structure

Detailed mapping, logging and section construction has shown that there are considerable
differences in stratal thicknesses and stratal
geometries within the Boixols thrust sheet (e.g.
Figs 4, 5, 7). The Lower Cretaceous section
shows a dramatic increase in thickness (in excess
of 2 km) from the Boixols anticline northwards
to Boumort (Figs 3, 7). This feature, together
with facies changes from platformal carbonates
to basinal marls (Fig. 4) and the identification of
palaeo-extensional fault scarps (Berastegui *et al.*
1990) indicate that the Lower Cretaceous megasequence 2 was deposited in an extensional basin
which had a northward-dipping basin margin
fault in the region of Boixols. This geometry
explains the reduced thicknesses and absence of
strata on the footwall of this basin-bounding
extensional fault (Fig. 5). Megasequence 3
(Upper Cretaceous, Fig. 4) is interpreted as
post-extensional in character and unconformably overlies megasequence 2 (Peybernés 1976;
Berastegui *et al.* 1990).

The geometries of megasequence 4 and in
particular that of the Arén sequence indicate
deposition during thrusting and folding of the
Boixols thrust sheet. Cross-sections in Fig. 7
show the general aspects of the onlap of the
Arén strata onto the Boixols anticline. Detailed
mapping of two areas of the frontal southern
limb of the Sant Corneli–Boixols–Nargo anticline, the Sant Corneli–Orcau area (Fig. 8a) and
the Sallent area (Fig. 8b), show that the Arén
sequence can be divided into three sub-sequen-
ces which clearly display local intraformational
unconformities and wedge geometries that thin
and onlap onto the forelimb of the anticlinal
structure and fan and thicken away from this
anticline (Fig. 8). The intraformational unconformities record progressive uplift and rotation
of a Maastrichtian fold limb during sedimentation. In these sequences, abrupt changes in
regional dip commonly follow a coarse clastic
influx/shallowing event within the sub-sequen-
ces. The present day location of the Arén growth
sequence in the footwall syncline to the Boixols
thrust indicates that the Sant Corneli–Boixols–
Nargo anticline may be interpreted as a fault-
propagation, growth anticline which initiated in
the Maastrichtian.

There is little evidence to indicate that the
Garumnian was deposited during active contraction and uplift of the Boixols thrust sheet, but
both the Maastrichtian strata and the Garumnian red beds have been subsequently folded
into the tight footwall Tremp–Sallent syncline
and the broad Isona anticline (Figs 3, 7).

R. M. G. BOND & K. R. McCLAY

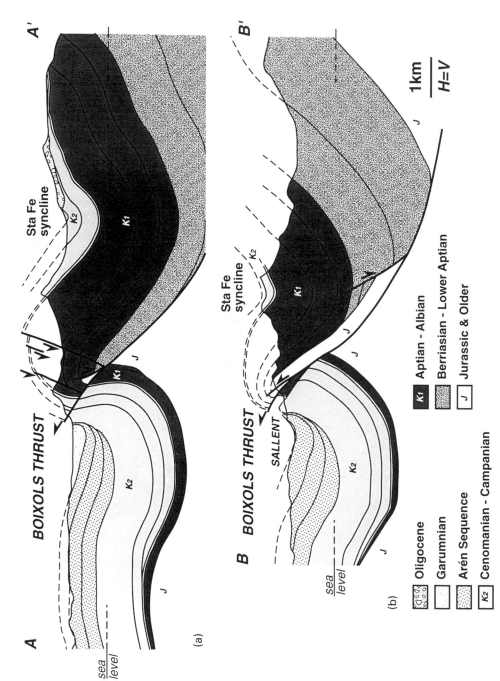

Fig. 7. Cross-sections through the Boixols thrust sheet: (**a**) section A–A' through the Abella de la Conca area (Fig. 3); (**b**) section B–B' though the Sallent area (Fig. 3).

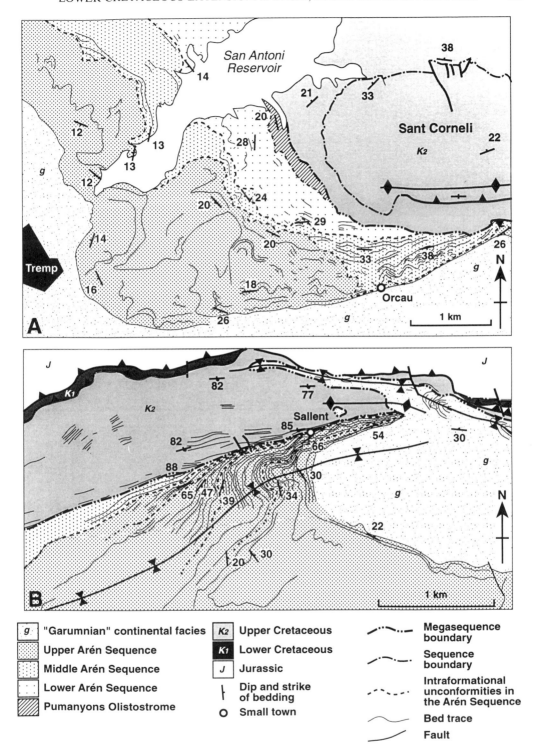

g	"Garumnian" continental facies	K2	Upper Cretaceous		Megasequence boundary
Upper Arén Sequence		K1	Lower Cretaceous		Sequence boundary
Middle Arén Sequence		J	Jurassic		Intraformational unconformities in the Arén Sequence
Lower Arén Sequence			Dip and strike of bedding		Bed trace
Pumanyons Olistostrome		O	Small town		Fault

Fig. 8. Detailed maps of the Campanian–Maastrichtian syn-inversion stratigraphic architecture at the leading edge of the Boixols thrust sheet. Sequence boundaries are based upon the work of P. Arbués. (**a**) Sant Corneli W-plunging anticline. (**b**) Sallent anticline–syncline growth structure.

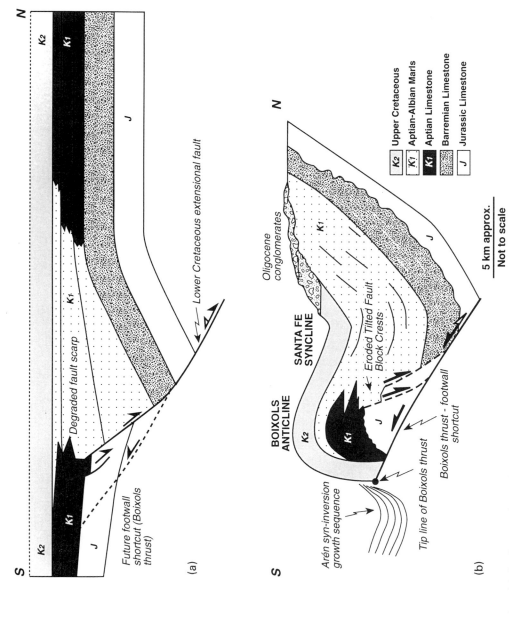

Fig. 9. Conceptual models for the extension and inversion of the Organyá basin in the Boixols thrust sheet. (**a**) Architecture at the end of the Cretaceous. (**b**) Inversion architecture.

Elsewhere in the Tremp–Graus, a tectonic control on sedimentary sequences within the Palaeogene strata has been postulated (Eichenseer *et al.* 1992). Contraction certainly continued into the Lower Oligocene because the lower sections of the Collegats conglomerate formation on the southern limb of the Boumort anticline are deformed in a growth fold associated with small-scale thrust imbrication. The stratal geometries within this growth fold suggest that a degree of detachment folding was possibly involved in the development of the Boumort anticline.

Discussion

The structural and stratigraphic relationships mapped within the Boixols thrust sheet have permitted five unconformity bounded megasequences to be identified (Fig. 4). These megasequences may be interpreted as the stratal response to phases of extension (megasequences 1 and 2), tectonic quiescence (megasequence 3) and phases of contraction (megasequences 4 and 5). A model for the extension of the Organyá basin and subsequent inversion and incorporation into the Boixols thrust sheet has been erected and is shown in Fig. 9.

During the Lower Cretaceous and deposition of megasequence 2, extension on a north-dipping extensional fault produced the half-graben Organyá basin with a depocentre to the north of the present day position of the Boixols thrust (Fig. 9a). A syn-extension growth wedge of Aptian–Albian marls was flanked by carbonate platforms to the south. During the Upper Campanian–Maastrichtian, the extensional fault system was reactivated in inversion, pushing the syn-extensional wedge out of the half-graben. The steeper sections of the extensional fault were not, however, reactivated and a footwall short-cut thrust developed (the upper part of the present day Boixols thrust) and part of the syn-extensional fault surface was preserved in the hangingwall of the thrust system (Fig. 9b). The inversion history and timing is indicated by the syn-inversion Arén growth sequence (megasequence 4) that is preserved in the footwall syncline (Tremp–Sallent syncline; Fig. 3) of the fault-propagation inversion anticline (Sant Corneli–Boixols–Nargo anticline; Figs 3, 7). After this initial contraction in the Late Cretaceous, a phase of relative tectonic quiescence followed in the Palaeocene, during which the Garumnian facies was deposited. The main phase of Pyrenean shortening occurred in the Eocene to Oligocene (Muñoz 1992) and the Boixols thrust sheet (and the Organyá basin)

were incorporated in the 'piggy-back' sequence of south-vergent, cover thrust sheets (e.g. Fig. 2). Late in the contractional history, the Upper Eocene–Oligocene Collegats conglomerates infilled the rugged topography of the eroded thrust wedge of the Pyrenees. The inversion history outlined above is in good agreement with the relative plate motions between Iberia and Europe (Roest & Srivastava 1991), whereby convergence began in the Late Campanian–Maastrichtian followed by a period of quiescence with renewed and more rapid convergence in the Eocene–Oligocene corresponding to the main phases of Pyrenean shortening (Muñoz 1992).

Analogue models of inverted extensional half-graben presented by Buchanan & McClay (1991) show many broad similarities with the geometrical model suggested here for the leading edge fold of the Boixols thrust sheet. In both the model and the natural example a broad anticline cored by syn-extensional deposits develops above the inverted basin bounding fault. Analogue models which have a rigid footwall block, however, cannot simulate the development of footwall shortcut thrust faults within the pre-rift section during inversion. In the analogue models (Buchanan and McClay 1991) syn-inversion strata were added at a rate sufficient to just bury the growing inversion anticline (i.e. 'fill to the top' of the growth structure). As a consequence, during inversion, the reactivated extensional fault propagated upwards at a steep angle into the syn-inversion strata. This contrasts with the Boixols example where the syn-extension strata onlap onto the growing anticline and hence the shortcut thrust is convex upwards, flattening towards the surface (Fig. 9b).

The evolution proposed above for the main Sant Corneli–Boixols–Nargo growth anticline has been forward modelled using a simple geometric model of dip-slip extension above a kinked planar fault followed by dip-slip inversion with the production of an anticlinal growth fold and a footwall short-cut fault (Figs 10, 11). Two forward models have been produced, the first with high syn-inversion sedimentation (Fig. 10), and the second with low syn-inversion sedimentation (Fig. 11). The initial half-graben geometry (Fig. 10a) was modelled using a vertical simple shear algorithm (Verrall 1981). The extensional fault had a 50° cut-off angle at the surface with an intermediate kink where the fault dip changed to 24° (Figs 10a, 11a). Syn-kinematic strata (represented by initially horizontal lines) were added incrementally during extension. The fault was

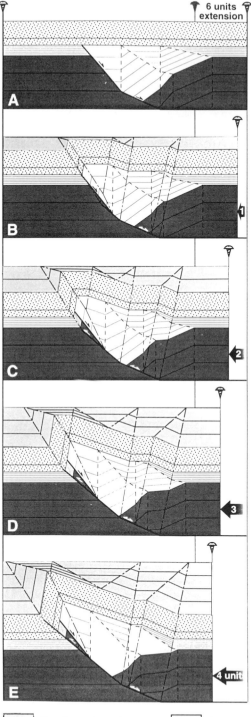

Fig. 10. Geometric model of the progressive inversion of an extensional basin over a propagating shortcut fault developing in the footwall. The rate of syn-inversion sedimentation is high, such that the inversion anticline is buried at all stages of inversion. (**a**) End extension architecture produced using a vertical simple shear algorithm over a 50°, 24°, 0° dipping extensional listric fault, showing a syn-extension growth sequence capped by a post-extension layer. (**b**) to (**e**) Progressive evolution of the structure during inversion over a 40° dipping shortcut to the 50° ramp of the initial extensional fault.

allowed to propagate up to the surface at a constant angle of 50° and at a rate sufficient to keep pace with sedimentation. The models include net regional subsidence such that syn-extension strata accumulated on the un-deformed footwall to the half-graben (Figs 10a, 11a).

The inversion of the extensional half-graben was modelled using the fixed-axis fault-propagation fold model of Suppe & Medwedeff (1990). This model, like the fault-bend model of Suppe (1983), assumes flexural slip along initially horizontal and parallel slip-planes (i.e. interbed slip). The inversion geometries shown in Figs 10b–e and 11b–e were calculated for the case of no overall layer parallel shear and with the addition of a 40° footwall shortcut fault propagating from the kink in the ramp of the extensional fault. The ramp to flat transition from a dip of 24° to a dip of 0° in the main fault surface was treated as a simple synclinal fault-bend fold (Suppe 1983). Horizontal syn-inversion growth strata were added above the developing fold for each increment of reverse slip on the fault. Complex internal geometries within the models (Figs 10b–e, 11b–e) result from the interference between the kink-band folds produced during extension and those produced during inversion.

In Fig. 10, the rate of syn-inversion sedimentation was high and continuous at 1.5 times the slip rate on the horizontal flat section of the master fault. This model produces a growth anticline with triangular dip domains bounded by kink band axial surfaces within the syn-inversion strata. The overall shape of this forward model as defined by the geometry of the post-extension strata (Fig. 10c–e) closely resembles the geometries of the Sant Corneli–

Syn-inversion strata   Syn-extension strata

Post-extension strata   Pre-extension strata

— 1 unit of displacement

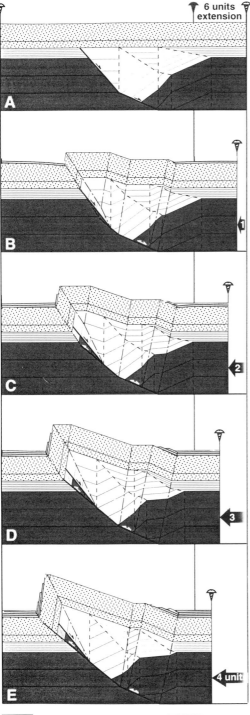

Fig. 11. A model with the same geometric evolution as the model in Fig. 10, but with a low syn-inversion sedimentation rate, such that the anticline is emergent at all stages of inversion. (**a**) End extension architecture produced using a vertical simple shear algorithm over a 50°, 24°, 0° dipping extensional listric fault, showing a syn-extension growth sequence capped by a post-extension layer. (**b**) to (**e**) Progressive evolution of the structure during inversion over a 40° dipping shortcut to the 50° ramp of the initial extensional fault. Growth sedimentation was modelled at the same slip-to-sedimentation ratio as for extension.

Boixols–Nargo anticline as shown in Fig. 7. However the syn-inversion Arén growth strata clearly onlap the Boixols structure indicating that the sedimentation rate did not keep pace with uplift produced by inversion. The second forward model in Fig. 11 shows the stratal architecture predicted by the fixed-axis fault-propagation model for a very slow rate of sedimentation ($0.2 \times$ the rate of slip on the horizontal flat section of the master fault). The complexities of the growth stratal geometries developed in the model in Fig. 10 are greatly reduced in Fig. 11 due to the lack of sedimentation on the fold crest. The terraces shown in Fig. 11b–e are geometric effects produced by the 'slip-then-deposit' nature of this modelling method.

Structural features not predicted by these forward models include the Santa Fe syncline (Fig. 7) and the lack of syn-inversion strata on the back limb of the major structure. These features may be attributed to post Oligocene tilting and uplift together with deformation associated with the Morreres backthrust that caused greater erosion of the northern section of the inversion structure.

The geometric model for the evolution of the Boixols thrust system (Fig. 9) offers an explanation for the structural and stratal relationships found in the Boixols thrust sheet. Careful analysis of structural and stratigraphic relationships has enabled syn-inversion growth geometries to be identified. Further research is needed to determine rates of growth folding and thrust slip in the Boixols thrust sheet. The preservation and identification of the footwall short-cut fault in the inverted half-graben system is particularly significant. Identification of similar geometries elsewhere may help to interpret

☐ Syn-inversion strata   ☐ Syn-extension strata

▨ Post-extension strata   ■ Pre-extension strata

— 1 unit of displacement

such complex structures at other localities in the Pyrenees and also in other thrust belts where basin inversion has preceded fully 'thin-skinned' thrust tectonics.

Conclusions

The structure of the Boixols thrust sheet is dominated by kilometric-scale E–W trending folds the most important of which is the Sant Corneli–Boixols–Nargo anticline at the leading edge of the thrust sheet. This anticline has been interpreted as a fault-propagation growth fold related to the inversion of a Lower Cretaceous extensional fault. Five unconformity bound megasequences have been identified within the Boixols thrust sheet: megasequence 1, Triassic–Jurassic; megasequence 2, Lower Cretaceous extension; megasequence 3, Upper Cretaceous tectonic quiescence; megasequence 4, Late Cretaceous inversion; and megasequence 5, Upper Eocene–Oligocene late-stage Pyrenean contraction. The Boixols thrust sheet incorporates the Lower Cretaceous Organyá extensional basin which underwent inversion in the Late Campanian–Maastrichtian and was incorporated into the Pyrenean south-vergent thrust system during the Eocene–Oligocene. The inversion history is recorded by the syn-inversion Arén growth sequence that is only preserved in the footwall syncline to the main fault-propagation growth fold.

During inversion a short-cut thrust fault developed in the footwall to the Cretaceous extensional fault system. This short-cut forms the present day Boixols thrust at the surface. The proposed inversion model for the Boixols thrust sheet and the Organyá basin has similar geometric features to analogue and simple geometric models of half-graben inversion structures. The inversion model for the Boixols thrust may be applied elsewhere in the Pyrenees and to structures in other foreland fold and thrust belts.

The research presented in this paper is part of the Fault Dynamics Project sponsored by ARCO British Ltd., BP Exploration, BRASOIL UK Ltd., CONOCO UK Ltd., MOBIL North Sea Ltd. and SUN International. We would also like to acknowledge the Spanish Ministry for Education and Science and the British Council (Acción Integrada Hispano-Británica 1992/94 no. 102 A and 45 B: Modelling of syn-sedimentary structures) for their financial support. P. Arbués is thanked for making unpublished stratigraphic maps and measured sections available, and for valuable discussion on the stratigraphy of the Arén sequence in the area. J. Poblet, J. A. Muñoz, I. Fitzsimons, S. Hardy, C. Puigdefabregas and G. Nichols are thanked for fruitful discussions and for suggestions that have greatly improved this manuscript. Fault Dynamics Publication No. 40.

References

BERASTEGUI, X., GARCIA-SENZ, J. M. & LOSANTOS, M. 1990. Tectonosedimentary evolution of the Organyá extensional basin (central south Pyrenean unit, Spain) during the Lower Cretaceous. *Bulletin de la Société Géologique de France, Paris*, **8 (VI)**, 251–264.

BUCHANAN, P. & MCCLAY, K. R. 1991. Analogue models of inversion structures. *Tectonophysics*, **188**, 97–115.

CHOUKROUNE, P., LE PICHON, X., SÉGURET, M. & SIBUET, J. C. 1973. Bay of Biscay and Pyrenees. *Earth and Planetary Science Letters*, **18**, 109–118.

EICHENSEER, H., LUTERBACHER, P. & LUTERBACHER, H. 1992. The Marine Paleogene of the Tremp Region (NE Spain) – Depositional Sequences, Facies History, Biostratigraphy and Controlling Factors. *Facies*, **27**, 119–152.

FISCHER, M. P., WOODWARD, N. B. & MITCHELL, M. M. 1992. The kinematics of break-thrust folds. *Journal of Structural Geology*, **14**, 451–460.

HUBBARD, R. J., PAPE, J. & ROBERTS, D. G. 1985. Depositional sequence mapping as a technique to establish tectonic and stratigraphic framework and evaluate hydrocarbon potential on a passive continental margin. *In*: BERG O. R. & WOOLVERTON D. G. (eds) *Seismic Stratigraphy II: an integrated approach*. American Association of Petroleum Geologists Memoir, **39**, 79–91.

KRAUS, S. 1990. Stratigraphy and Facies of the 'Garumnian' – Late Cretaceous to Early Palaeocene – in the Tremp Region, Central Southern Pyrenees. *Tübinger Geowissenschaftliche Abhandlungen, Reihe A, Geologie, Paläontologie, Stratigraphie*, **11**, 152.

LOSANTOS, M., BERASTEGUI, X., MUÑOZ, J. A. & PUIGDEFABREGAS, C. 1988. Corte geológico cortical del Pireneo Central (perfil ECORS): Evolución geodinámica de la cordillera Pirenaica. *In*: *Simposio sobre cinturones orogénicos, SGE, 1988*, II Congreso Geológico de España, Granada, 7–16.

MASSON, D. G. & MILES, P. R. 1984. Mesozoic seafloor spreading between Iberia, Europe and North America. *Marine Geology*, **56**, 279–287.

MELLERE, D. 1993. *Thrust-generated, back-fill stacking of alluvial fan sequences, south-central Pyrenees, Spain (La Pobla de Segur Conglomerates)*. International Association of Sedimentologists, Special Publication, **20**, 259–276.

MEY, P. H. W., NAGTEGAAL, P. J. C., ROBERTI, K. J. & HARTEVELT, J. J. A. 1968. Lithostratigraphic subdivision of Post-Hercynian deposits in the South-Central Pyrenees, Spain. *Leidse Geologische Mededelingen*, **41**, 221–228.

MUÑOZ, J. A. 1992. Evolution of a continental collision belt : ECORS – Pyrenees crustal Balanced cross-section. *In*: MCCLAY, K. R. (ed.) *Thrust Tectonics*. Chapman & Hall, London, 235–246.

NAGTEGAAL, P. J. C. 1972. Depositional history and clay minerals of the Upper Cretaceous Basin in the south-central Pyrenees, Spain. *Leidse Geologische Mededelingen*, **47**, 251–275.

PEYBERNÉS, B. 1976. Le Jurassique et al Crétacé inférieur des Pyrénées franco-espagnoles entre la Garonne et la Méditerranée. *Thèse Doct. Sci. Nat. Toulouse*.

—— & SOUQUET, P. 1984. Basement blocks and tecto-sedimentary evolution in the Pyrenees during Mesozoic times. *Geological Magazine*, **121**, 397–405.

PUIGDEFABREGAS, C., MUÑOZ, J. A. & VERGÉS, J. 1992. Thrusting and Foreland Basin evolution in the Southern Pyrenees. *In*: MCCLAY, K. R. (ed.) *Thrust Tectonics*, Chapman & Hall, London, 247–254.

—— & SOUQUET, P. 1986. Tecto-sedimentary cycles and depositional sequences of the Mesozoic and Tertiary from the Pyrenees. *Tectonophysics*, **129**, 173–203.

ROEST, W. R. & SRIVASTAVA, S. P. 1991. Kinematics of the plate boundaries between Eurasia, Iberia, and Africa in the North Atlantic from the Late Cretaceous to the present. *Geology*, **19**, 613–616.

SÉGURET, M. 1972. Étude tectonique des nappes et séries décollées de la partie centrale du versant sud des Pyrénées. *Publications de l'Université de Sciences et Techniques de Languedoc, série Geologie Structurale*, **2**, Montpellier.

SIMO, A. 1985. *Secuencias deposicionales del Cretácico superior de la Unidad del Montsec. (Pirineo Central)*. PhD thesis Universidad de Barcelona.

—— 1986. Carbonate platform depositional sequences, Upper Cretaceous, South Central Pyrenees (Spain). *Tectonophysics*, **129**, 205–231.

—— 1989. Upper Cretaceous platform-to-basin depositional-sequence development, Tremp Basin, South-Central Pyrenees, Spain. *In*: CREVELLO, P. D., WILSON, J. L., SARG, J. F. & READ, J. F. (eds) *Controls on Carbonate Platform and Basin Development*. Society of Economic Palaeontologists and Mineralogists, Special Publication, **44**, 365–378.

SUPPE, J. 1983. Geometry and Kinematics of Fault-Bend Folding. *American Journal of Science*, **283**, 684–721.

—— & MEDWEDEFF, D. A. 1984. Fault-propagation folding. *Geological Society of America Abstracts with Programs*, **16**, 670.

—— & —— 1990. Geometry and kinematics of Fault-Propagation folding. *Eclogae geologicae Helvetiae*, **83**, 409–454.

VAN HOORN, B. 1970. Sedimentology and Palaeogeography of an upper Cretaceous turbidite basin in the South-Central Pyrenees. Spain. *Leidse Geologische Mededelingen*, **45**, 73–154.

VERRALL, P. 1981. *Structural interpretation with applications to North Sea problems*. Joint Association of Petroleum Exploration Courses, London, Course Notes No. 3.

Inversion of an extensional-ramp basin by a newly formed thrust: the Cameros basin (N. Spain)

JOAN GUIMERÀ[1], ÁNGELA ALONSO[2] & JOSÉ RAMÓN MAS[2]

[1]*Departament de Geologia Dinàmica, Geofísica i Paleontologia, Facultat de Geologia, Universitat de Barcelona, Martí i Franqués s/n, 08028 Barcelona, Spain*
[2]*Departamento de Estratigrafía, Facultad de Ciencias Geológicas, Universidad Complutense de Madrid and Instituto de Geología Económica del C.S.I.C. Ciudad Universitaria, 28040 Madrid, Spain*

Abstract: The Cameros basin (Iberian Chain) was developed under a very subsident extensional regime during the latest Jurassic–Early Cretaceous. Its sedimentary fill constitutes a megasequence of more than 5000 m of vertical thickness which contains six depositional sequences. Most of the sediments are continental (alluvial and lacustrine) with only minor marine intercalations. A lateral accretionary geometry at the basin scale shows a SSW–NNE migration of the depocentre and an onlap to the north of the sedimentary sequences on the Mesozoic substratum of the northern boundary of the basin.

The overall Mesozoic structure of the basin, as seen from field studies and seismic profile interpretation, is a gentle WNW–ESE synclinorium 30–70 km wide and 150 km long. Extension in the pre-basin Mesozoic rocks is very small. The basin boundaries are defined by an unconformity of the basin filling rocks on the former Mesozoic substratum; it is bounded only locally by major faults at surface. The basin is interpreted as an extensional-ramp basin produced over an S-dipping ramp in a blind, flat-lying extensional fault some kilometres deep in the basement. The extensional displacement on the fault produced a synclinal basin which progressively widened with time. The depocentres of the successive depositional sequences were always located above the ramp and migrated to the north, inside the basin, as a result of the hangingwall displacement to the south. From a computer-modelled cross-section, we estimate the total displacement on the extensional fault to be about 33 km. This extension, in the hangingwall, would have taken place in the Pyrenean basin, north of the Cameros basin.

Tertiary compression (Palaeogene to Lower–Middle Miocene) inverted the basin by means of a newly formed E–W north-directed thrust. Overthrusting in the Tertiary Ebro basin fill, this new fault formed in the Keuper beds, in a weakness zone with uniform dip within the extensional hangingwall (about 30 km in cross-section). During the deformation it expanded to the north and south until it branched to the Iberian sole thrust, which might coincide with the Mesozoic extensional sole fault. Although this thrust produced a slight (2–3 km) inversion of the basement-Mesozoic cover unconformity, its maximum displacement was of about 28 km. Pre-basin Mesozoic rocks overlie the thrust surface and are overlain by the basin-filling rocks. A south-directed imbricate-fan thrust system developed in the southern margin. The total shortening was about 38 km, leading to the complete inversion of the basin.

The tectonic inversion of sedimentary basins is an issue of increasing interest in structural geology and in basin evolution studies (Butler 1989; Cartwright 1989; McClay *et al.* 1989; Roure *et al.* 1992). Usually the inversion occurs when the extensional faults that bound a basin formed during a previous extensional period invert their motion, becoming reverse-slip or thrust faults during a later compressional period, during which the whole basin may be compressed. A less studied way of basin inversion is by means of newly formed thrusts, which are facilitated by blocking, sealing or inappropriate orientation of the previous extensional faults with respect to the compression.

The aim of this paper is to study the inversion of the Cameros basin, a late Jurassic–early Cretaceous basin in the Iberian Chain, using data from field studies, seismic reflection profiles and oil exploration wells. Based on the results of this study, we propose a geometric and kinematic computer model for the Mesozoic extension and the Tertiary inversion of the basin, which is presented as an extensional-ramp basin inverted by a newly formed thrust.

From BUCHANAN, J. G. & BUCHANAN, P. G. (eds), 1995, *Basin Inversion*, Geological Society Special Publication No. 88, 433–453.

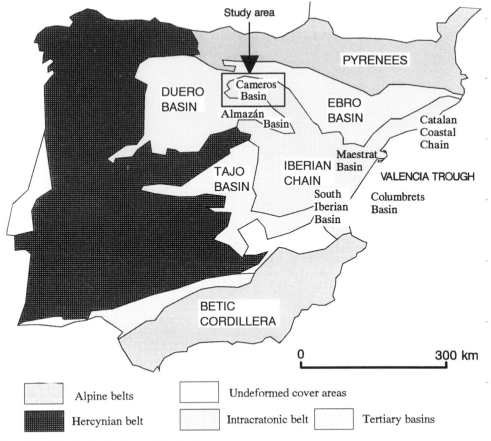

Fig. 1. Simplified geological map of the Iberian peninsula showing the location of the area studied.

Geological setting

The Iberian Mesozoic basins are located in the eastern part of the Iberian peninsula (they contain the present Iberian Chain, Catalan Coastal Chain and parts of the Ebro basin) and in the offshore Valencia trough, in the western Mediterranean (Fig. 1). They are composed of a sedimentary cover of Upper Permian and Triassic to Cretaceous age, containing carbonate and siliciclastic shallow marine and continental rocks, which lie over a widespread erosional surface which truncates the folded Palaeozoic rocks of the Hercynian basement. The thickness of this cover shows significant variations, from less than 1000 m to almost 7000 m.

From sequence stratigraphy, subsidence analysis and the integration of the intra-Mesozoic tectonic structures observed in the field, four evolutionary extensional stages can be recognized during the Mesozoic in the Iberian Chain (Álvaro *et al.* 1979; Vilas *et al.* 1983; Salas

& Casas 1993; Roca *et al.* 1994): 1) Triassic rift (Late Permian–Hettangian) which gave rise to the sedimentation from 0–2000 m of fluvial red beds with intercaled shallow marine carbonate, evaporite and salt accumulations; 2) Early and Middle Jurassic postrift (Sinemurian–Oxford-ian) recorded by shallow marine carbonates (up to 700 m) whose basal beds (carbonate breccia) overlie a partly to totally dismantled Triassic sequence (Esteban & Julià 1973; Giner 1978; Aurell *et al.* 1992; San Román & Aurell 1992); 3) Late Jurassic and Early Cretaceous rift (Kimmeridgian–Middle Albian), during which four more subsident basins (up to 4–5 km thick) developed in eastern Iberia (Fig. 1): the Cameros basin (Mas *et al.* 1993), the Maestrat basin (Salas 1987, 1989), the Columbrets basin (Roca & Guimerà 1992) and the South Iberian basin (Vilas *et al.* 1982); outside of these areas the rocks of this age lack or are only a few hundred metres thick; 4) Late Cretaceous postrift (Late Albian–Maastrichtian) recorded by up to 700 m of a basal terrigenous Utrillas Fm (late Albian), which lies on eroded rocks of the previous Mesozoic Sequences and even on the Hercynian basement, and marine carbonates which spread to all the Iberian basin showing little variations in thickness. Each of these stages produced a Depositional Supersequence.

Later, during the Eocene to Early Miocene, the Mesozoic basins were compressed and inverted, giving rise to the Iberian and Catalan Coastal chains, which overthrust the Ebro foreland basin and other Tertiary basins inside Iberia (Fig. 1).

The Iberian Chain is a thrust belt involving both the Hercynian basement and the Mesozoic cover (Guimerà 1984; Guimerà & Álvaro 1990). Folds and thrusts are developed both in the basement and in the cover and strike NW–SE and locally E–W (northern boundary of the Mesozoic Maestrat and Cameros basins) (Fig. 2). Two detachment levels can be deduced. One, located at the upper Triassic evaporitic levels, can be observed or deduced throughout most of the Iberian Chain, separating imbricate thrust systems in the cover from duplex thrust systems in the basement (Viallard 1989; Guimerà & Álvaro 1990). The other detachment is deduced from the existence of NW–SE-oriented thrusts in the basement; the depth of its sole thrust is not well established, and both thin-skinned (10–12 km deep sole thrust, Guimerà & Álvaro 1990) and thick-skinned (about 30 km, in the Moho, Salas & Casas 1993) have been proposed. The sense of displacement of the thrust sheets is to the NE or to the N in the northern and northeastern parts of the chain and to the SW in the southern and southwestern areas. In some of these NW–SE faults, a dextral component in the fault motions can also be deduced (Colomer & Santanach 1988; Guimerà 1988). Northeast of the Iberian Chain, the Catalan Coastal Chain displays a NE–SW sinistral-convergent strike-slip system (Guimerà 1984; Anadón *et al.* 1985). Motion on faults of all the orientations just described take place simultaneously at least during the Eocene and Oligocene, allowing a N010E compression direction to be deduced (Guimerà 1984; Guimerà & Álvaro 1990).

Stratigraphy of the Cameros area

Mesozoic pre-Cameros basin units: the Triassic and Early–Middle Jurassic megacycles

The Triassic record essentially corresponds to a megacycle bounded at the base and on top by widespread unconformities, the Earliest Triassic and Latest Triassic–Earliest Jurassic ones respectively. In the Cameros area, it displays the typical Bunstandstein, Muschelkalk and Keuper Germanic facies and lies unconformably over the Hercynian basement, which is made up of slightly metamorphic siliciclastic and carbonate rocks of Precambrian, Cambrian and Ordovician age.

The Buntsandstein is represented by conglomerates, sandstones and lutites which essentially correspond to alluvial systems; its thickness is very variable, ranging from 0–50 m in the western part to up to 500 m in the most eastern part. The Muschelkalk is represented by dolomites, limestones and marls which correspond to shallow carbonate platforms; its thickness varies from 0–20 m in the western part to 100 m in the eastern one. The Keuper is formed by clays, marls and evaporites deposited in coastal sabkha environments; its thickness is very variable (a mean of 30–40 m), reaching up to 100 m in the northern part.

The marine Jurassic record (which has a mean thickness of 600–700 m) essentially represents the evolution of carbonate ramps during a period of relative tectonic calm developed between two rifting phases, the Permian–Triassic and latest Jurassic–Early Cretaceous ones.

The Cameros basin fill: the Late Jurassic–Early Cretaceous megacycle

The sedimentary record in the Cameros basin is mainly formed by sediments of continental origin (alluvial and lacustrine systems) with very rare marine incursions (Gómez Fernández 1992; Alonso & Mas 1993). Chronostratigraphic data are not yet completed because of the generalized poor fossil record and local metamorphism. The main stratigraphic data are based on charophites, ostracods and palynologic associations for the continental record and some scarce marine calcareous algae for the rare marine incursions (Brenner 1976; Guiraud & Séguret 1985; Schudack 1987; Martin-Closas 1989; Clemente *et al.* 1991; Alonso & Mas 1993).

The basin infilling (Tithonian–Early Albian) corresponds to a large cycle or Megasequence bounded by two main unconformities at the base and the top (Mas *et al.* 1993). The stratigraphic gap which corresponds to the unconformity of the lower limit is more important in the Northwestern part of the basin (Early Tithonian–Barremian) than in the South and Southeastern sectors (part of the Early

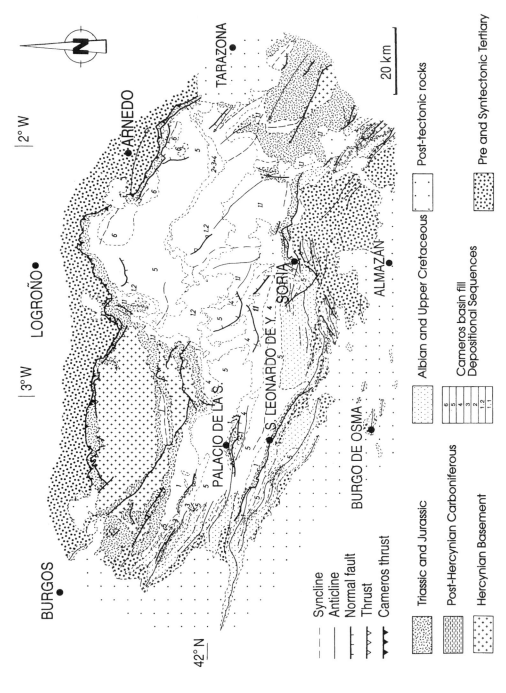

Fig. 2. Geological map of the Cameros basin and adjoining areas.

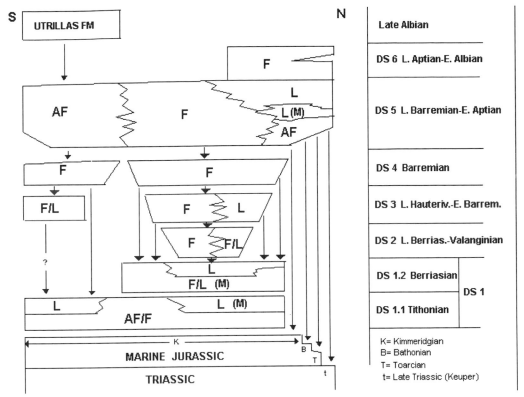

Fig. 3. Simplified stratigraphy of the Depositional Sequences (DS) which fill the basin. Depositional Systems: Alluvial fans (AF); Fluvial (F); Lacustrine (L); Minor Marine intercalations (M).

Tithonian). The upper limit is the unconformity (intra-Albian) that bounded the Upper Cretaceous Megacycle of the Iberian Ranges on its base.

The Megasequence can be divided into six Depositional Sequences (DS) bounded by unconformities (Mas *et al.* 1993), (Figs 3, 4):

DS1 (Tithonian–Berriasian). The lower limit is the Early Tithonian unconformity already mentioned. The upper limit is an unconformity (Latest Berriasian). The maximum thickness is reached in the eastern part of the basin (up to 3000 m). Several alluvial–lacustrine cycles are distinguished, three in the West and five in the East. In the eastern sector this sequence always overlies the Kimmeridgian marine sequence and in the western sector it overlies different Jurassic sequences such that the more westerly they are the older they are. This sequence can also be divided into two subsequences separated by a less important discontinuity: DS 1.1 (mainly Tithonian) and DS 1.2 (essentially Berriasian). In the eastern sector, during the second subsequence (DS 1.2), the first lacustrine episode is the only one which is characterized by the

development of carbonate–evaporitic playa-lakes (Fig. 3) whereas, throughout the basin, the other lacustrine episodes of the sequence essentially correspond to freshwater shallow carbonate lakes which only very occasionally display thin marine incursions (lagoonal sediments).

DS2 (Latest Berriasian–Valanginian). Its upper limit is an unconformity that includes part of the Valanginian and probably part of the Hauterivian stages. It is only represented at the western and eastern ends of the basin with a maximum thickness of 300 m and 200 m respectively. In the western sector it is represented by carbonate lacustrine deposits and in the eastern sector by siliciclastic and mixed siliciclastic–carbonate fluvial and lacustrine sediments.

DS3 (Late Hauterivian–Early Barremian). It is not represented in the central part of the basin, the thickness being up to 800 m in the western sector and up to 500 m in the eastern sector. Both in the western and in the eastern areas this sequence corresponds to siliciclastic–alluvial and carbonate–lacustrine units which display lateral changes of facies between them. The

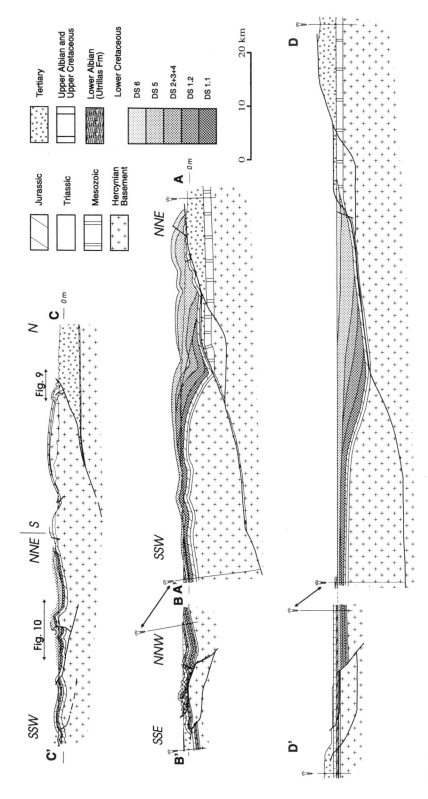

Fig. 4. A-A', B-B' and C-C': geological cross-sections through the Cameros basin. D-D': partial restoration of cross sections A-A' and B-B' at the state previous to the basin inversion. See Fig. 5 for location.

upper limit of the sequence is an unconformity (Early Barremian).

DS4 (Barremian). This sequence reaches a thickness of up to 900 m in the eastern sector of the basin where it is represented by two units, one of them siliciclastic fluvial in origin and the other one mixed siliciclastic–carbonate fluvial and lacustrine in origin. In the western sector it is characterized by fluvial deposits with a thickness being up to 800 m. The upper limit of this sequence is an important unconformity recognizable in all the Iberian Ranges (Late Barremian), which meant a generalized increment of subsidence and sedimentary accumulation rates in the basin.

DS5 (Late Barremian–Early Aptian). It is represented throughout the basin by a maximum thickness in the northeastern depocentral zone (up to 1900 m). Although this sequence is mainly represented by siliciclastic fluvial units, in the eastern sector it is also represented by a thick carbonate lacustrine unit (up to 1100 m) which presents some occasional episode of marine influence (lagoonal deposits). The upper limit of this sequence is an unconformity (intra-Aptian).

DS6 (Late Aptian–Early Albian). It is only present in the eastern sector and the maximum thickness preserved is 1500 m. This sequence corresponds to a unique alluvial unit with only rare thin interbedded carbonate lacustrine episodes. The upper limit is a large erosive unconformity (intra-Albian).

The Tithonian–Berriasian depositional sequence (DS 1) corresponds to the first rifting phase of the Latest Jurassic–Early Cretaceous Iberian Rifting (Mas et al. 1993). The sequences DS 2 and DS 3 (Latest Berriasian to Early Barremian) correspond to a stage of reduced rifting. And the sequences DS 4, DS 5 and DS 6 (Barremian to Early Albian) correspond to the second rifting phase of the Latest Jurassic–Early Cretaceous Iberian Rifting (Mas et al. 1993).

Mesozoic post-Cameros basin units: the Late Cretaceous Megasequence

The base of this Late Albian–Maastrichtian Megasequence, which in this area is up to 1000 m thick, is the intra-Albian unconformity which is present in large areas of the Iberian plate. From that moment the palaeogeographic configuration changed drastically and the Cameros Basin, like other latest Jurassic–early Cretaceous rift basins of Iberia, lost its identity. Large carbonate platforms occupied the Iberian

Seaway between the Iberian and Ebro Massifs during the Late Cretaceous in the Iberian basin realm (Alonso-Millán et al. 1989; Alonso et al. 1993).

The Megacycle starts over almost the whole area with siliciclastic continental deposits (Utrillas Fm) which overlie units ranging in age from the Lower Albian to the Palaeozoic. The Utrillas Fm itself is a very homogeneous, continuous but diachronous unit. Its age ranges from Mid–Late Albian to Turonian. Overlying these terrigenous rocks, a set of higher order cycles have been identified within this late Cretaceous Megasequence (Alonso et al. 1993).

Metamorphism affecting the Cameros basin

A low-grade metamorphism affected the Tithonian to lower Albian rocks filling the most subsident part of the Cameros basin (northwest of Tarazona, Fig. 2) (Guiraud 1983). The metamorphism conditions, areal extension and their relationship to the tectonic structures of the region are still under discussion. After Guiraud & Séguret (1985) it was dynamo-thermal, and Golberg et al. (1988) characterized it as reaching temperatures of 400° C in the most subsident part of the basin and an age of 99.5 ± 2.2 Ma (^{39}Ar–^{40}Ar method on newly formed phengites). More recently, Casquet et al. (1992) interpreted the metamorphism as hydro-thermal, with the maximum temperature at 350° C and an age of 108–86 Ma (Albian to Late Cretaceous) is deduced from K-Ar dating on authigenic illites. Both datings indicate that the metamorphism was produced just after the basin development (which was Tithonian–Early Albian); hence it does not appear to be related to the extensional tectonics which formed the basin. It would be related to the post-rift thermal evolution caused by the crustal thinning generated during the latest Jurassic–early Cretaceous basin formation.

Stratigraphy of the Tertiary basins

Two Tertiary basins bound the Cameros basin to the north and south. The northern one is the Ebro basin, which is the common foreland basin of the Pyrenees in the N, the Iberian Chain in the SW and the Catalan Coastal Chain in the SE. The southern one is the Almazán Basin, which is a WNW–ESE sub-basin of the larger Duero foreland basin (Figs 1, 2).

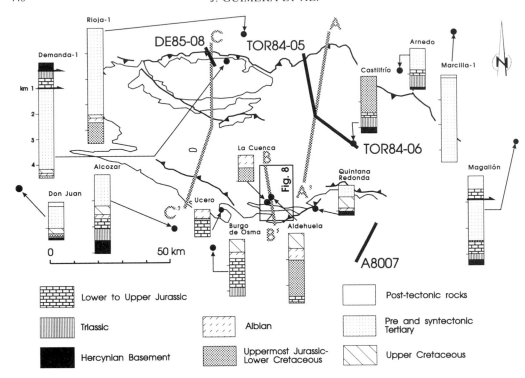

Fig. 5. Oil wells in the study area (modified from Lanaja 1987) and location of cross-sections of Fig. 4 and Fig. 7, seismic profiles of Fig. 6 and the map of Fig. 8.

The Ebro basin

North of the Cameros area more than 3000 m of terrigenous continental Tertiary rocks have been found in some oil-wells (see Demanda-1, Rioja-1 and Marcilla-1 oil-wells in Fig. 5) and deduced from seismic profiles (Fig. 6a, c) displaying a coarsening upwards succession (Demanda-1 oil-well; Lanaja 1987). Five sedimentary sequences separated by regionally widespread unconformities have been distinguished within this succession (Ramírez Merino *et al.* 1990; Muñoz Jiménez 1992). The first four sequences, with an age ranging from the Palaeogene to the Middle Miocene, are affected by the Cameros thrust; the fifth sequence is Late Miocene to Pliocene in age (Muñoz Jiménez 1992) and unconformably overlies the Cameros structures.

The Almazán basin

The Tertiary deposits which filled the basin are not well known, because the post-tectonic rocks cover most of the basin area. They can be grouped into three sequences or lithologic assemblages of mostly terrigenous continental rocks separated by regionally widespread unconformities (Navarro Vázquez *et al.* 1991a, b). According to these authors, the first and the second sequences are essentially considered Palaeogene in age; they would mainly correspond to the Eocene and Oligocene, and only the lower part of the first sequence could probably be Palaeocene in age (Navarro Vázquez *et al.* 1991a,b). These two sequences consist of more than 3000 m of sediment, calculated from seismic profiles (Fig. 6d), and are pre- and syn-tectonic respect to Alpine compressive deformation. The third sequence, considered Miocene–Pliocene in age, lies horizontally unconformable over different Palaeogene and Mesozoic units (Navarro Vázquez *et al.* 1991b). This clearly horizontal post-tectonic sequence is up to 200 m thick.

Structure of the Cameros basin and adjoining areas

The Cameros basin is located in the northernmost part of the Iberian Chain. It is a completely inverted basin included in an Alpine thrust-sheet (Cameros thrust-sheet) larger than the Cameros basin.

The Cameros thrust-sheet includes the uppermost Jurassic–lower Cretaceous basin fill and its Mesozoic substratum and the Hercynian basement, which crops out widely in the Sierra de la Demanda and surrounding areas as well as near the Moncayo peak (southwest of Tarazona; Fig. 2). It also includes the Almazán basin, which is then a piggy-back basin; the inverted Cameros basin overthrust, the Almazán basin along its southern boundary (Figs 2, 8). The thrust-sheet has a roughly E–W orientation and is at least 150 km in length and 80 km wide.

The northern border of this thrust-sheet is the Cameros sole thrust, which crops out for 120 km and has a roughly WNW–ESE trace. Emplaced synchronously with the sedimentation of the Palaeogene to Middle Miocene rocks of the Ebro basin, to the west and the southeast the Cameros thrust-sheet is overlain by post-tectonic upper Miocene rocks. Therefore, its actual length seems to be bigger, as deduced from the Magallón borehole (Fig. 5), where Jurassic rocks have been recognized thrusting over 1500 m of Tertiary rocks (Sánchez Navarro et al. 1990); this thrust may be the continuation of the Cameros thrust. The thrust trace draws three bows in map view; one, in the west, surrounding the Hercynian basement outcrop of the Demanda area, and the other two to the east, in the Cameros area.

To the southeast there is a transition between the Cameros and the Moncayo areas (Iberian Chain); the thickness of the basin fill reduces very quickly and NW–SE folds and faults appear, allowing the Mesozoic substratum of the basin and the Hercynian basement to crop out. Westwards, to the Sierra de la Demanda, the thickness of the Cameros basin fill also decreases, and WNW–ESE folds and thrusts then appear, involving the Mesozoic cover and the pre- and syn-tectonic Tertiary. These structures are unconformably overlain by the post-tectonic rocks of the Duero basin, preventing the observation of their continuation.

The southern border of the inverted Cameros basin is a roughly E–W-striking, south-directed thrust system with imbricate thrusts and thrust-propagation folds (Platt 1990; Casas Sainz 1992; Miegebielle et al. 1993), extending for about 150 km (Fig. 2). Thrusts and folds were synchronous with the sedimentation of the Tertiary rocks of the Almazán basin, its local foreland basin. Most of the rocks cropping out in the Almazán basin are post-tectonic Late Miocene and Pliocene rocks. Below them,

however, folds and thrusts affecting the Mesozoic and the pre- and syn-tectonic Tertiary rocks can be observed in seismic profiles (Fig. 6d) and in outcrops of Late Cretaceous rocks among the post-tectonic Tertiary (Fig. 2).

In order to describe more accurately the structure of the Cameros thrust-sheet and, especially that of the latest Jurassic–early Cretaceous Cameros basin, cross-sections from two transects running through the basin will be described, the Cameros and the Demanda transects.

The Cameros transect

Two geological cross-sections (Fig. 4 A–A' and B–B') and two seismic profiles (Fig. 6a, b) illustrate the most important geological features of this area.

The Mesozoic (Jurassic and, locally, upper Triassic) substratum of the basin fill is continuous throughout the basin, as can be deduced from the mapping (Fig. 2), seismic profiles (Fig. 6a, b) and the Castilfrío oil well (Fig. 5). This substratum underwent only minor extension, by normal faults of decametric to few hectometres of vertical slip (Guiraud 1983; Díaz Martínez 1988). There are, therefore, no major outcropping faults bounding the basin to the north.

Inside the basin, northward migration of the basin fill Depositional Sequences, and an onlap to the north of these sequences on the substratum, are also evident from surface (Fig. 2) and seismic (Fig. 6a; Fig. 4 A–A'). The basin fill reaches up to 5 km of vertical thickness and 9 km taking into account the north-migration of the Depositional Sequences.

The inversion of the basin was produced by a north-directed thrust located in the Upper Triassic (Keuper) beds. The displacement on this thrust is at least 23 km from the seismic profile (Fig. 6a), but may be estimated to be at least 28 km from the cross-section (Fig. 4 A–A'). North of the thrust, a short-cut in the thrust footwall has been recognized in the Arnedo area from surface structure, oil wells and seismic profiles (Ramírez Merino et al. 1990; Casas Sainz 1992); its continuation to the west could correspond to the horse or truncated thrust-sheet located in the Cameros thrust footwall of the seismic profile (a) in Fig. 6. The total shortening in cross-section A–A' (Fig. 4) is 33 km.

In the southern basin margin, a narrow band of E–W oriented folds and thrusts is observed, as shown in a more detailed map of the southern border of the Cameros unit and a geological cross-section through it (Figs 7, 8). The

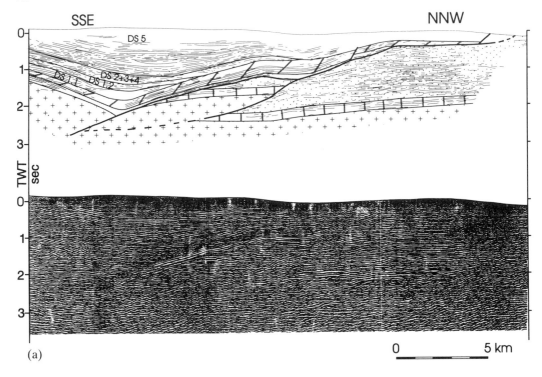

(a)

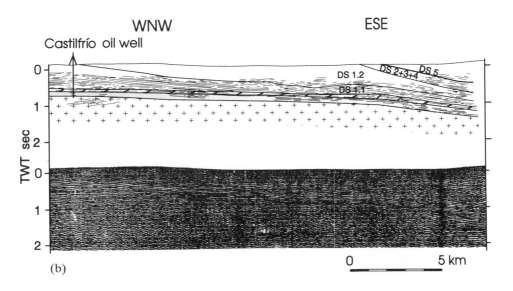

(b)

Fig. 6. Four seismic profiles in the area studied: (**a**) profile TOR84-05; (**b**) profile TOR84-06; (**c**) profile DE85-08; (**d**) profile A8007. For location see Fig. 5.

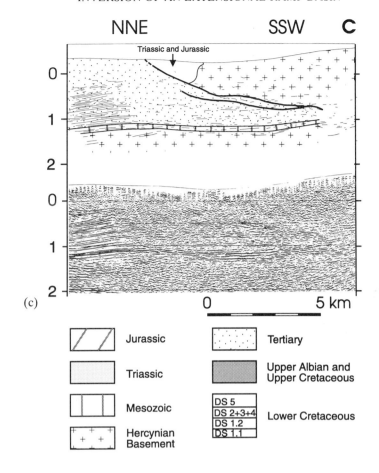

NNE   SSW   **C**

Triassic and Jurassic

(c)

0   5 km

	Jurassic		Tertiary
	Triassic		Upper Albian and Upper Cretaceous
	Mesozoic	DS 5 / DS 2+3+4 / DS 1.2 / DS 1.1	Lower Cretaceous
	Hercynian Basement		

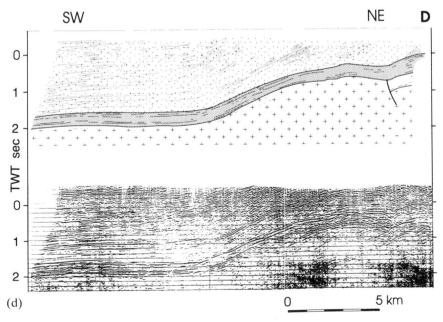

SW   NE   **D**

TWT sec

(d)

0   5 km

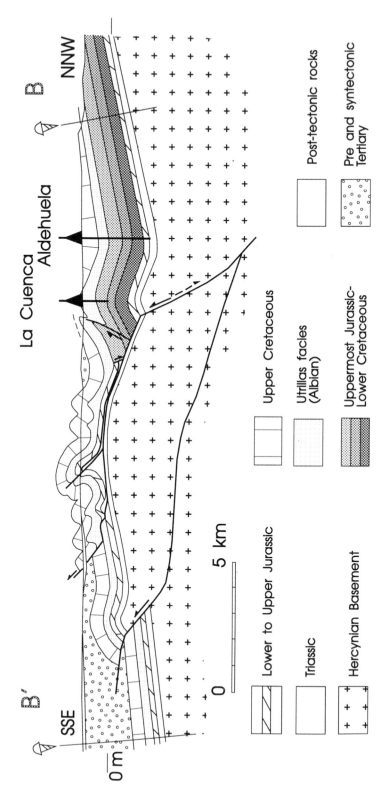

Fig. 7. Geological cross-section through the map of Fig. 8. For location see Figs 4 and 5.

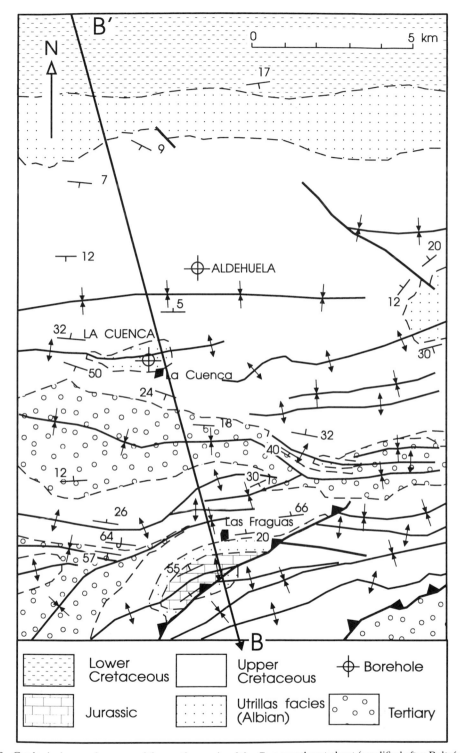

Fig. 8. Geological map of an area of the south margin of the Cameros thrust-sheet (modified after Beltrán Cabrera *et al*. 1980 and Valverde Hernández *et al*. 1991). The location of the cross-section of Fig. 7 is shown. For location see Fig. 5.

development of these compressional structures coincides with the disappearance of the Cameros basin fill, as deduced from field data and two oil wells. Thus, the Aldehuela oil well cut 1435 m of basin fill rocks, whereas 7 km to the south at Las Fraguas, the basin fill rocks are not found and the Albian overlies the pre-basin Jurassic.

This change is interpreted to be produced by a Cretaceous normal fault which, bounding the basin to the south (Meléndez & Vilas 1980; Clemente & Alonso 1990), is overlain by the Albian (Fig. 7). This fault was partially inverted and a short-cut formed at the Upper Triassic rocks, preventing the inversion of the upper segment of the Cretaceous normal fault. An imbricate thrust-system and associated folds developed in the Mesozoic cover south of the inverted fault. Another short-cut in the basement beneath the last structures is interpreted from the position of the Hercynian basement obtained in constructing the cross-section. The total shortening in this cross-section is 5.4 km.

Figure 4 D–D′ shows the partial restoration of cross-sections A–A′ and B–B′ before the Tertiary inversion. The composition of both restored sections gives an image of the geometry of the Cameros basin at the end of their development. Although a minor normal fault is developed in its southern margin, the basin appears to be a synclinal basin with no major fault bounding it to the north. The substratum of the basin fill is therefore always integrated by pre-basin Mesozoic rocks. The onlap of the basin fill depositional sequences to the north and the migration of their depocentres in that direction is evident.

The Demanda transect

One geological cross-section (Fig. 4C–C′) and one seismic profile (Fig. 6c) illustrate the structure through this transect, from the Ebro to the Almazán basins.

In this area the Cameros thrust has a bow-like pattern in map view. Hercynian basement and Mesozoic rocks thrust over 3000 m of Tertiary rocks, as seen in the Demanda borehole (Fig. 5). The Hercynian basement crops out at the core of an arch structure composed of two wide anticlines separated by a syncline, where Mesozoic rocks are present (Fig. 6c). A cover rim can be found around the limbs of the basement arch. In the north limb of the northern anticline, the Triassic to Jurassic beds dip from nearly vertical to reversed (Fig. 9 and Demanda-1 borehole in Fig. 5). Locally (Fig. 2), in the basement-cover contact of this north limb, a minor thrust

developed allowing the basement to thrust over the cover. The amount of displacement of the thrust-sheet is at least 9.5 km to the north, as deduced from the seismic profile at the central sector of the bow (Fig. 6c), but the cut-off of the footwall Mesozoic rocks cannot be clearly deduced in this profile.

At the west periclinal end of the arch, there are NW–SE-striking, near-vertical kilometric faults (Fig. 2). These faults seem to have been normal faults during the post-Hercynian Carboniferous sedimentation, giving rise to half-graben basins and producing significant variations in the thickness (Colchen 1974) of these Carboniferous rocks. Mesozoic activity of these faults is less clear, but they affect the Triassic, pre-basin Jurassic and the DS1 of the Cameros basin fill. These faults are cut by roughly E–W-oriented thrusts developed in the south limb of the southern Hercynian basement anticline (Fig. 2).

South of the Demanda arch, the western continuation of the Cameros basin is found, deformed by gentle ESE–WNW folds of kilometric size (Fig. 4c) and two major faults at surface.

The northern fault (Palacios de la Sierra fault) is an intra-Lower Cretaceous normal fault, as can be deduced from the lack of the DS4 in its footwall block and the existence of 400 m of rocks of this DS in the hangingwall (Fig. 10); this E–W fault is now folded and partially inverted, appearing as a reverse fault in the Jurassic and Lower Cretaceous rocks of its west end area.

The southern fault (the San Leonardo fault) is a WNW–ESE-striking south-directed thrust, cropping out for about 40 km. South of the San Leonardo fault, there are SSW-verging thrusts and folds, unconformably overlain by the post-tectonic Tertiary of the Almazán basin. The San Leonardo thrust and these SSW-verging compressional structures have been interpreted as a SSW-directed thrust system of imbricate thrusts and thrust propagation folds (Platt 1990).

Extension direction during the basin formation and transport direction of the thrust-sheet during the basin inversion

Guiraud & Seguret (1985, Fig. 5) estimate the extension direction during the basin formation from small quartz dykes and tension gashes filled of calcite affecting the basin fill in the northern Cameros area (Fig. 11). The extension direction deduced from these data is NE–SW around the areas where the present basin boundary is

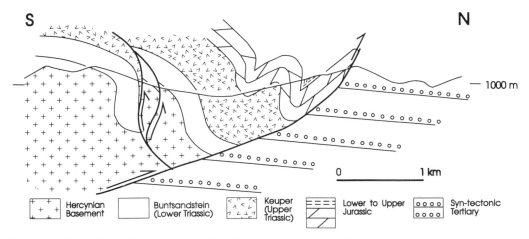

Fig. 9. Detailed section of the frontal part of the Cameros thrust-sheet in the Demanda area (northern portion of the C-C′ cross section). For location see Fig. 4.

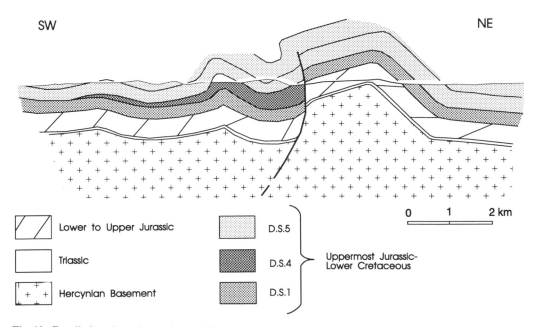

Fig. 10. Detailed section of central part of the Demanda (C-C′) cross section. For location see Fig. 4.

NW–SE (between Jubera and Baños, Fig. 11) and NW–SE where the present basin boundary is NE–SW (SSW of Soto, Fig. 11), south of these areas (south of Yanguas) the extension direction changes to NNE–SSW or almost N–S. Then, the extension directions deduced at the outcrop scale near the basin boundaries seem to be influenced by the local orientation of these, whereas towards the basin centre the extension

tends to be reorientated in a more NNE–SSW to N–S direction. From this we deduced a regional N–S to NNE–SSW extension direction during the basin formation.

To estimate the transport direction of the Cameros thrust-sheet during the basin inversion we apply the bow and arrow rule (Elliot 1976) to the sole thrust cropping out north of the sheet (Fig. 12); we assume that the arrow, along which

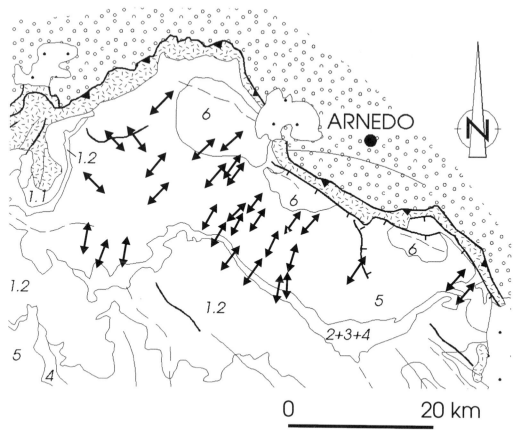

Fig. 11. Geological map of the northeastern part of the Cameros basin, showing the directions of extension at outcrop scale deduced from quartz dykes or calcite veins by Guiraud & Séguret (1985, fig. 5).

lies the maximum displacement, indicates the transport direction of the thrust-sheet. The arrow deduced in Fig. 12 is orientated towards the N–NNE. Nevertheless, the two ends of the bow chosen are not on the tip-line of the sole thrust, but they are the points where the sole thrust is covered by the post-tectonic Tertiary rocks. The thrust continues to the west and the southeast and, taking into account this continuation, we deduce that the arrow may have a more NNE orientation rather than a more N–S one.

In order to model the basin evolution, we assume the transport direction during the basin formation and inversion to be between N–S and NNE–SSW. During the basin formation it is assumed to be parallel to the extension direction previously deduced and during the basin inversion to be parallel to the arrow orientation

found applying the bow and arrow rule, also roughly between N–S and NNE–SSW.

Evolution of the Cameros basin

Basin formation

From what has been shown above, the Cameros basin is a synclinal basin with no major fault bounding it during its development (Fig. 4 D–D'). The N-migration of the basin fill Depositional Sequences and its lapping on the pre-basin Mesozoic substratum may be explained assuming that the basin formed over a roughly S-dipping ramp in a horizontal extensional fault, so it would be an extensional-ramp basin. Indeed, in scale models (McClay 1990, fig. 8; Roure *et al.* 1992, fig. 10), the same patterns in the units filling a synclinal basin have

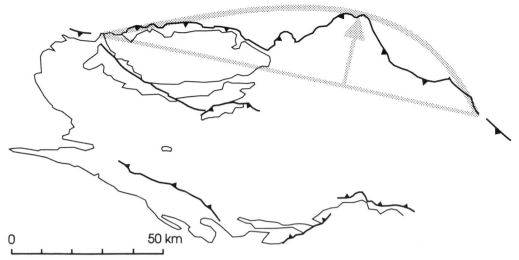

Fig. 12. Simplified map of the Cameros basin and the sole thrust croping out north of it. The application of the bow and arrow rule (Elliott 1976) is shown, in order to estimate the displacement direction of the Cameros thrust-sheet.

been obtained on horizontal extensional faults containing ramp segments over which the basin developed. In these models, the extension in the hangingwall rocks previous to the basin take place outside the synclinal basin, around the area where the extensional fault reaches the surface, so that the extension of the basin substratum is very small and their rock units remain continuous.

An important question is at what depth the extensional fault was located. According to Guiraud & Séguret (1985), Casas & Simón (1992) and Casas (1993) it would be located in the Keuper beds, so it would coincide with the thrust that inverted the basin (Fig. 4 D–D'). This hypothesis has a mechanical drawback which, in our opinion, makes it unlikely. To produce the basin, a slab of Jurassic rocks of about 500 m thick, several tens of kilometres wide (more than 30 km as we shall see later) and more than 100 km long had been to be pulled from the south without lack in its continuity nor formation of a fault over the ramp in the north basin boundary. Moreover, this had to be developed under subaerial conditions, as can be deduced from the continental nature of the rocks which filled the basin and from the lack of coeval rocks in many places north of it, which were being eroded. To avoid this problem we propose to locate the extensional fault in a deep position inside the Hercynian basement, at a depth of 7–11 km, coinciedent with the position of the sole thrust of the Iberian Chain proposed by Guimerà & Álvaro (1990). Then, the hang-

ingwall would be several kilometres thick and capable of allowing its displacement to the south without significant internal deformation. Summarizing, the Cameros basin would be produced by the motion of a deep extensional fault which contains two nearly horizontal sections (flats) separated by an intermediate section dipping to the south (ramp). We take a N 170 W sense of displacement for the hangingwall, parallel to the extension direction previously deduced.

To illustrate the hypothesis, we made a computer model using the program Fault! (Wilkerson & Usdansky 1989). This program constructs the fault-bend folds by vertical simple shear whereas the deformation mechanism observed in the studied area is bedding-parallel shear, and only one fault may be active at a time, so the modelling of coeval faults has to be performed in two successive stages. Despite these limitations, the computer model obtained is, in our opinion, a good representation of the evolution proposed for the Cameros basin.

Figure 13 shows the evolution obtained using Wilkerson & Usdansky's (1989) program for the Cameros cross-sections (Fig. 4 A–A', B–B'). Stage (a) shows the geometry of the extensional fault before the basin formation, and stages (b) to (f) the Tithonian to early Albian development and evolution of the basin as an extensional-ramp syncline as a result of the hangingwall displacement to the south. The width of the basin increases progressively until it reaches its final state. The depocentre at every stage is always located on the ramp and the former

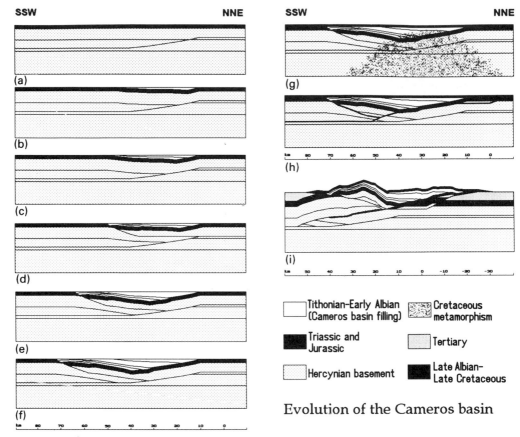

Fig. 13. Computer modelling of the Cameros basin formation and inversion, made using the program Fault!, by Wilkerson & Usdansky (1989). (**a**): pre-rift situation; (**b**): DS 1.1; (**c**) DS 1.2; (**d**) DS 2, 3 and 4; (**e**) DS 5; (**f**): DS 6; (**g**): late Cretaceous (metamorphism); (**h**): location of the faults which inverted the basin (before the basin inversion); (**i**) situation after the basin inversion (which took place during the Palaeogene and early–middle Miocene).

sedimentary units are displaced to the south during the sedimentation of the later units. The result is a migration to the north of the depocentres of every unit and an onlap to the north over the Mesozoic basin substratum.

To explain the secondary depocentres in the southern border of the Cameros basin, an antithetic fault is introduced in the model, at the southern border of the basin, during stages (d) and (e) (Fig. 13). This fault represents the major normal fault of cross-section B–B' in Figs 4 and 8. Similar faults have also been obtained in the scale models of McClay (1990) and Roure *et al.* (1992), as a result of the extension produced in the hangingwall when it passes from the ramp to the lower flat of the footwall.

The total extension obtained in this computer model is 33.3 km, and the maximum thickness of basin filling rocks is 5 km, determined by the differential depth between the two flats of the extensional fault (Fig. 13f).

Basin inversion

The basin inversion took place during the Palaeogene and the Early to Middle Miocene, as deduced from the age of the syn- and post-tectonic rocks filling the neighbouring Ebro and Almazán basins. The basin was inverted by a main north-directed thrust with a displacement of about 30 km in the north and a south-directed system of imbricate thrusts and thrust-propagation folds in the south, with total displacement of about 5 km. The northern main thrust is located at the upper Triassic (Keuper) beds in the north Cameros area and, towards the Demanda and the south part of the Cameros area, inside the Hercynian basement; joining both areas an oblique to frontal ramp could be deduced (Figs 2, 4 A–A' & C–C'). The hangingwall sense of displacement is assumed to be N 010 E, deduced from the application of the bow and arrow rule previously discussed and parallel to the sense of displacement assumed during the basin formation. This sense of displacement is roughly perpendicular to the Pyrenees and parallel to the compression direction inferred from the kinematics of the macrostructures in the Iberian Chain (Guimerà 1984 and Guimerà & Álvaro 1990).

Figure 13h shows the location of the Tertiary thrusts in relation to the previous Mesozoic structure. The northern main thrust is a newly formed fault. Its formation would have been facilitated by (1) the sealing on the Mesozoic extensional fault beneath the basin as a result of the late Cretaceous metamorphism (Fig. 10g; and (2) the existence, in the northern flank of the basin, of a potential weakness zone located in the Keuper beds, that shows a uniform dip (about 10° to the south) for about 30 km (Fig. 4 D–D'; Fig. 13h). From this weakness zone the new thrust might nucleate and spread to the north and to the south during the deformation, until it branched southwards to the lower flat of the Iberian sole thrust, which might coincide with the Mesozoic extensional sole fault. A short-cut formed in its footwall block. The thrust system of the southern basin margin developed from the inversion of the minor normal faults which bounded the Mesozoic basin to the south (Figs 8, 13h). The total shortening calculated for the Cameros cross-section (Fig. 4 A–A' & B–B') is 38.4 km and the one imposed on the model (Fig. 13i) is 35 km.

Therefore, the Cameros basin appears as a pop-up structure bounded by the Cameros thrust to the north and a less important back-thrust system to the south. In the model presented, the south-verging monocline located in the north of the Almazán basin (Figs 6d, 13i) could be generated as a result of the northwards displacement of the Cameros thrust-sheet along a south-dipping ramp. This implies that the Almazán Tertiary basin is a piggy-back basin inside the Iberian Chain.

Conclusions

The Cameros basin is an extensional-ramp basin, formed during the late Malm and early Cretaceous, over a roughly south-dipping ramp in a buried horizontal extensional fault several kilometres deep. The south-displacement of the hangingwall of this fault produced the basin over the ramp and increased the basin size progressively. The total extension during the basin formation was about 33 km. No main faults bound the basin-filling rocks to the north; they lie over the previous Mesozoic substratum.

The basin fill is grouped into six depositional sequences. The depocentres of these sequences were always located over the ramp and a progressive migration to the northeast of the successive depositional sequences was recorded. An onlap of the basin filling to the northeast on the pre-basin Mesozoic was formed as a result of this.

During the Palaeogene and early-middle Miocene the basin was inverted due to the Pyrenean compression. A new thrust formed at the potential weakness zone of the Upper Triassic (Keuper) beds in the north basin margin, where it had a uniform dip of about 10° to the south and 30 km wide. This new thrust may have expanded to the north and to the south during the deformation, until it branched to the Iberian horizontal sole thrust, which might coincide with the Mesozoic extensional fault. The north displacement of this thrust was about 30 km and generated the Almazán piggy-back basin. Beside this main thrust, an imbricate-thrust system developed on the south border of the Cameros basin. Therefore, the Cameros basin appears at surface as a pop-up structure about 80 km wide. The basin underwent a complete inversion. The shortening during Tertiary compression is similar to the lengthening during Mesozoic extension.

We thank E. Roca and F. Sàbat and an anonymous reviewer for their comments and suggestions. This work is a contribution to the projects PB89–0230 funded by the Comisión Interministerial de Ciencia y Tecnología, and PR179/91-3469 funded by the Universidad Complutense de Madrid.

References

ALONSO, A., FLOQUET, M., MAS, J. R. & MELÉNDEZ, A. 1993. Late Cretaceous Carbonate Platforms: Origin and Evolution. Iberian Range, Spain. *In*: TONI SIMO, J. A., SCOTT, R. W. & MASSE, J. P. (eds) *Cretaceous Carbonate Platforms*. American Association of Petroleum Geologists, Memoir, **56**, 297–316.

—— & MAS, J. R. in press. Control tectónico e influencia del eustatismo en la sedimentación del Cretácico inferior de la Cuenca de Los Cameros. España. *Cuadernos de Geología Ibérica*.

ALONSO-MILLÁN, A., FLOQUET, M., MAS, J. R. & MELÉNDEZ, A. 1989. Origin and evolution of an Epeiric Carbonate Platform. Upper Cretaceous. Spain. *XII Congreso Español de Sedimentología. Simposios*. Bilbao, 21–32.

ÁLVARO, M., CAPOTE, R. & VEGAS, R. 1979. Un modelo de evolución tectónica para la cadena celtibérica. Acta Geologica Hispanica, **14**, 172–177.

ANADÓN, P., CABRERA, L., GUIMERÀ, J. & SANTANACH, P. 1985. Palaeogene strike-slip deformation and sedimentation along the southeastern margin of the Ebro basin. *In*: BIDDLE, K. T. & CHRISTIE-BLICK, N. (eds) *Strike Slip Deformation,*

Basin Formation and Sedimentation. SEPM Special Publication, **37**, 303–318.

AURELL, M., MELÉNDEZ, A., SAN ROMÁN, J., GUIMERÀ, J., ROCA, E., SALAS, R., ALONSO, A. & MAS, R. 1992. Tectónica sinsedimentaria distensiva en el límite Triásico–Jurásico en la Cordillera Ibérica. *III Congreso Geológico de España, Salamanca 1992.* Actas, **1**, 50–54.

BELTRÁN CABRERA, F. J., RÍOS MITCHELL, J. M. & RÍOS ARAGÜÉS, L. M. 1980. *Mapa geológico de España.* Escala 1 : 50,000, hoja 349 (Cabrejas del Pinar). Instituto Tecnológico GeoMinero de España, Memoria.

BRENNER, P. 1976. Ostrakoden und Charophyten des spanischen Wealden (Systematik, Ökologie, Stratigraphie, Paläogeographie). *Palaeontographica.* (A) **152**, 113–201.

BUTLER, R. W. H. 1989. The influence of pre-existing basin structure on thrust system evolution in the Western Alps. *In*: COOPER, M. A. & WILLIAMS, G. D. (eds) *Inversion Tectonics.* Geological Society, Special Publication, **44**, 105–122.

CARTWRIGHT, J. A. 1989. The kinematics of inversion in the Danish Central Graben. *In*: COOPER, M. A. & WILLIAMS, G. D. (eds) *Inversion Tectonics.* Geological Society, Special Publication **44**, 153–175.

CASAS SAINZ, A. M. 1992. *El frente norte de las Sierras de Cameros: estructuras cabalgantes y campo de esfuerzos.* Zubía, monográfico, **4**, Instituto de Estudios Riojanos.

—— 1993. Oblique tectonic inversion and basement thrusting in the Cameros Massif (Northern Spain). *Geodinamica Acta,* **6**, 202–216.

—— & SIMÓN GÓMEZ, J. L. 1992. Stress field and thrust kinematics: a model for the tectonic inversion of the Cameros Massif (Spain). *Journal of Structural Geology,* **14**, 521–530.

CASQUET, C., GALINDO, C., GONZÁLEZ CASADO, J. M., ALONSO, A., MAS, J. R., RODAS, M., GARCÍA ROMERO, E. & BARRENECHEA, J. F. 1992. El metamorfismo en la Cuenca de Cameros. Geocronología e implicaciones tectónicas. *Geogaceta,* **11**, 22–25.

CLEMENTE, P. & ALONSO, A. 1990. Estratigrafía y sedimentología de las facies continentales del Cretácico inferior en el borde meridional de la Cuenca de Los Cameros. *Estudios Geológicos,* **45**, 90–109.

——, —— & PÉREZ-ARLUCEA, M. 1991. Secuencias de depósito en la parte occidental de la cuenca de los Cameros. Jurásico terminal–Cretácico inferior. *III Col. del Cret. de España. Resúmenes.* Universitat de Barcelona.

COLCHEN, M. 1974. Géologie de la Sierra de la Demanda. Burgos-Logroño (Espagne). *Memorias del Instituto Geológico y Minero de España,* **85**.

COLOMER, M. & SANTANACH, P. 1988. Estructura y evolución del borde sur-occidental de la Fosa de Calatayud-Daroca. *Geogaceta,* **4**, 29–31.

DÍAZ MARTÍNEZ, E. 1988. El Cretácico inferior del sector de Jubera (norte de la Sierra de Los Cameros, La Rioja): relación entre tectónica y sedimentación. *II Congreso Geológico de España, Comunicaciones.* Universidad de Granada **1**, 67–70.

ELLIOTT, D. 1976. Energy balance in thrust and deformation mechanisms of thrust sheets. Proceedings of the Royal Society, London, **A283**, 289–312.

ESTEBAN, M. & JULIÀ, R. 1973. Discordancias erosivas intrajurásicas en los Catalánides. *Acta Geologica Hispanica,* **8**, 153–157.

GINER, J. 1978. Origen y significado de las brechas del Lías de la Mesa de Prades (Tarragona). *Estudios Geológicos,* **34**, 529–533.

GOLBERG, J.-M., GUIRAUD, M., MALUSKI, H. & SÉGURET, M. 1988. Caractères pétrologiques et âge du métamorphisme en contexte distnsif du bassin sur décrochement de Soria (Crétacé inférieur, Nord Espagne). *Comtes Rendus de l'Academie des Sciences Paris,* Serie II, **307**, 521–527.

GÓMEZ FERNÁNDEZ, J. C. 1992. *Análisis de la Cuenca sedimentaria de los Cameros durante sus etapas iniciales de relleno en relación con su evolución paleogeográfica.* PhD thesis, Universidad Complutense, Madrid.

GUIMERÀ, J. 1984. Paleogene evolution of deformation in the northeastern Iberian Peninsula. *Geological Magazine,* **121**, 413–420.

—— 1988. *Estudi estructural de l'enllaç entre la Serralada Ibèrica i la Serralada Costanera Catalana.* PhD thesis, Universidad de Barcelona.

—— 1990. Structure et evolution de la compression alpine dans la Chaîne Ibérique et la Chaîne Cotiére Catalane (Espagne). *Bulletin de la Société Geologique de France,* 8ème série, **VI**, 339–340.

GUIRAUD, M. 1983. *Evolution tectono-sedimentaire du bassin wealdien (Cretacé inférieur) en relais de décrochement de Logroño–Soria (NW Espagne).* Thèse 3ème Cycle, Université du Languedoc, Montpellier.

——& SÉGURET, M. 1985. A releasing solitary overstep model for the Late Jurassic–Early Cretaceous (Wealdian) Soria strike-slip Basin (Northern Spain). **37**, 159–175.

LANAJA, J. M. 1987. Contribución de la explotación petrolífera al conocimiento de la geología de España. *I.G.M.E., Serv. Publ. Min. Indust. Energ.,* Madrid.

MCCLAY, K. R. 1990. Extensional fault systems in sedimentary basins: a review of analogue model studies. *Marine and Petroleum Geology,* **7**, 205–233.

——, INSLEY, M. W. & ANDERTON, R. 1989. Inversion of the Kechika Trough, Northeastern British Columbia, Canada. *In*: COOPER, M. A. & WILLIAMS, G. D. (eds) *Inversion Tectonics.* Geological Society, Special Publication **44**, 235–257.

MARTÍN-CLOSAS, C. 1989. *Els caròfits del Cretacé inferior de les Conques perifèriques del Bloc de L'Ebre.* PhD thesis, Universidad Barcelona.

MAS, J. R., ALONSO, A. & GUIMERÀ, J. 1993. Evolución tectonosedimentaria de una cuenca extensional intraplaca: La cuenca finijurásica–eocretácica de Los Cameros (La Rioja-Soria). *Revista de la Sociedad Geologica de España,* **6**, 129–144.

MELÉNDEZ, N. & VILAS, L. 1980. Las facies detríticas de la región de Picofrentes (Soria). *Boletin Real de la Sociedad Española de la Historia Naturalia (Geologia)*, **78**, 157–174.

MIEGEBIELLE, V., HERVOUET, Y. & XAVIER, J.-P. 1993. Analyse structurale de la partie méridionale du bassin de Soria (Espagne). *Bulletin des Centres de Recherches Exploration-Production Elf-Aquitaine*, **17**(1), 19–37.

MUÑOZ JIMÉNEZ, A. 1992. *Análisis tectosedimentario del Terciario del sector occidental de la Cuenca del Ebro (Comunidad de La Rioja)*. PhD thesis, Universidad de Zaragoza.

NAVARRO VÁZQUEZ, D., GRANADOS GRANADOS, L., MUÑOZ DEL REAL, J. K., GÓMEZ FERNÁNDEZ, J. J. et al. 1991a. *Mapa geológico de España*. Escala 1 : 50,000, hoja 350 (Soria). Instituto Tecnológico GeoMinero de España, Memoria.

——, ——, —— et al. 1991b. *Mapa geológico de España*. Escala 1 : 50,000, hoja 380 (Borobia). Instituto Tecnológico GeoMinero de España, Memoria.

PLATT, N. H. 1990. Basin evolution and fault reactivation in the western Cameros basin, Northern Spain. *Journal of the Geological Society*, London, **147**, 165–175.

RAMÍREZ MERINO, J. I., OLIVÉ, A., HERNÁNDEZ, A., ÁLVARO, M., AGUILAR, M. J., RAMÍREZ DEL POZO, J., ANADÓN, P., MOLINA, E., GALLARDO, J., GABALDÓN, V. & MARTÍN-SERRANO, A. 1990. *Mapa Geológico de España*. Escala 1 : 50,000, hoja 241 (Anguiano). Instituto Tecnológico Geo-Minero de España, Memoria.

ROCA, E. & GUIMERÀ, J. 1992. The Neogene structure of the eastern Iberian margin: structural constraints on the crustal evolution of the Valencia trough (western Mediterranean). *Tectonophysics*, **203**, 203–218.

——, —— & SALAS, R. 1994. Mesozoic extensional tectonics in the southeast Iberian Chain. *Geological Magazine* **131**, 155–168.

ROURE, F., BRUN, J.-P., COLETTA, B. & VAN DEN DRIESSCHE, J. 1992. Geometry and kinematics of estensional structures in the Alpine Foreland Basin of southeastern France. *Journal of Structural Geology*, **14**, 503–519.

SALAS, R. 1987. *El Malm i el Cretaci inferior entre el Massis de Garraf i la Serra d'Espadà*. PhD thesis, Universidad Barcelona.

—— 1989. Evolución estratigráfica secuencial y tipos de plataformas de carbonatos del intervalo Oxfordiense–Berriasiense en las Cordilleras Iberica oriental y Costero Catalana meridional. *Cuadernos de Geologia Ibérica*, **13**, 121–157.

—— & CASAS, A. 1993. Mesozoic extensional tectonics, stratigraphy and crustal evolution during the Alpine cycle of the eastern Iberian basin. *Tectonophysics*, **228**, 33–55.

SÁNCHEZ NAVARRO, J. A., SAN ROMÁN SALDAÑA, J., DE MIGUEL CABEZA, J. L. & MARTÍNEZ GIL, F. J. 1990. El drenaje subterráneo de la Cordillera Ibérica en la depresión del Ebro: aspectos geológicos. *Geogaceta*, **8**, 115–118.

SAN ROMÁN, J. & AURELL, M. 1992. Palaeogeographical significance of the Triassic–Jurassic unconformity in the north Iberian basins (Sierra del Moncayo, Spain). *Palaeogeography, Palaeoclimatology and Palaeoecology*, **99**, 101–117.

SCHUDACK, M. 1987. Charophytenflora und fazielle Entwicklung der Grenzschichten mariner Jura/Wealden in der Nordwestlichen Iberischen Ketten (mit Vergleichen zu Asturien und Kantabrien). *Paleontographica*, Abt. B., **204**, 1–6.

VALVERDE HERNÁNDEZ, M. 1991. *Mapa geológico de España*. Escala 1 : 50,000, hoja 378 (Quintana Redonda). Instituto Tecnológico GeoMinero de España, Memoria.

VIALLARD, P. 1989. Décollement de couverture et décollement médio-crustal dans une chaîne intraplaque: variations verticales du style tectonique des Ibérides (Espagne). *Bulletin de la Société Géologique de France*, 8ème série, **5**, 913–918.

VILAS, L., MAS, J. R., GARCÍA, A., ARIAS, C., ALONSO, A., MELÉNDEZ, N. & RINCÓN, R. 1982. Capítulo 8. Ibérica suroccidental. *In*: *El Cretácico de España*. Edicion de la Universidad Complutense, 457–514.

——, ALONSO, A., ARIAS, C., GARCÍA, A., MAS, J. R., RINCÓN, R. & MELÉNDEZ, N. 1983. The Cretaceous of the Southwestern Iberian Ranges (Spain). *Zitteliana*, **10**, 245–254.

WILKERSON, M. S. & USDANSKY, S. I. 1989. *Fault! A Cross Section Modelling Program for the IBM*. Unpublished.

Case studies: Asia

Structural development and stratigraphy of the Kyokpo Pull-Apart Basin, South Korea and tectonic implications for inverted extensional basins

JOSEPH. J. LAMBIASE[1] & WILLIAM P. BOSWORTH[2]

[1]*Amerada Hess Limited, London*

Present address: Department of Petroleum Geoscience, University of Brunei Darussalam, Gadong 3186, Negara Brunei Darussalam

[2]*Marathon Petroleum Egypt Ltd, Cairo*

Abstract: Field studies indicate that the Kyokpo Basin initially formed as a pull-apart during the Late Cretaceous Period. Syn-sedimentary normal and strike-slip faults suggest early transtensional movements. Fault-plane analysis indicates an approximately E–W extension direction for this early event. The basin filled with a succession of lacustrine deposits interbedded with abundant pyroclastic rocks and rare volcanic flows. Rapid initial subsidence is demonstrated by deep-water sediments that lie directly on pre-Late Cretaceous rocks. All exposed sedimentary facies were deposited in deep-water environments, indicating continued rapid subsidence during deposition of the entire sequence.

Laminated grey siltstones interbedded with thin, distal, sandy turbidites are the most abundant sediments. Channels filled with conglomeratic debris flow deposits (clasts as large as 1 m in diameter) cut into the siltstone beds. Massive and graded pebbly sandstone turbidites are also interbedded with siltstone. A thick accumulation of stacked, conglomeratic, debris flow deposits interbedded with sandstone turbidites forms a subaqueous fan complex at the southern part of the outcrop area.

Local compression caused up to 3 km of uplift and erosion following deposition of the Kyokpo strata. Thrust faults and small- and large-scale folds were formed by the W–E directed compression. Timing of the compression cannot be determined but probably correlates with a regional late Miocene transpressional event.

The Kyokpo Basin, located on the western side of the Korean Peninsula (Fig. 1), is part of a complex of small basins exposed in Korea and extending westward across the Yellow Sea into eastern China. The tectonic origin of these basins is uncertain; one possibility is that they originated by transtensional movements associated with the initial collision of India and Asia during the Late Cretaceous (He in press). Extensional half-graben and graben formed during this event continued to subside throughout the Palaeogene. Thick sedimentary sequences with predominately lacustrine and fluvial facies accumulated in the subsiding basins (Zhao *et al.* 1985; Yang in press). A subsequent regional transpressional event during the late Miocene is thought to be a product of the Himalayan orogen (Ma *et al.* 1982; He in press). This event caused widespread uplift, basin inversion and erosion, all of varied local intensity (Shen & Liang in press).

Field data were collected to analyse the detailed structural and stratigraphic history of the Kyokpo Basin. Outcrops were visited on the mainland north and south of the town of Kyokpo, and on the islands of Wido, Biando and Duksando (Fig. 1). In addition to structural and stratigraphic measurements, short cores were collected from the sedimentary rocks for subsequent geochemical, palynological and petrographic analysis. K–Ar dates were obtained from interbedded volcanic flows and underlying quartz monzonites to help constrain the timing of basin evolution.

Stratigraphy and depositional setting

Several hundred metres of basin-fill deposits crop out primarily as coastal exposures on the mainland and on several islands in the Kyokpo area (Fig. 1). The total thickness of strata could not be determined because of complex structural relationships (see the discussion below) and because the top of the sequence is not exposed. Measured stratigraphic sections indicate a stratigraphic thickness of at least 577 m (Fig. 2a), and

From BUCHANAN, J. G. & BUCHANAN, P. G. (eds), 1995, *Basin Inversion*,
Geological Society Special Publication No. 88, 457–471.

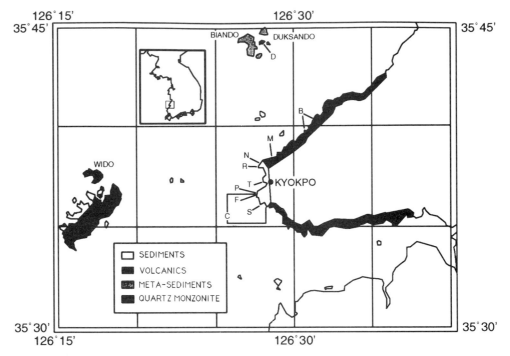

Fig. 1. Geological map of the Kyokpo Basin, South Korea. Inset shows location of map on the Korean peninsula. Outcrops on the mainland are restricted to coastal exposures; patterns indicate the lithology present at the adjacent coast in the areas visited. Letters designate locations of the following features: B, metamorphic basement; D, Duksando quartz monzonite porphyry; F, compressional folds on Wido and south of Kyokpo; M, dated quartz monzonite; N, northern measured stratigraphic section; P, dated rhyolite porphyry; R, dated rhyolite; S, southern measured stratigraphic section; T, Kyokpo Harbor duplex. The subaqueous fan complex occupies the bracketed area labelled C.

it is expected that the actual thickness may be much greater.

The basin-fill deposits consist of abundant volcanic rocks, a thick siltstone succession and coarse-grained clastics. All of the sediments were deposited in subaqueous environments, and palaeontological analysis indicates that it was a lacustrine basin with no evidence of marine incursions (Kaska pers. comm.). Volcanic debris is a major constituent of virtually all of the clastic sediments (Fig. 3a; Wegrzyn pers. comm.).

Volcanic rocks

The Kyokpo area has been recognized as a site of Cretaceous volcanism (Kim & Lee 1983; Won 1987), and acidic volcanic rocks dominate the exposed basin fill. Abundant pyroclastic rocks with subordinate rhyolite and rhyolite porphyry flows account for more than 50% of the observed outcrops, although some of the apparent dominance may be due to preferential exposure of the

relatively erosion-resistant volcanic rocks. Volcanic rocks occur at the top of the exposed section in some areas (Fig. 2a), at the top and bottom in others (Fig. 2b), and are interbedded with exposed sediments throughout the study area (Fig. 2a, b). Two of the interbedded flows yield a Late Cretaceous (Senonian) age for the basin fill (Table 1).

Siltstone succession

Laminated, grey siltstone is the most common lithology of sedimentary rock in the Kyokpo Basin (Fig. 3b), and it comprises almost the entire sequence in some areas. Few sedimentary structures were observed in the siltstone although slump-associated, soft sediment deformation affects some beds.

The siltstone is interbedded throughout with thin (0.1 m or less), laterally continuous, fine sandstone and, rarely, thin shale beds. The sandstone is abundant enough that individual siltstone beds can be as thin as a few centimetres

Table 1 *Results of K–Ar dating*

Location	Lithology	Mineral	Age (Ma)
*Duksando	Quartz Monzonite Porphyry	Biotite	87.2 ± 2.4
*Duksando	Quartz Monzonite Porphyry	K-Feldspar	83.9 ± 2.3
Kyokpo	Rhyolite	Whole Rock	83.5 ± 2.4
Kyokpo	Quartz Monzonite	K-Feldspar	128.0 ± 3.0
Kyokpo	Rhyolite Porphyry	K-Feldspar	76.1 ± 2.2

Sample locations are shown in Fig. 1. Age differences between the rhyolite and rhyolite porphyry are attributed to dating of different minerals. * Both Duksando dates are from the same intrusion. Age differences are attributed to dating of different minerals.

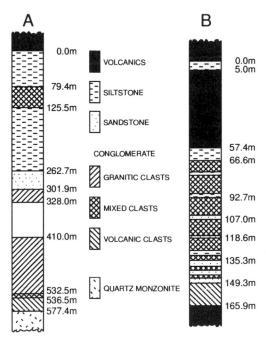

Fig. 2. Measured stratigraphic sections of Kyokpo sedimentary and volcanic units indicating dominant lithologies. (**a**) north of Kyokpo in a siltstone-dominated sequence; (**b**) south of Kyokpo in the subaqueous fan complex. Locations of the sections are shown on Fig. 1.

and they rarely exceed 1 m in thickness. In some areas, sandstone accounts for nearly 50% of the total thickness of the siltstone sequence. Many of the sandstone beds are visibly graded and small scours and flame structures are common. The sandstone beds are interpreted as distal turbidites, suggesting that the interbedded silt-stone was also deposited below wave base.

Coarse-grained clastics

Two coarse-grained clastic facies occur in outcrops in the Kyokpo Basin, including medium- to coarse-grained sandstone and boulder conglomerate. The conglomerate is much more abundant than the sandstone and is present throughout the area of sedimentary outcrops. It typically occurs as beds approximately 2 m thick that fill channels within the siltstone succession (Fig. 4a), although nearly the entire exposed sedimentary section on the headland south of Kyokpo is comprised of stacked conglomerate beds (Figs 1, 4b). The conglomerate consists mainly of clasts of the order of 0.3 m in diameter in a sand-sized matrix. In a few areas, clast diameters exceed 1 m and some reach 3 m.

Each conglomerate is interpreted as a subaqueous, matrix-supported debris flow. The basal bed surfaces often exhibit scour into underlying strata and many individual flows are inversely graded. A mixture of granitic and volcanic fragments comprise the large clasts in most flows, although some flows contain either volcanic or granitic clasts exclusively.

Outcrops of medium- and coarse-grained sandstone units are limited to the southern tip of the headland south of Kyokpo and the sedimentary section on Wido (Fig. 1). In those areas, the sandstone beds generally are 0.5–1 m thick. On Wido they are interbedded with siltstone as described above, while on the headland a succession of beds of similar thickness form a sandstone unit 39 m thick. Many of the coarse sandstone beds are normally graded and have erosional bases, and the interbedded siltstone units show evidence of loading. The sandstone beds are interpreted to be proximal turbidites.

Facies model

All of the outcropping facies in the Kyokpo Basin are indicative of deep-water sedimentary

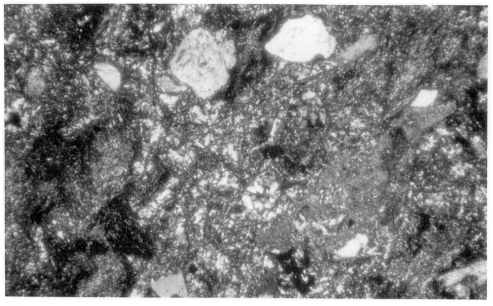

(a)

(b)

Fig. 3. (a) Photomicrograph (50×) of silicified, volcaniclastic sandstone. Silica and calcite have replaced the volcanic rock fragments and glass shards that are the dominant constituents; (b) laminated, grey siltstone at Kyokpo. Lens cap for scale.

(a)

(b)

Fig. 4. (a) Conglomerate-filled channels (lighter area) cutting into grey siltstone (darker areas) north of Kyokpo. At least 3 nested channels occur in the cliff and all are filled with debris flow deposits. (b) stacked, conglomeratic debris flow deposits in the axis of the subaqueous fan complex south of Kyokpo.

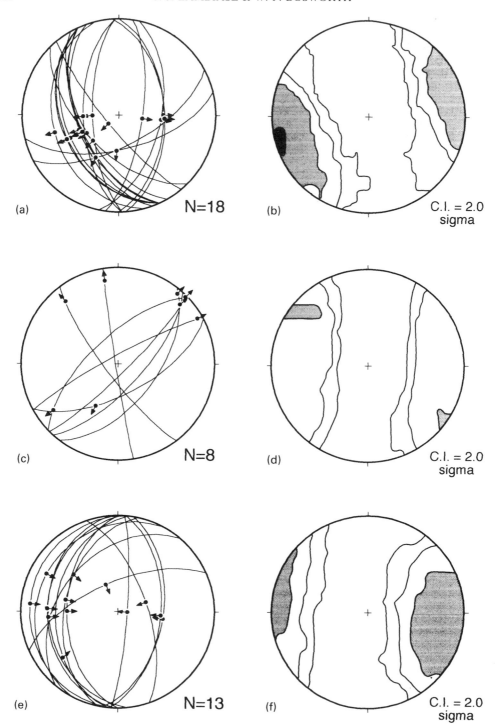

Fig. 5. Stereograms of structural data from the Kyokpo Basin. Equal area, lower hemisphere projections. (**a**) normal fault planes and striations; (**b**) normal fault extension axes; (**c**) strike-slip fault planes and striations; (**d**) strike-slip fault extension axes; (**e**) thrust fault planes and striations; (**f**) thrust fault shortening axes. Kinematic analysis according to the technique of Marrett & Allmendinger (1988, 1990).

environments. A deep lake must have occupied the basin immediately after its inception because deepwater deposits directly overlie older quartz monzonite (Fig. 2a; Table 1). All the siltstone in the Kyokpo Basin is a basinal facies, indicating that the lake persisted while the entire sequence was being deposited. Erosion of the abundant acidic volcanic rocks apparently introduced enormous volumes of silt into the basin, leading to deposition of siltstone in a setting that commonly accumulates claystone.

Each of the other three clastic facies reflects the introduction of coarse clastic material into a deep basinal environment by sediment gravity flows. They constitute a progression from proximal to distal, from debris flow conglomerates to coarse sandstone turbidites to thin, fine sandstone turbidites interbedded with siltstone. Geographically, their distribution suggests a subaqueous fan complex with its axis located on the headland south of Kyokpo (Fig. 1). Facies are apparently more distal to the north as shown by the decreasing amount of conglomerate relative to siltstone (Fig. 2). Wido represents a more distal facies than the mainland, although it is uncertain that the two sections are stratigraphically equivalent because a correlation could not be established.

The facies distribution suggests that all three facies may represent input from a single point source in close proximity to the headland south of Kyokpo, with a proximal to distal transition from south to north. However, this transition may be representative of a transect that is perpendicular to the axis of the subaqueous fan complex because palaeocurrent directions, although difficult to determine in the conglomerate facies, suggest E–W transport. The large volume of debris flow deposits and the coarse clast size required a steep slope and imply proximity to a high-relief fault scarp at the time of deposition. The incorporation of numerous granitic clasts into the debris flows indicates that this possibly was a basin-bounding fault.

The fill of the Kyokpo Basin includes volcanically-derived siltstone deposited in a rapidly subsiding, deep lacustrine basin that was fault-bounded at least to the east. At the same time, debris flows and turbidity currents constructed a large subaqueous fan complex from coarse sediment introduced via a point source along the basin's eastern margin. Syn-sedimentary acidic volcanism contributed a major proportion to the total basin fill.

Structure

Normal and strike-slip faulting

Sedimentary strata in the Kyokpo basin are cut by numerous small-scale normal and strike-slip faults. Large-scale growth faulting cannot be demonstrated in the limited exposures available for study. Field relationships suggest that the normal faults affected partially lithified sediments. Brecciation was minimal and sediment was often injected along fault planes. Ductile deformation occurred in strata adjacent to fault planes. The deformed beds do not show any structural fabric and deformation probably involved grain-boundary sliding mechanisms. High-angle, strike-slip faults have a similar appearance but are sometimes marked by thick breccias. These breccias often contain a fine-grained matrix of sandy or silty material supporting a mixture of angular and rounded clasts. The matrix commonly lacks through-going fractures. It is therefore not clear whether the breccias have a completely tectonic origin or involve sediments that partially filled fractures that were open to the surface during breccia formation.

Stepped, fibrous calcite mineralization is present on some of the normal and strike-slip faults. In other cases, grooves and mullions could be used to determine the direction of shear, but not its sense. Some of the faults occur as splays or in dense groups with internally consistent orientations and bedding offsets. In these areas bedding offsets could be used to determine reasonably that the same sense of shear had occurred on the associated fault surfaces once fibrous mineralization was found on one or more faults, although the precise direction of shear is unknown. Some early faults were rotated by later folding and thrust faulting, hence their orientations are not representative of the strain field at the time of their formation. Our kinematic analysis (Fig. 5) is based on 39 faults with complete slip data. However, 97 faults were measured in total and the orientations, except where folded, and partial slip data of the other 58 faults are consistent with Fig. 5.

Most normal faults strike N–S (Fig. 5a). Strike-slip faulting occurred as a predominantly NNW–SSE set and a sinistral NE–SW set (Figs 5c & 6). The outcrop-scale normal, oblique-slip and strike-slip faults are interpreted to have formed during initial development of the Kyokpo Basin and deposition of the Kyokpo sedimentary sequence. Where cross-cutting relationships were observed, strike-slip faults cut

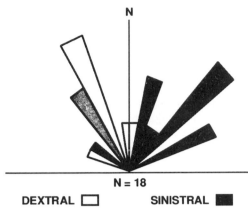

Fig. 6. Rose diagram of strike-slip fault orientations in the Kyokpo Basin where sense of shear is known. Intermediate shading indicates both sinistral and dextral orientations.

normal faults. Consequently, strike-slip faulting may represent a slightly later stage of this early deformation, although the number of examples is limited and a distinction may not be warranted.

Kinematic analysis of early faults

Families of small-scale faults with limited offsets, from small areas with uniform structural fabric, have been used to determine shortening and extension directions, and with further assumptions, palaeo-stress directions. This requires knowledge of fault plane orientation, slip direction, slip sense and timing of movement. Data suitable for this analysis were recorded at Kyokpo and Wido (Fig. 1).

Fault planes and slip directions for normal faults have been inverted to produce inferred extension axes (Fig. 5b) using the technique of Marrett & Allmendinger (1988, 1990). Similar data and contoured extension axes for strike-slip faults are given in Figure 5d. These results suggest that the early normal faults developed during an E–W to ENE–WSW extensional event, and that the strike-slip faults are compatible with the same stress field orientation.

Basin geometry

The basin-bounding faults of the Kyokpo sedimentary and volcanic sequence are not exposed. The contact with basement quartz monzonite north of Kyokpo is an unconformity. Further north, the contact with gneiss is a fault-reactivated unconformity. Elsewhere, the contacts are not exposed. The top of the upper

volcanic unit, for example, has not been observed in outcrop. The precise geometry of the Kyokpo Basin cannot, therefore, be determined without the addition of subsurface data.

Despite these limitations, several stratigraphic and structural observations constrain interpretations of the original form of the Kyokpo Basin. Units immediately adjacent to the basal unconformities of the sedimentary sequence are extremely coarse-grained and were deposited by subaqueous debris flows, implying both significant relief and deposition within a standing body of water. These debris flows are followed stratigraphically by intercalations of grey siltstone and turbidite sandstone, indicating continuous deepwater deposition as discussed above. Therefore, the stratigraphy indicates rapid generation of relief and subsidence, without an initial fluvial or sub-aerial phase of deposition as is commonly developed in extensional basins (Lambiase 1990; Schlische & Olsen 1990). Also, the topographic relief required to maintain the steep slopes necessary for debris flow transport implies proximity to a large fault. Inclusion of abundant granitic clasts in the debris flows indicates erosion on the footwall of rocks that pre-date extension. Coupled with the westerly transport direction of the debris flows, this suggests a basin-bounding fault to the east of the present outcrop.

The earliest phases of deformation are extensional and syn-sedimentary, and coeval with pervasive, complex strike-slip faulting. Fault patterns seen in outcrop at Kyokpo could represent an accommodation zone in an extensional rift setting unrelated to regional strike-slip faulting (Bosworth 1985; Rosendahl 1987). However, rapid subsidence and the predominance of strike-slip structures suggest that the Kyokpo Basin initially formed as a pull-apart.

Uplift and inversion

Thrusting and folding

Small-scale contractional features were observed throughout the Kyokpo area. These include numerous reverse faults with metre- or centimetre-scale offsets and folds of similar scale. In some cases these structures are associated with the strike-slip faults discussed above, and some folds and reverse faults may have been generated during the early transtensional basin-forming event. Several observations, however, argue for a later, independent tectonic episode that significantly shortened the basin, rotated early structures and inverted the large-scale

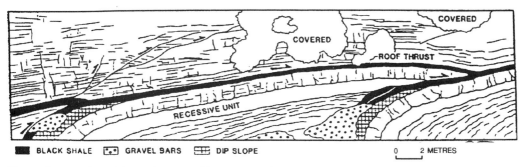

BLACK SHALE   GRAVEL BARS   DIP SLOPE   0   2 METRES

Fig. 7. Cross-section of the roof thrust and horses in the duplex at Kyokpo Harbour. The location of the duplex is shown on Fig. 1.

basin geometry. These relationships are that: 1) the geometries of a well-exposed duplex, described below, and of some folds in thin-bedded units indicate layer-parallel flexural slip, suggesting lithification prior to shortening; 2) thrust and reverse faults commonly cut normal and strike-slip faults, but the opposite is not observed; and 3) the shortening direction inferred from reverse faults is very consistent over the area of exposure. This suggests a regional cause rather than transpression along individual strike-slip faults with variable orientations.

Three large- or medium-scale 'compressional' features were observed in the Kyokpo area. The most intriguing is a series of horses and a roof fault exposed on the northern shore of Kyokpo Harbor, referred to as the Kyokpo Harbor duplex (Figs 1, 7). The repeated units include a distinctive black shale, well-jointed, pin-striped sandstone, and hummocky siltstone and sandstone. Portions of three horses are exposed, or two horses and a footwall ramp. The roof fault occurs at the top of the half-metre thick black shale. Slickenlines on small reverse fault splays from the roof fault indicate east-directed transport. Small-scale folds have approximately horizontal, N–S trending hinge lines.

South of Kyokpo, at the next major headland, the strata are involved in folding with a wavelength of several hundred metres and a minimum amplitude of a few tens of metres (Figs 1, 8). The fold axial plane at the headland dips approximately 60°W with a gently north-plunging hinge line. Bedding on the west limb is vertical to overturned and the fold appears to be very angular. No axial planar cleavage was observed in the vicinity of this folding, nor in any other Kyokpo exposures.

The third large compressional structure is located on Wido, on the south shore of a west-facing bay (Fig. 1). These outcrops preserve the core of a recumbent fold with amplitude and wavelength greater than 30 m (Fig. 9). The hinge of this fold trends approximately N 65°E, but the exposure is not complete enough to allow determination of the fold's vergence. As discussed below, the orientation of this structure is anomalous compared to those of the other compressional features. The fold may be detached from the underlying strata and may represent an unusual thrust geometry or possibly the slip surface for a large-scale slump feature. Bedding thickness around the fold is variable in some layers and roughly constant in others.

Kinematic analysis of thrust faults

Thrust and reverse fault planes and slip directions are given in Fig. 5e, including data collected at the Kyokpo Harbor duplex. Shortening axes derived from these data (Fig. 5f) suggest that the shortening direction was E–W to ESE to WNW, which is compatible with the large-scale folding at the headland south of Kyokpo, but is oriented approximately 45° counterclockwise from the shortening direction for the large recumbent fold on Wido.

The outcrop data do not constrain the total amount of shortening that has occurred across the Kyokpo Basin, although the observed large-scale folds and the duplex at Kyokpo Harbor indicate at least hundreds of metres of shortening. The scale and geometry of the folds and thrust faults are compatible with a major basin-inverting event. E–W shortening and resultant inversion of extensional structures

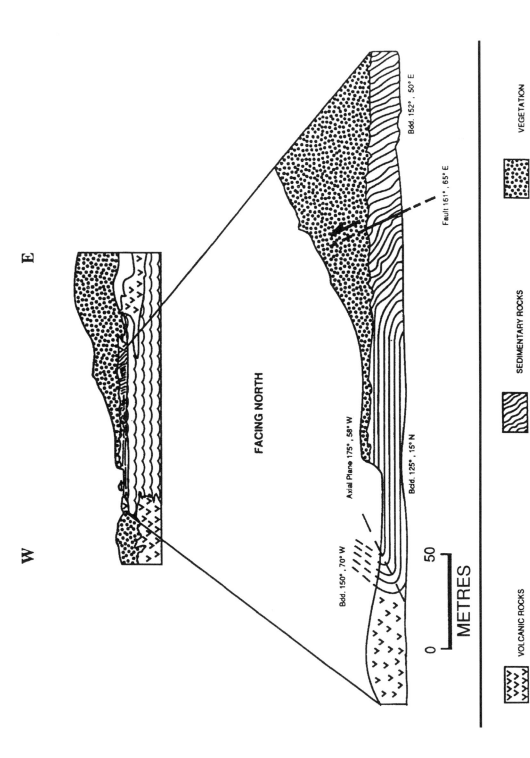

Fig. 8. Structural cross-section (from photographs) of the large-scale, overturned fold south of Kyokpo. The location of the fold is shown on Fig. 1.

W

E

FACING NORTH

Bdd. 152°, 50° E

Fault 161°, 65° E

Axial Plane 175°, 58° W

Bdd. 125°, 15° N

Bdd. 150°, 70° W

METRES

0 50

VOLCANIC ROCKS

SEDIMENTARY ROCKS

VEGETATION

Fig. 9. The recumbent fold on Wido. The fold may be detached from the underlying strata and may represent an unusual thrust geometry. It is approximately 10 m from the fold axis to the bottom of the photograph. The location of the fold is shown on Fig. 1.

have been interpreted from reflection seismic data offshore from the Kyokpo area (Fig. 10, Shoemaker pers. comm.). It is possible that N–S striking, basin-bounding faults in the coastal region also may have been reactivated as reverse faults to produce Sunda-style folds (Eubank & Makki 1981).

Depth of burial and rate of uplift

Precise depths of burial for the sedimentary section cannot be determined from the available data, but petrographical, palynological and geochemical evidence suggest a depth of several kilometres. Visual thin section examination reveals that the sediments are well compacted, and probably have been buried to approximately 3 km (Wegrzyn pers. comm.). High palaeo-temperatures are indicated by dark-coloured spores that have a thermal alteration index of approximately 4.5 (Edman pers. comm.). Vitrinite reflectance measurements range between 1.7 and 3.7% and corroborate the high degree of thermal maturity demonstrated by spore colour. These data suggest that an elevated heat flow associated with the syn-sedimentary volcanism has probably contributed significantly to the thermal maturity.

Static metamorphism has produced authigenic epidote and biotite in the Biando meta-sediments (Fig. 1), indicating a minimum of approximately 7.5 km of burial (Wegrzyn pers. comm.). The meta-sediments are intruded by the Duksando quartz monzonite which is age-equivalent to the syn-sedimentary rhyolite and rhyolite porphyry flows (Table 1). The presence and age relationship of the intrusion require a minimum burial depth of approximately 5 km for the meta-sediments at the same time that the Kyokpo sedimentary sequence was being deposited.

The timing of uplift and inversion of the Kyokpo Basin is not constrained by the field data, although it probably occurred during the late Miocene compressional event that affected the Yellow Sea region (He in press; Tong in press) and that is evident on offshore seismic data (Fig. 10). This implies rapid uplift of the Kyokpo strata, probably by as much as 3 km of uplift in a few million years. The Biando meta-sediments probably underwent even more rapid uplift. They were apparently transported to the surface from a depth of at least 5 km by the same event.

Discussion

Basin history

The structural and stratigraphic development of the Kyokpo Basin is compatible with the general regional history of the Yellow Sea region.

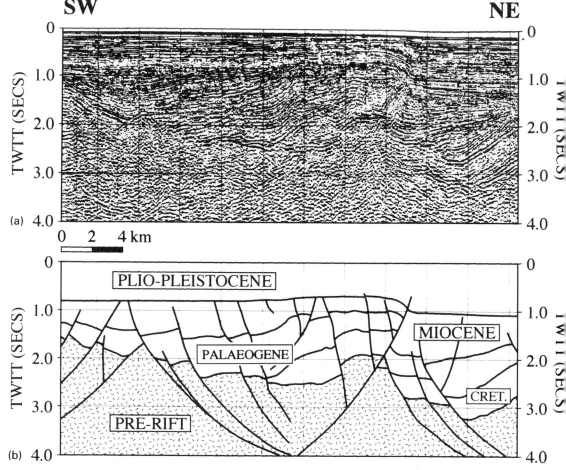

Fig. 10. Offshore seismic line from the eastern Yellow Sea. (**a**) migrated seismic data; (**b**) structural interpretation. Thickening of strata into normal faults is recorded in the Cretaceous to lower Miocene section. At some time during the Miocene, the faults were reactivated in a reverse sense, resulting in uplift and erosion. This continued into at least the early Pliocene as indicated by the geometry of the regional unconformity between the Miocene and Pliocene sections.

A Late Cretaceous age for the initial basin fill at Kyokpo makes it approximately age-equivalent to the oldest syn-extensional deposits in the western Yellow Sea (Shen & Liang in press). These strata consist of deepwater lacustrine deposits and, therefore, they also are facies equivalents of the Kyokpo sediments.

Palaeocene and Eocene sediments in the western Yellow Sea region coarsen upward from deepwater lacustrine rocks, through shallow water and deltaic environments, to fluvial deposits (Zhao *et al.* 1985; Tong in press). Together, the Late Cretaceous and Palaeogene formations comprise a basin-fill succession approximately 3 km thick. This succession is very

similar in character to the typical shallowing and coarsening upward sequence in nonmarine extensional basins identified by Lambiase (1990) from numerous examples worldwide.

The Kyokpo Basin may have had an equivalent Palaeogene history to the western Yellow Sea half-graben. Based on depth of burial estimates, subsidence resulting from extension allowed deposition of approximately 3 km of basin-filling strata, causing the compaction observed in outcrop. If this basin evolved as a typical nonmarine extensional basin, these sediments probably consisted of a shallowing- and coarsening-upward sequence similar to that elsewhere in the Yellow Sea. The Palaeogene rocks were subsequently eroded following late Miocene uplift, and the Kyokpo outcrops are the remnants of that sequence.

Tectonic implications

Field observations and regional considerations suggest that the structural development of the Kyokpo Basin spanned approximately 70 Ma and included two distinct tectonic events. Kinematic analyses accordingly indicate a major change in strain direction from early, partially syn-sedimentary E–W extension to late, west-over-east thrusting and shortening of well-lithified rocks. In a regional sense, both extensional and shortening features of the Yellow Sea basins have been linked to large-scale strike-slip faulting caused by the collision of India and Asia (Ma *et al.* 1982; He in press). This suggests that all the structural features observed in the Kyokpo Basin actually are the product of a complex, but single, regional tectonic event. Consequently, it is neither necessary nor desirable to invoke major changes in regional structural geometries to explain the evolution of this basin.

The above proposed tectonic history for the Kyokpo Basin appears to include a relatively common sequence of events that also occasionally occurs in what have been traditionally regarded as rift basins. Many rift basins originally formed by extensional tectonics were later subjected to compression, strike-slip faulting and/or basin inversion. Among these are the Karroo (Permo–Triassic) portion of the Morondava Basin in Madagascar, the Khartoum Basin in central Sudan, the Anza Rift in northeastern Kenya, the Benue Trough in Nigeria and an entire complex of basins on the Sunda Shelf in Indonesia.

In Sudan, the Khartoum Basin terminates at its northern end against the Central African Shear Zone (Browne & Fairhead 1983; Browne et al. 1985; Bosworth 1992). The Khartoum Basin underwent compression and inversion during the latest Cretaceous or Palaeogene, perhaps in response to continued slip on the Shear Zone under a different regional stress field. Folds of a similar age are present in outcrop in the Gebel Hegalu region of eastern Kenya and western Somalia (Joubert 1960) and in the subsurface of the Anza Rift in central eastern Kenya (Bosworth 1992; Bosworth & Morley 1994).

Several tectonic models have been proposed for the Eocene–Oligocene extensional event that initiated the Sunda Shelf basins (e.g. Eubank & Makki 1981; Moulds 1989). Strike-slip faulting associated with plate subduction during the Miocene caused the compression and uplift that produced the characteristic Sunda folds (Williams & Eubank 1994). Specific normal faults have been reactivated as reverse faults and the amount of compression and uplift varies considerably across the region.

Karroo (Permo–Triassic) sediments in the Morondava Basin were deposited in extensional half-graben (Nichols & Daly 1989) that formed part of a major rift system which pre-dated the breakup of Gondwanaland (Lambiase 1989). Syn-extensional deposition in the basin began in the Upper Permian with the Sakoa Formation and continued through deposition of the Sakamena Formation (Nichols & Daly 1989). The Sakamena was eroded to varying degrees prior to the deposition of the overlying Isalo Formation (Nichols & Daly 1989). Seismic data indicate that the Ilovo fault, which bounds the Karroo strata to the west, was reactivated as a reverse fault just before the Sakamena–Isalo erosional event and that age-equivalent flower structures occur in the basin (Hare pers. comm.). This suggests strike-slip induced inversion of the rift.

The Benue Trough has long been recognized as a rift basin that subsequently underwent compression. Extensional or transtensional basins' sub-basins originated in the Early Cretaceous and were filled with continental and marine clastics. A compressional event in the mid-Cretaceous that deformed the strata into folds of varying tightness resulted from strike-slip motion along several shear zones that cut the trough (Benkhelil et al. 1988).

The development of each of these basins is related, in varying degrees, to one or more major, regional strike-slip fault system. In all cases, the late compressional features are restricted to specific, reactivated normal faults or sub-basins. Shortening is, in general, not distributed throughout the original region of extension, as in the form of a fold-thrust belt. This localization of inversion and the regional association with long-lived strike-slip faults suggest that minor rotations of the far-field stress regime can account for the local changes from extension to shortening. By analogy with the Kyokpo Basin, some of these basins may have formed by transtension followed by transpression along a strike-slip fault system rather than by a regional change from extension to compression.

Conclusions

Rapid initial subsidence and early normal and strike-slip faulting indicate that the Kyokpo

Basin originated as a pull-apart with an E–W sense of extension during the Late Cretaceous. The basin began to fill in the Senonian with a sequence of deep-water lacustrine sediments and volcanic rocks. Subsidence and deposition apparently persisted through the Palaeogene, resulting in the deep burial of the now exposed strata.

Thrust faulting and folding were caused by west-over-east local compression, probably as part of a regional late Miocene transpressional event. At Kyokpo, 3 km of uplift and consequent erosion accompanied the compression; offshore, uplift probably exceeded 5 km. The same regional tectonics that generated the original transtensional movements may have been responsible for the later transpression and basin inversion, an occurrence that may be relatively common in inverted extensional basins.

We wish to thank Mike Stellas, Jim Shoemaker and John Sharp for valuable assistance in the field. Rich Wegrzyn did the petrographical analysis and Janell Edman did the geochemical analysis. Ken McMillen and Michele Bishop read an early version of the manuscript and made several helpful suggestions. We thank Marathon International Oil Company and Korea Petroleum Development Corporation for permission to publish this paper.

References

BENKHELIL, J., DAINELLI, P., PONSARD, J. F., POPOFF, M. & SAUGY, L. 1988. The Benue Trough: wrench-fault related basin on the border of the equatorial Atlantic. *In*: MANSPEIZER, W. (ed.) *Triassic-Jurassic rifting, continental breakup and the origin of the Atlantic Ocean and passive margins*. Developments in Geotectonics, **22**, part B, Elsevier, Amsterdam, 787–819.

BOSWORTH, W. 1985. Geometry of propagating continental rifts. *Nature*, **316**, 625–627.

—— 1992. Mesozoic and early Tertiary rift tectonics in East Africa. *Tectonophysics*, **209**, 115–137.

—— & MORLEY, C. K. 1994. Structural and stratigraphic evolution of the Anza rift, Kenya. *In*: PRODEHL, C., KELLER, G. R. & KHAN, M. A. (eds) Crustal and upper mantle structure of the Kenya Rift. *Tectonophysics*, **236**, 93–115.

BROWNE, S. E. & FAIRHEAD, J. D. 1983. Gravity study of the Central African Rift System: a model of continental disruption 1. The Nagoundere and Abu Gabra rifts. *Tectonophysics*, **94**, 187–203.

——, —— & MOHAMED, I. I. 1985. Gravity study of the White Nile rift, Sudan and its regional tectonic setting. *Tectonophysics*, **113**, 123–137.

EUBANK, R. T. & MAKKI, A. C. 1981. Structural geology of the central Sumatra back-arc basin. *Proceedings of the Indonesian Petroleum Association*, 10th Annual Convention. Indonesia Petroleum Association, Jakarta, 153–196.

HE, K. G. in press. Regional tectonic scenario and prospects of exploration for oil and gas in south Yellow Sea basins. *In*: *Proceedings of the Petroleum Geology Symposium, South Yellow Sea Basins, Shanghai*.

JOUBERT, P. 1960. Geology of the Mandera–Damassa area. *Geological Survey of Kenya, Report No. 48*.

KIM, O. J. & LEE, D. S. 1983. Summary of igneous activity in South Korea. *In*: RODDICK, J. A. (ed.) *Circum-Pacific Plutonic Terranes*. Geological Society of America, Memoir, **159**, 87–103.

LAMBIASE, J. J. 1989. The framework of African rifting during the Phanerozoic. *African Journal of Earth Sciences*, **8**, 183–190.

—— 1990. A model for tectonic control of lacustrine stratigraphic sequences in continental rift basins. *In*: KATZ, B. (ed.) *Lacustrine exploration: case studies and modern analogues*. American Association of Petroleum Geologists, Memoir, **50**, 265–276.

MA, X., DENG, Q., WANG, Y. & LIU, H. 1982. Cenozoic graben systems in N. China. *Z. Geomorphol., N. F., Suppl.*, **42**, 99–116.

MARRETT, R. & ALLMENDINGER, R. 1988. *Graphical and numerical kinematic analysis of fault slip data*. Geological Society, America Abstracts with Programs, **20**, 319.

—— & —— 1990. Kinematic analysis of fault slip data. *Journal of Structural Geology*, **12**, 973–986.

MOULDS, P. J. 1989. Development of the Bengkalis Depression, central Sumatra and its subsequent deformation – a model for other Sumatran grabens. *Proceedings of the Indonesian Petroleum Association, 18th Annual Convention*. Indonesia Petroleum Association, Jakarta, 217–247.

NICHOLS, G. J. & DALY, M. C. 1989. Sedimentation in an intracratonic extensional basin: the Karoo of the central Morondava Basin, Madagascar. *Geological Magazine*, **126**, 339–354.

ROSENDAHL, B. R. 1987. Architecture of continental rifts with special reference to east Africa. *Annual Reviews Earth & Planetary Science*, **15**, 445–503.

SCHLISCHE, R. W. & OLSEN, P. E. 1990. Quantitative filling model for continental extensional basins with applications to early Mesozoic rifts of eastern North America. *Journal of Geology*, **98**, 135–155.

SHEN, Z. & LIANG, G. in press. On regular patterns of hydrocarbon entrapment in Subei Basin. *In*: *Proceedings of the Petroleum Geology Symposium, South Yellow Sea Basins, Shanghai*.

TONG, B. R. in press. Petroleum geological features of northern basin in south Yellow Sea. *In*: *Proceedings of the Petroleum Geology Symposium, South Yellow Sea Basins, Shanghai*.

WILLIAMS, H. H. & EUBANK, R. T. 1994. Hydrocarbon habitat in the rift graben of the Central Sumatra Basin, Indonesia. *In*: LAMBIASE, J. J. (ed.) Hydrocarbon Habitat in Rift Basins. Geological Society, London, Special Publication, **80**, 331–372.

WON, C. K. 1987. Triassic to Paleogene igneous rocks. *In*: LEE, D.-S. (ed.) *Geology of Korea*. Kyohak-Sa Publishing Co., Seoul, 313–326.

YANG, R. Q. in press. Petroleum geological conditions

and assessment of southern basin in south Yellow Sea. *In: Proceedings of the Petroleum Geology Symposium, South Yellow Sea Basins, Shanghai.*

ZHAO, J., YOU, Y. & ZHOU, X. 1985. Subei-south Yellow Sea basin. *In: Stratigraphic correlation between sedimentary basins of the ESCAP region, Volume X, ESCAP Atlas of Stratigraphy IV.* People's Republic of China, Mineral Resources Development Series, **52**, 28–34.

Inversion-controlled uplift of an outer-arc ridge: Nias Island, offshore Sumatra

M. A. SAMUEL[1], N. A. HARBURY[1], M. E. JONES[1] & S. J. MATTHEWS[2]

[1]*Research School of Geological and Geophysical Sciences, Birkbeck College and University College London, Gower Street, London WC1E 6BT, UK*

[2]*BP Exploration, 4/5 Long Walk, Stockley Park, Middlesex UB11 1BP, UK*

Abstract: This study presents data from Nias Island which lies on an outerarc ridge 150 km to the west of mainland Sumatra. A new stratigraphy, geological map and cross-section of Nias have been produced from field data collected over all the island supplemented with aerial photography, LANDSAT and SAR (synthetic aperture radar) studies and these data form the basis for a new interpretation of the island. Where exposed the basement consists of ophiolitic rocks, originating from a variety of tectonic settings, and with a thin sedimentary cover. Three main sub-basins are identified on Nias. Late Paleogene–Neogene sedimentation in these basins was controlled primarily by major extensional faults though deposition within sub-basins was influenced by the presence of sub-basin transecting faults with significant vertical throws. Two phases of inversion are recognized: the first occurred during the Early Miocene, was limited to western regions and is identified by the presence of a localized unconformity and vitrinite and apatite fission track data. The second inversion initiated in Pliocene times and affected all the sub-basins on Nias. Latest Pliocene–Pleistocene rocks unconformably overlie the Miocene strata. Deformational styles within separate fault-bounded segments vary markedly; structural mapping demonstrates uplift and deformation has been largely controlled by reactivation of sub-basin extensional faults and oblique-slip movements on transecting faults. Diapiric melanges developed during inversion suggesting an intimate relationship between uplift and remobilization of overpressured mudrocks. This study concludes that the uplift of the sub-basins on Nias resulted from inversion of basin successions by reactivation of original major extensional faults rather than on thrust-slices developed in an accretionary prism. Nias does not form part of an accretionary complex; the accretionary prism has developed to the southwest of Nias.

The Sumatran Forearc, Indonesia forms part of the extensive Sunda subduction system which stretches from the island of Sumba in the east, to Burma in the north (Fig. 1). Oceanic lithosphere is presently being subducted along the 5600 km length of this subduction system and the overriding lithosphere changes from continental in Sumatra to oceanic in the region of Sumbawa and Flores (Hamilton 1988). Plate models predict convergence rates varying from 7.8 cm yr^{-1} near Sumbawa to 6 cm yr^{-1} near the Andaman Islands (Minster & Jordan 1978). Oblique convergence along the Sumatran sector is determined from slip vector studies of interplate earthquakes (Newcomb & McCann 1987; McCaffrey 1991) with associated right lateral movements on the Sumatran Fault Zone (Fitch 1972).

Nias Island forms part of a partially emergent, non-volcanic outerarc ridge, 110–150 km west of mainland Sumatra which separates the Sunda Trench and active accretionary prism from the forearc basin. The uplift of Nias has been explained by some authors as a consequence of the accretion of thick Bengal fan sediments which probably reached the area in Middle to Late Miocene times (e.g. Curray & Moore 1974).

Nias has classically been regarded as forming part of an uplifted subduction complex (Moore & Karig 1980; Leggett 1987; Rock 1987; Pickering *et al.* 1989; Allen & Allen 1990; Critelli 1993). Moore & Karig (1980) reported two main tectonostratigraphic units on the island. Their lowest unit, the 'Oyo Complex', a melange, contains angular inclusions immersed in a sheared matrix. They estimated that 70 percent of the inclusions were sandstones, shales and conglomerates. Other inclusions were mainly pillow basalts and cherts with subordinate mafic and ultramafic rocks. Moore & Karig suggested the 'Oyo Complex' was a tectonic melange of an accretionary complex comprised of deformed trench-fill turbidites and slices of oceanic crust and sediments accreted to the base of the inner trench slope. Neogene strata, termed the 'Nias

From BUCHANAN, J. G. & BUCHANAN, P. G. (eds), 1995, *Basin Inversion,*
Geological Society Special Publication No. 88, 473–492.

473

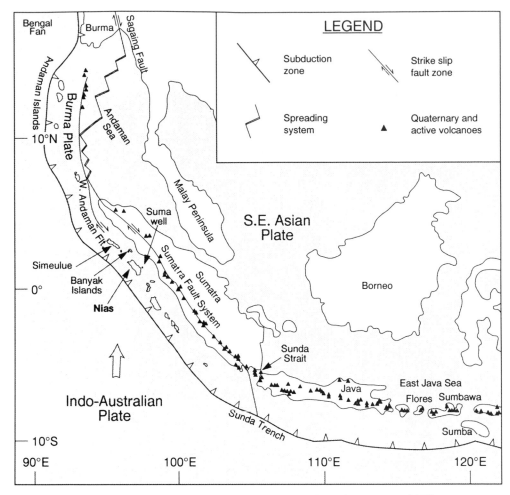

Fig. 1. Tectonic map of the Sunda Arc. Modified from Moore *et al.* (1980*b*) and Curray (1989).

beds', with an estimated thickness of just over 3000 m are documented overlying the 'Oyo Complex' (Moore *et al.* 1980*a*). The 'Nias beds' were interpreted as uplifted trench-slope basin deposits unconformably overlying the 'Oyo Complex'. The 'Nias beds' are reported as having been imbricated into an accretionary wedge during continuous deformation and uplift (Moore & Karig 1980).

In this study sedimentological and structural data, collected from all parts of Nias during four field seasons, combined with observations from unpublished seismic data, and LANDSAT and radar imagery, are presented. Stratigraphic data from over 20 river traverses have been recorded and dating of 350 samples has provided excellent biostratigraphic control. Fission track, vitrinite and XRD analyses provide thermo-tectonic constraints on the evolution of the island (Fig.

2). The new data presented in this paper lead us to the interpretation that the geology of Nias is best explained in terms of inversion of extensional sub-basins.

Nature of the basement

The oldest rocks exposed on Nias are a complex of ophiolitic rocks. These rocks occur in two distinct structural settings: as intact sections of an Ophiolite Complex and as blocks and clasts in melanges. The intact sections of ophiolite complex are exposed on Nias along parts of the west coast and on the north–eastern coast (Fig. 3). Large exposures of disrupted ophiolite occur on Simeulue and the Banyak Islands (Harbury & Kallagher 1991). Ophiolitic sections exhibit marked internal deformation, lithologies being faulted and sheared together. It is shown in this

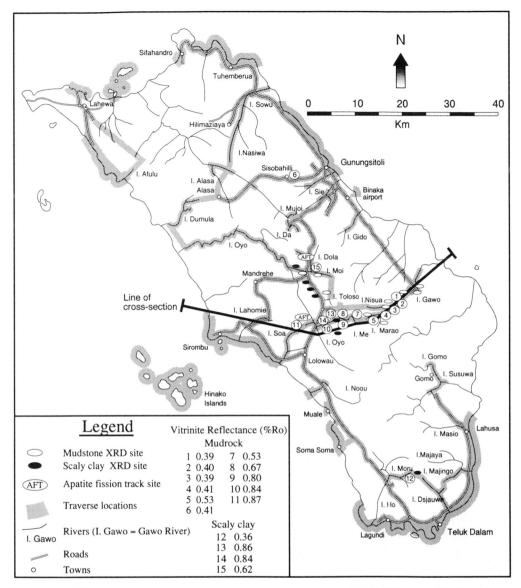

Fig. 2. Map of Nias showing the locations of road and river traverses completed in this study. The sites of samples analysed for vitrinite reflectance, apatite fission track ages and x-ray diffraction are indicated. The line of the cross-section in Fig. 5 is shown as a bold line.

paper that the melanges on Nias formed and incorporated rocks from parts of the Ophiolite Complex, after thick successions of late Palaeogene to Neogene sediments were unconformably deposited above the complex. The melanges do not therefore form basement on Nias.

A range of oceanic upper mantle, crustal, and pelagic rocks are represented in the Ophiolite Complex (Fig. 4). Peridotites are commonly serpentinized and rare bastite serpentinites

contain lineated bastites and olivines up to 50 cm in length. Plutonic rocks are mainly gabbroic although diorites and plagiogranites occur. Dolerites rarely contain olivine phenocrysts and a mixture of volcanic, submarine volcaniclastic and hydroclastic basaltic rocks have been identified. Sedimentary components of the Ophiolite Complex consist of cherts, pelagic limestones and ochres; these rocks are commonly intersheared.

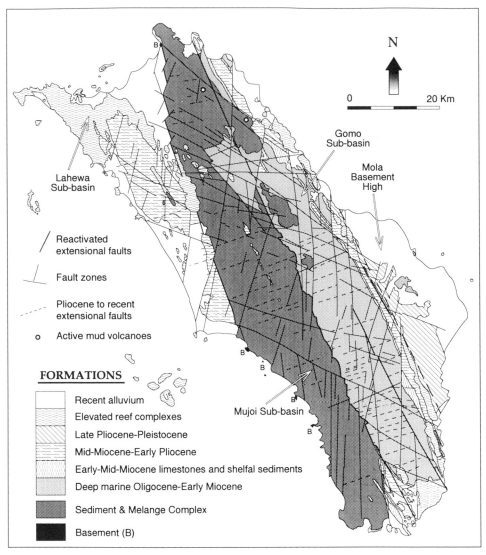

Fig. 3. Geological map of Nias compiled from field traverse data, aerial photographs, SAR and LANDSAT images.

The vast majority of ophiolitic material shows evidence of hydrothermal alteration with mineral assemblages typical of prehnite-actinolite and greenschist facies. There are rare examples of higher grade metamorphism: subordinate blueschists and metagreywackes have been identified in this study and Moore & Karig (1980) recorded garnet amphibolites.

It is difficult to fit the spectrum of rocks comprising the Ophiolite Complex into one original palaeogeographic setting. The lineated bastite serpentinites experienced deformation whilst they were still hot and able to deform in a ductile fashion, possibly in an oceanic transform zone. Diorites may have originally represented the plutonic portions of an oceanic island arc, whilst plagiogranites may represent differentiated intrusions at a mid-oceanic ridge. Pillow fragment breccias, the most common of the ophiolitic rocks on Nias, may form on the steep slopes of pillow volcanoes and large seamounts or accumulate at the foot of submarine fault scarps which are common in faulted mid-ocean ridges (Fisher & Schmincke 1984). The hydroclastic rocks on Nias may have formed as a result of explosive volcanism at shallow water depths.

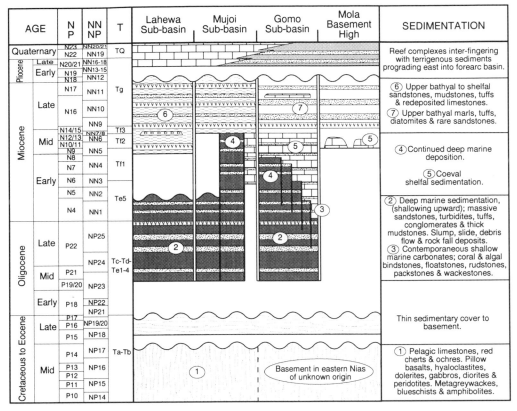

Fig. 4. Generalized chronostratigraphy for Nias. 'N' and 'P' denote planktic foraminiferal zones, 'NN' and 'NP' denote nannofossil zones and 'T' denotes far eastern letter stages. Sub-basin-bounding faults are marked by double vertical lines. The single vertical lines represent faults which transect the sub-basins. Stratigraphic variations on footwall and hangingwall sides of the faults results from marked vertical throws on these faults during sedimentation.

Analyses of basalts and gabbros from Nias show that they have geochemical characteristics of MORB type crust (Kallagher 1989). The range of ophiolitic material exposed on Nias appears to have been derived from oceanic crust, from oceanic islands and seamounts and from oceanic fracture zones.

The ophiolite complex is interpreted as comprising an amalgamated set of ophiolitic fragments. The amalgamation of distinct terranes is not uncommon at convergent plate boundaries (Hawkins *et al.* 1985; Blake *et al.* 1985). Continued subduction produces the opportunity for accretion of crustal and mantle fragments from a variety of different tectonic settings. Oblique subduction and transcurrent faulting can lead to the lateral displacement and emplacement of diverse fragments. Oceanic islands, seamounts, island arcs, transform fault zones and ridges are examples of topographic

highs that have a high chance of being incorporated into the Sumatran forearc as opposed to being subducted.

Sedimentary components of the Ophiolite Complex are generally recrystallised and dates are hard to obtain. Some of the pelagic limestones and cherts do however contain preserved fauna and the youngest age for the Ophiolite Complex, Mid-Eocene, comes from a red chert. The Ophiolite Complex on Nias is unconformably overlain by a thick sequence of Middle Oligocene to Miocene sedimentary rocks; the Complex was therefore in place after the Mid-Eocene but before the Mid-Oligocene. Whilst the evidence from western and northeastern Nias, Simeulue and parts of the Banyak Islands indicates that the basement is largely comprised of ophiolitic fragments, it is not easy to determine the nature of the basement beneath the present day forearc basin (Kieckhefer *et al.*

1980; Kieckhefer *et al.* 1981). The large gravity variations recorded over the forearc (Kieckhefer *et al.* 1981, Milsom *et al.* 1990) suggest that the basement of the forearc basin is extremely heterogeneous and is most unlikely to be composed solely of ophiolitic material.

Late Palaeogene-Neogene extension

Three main sub-basins can be identified on Nias (Fig. 3). From west to east they are the Lahewa, Mujoi and Gomo Sub-basins. The stratigraphic definition of these sub-basins and the structural controls on sub-basin formation and sedimentation are detailed below. To the northeast of the sub-basins, a basement high (the Mola Basement High), of unknown composition, is identified by interpretation of unpublished seismic sections (Fig. 5).

Sub-basin stratigraphy

The oldest sedimentary rocks recorded in all the sub-basins are of deep marine facies. They are characterized by massive micaceous fine grained sandstone beds which are in cases greater than 10 m thick. In well exposed sections, the massive sandstone beds are observed to be interbedded with thick units of mudstones, subordinate conglomerates and thinner sandstone and siltstone beds with well developed Bouma sequences. Mudstone samples are mostly barren of fauna indicating that deposition depths may have initially been close to or below the CCD (Moore *et al.* 1980a). The mudstone samples that do, however, contain small numbers of nannofossils and foraminifera give Middle and Upper Oligocene dates. Whilst Moore & Karig (1980) found Oligocene sedimentary rocks restricted only to the melanges on Nias and interpreted them as accreted trench deposits, some rivers in the central part of the island expose thick coherent sections of basal Oligocene sedimentary rocks which pass conformably up into similarly deformed Lower Miocene rocks (see below and Figs 8 & 11).

The Lower Miocene rocks are also of deep marine origin, with deposition largely by turbidity currents. Slide, slump and debris flow deposits are common in the successions. Bedded mudrocks contain increasing amounts of calcareous microfossils and studies of the smaller benthic foraminifera from our samples support the shallowing in depositional environment with time suggested by Moore *et al.* (1980a).

Whilst many areas of Nias record Lower Miocene turbidite deposition in lower to upper

bathyal environments, coeval limestones are located along strike from the areas of deep marine deposition in the Gomo Sub-basin and carbonates are also developed in southern fault bounded segments of the Mujoi Sub-basin (Fig. 3). A variety of shallow marine facies are represented, including reefal, carbonate mud bank and shoal. No upper Lower Miocene deposits have been recorded in the Lahewa Sub-basin and western parts of the Mujoi sub-basin (Fig. 4) indicating a distinct period of non-deposition and/or erosion.

Limestone deposition became predominant in the Gomo Sub-basin during the Mid-Miocene whilst deep marine sedimentation still continued in the central part of the Mujoi Sub-basin (Fig. 4). Seismic lines across eastern Nias, tied into the Suma well (Fig. 1), reveal earliest Middle Miocene reef build-ups resting unconformably on the Mola basement high (Fig. 5).

Middle Miocene limestones form a ridge running along the Gomo Sub-basin (Figs 3 & 7) and have been interpreted as deep marine by some workers on the basis of palaeobathymetric information collected from above and below the limestone sections (Moore *et al.* 1980a). Close examination of the sections reveals that indicative shallow marine facies including algal bindstones, tabular cross-bedded larger foraminiferal packstones, platy coral framestones and stick coral bafflestones are interbedded with calcarenites and calcirudites. These Middle Miocene shallow marine limestones are conformably overlain by Middle to Upper Miocene siliciclastic and graded limestone deposits. Palaeobathymetric data from these deposits indicate deposition in shelfal to upper bathyal water depths and the Upper Miocene graded limestones are interpreted as redeposited calcturbidites. The Gomo Sub-basin therefore records Oligocene to Mid-Miocene shallowing in environment of deposition with time, from at or near the CCD to shallow marine, with subsequent Late Miocene subsidence and upper bathyal deposition.

In the Lahewa and parts of the Mujoi Sub-basin, Middle to Upper Miocene rocks, predominantly marls and tuffs, unconformably overlie a 'sediment and melange complex' which is described and discussed in detail below. In the southern part of the Mujoi Sub-basin similar deposits rest conformably above the Middle Miocene limestones (Fig. 4) and indicate Mid-Miocene to Late Miocene deepening in environment of deposition; a similar gradation to that recorded for the Gomo Sub-basin. In the central parts of the Mujoi Sub-basin conformable sedimentation in upper bathyal water depths

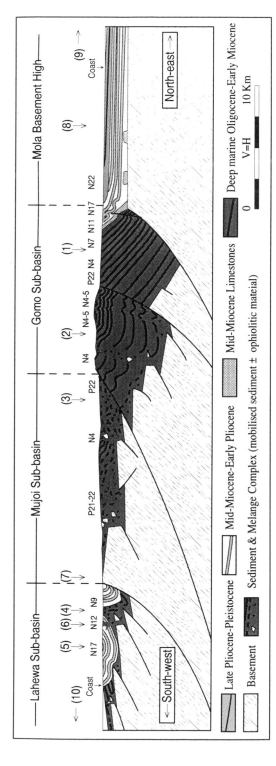

Fig. 5. Cross-section of Nias Island. The line of section is shown on Fig. 2. The section is drawn at true scale from continuous traverse information (including Figs 7, 10 & 11) and the position of critical dates is shown using the planktic foraminiferal zonation scheme (see Fig. 2). At depth the section is interpretative and schematic. The cross-section illustrates the following features: (1) Thick conformable succession of monoclinally-tilted Oligocene to Lower Pliocene sedimentary rocks in the north–eastern segment of the Gomo Sub-basin, unconformably overlain by Upper Pliocene and Pleistocene deposits. (2) Folding and thrusting in the south–western segment of the Gomo Sub-basin; Pliocene inversion has led to mudrock mobilization and present-day mud vulcanism. (3 and 4) Sediment and melange complex formed during Early Miocene inversion. (5 and 6) Middle Miocene to Lower Pliocene sedimentary rocks unconformably overlying the sediment and melange complex and intruded by it during Pliocene inversion (extrapolated from observations along-strike). (7) Disrupted ophiolitic material is exposed along-strike 7 km to the south of the section. (8) Flat-lying basement of unknown type at a depth of 2 km below easternmost Nias (unpublished seismic data, BP Jakarta, 1993, pers. comm. P. Bransden). (9) Extensional basins are developed to the east of Nias in the present-day forearc basin. (10) The modern accretionary prism lies to the south–west of Nias.

continued through the Mid-Miocene and Late Miocene.

Upper Pliocene to recent reef complexes and clastic deposits unconformably overlie Oligocene to Lower Pliocene sedimentary rocks and the ophiolite complex. Coarsening-up units of the terrigenous rocks prograde out towards the present day forearc basin.

Structural controls on sedimentation

Important stratigraphic variations exist not only between, but also within, the separate sub-basins. In particular marked along-strike sedimentological and thickness variations are recorded from relatively undeformed Oligocene to Lower Pliocene successions within the Gomo Sub-basin. Up to 6 km of sediment, predominately deep marine in nature, accumulated in the deepest part of the sub-basin whilst less than 2 km of sediment, including shallow marine limestones, were deposited on highs situated along strike.

Three major southeast striking sub-basin bounding faults have been identified (Fig. 3). The geological map has been produced from integration of field survey, aerial photographic and 1:100,000 scale LANDSAT and SAR (synthetic aperture radar) imagery data. Offsets of stratigraphy (Fig. 5) reveal that the sub-basins were originally extensional or transtensional in origin with a vertical throw component of over 5 km on the sub-basin bounding faults prior to reactivation. A fourth major southeast striking fault cuts the southern part of the Gomo Sub-basin.

The sub-basins are further transected by north and east–south–east striking fault zones (Fig. 3). These zones have been mapped from the SAR and LANDSAT images and aerial photographs. Many of the faults are evident in the field where they cross rivers as relatively narrow (up to 20 m wide) zones of intensely deformed rocks. In some areas straight river courses appear to run along the fault zones but exposures are very poor due to intense tropical weathering. Marked along strike variations in stratigraphy are found either side of the fault zones (Fig. 4). The sedimentological and thickness variations can be related to fault movements with thick deep marine sediments in hangingwall areas and thinner carbonate dominated sediments on uplifted footwalls (Fig. 6).

The majority of Oligocene and Lower Miocene rocks appear to have been derived from a mature continental provenance, probably Sumatran, as they contain an abundance of moderately to well rounded quartzose and

metamorphic fragments (Moore 1979). Distinctive conglomerates and rock fall deposits however indicate local structurally controlled sources for a proportion of the clastic lithologies. Some of the conglomerates, whilst containing quartzose and bioclastic material, are dominated by large clasts of ophiolitic material and this suggests that ophiolitic basement proximal to the sub-basins was being uplifted and eroded (Fig. 6). Dramatic rock fall deposits contain an angular assemblage of indurated and bioturbated siltstone clasts and ophiolitic basement lithologies. The siltstone clasts, which are up to 3 m in diameter and have well defined laminations where not destroyed by bioturbation, contain only minor amounts of fauna and may be deep marine turbiditic or contouritic deposits. The rock fall deposits are interpreted therefore as representing the rubble derived from fault scarps, developed in ophiolitic basement, capped with a thin cover (Fig. 6).

Pubellier *et al.* (1992) report 'a very thick clastic and limestone series bearing ages that range from Late Palaeocene–Early Eocene' overlying basement in southern and northern Nias. The sections they describe have been examined and sampled extensively during this study and by many previous workers (e.g. Douville 1912; Hopper 1940; Elber 1939; Burrough & Power 1968; Moore & Karig 1980; Moore *et al.* 1980a; Djamal *et al.* 1991; Harbury & Kallagher 1991) and the only Eocene material recovered is always identified as reworked. Faunas obtained during this study reliably date the sections as Oligocene and Early Miocene. It therefore appears that the only sedimentary cover the ophiolitic basement may have received prior to sub-basin formation was of fine grained deep marine clastics as typified by the indurated siltstone clasts.

Neogene inversion

Two phases of sub-basin inversion are recognized on Nias: the first inversion event occurred during the Early Miocene and was limited to certain areas whereas all the sub-basins were inverted during the second event which occurred during the Pliocene. These two events are described separately below although in areas that experienced both phases of inversion it is not always possible to distinguish the separate effects of deformation.

Early Miocene inversion

There is evidence, from some areas of Nias, for an aerially restricted but significant phase of

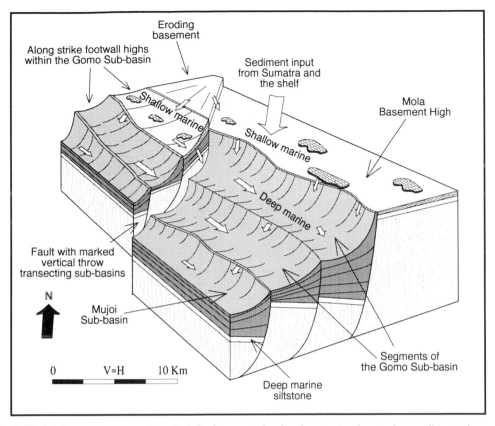

Fig. 6. Early Miocene reconstruction of sub-basin geometries showing structural controls on sedimentation.

inversion that occurred during the Early Miocene. The unconformity recorded in the Lahewa Sub-basin and western parts of the Mujoi Sub-basin indicates Early Miocene inversion in these areas, whilst sedimentation continued elsewhere. Middle Miocene sedimentary rocks unconformably overlie a 'sediment and melange complex' which consists of deformed Oligocene and lowest Miocene sedimentary successions intruded by melanges (Fig. 5). The unconformity is evident in the Oyo river and in the Dumula river of western Nias (Fig. 1) and can be mapped using aerial photographs. The Middle Miocene successions are gently folded whilst sedimentary sections within the 'sediment and melange complexes' are tightly folded and thrust. Since the matrix to the melanges contains clasts and blocks of Oligocene and lowest Miocene material, the formation of the melanges and the contractional structures in these areas must have occurred after the earliest Miocene but prior to the Mid-Miocene. Reworked Lower Miocene material in some of the Middle and Upper Miocene rocks on Nias

suggests that erosion of Lower Miocene sedimentary successions occurred in some areas.

Initial results from apatite fission track and vitrinite reflectance analyses on Oligocene and Miocene samples provide independent evidence for rapid cooling and presumably uplift of a part of the Mujoi sub-basin during the Early Miocene. The samples come from the Oyo and Dola rivers, within the Mujoi Sub-basin, central Nias (Fig. 2). Vitrinite values of Oligocene mudrock samples of between 0.8 and 0.9 %Ro suggest that the sediments were buried to a depth at which the apatites were completely reset. The sediments then passed rapidly up through the apatite annealing window (Green *et al.* 1989), as they give a tight cooling age of 20.1 ± 1.8 Ma. Vitrinite values for Lower Miocene samples suggest they were less deeply buried and the spread in the apatite grain ages indicates they were never completely annealed. An apatite age of 26.8 ± 4.8 therefore reflects, in part, the original apatite provenance.

There is no structural evidence for significant inversion of the Gomo sub-basin during the

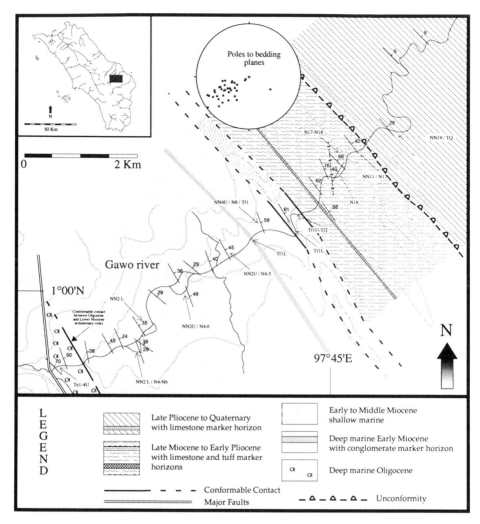

Fig. 7. Traverse map of the Gawo river in eastern Nias. The major deformational style has been tilting towards the northeast. Folding of Upper Miocene sedimentary rocks may be related to reactivation of the major sub-basin bounding fault (Fig. 5).

Early Miocene. The stratigraphic record indicates undisturbed deposition throughout the Miocene, although it is possible that small degrees of fault reactivation may not have much effect on sedimentation. It has not been possible to tightly define the area of the Mujoi Sub-basin that experienced inversion during the Early Miocene due to lack of exposures. However, the continuous stratigraphic and structural record within most of the Mujoi Sub-basin suggests that large areas of the Mujoi Sub-basin were not subject to marked inversion. It appears that inversion in Early Miocene times was therefore not of a regional nature, but was restricted to the Lahewa Sub-basin and the western Oyo region

of the Mujoi Sub-basin. It is likely that inverted segments were originally bound by southeast, north and east–south–east striking faults with contractional reactivation along local inversion axes.

Pliocene inversion

All the sub-basins on Nias have been subject to uplift and erosion during Pliocene to recent uplift and deformation. A major angular unconformity separates Upper Pliocene and Pleistocene sediments from Upper Miocene and lowest Pliocene rocks (Fig. 7). Clasts within the Upper Pliocene and Pleistocene terrigenous rocks are

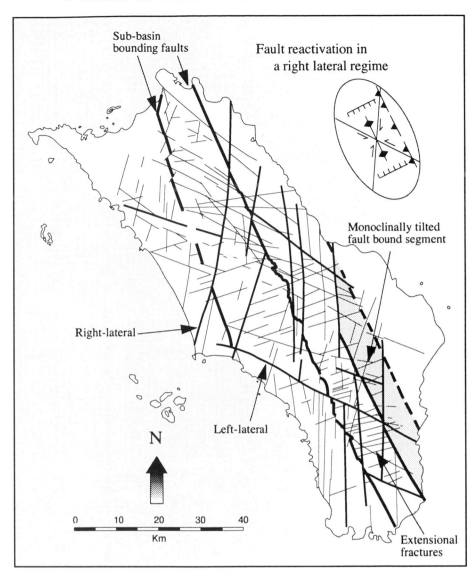

Fig. 8. Map of faults on Nias. Faults have been identified from SAR imagery and are confirmed by fieldwork where exposures allow. Sub-basin bounding faults and marked oblique-slip faults are shown in bold. These faults led to segmentation of the island and separate fault bounded segments experienced varying styles of deformation during inversion.

largely composed of ophiolitic, sandstone, limestone, mudstone and tuffaceous lithologies. Petrographic and microfossil studies indicate that the sedimentary clasts are sourced from the Oligocene to Early Pliocene sections with ophiolite clasts eroded from the basement.

The deep incision of many of the rivers on Nias (Moore & Karig 1980) and the presence of elevated Holocene reefs (Vita-Finzi & Situmorang 1989) indicates uplift is continuing to the present day; the largest river on the island, the Oyo (Fig. 1), is only 200 m above sea level where it runs through the central part of the island where peaks rise to over 800 m.

Structural mapping indicates that uplift and deformation have been largely controlled by contractional reactivation of the extensional faults and reactivated movements along the north and east–south–east striking fault zones. Fault analyses indicate that whilst compressional

Fig. 9. Examples of the variety of styles of deformation that have affected the rocks on Nias as seen in Oligocene and Lower Miocene sections: (**A**) A small-scale example of normal faulting illustrated by the offset of carbonaceous laminations. Normal faulting is the dominant deformational style in sedimentary rocks of the tilted segment of the Gomo Sub-basin. (**B**) Folds and thrusts cross-cutting earlier normal faults marked by calcite veins. (**C**) Scaly clay melange intruding through bedded sandstone. The melange matrix contains Oligocene and Lower Miocene sedimentary inclusions and also older ophiolitic inclusions. (**D**) Serpentinite (on right) thrust over sandstones. The thrust plane is marked by a 5 cm thick calcite vein.

(C)

(D)

reactivation occurred on the major southeast striking faults, extension parallel to the strike of the island has been taken up by oblique-slip right lateral movements on the north striking faults and left lateral movements on the east–south–east striking antithetic faults (Fig. 8). Kinematic indicators are rare in the field but senses of movement on the southeast faults are clear from offsets in the stratigraphy, well dated by biostratigraphic analyses (Fig. 5). Sharp changes in vitrinite reflectance values are also recorded across the faults. The small degree of offsets, caused by the oblique-slip faults, indicate that lateral movement has been limited and extension perpendicular to compression has also been accommodated on a well developed east–north–east striking network of fractures (Fig. 8). Whilst the stratigraphic evidence indicates that many of the north and east–south–east striking faults were active during late Palaeogene and Neogene extension and probably locally during Early Miocene inversion, some movements may have been initiated during Pliocene inversion.

Separate fault bounded segments have experienced varying degrees of deformation. A large segment of the Gomo sub-basin has suffered remarkably little deformation (Figs 7 & 8). Stratigraphic ridges run parallel to the strike of the island and the rocks are simply tilted with dips to the northeast, averaging 45°. There is no evidence for thrust deformation of the Lower Miocene section at the present erosion level in the Gawo river. Gentle folding of the Upper Miocene section can be related to minor reactivation of the main basin bounding fault (Figs 5 & 7). The dominant style of deformation observed in rocks of the tilted segment of the Gomo Sub-basin is small scale extensional faulting (Fig. 9A). Faults such as these are commonly observed in all sedimentary rocks on Nias including those of Pleistocene age suggesting that the faulting probably occurred during the Pliocene to recent inversion.

Many areas on Nias have experienced a greater degree of deformation than simple tilting during Pliocene inversion. Mapping upstream of the Gawo river shows that the rocks in this area of Nias have experienced tight folding and in cases thrusting. This segment of the Gomo sub-basin is separated from the segment to the northeast by a southeast striking contractional fault (Figs 8 & 10). Folds and thrusts generally verge towards the northeast but southwest verging compressional structures are also observed and in some examples thrusts and folds cross-cut extensional faults (Fig. 9B). These extensional faults developed prior to the compression, probably during original extension and sub-basin formation. They can be distinguished from the more recent normal faults as they are characteristically veined by calcite. The beds in the Nalomi river and upstream reaches of the Nisua rivers have a westerly strike as opposed to the northeast strike downstream (Fig. 10). This unusual strike may be a consequence of a broad fold pattern. Segments with atypical strikes have however been mapped in the field and are observed on aerial photographs, in many areas of Nias and the effects of oblique subduction along the Sumatran part of the Sunda Forearc are well documented in terms of right-lateral movements on faults parallel to the trench (e.g. the Sumatran Fault Zone, Fig. 1). Right-lateral movements on some of the major faults on Nias may be responsible for the changes in strike recorded in the field with independent rotations,

about vertical axes, of fault bound segments (Fig. 8).

Large areas of Nias consist of 'sediment and melange complex' (Fig. 3). Contacts between the sedimentary and melange sections are always intrusive. Where exposed the matrix to the melanges is a mudrock which always exhibits a well developed scaly foliation. Clasts and blocks supported within the matrix are angular to subangular and are predominantly of sedimentary origin. Large blocks are readily eroded out of the scaly clay matrix and rest in river beds. Two groups of melanges are recognized on Nias. 'Sediment only' melanges solely contain inclusions of sedimentary material. Inclusions have been examined petrographically and have been dated by microfossil analysis and are always recognizable as parts of the stratigraphic successions originally deposited in the sub-basins. 'Sediment and ophiolite' melanges contain, in addition, ophiolitic inclusions derived from the basement. In both melange groups at least 50%, and usually over 90% of inclusions are derived from the Oligocene and Lower Miocene deep marine sequences.

The scaly clay matrix to the melanges is commonly folded around and intruded into the clasts, blocks and deformed sedimentary sections (Fig. 9B). Melange intrusions occur on a variety of scales, from belts of melange up to 3½ km across, to thin (metre-scale) dykes and sills (Fig. 11). The intrusive contacts on Figure 10 are simplified and extrapolated, beyond the river courses where they are exposed, as straight lines in order to provide a graphic representation of the distribution of melanges. In reality the intricate nature of the intrusive contacts means that melange intrusions cannot be mapped with confidence beyond their exposures in river beds. Detailed mapping in rivers of central Nias indicates that the width of even the largest melange belts changes markedly along strike.

The range of vitrinite values for samples of scaly clay is comparable with that recorded for Oligocene and Lower Miocene mudrocks (Fig. 1). The observation that the major proportion of clasts within the matrix to the melanges are derived from the basal Oligocene and Lower Miocene successions and that the mudstone comprising the scaly clay experienced a similar thermal and burial history to the basal sedimentary sections indicates that the melanges are derived from these successions. XRD analyses of 7 scaly clay samples and 11 Oligocene and Lower Miocene mudrocks reveal that there is no mineralogical difference between the scaly clay and the sedimentary mudrocks and further

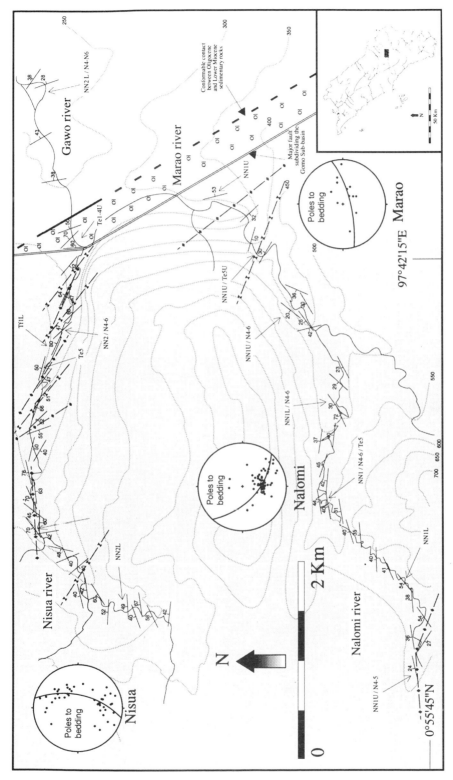

Fig. 10. Traverse map of rivers to the southwest of the Gawo river. Northeast verging compressional deformation is mapped in the lower reaches of the Nisua river and in the Marao river whilst upper reaches of the Nisua and Nalomi rivers show sections of rotated bedding with dips to the north.

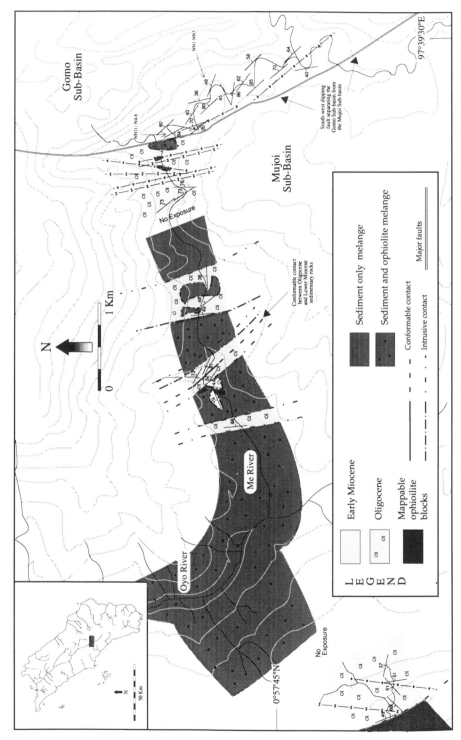

Fig. 11. Detailed map of 'sediment and melange complex' in the Mujoi Sub-basin, central Nias. The map records more detail than can be displayed on Fig. 3 and reveals sections of deformed sedimentary successions intruded by belts of melange. Intrusive contacts are mapped as straight lines to provide a graphic representation of the distribution of the melanges. Two types of melange are distinguished on the basis of their inclusions; both have an identical matrix of scaly clay.

suggest that the scaly clay is deformed deep marine mudrock.

The evidence already outlined from the Oyo and Dumula rivers shows that the melanges in some western parts of the island initially formed during the Early Miocene. In the Dola river (Fig. 2) a small sheet of serpentinite, in thrust contact with Lower Miocene sandstones, is exposed. The thrust plane dips towards the south–west and is marked by a calcite vein, 1 cm in thickness (Fig. 9D). The Oligocene and Lower Miocene sections in the river are heavily intruded by scaly clay and the melanges contain large amounts of serpentinite and other ophiolitic material. Thrusting occurred after deposition of the lowest Miocene sedimentary rocks and both the ophiolitic hangingwall material and the sedimentary footwall material have been subsequently intruded by melanges. Thrusting and melange formation could have occurred synchronously during either of the inversion events.

Melanges are however exposed in many areas of Nias where there is no evidence for Early Miocene inversion. Oligocene to Lower Pliocene successions have been intruded by, and incorporated into melanges exposed along strike of coherently deformed sections of the Gomo Sub-basin (Fig. 3). In the southern fault bounded segment of the Mujoi Sub-basin, melanges are observed where the scaly clay matrix has intruded Middle Miocene limestones. A large proportion of matrix inclusions are of Middle Miocene limestone and the scaly clay contains abundant calcareous microfossils derived from the limestone sections.

It is clear that large volumes of melange have been formed at a later stage than the Early Miocene inversion. The age relationships outlined above indicate that melange formation has occurred after Early Pliocene times during Pliocene to recent inversion. Melange formation continues to the present day with active mud volcanism in areas of the Gomo and Mujoi Sub-basins (Fig. 3). The mud volcanoes are poorly exposed in densely vegetated tropical jungle scarred by abundant landslides. The volcanoes cut through Upper Miocene to recent sediments and consist of small mounds, up to ½ m high, of a bubbling slurry of grey clay. This clay has the same distinctive grey colour as the scaly clay from the melanges on Nias. Large explosive eruptions of the volcanoes occurred in 1957 and 1976 (Hamidi 1976). Locals report earth tremors preceding the eruptions and during the eruptions a ½ m to 2 m thickness of rock detritus was spread aerially over a distance of up to 1 km from the centre of the volcanoes.

Gases from the eruptions ignited leading to widespread fires. During the eruption of the 20th August 1957, many people were badly injured and three villages had to be evacuated. Mudflows and widespread flooding of dammed river valleys led to the destruction of habitations and plantations.

Discussion

Contrary to previous interpretations, the rocks on Nias record a long history of deposition in extensional basins despite their proximity to the Sunda Trench and subduction zone (Fig. 12). Important questions are raised concerning the evolution of these basins, situated at the outer edge of the forearc. Firstly what were the reasons for the original sub-basin formation in the Nias area? Secondly what were the reasons for the periods of basin inversion? Thirdly what is the relationship between the melange formation and basin inversion?

The orientation of the original extensional and oblique slip faults on Nias is similar to that of faults bounding sub-basins within the present day forearc basin (Matson & Moore 1992). Basin formation may therefore have occurred at the same time over a large area of the Sumatran Forearc although the age of the oldest sediments within these basins remains undetermined (Matson & Moore 1992). The timing of major geological events in the Nias area can be related to first order changes in global plate motions. The Eocene collision between India and Eurasia corresponds in time with the formation of many of the basins of SE Asia (Daly et al. 1992). There are a number of possible causes for extension and transtension that can be related to the collision. Basin formation may have been related to sideways movements away from the zone of collision on crustal scale strike-slip faults (Tapponier et al. 1986) although the extent of motion on these faults is debated (Daly et al. 1992). A major effect of the collision was a marked decrease in the rate of convergence (Karig et al. (1979) and Daly (1992)) proposes oblique subduction and margin parallel extension as a major basin forming mechanism. Furthermore Matson & Moore (1992) propose that oblique motions of structural blocks of the overriding plate can account for basin formation.

Forearc areas are particularly sensitive to changes in subduction parameters (Daly 1989). The propagating collision of the Australian plate with the Sunda Arc during the Early Miocene is well documented in Eastern Indonesia (Hall et al. in press) and is probably responsible for the phase of inversion recorded in the East Java Sea

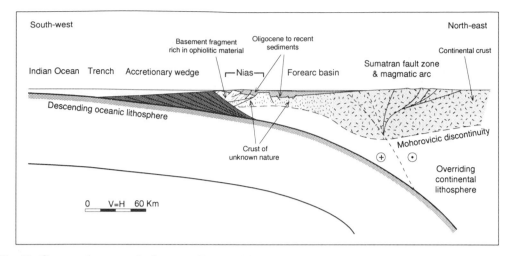

Fig. 12. Cross-sections over the Sumatran Forearc. The basins on Nias may have had a similar extensional history to basins identified by seismic lines across the present day forearc basin.

(Bransden & Matthews 1992). The effects of this collision on the Sumatran Forearc would be most marked at the outer edge and the collision may account for the Early Miocene inversion recorded in western parts of Nias.

An increase in subduction rate through the Late Miocene and Pliocene coupled with the accretion of thick Bengal fan sediments arriving in the area (Karig *et al.* 1979) led to an increase in compressional stresses at the outer margin of the Sumatran Forearc. The increase in stress may have driven fault reactivation on Nias.

The evidence on Nias indicates that melange formation occurred largely during the Pliocene inversion. The melanges formed by diapiric mobilization of the thick Oligocene and Lower Miocene mudrock dominated sedimentary successions. Diapiric melange formation is well documented from convergent margins and particularly accretionary prisms (e.g. Barber *et al.* 1986; Lash 1987; Brown & Orange 1993), however it is not typically associated with basin inversion. The basal sedimentary successions on Nias were not buried to great depths and the diapirism is a relatively near surface feature. Key parameters that appear to control whether the rocks reach shear failure and become mobilized are the rate of inversion and the degree of pore pressure elevation. Both of these parameters may have been atypically high on Nias; rapid sedimentation and swift burial would allow little time for the dissipation of pore pressures and Rose (1983) documents high pore pressures in wells drilled to the northeast of Nias in thick bathyal clay-dominated Neogene sedimentary sections.

Williams *et al.* (1984) has shown that basement material may be incorporated into diapiric melanges as it is thrust over sections of mobilized sedimentary rocks. This is a mechanism which is compatible with our observations from Nias.

Conclusions

The basement beneath most of Nias consists of a complex of ophiolitic fragments. Fragments from different tectonic settings were amalgamated, probably in an accretionary environment and the ophiolite complex collected a deep marine sedimentary cover. Extension of the basement began between Mid-Eocene and Mid-Oligocene times. Sub-basins were developed under transtension and deposition of thick sedimentary successions began. Vertical movements on faults transecting the sub-basins led to along strike topographic highs on which carbonate production occurred synchronously with deep marine sedimentation. Continued Neogene deposition was punctuated by a phase of localized Early Miocene inversion. By Mid-Miocene times the narrow and deep basins had become largely infilled and shallow marine sedimentation was widespread. Regional subsidence occurred during Middle to Early Pliocene times. All the sub-basins on Nias were uplifted, during a second phase of inversion that began during the Pliocene.

Deformation of the sedimentary successions occurred during the two phases of inversion. Separate parts of Nias experienced varying degrees of deformation. Extensional and oblique faults were reactivated in a contractional sense during inversion and controlled the deformation of the basinal successions. We suggest that it is not necessary to invoke accretionary prism processes to account for the late Paleogene and Neogene geology exposed on Nias. However, the development of a subduction related accretionary prism to the south–west of Nias has contributed to the observed fault reactivation during inversion.

We would like to thank B. Situmorang for his kind support of our field programmes in Indonesia, A. Bakri, F. Banner, J. Ling and Ludi who dated our samples and provided palaeobathymetric information, A. Carter and S. Hirons for fission track and XRD analyses, A. Barber, R. Hall, J. Milsom, B. Roberts and G. Roberts for helpful discussions, M. Cowgill for assistance in preparing the paper and D. Jackson for a very constructive review. This research was supported by British Petroleum and the University of London Geological Research in Southeast Asia Consortium and is publication no: 45 of the Research School of Geological and Geophysical Sciences, Birkbeck and University Colleges, London.

References

ALLEN, P. A. & ALLEN, J. R. 1990. *Basin analysis: principles and applications.* Blackwell Scientific publications, Oxford, 451pp.

BARBER, A. J., TJOKROSAPOETRO, S. & CHARLTON, T. R. 1986. Mud volcanoes, shale diapirs, wrench faults and melanges in accretionary complexes. *American Association of Petroleum Geologists Bulletin,* **70**, 1729–1741.

BLAKE, M. C. JR., ENGEBRETSON, D. C., JAYKO, A. S. & JONES, D. L. 1985. Tectonostratigraphic Terranes in Southwest Oregon. *In*: HOWELL, D. G. (ed) *Tectonostratigraphic terranes of the circum-Pacific region.* Circum-Pacific Council for Energy and Mineral Resources Earth Science Series, **1**, 147–157.

BRANSDEN, P. J. E. & MATTHEWS, S. J. 1992. Structural and stratigraphic evolution of the East Java Sea. *Proceedings Indonesian Petroleum Association,* **21**, 417–453.

BROWN, K. M. & ORANGE, D. L. 1993. Structural aspects of diapiric melange emplacement: the Duck Creek Diapir. *Journal of Structural Geology,* **15**, 831–847.

BURROUGH, H. C. & POWER, P. E. 1968. 1968 Field Survey, Southern Part of North West Sumatra Contract Area. *Union Oil Company of Indonesia,* report. **RGE 43** (unpublished).

CRITELLI, S. 1993. Sandstone detrital modes in the Palaeogene Liguride Complex, accretionary wedge of the southern Apennines (Italy). *Journal of Sedimentary Petrology,* **63**, 464–476.

CURRAY, J. R. & MOORE, G. F. 1974. Sedimentary and tectonic processes in the Bengal Deep-Sea Fan and geosyncline. *In*: BURK, C. A. & DRAKE, C. L. (eds) *The geology of continental margins.* Springer Verlag, New York, 617–627.

—— 1989. The Sunda Arc: A model for oblique plate convergence. *Netherlands Journal of Sea Research,* **24**, 131–140.

DALY, M. C. 1989. Correlations between Nazca/Farallon plate kinematics and forearc basin evolution in Ecuador. *Tectonics,* **8**, 769–790.

——, COOPER, M. A. & WILSON, I. 1992. Cenozoic plate tectonics and basin evolution in Indonesia. *Marine and Petroleum Geology,* **8**, 2–21.

DJAMAL, B., GUNAWAN, W. & SIMANDJUNTAK, T. O. 1991. Laporan Geologi Lembar Nias, Sumatera. Sekala 1 : 250,000. *Geological Research and Development Centre, Bandung.*

DOUVILLE, H. 1912. Les foramaniferes de l'ile de Nias. *Idem,* **9**.

ELBER, R. 1939. Geological review of the islands Simalur, Banjak Group, Nias, Batu Group, west coast of Sumatra. *BPM Report,* **888** (unpublished).

FISHER, R. V. & SCHMINCKE, H.-U. 1984. Pyroclastic rocks. *Springer Verlag, Berlin,* 472pp.

FITCH, T. J. 1972. Plate convergence, transcurrent faults and internal deformation adjacent to southeast Asia and the Western Pacific. *Journal of Geophysical Research,* **77**, 4432–4462.

GREEN, P. F., DUDDY, I. R., LASLETT, G. M., HEGARTY, K. A., GLEADOW, A. J. W. & LOVERING, J. F. 1989. Thermal annealing of fission tracks in apatite. 4. Quantitative modelling techniques and extension to geological time-scales. *Chemical Geology (Isotope Geoscience Section),* **79**, 155–182.

HALL, R., ALI, J. R., ANDERSON, C. D. & BAKER, S. in press. Origin and motion history of the Phillipine Sea Plate. *Tectonophysics.*

HAMIDI, S. 1976. 'Mud volcano' Hili Adulo, P. Nias. *Perpustakaan Direktorat Geologi,* **29**, (unpublished).

HAMILTON, W. B. 1988. Plate tectonics and island arcs. *Geological society of America Bulletin,* **100**, 1503–1527.

HARBURY, N. A. & KALLAGHER, H. J. 1991. The Sunda Outer-Arc Ridge, North Sumatra, Indonesia. *Journal of Southeast Asian Sciences,* **6**, 463–476.

HAWKINS, J. W., MOORE, G. F., VILLAMOR, C. & WRIGHT, E. 1985. Geology of the Composite Terranes of East and Central Mindanao. *In*: HOWELL, D. G. (ed) *Tectonostratigraphic terranes of the circum-Pacific region.* Circum-Pacific Council for Energy and Mineral Resources Earth Science Series, **1**, 437–463.

HOPPER, R. H. 1940. Geological reconnaissance in western and northern Nias. *NPPM. report (unpublished).*

KALLAGHER, H. J. 1989. *The structural and stratigraphic evolution of the Sunda Forearc Basin, North Sumatra, Indonesia.* PhD thesis, University of London, 387pp.

KARIG, D. E., SUPARKA, S., MOORE, G. F. & HEHANUSSA, P. E. 1979. Structure and Cenozoic evolution of the Sunda Forearc in the Central Sumatra region. *In*: WATKINS, J. S., MONTADERT, L. & WOOD-DICKERSON, P. (eds) *Geological and geophysical investigations of continental margins.*

American Association of Petroleum Geologists Memoir, **29**, 223–237.

KIECKHEFER, R. M., SHOR JR, G. G., CURRAY, J. R., SUGIARTA, W. & HEHUWAT, F. 1980. Seismic refraction studies of the Sunda Trench and forearc basin. *Journal of Geophysical Research*, **85**, 863–889.

——, MOORE, G. F., EMMEL, F. J. & SUGIARTA, W. 1981. Crustal structure of the Sunda forearc region west of central Sumatra from gravity data. *Journal of Geophysical Research*, **86**, 7003–7012.

LASH, G. G. 1987. Diverse melanges of an ancient subduction complex. *Geology*, **15**, 652–655.

LEGGETT, J. K. 1987. The Southern Uplands as an accretionary prism: the importance of analogues in reconstructing palaeogeography. *Journal of the Geological Society, London*, **144**, 737–752.

MCCAFFREY, R. 1991. Slip vectors and stretching of the Sumatran forearc. *Geology*, **19**, 881–884.

MATSON, R. G. & MOORE, G. F. 1992. Structural Influences on Neogene subsidence in the central Sumatra fore-arc basin. *American Association of Petroleum Geologists Memoir*, **53**, 157–181.

MILSOM, J., DIPOWIRJO, S., SAIN, B. & SIPAHUTAR, J. 1990. Gravity surveys in the North Sumatra Forearc. *United Nations CCOP Technical Bulletin*, **21**, 85–96.

MINSTER, J. B. & JORDAN, T. 1978. Present-day plate motions. *Journal of Geophysical Research*, **83**, 5331–5354.

MOORE, G. F. 1979. Petrography of subduction zone sandstones from Nias Island, Indonesia. *Journal of Sedimentary Petrology*, **49**, 71–84.

—— & KARIG, D. E. 1980. Structural geology of Nias Island, Indonesia: implications for subduction zone tectonics. *American Journal of Science*, **280**, 193–223.

——, BILLMAN, H. G., HEHANUSSA, P. E. & KARIG, D. E. 1980*a*. Sedimentology and paleobathymetry of Neogene trench-slope deposits, Nias Island, Indonesia. *Journal of Geology*, **88**, 161–180.

——, CURRAY, J. R., MOORE, D. G. & KARIG, D. E. 1980*b*. Variations in geologic structure along the Sunda Fore Arc, northeastern Indian Ocean. *In*: HAYES, D. E. (ed.) *The tectonic and geologic evolution of southeast Asian seas and islands*. American Geophysical Union Monograph, **23**, 145–160.

NEWCOMB, K. R & McCANN, W. R. 1987. Seismic history and seismotectonics of the Sunda Arc. *Journal of Geophysical Research*, **92**, 421–439.

PICKERING, K. T., HISCOTT, R. N. & HEIN, F. J. 1989. Deep-marine environments: clastic sedimentation and tectonics. Unwin Hyman, London, 416pp.

PUBELLIER, M., RANGIN, C., CADET, J-P., TJASHURI, I., BUTTERLIN, J. & MULLER, C. 1992. L'île de Nias, un edifice polyphase sur la bordure interne de la fosse de la Sonde (Archipel de Mentawai, Indonesie). *Comptes Rendues de l'Academie des sciences, Serie II*, **315**, 1019–1026.

ROCK, N. M. S. 1987. Discussion of a paper by J. K. Leggett. *Journal of the Geological Society, London*, **144**, 751–752.

ROSE, R. 1983. Miocene carbonate rocks of Sibolga basin, northwest Sumatra. *Proceedings of the Indonesian Petroleum Association*, **12**, 107–125.

TAPPONIER, P., PELTZER, G. & ARMIJO, R. 1986. On the mechanics of the collision between India and Asia. *In*: COWARD, M. P. & RIES, A. C. (eds) *Collision Tectonics*. Geological Society of London Special Publication, **19**, 115–157.

VITA-FINZI, C. & SITUMORANG, B. 1989. Holocene coastal deformation in Simeulue and Nias, Indonesia. *Marine Geology*, **89**, 153–161.

WILLIAMS, P. R., PIGRAM, C. J. & DOW, D. B. 1984. Melange production and the importance of shale diapirism in accretionary terrains. *Nature*, **309**, 145–146.

Fold growth during basin inversion – example from the East China Sea Basin

GUANG MING WANG[1,2]; MIKE P. COWARD[1], WENGUANG YUAN[3],
SHENSHU LIU[3] & WENQIANG WANG[3]

[1]*Department of Geology, Imperial College, London SW7 2BP, UK*
[2]*Present address: Primeline Petroleum Corporation, Parkview House, 14 South Audley Street, Mayfair, London W1Y 5DP, UK*
[3]*Shanghai Bureau of Marine Geological Survey, 526 Yan An Road W, Shanghai 200050, PR China*

Abstract: Understanding the progressive growth of fold structures during basin inversion is an important part of basin analysis. To achieve this, a complex inversion-related anticline in the East China Sea Basin has been studied to examine variations in the growth rate and growth kinematics during the Miocene inversion of the basin. Detailed studies of onlaps and truncations of seismic reflections, particularly within the syn-growth package, indicate that different parts of the anticlinal structure grew at slightly different times, i.e. the start and cessation of growth vary along and across the structure. Growth was related to the reactivation of the pre-existing, dominantly W-dipping early normal faults. The reactivation of these early normal faults and the development of new reverse faults progressed laterally and transversely across the structure. Sections have been depth-converted, back-stripped and reconstructed to restore the progressive development of the fold. The concept of this study may be applied to many other basins with a similar evolutionary history.

Positive basin inversion often reactivates early normal faults and develops associated hangingwall anticlines in syn-rift and particularly in post-rift sequences (Williams *et al.* 1989; Coward 1993; see other papers in Cooper & Williams 1989). The understanding of the geometries and kinematics of folds in an inverted basin is an important part of petroleum exploration. Folding during basin inversion often changes the geometry of structures forming hydrocarbon traps as well as the reservoir properties of the rocks and sealing properties of faults (Coward 1993). It is important to know the timing of fold growth relative to the burial and diagenetic history of reservoir rocks and maturation history of potential source rocks.

Dating fault movement by the study of growth sequences on the limbs of fault-related folds has been a very useful tool in structural interpretation (Cartwright 1992). Recently models have been produced predicting the progressive development of fault-bend-folds and fault-propagation folds (Suppe *et al.* 1992) and wedge-thrust related folds (Medwedeff 1992). These studies have focused on thin-skinned thrust-related folds. Many inversion structures and their related hangingwall folds, however, are thick-skinned. The post-rift sequences are shortened but also uplifted by shortening strains at depth. In the post-rift sequence, therefore, these folds may be considered as forced folds (Stearns 1978; Ameen 1988, 1991).

A grid of seismic profiles over a fold structure in the East China Sea Basin has been studied with the aim of determining its progressive development. Models are proposed for inversion-related fold growth, differing from the detachment-thrust related models. This paper aims to discuss the geometry of hangingwall folds associated with basement-involved tectonics, and to demonstrate the detailed growth history of the studied fold in the East China Sea Basin.

Different styles of inversion-related fold geometry and fold growth are discussed and the inversion structure in the East China Sea is placed in its regional tectonic setting. This exercise will have general application in the analysis of similar basins all over the world.

Overview of the basin development of the East China Sea

The East China Sea Basin is located between the SE coast of mainland China and the Ryukyu Islands of Japan, covering an off-shore area of 750 000 km^2 (Zhou *et al.* 1989). It comprises two

From BUCHANAN, J. G. & BUCHANAN, P. G. (eds), 1995, *Basin Inversion*,
Geological Society Special Publication No. 88, 493–522.

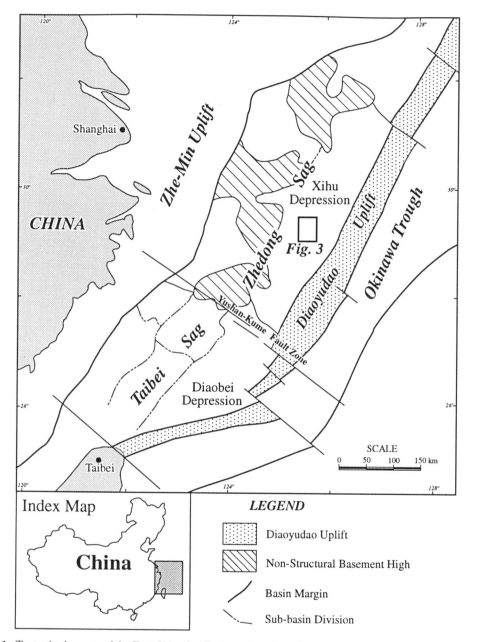

Fig. 1. Tectonic elements of the East China Sea Basin and location of the studied structure.

Mesozoic–Cenozoic basins: the East China Sea Shelf Basin (ECSSB) and the Okinawa Trough Basin (Fig. 1). The East China Sea Shelf Basin is a large Cretaceous–Tertiary basin infilled with Upper Cretaceous to Quaternary sediments, locally exceeding 10 km in thickness (Zhou *et al.* 1989; Yin & Liu 1990). It is bounded to the west by the Zhe–Min Uplift (Fig. 1) which is mainly a

zone of Mesozoic volcanic rocks underlain possibly by Proterozoic basement (Zhou *et al.* 1989; Wang 1990). The East China Sea Shelf Basin is bounded to the east by the Diaoyudao Uplift (or Fold Belt) (Fig. 1) (the Taiwan–Sinzi Belt of Letouzey & Kimura 1985), part of which has been recently re-interpreted as thrust belt (Wang *et al.* 1994). The East China Sea Shelf

Age	Formations	Facies	Tectonics
Quaternary	Donghai Group	*Marine*	
Pliocene	Santan Formation	*Paralic*	
Miocene	Liulang Formation	*Floodplain-Swamp*	
	Yuquan Formation	*Floodplain & Coastal Swamp*	*Inversion*
	Longjing Formation	*Lacustrine in N Fluvial in S*	
Oligocene	Huagang Formation	*Fluvial in N Paralic in S*	*Inversion*
Eocene	Pinghu /Oujiang Formation	*Coastal Swamp in N Marine in S*	*Rifting*
Palaeocene	Minyuefeng Formation	*Marine to Deltaic*	
	Lingfeng Formation	*Marine*	*Rifting*
	Shimentan Formation	*Terrestrial*	
Late Cretaceous	Late Cretaceous	*?*	*Rifting*

(right margin, rotated text: **Thermal Subsidence**)*

Fig. 2. Schematic stratigraphic and tectonic events in the East China Sea Shelf Basin.

Basin underwent a four-stage evolutionary history (see also Xu & Le, 1988; Wang 1990; Zhou *et al.* 1989) (Fig. 2).

(1) Extension began in the Late Cretaceous and continued until the Palaeocene/Eocene, probably related to back arc basin development above the subducting Pacific Plate/Philippines Sea Plate. Rifts, in the form of NE-trending half-graben, formed along the continental margin east of the Zhe–Min Uplift. A Late Cretaceous basaltic andesite (75 Ma) was penetrated by one well in the SW part of the basin (Tao *et al.* 1992). This type of back-arc related volcanic rock, however, is not widely observed, because the syn-rift sequence has been deeply buried in the main part of the basin.

(2) Extension was followed by subsidence, presumably related to cooling of the attenuated lithosphere. The resulting thermal subsidence is significant in the eastern part of the basin where there are locally over 5 seconds (TWT) (*c.* 9 km)

of post-rift sediments consisting of sandstones, shales and coal beds (Zhou, *et al.* 1989).

(3) Tectonic inversion occurred (i) during the Oligocene, uplifting the Diaoyudao Uplift in the eastern part of the ECSSB and (ii) during the Miocene, forming a series of NE-trending anticlinal structures in the centre and near the western margin of the ECSSB (Zhou *et al.* 1989; Yuan 1991). Miocene inversion was confined to the northern part of the ECSSB (the so-called Xihu Depression) (Fig. 1). In contrast, there are only a few gentle anticlines with no associated reverse faults developed in the southern part of the ECSSB (the so-called Diaobei Depression, see Fig. 1). This study deals with folds produced during the Miocene inversion in the centre of the Xihu Depression.

(4) The Plio-Pleistocene was a time of continued thermal subsidence without significant inversion or extension. In the northern part of the ECSSB, NE-trending reverse faults and

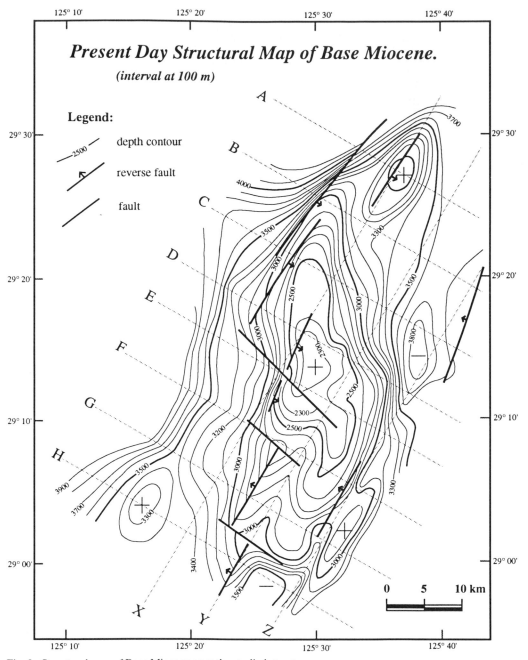

Fig. 3. Structural map of Base Miocene over the studied structure.

Fig. 4. Two seismic sections showing the change in polarity of the anticline axis and the general features of the structure (see Fig. 3 for location): (**a**) section F, in the south, with predominantly NW-dipping faults and fold axial planes; (**b**) section D, in the north, with predominantly SE-dipping faults and fold axial planes.

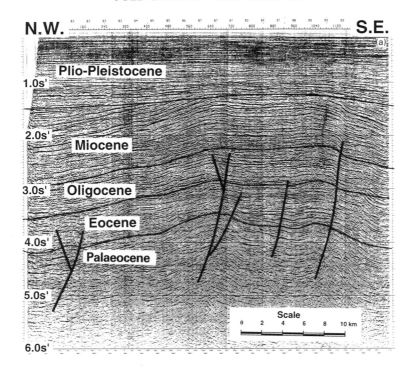

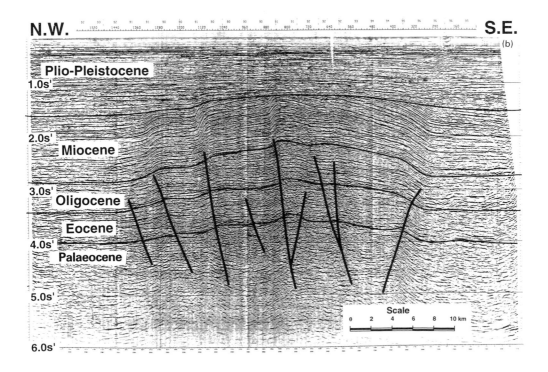

folds are truncated by an unconformity close to the base of the Pliocene. Occasionally NW-trending faults cut through the base Pliocene unconformity indicating that minor later tectonic movements have reactivated some earlier faults or generated late NW-trending structures in this basin. The Pliocene–Quaternary movements are dominantly normal or strike-slip along the NW-trending faults.

There are important variations in structure and structural development between different parts of the ECSSB. The most important variation is that between the northern and southern parts of the Basin, where the different sub-basins are separated by the NW-trending Yushan-Kume transform fault zone (Chai 1986) (Fig.1) (see also Zhou et al. 1989). NW-trending faults divide regions with different rift-related structures as well as different styles and intensities of inversion folds and faults (Yuan 1991; Zhou et al. 1989). The NW-trending faults are parallel to important structures which affect the basement onshore (Chai 1986; Lang & Yong 1993; Zhou et al. 1989).

General geometry of the studied structure

The structure studied is a complex anticline which grew in the central part of the Xihu Depression during the Miocene. This anticline is similar to other structures in the Xihu Depression; it is elongated in a NE–SW direction and is approximately 60 km long and 10–25 km wide. It has a complex detailed geometry as shown on the map of the Base Miocene (Fig. 3). Over 800 km of seismic data were studied on a grid spacing of c. 4 km. On Fig. 3 some of the seismic lines are shown and labelled for reference purposes.

Two sets of faults have developed: a NE–SW set and a NW–SE set (Fig. 3). The NE–SW set is parallel to the general trend of the basin and did not cut through the base Pliocene unconformity (Fig. 4); these faults are considered to predate the unconformity. The NE–SW set are strike faults and control the general geometry of the anticline (Figs 3, 4). During the regional study, it was found that these NE–SW trending reverse faults often appear to have normal displacement at depth, that is, some of the NE-SW set appear to be inverted normal faults. However because of the deep burial, the normal displacements are difficult to identify on seismic sections and hence impossible to map with any degree of accuracy in

the studied area. Subsequently details of the early normal fault geometry are not discussed in this paper.

Many of the NW–SE set of faults are truncated by the base Pliocene unconformity, but locally they cut through the unconformity and occasionally even displace the sediments at the sea bed. They have a long history of deformation throughout the basin evolution, and were also active from the Miocene to Plio-Pleistocene. The faults are steep and involve strike-slip displacement. The NW–SE set of faults seems to compartmentalize the fold. When mapped at different levels, the trend of this set of faults changes slightly from NW-trending at the Base Miocene level to WNW-trending at the Base Pliocene level. The NW-trending faults may be related to the basement transforms, e.g., the Yushan-Kume Fault Zone as mentioned above, as part of the NW-trending transform system.

Typical features of the anticline, as seen in cross section, are shown in Fig. 4. In the northern and central parts of the structure, the reverse faults are mainly SE-dipping with the overlying hangingwall anticlines having SE-dipping axial planes (Fig. 4b). In the south, the main axial planes are NW-dipping (Fig. 4a). Other features on the sections show that:

(i) reverse faults at depth pass up into folds at a high level, suggesting a thick-skinned forced fold model;

(ii) fold development involved rotation of both limbs during fold amplification; and

(iii) regional studies show that the folds are located above Late Cretaceous/Early Tertiary normal faults, whose geometry, however, is difficult to identify in detail.

Timing of the faulting and folding

The study examines the kinematics and growth history of the anticline by (i) identifying the syn-growth sequences on the flanks of the structure, and then (ii) back-stripping the growth sequence to produce a series of restored sections and maps.

The syn-growth sequences may be identified by basal onlaps and top truncations. Several horizons have been picked and tied between the dip and strike sections around the structure and hence the growth sequences may be correlated around the fold. The picked horizons cannot be dated as they are not intersected by any wells;

Fig. 5. Example of asymmetric pattern of growth sequences on two limbs of the structure (section D): (**a**) western limb, growth sequence: N–Q, and (**b**) eastern limb, growth sequence: P–Q.

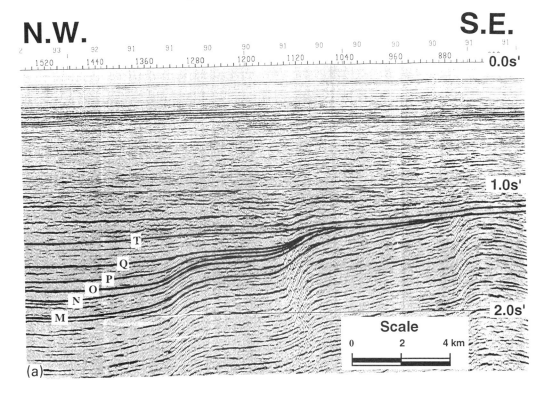

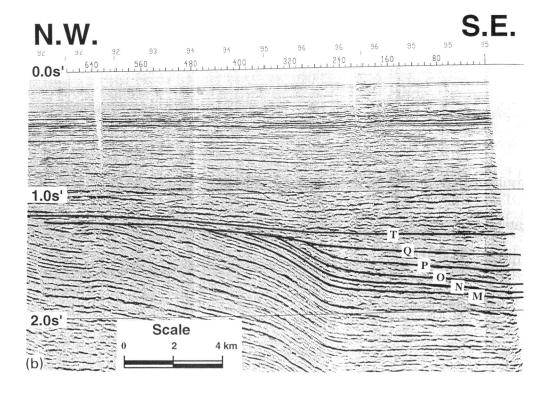

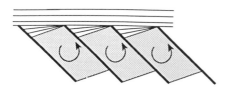 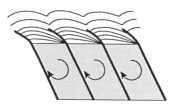

Rifting: Block Rotation

Pre-rift: block rotation and extension
Syn-rift: growth sedimentation

(a)

Inversion: Back Rotation

Pre-rift: reverse block rotation & contraction
Syn-rift: folding

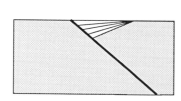 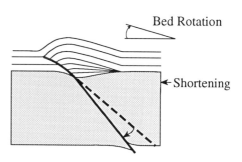

Rifting

Pre-rift: extension
Syn-rift: growth sedimentation

Inversion: Fault Block Flexural Cantilever

Pre-rift: contraction & cantilever flexure
Syn-rift: folding and rotation

(b)

Fig. 6. Possible models of back-limb rotation: (**a**) back-rotation of fault blocks: the syn-rift will be folded but there is no strain in the pre-rift; and (**b**) bed-rotation in the hangingwall: both syn-rift and pre-rift will be folded.

hence here are named by letters M, N, O, P, Q and T (in younging sequence). On the basis of the general assumption of seismic stratigraphy (Vail *et al.* 1977), that these picked seismic horizons represent chronological depositional surfaces, the growth of the fold can be dated in terms of this stratigraphic framework (M to T).

The geometry of the growth sequences illustrates that different parts of the structure developed at slightly different times. Some growth sequences have been folded (Fig. 5a), indicating that the fold growth history is not simple and that subsequent folding may obscure the early features. An important feature is that, in dip sections, the growth sequences have asymmetric patterns of growth on the two flanks of the structure (Fig. 5a, b). The same seismic

sequence may appear to be a growth sequence on one limb but be truncated on the other limb. The timing of tilting and uplift determined from the growth sequence appears to vary from one side of the anticline to the other (Fig. 5a, b).

Suppe *et al.* (1992) demonstrated that an asymmetric growth sequence may be developed on the two limbs of thin-skinned thrust-related folds. They described methods for analysing the growth of thin-skinned thrust-related folds assuming kink-band models where the growth sequences of the forelimb and back-limb are asymmetric but the forelimb and the back limb have constant dips. Though the geometric relationships between fold shape and ramp-flat geometry are valid for all dip-slip faults, a simple thin-skinned model is not directly applicable to

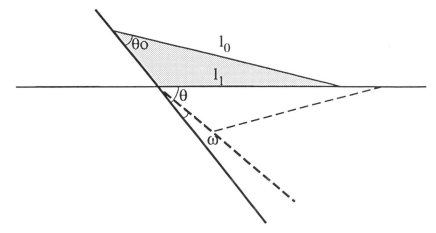

Fig. 7. Simplified geometry for fault block rotation.

the growth of the anticline in the East China Sea, where there is good evidence for displacement upon relatively steeply dipping faults and limb rotation during progressive development of the fold. Alternative models need to be considered for inversion-related folding.

Models for the development of hangingwall anticlines

The fold development above steeply-dipping inverted faults can be considered in terms of a combination of (i) forced folds, where the shape of the fold is controlled by the shape of some forcing member below and the beds change length during fold growth (Stearns 1978); and (ii) fault bend or fault propagation folds, where the shape of the fold is controlled by the curvature of the reverse fault as it propagates upwards through the post-rift sequence (Suppe 1983; Suppe & Medwedeff 1990).

In forced folds the forcing member can be an intrusive sill, batholith, faulted basement block or faulted massive sedimentary unit, imposing an upward force at a high angle to the overlying (cover) layers (Stearns 1978). Most discussions on forced folding have concentrated on a rigid basement with simple horizontally elevating hangingwall, associated with no tilting of the beds (e.g. Friedman *et al.* 1980; Ameen 1988; Mitra 1993). Examples of forced folds above inverted basement faults have been described by Ameen (1988, 1991) and Mitra (1993). The uplift of the basement block gives rise to bending of the cover sequence with complex strain development in the steep limb. Simple models (Ameen 1988, 1991; Mitra 1993) can be used to

illustrate the change in dip of the forelimb during inversion fold growth. However, as the models assume simple horizontal elevation with no tilting of the hangingwall, they cannot explain folds with back-limb rotation. In realistic geological environments, basement deformation will generally be more complicated than simple horizontal elevation and may involve rotation of fault blocks, or listric faulting, or folding of the basement itself. These complexities will cause the rotation of the back-limb of a hangingwall anticline.

The detailed geometries of folds produced during basin inversion will depend on the fault kinematics, that is, if the faults are listric, follow a domino rotation pattern, or act as single inverted faults. The folds develop as a result of the decrease in bed length due to (i) reversal in displacement up the fault and (ii) back-rotation of the faults.

Faults associated with block rotation

Block faults may back-rotate during inversion causing thrust sense reactivation along the fault plane, until their dip reaches the critical value to stop this reactivation (Coward 1993; Gillcrist *et al.* 1987). Back-rotation will be easier and more pronounced if the displacements are oblique; if they are highly oblique or strike-slip then the faults could rotate back to vertical. During back-rotation the syn-rift fill will be squeezed out into a series of folds and thrusts (Fig. 6). Note that the post-rift cover will be shortened in a series of folds or thrusts, some of which may detach close to the syn-rift/post-rift boundary and unless deep data are available, these folds

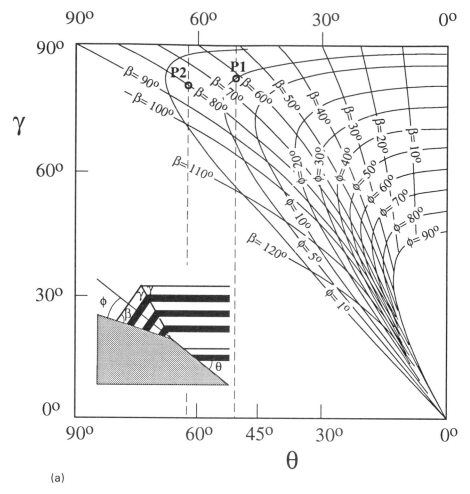

Fig. 8. Forelimb rotation: (**a**) graph showing the relationship between the change in fault dip (φ) and the new fault cut-off angle (β) (simplified from Suppe 1983); (**b**) and (**c**) convex-upwards flattening of an inverted fault can change the hangingwall cut off angle and develop a steep forelimb (see text for details).

and thrusts may be mistaken for thin-skinned detachment structures (Coward 1993).

Coward (1993) has shown examples of folds associated with rotated block faults from the North Sea and from experimental models. The folds (Coward 1993) are characterized by relatively planar back-limb dips and short hooked forelimbs. The faults may have extensional geometry at depth, passing through a null point (Williams *et al.* 1989) to a reverse fault at higher levels and then die out in a hooked tip beneath these forelimbs. A simple model for their development is shown in Fig. 7. The shortening in this rotational block model is given by:

$$\text{shortening} = l_1/l_0 = \sin\Theta_0/\sin\Theta = \sin\Theta_0/\sin(\Theta_0 + \omega)$$

where Θ_0 is the original fault dip angle and ω is the angle of rotation.

The analysis of kink-band folds by Suppe (1983) shows how the geometry of the hanging-wall changes as it moves over a fault with a change in dip. As the hangingwall moves over a curved fault plane which decreases in dip, the hangingwall cut-off angle will increase. The relationships between the original cut-off angle (Θ), the new cut-off angle (β), the change in dip of the fault (φ) and the fold interlimb angle (γ) have been calculated by Suppe (1983) and plotted graphically in Fig. 8a. Note that steeply dipping faults cannot be reactivated to form flat thrusts at high levels. There is no solution, assuming kink-band fold geometry, for a simple ramp-flat fault trajectory where the ramp dip exceeds 30°. Where a normal fault exceeds 30° in dip, then kink-band fault geometry allows for some decrease in dip of the fault upwards, but

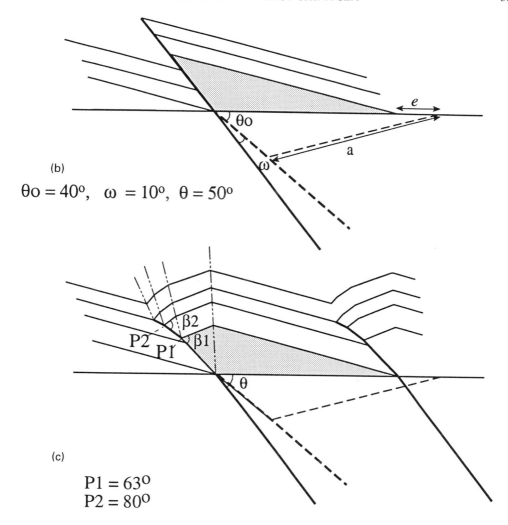

(b)

$\theta o = 40°, \quad \omega = 10°, \quad \theta = 50°$

(c)

$P1 = 63°$
$P2 = 80°$

always less than the original cut-off angle. The maximum change in dip of a fault (Θ_{max}) allowing kink-band geometry, is shown in Fig. 9.

Suppe's (1983) analysis of kink-band folds therefore can be applied to a limited range of fold structures above steeply dipping ramps. Assuming a model of rotated blocks during inversion (Fig. 8b), the back limb will change dip accompanying the rotation. The forelimb will change dip as the reactivated fault grows into the post-rift cover sequence. For a fault dipping at 50° following 10° rotation (Fig. 8b), that is, for a fault with an initial cut off angle (Θ) of 40°, if this fault grows into the post-rift sequence changing angle by 5° ($\phi = 5°$), the new hangingwall cut-off angle (β) will be 63° (P1 in Fig. 8a, c). The fault could change dip by another 2° ($\phi = 2°$) to produce a new cut-off angle (β) of 80° (P2 in Fig.

8a, c). As the fault changes dip its displacement decreases, so that it eventually dies out at a fault tip. This process of up-dip flattening of a fault will generate a steep forelimb with a gradual loss of fault slip (Fig. 8c). Note that the resulting fold can be analysed in terms of kink bands or considered as a smooth curve.

Single fault rotation

Single faults, or small groups of faults, which form the boundaries of extensional basins may invert by a process analogous to that of a flexural cantilever (Fig. 10). As the fault rotates, the hangingwall flexes upwards while the footwall flexes downwards. The resultant faults are characterized by long back limbs and steep fore-

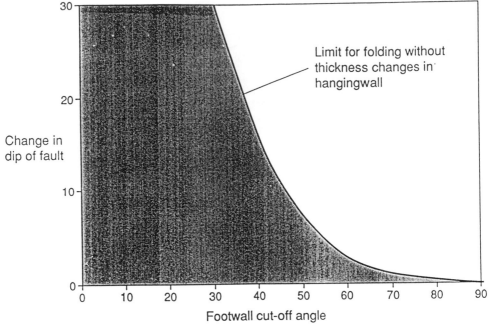

Fig. 9. Limit for change in fault shape with kink-band fault-bend folding. The shaded area shows the dip change possible with fault-bend folds.

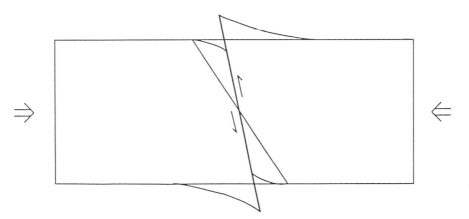

Fig. 10. Flexural cantilever model for fault reactivation during compression.

limbs as shown in Fig. 6B. During this type of inversion the shortening is given by (Fig. 11):

$$\text{Shortening} = l_1/l_0 = \sin(\Theta_0 + \omega - \delta)/\sin(\Theta_0 + \omega)$$

where Θ_0 is the original fault dip, δ the back limb dip and ω the rotation of the fault.

The final back limb dip depends on the initial dip of the fault and the angle of rotation. Steeper faults require more rotation and hence more shortening to develop the same back limb dip. The locked hangingwall fold may have a strained

steep limb or the reactivated fault may curve into the cover with a geometry similar to that described for rotated fault blocks in Fig. 8.

Keystone faults

Badley *et al.* (1989) showed a keystone block from the southern North Sea (Badley *et al.* 1989, fig. 6), uplifted on two faults with opposing dips. The faults in the basement flatten slightly into the post-rift sequence, causing the flat-topped

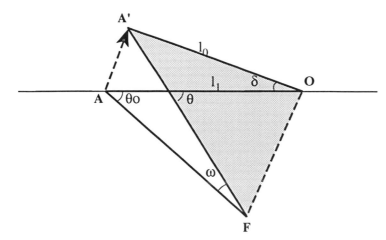

Fig. 11. Fault-bend fold relationship for bending associated with flexural cantilever model. AF, original fault plant; A'F, invested fault plane; A'O, rotated bed during inversion.

uplift to be bounded by two monoclines with opposing dips. The geometry of the monoclines can be examined assuming fault-bend fold models. Often the original normal faults form part of a symmetrical keystone graben; reactivation closes the graben uplifting the keystone block without change in dip.

Listric faults

Listric faults will reactivate easily at depth but the steeper parts of the faults will not reactivate, leading to the development of folds on the hangingwalls and 'short-cut' faults on the hangingwalls or footwalls (see Coward 1993, fig. 22). These footwall faults commonly lead to the development of 'floating islands' of pre-rift material, bounded by the original normal fault and by the shortcut. Isolated wedges of original footwall rock may be translated into the hangingwall of the thrust. The geometry of the hangingwall fold will depend on the change of dip of the shortcut structure. Where the listric faults are associated with rotation, several generations of shortcut structures may grow as the domino faults rotate. In experiments they are commonly convex-upwards and develop in a break-back sequence towards the hangingwall (Buchanan & McClay 1991). At small values of inversion they may only have limited displacement and form upward-fanning horsetail pattern. At higher values of contraction the footwall shortcuts may be responsible for generating a lower angle, more smoothly varying thrust trajectory.

Fold reconstruction in the East China Sea

The conceptual reconstruction models used in this study consider the fold to be developed as a thick-skinned inversion structure, possibly modelled as a cantilever with limb rotation during the development of the anticline (Fig. 12). The model envisages some folding of the syn-rift (syn-extension) sequence and forced folding of the post-rift sequence, associated with the propagation of the inverted fault to its tip (Fig. 12). It also envisages the fault to be straight and planar at depth, flattening upwards into the post-rift sequence. Horizontal shortening below the fault tip point is produced partly by fault displacement and partly by folding, but entirely by the folding above the tip point, resulting in equal shortening of the lower and upper sequences. Decoupling is not necessary.

This style of deformation will produce fold growth with limb rotation in both the back-limb and forelimb and a symmetric pattern of growth sequences, i.e. the timing of the folding worked out from growth sequences will be the same on both limbs (Fig. 12). Note however that the dips, thicknesses and extents of the growth sequences in two limbs are not necessarily exactly the same.

The seismic sections show that a slightly more complex model with multiple fault is possibly more realistic. Figure 13 shows a schematic model with two synthetic faults. In reality some of the faults may be antithetic. Where the faults are synthetic then rotation may be modelled in terms of a rotated block model, where each fault has a similar rotational axis, hence similar

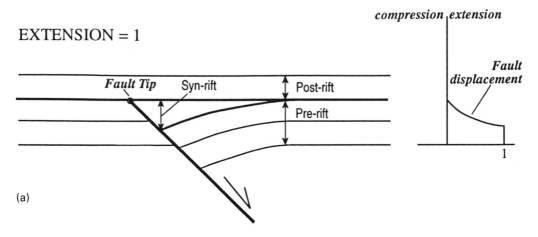

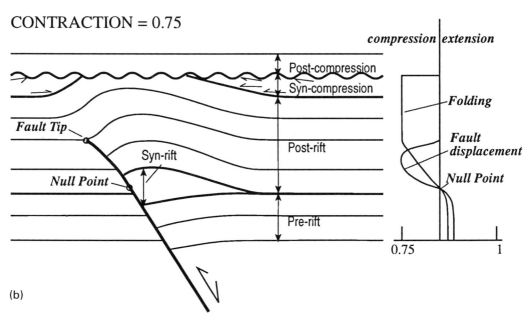

Fig. 12. Conceptual model of fold growth during inversion: (**a**) extension and the deposition of syn-rift sequence in the half graben; (**b**) inversion of the early normal fault after a thick post-rift sequence has been deposited. The syn-rift sequence is folded and the post-rift sequence is folded and faulted below the tip horizon. The syn-compressional growth sequences have basal on-laps, top truncations and an unconformity marks the start of deposition of the post-compression succession.

growth history. Alternatively the fault may be modelled in terms of a series of flexural cantilever faults, where each fault is reactivated independently.

When two (or more) faults grow individually by the cantilever model, complex asymmetric growth sequences can be developed (Fig. 13b, c). Early on-laps and truncations in the growth sequences can be eroded and the sequences appear as 'pre-growth sequences'. In these situations (Fig. 13b, c), the growth history will appear different on opposite limbs of the composite fold structure. Detailed analysis of the fold growth history can illustrate which structures were active at a particular time.

Restoration technique

Restoration of the structure has used a simple 2D vertical shear technique on the depth

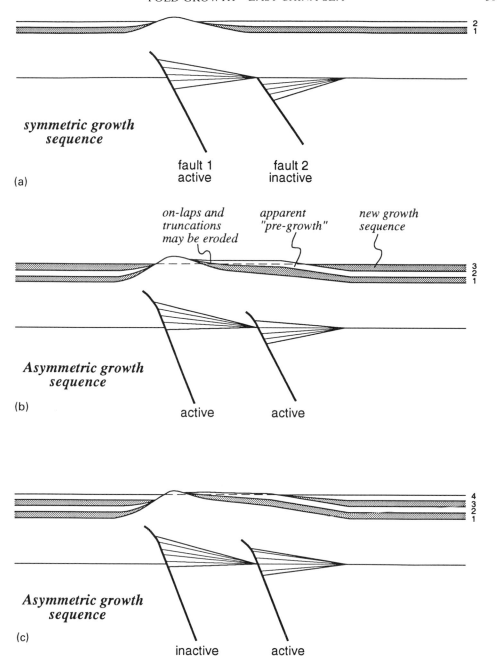

Fig.13. Multiple fault inversion fold growth model: (**a**) when fault 1 is active and fault 2 is inactive, a forced fold develops on fault 1 and symmetric growth sequences are deposited on both limbs (both 1-2); (**b**) following (a); faults 1 and 2 are both active, part of the right limb of the early fold is uplifted and new growth sequence is deposited further to the right. The growth sequences for the broader fold (comprising two forced folds) are asymmetric (left: 1–3 and right: 3); (**c**) following (b), if fault 1 now becomes inactive and fault 2 is still active, further asymmetry occurs in the growth sequences (left: 1–3 and right: 3–4).

sections. This gives errors in the restored line length but these errors are considered too small to affect the overall pattern. The picked horizons have been restored to a horizontal datum with zero water depth. Back-stripping of horizons outside the fold area restores the

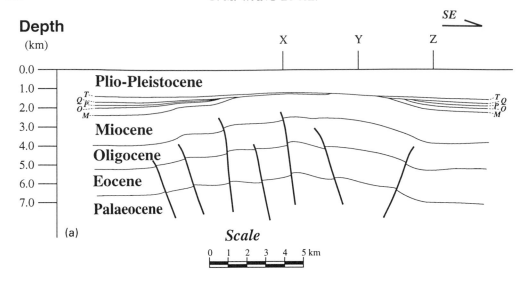

(a)

Scale

0 1 2 3 4 5 km

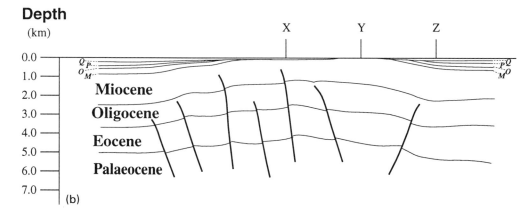

(b)

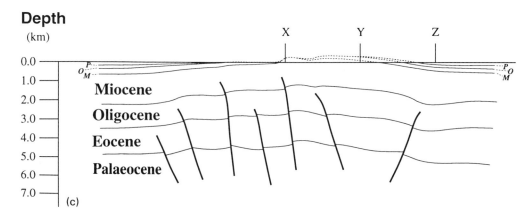

(c)

Fig. 14. Restoration of Section D. (X, Y and Z are the positions of the strike lines marked on Fig. 3).
(a) present-day depth section. Fault structure is simplified;(b) restored section at time T: the Plio-Pleistocene
sequence has been back-stripped; (c)restored section at time Q: restored to the upper most growth sequence.
The eroded pre-growth and early growth sequences have been extrapolated (dotted lines); (d) restored section

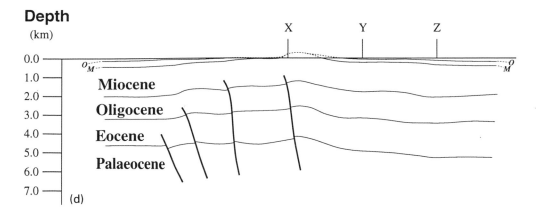

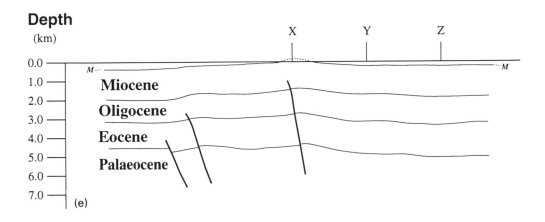

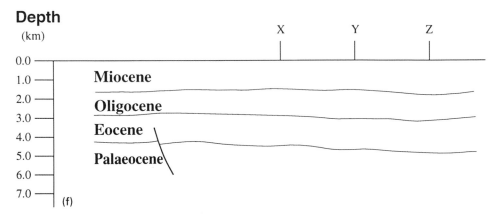

at time P: the faults which no longer have displacement have been removed; (e) restored section at time O: more faults are removed and the structure is now only confined to a small area. (f) restored section at time M: the fold disappears and the section displays a gently dipping horizon; a normal fault of the Palaeogene-Eocene age remained in the NW part of the section.

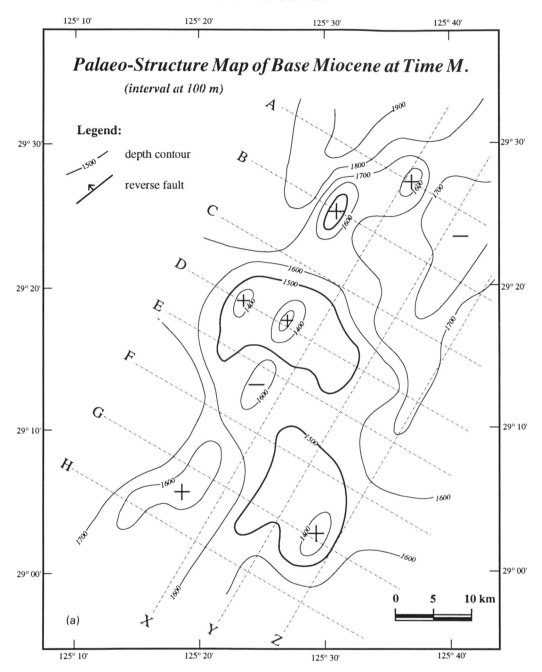

Fig. 15. Reconstructed palaeo-structural maps: (**a**) structural map of Base Miocene at time M; (**b**) structural map of Base Miocene at time O; (**c**) structural map of Base Miocene at time P; (**d**) structural map of Base Miocene at time Q;(**e**) structural map of Base Miocene at time T.

regional subsidence. In the folded area, the eroded parts of the earlier growth sequences as well as the pre-growth sequences have to be extrapolated. During restoration, faults are removed where the displacement becomes zero.

Other important assumptions have been made which could affect the shape of the fold: (i) sedimentation keeps pace with the regional

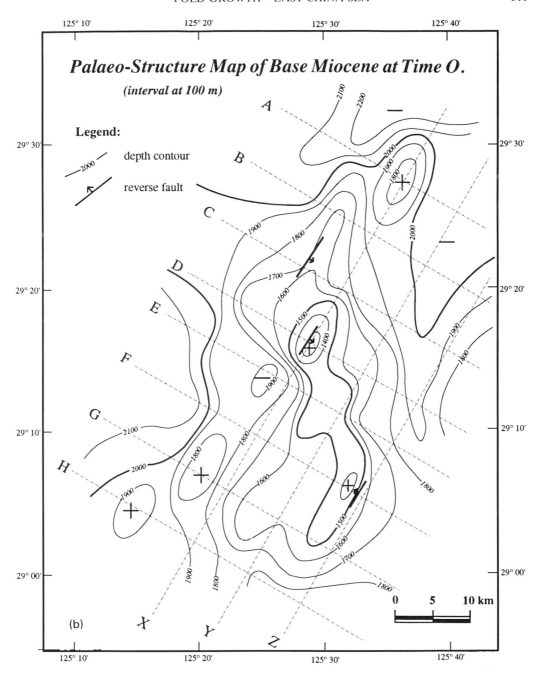

Palaeo-Structure Map of Base Miocene at Time O.
(interval at 100 m)

Legend:

——— depth contour

⤹ reverse fault

(b)

0   5   10 km

subsidence; (ii) regional sea level changes are ignored. There is no age data to correlate the picked horizons to regional sea level changes. Regional sea level changes will alter the water depth and erosion area and will produce regional progression/retrogression. They will therefore affect the size of the area of uplift, but not the variable pattern of uplift; (iii) erosion keeps pace with the fold growth; and (iv) differential compaction has been ignored. There is no data to permit the decompaction.

All the principle seismic sections were depth-converted using interval velocity data and well data provided by the SBMGS. The fault struc-

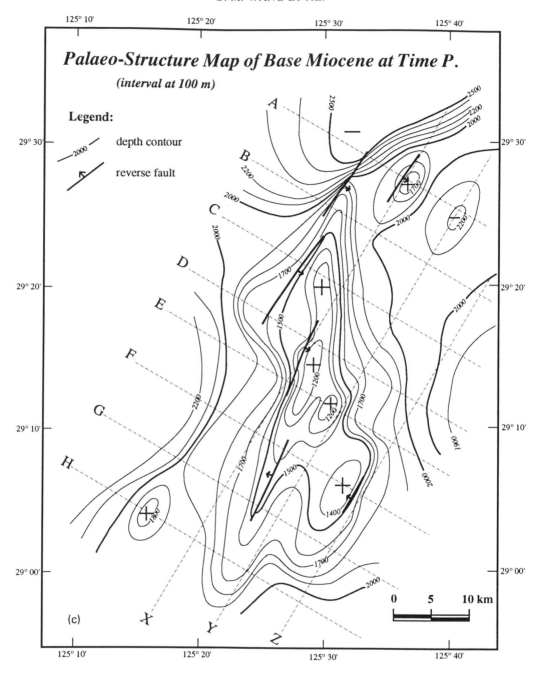

Palaeo-Structure Map of Base Miocene at Time P.

(interval at 100 m)

Legend:

— *2000* — depth contour

⤤ reverse fault

(c)

0   5   10 km

tures have been simplified to avoid unnecessary complications during restoration. The principle sections were restored to all six picks from T to M. Figure 14 shows an example of such an exercise in 5 restored sections. The restoration to pick N is very similar to that of pick O and thus omitted here.

During the reconstruction, strike-slip displacement on the NW/SE trending faults cannot be restored due to lack of sufficient data. These faults are simply removed during the restoration to pick T (back-stripping of Plio-Pleistocene sediments). The consequence of this assumption is discussed later.

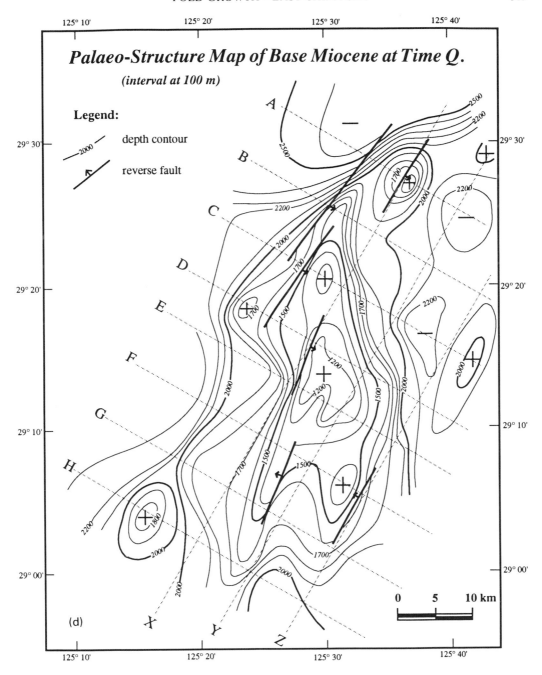

Palaeo-Structure Map of Base Miocene at Time Q.

(interval at 100 m)

Legend:

/ depth contour

↖ reverse fault

(d)

Results of the reconstruction

Palaeo-structural maps

The results of the restorations are shown in the palaeo-structural maps for the Base Miocene at the time of marker horizons M, O, P, Q, and T (Figs 15a–e). The map at the time N is not presented because it is very similar to that of at the time O. The structural evolution can be summarized as follows.

The growth of the anticline started from a very gentle feature at the time of M (Fig. 15a), growing into several domal structures at the time

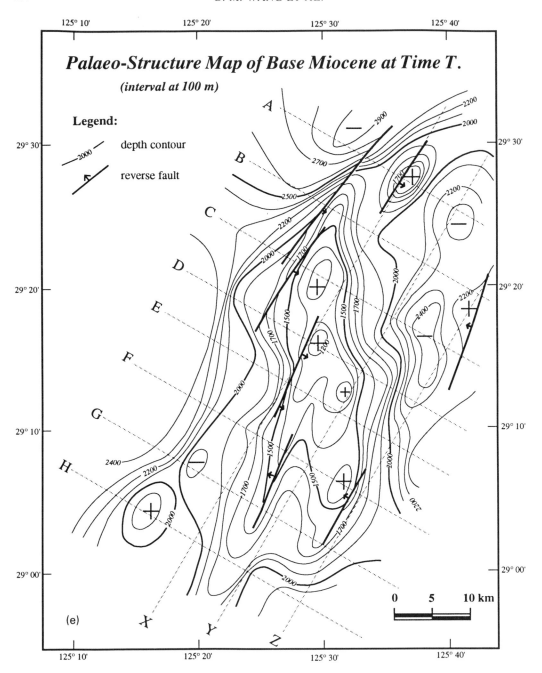

Palaeo-Structure Map of Base Miocene at Time T.

(interval at 100 m)

Legend:

— depth contour

↖ reverse fault

(e)

0   5   10 km

O (Fig. 15b). The domes are aligned in a right lateral en echelon pattern controlled by the inversion of a set of en echelon faults (Fig. 15b).

Subsequent growth of the anticline occurred both laterally and transversely (Fig. 15c–e) possibly controlled by the reactivation of older normal faults and development of new NE-trending reverse faults. There appears to have been a rapid fold growth phase between times O and P (Fig. 15c, d). The amplification rate of the fold varies and seems to be controlled by movement on individual inverted faults and possibly the development of new reverse faults. No overall relationship between the age of fold growth and amplification rate can be found. Local amplification rates vary significantly in

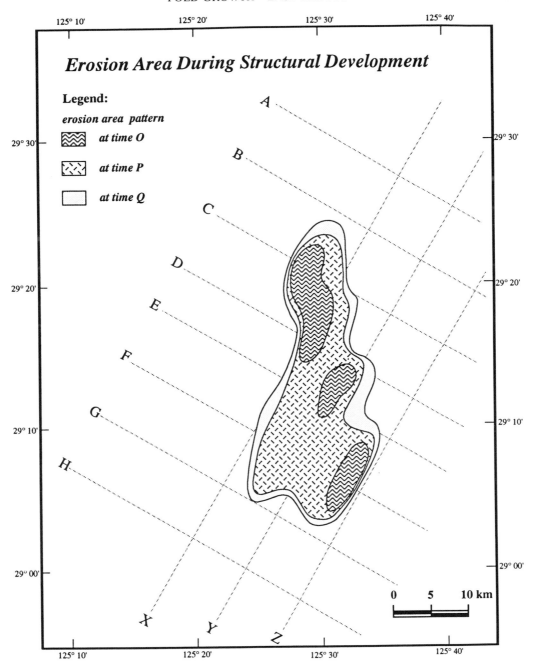

Fig. 16. Erosion area during structural development.

time and space, i.e. local amplifications of hangingwall folds above inverted faults can be accelerated, slowed down or stopped (Fig. 15c–e). Since reverse faults seems to propagate from a lower sequence (Palaeogene?) upward to the base of the Miocene, the important faults controlling the growth of folds may not appear on the map. Comparing Fig. 3 with Fig. 15, it is obvious that the present highs were not necessarily palaeo-highs during the fold growth.

The progressive fold growth can also be seen on maps showing areas of erosion (Fig. 16). At

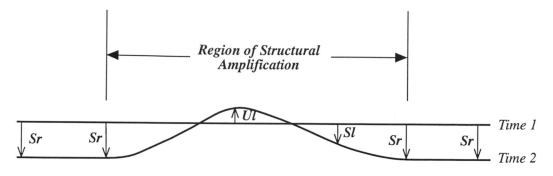

Fig. 17. Calculation of structural amplification (see text for details).

time O there are three isolated en echelon areas of erosion. These areas of erosion expanded to a much large area at time P and further modification occurs between times P and Q. Figure 16 shows the lateral and transverse growth of the fold.

Structural amplification

From the back-stripped/restored sections, structural amplification rates can be calculated for the time interval represented by any two picked horizons (Fig. 17). Assuming the shapes of a horizon at times 1 and 2 are as shown in Fig. 17, the structural amplification (AMs) of a particular point can then be calculated as:

AMs = Regional Subsidence (Sr)
    − Local Subsidence (Sl) or
AMs = Regional Subsidence (Sr)
    + Local Uplift (Ul)

The results are shown in three structural amplification maps of time M–O, time O–P and time P–Q (Fig. 18 a, b, c). These maps clearly illustrate the variations in displacement history throughout the fold and hence the variation in fault reactivation at depth. They also illustrate that the displacement history for each inverted fault is independent from that of its neighbours, i.e. individual faults can propagate (laterally and vertically), pause or further propagate (laterally and vertically) through time (Fig. 18 a, b, c).

Discussion

Errors

As stated above, this exercise aims to present a conceptual restoration of the fold/fault. Many inaccuracies may be introduced due to the assumptions and lack of data. More accurate analysis of fold growth requires a more detailed knowledge of the stratigraphy. Well data are required. The regional sea level changes and decompaction need to be considered to constrain a more accurate depth map of a horizon at a 'palaeo-time'. Both of these require additional stratigraphic and geochronological data. Further detailed depositional facies and palaeo water depth data will help to produce 'palaeo-geometry' to which horizons can be restored.

More accurate seismic (3D) data will allow more accurate fault/fold restoration. With detailed data, control line-length model, e.g. Suppe's kink-band model may be used to develop more accurate restorations of growth.

NW-trending faults

There are several NW-trending 'kinks' in the restored structural maps, particularly in the central part of the studied structure (Fig. 15b–e). These 'kinks' seem to be consistent with the NW-trending cross faults on the present day

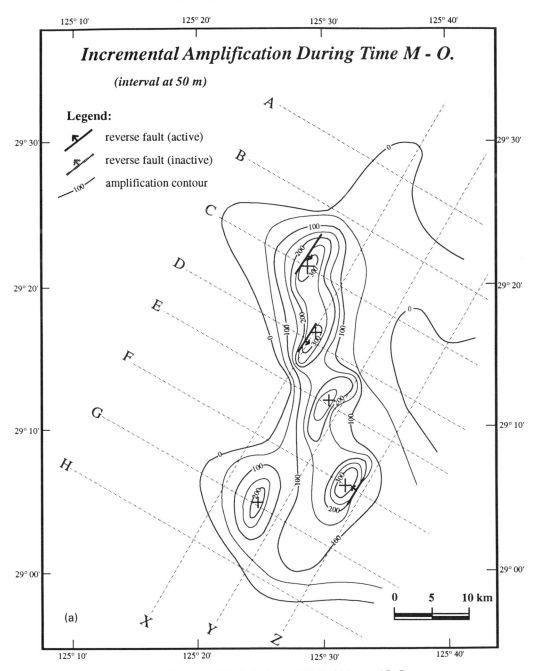

Fig. 18. Amplification maps for: (**a**) interval M–O; (**b**) interval O–P; (**c**) interval P–Q.

structural map (Fig. 3). The differences on opposite sides of the 'kink' (i.e. the dip of fault and geometry of the fold) cannot be resolved by a simple strike-slip displacement (Figs 3, 15). This suggests that the NW-trending faults may have played a role as transfer structures, active during fold growth. This suggests that the structure may be more segmented than shown on the maps.

The en echelon alignment of early inversion faults and resultant domes can be regarded as being caused by sinistral transpressional

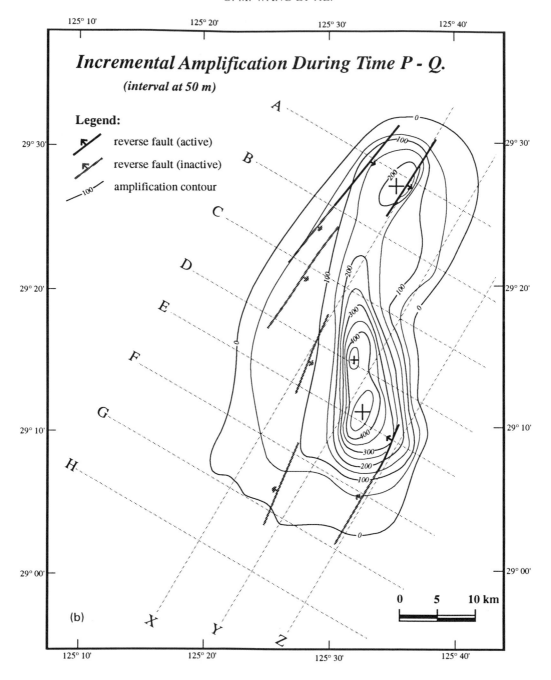

inversion of the basin (Fig. 19a) or as a set of push-up structures (Fig. 19b). Note that strike-slip/oblique slip inversion will produce synchronous growth of the fold along the fault strike. There is no evidence on either the palaeo-structural or the amplification rate maps (Figs 15, 18) to support this argument. From the palaeo-structural maps, it is evident that some of

the NW-trending faults were active during the growth of the folds (Fig. 15). This suggests that push up model is more realistic.

The presence of NW-trending faults as transfer structures enables the fault blocks to be inverted individually. The fold domes therefore were not 'displaced' by the strike-slip motion on the NW-trending faults, but rather are the

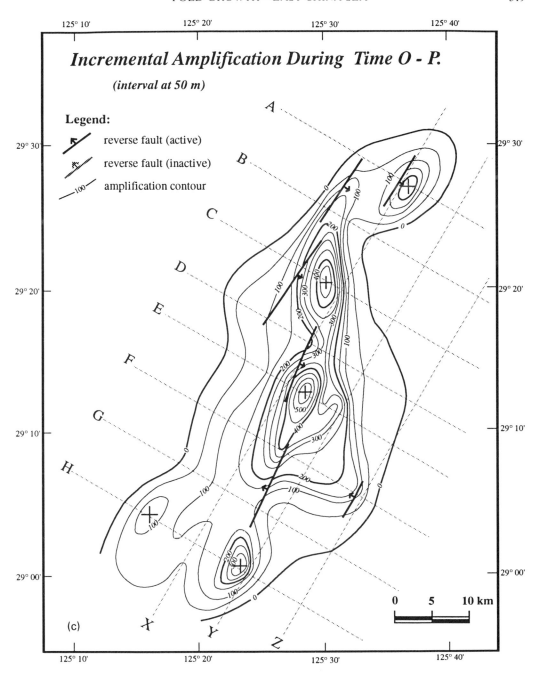

Incremental Amplification During Time O - P.

(interval at 50 m)

Legend:

reverse fault (active)

reverse fault (inactive)

amplification contour

(c)

localized 'push up' structures formed by the rotation of individual blocks between transfer zones.

Regional interpretation

The detailed interpretation and restoration given above have some bearing on the regional understanding of basin inversion. Sinistral transpresion along the strike of the faults is not favoured. Inversion appears to be related to push-up along NW-trending basement faults which were originally transfer faults during the rift phase. The relationship of these structures to regional fault pattern and the location of the push-up folds at the offset of these strike-slip

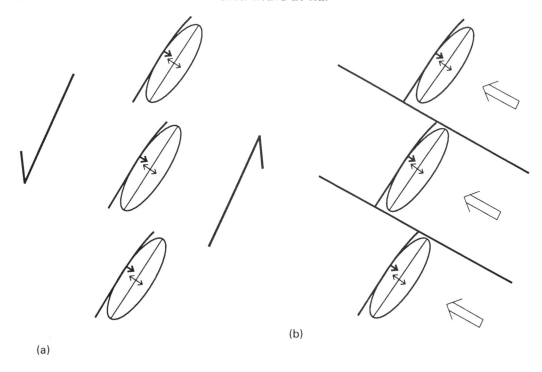

(b)

(a)

Fig. 19. Possible models for the development of the en echelon domal structures: (**a**) strike-slip related inversion fold growth, and (**b**) push-up structures where inversion is compartmentalized by transfer faults.

transfer structures, suggest the existence of a left-lateral shear couple across the northern part of the East China Sea Basin during the Miocene (Fig. 20). The shear couple may be related to the lateral expulsion of China away from the Himalayan Collision Zone and/or the variation in opening rate of the Okinawa Trough.

Conclusion

The following conclusions can be drawn from the current study on inversion kinematics.

(i) Fold growth in the Xihu Depression has resulted from the inversion of early normal faults and the development of new reverse faults during the basin inversion.

(ii) The fold began growth as a set of en echelon domes, related to inversion of an en echelon system of structures. Further fold development was accompanied by displacement along different faults. The folds are character- ized by gently dipping back-limbs and hooked fore-limbs. A rotational fault model is favoured involving the rotational reactivation of several faults, both synthetic and antithetic. The shape of the fold changes with time. The final geometry

of the fold is very different from that of the initial structure.

(iii) The fold grew laterally and transversely with a variation in amplification rate. The controlling reverse faults were activated at different times and have individual displacement histories. There is no simple relationship be- tween amplification rate and position in the structure.

(iv) The initial domes were 'offset' by NW- trending faults which possibly had an integral control on the structural development and became important as strike-slip or normal faults in the Plio-Pleistocene. The NW-trending faults compartmentalized the deformation in variably developed push up zones during fold growth.

(v) Note therefore that when modelling hydrocarbon migration, one has to take into account the complex fold history. The fracture patterns and the reservoir properties may vary through time. Structures have to be analysed individually, in particular the positions of palaeo-highs need to be matched with the timing of hydrocarbon maturation and migration. An important consideration is the compartmentaliz- ation of fold growth by transfer faults, and its

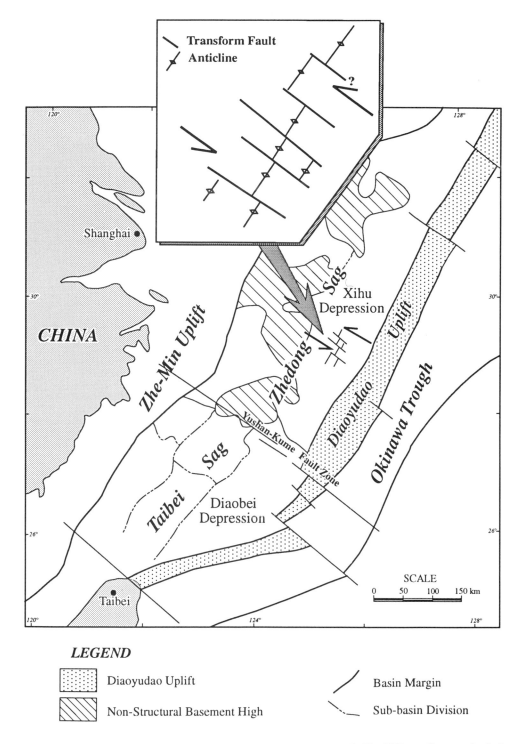

Fig. 20. Regional tectonic setting for a push-up interpretation of fold growth. The NW-trending transfer faults were active during the Miocene inversion, with a predominantly sinistral sense of movement. This implies that a left-lateral shear couple existed in the northern part of the East China Sea Basin during the Miocene.

associated implications for petroleum migration and reservoir continuity.

GMW and MPC were sponsored by Enterprise Oil plc. The Director of the Shanghai Bureau of Marine Geological Survey is thanked for permission to use data in this paper. The authors are grateful to Steven Knott & Schuman Wu for constructive comments on an early manuscripts which help to improve the paper.

References

AMEEN, M. S. 1988. *Folding of layered cover due to dip-slip basement faulting*. PhD thesis, Imperial College, London.

—— 1991. Possible forced folding in the Taurus-Zagros Belt of northern Iraq. *Geological Magazine*. **128**: **6**, 561–584.

—— 1992. Strain pattern in the Purbeck-Isle of Wight Monocline: a case study of folding due to dip-slip fault in the basement. *In*: BARTHOLOMEW, M. J. *et al.* (eds) *Basement Tectonics 8*. Kluwer Academic Publishers, 559–578.

BADLEY, M. E., PRICE, J. D. & BACKSHALL, L. C. 1989. Inversion, reactivated faults and related structures: seismic examples from the southern North Sea. *In*: COOPER, M. A. & WILLIAMS, G. D. (eds) *Inversion Tectonics*. Geological Society, London, Special Publication, **44**, 201–219

BUCHANAN, P. G. & McCLAY, K. R. 1991 Sandbox experiments of inverted listric and planar fault system. *Tectonophysics*, **188**, 97–115.

CARTWRIGHT, J. A. 1992. *Application of seismic stratigraphic methods in structural interpretation*. Joint Association for Petroleum Exploration Courses (JAPEC) (UK), **116**.

CHAI, LIGEN. 1986. The nature of the Yushan-Kume Fault belt and its controls over East China Sea. *Oil and Gas Geology*, **7**, 2, 107–115. [In Chinese]

COOPER, M. A. & WILLIAMS, G. D. 1989. *Inversion Tectonics*. Geological Society, London, Special Publication, **44**.

COWARD, M. P. 1993. Inversion Tectonics. *In*: HANCOCK, P. L. (ed.) *Continental Deformation*, Pergamon, 264–304.

FRIEDMAN, M., HUGMAN, R. H. H. & HANDIN, J. 1980. Experimental folding of rocks under confining pressure, Part VIII – forced folding of unconsolidated sand and of lubricated layers of limestone and sandstone. *Geological Society of America Bulletin. Part I*, **91**, 307–312.

GILLCRIST, R., COWARD, M. P. & MUGNIER, J. L. 1987. Structural inversion, examples from the Alpine Foreland and the French Alps. *Geodinimica Acta*, **1**, 5–34.

LANG, YIAOXIU & YONG, G. 1993. NW–trending structures in South China Block and their controls on mineralisation and seismisity. *Mineral Resources and Geology*, 7, No. 6: 390–395. [In Chinese.]

LETOUZEY, J. & KIMURA, M. 1985. Okinawa Trough genesis: structure and evolution of a back arc basin developed in a continent. *Marine and Petroleum Geology*, 2, 111–130.

MEDWEDEFF, D. A. 1992. Geometry and kinematics of an active, laterally propagating wedge thrust, Wheeler Ridge, California. *In*: MITRA, S. & FISHER, G. W. (eds) *Structural Geology of Fold and Thrust Belts*, 3–27.

MITRA, S. 1993. Geometry and kinematics evolution of inversion structures. *AAPG Bulletin*, **77(7)**, 1159–1191.

STEARNS, D. W. 1978. Faulting and forced folding in the Rocky Mountains foreland. *Geological Society of America Memoir*, **151**, 1–38.

SUPPE, J. 1983. Geometry and kinematics of fault–bend folding. *American Journal of Science*, **283**, 684–721.

——, CHOU, G. T. & HOOK, S. C. 1992. Rates of folding and faulting determined from growth strata. *In*: McCLAY, K. R. (ed.) *Thrust Tectonics*, Chapman & Hall, London, 105–123.

—— & MEDWEDEFF, D. A. 1990. Geometry and kinematics of fault-propagation folding. *Eclogae Geologicae Helvetiae*, **83**, 409–453.

TAO GUOBAO, LIANG, L & ZHU, P. 1992. Characteristics of Cenozoic geologic and tectonic development of the Oujiang Depression in Zhejiang – Fujian offshore area. *Scientia Geologica Sinica*, **1**, 1–9.

VAIL, P. R., TODD, R. G. & SANGREE, J. B. 1977. Chronostratigraphic significance of reflection. *AAPG Memoir*, **26**, 99–116.

WANG, GUOCHUN. 1990. Prospective petroliferous areas in the East China Sea Basin. *Journal of Petroleum Geology*, **13**, 71–78.

WANG, G. M., COWARD, M. P., YUAN, W. LIU, S. & WANG, W. 1994. The evolution of the East China Sea basin and its implication to the development of the west Pacific Marginal Seas. *AAPG Bulletin*, **78/7**, 1168.

WILLIAMS, G. D., POWELL, C. M. & COOPER, M. A. 1989. Geometry and kinematics of inversion tectonics. *In*: COOPER, M. A. & WILLIAMS, G. D. (eds) *Inversion Tectonics*. Geological Society, London, Special Publication, **44**, 3–15.

XU, WEILING & LE, JUNYING. 1988. Structural and tectonic evolution of the East China Sea Basin. *Marine and Quaternary Geology*, **8**, 1, 9–15 [In Chinese.]

YIN, PEILING & LIU, SHENSHU. 1990. Geological characteristics and hydrocarbon exploration of the East China Sea. *In: International Symposium on the Tectonic Evolution and Petroleum Potential of the East China Sea*. Stanford University, 80–100.

YUAN, WENGUANG. 1991. Petroleum exploration of the Continental Shelf Basin of East China Sea. *In: The Second International Symposium on the Tectonic Evolution and Petroleum Potential of the East China Sea*. Stanford University, 45–55.

ZHOU, Z., ZHAO, J. & YIN, P. 1989. Characteristics and tectonic evolution of the East China Sea. *In*: ZHU, X. (ed.) *Chinese Sedimentary Basins*. Elsevier Amsterdam. 165–71.

Case studies: Australasia

Inversion around the Bass Basin, SE Australia

KEVIN C. HILL[1], KATHY A. HILL[2], GARETH T. COOPER[2], ANDREA J. O'SULLIVAN[1], PAUL B. O'SULLIVAN[1] & M. JANE RICHARDSON[2]

[1]*Victorian Institute of Earth & Planetary Science, School of Earth Sciences, La Trobe University, Melbourne, Victoria 3083, Australia*

[2]*Victorian Institute of Earth & Planetary Science, Department of Earth Sciences, Monash University, Melbourne, Victoria 3168, Australia*

Abstract: Regional seismic reflection and fission track thermochronological studies of the Bass Basin area illustrate two types of inversion, both confined to the Bass failed rift. The first type involved 1–2 km of uplift, denudation and cooling of basement over more than 200 km along both margins of a failed rift, with lesser erosion within the rift basin. This occurred during renewed extension that bypassed the Bass Basin area leaving it as a failed rift. The uplift resulted from breaking of the lithosphere and may have been due, in part, to rebound following several kilometres of rapid sediment loading in the preceding 20 Ma. The second type comprised repeated structural inversion involving 1–3 km of uplift by compressional reactivation of extensional faults and along new reverse faults. These inversions occurred along the zone of maximum extension in a failed Mesozoic rift, such that Tasmania acted as a buttress to compression. The rifting and structural inversions are interpreted to have been controlled by a long-lived zone of weakness, perhaps overlying a Palaeozoic greenstone belt, with the maximum principal stress being roughly perpendicular to faults within the zone during inversion. Significantly, Miocene–Pliocene inversion resulted from the Australian craton being placed into compression following arc collision 3500 km to the north, emphasizing the importance of long distance transmission of compressional stresses through the lithosphere.

The aims of this paper are 1) to document episodes of structural inversion (fault reactivation) and periods of regional inversion (uplift/denudation) in the Bass Basin area and 2) to interpret the events in terms of plate interactions (Figs 1–3). The uplift and erosion events are analysed by a combination of seismic reflection interpretation and apatite fission track and vitrinite reflectance analyses. Seismic reflection data reveal the major unconformities and style of deformation, while apatite fission track and vitrinite reflectance analyses help constrain the maximum palaeotemperature and the timing and amounts of cooling and denudation. The latter analyses can reveal regional uplift not apparent from conventional seismic reflection data. The area is particularly suited to such analyses due to 1) the large volume of recently acquired high quality seismic reflection data, 2) the abundance of onshore Palaeozoic granites suitable for apatite fission track dating and 3) basin sediments rich in both apatite and plant fossils suitable for fission track and vitrinite reflectance measurements. This paper focuses on two areas, to the NW and SE of the Bass Basin, comprising the flanks of a failed rift, which record several uplift and denudation events.

Tectonic setting and sequence stratigraphy of the Bass Strait basins

The Bass Strait area lies between Tasmania and mainland Australia and contains the Otway, Bass and Gippsland basins (Figs 1 & 2). These basins were formed by two main periods of extension, Late Jurassic–Early Cretaceous intracontinental extension in all three basins and Late Cretaceous rifting which was focussed on the offshore Otway and Gippsland Basins (Fig. 3). During the latter phase, the Bass Basin was part of a failed rift and the margins, in particular were affected by several episodes of uplift, attributed to compression or wrenching (Fig. 3; Davidson & Morrison 1986). Both the extension and uplift events resulted from the Mesozoic and Tertiary separation of Australia from Antarctica, creating the Southern Ocean, and from the Late Cretaceous–Paleocene separation of Australia from New Zealand and the Lord Howe Rise, which opened the Tasman Sea (Figs 1 & 3;

From BUCHANAN, J. G. & BUCHANAN, P. G. (eds), 1995, *Basin Inversion*, Geological Society Special Publication No. 88, 525–547.

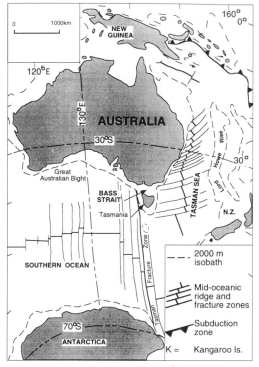

Fig. 1. Plate tectonic setting of Bass Strait, including the Bass Basin (Fig. 2), with oceanic crust to the west, south and east, due to the rapid Late Cretaceous spreading of the Tasman Sea and the slow Late Cretaceous and fast Tertiary spreading of the Southern Ocean. Present-day compressional plate margins are in New Zealand (N.Z.) and New Guinea.

Veevers *et al.* 1991). This paper is particularly concerned with the transition from an Early Cretaceous intracontinental rift to a Late Cretaceous failed rift in Bass Strait and with Neogene fault reactivation.

Intracontinental rift structures

In order to understand the causes of inversion, the orientation of the original (rift) extensional faults should be known. Etheridge *et al.* (1985) infer Early Cretaceous extension with NNE–SSW extension on NNW trending faults and orthogonal transfers, but Willcox *et al.* (1992) infer Late Jurassic–Early Cretaceous NNW–SSE extension on NNE trending faults, overprinted by Early Cretaceous NNW trending faults. In the onshore Otway Basin Perincek *et al.* (1994) show three Early Cretaceous syn-rift fault trends, E–W, NW–SE and NE–SW and Hill *et al.* (1994) suggest that Early Cretaceous extension may have been broadly N–S across an

area with pre-existing NE and NW basement fabrics.

In the study area syn-rift faulting was mainly down to the north or northwest resulting in a Late Jurassic–Barremian 'rift' sequence of thick alluvial and fluvial sediments. This sequence has a wedge-shaped geometry on seismic reflection data suggesting deposition during rotational normal faulting (Fig. 3 & for example Figs 4a/d & 4b/e). The Aptian–Albian sequence consists of several kilometres of widespread and more parallel-bedded volcanogenic, fluviatile sandstone throughout most of Bass Strait, with less evidence of syn-depositional faulting, suggesting regional subsidence (Fig. 3).

Transition to a failed rift

At *c.* 95 Ma, breakup occurred (Veevers 1986) and extension propagated along the west and east coasts of Tasmania, leaving Bass Strait as a failed rift with little Late Cretaceous sediment deposited. Renewed rifting to the west created a Late Cretaceous depocentre in the Otway Basin and ultimately lead to development of a passive margin. The boundary between passive margin and failed rift is marked by the Sorell Fault Zone linking to the Tasman Fracture Zone to the south and probably onshore to a tectonic lineament NW of Cape Otway (Figs 1 & 2; Foster & Gleadow 1992). This boundary is fundamental in separating the inverted area around the failed rift from that to the west which records only minor inversion.

West of the onshore lineament, fission track data suggest minimal regional uplift, but to the east the Palaeozoic margin of the Bass Strait failed rift records rapid cooling at 95–90 Ma due to regional uplift and denudation (Foster & Gleadow 1992; Dumitru *et al.* 1991; Duddy & Green 1992). Similarly, fission track analysis of Palaeozoic granites along Australia's east coast indicates significant uplift of the eastern margin at *c.* 80 Ma, associated with Tasman Sea rifting (Moore *et al.* 1986; Dumitru *et al.* 1991). Within the failed rift basin, Duddy (1994) reported mid-Cretaceous cooling ages from the Olangolah-1 well in the Otway Ranges (Fig. 2), inferring major folding of the Ranges at that time.

The 95 Ma breakup event resulted in a regional unconformity (Fig. 3) previously associated with wrench faulting and inversion around Bass Strait (Davidson & Morrison 1986; Willcox *et al.* 1992). Following 'breakup', renewed rifting in the Otway and Gippsland Basins and in the southern Bass Basin led to deposition of locally thick Late Cretaceous sequences (Figs 2 & 4c/f). However, Upper

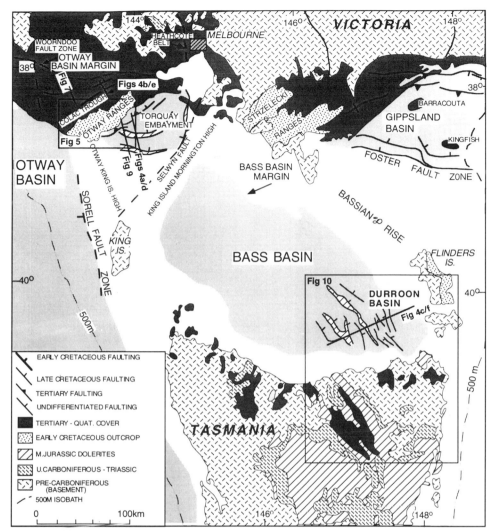

Fig. 2. Geology of the Bass Basin region showing 1) the main study areas; the Colac Trough, the Otway Ranges, the Torquay Embayment, the Durroon Basin and northeast Tasmania; 2) the location of seismic lines Figs 4a–f, 7 & 9; and 3) the main fault trends in the study areas. Most of the outcrop is Palaeozoic 'basement', Jurassic pre-rift dolerites and Tertiary basalts. The exceptions in the Otway and Strzelecki Ranges are the Lower Cretaceous syn-rift and Tertiary post-rift sequences representative of the thick Cretaceous and Tertiary sections in the Otway, Bass and Gippsland Basins. The Colac–Otway Ranges – Torquay area is normally considered to be part of the eastern Otway Basin although structurally distinct (see text) and separated from the main Otway Basin by the Otway–King Island High. Barracouta and Kingfish are two of the larger oil and gas fields of the Gippsland Basin.

Cretaceous section is rarely present in the Torquay Embayment and is absent in the onshore Otway Ranges and Colac Trough (Figs 2, 3, 4a/d & 4b/e). A regional Paleocene/ Maastrichtian peneplanation (e.g. Willcox *et al.* 1992; Duff *et al.* 1991) terminated the Late Cretaceous sequences and is overlain by a basal Tertiary sequence of coastal plain to nearshore barrier systems (Fig. 3; Bodard *et al.* 1985). The

peneplanation may have been associated with Late Cretaceous–Paleocene compressional deformation recorded in the Otway Basin, west of the Sorell Fault (Hall 1994; Tickell *et al.* 1992).

Tertiary thermal subsidence and uplift

Overall, the Tertiary sequence (Figs 3, 4a–f) was deposited during regional thermal subsidence

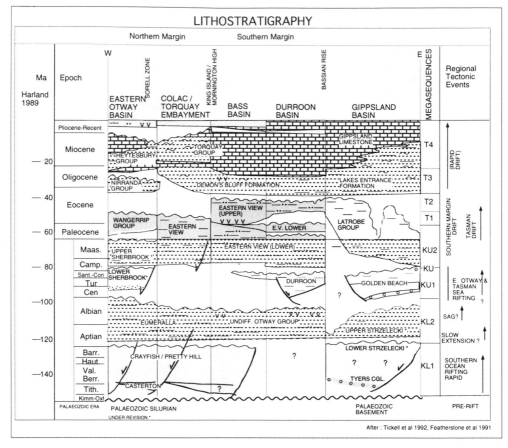

Fig. 3. The main stratigraphic sequences of Bass Strait, with emphasis on the Colac Trough, Torquay Embayment and Durroon Basin (Figs 1 & 2). The megasequences referenced throughout the text are shown on the left and a summary of the main tectonic events in the Bass Strait area is shown on the right. After Tickell *et al.* (1992), Featherstone *et al.* (1991), Hill & Durrand (1993), Veevers *et al.* (1991), Cande & Mutter (1982), Davidson & Morrison (1986), Willcox & Stagg (1990), Willcox *et al.* (1992) & Moore *et al.* (1992).

deposited during regional thermal subsidence (Hegarty *et al.* 1988) with increasing clastic starvation, which led to deposition of marls followed by a gradual transition to Neogene temperate climate carbonates. However, subsidence was interrupted by periodic uplift and erosion. The antiforms in the Gippsland Basin, containing large oil and gas accumulations, have been interpreted to have been formed by Eocene inversion of NE–SW trending transfer faults (Etheridge *et al.* 1985), whilst Reeckmann (1994) reports Oligocene uplift and erosion of the Otway Ranges. Young *et al.* (1991) state that in the Torquay Embayment, mid-Miocene and younger compression inverted existing NW-dipping Cretaceous normal faults with the degree and intensity of inversion decreasing to the NW and SE of the Otway Ranges. In the

Lower Cretaceous rocks exposed in the NE trending Strzelecki Ranges (Fig. 2) Constantine (1991) and Willcox *et al.* (1992) mapped NE, E and minor NW trending faults preserving Cretaceous extension, but with lineations showing reverse movement. The inversion occurred during the Tertiary to Recent as shown by the monoclines in the Tertiary cover.

Thermochronology

New apatite fission track and vitrinite reflectance analyses are reported here and integrated with seismic data in the next section in order to interpret the inversion events. To determine the timing and magnitude of uplift and erosion events, apatite fission track and vitrinite reflectance analyses were carried out in the Otway

Ranges and NE Tasmania (the northern and southern margin of the Bass failed rift, Fig. 2). Apatite fission track analysis can constrain the time of cooling from maximum palaeotemperatures, and the time–temperature history of rocks at temperatures up to c. 110°C for fluorapatite and up to 160°C for chlorapatite (Green 1989; Green *et al.* 1989). Vitrinite reflectance measurements (Rv max%) of surface samples can determine the maximum palaeotemperatures, and modelling of vitrinite reflectance data from boreholes constrains the palaeotemperature gradients (Burnham & Sweeney 1989).

To date, this study incorporates apatite fission track analysis of more than 30 samples from NE Tasmania and 10 samples from the Otway Ranges and more than 90 determinations of vitrinite reflectance (Rv max%) on surface samples from the Ranges. Apatite concentrates were separated from the samples using conventional magnetic and heavy liquid techniques and processed to analyse fission tracks following Naeser (1979) and Green *et al.* (1986). The principles used to interpret apatite fission track ages and confined track length distributions in terms of thermal history follow Gleadow *et al.* (1986), Green (1989) and Green *et al.* (1989). The data and analyses from the Tasmanian samples (O'Sullivan 1992) are to be presented in detail elsewhere, but the ages are summarized and discussed later in the text. The Otway Ranges data are discussed in more detail here and integrated with the seismic interpretation below.

Otway Ranges

Cooper has determined Rv max% values for over 90 surface samples of the Aptian–Albian Eumeralla Formation in the Otway Ranges. The contoured values show that the Ranges comprise a broad asymmetrical anticline (Fig. 5) with Rv max of >1.5% in the core indicating that samples currently at surface had previously attained temperatures of >180°C (modelling after Burnham & Sweeney 1989). Apatite fission track analysis was carried out on 9 new Otway Ranges surface samples of Eumeralla Formation, three from the coast and six from the core of the Ranges, along Wild Dog Creek (Tables 1 & 2). In addition, the central Otway Ranges sample (7842–194) of Gleadow & Duddy (1981) has been recounted. The samples from the core of the Ranges have Rv max values of >1.0% (Table 2) equating to maximum temperatures of >150°C, high enough to totally reset all but the most chlorine-rich apatites.

The stratigraphic age of the exposed Eumeralla sediments varies from c. 110 Ma along the coast to c. 125 Ma in the core of the Otway Ranges, yet six of the samples from the core of the Ranges yield apatite fission track ages of 97 ± 4, 96 ± 7, 83 ± 9, 82 ± 7, 87 ± 5 and 92 ± 6, with a mean age of 89 Ma (Table 2). These samples easily pass the chi-squared test at the 5% level (Galbraith 1981) and have long track lengths (14.1–14.8 μm), so are interpreted to represent fairly rapid cooling from temperatures of ≥150°C to temperatures of ≤60°C at c. 90 Ma.

Sample 7842–194 previously had a reported apatite fission track age of 118 ± 3 Ma (Gleadow & Duddy 1981) obtained by manual counting of only 10 grains prior to automation of microscope stages. Automated recounting of 25 grains yielded an age of 97 ± 4 Ma and a mean length of 14.5 μm, consistent with an interpretation of rapid cooling at 90–100 Ma. However, the low chi-squared probability suggests that this sample may have only just been overprinted, suggesting the presence of chlorine-rich apatites. Sample 2/30 from Wild Dog Creek, near the samples reported above, yielded an age of 130 ± 9 Ma,

Fig. 4. a/d: the mildly inverted extensional faults of the Torquay Embayment, showing thick Upper Jurassic to Lower Cretaceous rift sequences above NW dipping extensional faults. The monoclinal structures at the NW end of the section, deforming the Miocene section, are structurally similar to the adjacent Otway Ranges and are interpreted to have been formed by inversion of underlying Early Cretaceous extensional faults. See Figs 2 & 5 for location. **b/e**: NW-dipping Late Jurassic–Early Cretaceous extensional faults of the Colac Trough, similar to those across the Otway Ranges in the Torquay Embayment (**a/d**). The fault blocks have been planed-off so that the Upper Cretaceous sequence is absent. Minor Miocene–Pliocene inversion is apparent at SP 300 with significant Miocene–Pliocene inversion of the thickest part of the Lower Cretaceous section SE of SP 500 adjacent to the Otway Ranges. The syncline beneath the Palaeocene unconformity at SP 500 indicates mid-Late Cretaceous deformation of the Lower Cretaceous sequence prior to regional planation. **c/f**: extensional half-graben in the Durroon Basin with faults dipping down to the NE (Fig. 10). Note the 'blanket' deposit of Aptian–Albian (KL2) beds overlain by wedges of Upper Cretaceous sediment indicating Late Cretaceous fault development, related to opening of the Tasman Sea (Fig. 1). The Tertiary section records the post-rift thermal subsidence phase. The Aptian–Albian section has been eroded in the crests of the tilted fault blocks during initial faulting and is unconformably overlain by Turonian beds in the Durroon-1 well. The erosion may have been due to footwall uplift, but more likely was due to mid-Cretaceous regional uplift (see text).

NW **OS90A-11** 2000 1750

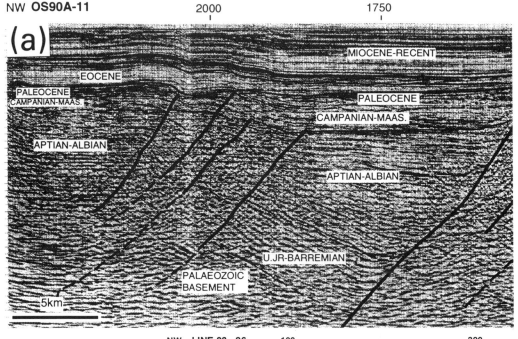

NW **LINE 83 - 06** 100 300

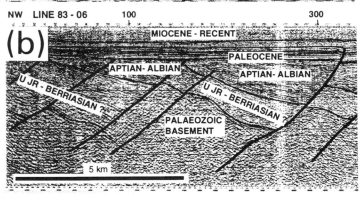

WSW **LINE BMR 88 - 306** 500

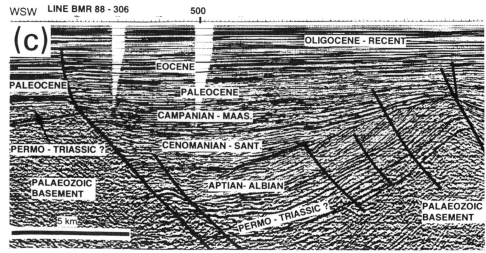

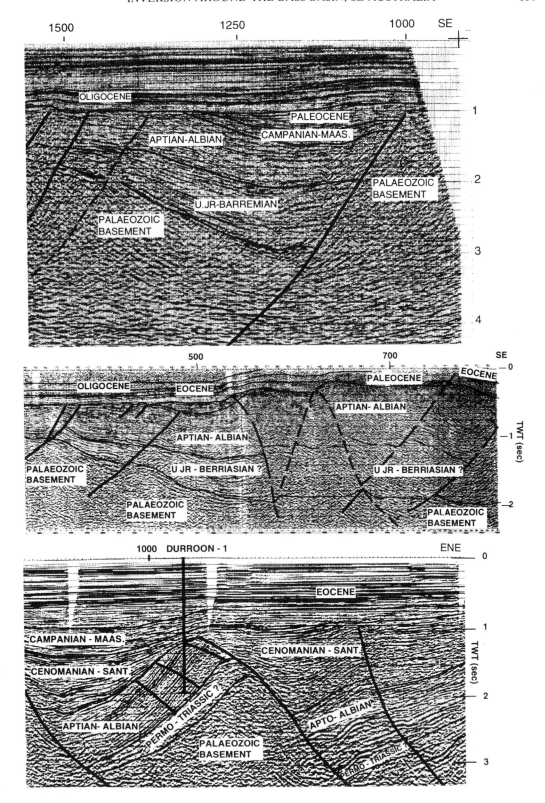

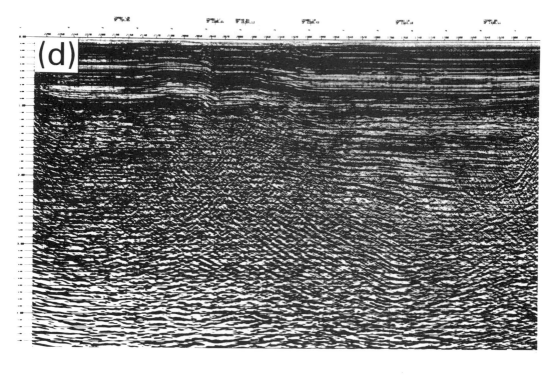

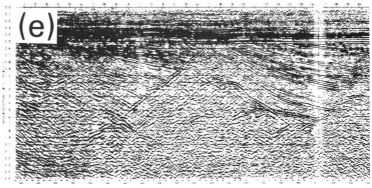

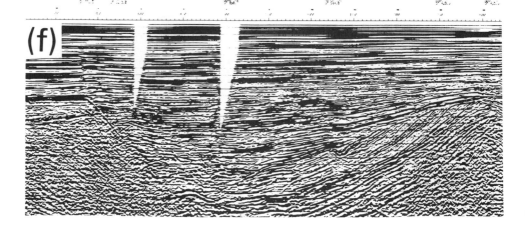

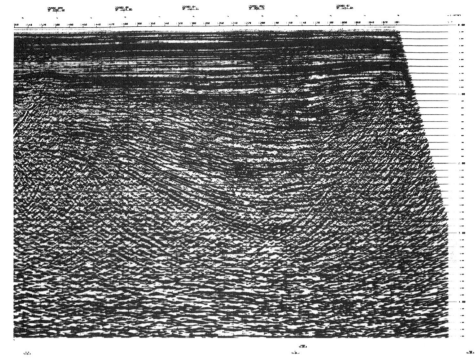

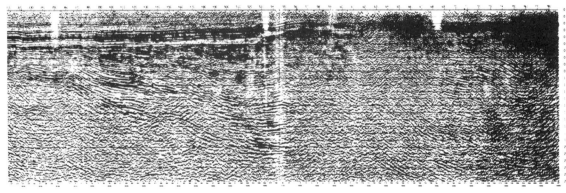

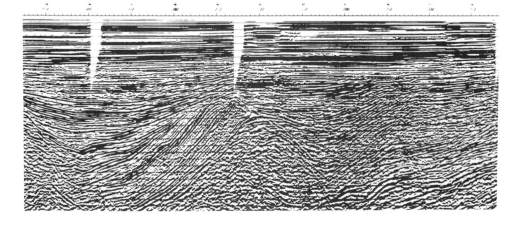

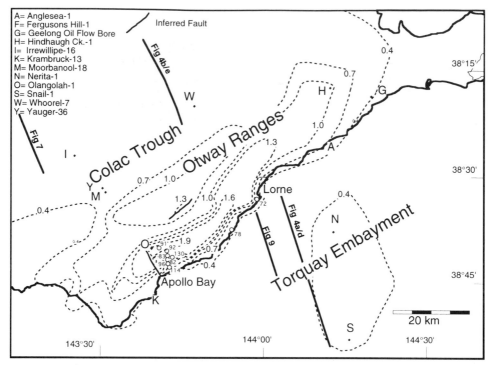

Fig. 5. Detailed map of the Otway Ranges area showing vitrinite reflectance (Rv max %) isoreflectors (after Cooper *et al.* 1993 and Struckmeyer 1988), the seismic line locations and the apatite fission track ages, mainly from Wild Dog Creek and the adjacent and parallel Skenes Creek Road, near Apollo Bay. Note that the isoreflectors show the broad, elongate, asymmetrical anticline structure of the Otway Ranges (Fig. 8).

Table 1. *Location of Otway Ranges fission track samples and of the nearest samples for vitrinite reflectance determinations. All samples are lithic sandstones and shales of the Eumeralla Formation, Aptian–Albian in age, mainly 120–105 Ma. Grid references are from the Lorne and Skenes Creek 1:25 000 map sheets*

| AFT sample | Apatite fission track sample | | | VR sample | | |
	Locality	Grid Ref	Elev. m	Nearest sample	Grid Ref
7842-194	(Gleadow & Duddy 1981)-inland				
1/10	Mariners Falls -inland	296128	120	3/22	313124
21/29	Wild Dog Ck -inland	316129	90	23/29	318139
12/28	Wild Dog Ck -inland	319107	40	11/28	319107
13/29	Wild Dog Ck -inland	328099	20	12/29	328099
3/28	Wild Dog Ck -inland	313124	75	1/30	335123
2/30	Wild Dog Ck -inland	335123	180	1/30	335123
1/28	Pt Grey Lorne -coast	603288	5	LF1	603288
L8-13	*c.* 5 km SW Lorne -coast	562253	0	T3	562253
19/29	Wild Dog Ck at coast	338092	5	18/29	338092

older than its stratigraphic age. This sample locality was >100 m higher than all other Wild Dog Creek samples, consistent with up to 10°C lower maximum palaeotemperatures, but still >130°C. Thus this sample is interpreted as being rich in chlorapatite, and as having attained lower maximum palaeotemperatures than its neighbours, so it was not completely overprinted and

Table 2. *Otway Ranges apatite fission track and nearest vitrinite reflectance analytical results and summary interpretation. The data are arranged in order of decreasing vitrinite reflectance (Rv max%) values. Typically, apatites from samples with a Rv max value of >1.0%, corresponding to a temperature of >150°C, have been completely annealed, so record the age of cooling from those high temperatures. Sample 2/30 is an exception, probably due to the presence of chlorapatite grains resistant to annealing even at such elevated temperatures. The upper seven samples are from the core of the Otway Ranges whilst the lower three are from the coast (Table 1).*

Sample Number	No. of grains	Standard track density ×10⁶cm⁻²	Fossil track density ×10⁵cm⁻²	Induced track density ×10⁶cm⁻²	Chi square prob %	Fission track age Ma	U ppm	Mean track length μm	SD μm	Vitrinite reflect Rv max %	VR Range Rv max %	VR Counts n	Interpretation of Fission Track Age
7842-194	25	1.431 (3031)	9.405 (1390)	2.430 (3592)	5.7	96.9 ± 3.6	20.9	14.54 ± 0.15 (100)	1.58				Denudation/cooling age
1/10	25	1.484 (3711)	0.346 (472)	1.039 (1416)	24.8	86.9 ± 4.9	8.6	14.58 ± 0.12 (109)	1.18	**1.88**	1.57–2.22	16	Denudation/cooling age
21/29	20	1.425 (3070)	2.672 (329)	72.43 (892)	71.8	92.0 ± 6.2	6.3	14.14 ± 0.18 (100)	1.75	**1.52**	1.33–1.70	30	Denudation/cooling age
12/28	20	1.425 (3070)	2.613 (118)	79.06 (357)	73.2	82.5 ± 8.9	6.8	14.54 ± 0.14 (50)	0.96	**1.42**	1.24–1.63	30	Denudation/cooling age
13/29	20	1.425 (3070)	3.232 (332)	84.20 (865)	39.7	95.7 ± 6.5	7.3	14.77 ± 0.18 (100)	1.80	**1.09**	0.99–1.23	31	Denudation/cooling age
3/28	20	1.425 (3070)	2.480 (168)	75.73 (513)	99.8	81.8 ± 7.4	6.5	14.43 ± 0.19 (102)	1.93	**1.04**	0.82–1.18	24	Denudation/cooling age
2/30	20	1.425 (3070)	4.548 (379)	87.00 (725)	49.0	130.0 ± 8.6	7.5	13.88 ± 0.24 (100)	2.36	**1.04**	0.82–1.18	24	Partially annealed, some old chlorapatite grains.
1/28	20	1.425 (3070)	4.628 (140)	1.488 (450)	98.8	77.7 ± 7.7	12.8	13.87 ± 0.19 (102)	1.88	**0.68**	0.65–0.73	13	Late Cretaceous–Early Tertiary cooling
L8-13	20	1.620 (6887)	0.314 (366)	1.240 (1446)	6.9	71.8 ± 4.4	10.0	14.6 ± 0.2 (98)		**0.68**	0.52–0.86	30	Late Cretaceous–Early Tertiary cooling
19/29	20	1.425 (3070)	5.653 (645)	1.231 (1405)	41.9	114.3 ± 5.9	10.6	14.81 ± 0.14 (101)	1.35	**0.4**	0.36–0.57	30	Stratigraphic age

Note: Brackets show number of tracks counted. Standard and induced track densities measured on mica external detectors (g = 0.5), and fossil track densities on internal mineral surfaces. Errors quoted at ± 1σ. Ages for samples calculated using ζ = 352.7 for dosimeter glass SRM612

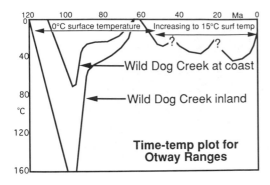

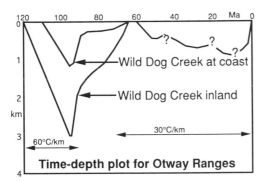

Fig. 6. Inferred time–temperature and time–depth geohistory diagrams for the Otway Ranges, based upon fission track and adjacent seismic reflection data. A 0°C surface temperature is assumed for the Cretaceous as Bass Strait was in high latitudes, warming up in the Tertiary. The depth conversion assumes that the high temperature gradients indicated by vitrinite reflectance data from wells were present in the Early Cretaceous, but dropped back to 'normal' following the mid-Cretaceous reduction in extension. The Palaeocene and Pliocene denudation events are inferred from seismic data.

temperature histories have been constructed, modelled after Laslett *et al.* (1987; Fig. 6). These histories incorporate cooling from elevated Early Cretaceous temperature gradients of *c.* 60°C km^{-1} (Cooper *et al.* 1993; Duddy 1994). Figure 6 shows an inferred 2–3 km of erosion of the Ranges from 95–90 Ma and subsequent Late Cretaceous cooling, consistent with fission track data from the Lorne area. Previously reported Tertiary events are shown schematically, except for Late Miocene–Pliocene cooling, which is consistent with inversion seen on seismic reflection data (below) and has been reported from vitrinite reflectance analysis of the Nerita-1 and Anglesea-1 wells (using BasinMod^{TM}; Cooper *et al.* 1993; Duddy 1994).

Seismic reflection interpretation

Although the cooling and denudation history can be inferred from the thermochronology, the style of deformation, particularly evidence for structural inversion, can only be determined from seismic reflection data. As part of a regional Bass Strait tectonic study, approximately 5500 km of seismic data have been examined from the Torquay Embayment and Colac Trough, on either side of the Otway Ranges and from the Durroon Basin (Fig. 2). The data range from fair quality 1973 onshore regional stack sections to very good 1990 offshore 60 fold migrated 2D data in the Torquay Embayment. Five seismic reflection lines are presented (Figs 4a–c, 7 & 11) to illustrate the style and timing of extension and of subsequent inversion in the Torquay, Colac and Durroon basins. Well coverage includes Palaeozoic basement penetrations onshore with section as old as Aptian being penetrated offshore within the area of study.

Inversion of the margins of the Bass failed rift

In this section, the results of seismic reflection data interpretation are integrated with thermochronological analyses to determine inversion of the margins of the Bass failed rift. The Torquay Embayment, Otway Ranges and Colac Trough are part of the northern margin of the Bass failed rift and the Durroon Basin and NE Tasmania are part of the southern margin (Fig. 2). Inversion of the northern and southern margins is considered separately below and then contrasted. The events interpreted from seismic reflection and thermochronological data are summarized in Table 3.

retained some old single grain ages. However, like its neighbours, it cooled rapidly at *c.* 90 Ma.

The coastal sample 1/29 yielded an age of 114 ± 6 Ma, interpreted as a stratigraphic age as Rv max values are 0.4% indicating relatively little post-depositional heating (Tmax *c.* 70°C). However, samples 1/28 and L8–13 along the coast to the NE near Lorne yielded ages of 78 ± 8 Ma and 72 ± 4 Ma (Fig. 5 and Tables 1 & 2). These samples record a Late Cretaceous to Early Tertiary (?70–60 Ma) cooling event from temperatures in the *c.* 80–100°C range (Rv max values of 0.68%).

Compelling evidence for mid-Cretaceous cooling and denudation comes from the thermochronological analyses of the Otway Ranges. Combining the vitrinite reflectance and apatite fission track analysis data, time- and depth-

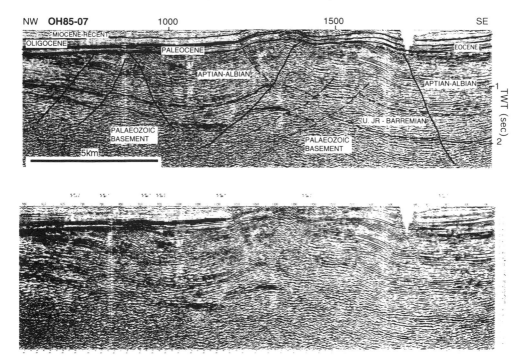

Fig. 7. Aptian–Albian (KL2) beds in the Colac Trough (Fig. 5) were deformed into an anticline then planed off leaving the syncline at SP 1000 unconformably below Palaeocene strata. The structure was re-inverted in the Pliocene as a pop-up structure. A small extensional fault appears to have been reactivated (c. SP 1300), but the consistent regional dip and elevation of basement on either side of the pop-up structure suggest minimal Cretaceous extensional faulting there. Thus the mid-Cretaceous deformation was accommodated by newly formed compressional/ transpressional faults as well as inversion of extensional faults.

Northern margin of the Bass failed rift

(i) Extension Within the Torquay Embayment and Colac Trough, regional seismic data clearly show extension across major NW to NNW-dipping, basin bounding faults (Figs 5, 4a/d [SP 1200] & 4b/e), giving rise to characteristic wedges of Late Jurassic–Barremian (Crayfish Group) sediment up to 4 km thick. The overlying Aptian–Albian megasequence (KL2, Eumeralla/Otway Group) is generally more parallel bedded and less faulted (Figs 4a/d & 4b/e). However, in the Torquay Embayment and probably in the Otway Ranges, extensional faulting persisted into the Aptian–Albian as indicated by growth in that sequence (Fig. 4a/d).

(ii) Mid-Cretaceous inversion Palaeozoic basement along the northern margin of the Bass failed rift was regionally uplifted and denuded in the mid-Cretaceous (e.g. Foster & Gleadow 1992; Dumitru *et al.* 1991; Duddy & Green 1992). Thermochronological analyses of the Otway Ranges also show significant uplift and denudation from 95–90 Ma (Fig. 6 & Tables 2–3), but the homogeneity of the Eumeralla Formation exposed throughout the Ranges makes it difficult to determine the structural style from field mapping. However, the deformation is clearly apparent from seismic reflection data from the adjacent Colac Trough and Torquay Embayment (Figs 2 & 4).

In the Colac Trough, the Aptian–Albian sequence in the tops of tilted fault blocks is clearly planed off and overlain by Lower Paleocene beds (Fig. 4b/e). The peneplanation is interpreted to be synchronous with the mid-Cretaceous denudation of the Otway Ranges and of Palaeozoic basement to the north. Importantly, adjacent to the Otway Ranges, around SP 500 on Fig. 4b/e and SE of SP 1000 on Fig. 7, the Aptian–Albian sequence is folded and then truncated at the unconformity, indicating significant mid-Cretaceous structural deformation, as well as regional uplift and denudation. The folding coincides with the thickest Early Cretaceous section on both lines suggesting inversion of a half-graben structure.

Table 3. *A summary of the seismic reflection and thermochronological observations, and interpretation from this study for the northern and southern margin of the Bass failed rift. See Fig. 2 for locations.*

Interval	Northern margin of Bass Failed Rift (Torquay Embayment, Colac Trough, Otway Ranges/ E. Otway Basin)		Southern margin of Bass Failed Rift (Durroon Basin, Tasmania, Bass Basin)		Inferred cause
	Thermochronology	Seismic	Thermochronology	Seismic	
Miocene-Recent	Cooling in Otway Ranges & Torquay wells	Inversion of extensional faults ←→ Regional subsidence — Torquay — Colac		←→ Regional subsidence ←→	Collision of Australia's northern margin with island arc in New Guinea
Oligocene					Commence Southern Ocean rapid spreading
Eocene	Significant cooling in the Lorne area	Erosion/Torquay — Minor basement extensional faulting in Torquay Embayment		Erosion	Pacific Plate moves west? — End of Tasman Drift
Paleocene		Erosion in Torquay Embayment — Extensional faulting Torquay Embayment	Significant cooling of local 'blocks' in NE Tasmania - faulting?	Erosion in Durroon Basin	Thermal subsidence Onset of Tasman Drift
Campanian - Maastrichtian		Extensional faulting Otway Basin		Erosion in Durroon Basin	Tasman Sea extension
Cenomanian-Santonian		Inverted extensional faults & thrusts in Otway Ranges then non-deposition/erosion in Ranges & Colac Trough ←→		Extensional faulting & half graben fill in Durroon & Gippsland basins	
"Mid" Cretaceous (ca 95 Ma)	Rapid cooling, regional denudation on and north of Otway Ranges, including basement in basin margin		Rapid regional cooling of basement in the basin margin (NE Tasmania), indicating significant denudation	←→ Non deposition/erosion ←→	Thermal uplift associated with Breakup; Local compressional inversion of faults (Otway Ranges)
Early Cretaceous (97.5 Ma-125Ma)		Late stage extensional faulting; Widespread deposition of Eumeralla		Widespread deposition of Eumeralla	Slow late-stage extension and major sediment loading
Late Jurassic - Early Cretaceous (125 Ma-150Ma)		Extensional faulting & half graben fill		Extensional faulting & half graben fill in Bass Basin & non deposition -erosion in Durroon Basin	Extension of Australia's Southern Margin

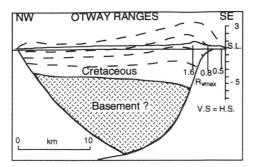

Fig. 8. Sketch section of the structure of the Otway Ranges. The amount denuded is indicated by the vitrinite reflectance values (the upper dotted line being Rv max of 0.5) and fission track data indicate mid Cretaceous denudation. The structural style is determined from field mapping and seismic data (Figs 4a/d, 4b/e & 9).

However, the basement and sedimentary section appear to have a consistent regional dip and elevation on either side of the anticline (Fig. 7) indicating that the mid-Cretaceous deformation also took place along newly formed compressional/transpressional faults as well as reactivation of existing extensional faults. The inversion of the Otway Ranges is shown schematically on Fig. 8, discussed below.

In the Torquay Embayment Neogene inversion is evident (Fig. 9), but there is less evidence for mid-Cretaceous inversion. A c. 110 to 80 Ma gap in stratigraphy is documented in the Nerita-1 well (Fig. 5; MacPhail 1989), but there is only minor angularity across the mid-Cretaceous unconformity (Fig. 4a/d) suggesting uplift over a large area. However, within the hangingwall of the basin-bounding extensional fault, the Aptian–Albian sequence is interpreted to be increasingly truncated to the SE, (Fig. 4a/d, SP 1100–1400). Assuming that the Aptian–Albian sequence was planed off to horizontal, and is now overlain by Campanian or older beds (Fig. 4a/d), then minor mid-Cretaceous inversion of the basin-bounding extensional fault is implied.

(iii) Late Cretaceous and Palaeogene The Late Cretaceous sequences are not preserved in the Colac Trough, but remnants in the Torquay Embayment show minor growth into rejuvenated extensional faults (Fig. 4a/d). Palaeocene to Recent sediments blanket the Bass Strait area, but Fig. 9 (SP 2100–2300) and Fig. 4a/d (SP 1100) show thickening of the Tertiary section in the Torquay sub-basin suggesting reactivated extensional faulting.

A Late Cretaceous event is indicated by 78 ± 8 Ma and 72 ± 4 Ma fission track cooling ages of samples 1/28 and L8–13 from near Lorne (Fig. 5 and Tables 1 & 2) perhaps associated with Late Cretaceous compression recorded in the Otway Basin to the west (Hall 1994). Alternatively, local intrusions are possible and this event needs further investigation. Offshore, in the Torquay Embayment, the Cretaceous and Palaeocene section is truncated in the highs of tilted fault blocks and is overlain by lowermost Eocene section (Fig. 4a/d, SP 1400–1500). However, on the sections available, this denudation event does not appear to be due to inversion.

(iv) Miocene to Recent The Otway Ranges form prominent hills up to 600 m high and are characterized by oversteepened creek profiles and mass movement suggesting recent uplift. The seismic data from both sides of the Ranges reveal the nature of the uplift. For instance minor Neogene inversion of Early Cretaceous extensional faults in the Colac Trough is clearly shown on Fig. 4b/e, SP 300. Also in the Colac Trough 5–10 km wide 'pop-up' anticlines deform the Oligocene and probably the Miocene section (Figs 4b/e & 7). These anticlines are superimposed on the anticlines formed there in the mid-Cretaceous and probably reactivate the same compressional/transpressional faults.

In the Torquay Embayment three SE-verging monoclinal structures, each 2 km wide with c. 200 m of relief at the top Oligocene level are underlain by steeply NW-dipping reverse faults (Fig. 9, SP 2040–2280). The monoclines deform the Miocene–Recent section but, internally, show growth of the Oligocene–Eocene and Paleocene section into the faults, suggesting inversion of faults that were extensional in the Paleogene. By analogy with the half-graben to the SE (e.g. Fig. 4a/d, SP 1600) the faults beneath the monoclines are interpreted to have also been extensional during the Cretaceous.

Although smaller, the monoclines on Fig. 9 are analogous to the overall structure of the Otway Ranges, which comprises a large SE-verging monocline, up to 600 m high (Figs 5 & 8). Thus the present topographical expression of the Ranges is considered to be due to Miocene–Pliocene inversion of Cretaceous and ?Palaeogene extensional faults, illustrated schematically on Fig. 8. However, the Miocene–Pliocene event was not recorded by the thermochronological data (although modelled by Cooper *et al.* 1993) indicating that the cooling was insignificant compared to the mid-Cretaceous event. The latter is shown on Fig. 8 by the dashed lines illustrating the amount of section eroded in the

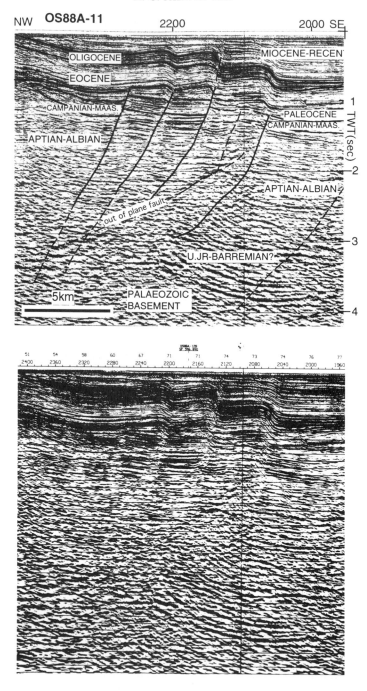

Fig. 9. Three SE-verging monoclinal structures in the Torquay Embayment, immediately offshore from the Otway Ranges and similar in structural style to that of the SE Otway Ranges (Figs 5 & 8). Each monocline is 2 km wide with *c.* 200 m of relief at the top Oligocene level and underlain by steeply NW-dipping reverse faults. The monoclines deform the Miocene Peubla Formation and show hangingwall growth of the Oligocene–Eocene section and wedging of the underlying Palaeocene section, demonstrating Miocene–Pliocene inversion of faults that were extensional in the Palaeogene. By analogy with the half-graben to the SE (e.g. Fig. 4a/d, SP 1600) the faults beneath the monoclines are interpreted also to have been extensional during the Cretaceous.

mid-Cretaceous, constrained by vitrinite reflectance data. Therefore the Otway Ranges are inferred to have undergone major inversion, denudation and cooling in the mid-Cretaceous and minor inversion in the Miocene–Pliocene with as yet little denudation and cooling. Both events involved compressional/transpressional faults and reactivation of extensional faults.

Southern margin of the Bass failed rift

A composite grid of 2000 km of 1982–1991 fair to very good quality seismic reflection data from the Durroon Basin (Figs 2 & 10) has been interpreted and the periods of uplift and sedimentation inferred from the data have been related to apatite fission track analyses from NE Tasmania (O'Sullivan 1992). The data and analyses from the Tasmanian samples (O'Sullivan 1992) are to be presented in detail elsewhere, but the ages are summarized on Fig. 10. The samples are from Devonian granites, Permo–Triassic sediments and Jurassic dolerites such that ages younger than 160 Ma record one or more post crystallisation cooling event(s). The 183–155 Ma ages from the NE coast are provenance ages of Cretaceous sediments within the onshore extension of the Durroon Basin.

(i) Intracontinental extension In contrast with the northern margin of Bass Strait, the Late Jurassic to Barremian sequence is thin to absent in the Durroon Basin and there is no evidence for Late Jurassic to Barremian extensional faults (Fig. 3 & Table 3). This indicates that intracontinental rifting was concentrated along the northern margin, the site of the mid-Cretaceous and Miocene–Pliocene inversions outlined above. The Durroon Basin is floored by largely parallel bedded Aptian–Albian sequence (Fig. 4c/f), demonstrating that it was subject to regional subsidence probably following the main Late Jurassic–Barremian rift phase.

(ii) Mid-Cretaceous inversion There are no Early Cretaceous extensional faults in the Durroon Basin and there is very little mid-Cretaceous structural inversion. On the contrary, NW–SE trending extensional faults developed in the Cenomanian (Figs 10 & 4c/f), orthogonal to the Early Cretaceous Torquay Embayment faults and consistent with Late Cretaceous extension in the Gippsland Basin to the east and the Otway Basin to the west. However, regional uplift in the early Late Cretaceous is suggested by the erosional truncation of the Aptian–Albian sequence on footwall highs (Fig. 4c/f, SP 750 and SP 1050), during

deposition of the basal Cenomanian–Santonian sequence in the adjacent troughs. At the Durroon-1 well location (Fig. 4c/f) the Aptian–Albian sequence is unconformably overlain by Turonian sediments (c. 90 Ma), restricting the uplift with erosion to the Cenomanian. Such regional uplift is consistent with the gap in time represented by the unconformity in Nerita-1 in the Torquay Embayment.

The main evidence for mid-Cretaceous uplift and denudation comes from the basement rocks of the basin margin in NE Tasmania. Surface samples from the elevated areas (c. 1000 m) around Poatina and Ben Nevis yield apatite fission track ages of 90–100 Ma consistent with >60°C of mid-Cretaceous cooling (Fig. 10). The Poatina samples were clearly uplifted by faulting as they are juxtaposed against similar rocks yielding much older ages. Similarly, the 89 Ma age on the NE coast near a 135 Ma age suggests fault controlled mid-Cretaceous uplift and erosion (1–2 km) at the southern margin of the Durroon Basin. However, the widespread granites of the Ben Nevis area appear to have undergone regional uplift and denudation in the mid-Cretaceous. The regional extent of the 90–100 Ma ages indicates mid-Cretaceous uplift and erosion of basement along the southern margin of the failed rift as documented in basement along the northern margin (Table 3; Foster & Gleadow 1992; Dumitru *et al.* 1991; Duddy & Green 1992).

(iii) Late Cretaceous and Paleogene Significant growth on NE-dipping extensional faults occurred in the Durroon Basin in the early Late Cretaceous giving rise to wedges of Cenomanian–Santonian sediments up to 2.5 km thick unconformably overlain by a Campanian–Maastrichtian 'sag' sequence. However, inversion of these sequences is minimal and restricted to the few examples seen along the Tasmanian margin where Late Cretaceous and Palaeocene beds have been uplifted, planed off and unconformably overlain by Eocene sediments. There, inversion occurred episodically through the Late Cretaceous–Palaeocene but the main inversion was probably in the Late Palaeocene.

Similar age denudation is recorded in NE Tasmania, in that apatite fission track analysis of a granite sample from the east coast yields a cooling age of 56 Ma (Fig. 10), indicating >50°C of cooling at the end of the Palaeocene. This is consistent with analysis of the lowland Jurassic dolerites near Poatina and Launceston which yield partially reset ages of 64 and 68 Ma, indicating Tertiary cooling. The 56 Ma age is

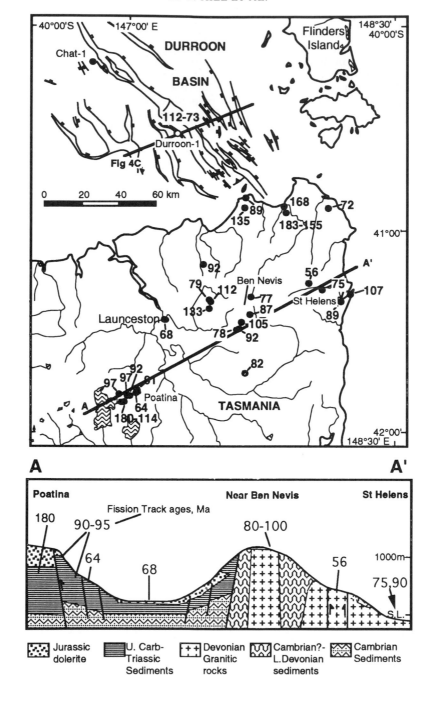

Fig. 10. Apatite fission track ages on Devonian granites and Jurassic dolerites of NE Tasmania (after O'Sullivan 1992) and trends of the Late Cretaceous extensional faults within the Durroon Basin, illustrated on Fig. 4c/f. The higher peaks of NE Tasmania typically record rapid-cooling ages of 90–100 Ma, whilst samples from the valleys record younger partially reset ages. The 56 Ma age near St Helens is significant as it has been reset indicating rapid cooling at *c.* 60–50 Ma. The older ages juxtaposed against the 97 Ma ages at Poatina indicate ?local fault offset.

probably related to fault uplift as it is adjacent to older and therefore less cooled samples.

(iv) Miocene to Recent In contrast to the northern margin of the Bass failed rift, there is no evidence of Miocene–Pliocene inversion along the southern margin (Table 3).

Summary of the inversion events of the Bass Strait region

The main events are shown on Table 3 and are summarized here. Within Bass Strait, intracontinental, Late Jurassic–Barremian rifting was along the northern margin only, with faults down to the NNW. However, Aptian–Albian regional subsidence and deposition of 1–5 km of volcanogenic sediment was over a much wider area, including the southern margin. Mid-Cretaceous structural inversion (due to compression) was largely confined to the Otway Ranges, along the northern margin of the newly developed failed rift, resulting in up to 3 km of denudation. Otherwise there was minor regional uplift and denudation basinwide, but importantly, 1–2 km of mid-Cretaceous regional uplift and denudation of basement in both northern and southern margins of the failed rift.

Rare Palaeocene inversion in the southern Durroon Basin is observed and significant local Palaeocene cooling in NE Tasmania is recorded. Tertiary events have been recorded previously in the Otway Ranges along the northern margin, but probably the most significant event there is compressional fault reactivation in the Miocene–Pliocene to form the present day Ranges.

Discussion

A number of questions are raised from this study concerning the mechanisms governing inversion of sedimentary basins, two of which are addressed here.

1) What is the nature of the widespread, rapid Aptian–Albian subsidence and of the subsequent *c.* 95 Ma rapid uplift and denudation around Bass Strait with compression in the Otway Ranges?

In the Aptian–Albian, following Late Jurassic–Barremian intracontinental rifting, a 1–?5 km thick 'blanket' of dominantly terrigenous volcaniclastic sediment was deposited throughout the Bass Strait region, an area *c.* 1000 km long and >300 km wide (Fig. 11). Hill *et al.* (1994) point out that the volcaniclastic sediment is andesitic (Gleadow & Duddy 1981)

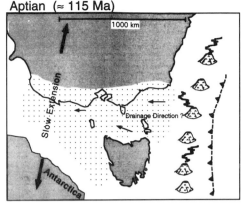

Aptian (≈ 115 Ma)

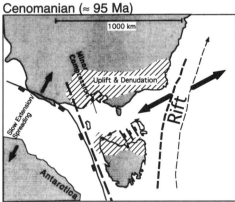

Cenomanian (≈ 95 Ma)

Fig. 11. Tectonic model for uplift, denudation and inversion of the Bass Strait area associated with 95 Ma breakup. Following Neocomian extension creating a rift graben, a volcanic arc developed along the eastern Australian margin in the Aptian–Albian, flooding the rift with volcanogenic sediments, causing regional subsidence. At *c.* 95 Ma renewed extension caused Tasman Sea rifting, ending the volcanogenic sediment supply. There was also renewed extension in the main part of the Otway Basin. We suggest that the lithosphere was broken at this time allowing rebound around the Bass failed rift causing uplift and denudation of the margins. In addition, there was *c.* 5% shortening concentrated on the previous depocentre, the Otway Ranges, causing inversion.

and probably arc-derived and that palaeocurrents indicate derivation from the east (Constantine 1992) where Veevers *et al.* (1982) infer an Aptian–Albian volcanic arc along the eastern margin of Australia (Fig. 11).

We suggest that at this stage the lithosphere had been extended to form an intracontinental rift, but that it had not been broken so that post-rift subsidence in the Aptian–Albian was regional in nature. The subsidence was partly thermally induced, but was largely driven by the influx of >500 000 km³ of fluvial sediment, mainly derived from a volcanic arc to the east. Subsidence and sedimentation may have occurred over a wider area than is now preserved, extending across basement that is now exposed in the margins of the present basins.

At *c.* 95 Ma, renewed extension caused the lithosphere to break-up (Veevers *et al.* 1991) but not along the lines of the former intracontinental rift. Instead the Tasman Fracture Zone, the Sorell Fault and its extension into an onshore lineament (Figs 1 & 2, Foster & Gleadow 1992) became a fundamental break, separating an area of renewed subsidence in the Otway Basin from the Bass failed rift. A similar break is inferred along the east coast of Tasmania, due to initial rifting prior to opening of the Tasman Sea. This terminated the supply of volcaniclastic sediment (Fig. 11; Gleadow & Duddy 1981).

We propose that the breaking of the lithosphere and the end of sedimentary input allowed uplift of the margins of the newly-formed Otway and Tasman rifts. Within the Bass failed rift, there was relatively minor uplift of the basin, but significant uplift of the margins, perhaps in part due to rebound following Aptian–Albian sediment loading. The new stress regime that initiated Late Cretaceous rifting allowed minor NW–SE compression. This caused inversion of the Otway Ranges extensional faults as they were oriented perpendicular to the compression. In addition, comparison of Torquay Embayment and Colac Trough data (Figs 4a/d & 4b/e) with the Australian Geological Survey Organisation deep seismic line across the Otway Ranges (AGSO 1992–01; Finlayson *et al.* 1994) indicates that the Otway Ranges was probably the most extended area in the Early Cretaceous. Thus it seems likely that the area most extended became the focus of subsequent inversion.

Clearly this rebound model is speculative at present. Further analysis of deep seismic reflection data, subsidence curves and lithospheric modelling is necessary to test these hypotheses.

2) Why did the Otway Ranges area invert in the Miocene–Pliocene when SE Australia was entirely bound by oceanic crust (Fig. 1)?

During the Late Miocene–Pliocene the Australian continent was subjected to compression between the colliding Melanesian and Timor arcs to the north and Southern Ocean ridge-push forces to the south (Denham & Windsor 1991; Cooper & Taylor 1987; Hill 1991; Etheridge *et al.* 1991, e.g. Bott 1991). The compression caused widespread deformation in New Guinea and minor inversion in Australia's interior (Etheridge *et al.* 1991) and propagated to the southern margin. Thus Miocene–Pliocene

structural inversion of the Otway Ranges is interpreted to result from compression due to arc collision 3500 km to the north. In the Otway Ranges area the maximum principal stress was approximately perpendicular to the Early Cretaceous extensional faults, as it is today (Denham & Windsor 1991), causing their reactivation to build the Ranges.

Along the southern margin significant inversion seems largely to be restricted to the northern margin of the Bass failed rift. This suggests that Tasmania acted as a buttress, bound by oceanic lithosphere to the south (Fig. 1) and/or that the northern margin is a fundamental zone of weakness probably having been the focus of maximum extension.

Basement fabric

The question remains, why was the Otway Ranges area a zone of weakness such that it was the focus of extension and then inversion? To the NE of the Ranges the Palaeozoic Heathcote Greenstone Belt trends N–S. It has previously been suggested that the Greenstone Belt continues to the southwest beneath the Ranges (Fig. 2; Harrington *et al.* 1973). This southwest fabric could account for the location of the rift through the Otway Ranges area and the NE–SW orientation of the rift faults, oblique to the general E–W trend. We are currently studying gravity and magnetic data in conjunction with the AGSO deep seismic line to investigate the possibility that the Greenstone Belt controlled both the original extension and subsequent inversion.

Conclusions

Following our seismic reflection and fission track thermochronology study of the Bass Basin area, several significant conclusions may be drawn concerning the types and controlling mechanisms of inversion.

Two types of inversion are recorded:

- repeated structural inversion involving 1–3 km of uplift by compressional reactivation of extensional faults and along new reverse faults.
- 1–2 km of regional uplift, denudation and cooling of basement over more than 200 km along both margins of a failed rift, with lesser erosion or non-deposition within the rift basin.

The structural inversions occurred along the site of a failed Mesozoic rift and were probably focussed on the site of maximum extension, interpreted as a long-lived zone of weakness. The location and orientation of the rift may have been controlled by an underlying Palaeozoic greenstone belt. The structural inversions are interpreted to have been controlled by three factors:

- the long-lived zone of weakness;
- the maximum principal stress being roughly perpendicular to faults within the zone during inversion; and
- failure of the rift, providing a buttress (Tasmania) to compression not present in the adjacent passive margins.

Significantly, the most recent inversion resulted from the Australian craton being placed into compression following arc collision 3500 km to the north of the area. This emphasizes the importance of the transmission of compressional stresses over long distances through the lithosphere.

The regional uplift of the margins of the failed rift occurred as it failed during renewed extension, c. 50 Ma after the start of the original rifting. We suggest that the lithosphere was broken during renewed extension allowing uplift of the margins of the failed rift. The uplift may have been due in part to rebound following several kilometres of rapid sediment loading in the 20 Ma prior to renewed rifting.

The Australian Geological Survey Organisation, BHP Petroleum, Bridge Oil, Esso, Gas & Fuel, Geotrack International, SAGASCO and the Shell Company of Australia all supplied seismic reflection data and/or funding for this study. This work is part of a regional Bass Strait study funded by an Australian Research Council grant with neutron irradiations supported by the Australian Nuclear Science and Technology Organisation. The study contributes towards the Otway National Geoscience Mapping Accord.

References

BODARD J. M., WALL V. J. & KANEN R. A. 1985. Lithostratigraphic and depositional architecture of the Latrobe Group, offshore Gippsland Basin. *In*: GLENIE, R. C. (ed) *Second South-Eastern Australia Oil Exploration Symposium*. Petroleum Exploration Society of Australia, 113–136.

BOTT, M. H. P. 1991. Ridge push and associated plate interior stress in normal and hot spot regions. *Tectonophysics*, **200**, 17–32.

BURNHAM, A. K. & SWEENEY, J. J. 1989. A chemical kinetic model of vitrinite maturation and reflectance. *Geochemica et Cosmochimica Acta*, **53**, 2649–2657.

CANDE, S. C. & MUTTER, J. C. 1982. A revised identification of the oldest sea-floor spreading anomalies between Australia and Antarctica. *Earth and Planetary Science Letters* **58**, 151–160.

CONSTANTINE A. E. 1991. Structural history of the Late Jurassic–Early Cretaceous Strzelecki Group, onshore Gippsland Basin. Abstract, *Fifth Victorian Universities Geology Conference*.

—— 1992. Fluvial architecture and sedimentology of the Late Jurassic–Early Cretaceous Strzelecki and Otway Groups. *Geological Society of Australia Abstracts* **32**, 145.

COOPER, G. T., HILL, K. C. & WLASENKO, M. 1993. Thermal modelling in the eastern Otway Basin. *The APEA Journal* (Australian Petroleum Exploration Association), **33**, 205–213.

COOPER, P. & TAYLOR, B. 1987. Seismotectonics of New Guinea: a model for arc reversal following arc-continent collision. *Tectonics*, **6**, 53–67.

DAVIDSON, J. K. & MORRISON, K. C. 1986. A comparison of Hydrocarbon Plays in the Bass, Gippsland, Otway and Taranaki Basins. *In*: GLENIE, R. C. (ed.) *Second South-Eastern Australia Oil Exploration Symposium*. Petroleum Exploration Society of Australia, 365–374.

DENHAM D. & WINDSOR C. R. 1991. The crustal stress pattern in the Australian continent. *Exploration Geophysics*, **22**, 101–106.

DUDDY, I. R. 1994. The Otway Basin: thermal, structural, tectonic and hydrocarbon generation histories. *In*: FINLAYSON, D. M. (compiler) 1994. *NGMA/PESA Otway Basin Symposium, Melbourne, 20 April 1994: Extended abstracts*. Australian Geological Survey Organisation, Record **1994/14**, 35–42.

—— & GREEN, P. F. 1992. Tectonic Development of the Gippsland Basin and Environs: Identification of Key Episodes Using Apatite Fission Track Analysis. *In*: BARTON, C. (ed.) *Energy, Economics and Environment, Gippsland Basin Symposium, Melbourne June 1992*. Australian Institute of Mining and Metallurgy, 111–120.

DUFF, B. A., GROLLMAN, N. G., MASON, D. J., QUESTIAUX, J. M., ORMEROD, D. S. & LAYS P. 1991. Tectonostratigraphic evolution of the south east Gippsland Basin. *The APEA Journal* (Australian Petroleum Exploration Association), **31**, 116–130.

DUMITRU, T. A., HILL, K. C., COYLE, D. A., DUDDY, I. R., FOSTER, D. A., GLEADOW, A. J. W., GREEN, P. F., LASLETT, G. M., KOHN, B. P. & SULLIVAN, A. B. 1991. Fission Track Thermochronology: Application to Continental Rifting of Southeastern Australia. *The APEA Journal* (Australian Petroleum Exploration Association) **31**, 131–142.

ETHERIDGE, M. A., BRANSON, J. C. & STUART-SMITH, P. G. 1985. Extensional Basin – Forming Structures in Bass Strait and their importance for hydrocarbon exploration, *The APEA Journal*, (Australian Petroleum Exploration Association), **25**, 344–61.

——, McQUEEN, H. & LAMBECK K. 1991. The role of intraplate stress in Tertiary (and Mesozoic) deformation of the Australian continent and its margins: a key factor in petroleum trap formation. *Exploration Geophysics*, **22**, 123–128.

FEATHERSTONE, P., AIGNER, T., BROWN, L., KING, M.

& LEU, W. 1991. Stratigraphic Modelling of the Gippsland Basin. *The APEA Journal*, (Australian Petroleum Exploration Association), **31**, 105–115.

FINLAYSON, D. M., JOHNSTONE, D.W., OWEN, A. J. & WAKE-DYSTER, K.D. 1994. Deep seismic profiling: basement controls on Otway Basin development, *In*: FINLAYSON, D. M. (compiler) 1994. *NGMA/PESA Otway Basin Symposium, Melbourne, 20 April 1994: extended abstracts.* Australian Geological Survey Organisation, Record **1994/14**, 13–18.

FOSTER, D. A. & GLEADOW, A. J. W. 1992. Reactivated tectonic boundaries and implications for the reconstruction of southeastern Australia and northern Victorialand, Antarctica. *Geology,* **20**, 267–270.

GALBRAITH, R. F. 1981. On statistical models for fission track counts. *Mathematical Geology,* **13**, 471–488.

GLEADOW A. J. W. & DUDDY, I. R. 1981. Early Cretaceous volcanism and the early breakup history of southeastern Australia: Evidence from fission track dating of volcaniclastic sediments. *In*: CRESSWELL M. M. & VELLA P. (eds) *Fifth International Gondwana Symposium, Wellington, New Zealand.* A. A. Balkema, Rotterdam, 295–300.

——, ——, GREEN, P. F. & HEGARTY, K. A. 1986. Fission track lengths in the apatite annealing zone and the interpretation of mixed ages. *Earth and Planetary Science Letters* **78**, 245–254.

GREEN, P. F. 1989. Thermal and tectonic history of the East Midlands shelf (onshore UK), and surrounding regions assessed by apatite fission track analysis. *Journal of the Geological Society*, London **146**, 755–773.

——, DUDDY, I. R., GLEADOW, A. J. W., TINGATE, P. R. & LASLETT, G. M. 1986. Thermal annealing of fission tracks in apatite, 1. A qualitative Description. *Chemical Geology* **59**, 237–253.

——, ——, LASLETT, G. M., HEGARTY, K. A., GLEADOW, A. J. W. & LOVERING, J. F. 1989. Thermal annealing of fission tracks in apatite: 4 – Quantitative modelling techniques and extension to geological time scales. *Isotope Geosci.* **79**, 155–182.

HALL, M. 1994. Towards a structural history for the Otway Basin; the view from the southeast. *In*: FINLAYSON, D. M. (compiler) 1994.*NGMA/PESA Otway Basin Symposium, Melbourne, 20 April 1994: extended abstracts.* Australian Geological Survey Organisation, Record **1994/14**, 55–58.

HARRINGTON, H. J., BURNS, K. L. & THOMPSON B. R. 1973. Gambier–Beaconsfield and Gambier–Sorell Fracture Zones and the Movement of Plates in the Australia–Antarctica–New Zealand Region. *Nature Physical Science,* **245**, 109–112.

HEGARTY, K. A., WEISSEL, J. K. & MUTTER, J. C. 1988. Subsidence history of Australia's southern margin: constraints on basin models. *American Association of Petroleum Geologists Bulletin* **72**, 615–633.

HILL, K. A. & DURRAND, C. 1993. The western Otway Basin – an overview of the rift and drift history using serial composite seismic profiles. *PESA Journal* (Petroleum Exploration Society of Australia) **21**, 64–78.

——, FINLAYSON D. M., HILL K. C., PERINCEK D. & FINLAYSON B. 1994. The Otway Basin: pre-drift tectonics, *In*: FINLAYSON, D. M. (compiler) *NGMA/PESA Otway Basin Symposium, Melbourne, 20 April 1994: extended abstracts.* Australian Geological Survey Organisation, Record **1994/14**, 43–48.

HILL, K. C. 1991. Structure of the Papuan Fold Belt, Papua New Guinea. *American Association of Petroleum Geologists Bulletin* **75**, 857–872.

LASLETT, G. M., GREEN, P. F., DUDDY, I. R. & GLEADOW, A. J. W. 1987. Thermal annealing of fission tracks in apatite, 2. A Quantitative Analysis. *Chemical Geology* **59**, 237–253.

MACPHAIL, M. K. 1989. Palynological analysis of samples from Nerita-1A, Torquay Sub-Basin. Unpublished report for Shell Company of Australia.

MOORE, A. M. G., WILLCOX, J. B., EXON, N. F. & O'BRIEN, G. W. 1992. Continental shelf basins on the west Tasmania margin. *The APEA Journal* (Australian Petroleum Exploration Association), **32**, 231–250.

MOORE, E. M., GLEADOW, A. J. W. & LOVERING, J. F. 1986. Thermal evolution of rifted continental margins: new evidence from fission tracks in basement apatites from southeastern Australia. *Earth and Planetary Science Letters,* **78**, 255–270.

NAESER, C. W. 1979. Fission track dating and geologic annealing of fission tracks. *In*: JAGER, E. & HUNZIKER, J. C. (eds) *Lectures in Isotope Geology.* Springer Verlag, New York, 154–169.

O'SULLIVAN, A. J. 1992. Thermochronology of Tasmania and the southern Bass Basin, Australia. Abstract, *Nuclear Tracks and Radiation Measurements,* **21**, 584 (Special issue). 7th International Workshop on Fission Track Thermochronology, July 13–17, 1992. Philadelphia, USA.

PERINCEK, D., COCKSHELL, C. D., FINLAYSON, D. M. & HILL K. A. 1994. The Otway Basin: Early Cretaceous rifting to Miocene strike-slip, *In*: FINLAYSON, D. M. (compiler) *NGMA/PESA Otway Basin Symposium, Melbourne, 20 April 1994: extended abstracts.* Australian Geological Survey Organisation, Record **1994/14**, 27–33.

REECKMANN, S. A. 1994. Geology of the onshore Torquay Sub-Basin: a sequence stratigraphic approach. *In*: FINLAYSON, D. M. (compiler) 1994. *NGMA/PESA Otway Basin Symposium, Melbourne, 20 April 1994: extended abstracts.* Australian Geological Survey Organisation, Record **1994/14**, 3–6.

STRUCKMEYER, H. I. M. 1988. Source rock and maturation characteristics of the sedimentary sequences of the Otway Basin. PhD thesis, University of Wollongong, NSW.

TICKELL, S. J., EDWARDS J. & ABELE, C. 1992. Port Campbell Embayment 1 : 100,000 map geological report. *Geological Survey of Victoria Report* **95**, 97p.

VEEVERS, J. J. 1986. Breakup of Australia and Antarctica estimated as mid-Cretaceous (95+/−5 Ma) from magnetic and seismic data at the continental margin. *Earth and Planetary Science Letters* **77**, 91–9.

——, JONES, J. G. & POWELL, C. M. 1982. Tectonic framework of Australia's sedimentary basins. *The APEA Journal* (Australian Petroleum Exploration Association), **22**, 282–300.

——, POWELL, C. M. & ROOTS, S. R. 1991. Review of Sea floor spreading around Australia I. Synthesis of the patterns of spreading. *Australian Journal of Earth Sciences* **38**, 373–98.

WILLCOX, J. B., COLWELL, J. B. & CONSTANTINE, A. E. 1992. New ideas on Gippsland Basin Regional Tectonics. *In*: BARTON, C. (ed.) *Energy, Economics and Environment, Gippsland Basin Symposium, Melbourne June 1992.* Australian Institute of Mining and Metallurgy, 93–110.

—— & STAGG, H. M. J. 1990. Australia's southern margin: a product of oblique extension. *Tectonophysics* **173**, 269–281.

YOUNG, I. M., TRUPP, M. A. & GIDDING M. J. 1991. Tectonic Evolution of Bass Strait-Origins of Tertiary Inversion. *Exploration Geophysics,* **22**, 465–468.

Development of structurally inverted basins: a case study from the West Coast, South Island, New Zealand

DANIEL J. BISHOP,[1] & PETER G. BUCHANAN[2,3]

[1]*Department of Geology, Victoria University of Wellington, PO Box 600, Wellington, New Zealand*

[2]*CogniSeis Development Inc., Stanley House, Kelvin Way, Crawley, Sussex RH10 2SX, UK*

[3]*Present address: Oil Search Ltd., NIC Haus, PO Box 1031, Champion Parade, Port Moresby, Papua New Guinea*

Abstract: Cretaceous–Palaeogene basins in the West Coast of South Island, New Zealand, underwent inversion during the Neogene. Initial extension was oriented WNW–ESE and had a typical magnitude of *c.* 5%, whilst subsequent shortening in the same direction was typically *c.* 3–4%. This deformation produced a NNE–SSW trending basin and range province. Shortening was commonly partitioned into narrow zones along basin and range margins, leaving unreactivated extensional structures inbetween. Such margins are characterized by subparallel and en echelon Cretaceous – Palaeogene faults, which have been reactivated as predominantly reverse faults during the Neogene. Reverse faults propagated into the syn-inversion sedimentary sequence with increasingly shallow dip. Inversion caused local erosion of the original rift deposits. New back-thrusts, low-angle thrusts and folds developed during the late Neogene. Locally, incompetent Eocene mudstones deformed in a quasi-plastic manner, so that reverse faults at deeper structural levels became decoupled from thrust faults at shallower levels. Computer-aided section restoration has allowed validation of the interpretations, provided estimates of the timing and magnitude of deformation, and given a better understanding of the sequential development of inversion structures. Natural inversion structures show both contrasts and similarities to those produced in sand-box experiments.

Regional setting

The New Zealand region presently encompasses part of the convergent plate boundary at the junction of the Australian and Pacific plates (Fig. 1). Offshore eastern North Island, a major subduction accretion complex is developed, with associated magmatic arc and back-arc rifting onshore. The character of the plate boundary changes abruptly southwards, in the region of the Cook Strait, which separates North Island from South Island (Fig. 1). Whilst to the north of the Cook Strait mainly oceanic crust of the Pacific Plate subducts beneath the mainly continental crust of North Island, to the south of the Cook Strait continental crust of the Pacific Plate is in collision with continental crust of the Australian Plate, which is causing the present orogeny in South Island, known as the Kaikoura Orogeny (Nathan *et al.* 1986). By a mechanism not yet fully understood, the subduction zone in the north is connected to a series of major right-lateral transpressional faults, which in turn are replaced southwards by the transpressional Alpine Fault (Fig. 1). The Alpine Fault effectively divides South Island into two major tectonostratigraphic domains. Basin inversion that has occurred in parts of the West Coast region forms the subject of this paper.

It is now widely accepted that there has been major right-lateral strike-slip movement across the Alpine Fault, as evidenced by the offset of distinctive Permian lithological units by *c.* 480 km. However, when abouts strike-slip movement began has remained disputed, since there are no recognized reference features younger than the Permian on opposite sides of the Alpine fault, apart from those of late Quaternary age. Whilst some authors have preferred a Late Mesozoic age (e.g. Suggate 1963, 1979; Grindley & Oliver 1979), the consensus of opinion is now that most of the strike-slip movement was Neogene, based on evidence from plate tectonic reconstructions (e.g. Walcott 1978, 1987; Stock & Molnar 1987), and the geology of Cenozoic basins in South Island (Carter & Norris 1978; Kamp 1986*a,b*; Nathan *et al.* 1986), which suggests that inception of right-lateral strike-slip movement began in the Miocene. This concurs with independent evidence from radiometric dating of lamprophyre dykes close to the Alpine Fault that have

From BUCHANAN, J. G. & BUCHANAN, P. G. (eds), 1995, *Basin Inversion*, Geological Society Special Publication No. 88, 549–585.

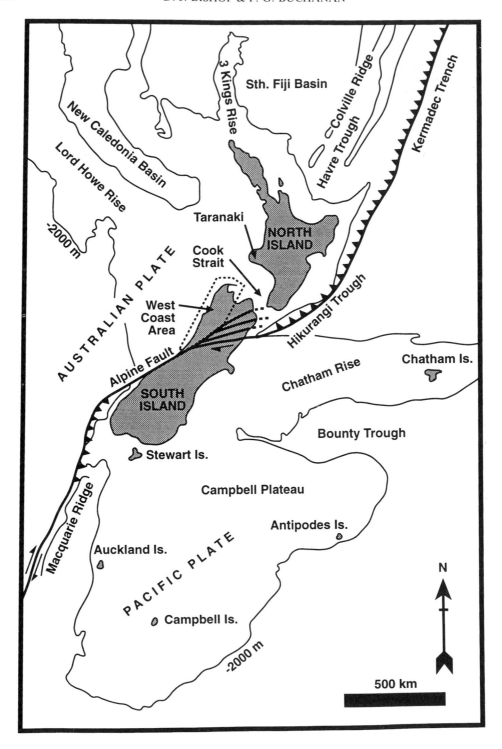

Fig. 1. Location of the West Coast area of South Island with respect to the present plate tectonic situation of New Zealand.

been dated at c.23 Ma, the timing of which is thought to correspond to initiation of the Alpine Fault (Cooper et al. 1987).

However, it seems likely that the Alpine Fault was a major structural feature prior to the Neogene, as shown by plate tectonic reconstructions (Grindley & Davey 1982; Bishop 1992b). To understand Cretaceous–Cenozoic deformation in the study area, it is important to realize the existence of earlier geological structures, which seem to have played an important part in controlling subsequent deformation. The basement rocks of the South Island can be subdivided into at least nine tectono-stratigraphic terranes (Bishop et al. 1985) comprising incomplete remnants of magmatic arcs, forearc basins, trench slope basins and accretionary complexes, which accreted to the Gondwana margin during the Palaeozoic and Mesozoic and produced a strong NNE–SSW structural grain in the West Coast area (Bishop et al. 1985; Cooper 1986). This collisional event is known as the Rangitata Orogeny (Nathan et al. 1986). In the West Coast, the two major terranes have been described by Cooper (1986) as the western 'Buller Terrane', and the eastern 'Takaka Terrane' (Fig. 3), and described by Bishop et al. (1985) as the 'Karamea Terrane' and the 'Golden Bay Terrane' respectively. The two terranes are separated by the Anatoki Thrust fault (Fig. 3). NNE–SSW trending structures have been reactivated at different times during the middle Cretaceous–Cenozoic, as described by Nathan et al. (1986) and Bishop (1992a,d).

No Jurassic or Early Cretaceous strata older than late Albian have been identified in the study area, probably owing to erosion during the closing phases of the Rangitata Orogeny (Nathan et al. 1986). Pororari Group red-beds, coarse alluvial sediments and lacustrine sediments were deposited during the middle–Late Cretaceous, outcrop in the onshore region and are believed to have been deposited in WNW–ESE trending rift basins (Gage 1952; Newman 1985; Bishop 1992b). This WNW–ESE rift orientation can be recognized throughout the New Zealand sector of Gondwana (Bishop 1992b).

A regional unconformity of late Campanian age in the New Zealand region (Nathan et al. 1986) is penecontemporaneous with the onset of seafloor spreading in the Tasman Sea (Weissel & Hayes 1977) and is thought therefore to represent a break-up unconformity (Cook & Beggs 1990; Bishop 1992b) of the type originally described by Falvey (1974). This regional unconformity separates all Pororari Group strata with their WNW–ESE depositional trend, from younger Paparoa Coal Measures and Pakawau Group strata with their mainly NNE–SSW depositional trend (Nathan et al. 1986). The Paparoa Coal Measures and Pakawau Group sediments are mainly terrestrial in origin, comprising coal beds, alluvial and lacustrine deposits, and were deposited during the Late Cretaceous and Palaeocene (Nathan et al. 1986). These younger rift sediments were deposited in a major rift system which connected the eastern end of the New Caledonia Basin with the Tasman Sea, via Taranaki (Fig. 1) and the West Coast (Kamp 1986a; King 1990; Bishop 1992a).

Seafloor spreading in the Tasman Sea ceased at about 60 Ma (Weissel & Hayes 1977). The resultant tectonic quiescence allowed erosion to reduce the West Coast region to a virtual peneplain in the earliest Eocene, as described by Cotton (1916) and Nathan et al. (1986).

Renewed extension occurred during the Eocene and Oligocene in the West Coast region, probably as a result of the propagation of a rift zone northwards from the Southeast Indian Ridge into the South Island, as proposed by Kamp (1986a) on the basis of plate tectonic reconstructions. This period saw a gradual transgression until marine carbonate deposition was taking place over most of the New Zealand region by the end of the Oligocene (Carter & Norris 1976; Nathan et al. 1986).

Deposition of limestone rocks in the West Coast was abruptly halted at the beginning of the Miocene as a result of the major influx of terrigenous clastic sediments into basinal areas, which occurred at that time. This was due to uplift and erosion of the Southern Alps (Fig. 2), as the transpressional Alpine Fault formed and continental collision began along the Australia–Pacific plate boundary (Fig. 1). Since then some of the major pre-existent extensional faults in the West Coast have undergone reactivation as reverse and strike-slip faults. Indeed, many structures are still actively developing in the West Coast region, as indicated by folding and faulting of Quaternary river terraces (Yeats 1985; Suggate 1987; Roder & Suggate 1990), as well as by seismicity (Fyfe 1929; Lensen & Otway 1971; Holt 1991). The axis of maximum shortening is orientated approximately WNW–ESE and has been so since the late Neogene (Bishop 1992c; Holt 1991). West of the Alpine Fault the reverse component of Neogene fault movements is nearly always much greater than the strike-slip component. Whereas reverse faults have throws of up to c. 3–4km (Bishop 1992a), strike-slip motion has not been demonstrated on any fault to be more than c. 300 m (Barry & MacFarlan 1988).

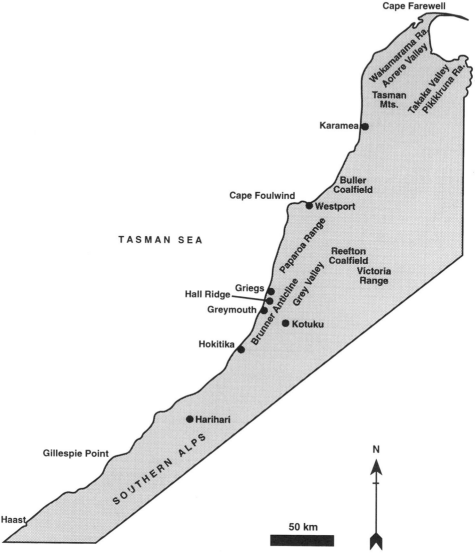

Fig. 2. Map of the West Coast region of South Island, showing the main towns and other localities mentioned in the text.

Major Neogene reverse faults bound uplifted and partly eroded Cretaceous–Palaeogene basins, and separate these structural blocks from adjacent contractional Neogene basins (Fig. 4). The uplift and structural inversion of Cretaceous–Palaeogene basins during the Neogene was first documented by Wellman (1948), and in a series of manuscripts by Wellman in 1950, which are now lodged at the New Zealand Institute of Geological and Nuclear Sciences. Although they were never published, these manuscripts remain of very great historical importance, since they provided the essential preliminary groundwork for much of the mapping since carried out in the West Coast by government geoscientists (e.g. Nathan *et al.* 1986). Wellman (1948) coined the term 'eversion' to describe the observed uplift and erosion ('turning inside out') of a basin, such that the deepest parts become the crest of a mountain range, whilst the margins of the basin become the margins of the mountain range. This term is still occasionally used (e.g. Sissons & Walsh 1986). The term 'inversion' was applied to

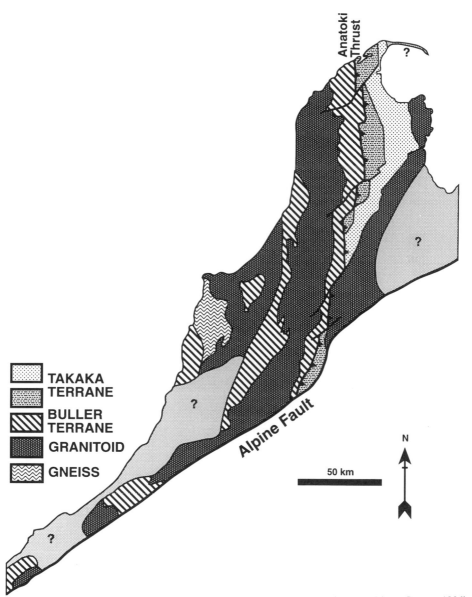

Fig. 3. Map showing basement terranes west of the Alpine Fault, South Island (modified from Cooper 1986).

similar structures in New Zealand by Cotton (1958). Since then this has become the conventional term to describe a switch in tectonic mode from extension to compression, such that extensional basins are contracted and become regions of positive relief, and pre-existing extensional faults are reactivated such that they undergo reverse slip and may eventually become thrust faults (Cooper & Williams 1989; Williams *et al*. 1989).

The major inversion structure known as the Brunner Anticline (Fig. 2) was mapped in detail by Gage (1952) and Nathan (1978); inversion there and to the north was further described by Laird (1968) and Laird & Hope (1968). Together with Wellman's pioneering work, these papers were seminal to the authors' understanding of basin inversion in the West Coast. However, early workers were hampered by the dense vegetative cover in many parts of the

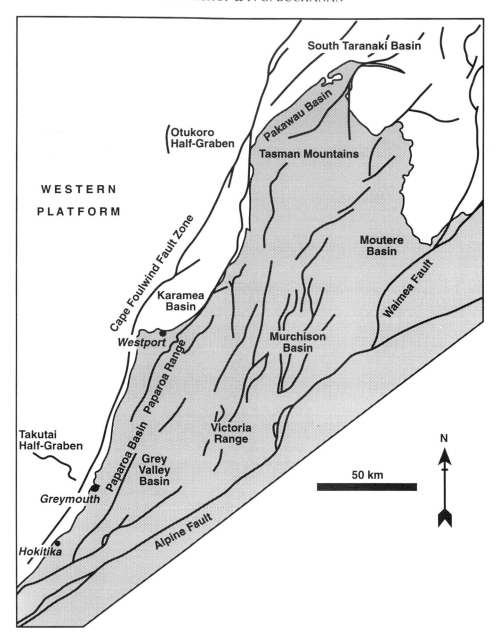

Fig. 4. Map showing the main faults that have been active intermittently during the Cretaceous and Cenozoic, and the location of major sedimentary basins and mountain ranges (modified from Nathan *et al*. 1986).

region, and consequent lack of exposure. In some cases this has led to an incomplete understanding of the structural complexities, particularly the geometry of faults in the region. Many published cross-sections show most faults as being vertical (e.g. Suggate 1984; Nathan *et al*. 1986), when in fact all available field evidence shows that most faults dip steeply at 60–85° and are often not single discrete fault planes, but rather complex zones of faulting and folding (Gage 1952; Ravens 1990; Bishop 1992*a*). Only with the release of commercial seismic reflection data into the public domain has it been possible to delineate subsurface structures with more confidence, as will be described in this paper.

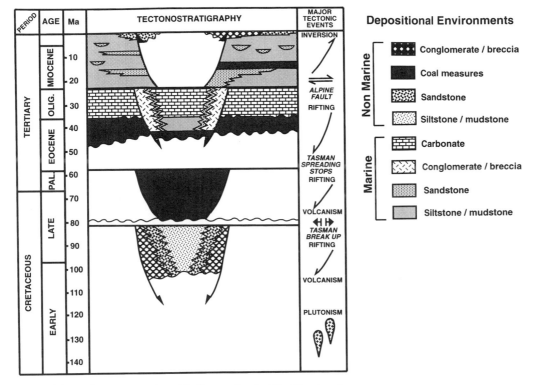

Fig. 5. Schematic tectonostratigraphic diagram for the West Coast region.

Stratigraphy

The stratigraphic nomenclature in the West Coast is complex, but an excellent review has been provided by Nathan *et al.* (1986). A very simplified tectonostratigraphic scheme for the West Coast is presented in this paper (Fig. 5) which represents a synthesis of a very large amount of published and unpublished literature (see Nathan *et al.* 1986 and references therein; Bishop 1992 *a,b,d*). A regional unconformity of approximately middle Albian age is developed across the West Coast region, truncating igneous and metamorphic rocks of Precambrian–Triassic age (Adams 1975; Raine 1980) that are generally referred to as 'basement' by New Zealand geologists. In contrast to the overlying strata (or 'cover'), the basement is either crystalline or very well lithified.

The 'cover' comprises strata of middle Cretaceous–Recent age (Fig. 5). Middle–Late Cretaceous rocks belonging to the Pororari Group consist of red-beds, alluvial and lacustrine deposits, together with some tuffs (Gage 1952; Newman 1985; Nathan *et al.* 1986). Late Cretaceous–Palaeocene rocks of the Paparoa Coal Measures and Pakawau Group comprise a mixed volcano-sedimentary succession, consisting primarily of coal beds, alluvial and lacustrine deposits and basic volcanic rocks (Gage 1952; Newman 1985; Nathan *et al.* 1986). More recent work has shown also that there were marine incursions in the northern part of the West Coast, during the Cretaceous (Thrasher 1992*b*). Widespread erosion and peneplanation across the West Coast during the late Palaeocene and early Eocene was followed by deposition of further coal measures, which were in turn overlain in many places by sandstones and shales deposited in restricted anoxic marine basins (Nathan *et al.* 1986). Widespread deposition of marine limestone began at the start of the Oligocene, followed in the Miocene and Pliocene by marine shales and intercalated marine sandstones (Nathan *et al.* 1986). Alluvial deposition recommenced in many areas during the Pliocene and continues to the present (Nathan *et al.* 1986).

The Cretaceous–Cenozoic sedimentary cover is unmetamorphosed and many of the rocks are soft and friable, especially the marine mudstones. At certain stratigraphic levels there are strong contrasts in rheological strength between successive sedimentary units, e.g. there are

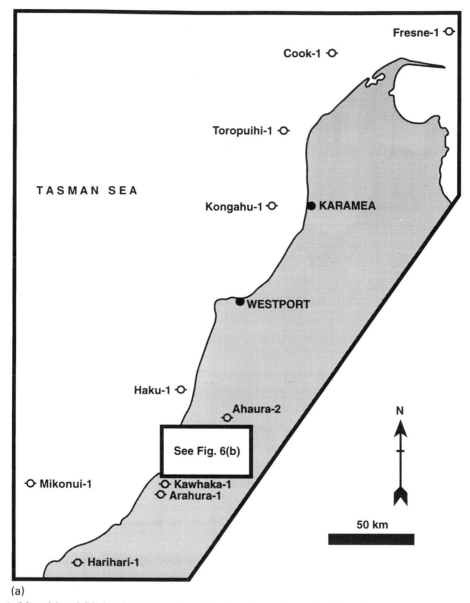

(a)

Fig. 6. Maps (a) and (b) showing the locations of exploration wells used in this study.

strong competence contrasts between the Oligo-cene limestone, underlying Eocene mudstone and overlying Miocene mudstone. These rheo-logical factors will be demonstrated in this paper to be of importance in controlling the style of basin inversion in the West Coast.

The stratigraphic column for the West Coast (Fig. 5) records an overall transgression during the Eocene, owing to repeated rifting and subsidence since the middle Cretaceous, and regression during the Pliocene, owing to acceler-ation of the Kaikoura Orogeny which had begun at the start of the Miocene. In this paper, the pre-rift sequence comprises the 'basement' rocks, the syn-rift sequence comprises middle Cretaceous–Oligocene rocks, and the syn-inversion sequence comprises Miocene–Recent rocks.

Since the beginning of this century, a con-siderable number of exploration wells have been

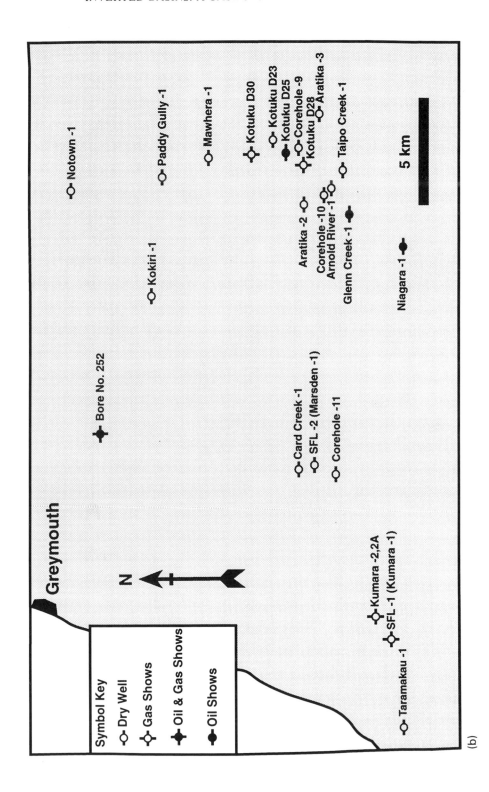

(b)

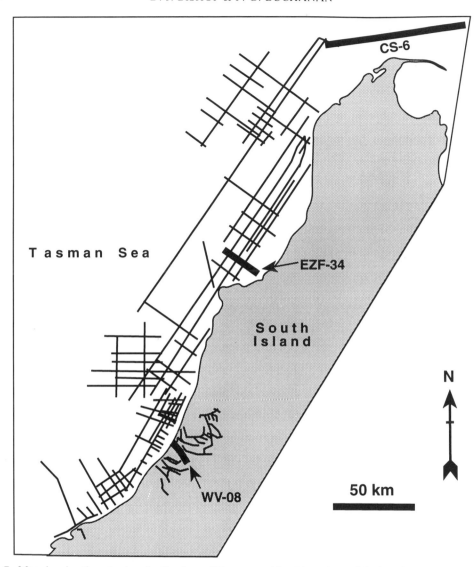

Fig. 7. Map showing the seismic reflection line grid interpreted for this study, and the location of lines WV-08, EZF-34 and CS-6, which are described and restored in this paper.

drilled in the West Coast, offshore and onshore (Ministry of Energy 1989). The presence of oil seeps, such as the one at Kotuku (Fig. 2) described by Morgan (1911), encouraged the belief that reservoirs of hydrocarbons exist in the West Coast region which might be economic to exploit. As yet this has not come to pass, although several wells have produced oil and gas shows, e.g. Niagara-1 (Matthews 1990). Well data for this paper was collated from publications by Wellman (1971), Nathan *et al.* (1986), the Ministry of Energy (1989) and unpublished well completion and petroleum reports lodged at

the New Zealand Institute of Geological and Nuclear Sciences. Altogether 38 wells (Fig. 6) have been tied to a seismic reflection line grid, which extends along the most of the offshore West Coast and onshore in the region of Greymouth (Fig. 7).

Seismic reflection data

Since the 1960's, a number of seismic reflection surveys have been conducted in the West Coast. Coverage by seismic reflection surveys has been limited to the onshore region, and offshore

Table 1 *Average interval velocities and lithologies used for depth conversion and decompaction*

Age of Unit	Average Velocity (m/s)	Average Lithology
Pliocene–Recent	2440	shale + sandstone
Miocene	2960	shale
[1]Miocene–Recent	2700	shale
Eocene–Oligocene	3310	shale
[2]Eocene–Oligocene	3900	sandstone
Cretaceous–Palaeocene	3900	sandstone
[1]Cretaceous–Oligocene	3900	sandstone

Notes: [1]Line CS-6 only. [2]Line EZF-34 only.

within 50 km of the coast. Survey data extends from offshore Haast (Fig. 2), northward along the coast and into Taranaki (Nathan *et al*. 1986; Thrasher & Cahill 1990). Nathan *et al*. (1986) also catalogued the onshore surveys that have been carried out in the Murchison and Moutere basins (Fig. 4), in the Grey Valley and between Greymouth and Gillespie Point (Fig. 2). In this paper, data used ranges in age from 1969 to 1985, and are all in the public domain. Seismic reflection data interpreted for this study cover the offshore area from Harihari to Cape Fare-well and the onshore area from Harihari to just northeast of Greymouth (Figs 2 & 7).

A total of 179 seismic reflection lines selected from 11 different surveys were interpreted amounting to over 2700 line-km in length. The data are from conventional 2-D surveys and range from 12 to 60 fold. Most lines used were stacked unmigrated lines. A few lines in complex structural areas were migrated. As can be seen from the following examples, data quality is very variable and generally not good, owing in part to the antiquity of the lines and in part to the physical problems encountered whilst shooting the onshore surveys. Despite these limitations, the large scale geometric features of inverted basins can be recognized on the seismic reflection lines.

Section restoration methodology

In this paper, Neogene inversion structures have been interpreted from the seismic reflection data and restored using specialized computer soft-ware. Recent advances in computer hardware and software technology now allow full resto-ration of geological and seismic sections using a computer, which offers significant advantages over traditional manual methods. Use of a computer increases speed and accuracy, and provides the ability to test and modify interpre-tations rapidly and interactively. In this paper, a commercial program Geosec was used (Kligfield *et al*. 1986; Geiser *et al*. 1988; Geiser *et al*. 1991), although similar results could have been ob-tained using other commercially available resto-ration packages such as Locace (Moretti & Larrére 1989) and Restore (Schultz-Ela 1991). Geosec was used to depth convert, retrodeform and decompact seismic sections, selected from the interpreted regional seismic reflection line and well database, described above.

The original time interpretations were digi-tized and then depth converted using the Geosec program. Meaningful restorations can not be done on sections for which the vertical axis is time. To depth convert the sections, average interval velocities for each stratigraphic unit (Table 1) were calculated from the well log data. These velocities were used to convert the time sections to the depth sections.

It should be stressed that these interval vel-ocities only crudely approximate the true sub-surface velocity structure. Since there has been widespread inversion, there cannot be any simple relationship between the present depth of a unit and its interval velocity, as there often is in sedimentary basins which have not under-gone inversion. Unfortunately, there is insuf-ficient well control in the region to enable the effect of uplift on interval velocities to be modelled. Therefore, for simplicity, the inter-val velocities within each unit were assumed to be invariant with depth. Local lateral facies changes also affect interval velocities and some account of this is taken (Table 1) by using different velocities for depth conversion of dif-ferent lines. Although the interval velocities used here are only crude approximations of the true velocity structure, the isopach and depth maps produced using these interval velocities (Bishop 1992*a*) were in close agreement with the depths and thicknesses recorded in explor-ation wells. Furthermore, the depth converted sections were successfully restored, as will be shown in this paper, which indicates that the interval velocities used were satisfactory.

The restoration methodology used is approximately the same as that outlined by Kligfield *et al.* (1986), Geiser *et al.* (1988) and Geiser *et al.* (1991), and described in the Geosec user's manual. To restore the depth sections, they were first subdivided into fault- and bedding-bounded regions that could be treated as discrete structural entities or 'components'. These 'components' were then successively restored to a horizontal datum using appropriate deformation mechanisms. The mechanisms were chosen such that no space problems were created on the restored section, cross-sectional area balance was maintained, and the restored section was geologically plausible. If no mechanism could be found to meet these demands, then there had to be an error in the original interpretation or the interval velocity model. Only minor modifications of the original interpretations were found to be necessary in order to produce valid restorations.

Insufficient data were available to enable precise palaeobathymetric profiles to be constructed along the length of each seismic section. Therefore the sections could not be restored to such profiles. Instead, at each stage of restoration, the overburden layer was removed and underlying layers restored to an arbitrary horizontal datum.

During restoration of these sections it was found that in most cases the mechanism of vertical simple shear allowed construction of properly balanced restored cross-sections. The simple shear method was originally developed by the Chevron Company and described by Verrall (1981), since when it has been widely used in section restoration work. The application of simple shear during restoration maintains cross-sectional area, and by implication, conserves rock mass. However, bed length is not necessarily conserved.

During restoration of line WV-08, steep antithetic shear was also used with good results. A shear angle of 65° to the horizontal was found to be the optimum angle. This angle is comparable to the dips of many of the faults in the area (Gage 1952; Bishop 1992*a*), and there is also abundant field evidence of steep shear zones in the basement, as described later in this paper. Steeply inclined or vertical shear is thought to be an important deformation mechanism in the West Coast. Occasionally, small components were best restored using flexural slip as a mechanism. This is also considered to be an appropriate mechanism, since flexural slip between bedding planes has been observed in the

field, as described by Yeats (1985) and later in this paper.

At each stage of a sequential restoration, the effects of sedimentary compaction by the overburden were accounted for. In areas of uplift and erosion the sections were not decompacted, but elsewhere it was assumed that normal decompaction could be applied. Several workers have published relationships which describe how different lithologies compact with increasing burial (e.g. Sclater & Christie 1980). Such relationships can be programmed as algorithms, which can be used to calculate how much decompaction should occur as layers are stripped off (Geiser *et al.* 1991). To decompact the sections presented here, each stratigraphic interval was assumed to consist of either sandstone, shale or a 1:1 mixture of each (Table 1). Computer algorithms based on the empirically derived compaction curves for shale and sandstone described by Sclater & Christie (1980) were used to decompact the sections presented here. As described by Geiser *et al.* (1991), the Geosec program compares the positions of all corresponding points on the deformed and undeformed state sections, thus enabling a truly 2-dimensional decompaction of the sections, rather than a 1-dimensional decompaction as we might carry out for a well, for example.

The Geosec program decompacts from the topmost horizon downwards ('top down'). It has been suggested (A. Welbon pers. comm.) that this approach is not geologically realistic, and that the decompaction should be carried out from the bottommost horizon upwards ('bottom up'). However, it is the authors opinion that using neither the topmost nor the bottommost horizon as a datum for decompaction can give a truly geologically realistic result. Decompaction from the top down causes cumulative errors in decompaction to appear at deeper structural levels, whereas decompaction from the bottom up causes cumulative errors to appear at shallower structural levels. However, decompaction from the bottom up can generate a considerable surface topography, which cannot be justified or supported from other geological evidence. Indeed, such a topography would mainly be the product of inaccuracies in the decompaction method, and it is unlikely that the topography would reflect reality. In this case study the structure close to the surface is of prime interest, not the basement, hence a top down decompaction is preferred. Any inaccuracies inherent in the decompaction method are transmitted into the basement, a region which does not concern this paper.

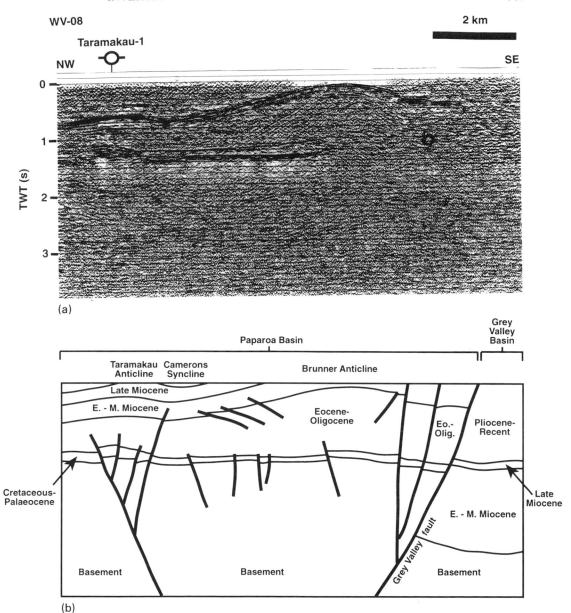

Fig. 8. (a) Line WV-08 and (b) interpretation.

Basin inversion structures

Interpretations of three key seismic reflection lines are presented below. These lines show the effects of Neogene basin inversion in the West Coast. Full sequential structural restorations of each section describe the evolution of these inversion structures.

Line WV-08

Seismic reflection line WV-08 (Figs 7 & 8) was shot by Hosking Geophysical Corporation for Petrocorp in 1983 and processed in 1984. A Vibroseis source was used and the section presented here (Fig. 8) is a migrated 24 fold CDP stack time section.

To the left (west) and centre of line WV-08 is a Late Cretaceous–Oligocene basin known as the Paparoa Basin (Figs 4 & 8), which has undergone inversion during the Neogene. At the right end (east) of the section is a syn-inversion contractional basin known as the Grey Valley Basin (Figs 4 & 8). Three strands of a major reverse fault known as the Grey Valley fault (Fig. 8) separate the inverted extensional basin from the contractional basin. The geometry of faults within the basement is not well imaged on the seismic data, so some reliance has been placed on downward extrapolation of fault dips in the interpretation of the fault systems. The syn-rift section thickens toward the Grey Valley fault, indicating that the fault was an active growth fault during rifting. This is further evidenced by outcropping Eocene breccia beds north of this section line, which thin and fine away from the eastern margin of the basin and are thought to have been derived from an active submarine fault scarp along the eastern margin of the basin during extension in the Eocene (Laird 1968; Nathan 1978).

Syn-rift strata are not interpreted on the seismic section to the right (east) of the Grey Valley fault (Fig. 8). The lowest reflector is interpreted to be at the base of the syn-inversion sequence, within a few metres of the top of the basement. Seismic reflection lines from another survey show that to the east of the Grey Valley fault, syn-rift strata may be locally thick, but are more often extremely thin (below seismic reflection resolution) or absent. Line WV-08 has been processed to resolve the reflectors in the upper part of the section, hence the difficulty in resolving the top of the basement. Also, there is a shadow zone beneath the Grey Valley fault hangingwall block, which makes it hard to pick reflectors in the footwall. However, the top basement pick is consistent with other lines in the seismic reflection line grid where the top basement reflector is better resolved.

Inversion has resulted in reverse movement on the Grey Valley fault, uplift and erosion on the left of the section, and major subsidence on the right of the section. The original Paparoa Basin floor was concave-up, but is now nearly flat-lying, and the younger strata which were originally nearly flat-lying are now folded into a broad anticline known as the Brunner Anticline (Figs 2 & 8). Both Miocene and Pliocene strata have been eroded from the crest of the anticline.

The Grey Valley fault comprises three discrete steep reverse faults which converge at depth and define downward-tapering wedge-shaped blocks. The faults clearly cut the hangingwall and are not comparable with the footwall shortcut faults which formed in the sand-box models of basin inversion described by Buchanan & McClay (1991, 1992). The three separate faults are thought to have been original extensional faults at the eastern margin of the Paparoa Basin that have all been reactivated in a reverse sense. Buchanan & McClay (1991, 1992) used only single basin boundary faults for their extension and inversion experiments, whereas in reality basins are commonly bounded by more than one discrete fault. On inversion, extensional fault zones become reverse fault zones. Comparable examples of this phenomenon have been documented elsewhere in the World, e.g. the Eakring–Foston fault in England (Fraser et al. 1990), and in the Middle East (Koopman et al. 1987).

A secondary inversion structure is developed at the left end of line WV-08 (Fig. 8) which is called the Taramakau Anticline. This 'pop-up' is due to reactivation of steep reverse faults on the western side of the Paparoa Basin. 'Pop-up' structures are common to inversion settings elsewhere, e.g. the Caunton Oilfield in England (Fraser et al. 1990). The main reverse fault bounding the 'pop-up' is interpreted to dip to the right, towards the Paparoa Basin axis. The 'pop-up' is also bounded by reverse faults which are antithetic to the main reverse fault. Similar 'pop-ups' have been modelled in sand-box experiments (Buchanan & McClay 1991).

Inversion has caused gentle folding of the syn-inversion sequence: the Camerons Syncline separates the Taramakau Anticline from the Brunner Anticline (Wellman – unpublished data). The folding is shown well by the strong reflectors on the Oligocene limestone (Fig. 8).

The limestone layer imaged on line WV-08 is cut by several thrusts, which cannot be traced into either the Eocene mudstone below or the Miocene mudstone above. Thus these thrust faults are effectively 'decoupled' from basement faults by the layer of Eocene mudstone. In the field, the limestone is brittle and generally competent, in contrast to the mudstone which is soft, friable and generally very incompetent. Whilst faults have been mapped cutting the limestone, few have been mapped cutting the mudstone, as can be seen on maps by Nathan (1978) and Bishop (1992a). So, although faults can be recognized in the field and on seismic reflection data, cutting brittle units such as the crystalline basement, coal measures and limestone, there is little evidence that similar amounts of faulting occur within the mudstone units. Since deformation associated with basin inversion has been transmitted from the basement into higher stratigraphic levels, it follows that although the mudstones do not exhibit obvious signs of deformation such as faulting, they clearly must have undergone deformation of some sort. It is suggested that the mudstone units may deform in a quasi-plastic manner, rather than in a brittle manner.

Between the Grey Valley fault to the east (right) and the Taramakau Anticline to the west (left), there are several normal faults with very small throws which cut the Cretaceous–Palaeocene coal measures (Fig. 8). Normal faults like these have been mapped by many workers in the West Coast and are thought to be of Cretaceous–Palaeogene age (e.g. Gage 1952; Nathan 1978). They are preserved away from the margins of the major structural blocks which are commonly about 8 km across (Bishop 1992a). Regional shortening has resulted in reactivation of normal faults at the margins of these blocks as reverse faults, whilst normal faults within the blocks have escaped reactivation. Fieldwork has shown (Bishop 1992a) that this phenomenon of strain partitioning becomes particularly apparent closer to the Alpine Fault, where shortening is greatest, such that completely unreactivated normal faults are found close to intense folding and thrusting, e.g. in the Reefton Coalfield area (Fig. 2) described later in this paper.

A full restoration of line WV-08 illustrates the formation of the southern part of the Paparoa Basin as a half-graben, and its subsequent inversion to form an asymmetric anticlinal and reverse fault bounded structure (Fig. 9). The Top Palaeocene restoration shows a basin bounded by a major fault to the east (right). East of this fault there may have been a highland region during the Cretaceous–Palaeocene which suffered erosion and supplied a lot of the course terrigenous clastic sediment which now comprises the Cretaceous–Palaeocene beds. Other intra-basin normal faults were active at this time also. Major subsidence of the basin and extension across the half-graben boundary fault took place in the Eocene–Oligocene. Normal faults close to the basin margins continued to be active. It was during this period that sedimentary breccias were shed into the basin along the eastern margin.

The asymmetric nature of the inversion structure was inherited from the pre-existent asymmetrical half-graben. The originally asymmetrical nature of the Cretaceous–Palaeogene basins in the West Coast region has been recognized from areas where there has been no inversion, as exemplified by the middle Cretaceous Takutai Half-Graben offshore Greymouth, and the Palaeogene Otukoro Half-Graben offshore Karamea (Fig. 4), and described by Bishop (1992 a, b). It has also been recognized indirectly from the analysis of vitrinite reflectance data (Suggate 1959; Nathan et al. 1986). Increasing vitrinite reflectance indicates increasing depth of burial, and when the vitrinite reflectance data for the Cretaceous–Eocene coal measure basins are mapped, it is clear that burial in the basins was asymmetrical about the long axes of the basins, with greatest burial being closest to the normal boundary faults (Suggate 1959; Nathan et al. 1986).

For this set of restorations (Fig. 9), it was assumed that the presently eroded Miocene units originally extended across the entire length of the section. Thus, for the purposes of the restoration, they were projected across the main inversion axis, and above the regional datum that is close to mean sea-level. The restoration presented here must be considered with this assumption in mind.

The Top Middle Miocene restoration (Fig. 9) shows how the basin bounding normal faults, and the Grey Valley fault splay in particular, became reactivated as reverse faults. The inversion was synchronous with regional subsidence and deposition, particularly in the contractional Grey Valley Basin which formed east of the Grey Valley fault splay.

If the Late Miocene thickness is projected across the section as described above, then the Base Pliocene restoration shows that regional subsidence and sedimentation took place in the late Miocene, but that there was little reverse fault movement. An alternative to this hypothesis is that if the Late Miocene sequence thinned or became absent across the main inversion axis, then some reverse movement on the faults would be predicted during the Late Miocene. This problem illustrates how missing stratigraphy in an area of inversion tectonics can severely hamper structural restoration.

Comparison of the Base Pliocene restoration and the present depth interpretation shows how very little Pliocene sedimentation took place on and around the inversion structure, whilst a considerable thickness accumulated in the contractional basin to the east. The Grey Valley fault splay was clearly very active in the Pliocene.

Line EZF-34

Seismic reflection line EZF-34 (Figs 7 & 10) was acquired by Western Geophysical for Esso and processed in 1969. An aquapulse source was used and the section presented here (Fig. 10) is an unmigrated 12 fold CDP stack time section.

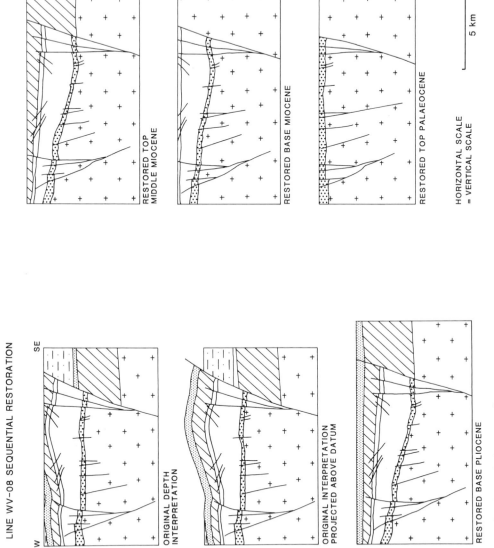

Fig. 9. Restoration of line WV-08. No vertical exaggeration.

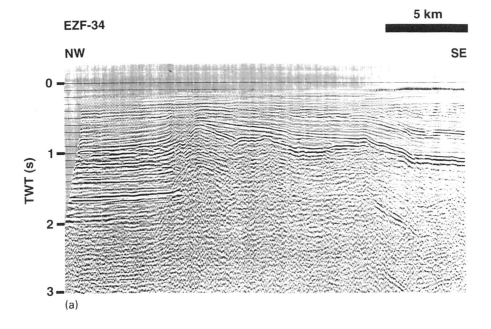

EZF-34

5 km

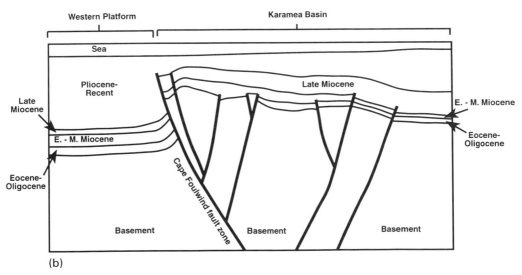

(b)

Fig. 10. (a) Line EZF-34 and (b) interpretation.

Line EZF-34 shows an inversion structure in the Karamea Basin (Fig. 4). The main reverse fault bounding the uplifted block is part of the Cape Foulwind fault zone (Nathan *et al*. 1986). This together with two major antithetic reverse faults bounds a 'pop-up' block, which is the locus of greatest uplift and where a base Miocene unconformity has cut down to the level of basement. A thin remnant of the syn-rift layer remains on the right (east) of the section.

The full structural restoration of line EZF-34 shows how the inversion structure developed (Fig. 11). The first Base Miocene restoration shows how all the reverse faults were restored so that no offset remained on any of them. In this restoration, it was assumed that rift sediments of Eocene–Oligocene age originally existed along the entire length of the section although this may not have been the case. Mapping of Eocene–Oligocene strata west of the Cape Foulwind fault

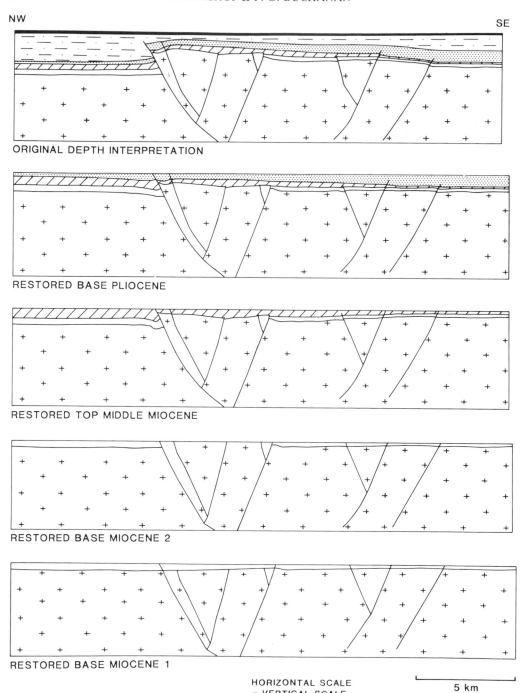

NW

SE

ORIGINAL DEPTH INTERPRETATION

RESTORED BASE PLIOCENE

RESTORED TOP MIDDLE MIOCENE

RESTORED BASE MIOCENE 2

RESTORED BASE MIOCENE 1

HORIZONTAL SCALE
= VERTICAL SCALE

5 km

Fig. 11. Restoration of line EZF-34.

zone showed that thickness decreases westward, and eventually their occurrence becomes patchy and hard to recognize from seismic reflection data (Bishop 1992a). There is little evidence on this line of extensional faulting during the Eocene–Oligocene. Instead the line shows that regional subsidence was more important. Bishop (1992a) observed that in the West Coast region as a whole, the Eocene–Oligocene was characterized by development of local fault bounded basins, as seen on line WV-08 (Fig. 8), separated by broader areas of subsidence and crustal 'sag', as seen on line EZF-34.

By comparing the first and second Base Miocene restorations, it can be seen how the first inversion feature to develop was a local 'pop-up' structure, with associated erosion of Eocene–Oligocene sediments. Later restorations show how the activation of reverse faults spread outwards from this initial pop-up and how uplift also caused thinning and gentle folding of the syn-inversion sedimentary layers.

The main syn-inversion depocentres on line EZF-34 have been on either side of the central inversion structure. Reverse movement on the Cape Foulwind fault zone has been considerable since the beginning of the Pliocene, with the development of a thick Pliocene–Recent sequence in the footwall area and further offshore on the Western Platform (Bishop 1992a).

Line CS-6

Seismic reflection line CS-6 (Figs 7 & 12) was shot by Geophysical Services Incorporated in 1970, but was reprocessed by Geco (NZ) in 1985. An airgun source was used and the section presented here (Fig. 12) is a migrated 24 fold CDP stack time section.

Although showing similarities, the structural interpretation and restorations presented here (Fig. 12) are much more detailed than those presented by Sissons & Walsh (1986) and Thrasher (1992a). The 'Mid Late Cretaceous' horizon (Thrasher 1992a) or 'Top Lower Pakawau' horizon (Sissons & Walsh 1986), which lies at the top of the Rakopi Formation (Thrasher 1992b), has been picked on the interpretation of CS-6 presented here, separating undifferentiated Cretaceous from the Late Cretaceous. Sissons & Walsh (1986) also pick 'Top Pakawau' on their interpretation, which equates to the 'Top Cretaceous' pick of Thrasher (1992a,b). Following Thrasher (1992a,b) we recognize this horizon as being 'Top Cretaceous' and have picked it on our interpretation of line CS-6 (Fig. 12).

The high amplitude reflector picked by Sissons & Walsh (1986) as 'Base Cobden' and by Thrasher (1992a) as 'Top Eocene' is thought by these authors to be the result of a velocity–density contrast at the base of the Oligocene limestone, based on synthetic seismogram studies farther north in the basin (Thrasher pers. comm.). However, to the south of line CS-6 the transition from the Eocene siliciclastic beds to Oligocene limestone is often quite gradual and marked by semi-marine and marine sandstones (Nathan et al. 1986). In contrast, the contacts between Oligocene limestone and overlying Miocene marine shales and sandstones are usually relatively sharp, and it is proposed therefore that the top of the limestone is more likely to be the cause of the high amplitude reflector, picked on line CS-6 as the base of the Miocene. In any case, the limestone in this region is generally less than 60 m thick (Nathan et al. 1986), often less than one seismic reflection wavelet, and hence it is difficult to distinguish between the base and the top of the limestone on the seismic reflection lines. What is important is that the reflector picked is close to (within 60 m) of the base of the Miocene, which is the base of the syn-inversion sequence.

The high on the west (left) of our interpretation (Fig. 12) is a 'pop-up' bounded on the western side by a major reverse fault zone named the Kahurangi fault zone (Sissons & Walsh 1986). This high has been drilled by the well Cook-1 and is commonly known by this name. A complex of minor steep reverse faults has formed on the eastern (right) side of the 'pop-up'. Another smaller 'pop-up' has formed near the centre of the line.

At the eastern end of the line is a major asymmetric anticline, named the Wakamarama Anticline, which can be traced southward into the onshore region (Bishop 1971). This anticline was drilled by the well Fresne-1. The crest of the anticline coincides with the thickest part of the syn-rift sequence. The Wakamarama fault bounds this major inversion structure on the eastern (right) side. A thin syn-rift sequence occurs to the east of the Wakamarama fault and a thick sequence to the west, producing a classic 'harpoon' or 'arrow-head' structure and a 'null point' of the type described by Badley et al. (1989), McClay & Buchanan (1991) and Williams et al. (1989).

The restoration of line CS-6 illustrates the evolution of the extensional basin and the subsequent basin inversion (Fig. 13). Initially a major graben formed during the Cretaceous, bounded by the Kahurangi and Wakamarama faults. This basin was filled mainly with

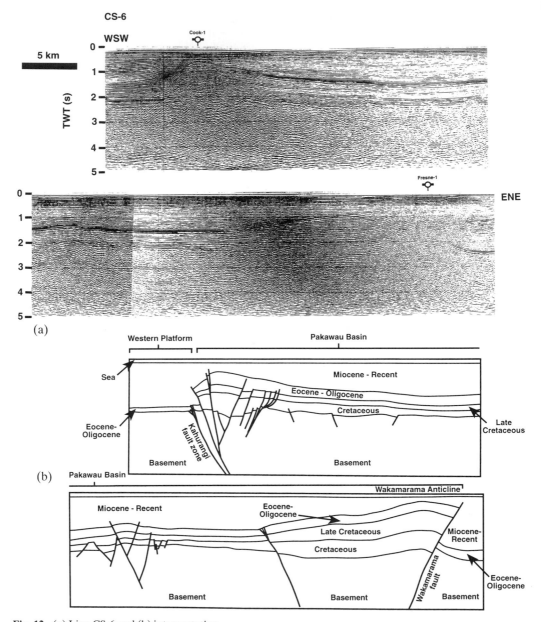

Fig. 12. (a) Line CS-6 and (b) interpretation.

terrestrial clastic sediments and coal measures and represents the offshore northern continuation of the Pakawau Basin (Nathan *et al.* 1986). The basement footwall highs on either side of the basin were probably uplifted highland areas and provided a source for the sedimentary fill of the basin. The deepest parts of the graben were adjacent to the two main boundary faults. A number of intra-graben normal faults also existed, particularly near the Kahurangi fault.

Extension and subsidence continued into the Palaeocene, particularly in the centre and east of the section, with the activation of several normal faults. As shown by the Base Miocene restoration, extension and subsidence continued into the Eocene and Oligocene, although it had a rather different character, with growth faults being less important than regional subsidence. Subsidence caused the footwall blocks of both the Kahurangi and Wakamarama faults to be

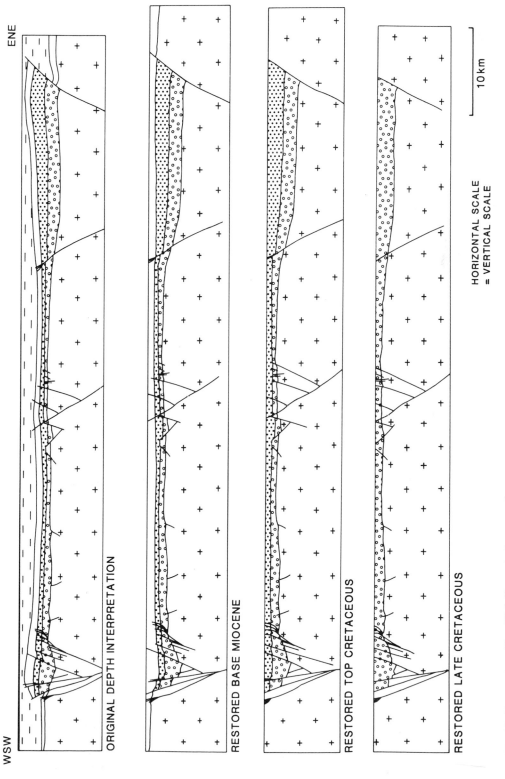

WSW

ENE

ORIGINAL DEPTH INTERPRETATION

RESTORED BASE MIOCENE

RESTORED TOP CRETACEOUS

RESTORED LATE CRETACEOUS

10 km

HORIZONTAL SCALE
= VERTICAL SCALE

Fig. 13. Restoration of line CS-6. No vertical exaggeration.

overtopped by Eocene–Oligocene deposition. This line shows both the effects of extensional faulting of the kind observed on line WV-08 and also the effects of regional sag and subsidence like that observed on line EZF-34.

The Base Miocene restoration can be compared with the original depth section (Fig. 13) to see the effects of basin inversion. The Wakamarama fault has undergone a considerable amount of reverse offset, whilst inversion at the western end of the line has been partitioned amongst several pre-existing faults. The several normal faults which bounded the western side of the graben have all been reactivated to form a reverse fault splay. The pop-up structure drilled by Cook-1 is also bounded by a number of other reverse faults on the eastern side. The reverse faults generally become more shallow dipping in the syn-inversion sequence. However, not all the normal faults have become reactivated as reverse faults, as has also been observed on line WV-08. The syn-inversion sedimentary sequence thins over the main inversion structures and is thickest in the footwall blocks of the Wakamarama and Kahurangi reverse faults.

Timing and magnitudes of extension and shortening

The percentage extension and shortening shown by the restored sections has been plotted versus time (Fig. 14), using the New Zealand chronostratigraphic timescale of Edwards *et al.* (1988). Extension is shown as a positive percentage change in section length and shortening as a negative percentage change. The values were calculated as a percentage of the original section length. In the case of line EZF-34 the original section was taken to be the Base Miocene restoration. For the other lines, the total length of a flattened top basement surface was measured, without faults, and this figure used as the original section length.

Hitherto it has been difficult to estimate the amount of extension in the West Coast region during the Cretaceous–Oligocene period, owing to the structural overprint of Neogene inversion. Previous estimates have come from areas where little Neogene deformation has taken place, such as in the stable Western Platform area outwith the Neogene deformation front, and within large structural blocks which have resisted inversion, such as those which occur within the Moutere area (Fig. 4). The Otukoro Half-Graben (Fig. 4) has an orientation parallel to most of the structures described in this paper (i.e. NNE–SSW), and is thought to be of

Eocene–Oligocene age (Bishop 1992*a*), i.e. it formed penecontemporaneously with the extensional basins described in this paper. Extension of *c.* 5% (*c.* 0.5 km) occurred in the Otukoro Half-Graben. This is comparable with the half-grabens of probable Palaeocene–Eocene age buried beneath the Neogene Moutere Basin (Fig. 4), which show extension of *c.* 6–7% (Lihou 1992). The restorations presented in this paper provide similar estimates of extension: line CS-6 shows *c.* 5.4% (4.6 km), and line WV-08 shows *c.* 5.2% (0.7 km). The amount of extension which caused subsidence along line EZF-34 could not be determined. In conclusion, total extension during the Late Cretaceous–Oligocene period was typically *c.* 5% in the West Coast region.

The magnitude of basin inversion shows more variability. Bishop (1992*a*) constructed regional structural cross-sections through the entire West Coast province, which gave estimates of *c.* 3–4% shortening, with a maximum shortening orientation approximately parallel to the previous extensional direction (i.e. WNW–ESE). This is comparable with estimates of *c.* 5% for both the Takaka Valley (Fig. 2) calculated by Judd (unpublished data) and the Moutere Basin (Lihou 1992). Restoration of lines CS-6 and EZF-34 produces estimates of *c.* 4.0% (3.6 km) and *c.* 3.3% (0.8 km) shortening respectively. These are also probably reasonable estimates for total Neogene regional shortening in the West Coast of South Island.

However, line WV-08 shows a much greater shortening of *c.* 16.2% (2.1 km). This can be explained by the fact that line WV-08 is much closer to the Alpine Fault plate boundary than the other lines, and hence has been subjected to much more intense deformation. Also, WV-08 is shorter than the others presented in this paper, and traverses the inversion structure only. Hence the shortening measured from the restoration of WV-08 should not be regarded as a regional value.

Even more extreme shortening has been measured in the Murchison Basin (Fig. 4) by Lihou (1993), closer still to the plate boundary. Lihou calculated *c.* 50% shortening across the Murchison Basin. In conclusion, the measured magnitude of shortening depends on proximity to the plate boundary and the length of the section baseline. However, the original estimates of 3–4% total regional shortening by Bishop (1992*a*) are probably typical of the West Coast as a whole.

Extension in the West Coast is thought to have begun in the middle Cretaceous, about 110 Ma ago (Nathan *et al.* 1986; Bishop 1992*a,b*), and

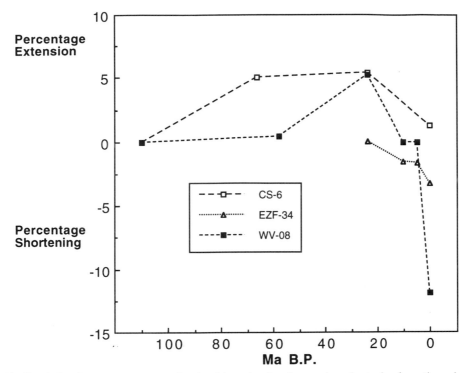

Fig. 14. Graph showing percentage extension (positive values) and percentage shortening (negative values) for each section through time.

this is shown in the graphical representation of extension and shortening since that time (Fig. 14). Extension generally accelerated during the Palaeocene–Oligocene. Shortening and inversion began at the beginning of the Miocene and has been on-going since then. Line EZF-34 shows the least inversion, due to its distance from the plate boundary. Lines EZF-34 and WV-08 show that inversion during the Neogene was punctuated in places by periods of tectonic quiescence in the Late Miocene. This may have been due to local tectonic block reorganizations as the azimuth of plate convergence rotated during the Neogene (Walcott 1987). Lines WV-08 and CS-6 show dramatic inversion during the Neogene, with the amount of shortening along line WV-08 exceeding the amount of original extension by the Middle Miocene.

Appearance of basin inversion structures in map view

The West Coast seismic reflection and well data set allow comparison of inversion structures in the onshore and offshore domains. The inversion structures described here have been mapped in both onshore and offshore regions by Bishop (1992a) using the seismic reflection grid presented in this paper (Fig. 7). It was shown by Bishop (1992a) that most of the major structural elements can easily be traced across the present day coastline. The structural relief created by the inversion of basins in the West Coast is exemplified in many areas by the relief of the deformed Oligocene limestone. The top of the Oligocene limestone is an effective strain marker, since it represents the end of rifting and regional subsidence, and the onset of inversion tectonics.

A contour map of the top Oligocene limestone reflector, at the southern end of the inverted Paparoa Basin, shows how several of the above-mentioned features of structural inversion appear in map view (Fig. 15). The region is divided into three regions by two major faults which are the Cape Foulwind fault zone in the west and the Grey Valley fault in the east, both of which trend NNE-SSW. The western province is relatively unfolded or faulted and represents the foreland region, furthest from the plate boundary. The central region is the southern end of the Paparoa Basin which has undergone inversion. The eastern region is the

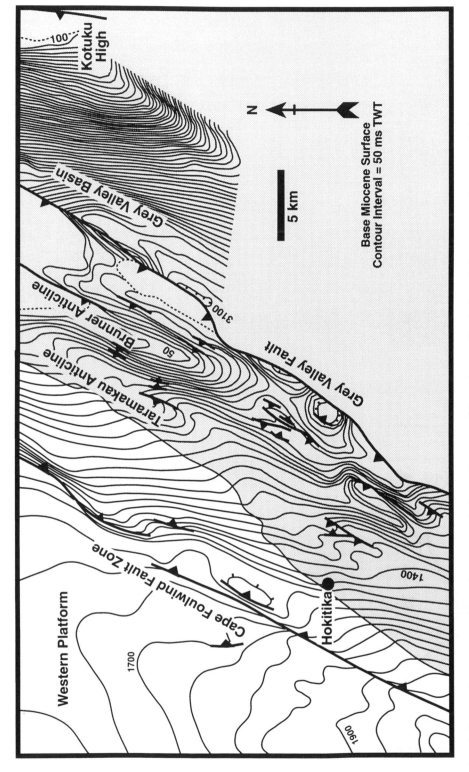

Fig. 15. Structural contour map of inversion structures in the southern part of the study area. For clarity, only key contour values are marked.

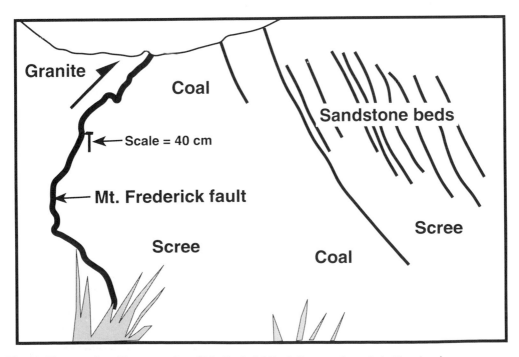

Fig. 16. Photograph and interpretation of Mt. Frederick Fault (hammer for scale is 40 cm long).

Grey Valley Basin which has undergone subsidence synchronously with inversion. The Grey Valley fault was the original main boundary normal fault to this part of the Paparoa Basin, dipping steeply to the WNW. It has since been reactivated as a steep reverse fault, which bounds the deepest part of the Grey Valley Basin. The Cape Foulwind fault zone comprises several en echelon faults, which were originally normal faults stepping down into the Paparoa Basin, but have been reactivated as reverse faults dipping steeply to the ESE. The axis of the basin lies just west of the Grey Valley fault and owing to inversion forms a broadly anticlinal feature, which is the Brunner Anticline, comprising local highs separated by saddles, plunging SSW at about 10°. The anticline is strongly asymmetric with a gently dipping limb to the west and a steeply dipping limb to the east. As shown by the interpreted seismic sections and the section restorations presented above, this asymmetry is due to the inversion of an originally asymmetric half-graben. The anticline is disrupted by a number of thrust faults which verge both WNW and ESE. Most of these faults occur in the region of greatest uplift near the crest of the Brunner Anticline. As mentioned earlier, these thrusts are decoupled from basement structures and were formed entirely during basin inversion, rather than being reactivated basement faults.

Field examples of inversion structures

Extensive fieldwork was carried out by Bishop (1992a) to examine how the inversion structures mapped from seismic reflection data compared with outcrop scale inversion structures. If we are to understand the mechanisms of basin inversion in a region, it is important to examine its effects at all available scales of observation. The quality of the seismic reflection data available for this study was not good, as has already been shown, so field investigation provided important supplementary data. Even had the seismic reflection data been better, it would still have been prudent to 'ground-truth' the seismic interpretations wherever possible.

Style of faulting

Examples of both low angle thrusts and high angle reverse faults can be found in the field, as well as on seismic reflection data. There is also evidence of flexural-slip being an important deformation mechanism.

In the Buller Coalfield region of South Island (Fig. 2), the Mt. Frederick fault is a good

example of a high angle reverse fault (Fig. 16). The hangingwall comprises granitoid rocks and Lower Palaeozoic metasediments, which are 'basement' in this part of the West Coast. The footwall comprises Eocene coal measures. The fault plane dips at about 56°. Granite in the hangingwall and coal in the footwall are both extremely brecciated, although the fault contact is quite sharp (Fig. 16). The contact is marked by clay gouge a few centimetres thick. Steeply dipping sandstone beds in the limb of a footwall syncline exhibit bedding plane slickensides parallel to bedding dip, indicating that folding has occurred by slip between beds. Yeats (1985) also described examples of flexural slip in the West Coast. Thus, the use of the flexural slip mechanism during section restoration can be justified on the basis of field evidence.

In some places low-angle thrusts can be observed which deform the cover sequence. North of Greymouth, thrusts occur within Late Cretaceous coal measures. In the example at Griegs beach (Figs 2 & 17), the main thrust has a ramp-flat geometry, and offsets sandstone, conglomerate and coal beds by about 0.5 m. More minor synthetic and antithetic thrusts occur in the footwall block. Fault surfaces are polished and slickensides indicate movement parallel to the dip of the fault. Some brecciation is observed, but as is the case with other low-angle faults in the West Coast, there is much less fault breccia and gouge than there is generally associated with high-angle faults.

In some places there has been a complex interaction between fault movement and flexural slip. There is a good example within the Eocene coal measures on Hall Ridge, near Greymouth (Figs 2 & 18). A three phase deformation history is proposed to explain this structure: (1) a fault initially truncated the coal bed; (2) bedding plane slip along the base of the coal bed and the juxtaposed sandstone bed caused the fault to become offset, and also the coal bed to be folded; (3) the lower portion of the fault was reactivated as a reverse fault, accentuating the fold in the coal bed above.

Fault splays

Probably one of the best field examples of a splayed reverse fault is the Pikikiruna fault, which trends N–S to NE–SW, along the western flanks of the Pikikiruna Range and the eastern margin of the Takaka Valley (Figs 2 & 19). Detailed mapping of the contact between the basement rocks of the Pikikiruna Range and Tertiary sediments indicates that steep faults are present (Grindley 1971; Judd, unpublished

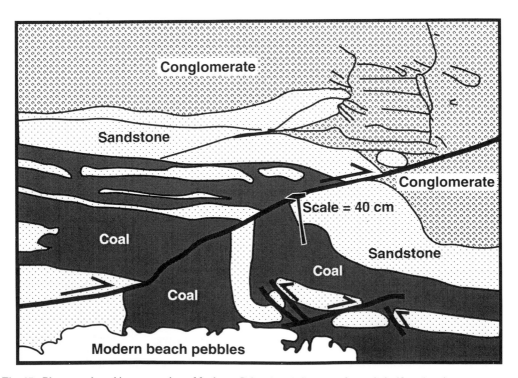

Fig. 17. Photograph and interpretation of faults at Griegs beach (hammer for scale is 40 cm long).

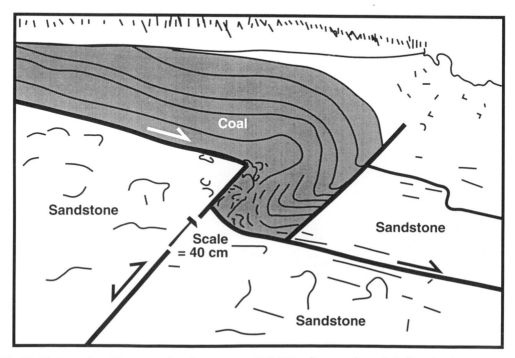

Fig. 18. Photograph and interpretation of structure on Hall Ridge (hammer for scale is 40 cm long).

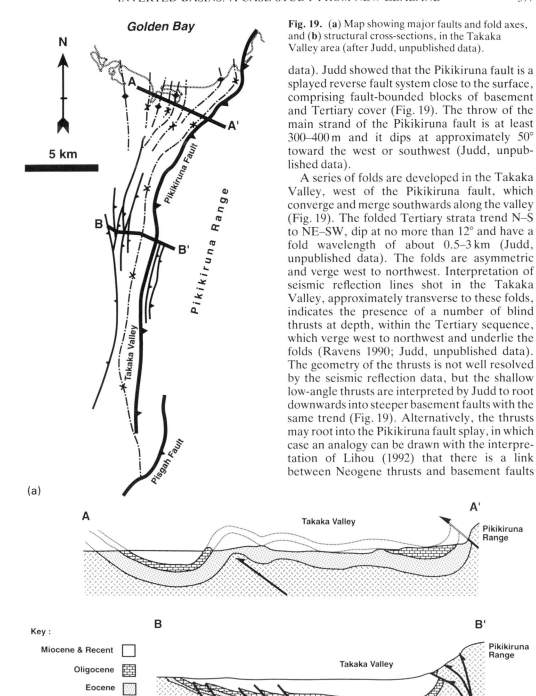

(a)

Fig. 19. (a) Map showing major faults and fold axes, and (b) structural cross-sections, in the Takaka Valley area (after Judd, unpublished data).

data). Judd showed that the Pikikiruna fault is a splayed reverse fault system close to the surface, comprising fault-bounded blocks of basement and Tertiary cover (Fig. 19). The throw of the main strand of the Pikikiruna fault is at least 300–400 m and it dips at approximately 50° toward the west or southwest (Judd, unpublished data).

A series of folds are developed in the Takaka Valley, west of the Pikikiruna fault, which converge and merge southwards along the valley (Fig. 19). The folded Tertiary strata trend N–S to NE–SW, dip at no more than 12° and have a fold wavelength of about 0.5–3 km (Judd, unpublished data). The folds are asymmetric and verge west to northwest. Interpretation of seismic reflection lines shot in the Takaka Valley, approximately transverse to these folds, indicates the presence of a number of blind thrusts at depth, within the Tertiary sequence, which verge west to northwest and underlie the folds (Ravens 1990; Judd, unpublished data). The geometry of the thrusts is not well resolved by the seismic reflection data, but the shallow low-angle thrusts are interpreted by Judd to root downwards into steeper basement faults with the same trend (Fig. 19). Alternatively, the thrusts may root into the Pikikiruna fault splay, in which case an analogy can be drawn with the interpretation of Lihou (1992) that there is a link between Neogene thrusts and basement faults

(b)

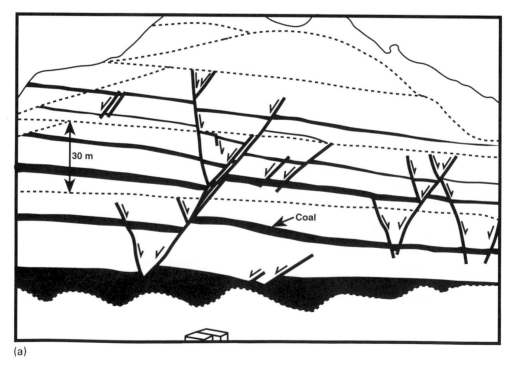

(a)

Fig. 20. Photographs and interpretations of (a) extensional faults (each bench 30 m high) and (b) contractional folds at the Garvey Creek Mine, Reefton Coalfield (hammer for scale is 40 cm long).

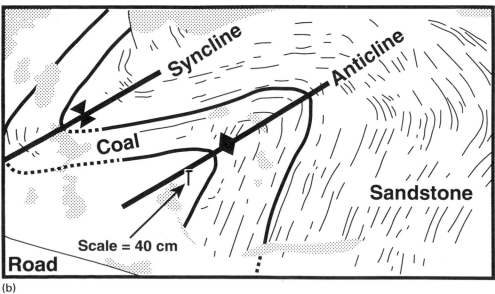

(b)

within the Kawatiri fault zone of the Moutere Basin, which lies to the east of the Pikikiruna Range.

Partitioning of deformation

As shown by the seismic interpretations presented above, the deformation due to contraction and inversion is partitioned into narrow zones along the margins of major structural blocks, and within those blocks there may be preserved extensional features which have undergone no inversion. The juxtaposition of extensional and contractional structures can also be seen in the field, in the Reefton Coalfield, where normally faulted Eocene coal measures are to be found close to severely faulted and folded coal measures. A good example is found at the Garvey Creek Mine, in the Reefton Coalfield, where extensional structures are exposed in the main highwall, whilst less than 0.5 km down the road from the mine there are tightly folded coal measures and recumbent folds exposed in road cuttings (Fig. 20).

Comparisons and contrasts with a sand-box model

The effects of basin inversion were modelled by Buchanan (1991) using sand-box experiments, and further described and explained by Buchanan & McClay (1991, 1992). In these experiments the basement was represented by rigid base-plates at the bottom of the sand-box experiment, and the deformable cover was represented by sand. In the sand-box experiments, the sand responded passively to deformation caused by movement of the base-plates. If the base-plates are used as analogues for basement blocks in the West Coast, then the model is accurate only if the basement in the West Coast behaved rigidly with respect to the deforming sediments above.

However, in the West Coast there is evidence that, although the basement is rheologically stronger than the Cretaceous–Cenozoic cover, the basement fault blocks have been deformed internally. It has been observed in the field that basement rocks are sometimes very fractured and broad folds may be formed (Wellman (unpublished data); Bishop 1992a; Lihou 1993). Multiple steeply dipping shear planes can commonly be seen within the basement. It is envisaged that minor faulting, brecciation and cataclasis in the basement enabled the generation of 'pseudo-folded' basement, as has been described in the Rocky Mountain foreland of the western U.S.A. by Blackstone (1983) and Cook

(1988). Hence, the results of the sand-box experiments may well differ from what we see in the West Coast, because in the sand-box experiments the basement analogue is rigid, whereas in reality the basement has been deformed.

An example of one of the sand-box experiments conducted by Buchanan (1991) is reproduced here (Fig. 21) to enable comparison with the West Coast inversion structures. Initial extension of the sand-box resulted in development of a graben, bounded by major normal faults and some intra-graben normal faults. During the early stages of inversion, the main boundary normal fault was reactivated as a reverse fault. Reverse faults propagated into the syn-inversion sequence with progressively shallower dip. Continued inversion resulted in the formation of reverse fault splays at the inverted basin margins, new thrust faults at high structural levels, cross-cutting footwall shortcut faults and significant folding of both the syn-rift and syn-inversion sequences.

Some comparisons with the West Coast can easily be made. Some of the early normal faults within the sand-box model basin never became reactivated as reverse faults, and this is true of many of the West Coast inverted basins. Also, in both the sand-box model and the West Coast, shortening was not evenly partitioned, but was concentrated across certain faults, usually the original basin boundary faults.

In the West Coast and in the sand-box model, reverse faults which propagate into the syn-inversion strata show progressively shallower dips toward fault tips. They become more thrust-like in nature as inversion progresses. The result of the propagation of reverse faults into the syn-inversion sequences with increasingly shallow dip is that the faults develop a distinctive convex-up geometry. This geometry can be seen on the interpreted seismic lines and cross sections presented in this paper, as well as in the sand-box model example.

Furthermore, in both the West Coast and the sand-box model, new thrust faults are developed during inversion, at higher structural levels, within the syn-inversion strata. These thrust faults are not necessarily connected with deeper basement faults either, as observed from the data and interpretations presented in this paper. The lack of direct connection between faults at basement level and faults within the inversion sequences suggests that non-brittle deformation takes place during inversion, in both the West Coast and the sand-box model analogue. Non-brittle deformation, or 'quasi plastic' deformation, would allow strain to be transmitted from lower to higher structural levels.

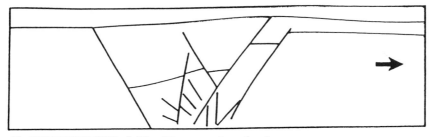

A) EXPERIMENT I-50 (7 cm extension)

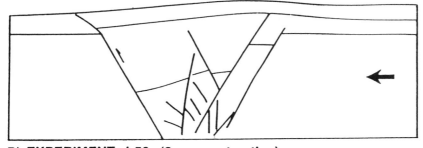

B) EXPERIMENT I-50 (2 cm contraction)

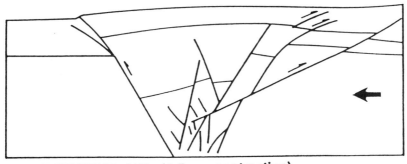

C) EXPERIMENT I-50 (4.5 cm contraction)

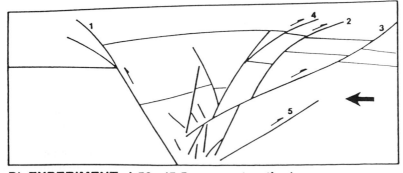

D) EXPERIMENT I-50 (5.5 cm contraction)

Fig. 21. Line drawing of sand-box model, showing development of inversion structures (from Buchanan 1991).

Despite these similarities, there are also some strong contrasts which need some explanation. There are two major differences between the analogue model shown here and the West Coast situation. One difference is that the analogue model does not model the effects of multiple normal boundary faults which have been observed in the West Coast. This is simply a limitation of the experiment design.

The other main difference is that cross-cutting footwall shortcuts have not been observed in the West Coast. In line CS-6, fault splays at basement level either side of the Cook-1 structure may be interpreted as footwall shortcut faults. However, this is not a feature observed elsewhere, there are no cross-cutting relationships, and nothing has been observed that is comparable with the major cross-cutting footwall shortcut faults exhibited by the sand-box model. This difference between the West Coast examples and the sand-box model may be due to the strength of faults relative to the material in which they are developed. In the sand-box model, the faults do not show significantly less cohesion than the surrounding material. Hence during inversion the shortening can readily be accommodated by the development of new faults at more appropriate angles. In the real geological situation, faults are major planes of weakness and show much less cohesion than the surrounding rock. Because it is relatively difficult to develop new faults, the pre-existing faults are utilised, even if they may not have the most appropriate orientation with respect to the shortening direction.

Summary

In this paper, inversion structures have been described using both subsurface and field data, and despite differences of scale, there is consistency of interpretation and structural styles. Periods of rifting affected the entire West Coast region during the Cretaceous–Oligocene period, resulting in about 5% regional extension. Rifting was characterized by locally deep extensional basins, interspersed with broader areas of regional crustal subsidence. Normal faults were steep and followed the pre-existing structural grain and lines of weakness inherited from the Late Mesozoic Rangitata Orogeny.

Shortening began in the earliest Miocene, at about the same time as the Alpine Fault became active as a transpressional boundary between the Australian and Pacific plates. Transpressional deformation caused the reactivation of major basin boundary normal faults as predominantly reverse faults, and the local elevation of pop-up structures. Continued shortening during the Miocene produced reactivation of more minor faults as antithetic reverse faults and backthrusts, so that areas affected by inversion increased in size. Uplift of basins and local erosion of rift strata on structural highs were accompanied by rapid subsidence of adjacent blocks. Basins undergoing inversion were separated from the new contractional basins by basin margin faults reactivated as steep reverse faults. Some of these major reverse faults formed splays, with propagation of reverse faults into the syn-inversion sedimentary sequence at progressively shallower angles. Although inversion resulted in major local uplift, the West Coast was inundated with sediment derived from erosion of the Southern Alps. As a result, inversion structures suffered periods of sedimentary burial as well as periods of erosion.

In some areas there was a temporary lull in inversion during the Late Miocene, the cause of which has yet to be explained, but may well relate to some change in the rate or direction of convergence across the Australian–Pacific plate boundary. Inversion was active in most areas during the Pliocene–Recent, with the development of new low-angle thrusts and folds, and the dramatic subsidence and infilling of contractional basins. Structural highs became eroded or developed particularly thin syn-inversion sequences. Many of the thrust faults developed at higher structural levels were structurally decoupled from basement faults by quasi-plastically deformed Eocene mudstones. In some areas shortening resulted in the development of fold and thrust belts, such as in the Takaka Valley.

Total Neogene shortening in the region was typically c. 3–4% and directed WNW–ESE, although shortening was much greater closer to the Alpine Fault plate boundary, and locally greater around inversion structures. Despite this, many extensional faults have escaped inversion entirely, and there are dramatic juxtapositions of extensional and contractional structures to be found in the field, particularly close to the Alpine Fault. Many of the features observed on seismic reflection lines and in the field can be recognized in sand-box models of basin inversion, although multiple normal basin boundary faults have not been modelled, and cross-cutting footwall shortcut faults have not been recognized in the West Coast.

Restoration of the sections presented in this paper, together with field evidence, indicates that vertical and steeply inclined shear and flexural slip may all be deformation mechanisms in the West Coast.

This paper illustrates the insights which may be gained from studying both the surface and subsurface expression of basin inversion within a region, and the benefits of carrying out full structural restoration of interpreted seismic reflection lines.

The authors would like to thank R. I. Walcott, D. J. May, A. S. Jayko, G. P. Thrasher, G. Rait, J. C. Lihou, M. J. Judd and M. Mueller for many useful discussions regarding the field geology and seismic reflection data. A. Welbon and one anonymous reviewer are thanked for their helpful reviews of the manuscript. M. J. Judd is thanked for giving permission to reproduce his unpublished maps and structural cross-sections of the Takaka Valley. L. Ellis and the Ministry of Commerce, Wellington, are thanked for reproducing the public-domain seismic reflection lines presented in this paper. Fieldwork and initial seismic reflection interpretation was carried out as part of the first author's PhD project in New Zealand, which was funded by Victoria University of Wellington and the New Zealand Vice Chancellors Committee. The remainder of the work was carried out jointly by both authors at The University of Edinburgh and CogniSeis Development Inc., Crawley. The University of Edinburgh is thanked for providing material and financial support during the writing of the paper. CogniSeis Development Inc., are thanked for allowing us to use their software package Geosec.

References

ADAMS, C. J. D. 1975. Discovery of Precambrian rocks in New Zealand: age relations of the Greenland Group and the Constant Gneiss, South Island. *Earth and Planetary Science Letters*, **28**, 98–104.

BADLEY, M. E., PRICE, J. D. & BLACKSHALL, L. C. 1989. Inversion, reactivated faults and related structures – seismic examples from the Southern North Sea. *In*: COOPER, M. A. & WILLIAMS, G. A. (eds) *Inversion Tectonics*. Geological Society, London, Special Publication, **44**, 201–219.

BARRY J. & MACFARLAN, D. 1988. Geology and coal resources of the Upper Waimangaroa and Mt. William South sectors, Buller Coalfield. *Market Information and Analysis Coal Geology Report 13*. Ministry of Energy, Wellington.

BISHOP, D. G. 1971. *Sheet S1 and S3 – Farewell and Collingwood. Geological map of New Zealand. 1:63.360*. Department of Scientific and Industrial Research, Wellington.

——, BRADSHAW, J. D. & LANDIS, C. A. 1985. Provisional terrane map of South Island, New Zealand. *In*: HOWELL, D. G. (ed.) *Tectonostratigraphic terranes of the circum-Pacific region*. Circum-Pacific Council for Energy and Mineral Resources, Earth Science Series, **1**, 515–521.

BISHOP, D. J. 1992a. *Middle Cretaceous–Tertiary tectonics and seismic interpretation of North Westland and Northwest Nelson, New Zealand*. PhD thesis, Victoria University of Wellington.

—— 1992b. Extensional tectonism and magmatism during the middle Cretaceous to Paleocene, North Westland. *New Zealand Journal of Geology and Geophysics*, **35**, 81–91.

—— 1992c. Neogene deformation in part of the Buller Coalfield, Westland, South Island, New Zealand. *New Zealand Journal of Geology and Geophysics*, **35**, 249–258.

—— 1992d. Cretaceous and Cenozoic tectonics of the West Coast region of the South Island. *1991 New Zealand Oil Exploration Conference Proceedings*. Energy and Resources Division, Ministry of Commerce, Wellington, 122–133.

BLACKSTONE, D. L. 1983. Laramide compressional tectonics, southeastern Wyoming. *University of Wyoming Contributions to Geology*, **22**, 1–38.

BUCHANAN, P. G. 1991. *Geometries and kinematic analysis of inversion tectonics from analogue model studies*. PhD thesis, University of London.

—— & MCCLAY, K. R. 1991. Sandbox experiments of inverted listric and planar fault systems. *Tectonophysics*, **188**, 97–115.

—— & —— 1992. Experiments on basin inversion above reactivated domino faults. *Marine and Petroleum Geology*, **9**, 486–500.

CARTER, R. M. & NORRIS, R. J. 1976. Cainozoic history of southern New Zealand: an accord between geological observations and plate tectonic predictions. *Earth and Planetary Science Letters*, **31**, 85–94.

COOK, D. G. 1988. Balancing basement-cored folds of the Rocky Mountain foreland. *In*: SCHMIDT, C. J. & PERRY, W. J. (eds) *Interaction of the Rocky Mountain Foreland and Cordilleran Thrust Belt*. Geological Society of America, Memoir, **171**, 53–64.

COOK, R. A. & BEGGS, J. M. 1990. The exploration potential of the Great South Basin. *1989 New Zealand Oil Exploration Conference Proceedings*. Energy and Resources Division, Ministry of Commerce, Wellington, 55–61.

COOPER, A. F., BARREIRO, B. A., KIMBROUGH, D. L. & MATTINSON, J. M. 1987. Lamprophyre dike intrusion and the age of the Alpine Fault, New Zealand. *Geology*, **15**, 941–944.

COOPER, M. A. & WILLIAMS, G. D. 1989. (eds) *Inversion Tectonics*. Geological Society, London, Special Publication, **44**.

COOPER, R. A. 1986. A terrane interpretation of the early evolution of New Zealand. *Abstracts from the Lower Palaeozoic workshop, University of Canterbury*. Geological Society of New Zealand Miscellaneous Publication, **34**, 14–16.

COTTON, C. A. 1916. Block mountains and a 'fossil' denudation plain in northern Nelson. *Transactions of the New Zealand Institute*, **48**, 59–75.

—— 1958. *Geomorphology: an introduction to the study of landforms*. Whitcombe and Tombs, Christchurch.

EDWARDS, A. R., HORNIBROOK, N. DE B., RAINE, J. I., SCOTT, G. H., STEVENS, G. R., STRONG, C. P. & WILSON, G. J. 1988. A New Zealand Cretaceous–Cenozoic Geological Time Scale. *New Zealand Geological Survey Record*, **35**, 135–149.

FALVEY, D. A. 1974. The development of continental margins in plate tectonic theory. *Australian*

Petroleum Exploration Association Journal, **14**, 95–106.

FRASER, A. J., NASH, D. F., STEELE, R. P. & EBDON, C. C. 1990. A regional assessment of the intra-Carboniferous play of Northern England. *In*: BROOKS, J. (ed.) *Classic Petroleum Provinces*. Geological Society, London, Special Publication, **50**, 417–440.

FYFE, H. E. 1929. Movement on the White Creek Fault, New Zealand, during the Murchison Earthquake of 17 June 1929. *New Zealand Journal of Science and Technology*, **11**, 192–197.

GACE, M. 1952. The Greymouth Coalfield. New Zealand Geological Survey Bulletin, **45**.

GEISER, J., GEISER, P. A., KLIGFIELD, R., RATLIFF, R. & ROWAN, M. 1988. New applications of computer-based section construction: strain analysis, local balancing, and subsurface fault prediction. *The Mountain Geologist*, **25**, 47–59.

—— RATLIFF, R., KLIGFIELD, R. & MORRIS, A. 1991. Simultaneous decompaction and restoration: an improved method for basin modeling. *American Association of Petroleum Geologists Bulletin*, **75**, 579.

GRINDLEY, G. W. 1971. *Sheet S8 – Takaka. Geological map of New Zealand 1:63360*. Department of Scientific and Industrial Research, Wellington.

—— & DAVEY, F. J. 1982. The reconstruction of New Zealand, Australia and Antarctica. *In*: CRADDOCK, C. (ed.) *Antarctic Geoscience*. University of Wisconsin Press, Madison, 15–29.

—— & OLIVER, P. J. 1979. Paleomagnetism of Upper Cretaceous dikes, Buller Gorge, North Westland, in relationship to the bending of the New Zealand orocline. *In*: WALCOTT, R. I. & CRESSWELL, M. M. (eds) *The Origin of the Southern Alps*. Royal Society of New Zealand Bulletin, **18**, 131–147.

HOLT, W. E. 1991. Distributed faulting, relative velocities, and block rotation in Northwest Nelson. *Neotectonics of the Buller region: workshop abstracts volume*. Department of Scientific and Industrial Research, Wellington.

KAMP, P. J. J. 1986a. Late Cretaceous–Cenozoic tectonic development of the southwest Pacific region. *Tectonophysics*, **121**, 225–251.

—— 1986b. The mid-Cenozoic Challenger Rift System of western New Zealand and its implications for the age of Alpine fault inception. *Geological Society of America Bulletin*, **97**, 255–281.

KING, P. R. 1990. Polyphase evolution of the Taranaki Basin, New Zealand: changes in sedimentary and structural style. *1989 New Zealand Oil Exploration Conference Proceedings*. Energy and Resources Division, Ministry of Commerce, Wellington, 134–150.

KLIGFIELD, R., GEISER, P. & GEISER, J. 1986. Construction of geologic cross sections using microcomputer systems. *Geobyte*, **1**, 60–69.

KOOPMAN, A., SPEKSNIJDER, A. & HORSFIELD, W. T. 1987. Sandbox model studies of inversion tectonics. *Tectonophysics*, **137**, 379–388.

LAIRD, M. G. 1968. The Paparoa Tectonic Zone. *New Zealand Journal of Geology and Geophysics*, **11**, 435–454.

—— & HOPE, J. M. 1968. The Torea Breccia and the Papahaua Overfold. *New Zealand Journal of Geology and Geophysics*, **11**, 418–434.

LIHOU, J. C. 1992. Reinterpretation of seismic reflection data from the Moutere Depression, Nelson region, South Island, New Zealand. *New Zealand Journal of Geology and Geophysics*, **35**, 477–490.

—— 1993. The structure and deformation of the Murchison Basin, South Island, New Zealand. *New Zealand Journal of Geology and Geophysics*, **36**, 95–106.

LENSEN, G. J. & OTWAY, P. M. 1971. Earthshift and post-earthshift deformation associated with the May 1968 Inangahua earthquake, New Zealand. *Royal Society of New Zealand Bulletin*, **9**, 107–116.

MATTHEWS, E. R. 1990. Exploration in the onshore Westland basin. *In*: *1989 New Zealand Oil Exploration Conference Proceedings*. Energy and Resources Division, Ministry of Commerce, 62–70.

MINISTRY OF ENERGY 1989. *Petroleum wells in New Zealand 1865–1989*. Ministry of Energy, Wellington.

MORETTI, I. & LARRÉRE, M. 1989. Computer-aided construction of balanced geological cross sections. *Geobyte*, **4**, 16–24.

MORGAN, P. G. 1911. *The geology of the Greymouth subdivision, North Westland*. New Zealand Geological Survey Bulletin, **13**.

NATHAN, S. 1978. *Sheet S44 – Greymouth. Geological map of New Zealand, 1:63, 360*. Department of Scientific and Industrial Research, Wellington.

——, ANDERSON, H. J., COOK, R. A., HERZER, R. H., HOSKINS, R. H., RAINE, J. I. & SMALE, D. 1986. *Cretaceous and Cenozoic sedimentary basins of the West Coast region, South Island, New Zealand*. New Zealand Geological Survey Basin Studies, **1**, Department of Scientific and Industrial Research, Wellington.

NEWMAN, J. 1985. *Paleoenvironments, coal properties and their interrelationships in Paparoa and selected Brunner Coal Measures on the West Coast of the South Island*. PhD thesis, University of Canterbury.

RAINE J. I. 1980. Palynology of the Triassic Topfer Formation, Reefton, South Island, New Zealand. *Abstracts volume, 5th Gondwana Conference*, Victoria University of Wellington.

RAVENS, J. M. 1990. Shallow seismic reflection surveys in the Takaka Valley, Northwest Nelson. *New Zealand Journal of Geology and Geophysics*, **33**, 23–28.

ROBINSON, R. & ARABASZ, W .J. 1975. Microearthquakes in the North-West Nelson region, New Zealand. *New Zealand Journal of Geology and Geophysics*, **18**, 83–91.

RODER, G. H. & SUGGATE, R. P. 1990. *Sheet L29BD – Upper Buller Gorge, Geological map of New Zealand 1:50.000*. Department of Scientific and Industrial Research, Wellington.

SCHULTZ-ELA, D. D. 1991. Practical restoration of extensional cross sections. *Geobyte*, **6**, 14–23.

SCLATER, J. G. & CHRISTIE, P. A. F. 1980. Continental stretching: an explanation of the post-mid-

Cretaceous subsidence of the Central North Sea basin. *Journal of Geophysical Research*, **85**, 3711–3739.

SISSONS, B. A. & WALSH, R. D. 1986. Establishing the elusive paleo-play – a case study of the Kahurangi Fault Zone, Southern Taranaki, New Zealand. *In*: GLENIE, R. C. (ed.) *Second South-Eastern Australia Oil Exploration Symposium*. Petroleum Exploration Society of Australia, 331–336.

STOCK, J. M. & MOLNAR, P. 1987. Revised history of early Tertiary plate motion in the south-west Pacific. *Nature*, **325**, 495–499.

SUGGATE, R. P. 1959. *New Zealand Coals: their geological setting and its influence on their properties*. Department of Scientific and Industrial Research, Wellington, Bulletin **134**.

—— 1963. The Alpine Fault. *Transactions of the Royal Society of New Zealand (geology)*, **2**, 105–129.

—— 1979. The Alpine Fault bends and the Marlborough Faults. *Royal Society of New Zealand Bulletin*, **18**, 67–72.

—— 1984. *Sheet M29 AC-Mangles Valley. Geological Map of New Zealand, 1:50.000*. Department of Scientific and Industrial Research, Wellington.

—— 1987. Active folding in North Westland, New Zealand. *New Zealand Journal of Geology and Geophysics, 30*, 169–174.

THRASHER, G. P. 1992a. Late Cretaceous source rocks of Taranaki Basin. *1991 New Zealand Oil Exploration Conference Proceedings*.Energy and Resources Division, Ministry of Commerce, Wellington, 147–154.

—— 1992b. *Late Cretaceous geology of Taranaki basin, New Zealand*. PhD thesis, Victoria University of Wellington.

—— & CAHILL, J. P. 1990. *Subsurface maps of the Taranaki Basin region, New Zealand*. New Zealand Geological Survey Report, **G142**.

VERRALL, P. 1981. Structural interpretation with application to North Sea problems. *Course notes no. 3*. Joint Association of Petroleum Exploration Courses (UK).

WALCOTT, R. I. 1978. Present tectonics and Late Cenozoic evolution of New Zealand. *Geophysical Journal of the Royal Astronomical Society*, **52**, 137–164.

—— 1987. Geodetic strain and the deformational history of the North Island of New Zealand during the late Cenozoic. *Philosophical Transactions of the Royal Society of London*, **A321**, 163–181.

WEISSEL, J. K. & HAYES, D. E. 1977. Evolution of the Tasman Sea reappraised. *Earth and Planetary Science Letters*, **36**, 77–84.

WELLMAN, H. 1948. Geology of the Pike River Coalfield, North Westland. *New Zealand Journal of Science and Technology*, **B30**, 84–95.

—— 1971. Geology of the Kotuku oilfield, Westland. *New Zealand Geological Survey Report*, **48**.

WILLIAMS, G. D., POWELL, C. M. & COOPER, M. A. 1989. Geometry and kinematics of inversion tectonics. *In*: COOPER, M. A. & WILLIAMS, G. A. (eds) *Inversion Tectonics*. Geological Society, London, Special Publication, **44**, 3–15.

YEATS, R. S. 1985. Flexural-slip faulting in Grey-Inangahua Depression, South Island: field guide to Giles Creek and Blackball scarps. *New Zealand Geological Survey Report*, **EDS99**.

Tertiary uplift of a deep rift-sag basin, Cardigan Bay, offshore Wales, UK

JONATHAN P. TURNER[1] & STEVE G. CORBIN[2]

[1]University of Birmingham, School of Earth Sciences, Birmingham B15 2TT, UK

[2]LASMO North Sea plc, Adelaide House, Chiswick High Road, London W4 5RS, UK

Seismic mapping in the Cardigan Bay basin (CBB) shows that very deep Mesozoic subsidence (post-Carboniferous succession c. 6 km) exerted a significant influence over both the mode and scale of Late Cretaceous–Tertiary uplift. The basin comprises a thick sequence of Mesozoic sediment (including the thickest and most complete Liassic succession in the UK), that fills a central, fault-bounded 'sag' depression, expressed as near-circular isopachs. Inversion, in the sense of reverse reactivation of initially normal faults, is most clearly manifest on seismic data along the northwest and southern margins of the basin, where the Mesozoic succession is thinnest. From integration of seismic data with regional results from apatite fission track analysis (e.g. Lewis et al. 1992), however, it is increasingly clear that the basal Tertiary unconformity records substantial uplift and erosion across the entire basin.

Gravity modelling-derived depth to the moho beneath the CBB suggests it is depressed by up to 10 km, from a regional depth of c. 20 km (Meisner et al. 1986). On the BIRPS deep seismic profile, SWAT 2 & 3, (Klemperer & Hobbs 1991), the CBB is interpreted as a Mesozoic half-graben, located immediately foreward of the Variscan Front. Consequently its thickened crust cannot easily be interpreted to have undergone thickening during Variscan thrusting. We assign this crustal thickening to Alpine reactivation of deep, Variscan and/or Caledonian structures. Such an interpretation leads to an additional mechanism for regional Late Cretaceous–Early Tertiary uplift of the CBB, whereby isostatic uplift occurs in response to the formation of a relatively buoyant, deep crustal root.

The structural evolution of the basin is complicated by the fact that it is affected by three major, basin-bounding faults related to Caledonian, Variscan and Alpine (Laramide) stress. Structural interpretation, with emphasis on the areas in which these lineaments interact with each other, allows the following summary of CBB structural evolution.

Stephanian (Variscan): crustal thickening, uplift and foreland basin subsidence.

Permo-Triassic: intracontinental rifting and subsidence.

Lower Jurassic: rapid subsidence and eustatic sealevel rise produce a significant bathymetric deep. Gravitational collapse of basin margin faults in response to unfilled bathymetry. Initial subsidence trigger not yet known.

Upper Jurassic: shallowing bathymetry in response to rapid Lower Jurassic sedimentation.

Palaeocene (Laramide). Reactivation of deep basement faults, crustal thickening and isostatic uplift. Erosion of the Cretaceous succession above the locus of Jurassic subsidence; inversion of suitably orientated segments of the basin margin.

References

LEWIS, C. L. E., GREEN, P. F., CARTER, A. & HURFORD, A. J. 1992. Elevated K/T paleotemperatures throughout Northwest England: three kilometres of Tertiary erosion? *Earth and Planetary Science Letters*, **112**, 131–145.

KLEMPERER, S. & HOBBS, R. 1991. *The BIRPS Atlas – Deep Seismic Profiles Around the British Isles*. Cambridge University Press.

MEISNER, R., MATTHEWS, D. & WEVER, TH. 1986. The 'Moho' in and around Great Britain. *Annales Geophysicae*, **4B**, 659–644.

From BUCHANAN, J. G. & BUCHANAN, P. G. (eds), 1995, *Basin Inversion*, Geological Society Special Publication No. 88, 587.

587

Index

Page numbers in *italics* refer to Figures or Tables